Lecture Notes in Electrical Engineering

Volume 210

For further volumes:
http://www.springer.com/series/7818

Wei Lu · Guoqiang Cai
Weibin Liu · Weiwei Xing
Editors

Proceedings of the 2012 International Conference on Information Technology and Software Engineering

Information Technology

 Springer

Editors

Wei Lu
Beijing Jiaotong University
Beijing
People's Republic of China

Guoqiang Cai
Beijing Jiaotong University
Beijing
People's Republic of China

Weibin Liu
Beijing Jiaotong University
Beijing
People's Republic of China

Weiwei Xing
Beijing Jiaotong University
Beijing
People's Republic of China

ISSN 1876-1100   ISSN 1876-1119   (electronic)
ISBN 978-3-662-51184-8   ISBN 978-3-642-34528-9   (eBook)
DOI 10.1007/978-3-642-34528-9
Springer Heidelberg New York Dordrecht London

Committees

Honorary Chair
Yaoxue Zhang, Central South University, China

General Chairs
Wei Lu, Beijing Jiaotong University, China
Jianhua Ma, Hosei University, Japan

Steering Committee Chairs
Zengqi Sun, Tsinghua University, China
Shi-Kuo Chang, University of Pittsburgh, Knowledge Systems Institute, USA
Mirko Novak, Institute of Computer Science, Czech Academy of Sciences, Czech Republic

Program Committee Chairs
Guoqiang Cai, Beijing Jiaotong University, China
Weibin Liu, Beijing Jiaotong University, China

Organizing Committee Chairs
Weiwei Xing, Beijing Jiaotong University, China
Qingshan Jiang, Shenzhen Institute of Advanced Technology, Chinese Academy of Science, China
Kin Fun Li, University of Victoria, Canada
Tole Sutikno, Ahmad Dahlan University, Indonesia

Technical Program Committee Members
Bin Luo, Nanjing University, China
Charles Clarke, University of Waterloo, Canada
Chih Lai, University of St.Thomas, USA
Chris Price, Aberystwyth University, England
David Levy, University of Sydney, Australia
Hong-Chuan Yang, University of Victoria, Canada

Preface

The 2012 International Conference on Information Technology and Software Engineering (ITSE2012) was held by Beijing Jiaotong University in Beijing on December 8–10, 2012. The objective of ITSE2012 was to provide a platform for researchers, engineers, academicians as well as industrial professionals from all over the world to present their research results and development activities in Information Technology and Software Engineering. This conference provided opportunities for the delegates to exchange new ideas and application experiences face-to-face, to establish research or business relations and to find global partners for future collaboration.

We have received more than 1,300 papers covering most of the aspects in Information Technology, Software Engineering, Computing Intelligence, Digital Media Technology, Visual Languages and Computing, etc. All submitted papers have been subject to a strict peer-review process; about 300 of them were selected for presentation at the conference and included in the ITSE2012 proceedings. We believe the proceedings can provide the readers a broad overview of the latest advances in the fields of Information Technology and Software Engineering.

ITSE2012 has been supported by many professors (see the name list of the committees); we would like to take this opportunity to express our sincere gratitude and highest respect to them. At the same time, we also express our sincere thanks for the support of every delegate.

Wei Lu
Chair of ITSE2012

Contents

Chapter 1
Generalized Graph Regularized Non-negative Matrix Factorization for Data Representation

Yang Hao, Congying Han, Guangqi Shao and Tiande Guo

Abstract In this paper, a novel method called Generalized Graph Regularized Non-Negative Matrix Factorization (GGNMF) for data representation is proposed. GGNMF is a part-based data representation which incorporates generalized geometrically-based regularizer. New updating rules are adopted for this method, and the new method convergence is proved under some specific conditions. In our experiments, we evaluated the performance of GGNMF on image clustering problems. The results show that, with the guarantee of the convergence, the proposed updating rules can achieve even better performance.

Keywords GGNMF · Asymptotic convergence · Stationary · Image clustering

1.1 Introduction

Non-negative matrix factorization (NMF) is a useful part-based method for data representation. Since Lee and Seung [1] proposed the specific multiplicative updating rules in 1999, NMF has been widely used in pattern recognition and data mining. It aims to find two non-negative matrices whose product can well approximate the initial non-negative data matrix. In recent years, many variants of NMF have been proposed. Li [2] proposed the Local-NMF (LNMF) which imposes a spatially localized constraint on the bases. Hoyer [3] added sparseness constraint to the original NMF model. Ding [4] proposed a semi-NMF method

Y. Hao · C. Han (✉) · G. Shao · T. Guo
School of Mathematical Sciences, Graduate University of Chinese Academy of Science, Beijing, China
e-mail: hancy@gucas.ac.cn

W. Lu et al. (eds.), *Proceedings of the 2012 International Conference on Information Technology and Software Engineering*, Lecture Notes in Electrical Engineering 210, DOI: 10.1007/978-3-642-34528-9_1, © Springer-Verlag Berlin Heidelberg 2013

which relaxes the nonnegative constraint on the base matrix. Daoqiang Zhang [5] proposed a two-dimensional NMF method which aims to preserve spatial structure of the image data. In addition, Wang [6] and Zafeiriou [7] proposed independently a discriminative NMF, which minimizes the within-class scatter and maximizes the between class scatter in the subspace spanned by the bases.

Deng Cai [8] has proposed a novel algorithm called graph-regularized non-negative matrix factorization (GNMF) which is motivated by recent progresses in matrix factorization and manifold learning. This method explicitly considers the local invariance and geometrical information of the data space by constructing a nearest neighbor graph. However, when we consider applying NMF to clustering, the algorithm should exploit the intrinsic geometry of data distribution on mani-fold. As some papers discussed previously [8, 9], NMF is an optimization of convex cone structure. Instead of preserving the locality of points in a Euclidean manner, it is more objective to use some more reasonable manners to measure the similarity or distance of the data. In our paper, we develop a general model in which we can use different manners to measure the distance of different source data and related coefficients. We also proposed slightly modifications of the existing multiplicative algorithms and proved their convergence to a stationary point using the similar techniques invented by Chih-Jen Lin [10]. In our experi-ment, we implement the proposed algorithm and test its performance on image clustering. It showed that the more number of classes, the better performance our modified updating rules would achieve on compared with the GNMF method. Obviously, we have to pay some acceptable cost of speed for the convergence.

The rest of this paper is organized as follows: In Sect. 1.2, we give a brief review of the related work of NMF, including the original NMF and GNMF. Section 1.3 presents the GGNMF model and the modified updating rules which could be proven to converge to a stationary point with some specific conditions. Section 1.4 shows experimental results about the performance of our modified algorithm on image clustering.

1.2 Related Work

1.2.1 Non-negative Matrix Factorization

The original model of NMF proposed by Lee and Seung [1] is:

$$\min \frac{1}{2} \|V - WH\|_F^2$$
$$s.t. \ W \in R_+^{m \times r}, H \in R_+^{r \times n}$$

where $V \in R_+^{m \times n}$ is the data matrix, such as images, signals, spectrums. In reality, we have $r < \min(m, n)$, which makes NMF essentially find a compressed approximation of the original data matrix.

There are two commonly used cost functions which quantify the quality of the approximation. The first one is Euclidean distance between two matrices:

$$f_1 = \|V - WH\|_F^2 = \sum_{i,j}\left(V_{ij} - \sum_k W_{ik}H_{kj}\right)^2$$

The second one is the generalized K–L divergence between two matrices:

$$f_2 = D\left(V\|WH\right) = \sum_{i,j}\left(V_{ij}\log\frac{V_{ij}}{(WH)_{ij}} - V_{ij} + (WH)_{ij}\right)^2$$

The objective functions are both convex in $\mathbf{U}$ only or $\mathbf{V}$ only, and they are not convex in both variables together. The multiplicative updating rules presented by Lee and Seung minimizing f_1 is:

$$W_{ia}^{k+1} = W_{ia}^k - \eta_{ia}\cdot\nabla_W f_1(W^k, H^k)_{ia};$$
$$H_{bj}^{k+1} = H_{bj}^k - \delta_{bj}\cdot\nabla_H f_1(W^{k+1}, H^k)_{bj}$$

where $\nabla_W f_1\left(W^k, H^k\right)_{ia}$ and $\nabla_H f_1\left(W^{k+1}, H^k\right)_{bj}$ are the partial derivatives of f_1 with respect to W_{ia} and H_{bj}:

$$\nabla_W f_1(W^k, H^k)_{ia} = ((WH - V)H^T)_{ia},$$
$$\nabla_H f_1(W^{k+1}, H^k)_{bj} = (W^T(WH - V))_{bj}.$$

The η_{ia} and δ_{bj} are usually referred as step size parameters. Lee and Seung set the step sizes as:

$$\eta_{ia} = \frac{W_{ia}^k}{(W^k H^k (H^k)^T)_{ia}} \quad \text{and} \quad \delta_{bj} = \frac{H_{bj}^k}{((W^{k+1})^T W^{k+1} H^k)_{bj}} \tag{1.1}$$

This algorithm is thus a gradient-descent method. Lee and Seung also proposed a multiplicative algorithm to minimize f_2. Because algorithms for these two methods are the same in essence, we will focus on incorporating f_1 with the generalized graph regularizer in Sect. 1.3.

1.2.2 Graph Regularized Non-negative Matrix Factorization

As Deng Cai discussed in his article [9], NMF performs the learning in the Euclidean space which covers the intrinsic geometrical and discriminating structure of the data space. He introduced GNMF which tries to avoid this limitation by incorporating with a geometrically based regularizer. This method based on a natural local invariance assumption about the intrinsic geometry of the data distribution: if two data points v_1 and v_2 are "close" in the intrinsic geometry of the data distribution, then h_1 and h_2, which are the representations of the two points in the basis space, also should be "close". For this reason, GNMF aims to preserve the consistency of geometrical structure in its non-negative matrix factorization.

Similar to NMF, here is objective function in the Euclidean form:

$$f = \|V - WH\|_F^2 + \frac{\lambda}{2} \sum_{i,j}^{n} \|h_i - h_j\|_2^2 C_{ij}$$

where $C = [C_{ij}]$ is the weighting matrix, and the parameter λ controls the smoothness of the new representation. C_{ij} negatively correlated with the closeness of two original points v_i and v_j. We usually uses 0–1, heat kernel, or dot-product weightings to measure the closeness of the two data vectors [8].

The algorithm minimizing f is:

$$W_{ia}^{k+1} = W_{ia}^k \frac{(V(H^k)^T)_{ia}}{(W^k H^k (H^k)^T)_{ia}}, \quad \text{and } H_{bj}^{k+1} = H_{bj}^k \frac{((W^{k+1})^T V + \lambda H C^T)_{bj}}{(((W^{k+1})^T W^{k+1} + 2\lambda D)H^k)_{bj}}$$

where: C is the weight matrix, D is a diagonal matrix whose entries are $D_{jj} = \sum_l C_{jl}$.

1.3 Generalized Graph Regularized Non-negative Matrix Factorization

By incorporating with a geometrically based regularizer, GNMF performs its learning with manifold regularization. However, NMF could also be analyzed as an optimization of convex cone structure [11]. Instead of preserving the locality of points in a Euclidean manner, we should be free to consider a more general form to describe the geometrical structure of the original data, such as images, signals and documents. Then we proposed the GGNMF model.

1.3.1 Description of GGNMF

As we discussed above, we would consider the locality and similarity of the original data matrix more efficiently. We propose a more general model in which we used $G(v_i, v_j)$ to substitute for the weighting matrix $C = [C_{ij}]$, and $K(h_i, h_j)$ to represent the "closeness" of relevant coefficient vectors h_i and h_j.

With the generalized forms, our GGNMF model is:

$$\min \frac{1}{2} \|V - WH\|_F^2 + \frac{\lambda}{2} \sum_{i,j}^{n} K(h_i, h_j) G(v_i, v_j) \tag{1.2}$$

$$s.t. \quad W \in R_+^{m \times r}, H \in R_+^{r \times n}$$

We believe that there must be a variety of forms of $K(h_i, h_j)$ and $G(v_i, v_j)$ for different original data sets. Because different kinds of original data have different characteristics, we can choose different distances to improve the efficiency, accuracy, and even convergence of this model. For example, given geometrical structure of cone, we can use angle to measure the "closeness" of h_i and h_j.

1.3.2 Multiplicative Updating Rules for GGNMF

In this section, we proposed the multiplicative updating rules for GGNMF. In order to ensure the convergence with some conditions of $K(h_i, h_j)$ and $G(v_i, v_j)$, we slightly modified the existing rules. In the next subsection, we will show the necessary conditions of $K(h_i, h_j)$ and $G(v_i, v_j)$ to guarantee algorithm's convergence.

The idea is based on the techniques invented by Chih-Jen Lin, who modified the updating rules of Lee and Seung and proved that there is at least one limit stationary point. We focused on the objective function:

$$F(W, H) = \frac{1}{2}\|V - WH\|_F^2 + \frac{\lambda}{2}\sum_{i,j}^{n} K(h_i, h_j)\, G(v_i, v_j) \qquad (1.3)$$

A natural idea is alternative descent method. Based on this method, we usually update H and W alternatively using gradient-descent method. Then we can obtain its first-order gradient information respect to H_{bj} and W_{ia}:

$$\nabla_H F(W, H)_{bj} = (W^T WH - W^T V)_{bj} + \lambda \sum_{i,j=1}^{n} \nabla_H K(h_i, h_j)_{bj} \cdot G(v_i, v_j)$$

$$\nabla_W F(W, H)_{ia} = (WHH^T - VH^T)_{ia}$$

As discussed by Chih-Jen Lin [10], we could know that there are two difficulties raised in the Eq. (1.1):

1. If the numerator of the step size is zero but the gradient still less than zero, however H_{bj}^{k+1} or W_{ia}^{k+1} can not be further improved by the rule.
2. The denominator of the step size may also be zero which can not be reasonable.

Hence we modify the step size using the techniques proposed by Chih-Jen Lin:

$$\bar{\delta}_{bj} = \frac{\overline{H}_{bj}^k}{(\nabla_H^2 F(W^k, H^k) \cdot \overline{H}^k)_{bj} + \tau}$$

where:

$$\bar{H}_{bj}^k \equiv \begin{cases} H_{bj}^k & \text{if } \nabla_H F(W^k, H^k)_{bj} \geq 0, \\ \max(H_{bj}^k, \sigma) & \text{if } \nabla_H F(W^k, H^k)_{bj} < 0. \end{cases}$$

Both τ and σ are appropriate small positive numbers. Similarly, we can define $\bar{\eta}_{ia}$ for the iteration of $\mathbf{W}$.

Based on the modifications of the step sizes, the algorithm for **GGNMF** is as follows:

Multiplicative updating rules for GGNMF

Given $\tau > 0$ and $\sigma > 0$. Initialize $W_{ia}^1 \geq 0$, $H_{bj}^1 \geq 0$, for k = 1, 2…:

1. If (W^k, H^k) is stationary, stop.

Else

$$H_{bj}^{k,n} = H_{bj}^k - \alpha_{bj} \cdot \bar{\delta}_{bj} \cdot \nabla_H F(W^k, H^k)_{bj},$$

$$W_{ia}^{k,n} = W_{ia}^k - \bar{\eta}_{ia} \cdot \nabla_W F(W^k, H^{k,n})_{ia}$$

where: $\alpha_{bj} \in [0, 1]$ is a coefficient which could be helpful in the prove of convergence.

2. Normalize $W^{k,n}, H^{k,n}$ to W^{k+1}, H^{k+1} (If the whole column of $W^{k,n}$ is zero, then the corresponding row in H^{k+1} are unchanged).

We denote W_{ia}^k as the ath element of the ith row of the basis matrix W^k, and H_{bj}^k as the bth element of a certain column of the coefficient matrix H^k. We also denote $W^{k,n}$ and $H^{k,n}$ as the intermediate matrices before normalization, and impose the normalization operation in order to make $\{W^{k,n}, H^{k,n}\}$ in a bounded set which is a critical step in the proof of the convergence.

1.3.3 Convergence Analysis

With the character of alternative descent method, the objective function value could not be increased every iteration. Generally speaking, it is relatively difficult to prove the convergence, especially for the updating rules of the generalized model. Before discussing the convergence of our algorithm, we have to preset some necessary requirements for $K(h_i, h_j)$ and $G(v_i, v_j)$:

1. $K(h_i, h_j)$ and $G(v_i, v_j)$ should be symmetric and non-negative functions.
2. $K(h_i, h_j)$ should be second-order differentiable, convex function for each variables, and $\nabla_{h_i}^2 K(h_i, h_j) \geq 0$.

With the above conditions for $K(h_i, h_j)$ and $G(v_i, v_j)$, we could prove the algorithm converging to a stationary point using the techniques invented by Chih-Jen Lin.

We begin the convergence analysis by showing that from H^k to $H^{k,n}$. When W^k is fixed, the function (1.3) is the sum of n functions, each of which relates to only one column of H. Thus, it is enough to consider any column $\mathbf{h}$ and discuss the function:

$$\bar{f}(W, h) = \frac{1}{2} \|v - Wh\|_F^2 + \lambda \sum_{j}^{n} F(h, h_j) G(v, v_j) \tag{1.4}$$

We can obtain its first-order, and second-order gradient information respect to h:

$$\nabla_h \bar{f}(W,h) = W^T W h - W^T v + \lambda \sum_{j=1}^{n} \nabla_h K(h,h_j) \cdot G(v,v_j)$$

$$\nabla_h^2 \bar{f}(W,h) = W^T W + \lambda \sum_{j=1}^{n} \nabla_h^2 K(h,h_j) \cdot G(v,v_j)$$

According to the idea of Chih-Jen Lin, we can define a similar auxiliary function on non-KKT indices:

$$A(h,h^k) \equiv \bar{f}(h^k) + (h - h^k)_I^T \nabla \bar{f}(h^k)_I + \frac{1}{2}(h - h^k)_I^T \overline{D}_{II}(h - h^k)_I$$

where

$$\begin{aligned} I &\equiv \{b | h_b^k > 0, \nabla \bar{f}(h^k)_b \neq 0 \text{ or } h_b^k = 0, \nabla \bar{f}(h^k)_b < 0 \} \\ &= \{b | \bar{h}_b^k > 0, \nabla \bar{f}(h^k)_b \neq 0 \} \end{aligned} \tag{1.5}$$

and $\overline{D}_{II}$ is the sub-matrix of a diagonal matrix $\overline{D}$ with elements:

$$\overline{D}_{bb} = \begin{cases} \frac{(\nabla_h^2 \bar{f}(W^k,h^k)\bar{h}^k)_b + \tau}{\bar{h}_b^k}, & \text{if } b \in I \\ 0, & \text{if } \notin I \end{cases} \tag{1.6}$$

Our new auxiliary function looks like a straightforward extension of that designed by Lee and Seung, but this modification is not trivial. While we define $A(h,h^k)$ like this, the indices satisfying $h_b^k = 0$ and $\nabla \bar{f}(h^k)_b < 0$ are taken care of. We also need that $A(h,h^k)$ leads to the non-increasing property, that is: $\bar{f}(h) \leq A(h,h^k) \leq A(h^k,h^k) = \bar{f}(h^k)$. By the following theorems we know GGNMF algorithm is valid and reliability.

Theorem 1 [10] *Let σ and τ be pre-defined small positive numbers and h^k be any column of H^k. Let I and $\overline{D}$ be defined as in (1.5) and (1.6). Let $I' \equiv \{1,\ldots,r\}\backslash I$, and use $\otimes$ to denote element-wise product of two vectors or matrices. If $h_I^{k,n} = h_I^k - \alpha_I \otimes (\overline{D}_{II}^{-1} \cdot \nabla_h f(h^k)_I)$, and $h_{I'}^{k,n} = h_{I'}^{k,n}$, we have:*

$$\bar{f}(h) \leq A(h,h^k) \leq A(h^k,h^k) = \bar{f}(h^k) \tag{1.7}$$

Moreover, if $h^{k,n} \neq h^k$, then $\nabla \bar{f}(h^k)_I \neq 0$ and both inequalities in (1.7) are strict. Theorem 1 immediately implies that the function value is strictly decreasing:

Theorem 2 [10] *If the rules generate an infinite sequence $\{W^k, H^k\}$, then we have:*

$$F(W^{k+1}, H^{k+1}) = F(W^{k,n}, H^{k,n}) \leq F(W^k, H^{k,n}) \leq F(W^k, H^k)$$

Moreover, one of the two inequalities is strict.

Next, we modify the proof proposed by Chih-Jen Lin [10] to show that H^k and $H^{k,n}$ converge to the same point.

Theorem 3 *Assume $\{H^k\}, k \in \kappa$ is a convergent sub-sequence and*

$$\lim_{k \in \kappa, k \to \infty} H^k = H^*.$$

Then we have $\lim_{k \in \kappa, k \to \infty} H^{k,n} = H^*$.

Then we are ready to prove that at any limit point (W^*, H^*), the matrix H^* satisfies KKT-conditions:

Theorem 4 *Assume $\{W^k, H^k\}, k \in \kappa$ is a convergent sub-sequence and:*

$$\lim_{k \in \kappa, k \to \infty} (W^k, H^k) = (W^*, H^*)$$

We have that: if $h_b^* > 0$ then $\nabla \bar{f}(h^*)_b = 0$; if $h_b^* = 0$,then $\nabla \bar{f}(h^*)_b \geq 0$.
The main convergence statement is Theorem 5.

Theorem 5 *Any limit point of the sequence $\{W^k, H^k\}$ is a stationary point of this optimal problem.*

The remaining task is to prove that at least one limit point exists.

Theorem 6 *The sequence $\{W^k, H^k\}$ has at least one limit point.*

We had finished the whole proof, but it is too long to be involved in this paper. We only presented some important results.

1.4 Experimental Results

Though the modified algorithm has the same computational complexity per iteration, it is important to check its practical performance. Previous studies showed that NMF is very powerful for clustering, especially in the document and image clustering tasks [12–14]. However, we could focus on image clustering and make some basic assumptions such as:

$$K(h_i, h_j) = \left\| h_i - h_j \right\|_2^2$$

$$G(v_i, v_j) = \begin{cases} 1 & if \ v_i \in N_p(v_j) \quad or \quad v_j \in N_p(v_i) \\ 0 & otherwise \end{cases},$$

Table 1.1 Statistics of the two data sets

Dataset	Size	Dimensionality	Classes
COIL20	1,440	1,024	20
PIE	2,856	1,024	68

where v_i represents the ith image which is the ith column of V, and $N_p(v)$ denotes the p-nearest neighbors of v in Euclidean distance. It is obvious that under these settlements our model is the same with Deng Cai's, where $K(h_i, h_j)$ and $G(v_i, v_j)$ could satisfy the necessary conditions for the proof of convergence. Then we could see the tradeoff of the performance in clustering and the cost of accuracy and speed for the convergence.

1.4.1 Data Sets

Two image data sets are used in this experiment. These data sets are summarized in (Table 1.1)

- The first data set is the COIL20, which contains 32×32 gray scale images of 20 objects viewed from varying angles.
- The second data set is the CMU PIE face database, which contains 32×32 gray scale face images of 68 individuals. Each has 42 facial images under different conditions.

Experiments were implemented in MATLAB on an Intel(R) Core(TM) 2 Quad 2.40 GHz computer.

1.4.2 Clustering Results

The clustering results are evaluated by comparing the obtained label of each sample with the label provided by the data set. The normalized mutual information metric (NMI) and Accuracy (AC) are used to measure the clustering performance [14]. Tables 1.2 and 1.3 show the clustering results on COIL20 and PIE data sets using original NMF, GNMF, and GGNMF algorithms.

From the Tables 1.2 and 1.3, we can know that:

1. No matter how many classes need to be clustered, the performance of GGNMF is better than the original NMF.
2. When the number of classes is relatively small, the cost of convergence of GGNMF is obvious. However along with the growth of the classes, the performance of GGNMF is getting better.

Table 1.2 Clustering performance on COIL20

K	Normalized Mutual information (%)			Accuracy (%)		
	NMF	GNMF	GGNMF	NMF	GNMF	GGNMF
4	71.8 ± 18.4	90.9 ± 12.7	88.9 ± 12.1	81.0 ± 14.2	93.5 ± 10.1	94.2 ± 8.5
8	71.0 ± 7.4	89.0 ± 6.5	88.8 ± 6.5	69.3 ± 8.6	84.0 ± 9.6	86.4 ± 8.8
12	73.3 ± 5.5	88.0 ± 4.9	88.1 ± 4.5	69.0 ± 6.3	81.0 ± 8.3	85.9 ± 6.5
14	73.8 ± 4.6	87.3 ± 3.0	88.4 ± 2.8	67.6 ± 5.6	79.2 ± 5.2	81.6 ± 5.9
16	73.4 ± 4.2	86.5 ± 2.0	88.5 ± 2.4	66.0 ± 6.0	76.8 ± 4.1	78.4 ± 4.5
18	72.4 ± 2.4	85.8 ± 1.8	88.9 ± 2.0	62.8 ± 3.7	76.0 ± 3.0	76.5 ± 3.0
20	72.5	87.5	88.1	60.5	75.3	77.2
Avg	72.7	88.4	88.3	68.1	80.8	82.9

Table 1.3 Clustering performance on PIE

K	Normalized mutual information (%)			Accuracy (%)		
	NMF	GNMF	GGNMF	NMF	GNMF	GGNMF
4	66.2 ± 4.0	86.1 ± 5.5	87.1 ± 4.6	57.8 ± 6.3	80.3 ± 8.7	83.1 ± 11.2
8	77.2 ± 1.7	88.0 ± 2.8	89.7 ± 2.1	62.0 ± 3.5	79.5 ± 5.2	81.5 ± 7.8
12	80.4 ± 1.1	89.1 ± 1.6	89.4 ± 1.7	63.3 ± 3.7	78.9 ± 4.5	80.1 ± 6.5
14	82.0 ± 1.1	88.6 ± 1.2	89.6 ± 1.3	63.7 ± 2.4	77.1 ± 3.2	79.3 ± 4.6
16	83.4 ± 0.9	88.8 ± 1.1	89.3 ± 1.1	65.2 ± 2.9	75.7 ± 3.0	77.6 ± 3.9
18	84.1 ± 0.5	88.7 ± 0.9	88.9 ± 0.7	65.1 ± 1.4	74.6 ± 2.7	75.9 ± 3.2
20	85.8	88.6	89.5	66.2	75.4	76.3
Avg	79.9	88.3	89.0	63.3	77.4	79.1

1.4.3 Efficiency Comparison

Table 1.4 report the average iteration speed of original NMF, GNMF and GGNMF
on COIL-20 and PIE. Clearly, the modified GGNMF is slower than NMF and
GNMF. But with the classes increasing, the GGNMF iteration time becomes closer
and closer to NMF and GNMF. Mover, GGNMF's convergence is guaranteed with
no flows. However, the cost of speed is still acceptable. Hence, in practice if one
would like to safely use GGNMF to clustering, our modification is a possible
choice.

1.4.4 Sparseness Study

Previous researchers [3] show that NMF codes naturally favor sparse and parts-
based representations which in pattern recognition can be more robust than non-
sparse global features. The additive combinations of the basis vectors seems to
make sure this quality of NMF, however recent studies show that NMF does not

Table 1.4 Average iteration time on the image sets

K	Average iteration time COIL20 (s)			K	Average iteration time PIE (s)		
	NMF	GNMF	GGNMF		NMF	GNMF	GGNMF
4	0.481	0.498	0.744	10	1.341	1.424	1.953
8	1.839	1.709	2.458	20	3.369	3.661	5.131
12	2.997	3.155	4.263	30	6.150	7.002	9.318
14	3.724	3.925	4.994	40	10.101	11.215	14.771
16	4.276	4.538	5.996	50	15.447	15.997	22.285
18	5.372	5.655	6.962	60	21.563	20.960	30.084
20	6.128	6.533	7.951	68	27.072	26.082	38.355

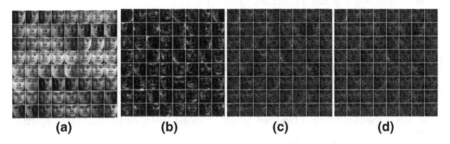

Fig. 1.1 **a** Original images. **b** Basis vectors learned by NMF. **c** Basis vectors learned by GNMF. **d** Basis vectors learned by GGNMF

always result in parts-based representations. As discussed by Deng Cai [8], GNMF can efficiently improve the sparseness of the basis vector (column vectors of U).

In this section, we investigated the sparseness of the basis vectors learned in GGNMF. Figure 1.1 shows the original images, and basis vectors learned by NMF, GNMF, and GGNMF in the PIE data sets respectively. Our experiment results suggest that, there is no considerable difference between the sparseness of the basis vectors obtained by GNMF and modified GGNMF algorithms, both of them can learn a better parts-based representation than NMF.

1.5 Conclusion

We have presented a general model based on GNMF, and proposed a generalized algorithm which could converge to a stationary point under some specific conditions. In a short, the GGNMF model is a framework for optimization of data representation. We could work on finding more efficient manners for different data sets.

How to choose appropriate manners for different kinds of data remains to be investigated in our future work. In order to factorize the data matrix into the product of basis matrix W and coefficient matrix H more accurately, we should do much more research to discover the geographic structures of different data sets.

Acknowledgments This work was supported by the National Natural Science Foundation of China (11101420, 10831006). In addition, I am truly grateful for the inspiration and help of Deng Cai. Any opinions, findings, and conclusions expressed here are those of the authors' and do not necessarily reflect the views of the funding agencies.

References

1. Lee DD, Seung HS (2001) Algorithms for non-negative matrix factorization Advances in neural information processing systems 13, MIT Press
2. Li SZ, Hou XW, Zhang HJ, Cheng QS (2001) Learning spatially localized, parts-based representation. In proceedings of CVPR, vol 1:207–212
3. Hoyer PO (2004) Non-negative matrix factorization with sparseness constraints. J Mach Learning Res 5:1457–1469
4. Ding C, Li T, Jordan MI (2010) Convex and semi-nonnegative matrix factorizations. IEEE Trans Pattern Analysis Mach Intelligence 32(1):45–55
5. Zhang D, Zhou Z, Chen S (2005) Two-dimensional non-negative matrix factorization for face representation and recognition, AMFG'05. Proceedings of the 2nd international conference on analysis and modelling of faces and gestures, Springer-Verlag Berlin, Heidelberg, pp 350–363
6. Wang Y, Jia Y, Hu C, Turk M (2004) Fisher non-negative matrix factorization for learning local features. ACCV
7. Zafeiriou S, Tefas A, Buciu Ioan, Pitas I (2006) Exploiting discriminant information in nonnegative matrix factorization with application to frontal face verification. IEEE Trans Neural Networks 17(3):683–695
8. Cai D, He X, Han J, Huang T (2011) Graph regularized non-negative matrix factorization for data representation. IEEE Trans Pattern Anal Mach Intell 33(8):1548–1560
9. Cai D, He X, Wu X, Han J (2008) Non-negative matrix factorization on manifold. Proceedings of the 2008 International Conference on Data Mining (ICDM'08), Pisa, Italy
10. Lin CJ (2007) On the convergence of multplicative update algorithms for nonnegative matrix factorization. IEEE Trans Neural Networks. vol 18(6)
11. Donoho D, Stodden V (2003) When does non-negative matrix factorization give a correct decomposition into parts? Advances in Neural Information Processing Systems 16, MIT Press
12. Cai D, He X, Han J (2005) Document clustering using locality preserving indexing. IEEE Trans. Knowledge and Data Eng 17(12):1624–1637
13. Xu W, Liu X, Gong Y (2003) Document clustering based on non-negative matrix factorization," Proc. Ann. Int'l ACM SIGIR conf. Research and Development in Information Retrieval, pp 267–273
14. Shahnaza F, Berrya MW, Paucab V, Plemmonsb RJ (2006) Document clustering using non-negative matrix factorization. Inf Process Manage 42(2):373–3863

Chapter 2
Research on Conformal Phased Array Antenna Pattern Synthesis

Guoqi Zeng, Siyin Li and Zhimian Wei

Abstract Phased array antenna has many technical advantages: high power efficiency, shaped beam, fast tracking by electric scanning, high stealth performance. If the array elements distribute on the airframe surface, the shape of the array is the same as the airframe contour, then the array is a conformal phased array. Conformal phased array has many advantages compared to planar phased array: smaller volume, no effect to aerodynamic performance of aircrafts, wider scanning range. In a conformal array, antenna elements are not placed in one plane because of its conformal structure. So the array elements and array factors cannot be separated, and the polarization direction of each element is different from each other, which will cause severe cross-polarization component. A field vector synthesis method is used in this paper to analyze the pattern of conformal phased array antenna, which can avoid dealing with the array element and array factor. This method is suitable for conformal array of any shape. This method is verified by calculating the pattern of a conformal array of truncated cone shape.

Keywords Conformal phased array · Pattern synthesis · Coordinate transformation

G. Zeng (✉) · Z. Wei
UAV Research Academy, Beihang University, Beijing 100191, China
e-mail: zengguoqi@sina.com

Z. Wei
e-mail: xiangjwxj@126.com

S. Li
School of Electronic and Information Engineering, Beihang University,
Beijing 100191, China
e-mail: lisiyin_staff@sina.com

W. Lu et al. (eds.), *Proceedings of the 2012 International Conference on Information Technology and Software Engineering*, Lecture Notes in Electrical Engineering 210, DOI: 10.1007/978-3-642-34528-9_2, © Springer-Verlag Berlin Heidelberg 2013

2.1 Introduction

Due to the importance of conformal phased array, USA, Japan and European Countries attach great importance to the development in this area. In recent years, conformal phased array antenna has been more and more widely applied due to the requirement of kinds of advanced aircrafts. The American "Advanced Anti-Radiation Guided Missile (AARGM)" Project focuses on the research on the technique of wide band passive receiving conformal phased array [1]. There is a special academic annual conference in Europe every other year since 1999 [2]. The Raytheon company promote a conformal phased array scheme used in a dual mode seeker, which contains a dielectric frustum radome structure (the open slot on it provides the space to place the infrared sensor), and an electronically steerable slot array in the radio-frequency range [3].

As for conformal phased array antenna, antenna elements are not placed in one plane because of its conformal structure. So the array elements and array factors cannot be separated, and the polarization direction of each element is different from each other, which will cause severe cross-polarization component. Thus, we have to find a new method to solve the pattern synthesis problem of a conformal structure.

Pattern synthesis of conformal phased array has been widely researched around the world. Reference [4] uses moment method, Ref. [5] uses numerical methods such as finite element method to analyze pattern of conformal phased array. Reference [6] uses the least squares method to optimize the in-phase polarization component and cross-polarization component of the spherical conformal array pattern. Reference [7] uses the iterative least square method to analyze pattern of spherical conformal array by change the incentive phase only. Reference [8] uses the least squares method and changes the phase only to analyze pattern of parabolic conformal structure. The numerical calculating method can only calculate a certain size conformal array, and computation speed is very slow. The methods used in Ref. [6–8] are only suitable for specific conformal structure, don't have versatility. In this paper, the field vector synthesis method is used to analyze pattern of conformal array. This method can be used in any shape of conformal structure and time cost is acceptable.

2.2 Planar Array Pattern Synthesis

$$\overrightarrow{E(\theta, \varphi)} = \sum_{n=0}^{N-1} A_n e^{j\varphi_n + jkl\cos\theta} F(\theta, \varphi) = F(\theta, \varphi) \sum_{n=0}^{N-1} A_n e^{j\varphi_n + jkl\cos\theta} = F(\theta, \varphi) f(\theta, \varphi)$$

$$(2.1)$$

Fig. 2.1 One-dimensional
antenna array

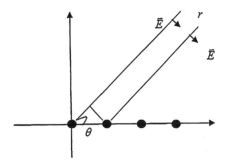

$$\overrightarrow{E(\theta, \phi)} = \sum_{m=0}^{M-1} \sum_{n=0}^{N-1} A_{m,n} e^{j\varphi_{m,n} + jk(ml \sin\theta \cos\phi + nl \sin\theta \sin\phi)} F(\theta, \phi)$$

$$= F(\theta, \phi) \sum_{m=0}^{M-1} \sum_{n=0}^{N-1} A_{m,n} e^{j\varphi_{m,n} + jk(ml \sin\theta \cos\phi + nl \sin\theta \sin\phi)} = F(\theta, \phi) * f(\theta, \phi)$$

$$(2.2)$$

From Eqs. (2.1) and (2.2), it can be found that all the elements are in one plane. So the electromagnetic field in the far field region in direction (θ, ϕ) is the superposition of field of each element in direction (θ, ϕ). The right side of the equations above can be divided into two parts, the field of each element in direction (θ, ϕ) is called the element component and the rest is called array factor (Figs 2.1, 2.2).

2.3 Spatial Field Calculation of Conformal Structure

According to the pattern synthesis method of a planar array, total field can be calculated by superimposing the contribution field of each element. We have to transform the pattern in local coordinate system of each element into global coordinate system by coordinate transformation. The field in direction (θ, φ) in global coordinate system is the superposition of the field $\overrightarrow{E}(\theta_{m,n}, \varphi_{m,n})$ of element (m, n) at the coordinate $(\theta_{m,n}, \varphi_{m,n})$ in local coordinate system. (Fig. 2.3)

$$\vec{F}_W(\theta, \varphi) = Z_L^W \vec{F}_L(\theta', \varphi')$$

$$\vec{E}(\theta, \varphi) = \sum_{n=0}^{N-1} \sum_{m=0}^{M-1} A_{m,n} e^{j\varphi_{m,n} + jk(\vec{r}' \cdot \hat{r})} [\vec{F}_W(\theta, \varphi)]$$

$$= \sum_{n=0}^{N-1} \sum_{m=0}^{M-1} A_{m,n} e^{j\varphi_{m,n} + jk(\vec{r}' \cdot \hat{r})} [Z_L^W \vec{F}_L(\theta'_{mn}, \varphi'_{mn})]$$

$$(2.3)$$

Fig. 2.2 Two-dimensional
antenna array

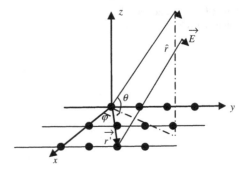

Fig. 2.3 Element
distribution of conformal
array antenna

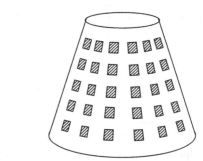

In which, Z_L^W is the transform matrix from local coordinate system to global coordinate system, $\vec{F}_L(\theta', \varphi')$ is the pattern of each element in local coordinate system, $\vec{F}_W(\theta, \varphi)$ is the pattern of each element in global coordinate system, $\vec{E}(\theta, \varphi)$ is the field in global coordinate system, $A_{m,n}$ is the amplitude of element (m, n), $\varphi_{m,n}$ is the phase of element (m, n).

Different from the planar array, $\vec{F}_W(\theta, \varphi)$ in Eq. (2.3) is various. Thus, the pattern cannot be divided into two parts just like Eqs. (2.1) and (2.2). In Eq. (2.3), $\vec{F}_L(\theta'_{mn}, \varphi'_{mn})$ is the known component, Z_L^W and $(\theta'_{mn}, \varphi'_{mn})$ are unknown. And $(\theta'_{mn}, \varphi'_{mn})$ in local coordinate system is corresponding to (θ, φ) in global coordinate system.

2.3.1 Spatial Field Calculation of Truncated Cone Shape Conformal Array

As showed in Fig. 2.4, r is the radius of top surface of truncated cone and R is the radius of the bottom surface, h is height. So the dip angle is $\theta_0 = a \tan\left(\frac{h}{R-r}\right)$. In Fig. 2.4, xyz is the global coordinate system and $x''y''z''$ is the local coordinate system.

Fig. 2.4 Coordinate system of truncated cone shape conformal structure

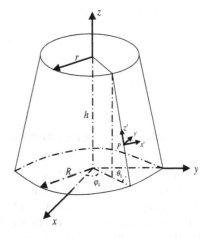

The transform process from xyz to $x''y''z''$ is: firstly, rotate xyz around z axis by φ_0 to get $x'y'z'$, then rotate $x'y'z'$ around y' axis by θ_0 to get the $x''y''z''$ coordinate system. Oppositely, rotate $x''y''z''$ around y'' axis by $-\theta_0$ to get $x'y'z'$, and then rotate around z' axis by $-\varphi_0$ to get the xyz coordinate system.

The transformation matrix from global coordinate system to local coordinate system is:

$$Z_L^{W'} = Z_W^L = Z_Y Z_Z$$

$$= \begin{bmatrix} \cos\theta_0 & 0 & -\sin\theta_0 \\ 0 & 1 & 0 \\ \sin\theta_0 & 0 & \cos\theta_0 \end{bmatrix} \begin{bmatrix} \cos\varphi_0 & \sin\varphi_0 & 0 \\ -\sin\varphi_0 & \cos\varphi_0 & 0 \\ 0 & 0 & 1 \end{bmatrix} \quad (2.4)$$

$$= \begin{bmatrix} \cos\theta_0\cos\varphi_0 & \cos\theta_0\sin\varphi_0 & -\sin\theta_0 \\ -\sin\varphi_0 & \cos\varphi_0 & 0 \\ \sin\theta_0\cos\varphi_0 & \sin\theta_0\sin\varphi_0 & \cos\theta_0 \end{bmatrix}$$

Assume that (θ, φ) is the angle in global coordinate system, and (θ', φ') is the corresponding angle in local coordinate system of a certain array element.

$$\begin{Bmatrix} x_L \\ y_L \\ z_L \end{Bmatrix} = Z_W^L \begin{Bmatrix} x_W \\ y_W \\ z_W \end{Bmatrix} = \begin{bmatrix} \cos\theta_0\cos\varphi_0 & \cos\theta_0\sin\varphi_0 & -\sin\theta_0 \\ -\sin\varphi_0 & \cos\varphi_0 & 0 \\ \sin\theta_0\cos\varphi_0 & \sin\theta_0\sin\varphi_0 & \cos\theta_0 \end{bmatrix} \begin{bmatrix} \sin\theta\cos\varphi \\ \sin\theta\sin\varphi \\ \cos\theta \end{bmatrix}$$

So,

$$\varphi' = a\tan\left(\frac{\sin\theta\sin(\varphi - \varphi_0)}{\cos\theta_0\sin\theta\cos(\varphi - \varphi_0) - \sin\theta_0\cos\theta}\right) \quad (2.5)$$

$$\theta' = a\cos(\sin\theta_0\sin\theta\cos(\varphi_0 - \varphi) + \cos\theta_0\cos\theta)$$

For the circular column structure, $\theta_0 = 90°$. Then,

$$\varphi' = a\tan(\tan\theta\sin(\varphi_0 - \varphi)) \tag{2.6}$$

$$\theta' = a\cos(\sin\theta\cos(\varphi_0 - \varphi))$$

2.3.2 Direction Determination of Spatial Field

In a planar array, in far field region in direction (θ, φ), the direction of pattern of each element is also (θ, φ). And the field direction of each element is the same. So the amplitude of each element can be added together directly. But in a conformal phased array, elements are not in one plane, thus the field direction of each element is different from each other in far field region in direction (θ, φ). So the vector superposition method has to be taken.

Each element radiates and produces spatial field. The direction of spatial field is $\hat{\theta}$ when field is vertical polarization. And its direction is $\hat{\varphi}$ for horizontal polarization. For circular polarization field, the amplitude of each field can be added together directly.

The transform formula from rectangular coordinate to spherical coordinate is as follows:

$$\begin{cases} \vec{i_{rs}} = \vec{i_x}\sin\theta\cos\varphi + \vec{i_y}\sin\theta\sin\varphi + \vec{i_z}\cos\theta \\ \vec{i_\theta} = \vec{i_x}\cos\theta\cos\varphi + \vec{i_y}\cos\theta\sin\varphi - \vec{i_z}\sin\theta \\ \vec{i_\varphi} = -\vec{i_x}\sin\varphi + \vec{i_y}\cos\varphi \end{cases} \quad \begin{cases} \vec{i_x} = \vec{i_{rs}}\sin\theta\cos\varphi + \vec{i_\theta}\cos\theta\cos\varphi - \vec{i_\varphi}\sin\varphi \\ \vec{i_y} = \vec{i_{rs}}\sin\theta\sin\varphi + \vec{i_\theta}\cos\theta\sin\varphi + \vec{i_\varphi}\cos\varphi \\ \vec{i_z} = \vec{i_{rs}}\cos\theta - \vec{i_\theta}\sin\theta \end{cases}$$

2.4 Example of Field Calculating

The longitudinal section of a truncated cone shape conformal structure is as shown in Fig. 2.5. The radius of top line is 40 mm, bottom line is 80 mm and the height is 100 mm. Set frequency as 35 GHz, thus the wave length is 8 mm. Assume that the pattern of array element $\vec{F}_L(\theta'_{mn}, \varphi'_{mn})$ in Eq. (2.3) is as shown in Fig. 2.6. So just the scalar sum of fields should be considered in calculation, and the summation is the pattern of the conformal array. Consider two kinds of uniform distribution of

Fig. 2.5 Longitudinal section of truncated cone shape conformal structure

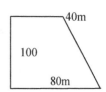

Fig. 2.6 Pattern of array element

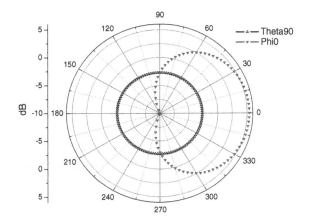

Fig. 2.7 Elements distribution of truncated cone conformal array and its 3-D pattern (36 × 20 elements)

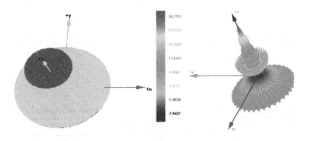

Fig. 2.8 Elements distribution of truncated cone conformal array and its 3-D pattern (8 × 8 elements)

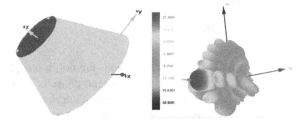

array elements with equal angle difference. The first one, elements cover the whole inclined plane of the truncated cone. And the next one, elements cover part of the inclined plane. Assume that all elements have the same amplitude in calculation. The synthesized pattern are as shown in Figs. 2.7, 2.8 and 2.9 (Table 2.1).

As shown in Fig. 2.7, the number of element is 36 × 20 = 720, and the output calculation result has 361 points in azimuth and 361 points in elevation; it takes 240 s to calculate, totally. As for Fig. 2.8, with the number of element 8 × 8 = 64, it takes 22 s to calculate. And the computer configuration is: IBM T61: CPU Intel Core2 Duo 2.2 GHz, RAM 1.96 GHz.

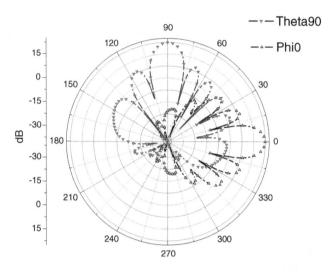

Fig. 2.9 Pattern section in main lobe direction $\theta = 90°$ and $\varphi = 0°$

Table 2.1 Pattern parameters of truncated cone conformal structure

Section	Main lobe (dB)	3 dB band-width (°)	First side lobe (dB)	Maximum side lobe (dB)
$\varphi = 0°$	22.30	13.72	9.79	9.79
$\theta = 90°$	22.30	13.10	9.16	9.16

2.5 Conclusions

In a conformal array, all the elements are not in one plane. When synthesizing the far field pattern, the field amplitude and phase of each element are different from each other. So the elements and array factors cannot be separated. At the same time the polarization direction of each element is different from each other, which will cause severe cross-polarization component. In this paper, the field vector synthesis method is introduced; establish global coordinate system based on pattern of the conformal array and establish local coordinate system based on the location of each element. According to the relationship between global and local coordinate system, there is the field in direction $(\theta_{mn}, \varphi_{mn})$ in local coordinate system of each element responding to the field in direction (θ, φ) in global coordinate system, from which the amplitude and phase contribution to the total field by each element can be calculated. And then the pattern of the array can be synthesized. In this paper the pattern of a truncated cone conformal array with different element distribution situation is analyzed: all covered by array element and partly covered by array element.

However, when synthesizing the pattern of a real conformal array antenna, the mutual coupling among elements should be taken into consideration. Also the amplitude or phase of each element should be optimized to synthesis the pattern, which needs further studying.

References

1. Cong min (1999) The upgrade and improve plan of HARM missile. Winged Missiles J (3) (in Chinese)
2. Josefsson L, Patrik P (2006) Conformal array antenna theory and design [M]. Willey-Inter Science Publication, New York
3. Park PK, Robertson RS (2003) Electronically scanned dielectric continuous slot antenna conformal to the cone for dual mode seeker: USP, 6653984 [P]. 25 Nov 2003
4. Raffaelli S, Zvonimir S, Per-Simon K (2005) Analysis and measurements of conformal patch array antennas on multilayer circular cylinder. IEEE Trans Antennas Propag 53(3):1105–1113
5. Charles AM, Kempel LC, Schneider SW (2004) Modeling conformal antennas on metallic prolate spheroid surfaces using a hybrid finite element method. IEEE Trans Antennas Propag 52(3):750–758
6. Vaskelainen LI (1997) Iterative least-squares synthesis methods for conformal array antennas with optimized polarization and frequency properties. IEEE Trans Antennas Propag 45(7):1179–1185
7. Vaskelainen LI (2000) Phase synthesis of conformal array antennas. IEEE Trans Antennas Propag 48(6):987–991
8. Dinnichert M (2000) Full polarimetric pattern synthesis for an active conformal array. In proceedings of IEEE international conference on phased array systems and technology, pp 415–419 May 2000

Chapter 3
Research on the Optimization of Wireless Sensor Network Localization Based on Real-Time Estimation

Xiancun Zhou, Mingxi Li and Fugui He

Abstract Range-based localization of wireless sensor networks is to obtain sensory data for the calculation of the distance to the target. The positioning accuracy is directly affected by environmental factors. The major localization error of the traditional methods will be made by the use of fixed parameters for the measurement of distance. In this chapter, we proposed a RSSI-based cooperative localization algorithm by adjusting the signal propagation model dynamically. The localization method was optimized and the triangular range localization method was taken as an example to verify the optimization result. From the simulation and experiment, we verify that the proposed algorithm can provide higher localization accuracy.

Keywords Sensor network · Range-based localization · RSSI

3.1 Introduction

Target localization is one of the key application areas of sensor networks. For the application with strict requirements on target localization accuracy, range-based localization methods are practically adopted. The range-based localization methods commonly used include: RSSI (Received Signal Strength Indicator) [1], TOA (Time

X. Zhou (✉) · F. He
School of Information Engineering, West Anhui University, lu'an, Anhui, China
e-mail: zhouxcun@mail.ustc.edu.cn

M. Li
School of Computer Science, University of Science and Technology of China,
Hefei, China
e-mail: mxli@mail.ustc.edu.cn

W. Lu et al. (eds.), *Proceedings of the 2012 International Conference on Information Technology and Software Engineering*, Lecture Notes in Electrical Engineering 210, DOI: 10.1007/978-3-642-34528-9_3, © Springer-Verlag Berlin Heidelberg 2013

Of Arrival) [2], TDOA (Time Difference Of Arrival) [3], AOA (Angle Of Arrival) [4] and so on. Compared with other range-based localization methods, the localization methods based RSSI can simultaneously broadcast their own coordinates at beacon nodes and complete by the RSSI measurement. With the known transmit strength in RSSI, the receiving node calculates the signal losses in the communication process according to the received signal strength, and then transforms the propagation loss into distance with the theoretical or empirical signal propagation model. Currently, RSSI-based wireless sensor network localization algorithm provides various approaches to improve the accuracy of localization. In the literature [5], an algorithm proposed based on extended Kalman filter (EKF) to reduce error of target localiza- tion, but the method brings a lot of communication and computing cost. A RSSI-based self-localization algorithm is presented in the literature [6]. The algorithm performs well in localization accuracy when anchors are at the fringe of the networks. However, it increases work complexity of the anchor nodes layout and the burden of communication. Online correction method [7] can improve the environmental adaptability of the sensor network and estimate the path loss exponent n in the application environment of the sensor network with precise location information of the network anchors through real-time estimation. Different from the above opti- mization methods, in our research, based on the principle of online correction, the localization optimization method based on real-time estimation can be applied to reduce the impact of environmental complexity on the measured data.

3.2 RSSI Range-Based Localization Method

3.2.1 Trilateral Range-Based Localization Method

Based on RSSI range-based localization, the most representative one is trilatera- tion. As shown in Fig. 3.1, it is known that the coordinates of three nodes N1, N2 and N3 are respectively (x_1, y_1), (x_2, y_2) and (x_3, y_3), and their distances to the unknown target node are d_1, d_2 and d_3. Assume that the coordinates of the target nodes are (x, y).

According to the Pythagorean Theorem, the following formula is established:

$$\begin{bmatrix} (x - x_1)^2 + (y - y_1)^2 \\ (x - x_2)^2 + (y - y_2)^2 \\ (x - x_3)^2 + (y - y_3)^2 \end{bmatrix} = \begin{bmatrix} d_1^2 \\ d_2^2 \\ d_3^2 \end{bmatrix} \tag{3.1}$$

Appropriate ranging model was selected to process the collected RSSI data and d_1, d_2 and d_3 can be obtained, and then substituted to the Eq. 3.2 to calculate the coordinates of the target. Therefore, the selection of an appropriate RSSI ranging model is one key part to obtain accurate distance information in RSSI ranging localization.

Fig. 3.1 Trilateration

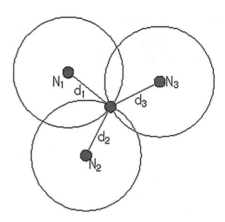

3.2.2 Ranging Model

In fact, an explicit analytic expression between the path loss of the distance and field strength values is generally unavailable. Massive field measurement data present that the common signal attenuation ranging model complies with the following equation:

$$PL(d) = PL(d_0) - 10n \lg(d/d_0) - X_\sigma \qquad (3.2)$$

Where d stands for the distance between the transmitter and the receiver, and d_0 is the reference distance; n is the path loss exponent which depends on the surroundings and building types; X_σ refers to the normal random variable of the standard deviation of σ, which is mainly caused by the sensor sensing accuracy; $PL(d_0)$ refers to the received signal strength at the corresponding meters node d_0. It can be found from the above equation that whether the path loss exponent n is consistent with the application environment is of great importance for the ranging results, which also directly affects the localization accuracy of RSSI ranging.

3.3 The Optimization Method of Real-Time Estimation Range-Based Localization

3.3.1 The Optimization Model of Real-Time Estimation Range-Based Localization

To improve the ranging accuracy from the target point to three anchor points is the fundamental way of improving the localization accuracy of trilateration, and how to get the path loss exponent n that meets the requirements of the actual environment is the key to improve ranging accuracy. The principle of the optimization

Fig. 3.2 The optimized
trilateral localization
measurement based on real-
time estimation

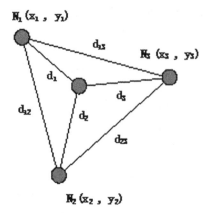

method in the trilateral localization measurement based on real-time estimation is to compare the relationship between the actual distance and the measured distance of three anchors, obtaining the path loss exponent n of communication lines among the anchors. In this way, the propagation path loss exponent n between three anchor nodes and the target node can be calculated.

As shown in Fig. 3.2, distances between anchors are respectively d_{12}, d_{23} and d_{13}, and the measured distances are respectively d'_{12}, d'_{23} and d'_{13}.

According to RSSI range-based model, the following equation set can be established:

$$n_{ij} = \frac{PL(d_{ij}) - PL(d'_{ij})}{10\lg(d'_{ij}/d_{ij})} \quad (i \neq j \in \{1, 2, 3\}) \quad (3.3)$$

From Eq. 3.3, we can get the estimation values of the attenuation coefficients between anchors as n_{12}, n_{23} and n_{13}. The estimated values of real-time attenuation coefficients of the target node to three anchors are calculated as n_1, n_2 and n_3. The easiest method is to calculate the average value of trilateral attenuation coefficients, as Eq. 3.4:

$$\begin{cases} n_1 = \frac{1}{2}(n_{13} + n_{12}) \\ n_2 = \frac{1}{2}(n_{23} + n_{12}) \\ n_3 = \frac{1}{2}(n_{13} + n_{23}) \end{cases} \quad (3.4)$$

We substitute Eq. 3.3 in Eq. 3.4 and solve the equation set to obtain n_1, n_2 and n_3 as Eq. 3.5:

$$n_1 = \frac{1}{2} \left(\frac{PL(d_{12}) - PL(d'_{12})}{10 \lg \left(d'_{12}/d_{12} \right)} + \frac{PL(d_{13}) - PL(d'_{13})}{10 \lg \left(d'_{13}/d_{13} \right)} \right)$$

$$n_2 = \frac{1}{2} \left(\frac{PL(d_{12}) - PL(d'_{12})}{10 \lg \left(d'_{12}/d_{12} \right)} + \frac{PL(d_{23}) - PL(d'_{23})}{10 \lg \left(d'_{23}/d_{23} \right)} \right) \qquad (3.5)$$

$$n_3 = \frac{1}{2} \left(\frac{PL(d_{13}) - PL(d'_{13})}{10 \lg \left(d'_{13}/d_{13} \right)} + \frac{PL(d_{23}) - PL(d'_{23})}{10 \lg \left(d'_{23}/d_{23} \right)} \right)$$

The obtained path loss coefficients were substituted into Eq. 3.1 respectively to calculate the RSSI ranging estimation values d_1, d_2 and d_3. which were from substituted into Eq. 3.2 for triangular localization model to calculate the relative localization coordinates of the target nodes (x_1, y_1), (x_2, y_2) and (x_3, y_3).

3.3.2 The Improvement of Trilateral Ranging Method

As the trilateral ranging method obtains the position coordinates of the target node with the reference parameters of distances between the target and three anchors respectively, the precise node location coordinates can only be found under the ideal circumstance, that is, the appropriate path loss index n and the normal random error X_σ are found. Yet, this distance measurement inevitably leads to the errors of measurement data (only the internal triangle formed by three anchors is considered), resulting in incorrect location coordinates of the nodes where $n > n'$ and $n < n'$ without solutions, as shown in (Fig. 3.3).

Under these two situations, Eq. 3.2 can be changed into the following equation for solutions:

$$\min \sum_{i=1}^{3} \left| \sqrt{(x_i - x)^2 + (y_i - y)^2} - d'_i \right| \qquad (3.6)$$

In Eq. 3.6, (x_i, y_i) are the coordinates of the anchor N_i and d'_i is the measured distance of N_i to the target node. Solving Eq. 3.6, we can obtain the approximate optimal solution of location coordinates of the target node.

3.4 Experimental Verification and Result Analysis

The verification experiment adopted the independently developed sensor nodes and data analysis software, and the experimental environment was the open playground. In the experiment, three anchor nodes were set with the relative

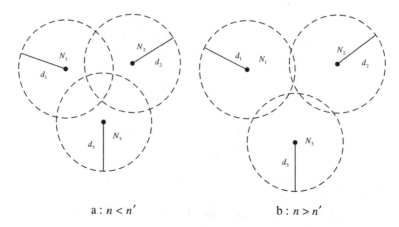

Fig. 3.3 Two situations of trilateration localization method without solutions

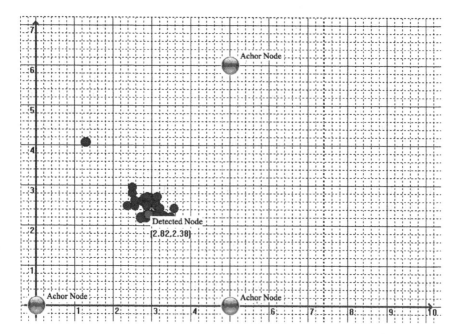

Fig. 3.4 Experimental results without optimization method

position coordinates of (0, 0), (5, 0) and (5, 6) with the unit of meters, that is, three anchor nodes were connected into a right triangle. The relative position coordinates of a node to be positioned was set as (2, 2). In the laboratory, different radio disturbance and communication separation environments were set and the localization experiments were conducted with the ordinary RSSI range-based

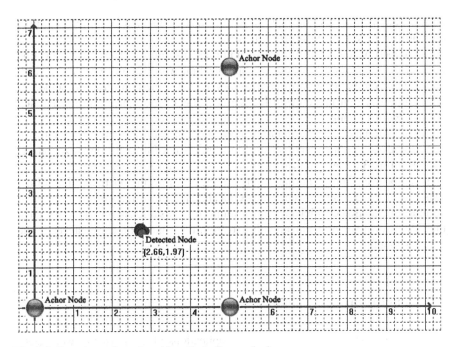

Fig. 3.5 Experimental results with optimization method

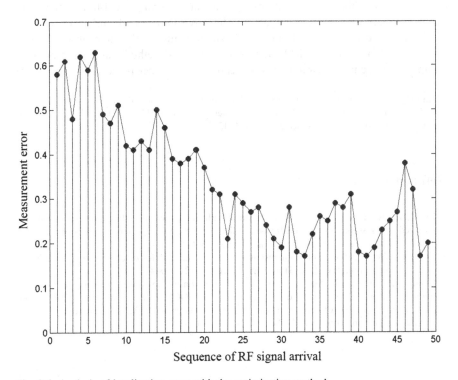

Fig. 3.6 Analysis of localization error with the optimization method

localization method and the optimized localization method based on real-time estimation respectively to observe the results.

The experimental results were shown in Figs. 3.4 and 3.5. The average location error of the static target without the optimization was 1.26 m. After the optimization, the average location error was 0.48 m. Thus, the optimization method of real-time estimation range-based localization effectively reduces the localization error of RSSI ranging.

After the analysis of the obtained node location data, the maximum error of 0.63 m and the minimum error of 0.17 m were obtained after the optimization. As shown in Fig. 3.6, the localization error is less volatile:

It can be seen from the trend of error changes that as the operation time of the sensor network increases, values of the localization error decrease because more accurate attenuation coefficient can be estimated with sufficient measurement data, thus identifying signal attenuation due to from environmental disturbances in the wireless ranging.

3.5 Conclusion

The target ranging is one of the typical applications of the sensor network technology. In this research, errors in the ranging process were analyzed. The proposed optimization method of sensor network localization through real-time estimation and correction weakens the impact of the environmental disturbances on the localization accuracy of RSSI ranging. Since the nature of the attenuation of most sensor signals is similar to RSSI RF attenuation, this method can also be used to optimize the accuracy of range-based localization of other perceived signals.

Acknowledgments The work presented in this chapter is supported by the Nature Science Foundation of Anhui Education Department under grant No. KJ2012B212 and the National Science Foundation of China under grant No. 61170233.

References

1. Liu C, Wu K, He T (2004) Sensor localization with ring overlapping based on comparison of received signal strength indicator. In: Proceedings of the 1st IEEE international conference on mobile Ad-hoc and sensor systems, Los Alamitos, 516–518
2. Cheung K, So H (2005) A multidimensional scaling framework for mobile location using time-of-arrival measurements. IEEE Trans Sign Process 53(2):460–470
3. Cheng XZ, Andrew T, Xue G (2004) TPS: a time-based positioning scheme for outdoor wireless sensor net works. In: Proceedings of IEEE INFOCOM. IEEE, New York, 2685–2696
4. Ni C, Nath B (2003) Ad hoc positioning system (APS) using AoA. In: IEEE conference on computer communications. IEEE, New York, 1734–1743

5. Caltabiano D, Muscato G, Russo F (2004) Localization and self-calibrartion of a robot for volcano exploration. In: Proceedings of the 2004 IEEE ICRA. IEEE robotics and automation society, New Orleans, 1:586–591
6. Wang S, Yin J, Cai Z et al (2008) A RSSI-based Self-localization algorithm for wireless sensor networks. J Comput Res Dev 45(Suppl.):385–388 (in Chinese)
7. Feng J, Megerian S, Potkonjak M (2003) Model-based calibration for sensor networks. In: IEEE, 737–742

Chapter 4
Process Calculus for Cost Analysis of Process Creation

Shin-ya Nishizaki, Mizuki Fujii and Ritsuya Ikeda

Abstract Many researchers have proposed formal frameworks for analyzing cryptographic and communication protocols and have studied the theoretical properties of the protocols, such as authenticity and secrecy. The resistance of denial-of-service attacks is one of the most important issues in network security. Several researchers have applied process calculi to security issues. One of the most remarkable of these studies is Abadi and Gordon's, based on Milner's pi-calculus. For the denial-of-service attack, Medow proposed a formal framework based on the Alice-and-Bob notation, and Tomioka et al. proposed a process calculus based on Milner's pi-calculus, the Spice-calculus. Using the Spice-calculus, we can evaluate the computational cost of executing processes. In our previous works, the Spice-calculus could analyze computational costs such as arithmetic operations, hash computation, and message transmission. However, the cost of process creation was disregarded. In this paper, we improve the Spice-calculus through adding cost evaluation of process creations. We extend the syntax of the cost in the Spice-calculus and operational semantics of the Spice-calculus. We then present an example of the improved Spice-calculus.

Keywords Denial-of-service · Attack · Process calculus · Pi-calculus

S. Nishizaki (✉) · M. Fujii · R. Ikeda
Department of Computer Science, Tokyo Institute of Technology, 2-12-1-W8-69, Ookayama, Meguro-ku, Tokyo 152-8552, Japan
e-mail: nisizaki@cs.titech.ac.jp

M. Fujii
e-mail: mizuki.fujii@lambda.cs.titech.ac.jp

R. Ikeda
e-mail: ritsuya.ikeda@lambda.cs.titech.ac.jp

W. Lu et al. (eds.), *Proceedings of the 2012 International Conference on Information Technology and Software Engineering*, Lecture Notes in Electrical Engineering 210, DOI: 10.1007/978-3-642-34528-9_4, © Springer-Verlag Berlin Heidelberg 2013

4.1 Introduction

Vulnerability of communication protocols can cause various kinds of attacks, which cause substantial damage to systems connected to the Internet. A *denial-of-service attack* is one of the constant concerns of computer security. It can make the usual services of a computer or network inaccessible to the regular users. An archetypal example of DoS attack is a *SYN flooding attack* [1] on the Transmission Control Protocol (TCP) (Fig. 4.1).

Before sending data from a source host S to a destination host D, the hosts have to establish a connection between S and D, called *three-way handshake*. The host S begins by sending a SYN packet including an initial sequence number x. The host D then replies to S with a message in which the SYN and ACK flags are set, indicating that D acknowledges the SYN packet from the S. The message includes D's sequence number y and incremented S's sequence number $(x + 1)$ as an ACK number. The host S sends a message with an ACK bit, a SEQ number $(x + 1)$, and an ACK number $(y + 1)$.

Consider a situation that lasts for a short period during which an attacker, A, sends an enormous number of connection requests with spoofed source IP-addresses to a victim host, D. The number of actual implementations of half-opened connections per port is limited since the memory allocation of data structures for such a connection is limited (Fig. 4.2).

Recently, several researchers have studied DoS attacks from various viewpoints, such as protection from DoS attacks using resource monitoring [2] and the development of DoS resistance analysis [3, 4]. She extended the Alice-and-Bob notation by attaching an atomic procedure annotation to each communication.

Fig. 4.1 Three-way handshake of TCP

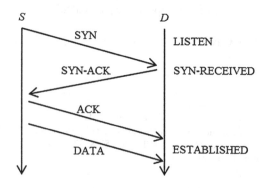

Fig. 4.2 SYN-flood attack

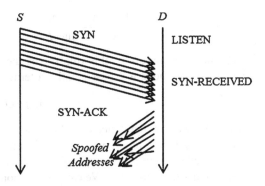

4.2 Process Calculus for Analyzing Denial-of-Service Attacks

4.2.1 The Spice Calculus

Tomioka et al. [5] proposed a formal system for DoS resistance analysis, the Spice calculus, which is based on Milner's process calculus [6] and its extension for secure computation [7]. A characteristic feature of the calculus is a type system which enables us to track the computational cost of each computer node. The type of the Spice calculus represents the configuration of a distributed system.

For example, a typing of the Spice-calculus

$$\mathcal{A}|(\mathcal{B}|\mathcal{C}) \triangleright P_1|(Q|P_2)$$

means that the process P_1 has the type on the left hand side. This notation means that $\mathcal{A}, \mathcal{B}, \mathcal{C}$

P_1, Q, P_2 has type $\mathcal{A}, \mathcal{B}, \mathcal{C}$, respectively; more intuitively, this means that processes P_1, Q, P_2 are executed on machines $\mathcal{A}, \mathcal{B}, \mathcal{C}$, respectively. The type $\mathcal{A}|(\mathcal{B}|\mathcal{C})$ means a distributed system which consists of $\mathcal{A}, \mathcal{B}, \mathcal{C}$. The types of Spice-calculus [5] are defined by the following grammar.

$$\mathcal{A} ::= \mathbf{a} :: \{x_1, \ldots, x_n\} \mid (\mathcal{A}|\mathcal{B})$$

The typing of the Spice-calculus contributes to the following:

- identifying the origins of computational costs and
- formalizing current memory usage during execution of processes.

The type $\mathbf{a} :: \{x_1, \ldots, x_n\}$ means that in the single computer node $\mathbf{a}$, memory cells $x_1, \ldots, x_n$ are occupied.

In order to track memory usage accurately, stricter restriction is imposed in the Spice-calculus than in Milner's original process calculus, the pi-calculus. For example, a typing rule [5]

$$\frac{\mathbf{a} :: \mathcal{V} \triangleright P \quad \mathcal{V} \supseteq fv(M) \quad \mathcal{V} \supseteq fv(N)}{\mathbf{a} :: \mathcal{V} \triangleright \operatorname{out} M\langle N\rangle; P} \ \mathbf{TypeOut}$$

The first assumption of this rule forces the process P to be executed on the *single* computer node **a**. Consequently, a term

$$\operatorname{inp} n(x); (p \mid Q)$$

is not allowed in the Spice-calculus, since $(P \mid Q)\$$ must be executed on a distributed system composed of multiple computer nodes.

The Spice-calculus is given operational semantics as a transition relation, in other words, small-step semantics. We can find the computational cost consumed in each transition step of the semantics. Moreover, we can identify not only the level of the computational cost but also the origin of this cost. For example, [5], a rule

$$\frac{M \downarrow V : c}{\operatorname{hash}(M) \downarrow hash_V : c + hash}$$

formulate hash-value computation. The cost c is incurred for evaluation of M and *hash* for the hash-value computation. There are two kinds of transition relation between processes giving the operational semantics of the Spice-calculus: one is a reduction relation and another commitment relation. *Inner-process computation* is formalized as the reduction relation and *inter-process computation* as the commitment relation. The following is one of the rules defining the reduction relation:

$$\frac{M \downarrow V : c \quad N \downarrow V : d}{(\operatorname{match} M \operatorname{is} N \operatorname{err} \{P\}; Q) \ > \ Q : \mathbf{a} \cdot (c + d + match)} \ \mathbf{RedMatch}$$

The costs for evaluating M and N are c and d, respectively. The resulting value of M and N is V. The following is one of the rules defining the commitment relation:

$$\frac{}{\mathbf{a} :: \mathcal{V} \vdash \operatorname{inp} n(x); P \xrightarrow{n} (x)P : \{\mathbf{a} \cdot store_x\}} \ \mathbf{CommIn}$$

In this rule, the process inp $n(x); P$ is transit to an intermediate form $(x)P$, which is waiting for arrival of a message at the port n. The cost $store_x$ occurs at the node **a**.

Recently, Cervesato proposed another formal approach to quantitative analysis of secure protocols [8].

4.2.2 Formalization of Process Creation

In the existing version of the Spice-calculus [5], the costs of memory and computation, such as hashing and arithmetic operations, are formalized. However, the cost of process creation is overlooked. In this paper, we improve the Spice-calculus by adding the costs of process creation.

Actually, process creation entails a certain amount of cost. For example, in a variation of the Unix system, a Process Control Block (PCB) is allocated in a kernel memory space to save the process context when the process is created. Hence process-creating is considered as expensive. In order to formulate process creation in the Spice-calculus, we introduce several new rules for the operational semantics. In the reduction relation between processes, the process replication is represented as a rule

$$\text{repeat } P > P \,|\, (\text{repeat } P)$$

The cost of creating a process is represented as *process* and is incorporated into the rule as follows.

$$\text{repeat } P > P \,|\, (\text{repeat } P) : \mathbf{a} \cdot process$$

The terminated process is formalized as the nil stop. In order to formalize elimination of the terminated processes precisely, we introduce another kind of terminated process end and we add a new reduction rule in which stop is transit to end as follows.

$$\text{stop} > \text{end} : \mathbf{a} \cdot -process$$

When a process is terminated and eliminated, a cost *process* is deducted.

4.3 Example of Formalization: Three-Way Handshake

TCP's three-way handshake is described as follows in the Alice-and-Bob notation.

$$A \rightarrow B : A, B, S_A$$
$$B \rightarrow A : B, A, S_B, S_A + 1$$
$$A \rightarrow B : A, B, S_A + 1, S_B + 1$$

In the Spice-calculus, we can write the protocol as the following process expressions.

$$
\begin{aligned}
P_A \stackrel{def}{=} \quad & \text{new}(S_A); \\
& \text{store } x_{sa} = S_A; \\
& \text{out } c\langle (A,\ B,\ x_{sa}) \rangle; \\
& \text{inp } c\ (p); \\
& \text{split } [x'_b,\ x'_a,\ x'_{sb},\ x'_{sa+1}] \text{ is } p \text{ err}\{\text{free } x_{sa},\ p\}; \\
& \text{free } p; \\
& \text{match } x'_{sa} \text{ is } A \text{ and } x'_{sb} \text{ is } B \text{ and } x'_{sa+1} \text{ is succ}(x_{sa}) \\
& \quad \text{err}\{\text{free } x_{sa},\ x'_b,\ x'_a,\ x'_{sb},\ x'_{sa+1}\}; \\
& \text{free } x'_b,\ x'_a,\ x'_{sa+1}; \\
& \text{out } c\ \langle (A,\ B\ \text{succ}\ (x_{sa}),\ \text{succ}\ (x'_{sb})) \rangle; \\
& P'_A
\end{aligned}
$$

$P_B \overset{def}{=}$ new(S_B);
        inp c (q_1);
        split $[y_a, y_b, y_{sa}]$ is q_1 err$\{$free $q_1\}$;
        free q_1;
        match y_b is B err$\{$free $y_a, y_b, y_{sa}\}$;
        free y_b;
        store $y_{sb} = S_B$;
        out $c\langle(B, y_a, y_{sb}, \text{succ}(y_{sa}))\rangle$;
        inp c (q_2);
        split $[y'_b, y'_a, y'_{sa+1}, y'_{sb+1}]$ is q_2 err$\{$free $y_a, y_{sa}, y_{sb}, q_2\}$;
        free q_2;
        match y'_b is B and y'_a is y_a and y'_{sa+1} is $\text{succ}(y_a)$
        and y'_{sb+1} is succ (y_{sb});
          err$\{$free $y_a, y_{sa}, y_{sb}, y'_b, y'_a, y'_{sa+1}, y'_{sb+1}\}$;
        free $y'_b, y'_a, y'_{sa+1}, y'_{sb+1}$;
        P'_B

Then a normal situation of communication between these two processes is represented as *NormalConfig*:

$$NormalConfig \overset{def}{=} (\text{repeat } P_A \mid \text{repeat } P_B).$$

On the other hand, a situation of a SYN-flood attack [9] on the three-way handshake protocol is described in Alice-and-Bob notation as follows.

$$I \rightarrow B : I_i, B, S_I$$
$$B \rightarrow I : B, A, S_B, S_{I_i} + 1$$
$$(i = 1, 2, \ldots)$$

The intruder in this attack is written as a process P_I as follows.

$$P_I \overset{def}{=} \text{new}(i); \text{ new}(s); \text{out } c\langle(i, B, s)\rangle; \text{stop}.$$

Since the intruder spoofs the senders' IP-addresses, the responding packets from the victim B are lost. In order to represent such lost packets, we introduce a process P_N which formulates the network:

$$P_N \overset{def}{=} \text{inp } c(r); \text{stop}.$$

The attacking situation is written as a process *AttackConfig*:

$$AttackConfig \overset{def}{=} (\text{repeat } p_I \mid \text{repeat } p_B \mid \text{repeat } p_N).$$

Then the process is typed as

$$AttackConfig : (\mathbf{i}|\mathbf{b}|\mathbf{n}) \triangleright (\text{repeat } p_I \mid \text{repeat } p_B \mid \text{repeat } p_N).$$

Reducing the process *AttackConfig*, we know the level of costs consumed during execution.

$$
\begin{aligned}
&AttackConfig \\
\rightarrow\rightarrow\ &new(S_B); (\text{stop} \mid \text{repeat}\, P_A \mid (\text{inp}\, c\, (q_2); \cdots) \\
&\mid\text{repeat}\, P_B \mid \text{stop} \mid \text{repeat}\, P_N) \\
&: \{\mathbf{i} \cdot proces,\ \mathbf{b} \cdot proces,\ \mathbf{n} \cdot proces\} \\
&+\{\mathbf{i} \cdot pair,\ \mathbf{b} \cdot store\} \\
&+\{\mathbf{i} \cdot -proces,\ \mathbf{b} \cdot 2store,\ \mathbf{b} \cdot match\} \\
&+\{\mathbf{b} \cdot pair,\ \mathbf{n} \cdot store\} \\
&+\{\mathbf{n} \cdot -proces\} \\
&= \{\mathbf{i} \cdot pair,\ \mathbf{b} \cdot proces,\ \mathbf{b} \cdot pair, \\
&\qquad \mathbf{b} \cdot 3store,\ \mathbf{b} \cdot match,\ \mathbf{n} \cdot store\}
\end{aligned}
$$

Here we know that not only memory cost but also process creation is consumed by the SYN-flood attack.

4.4 Conclusion

In our previous works, the Spice-calculus could analyze computational costs such as arithmetic operations, hash computation, and message transmission. However, the cost of process creation was disregarded. In this paper, we improved the Spice-calculus through adding cost evaluation of process creation and elimination. We extended the syntax of the cost in the Spice-calculus and operational semantics of the Spice-calculus. We then described TCP's three-way handshake and the SYN-flood attack as examples of our new calculus.

As well as the work presented in this paper, we have several other subjects to study in future. One of the most important ones is formalizing and analyzing the SYN-cookie method against the SYN-flood attack and other DoS-resistant methods [10, 11].

Acknowledgments This work was supported by Grants-in-Aid for Scientific Research (C) (24500009).

References

1. Schuba CL, Krsul IV, Kuhn MG, Spafford EH, Sundaram A, Zamboni D (1997) Analysis of a denial of service attack on TCP. In: Proceedings of the 1997 IEEE symposium on security and privacy, pp 208–223. IEEE Computer Society, IEEE Computer Society Press
2. Millen JK (1993) A resource allocation model for denial of service protection. J Comput Secur 2(2/3):89–106
3. Meadows C (1999) A formal framework and evaluation method for network denial of service. In: Proceeding of the 12th IEEE computer security foundations workshop, pp 4–13

4. Meadows C (2001) A cost-based framework for analysis of denial of service networks. J Comput Secur 9(1/2):143–164
5. Tomioka D, Nishizaki S, Ikeda R (2004) A cost estimation calculus for analyzing the resistance to denial-of-service attack. In: Software security—theories and systems. Lecture Notes in Computer Science, vol 3233. Springer, New York, pp 25–44
6. Milner R, Parrow J, Walker D (1992) A culculus of mobile processes, part i and part ii. Inf Comput 100(1):1–77
7. Abadi M, Gordon AD (1997) A calculus for cryptographic protocols: the spi calculus. In: Fourth ACM conference on computer and communication security, pp 36–47. ACM Press, New York
8. Cervesato I (2006) Towards a notion of quantitative security analysis. In: Gollmann D, Massacci F, Yautsiukhin A (eds) Quality of protection: security measurements and metrics—QoP'05, pp 131–144. Springer advances in information security 23
9. TCP SYN Flooding and IP spoofing attacks (1996), CA-1996-21
10. Aura T, Nikander P (1997) Stateless connections. In: International conference on information and communications security ICICS'97. Lecture notes in computer science, vol 1334, pp 87–97. Springer, Berlin
11. Aura T, Nikander P, Leiwo J (2001) DOS-resistant authentication with client puzzles. In: Security protocols, 8th international workshop. Lecture notes in computer science, vol 2133, pp 170–177. Springer, Berlin

Chapter 5
Parameter Estimation of LFM Signal by Direct and Spline Interpolation Based on FrFT

Jun Song and Yunfei Liu

Abstract An important cause of performance loss in parameter estimation of linear frequency modulation (LFM) signal based on Fractional Fourier Transform (FrFT) is the "picket fence effect" because of the discretization of FrFT. In this work, a direct and spline interpolation method, along with a coarse search, is investigated and is used for parameter estimation of LFM signal based on FrFT. In FrFT domain, the parameter estimation of a LFM signal is equivalent to search the peak of the FrFT coefficients. The direct and spline interpolation method, instead of a conventional fine search algorithm, can not only improve the estimation accuracy of the peak coordinate, which leads to a better performance of the LFM parameter estimation, but also save computation load. Performance of the proposed method is analyzed by digital simulations at different signal to noise ratio (SNR) levels, and good performance is achieved.

Keywords LFM signal · Parameter estimation · FrFT · Spline interpolation

5.1 Introduction

In this work, we consider the parameter estimation of Linear frequency modulation (LFM) signal. In the recent decades, the fractional Fourier transform (FrFT) is widely used in the parameter estimation for LFM signal because the FrFT can be interpreted as a decomposition to a basis formed by the orthogonal LFM functions.

J. Song (✉) · Y. Liu
Department of Information Science and Technology, Nanjing Forestry University,
No.159 Longpan Road, Nanjing, China
e-mail: georgecumt@163.com

W. Lu et al. (eds.), *Proceedings of the 2012 International Conference on Information Technology and Software Engineering*, Lecture Notes in Electrical Engineering 210, DOI: 10.1007/978-3-642-34528-9_5, © Springer-Verlag Berlin Heidelberg 2013

Digital FrFT (DFrFT) and Radon-ambiguity transform are employed in [1–3] to estimate the parameters of LFM signal,but the precision of estimation is limited because of the discretization of FrFT. A direct interpolation method based on FrFT is used in [4], and the estimation precision of part parameters of LFM signal is improved at proper signal to noise ratio (SNR) levels, but that of chirp rate can not be improved. DFrFT and quasi-Newton method is introduced in [5], and the precision is perfect. In substance, the method of [5] is based on a coarse search and a nonlinear optimization fine search, so it needs many extra calculations.

This paper introduces a direct and spline interpolation method, along with a coarse search, for parameter estimation of LFM signal based on FrFT. The coarse search returns a coarse estimate of the parameters by searching for the maximum DFrFT coefficient of the LFM signal. This coarse estimate can be improved by a fine search algorithm, i.e. that of [5], and some interpolation method. Compare with the fine search algorithm, interpolation method is usually more efficient since they use prior knowledge about the signal. According to this enlightenment, a direct and spline interpolation method is investigated to overcome the performance loss caused by "picket fence effect" because of discretization of FrFT. Simulation result indicates that the root mean square error (RMSE) is marginally above CRB (Cramer-Rao Bound) under proper low SNR levels.

5.2 FrFT and DFrFT

The definition of FrFT for signal $x(t)$ is as follows

$$X_\alpha(u) = F^p[x(t)] = \int_{-\infty}^{+\infty} x(t)\tilde{K}_p(u,t)dt, \tag{5.1}$$

where, $\alpha = p \cdot \pi/2$, p is a real number and referred to the fractional order, $F^p[g] = F^\alpha[g]$ denotes the FrFT operator, and $\tilde{K}_p(u,t)$ is the kernel of FrFT:

$$\tilde{K}_p(u,t) = \begin{cases} C_\alpha \cdot e^{j\pi(u^2 \cot\alpha - 2ut\csc\alpha + t^2\cot\alpha)} & \alpha \neq m\pi \\ \delta(t-u) & \alpha = 2m\pi \\ \delta(t+u) & \alpha = (2m\pm 1)\pi \end{cases}, \tag{5.2}$$

where $C_\alpha = \sqrt{1 - j\cot\alpha}$, and m is an integer.

Ozaktas proposed a decomposition algorithm for the DFrFT in an order of α [6], and that is of the form as

$$X_\alpha(U) = \frac{C_\alpha}{2L} \sum_{n=-N}^{N} \exp\left[j\pi\left(\lambda_1\left(\frac{U}{2L}\right)^2 - 2\lambda_2\frac{nU}{(2L)^2} + \lambda_1\left(\frac{n}{2L}\right)^2\right)\right]x(n), \tag{5.3}$$

with $\lambda_1 = -\cot(\alpha)$, $\lambda_2 = \csc(\alpha)$, $L = \sqrt{N}$, $U = u/L$ and $x(n)$ is samples of $x(t)$.

5.3 Interpolation Algorithm

A noisy monocomponent LFM signal under consideration can be expressed as

$$
\begin{aligned}
x(n) = s(n) + w(n) &= b_0 \cdot \exp(j\theta_0 + j2\pi f_0 \cdot n \cdot \Delta + j\pi k_0 \cdot n^2 \cdot \Delta^2) + w(n), \\
&- (N-1)/2 \le n \le (N-1)/2
\end{aligned}
\tag{5.4}
$$

where Δ is sampling interval, b_0 is the amplitude, θ_0 is the initial phase, f_0 is the initial frequency, k_0 is the chirp rate, and $w(n)$ is additive Gaussian noise with zero mean and variance of $2\sigma^2$.

According to the definition of FrFT and (5.4), a LFM signal is turned into an impulse only for a particular fractional order, but the Gaussian white noise will be distributed evenly in time–frequency plane [7]. Therefore, for a given LFM signal as in (5.4), the DFrFT can be computed for a given α and an energy impulse peak $|X_\alpha(U)|$ can be searched on the 2-D plane of (α, U), then we can get the location information of the peak coordinate $(\widehat{\alpha}_0, \widehat{U}_0)$. When the peak coordinate $(\widehat{\alpha}_0, \widehat{U}_0)$ is identified, the four parameters of the LFM signal are estimated by using the following relations [5]:

$$
\begin{cases}
\widehat{k}_0 &= -\cot(\widehat{\alpha}_0) \cdot \frac{f_s^2}{N}; \\
\widehat{f}_0 &= \widehat{U}_0 \cdot \frac{f_s}{N} \csc(\widehat{\alpha}_0); \\
\widehat{b}_0 &= \dfrac{\left| X_{\widehat{\alpha}_0}(\widehat{U}_0) \right|}{L \cdot \left| C_{\widehat{\alpha}_0} \right|}; \\
\widehat{\theta}_0 &= \arg\left[\dfrac{X_{\widehat{\alpha}_0}(\widehat{U}_0)}{C_{\widehat{\alpha}_0} \exp(j\pi \widehat{U}_0^2 \cot(\widehat{\alpha}_0)/N)} \right].
\end{cases}
\tag{5.5}
$$

It follows from (5.5) that the parameter estimation of the LFM signal is equivalent to the estimation of the peak coordinate $(\widehat{\alpha}_0, \widehat{U}_0)$.

Due to the restriction of search step size of α and discretization of u, there would be some deviation between the searched peak location (α_k, U_k) referred to as the quasi-peak and the real peak location (α_0, U_0) indeed. The deviation between the quasi-peak (α_k, U_k) and the real peak (α_0, U_0) is the decisive factor that affects precision of parameter estimation.

5.3.1 Interpolation of α

Theoretically, if the search step size of α is sufficiently small, the α_k would be enough close to α_0, and at the same time, it requires tremendous computation. The computational load of the search algorithm can be reduced observably by inviting a larger search step size and combining it with an interpolation technique.

According to (5.1), the FrFT of a LFM signal is of

$$X_\alpha(u) = Co(\alpha, u) \int_{-L/2}^{L/2} e^{j2\pi(f_0 - u \cdot \csc \alpha)v} \cdot e^{j\pi(k_0 + \cot \alpha)v^2} \, dv, \tag{5.6}$$

with $Co(\alpha, u) = \sqrt{1 - j \cot \alpha} \cdot b_0 \cdot e^{j\pi u^2 \cdot \cot \alpha + j\theta_0}$.

If we ignore the influence of u at the point of quasi-peak, (5.6) can be represented as:

$$X_\alpha(\widehat{u}_0) = Co(\alpha, \widehat{u}_0) \int_{-L/2}^{L/2} e^{j\pi(k_0 + \cot \alpha)v^2} \, dv = 2Co(\alpha, u) \int_0^{L/2} e^{j\pi qv^2} \, dv, \tag{5.7}$$

where $q = k_0 + \cot \alpha = \cot \alpha - \cot \alpha_0$.

If α fulfils function $\cot \alpha = -k_0$, the amplitude of (5.7) reaches its maximum [8]:

$$|X_{\max}| = \frac{b_0 \cdot L}{\sqrt{|\sin \alpha|}}. \tag{5.8}$$

In engineering applications, the α_k of quasi-peak could hardly fulfill the equation $\cot \alpha_k = \cot \alpha_0 = -k_0$ because of a restriction of the search step size. But, the deviation between α_k and α_0 would not exceed half of a search step. In an effort to obtain the estimation of α_0 (presented as $\widehat{\alpha}_0$), we can employ the α_k and its neighbor $\alpha_{k\pm1}$ to get an approach of $\widehat{\alpha}_0$ by method of interpolation. Therefore, the DFrFT coefficients those correspond to α_k and $\alpha_{k\pm1}$ should be calculated.

We rearrange formula (5.7), and then get

$$X_\alpha(\widehat{u}_0) = \frac{L \cdot Co(\alpha, \widehat{u}_0)}{R} \cdot \int_0^R e^{jv^2} \, dv, \tag{5.9}$$

where

$$R = \frac{\sqrt{\pi \cdot L^2}}{2} \sqrt{|q|} = \frac{\sqrt{\pi \cdot L^2}}{2} \sqrt{|\cot \alpha - \cot \alpha_0|}. \tag{5.10}$$

According to Fresnel integral, there is no analytic representation for (5.9), and the $X_\alpha(\widehat{u}_0)$ is a nonlinear function with variable α. As a result, cubic splines are particularly convenient for determining a approximate representation of $X_\alpha(\widehat{u}_0)$, though a variety of other ones are possible. The piecewise continuous polynomial representation of $X_\alpha(\widehat{U}_0)$ can be expressed as shown in (5.11):

$$X_\alpha(\widehat{U}_0) = \begin{cases} a_1\alpha^3 + b_1\alpha^2 + c_1\alpha + d_1 & \alpha_1 \le \alpha \le \alpha_2 \\ \quad \vdots & \quad \vdots \\ a_i\alpha^3 + b_i\alpha^2 + c_i\alpha + d_i & \alpha_i \le \alpha \le \alpha_{i+1} \\ \quad \vdots & \quad \vdots \\ a_{N-1}\alpha^3 + b_{N-1}\alpha^2 + c_{N-1}\alpha + d_{N-1} & \alpha_{N-1} \le \alpha \le \alpha_N \end{cases} \tag{5.11}$$

where α_i is the ith moment of searching α, and a_i, b_i, c_i, d_i are spline weighting coefficients which can be found through a variety of numerical methods, including matrix pseudo inverses [9].

In order to get an accurate estimation of $\widehat{\alpha}_0$, we now compute pattern-matching interpolations between $X_{\alpha_{k-1}}(\widehat{U}_0)$ and $X_{\alpha_k}(\widehat{U}_0)$ together with that between $X_{\alpha_k}(\widehat{U}_0)$ and $X_{\alpha_{k+1}}(\widehat{U}_0)$. Assuming that there are interpolations with number of Q in each section, the total number of interpolations is 2Q in the two sections above. Then the α corresponding to interpolation with maximum amplitude can be searched and it is determined to be the estimation of $\widehat{\alpha}_0$.

5.3.2 Interpolation of U

In the deviation of quasi-peak (α_k, U_k) to the real peak (α_0, U_0), that of U_k to U_0 is mainly attributable to the discretization of u, so it also needs to modify by interpolation.

According to DFrFT, we apply the expression of the LFM signal:

$$X_\alpha(U) = \frac{C_\alpha \cdot b_0}{L} \exp\left(-j\pi \cot\alpha \left(\frac{U}{2L}\right)^2\right) \sum_{n=-N}^{N} e^{j2\pi n\left(\frac{f_0}{f_s} - \frac{U\csc\alpha}{L^2}\right) + j\pi n^2\left(\frac{k_0}{f_s^2} + \frac{\cot\alpha}{L^2}\right)}. \quad (5.12)$$

If α meets the function $\cot\alpha \approx -k_0 L^2/f_s^2$, the expression in (5.12) thus can be simplified to:

$$X_{\widehat{\alpha}_0}(U) \doteq \frac{C_{\widehat{\alpha}_0} \cdot b_0}{L} \exp\left(-j\pi \cot\widehat{\alpha}_0 \left(\frac{U}{2L}\right)^2\right) \sum_{n=-N}^{N} \exp\left[j2\pi n\left(\frac{f_0}{f_s} - \frac{U \cdot \csc\widehat{\alpha}_0}{N}\right)\right]. \quad (5.13)$$

According (5.5), we know that at the real peak, the U_0 meets the function $f_0 = U_0 \cdot \frac{f_s}{N}\csc(\widehat{\alpha}_0)$.

But the discrete U_k would a little apart from U_0 inevitably. In order to make up the deviation, firstly, we suppose that $U_0 = U_k + \eta, (|\eta| < 0.5)$ without loss of generality, and there will be of the form

$$f_0 \doteq (U_k + \eta) \cdot \frac{f_s}{N}\csc(\widehat{\alpha}_0). \quad (5.14)$$

According to formula (5.13) and (5.14), we arrive at

$$X_{\widehat{\alpha}_0}(U_k) = \frac{C_{\widehat{\alpha}_0} \cdot b_0}{L} \exp\left(-j\pi \cot\widehat{\alpha}_0 \left(\frac{U_k}{2L}\right)^2\right) \sum_{n=-N}^{N} \exp\left[j2\pi n\left(\frac{\eta \cdot \csc\widehat{\alpha}_0}{N}\right)\right]. \quad (5.15)$$

In the same way, the DFrFT coefficient at point of $U_k + pn; (pn = \pm 0.5)$ can be computed

$$X_{\widehat{\alpha}_0}(U_k + pn) = \widetilde{C}_{\widehat{\alpha}_{0,U_k}} \cdot \frac{1 + e^{j2\pi\eta \cdot \csc \widehat{\alpha}_0}}{1 - e^{j2\pi(pn+\eta)\csc \widehat{\alpha}_0/N}} \doteq \widetilde{\widetilde{C}}_{\widehat{\alpha}_{0,U_k}} \cdot \frac{\eta}{(pn + \eta) \cdot \csc \widehat{\alpha}_0}, \quad (5.16)$$

where

$$\widetilde{C}_{\widehat{\alpha}_{0,U_k}} = \frac{C_{\widehat{\alpha}_0} \cdot b_0}{L} \exp\left(-j\pi \cot \widehat{\alpha}_0 \left(\frac{U_k + pn}{2L}\right)^2\right), \text{ and } \widetilde{\widetilde{C}}_{\widehat{\alpha}_{0,U_k}}$$

$$= \widetilde{C}_{\widehat{\alpha}_{0,U_k}} \cdot \frac{1 + e^{j2\pi\eta \cdot \csc \widehat{\alpha}_0}}{-j2\pi\eta/N}.$$

According to (5.16), the deviation η can be calculated by

$$\frac{X_{0.5} - X_{-0.5}}{X_{0.5} + X_{-0.5}} = \frac{\left|\widetilde{\widetilde{C}}_{\widehat{\alpha}_{0,U_k}}\right| \cdot \frac{\eta}{(0.5-\eta)\csc \widehat{\alpha}_0} - \left|\widetilde{\widetilde{C}}_{\widehat{\alpha}_{0,U_k}}\right| \cdot \frac{\eta}{(\eta+0.5)\csc \widehat{\alpha}_0}}{\left|\widetilde{\widetilde{C}}_{\widehat{\alpha}_{0,U_k}}\right| \cdot \frac{\eta}{(0.5-\eta)\csc \widehat{\alpha}_0} + \left|\widetilde{\widetilde{C}}_{\widehat{\alpha}_{0,U_k}}\right| \cdot \frac{\eta}{(\eta+0.5)\csc \widehat{\alpha}_0}} = 2\eta \quad (5.17)$$

where $X_{0.5} = \left|X_{\widehat{\alpha}_0}(U_k + 0.5)\right|; X_{-0.5} = \left|X_{\widehat{\alpha}_0}(U_k - 0.5)\right|.$

So, we can obtain the estimation of U_0 at point of real peak by

$$\widehat{U}_0 = U_k + \eta = U_k + \frac{1}{2} \cdot \frac{X_{0.5} - X_{-0.5}}{(X_{0.5} + X_{-0.5})}. \quad (5.18)$$

5.3.3 The Algorithm Flow

According to subsections above, the estimator by direct and spline interpolation based on FrFT is summarized as follows:

1. Let: $X_\alpha(U) = |DFrFT(x(n))|$, $U = 1, 2, \ldots N - 1, \alpha = 0 \sim \pi$.
2. Find: $\{\alpha_k, U_k\} = \arg \underset{\alpha,U}{\text{Max}}(X_\alpha(U))$.
3. Set $\widehat{U}_0 = U_k, \widehat{\alpha}_0 = \alpha_k$.
4. Calculate the spline coefficients a_i, b_i, c_i, d_i from (5.11) where corresponding weighting coefficients are obtained by numerical method.
5. Compute pattern-matching interpolations between $X_{\alpha_{k-1}}(\widehat{U}_0)$ and $X_{\alpha_k}(\widehat{U}_0)$ together with that between $X_{\alpha_k}(\widehat{U}_0)$ and $X_{\alpha_{k+1}}(\widehat{U}_0)$, and then search the α corresponding to interpolation with maximum amplitude to get the estimation of $\widehat{\alpha}_0$.
6. Calculate $X_{\pm 0.5} = \left|X_{\widehat{\alpha}_0}(\widehat{U}_0 \pm 0.5)\right|$ as in (5.3)

   then $\eta = \frac{1}{2}\frac{X_{0.5} - X_{-0.5}}{X_{0.5} + X_{-0.5}}$,

   and modify $\widehat{U}_0$ as $\widehat{U}_0 = \widehat{U}_0 + \eta$;
7. Finally: Substitute $\left(\widehat{\alpha}_0, \widehat{U}_0\right)$ to (5.5) to get the approach of parameters.

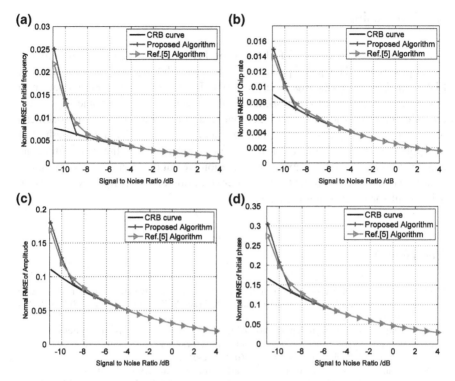

Fig. 5.1 Normal root mean square errors (RMSE) of the parameters estimated using the proposed algorithm **a** Normal RMSE of initial frequency f_0, **b** Normal RMSE of chirp rate k_0, **c** Normal RMSE of amplitude b_0, and **d** Normal RMSE of initial phase θ_0

5.4 Performance Analysis

In order to show the validity of the proposed algorithm, the results of Monte Carlo experiments are presented graphically. In the experiments, a monocomponent LFM signal with length of N = 513 is chosen, and is modeled as: $k_0 = 800\,\text{Hz/s}, f_0 = 30\,\text{Hz}, \theta_0 = 0.5\,\text{rad}$. The sampling frequency is $f_s = 2{,}000$ Hz. The scope of fractional order p is [0.5,1.5] with search step of 0.01 [5]. The results are plotted as functions of SNR varied from -11 to 4 dB. At each level of SNR, we run 200 Monte Carlo simulations, and the normal RMSEs(root mean square error) obtained are shown in Fig. 5.1. For comparison, the CRB in [8] and normal RMSE by method of [5] are also plotted.

As shown in Fig. 5.1, the proposed interpolation algorithm can do as well as that of [5], and the curve of proposed algorithm is marginally above CRB when input SNR is higher than -9 dB. What's more, the computation load of the proposed algorithm is lighter than that of [5]. If the input SNR decreases, the performance of our algorithm declines rapidly. Obviously, the reason is that at low SNR levels, there would be some reverse interpolation due to the interference of noise.

5.5 Conclusion

The parameter estimation of LFM based on FrFT is becoming very attractive in signal processing. This paper presents a new approach for parameter estimation of LFM signal by interpolations based on FrFT. In fact, the parameter estimation in this paper is equivalent to the estimation of the peak on the 2-D plane (α, U). On condition that the search step is bigger than that of previous, the new approach can obtain an accurate estimation of peak in fractional Fourier domain by method of interpolation, then parameter estimation are carried out. At the same time, the computation load and complexity of estimation are cut down, and the performance is well matched the theoretical curve under condition of proper low SNR levels, so it is useful for engineering applications.

References

1. Zhao X, Tao R, Zhou S (2003) Chirp signal detection and multiple parameter estimation using radon-ambiguity and fractional fourier transform. Trans Beijing Inst Technol 23(3):371–374 (in Chinese)
2. Guo B, Zhang H (2007) The application of iterative algorithm to chirp signal detection and parameter estimation using radon-ambiguity transform and fractional fourier transform. J Electron Inf Technol 29(12):3024–3026 (in Chinese)
3. Du D, Tang B (2004) An approach for estimating parameters of LFM signal based on fractional fourier transform and subspace orthogonality. J UEST China 33(3):247–249 (in Chinese)
4. Yuan Z, Hu W, Yu W (2009) Parameter estimation of LFM signals using FrFT interpolation. Sig Process 25(11):1726–1731 (in Chinese)
5. Qi L, Tao R, Zhou S et al (2004) Detection and parameter estimation of multicomponent LFM signal based on the fractional Fourier transform. Sci China: Ser. F Inf Sci 47(2):184–198
6. Ozaktas HM, Arikanet O, Kutay A (1996) Digital computation of the fractional fourier transform. IEEE Trans Sig Process 44(9):2141–2150
7. Almeida LB (1994) The fractional Fourier transform and time frequency representations. IEEE Trans Signal Process 42(11):3084–3091
8. Ristic B, Boadshash B (1998) Comments on 'The Cramer-Rao lower bounds for signals with constant amplitutde and polynomial phase'. IEEE Trans Signal Process 46(6):1708–1709
9. Hui F, Shan Y, Li L et al (2010) Spline based nonparametric estimation of the altimeter sea state bias correction. IEEE Geosci Remote Sens Lett 7(3):577–581

Chapter 6
An Analysis Method for Network Status of Open-Pit WLAN Based on Latency Time

Jiusheng Du, Yijin Chen, Zheng Hou and Xiaona Wu

Abstract Network status of the wireless local area network in open-pit has a major impact on production. Traditionally, signal strength reflects the status of a WLAN; nevertheless, for non-standard wireless adapter, this is inaccurate. Combined with spatial analysis, this paper adopts the spatial characters of latency time, namely date transmission time from clients to the server, to reflect network status. Based on an open-pit dispatch system, the experiment showed this method is more suitable.

Keywords Latency time · Kriging interpolation · Signal strength · Open-pit

6.1 Introduction

The recent decade has witnessed a rapid growth in wireless communications technology. Many public places such as hotels, cafes, restaurants, airports, train stations, and university campuses are offering WLANs based on the IEEE 802.11b/g standards for accessing the Internet remotely [1]. Nowadays the development of modern opencast mining is on the trend to be large-scale production, informationized management, scientific technology, and environmental friendly [2]. The main communication ways in current coal mine are PHS,

J. Du (✉) · Y. Chen · X. Wu
609, Zonghe Building, College of Geoscience and Surveying Engineering, China University of Mining and Technology (Beijing), No.Ding 11, Xueyuan Road, 100083 Haidian District, Beijing, People's Republic of China
e-mail: dujiush82@163.com

Z. Hou
Mining Department, Henan Engineering Technical School, No. 801, Bilian Road,, Jiaozuo 454000, People's Republic of China

W. Lu et al. (eds.), *Proceedings of the 2012 International Conference on Information Technology and Software Engineering*, Lecture Notes in Electrical Engineering 210, DOI: 10.1007/978-3-642-34528-9_6, © Springer-Verlag Berlin Heidelberg 2013

SCDMA, wireless digital interphone, WLAN and Next Generation advanced PHS. Among these, WLAN especially may be the main communication method of mine enterprise in the future [3]. Therefore how we obtain the network status of the wireless local area network in open-pit is very meaningful.

In theory, by means of the open-pit topography and the first Fresnel zone formula, the signal strength of each point in the path will be calculated out. However, this method is not applied for dynamic environments, such as open-pit mines. Be-cause the calculation relies on terrain, with mining processes, the first Fresnel zone based on the analysis will become invalid, then it needs to re-topographic survey and complexly calculate again.

Therefore, in this paper, we devote ourselves to find a practicable way to obtain the network status with focus on spatial analysis.

6.2 Problem Statements

6.2.1 System Architecture

In this paper, we take an open-pit dispatch system as an example. The system mainly includes three parts:

- For communication, wireless signal towers and signal vehicles are deployed. In the first case, the wireless signal towers in the mining area are equipped with: 5.8 GHz antenna for signal directional transmission of signals between towers; 2.4 GHz antenna for signal coverage. In the second case, signal vehicles are equipped with antennas to avoid signal dead, and their roles are similar to the signal tower.
- Dispatching centre is undoubtedly the core of the overall system. By communication, all information of vehicles is stored in the database of dispatching centre; on the other hand, dispatching centre sends dispatch order to vehicles.
- Vehicles are equipped with GPS antenna and Wi-Fi antenna, so they can send their status information (ID, driver name, GPS coordinates, load status, sending time) to the dispatching centre. They are utilized as the data terminal.

6.2.2 Wireless Adaptor

Wireless adaptors are installed on vehicle equipment, so the vehicle can have access to APs, and then communicate with the dispatching centre. Unlike standard adapters such as Atheros [4], the wireless adaptors used in this system are non-standard. However, this exactly provides convenience for users. We had developed a driver for this adaptor. Based on this driver, the wireless can work based on TCP/IP protocol or UDP protocol; what's more, it re-associates with WLAN at regular intervals in order to fit vehicle's state of motion.

6.2.3 Latency

Although the signal coverage has been fully taken into consideration for the selection of wireless signal tower's location and the installation of signal tower antenna's direction, the open-pit landform is complex, and vehicles are in motion, moreover, vehicle terminals connected to a wireless network need a certain period of time, so all of the above result in vehicles cannot be always online.

What be worth to be carried more is that, as stated in front, the driver for wireless adapter used in this system is self-developed. It defines that by scanning all signal towers at regular intervals, vehicle terminals select the tower which with stronger signal strength to access to WLAN, while vehicles are in motion. It also provides that in no signal or weak signal area, the generated data in vehicle be cached until next scanning, and the date will be send to dispatch centre when the vehicle is re-associated. Obviously, caching results in that the transmission time between vehicles and dispatch centre may be long. This is the main reason for latency.

6.3 The Analysis Method

6.3.1 Initial Data

Until next scanning, the 'SSID', 'rssi' are still using the last scanning's result. Therefore, only the first record generated after each scan can accurately reflect the status information of the point, and these are valid basis for analysis.

By analysing initial data as shown in Table 6.1, we can obtain the table of every truck for per day as shown in Table 6.2. If the vehicle terminal's system time is inconsistent with the server system time, it will cause systematic error. Table 6.1 shows an extreme example: at first sight, it means that receiving time is earlier than sending time.

6.3.2 Spatial Analysis Method

As mentioned in Sect. 6.1, making use of the open-pit topographic and the first Fresnel zone calculation formula to obtain network status is not suitable for dynamic environment. We discuss another way which uses spatial analysis method to solve this problem.

Kriging interpolation is regarded as one of the main method of geological statistics. It fully absorb the idea of space statistics, think any space of continuous change attributes is very irregular, cannot imitation by simple smooth mathematical functions, but can described by random surface function. From the angle of interpolation Kriging interpolation is one of the methods that is linear interpolation

Table 6.1 Shows partial initial data

Now position	Now time	Submit time	SSID	rssi
0xE6100000010DE09C......	20:27:05	19:28:24	CSH-Tower-2	40
0xE6100000010D96C0......	20:27:10	19:28:24	CSH-Tower-2	40
0xE6100000010D190E......	20:27:15	19:28:24	CSH-Tower-2	48

'Now Time' is the time of record sending, and it uses vehicle' system time. 'Submit Time' is the time of record receiving, and it uses database server's system time. 'SSID' field is the wireless tower's name, and 'rssi' field is the signal strength

Table 6.2 Data structure of the table of every truck for per day

Field name	Type	Length	Note
Now position	Geography		Point coordinates
Span	int		The initial latency time
Newspan	int		The latency time without system error
SSID	nvarchar	15	Access AP's name
rssi	nvarchar	3	Wireless signal strength

This table is stored in SQL Server 2008, which has specialized spatial data types: geography, geometry

optimal, unbiased estimation for directional distribution data [5]. Kriging interpolation is defined as formula (6.1):

$$Z(x_0) = \sum_{i}^{n} \lambda_i Z(x_i) \tag{6.1}$$

Here $Z(x_i)$ is the observed value of i point, $Z(x_0)$ is the value of estimation point, λ_i is the power, and we have formula (6.2):

$$\sum_{i}^{n} \lambda_i = 1 \tag{6.2}$$

Network status is influenced by terrain and the spatial location, so when evaluating the attribute value of a point, Kriging interpolation can be chosen. Since this investigation is an initial study, we only use ordinary Kriging interpolation, without considering the characteristic of antenna theory.

6.4 Preferences

As mentioned in Table 6.2, both signal strength ('rssi' field) and latency time ('newspan' field) were obtained. The former one is the refection of electromagnetic wave energy; the latter one is the refection of system performance. In theory, the spatial distribution of each of them will be in accordance with the network status of the open-pit's WLAN. In fact, experiments show that was not the case.

Fig. 6.1 Is the spatial distribution of 'rssi' by Kriging interpolation, and the smaller the value, the better the signal strength

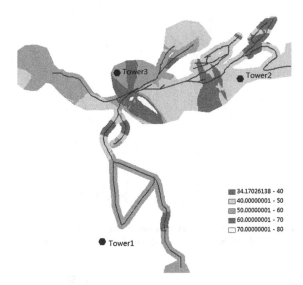

6.4.1 Signal Strength

WLAN vendors can define RSSI metrics in a proprietary manner [6]. For convenience, the measured RSSI of every point had been processed by formula (6.3), and 'rssi' is the processed value, and 'RSSI' is the actual received signal strength. This process only changes the value, and do not affects the spatial distribution.

$$rssi = RSSI + 120 \qquad (6.3)$$

6.4.1.1 The Distribution of Signal Strength

Wireless signal tower's location and the installation of signal tower antenna's direction were considered for the network performance. But from Fig. 6.1, the better performance area are so less, what' more, the tower's locations are not affected.

6.4.1.2 The Distance Between Wireless Adapter and Wireless Tower

According to the first Fresnel zone calculation formula, the distance between transmitting antenna and receiving antenna influences the receiving signal strength. From Fig. 6.2, there is no correlation between this 'rssi' value and distance, in range from 0 to 2,350 m.

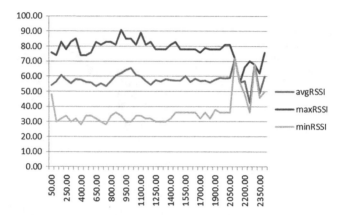

Fig. 6.2 Indicates the relationship between 'rssi' and distance. The *vertical* axis shows 'rssi' value, and the *horizontal* axis shows the distance (meter)

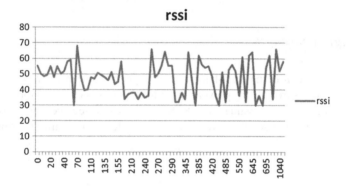

Fig. 6.3 Indicates the relationship between 'rssi' and latency time. The *vertical* axis shows 'rssi' value, and the *horizontal* axis shows the latency time (second)

6.4.1.3 The Correlation Between 'rssi' and Latency Time

From Fig. 6.3, there is also no correlation between this 'rssi' value and latency time.

6.4.1.4 Reasons

Through above analysis, the obtained signal strength is not a representative symbol of network status. There are three main reasons:

1. The signal strength acquisition time is inconsistent with the GPS data acquisition time, that results in spatial location do not match the 'rssi' value.
2. Scan time of the wireless signal is inconsistent with the data sending time.
3. The wireless signal values collected by the hardware are inaccurate.

Fig. 6.4 Is the spatial distribution of latency time by Kriging interpolation, and the smaller the value, the shorter the latency time

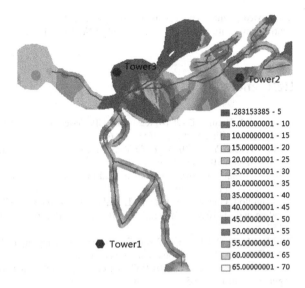

6.4.2 Latency Time

6.4.2.1 Data Pre-Processing

If the network status is too bad for vehicles to re-associate, the latency time will be very large, and it is meaningless. Analysis shows that higher than 80 % of the latency time was no more than 70 s, so 70 s could be selected as the threshold of the 'newspan' field. If the value is larger than 70 s, it is set to 70 s.

6.4.2.2 Spatial Interpolation

From Fig. 6.4, the spatial distribution of latency time mainly succeeds to match reality. Because it is in accord with wireless signal tower's location and antenna's direction, and it is in keeping with system performance and users' feelings.

6.5 Conclusion

For open-pit's dynamic environments, ordinary theoretical method is not practicable for analysis wireless network status. This paper focuses on spatial analysis, and proposes a novel method based on latency time. Compared with signal strength, latency time is more effective. Statistics of latency time of a period has been analysed. The open-pit wireless network signal quality has been represented by Kriging interpolation; both professionals and non-professionals will intuitively know the wireless network condition within this region.

Acknowledgments This research was supported by the Fundamental Research Funds for the Central Universities under the grant number No.2010YD06.

References

1. Ibrahim A, Ibrahim D (2010) Real-time GPS based outdoor WiFi localization system with map display. Adv Eng Softw 41:1080–1086
2. Zhou Y, Cao Y (2010) A brief talk on open-pit mine production scheduling. Opencast Min Technol (S1):108. (in Chinese)
3. Wang P, Jiang Q (2010) Analysis of coal mine wireless communication technology. Saf Coal Min 41(11):111. (in Chinese)
4. Tinnirello I, Giustiniano D, Scalia L et al (2009) On the side-effects of proprietary solutions for fading and interference mitigation in IEEE 802.11b/g outdoor links. Adv Eng Softw 53:141–152
5. Li DL, Liu YD, Chen YY (2011) Computer and computing technologies in agriculture IV. In: 4th IFIP TC 12 international conference. Springer, Berlin
6. Coleman D, Westcott D (2012) Cwna: certified wireless network administrator official study guide: exam Pw0-105. Wiley, England

Chapter 7
Multi-Source Scheme for Reprogramming in Wireless Sensor Networks

Weihong Wang, Shusheng Zheng and Chunping Wang

Abstract Reprogramming may become inefficient with the increase of network scale, we proposed a multi-source scheme which can efficiently disseminate code image, and several approaches of the source node placement were discussed. The scheme with these approaches was evaluated in real applications and simulation experiments. Experimentation results confirmed that multi-source scheme achieves substantial improvement in both time to completion and energy consumption in large scale networks. The time to completion was reduced approximately 44.4 % while the node spacing was 6 m. Simulation experiments showed the segmentation was the most efficient in grid networks, maximum in minimal dominating set worked best in random networks.

Keywords Reprogramming · Multi-source · Time to completion · Energy consumption

7.1 Introduction

Wireless sensor networks are essential for many applications, such as environment monitoring, rescue missions and battlefield surveillance. Most of these applications need to reprogram the code running on the nodes to fix bugs, insert new applications or delete useless codes.

W. Wang (✉) · S. Zheng · C. Wang
College of Compute Science and Technology, Zhejiang University of Technology,
No 288 Liuhe Road, Hangzhou, China
e-mail: wwh@zjut.edu.cn

W. Lu et al. (eds.), *Proceedings of the 2012 International Conference on Information Technology and Software Engineering*, Lecture Notes in Electrical Engineering 210, DOI: 10.1007/978-3-642-34528-9_7, © Springer-Verlag Berlin Heidelberg 2013

XNP [1] is the first method implementing basic network reprogramming function, in which the code image is propagated to nodes situated within by one hop.

Multihop Over-the-air Programming (MOAP) proposed a more comprehensive approach for network reprogramming [2]. It extends XNP to operate over multiple hops. In the MOAP, the code image is divided up into segments, but spatial multiplexing is not considered to optimize the reprogramming either.

Deluge [3], a reliable reprogramming protocol for propagating large data objects, is a protocol based on Trickle. Deluge uses a monotonically increasing version number, segments the binary code image into a set of fixed-size pages. It allows for spatial multiplexing and supports efficient incremental upgrades.

Multi-hop network reprogramming (MNP) [4] extends Deluge to propose a sender selection mechanism, in which senders compete with each other based on the number of distinct requests they have received. MNP reduces the active radio time of a node by putting the node into "sleep" while it not participate data transmission, and the sender selection mechanism reduces the issue of message collisions and hidden terminal problems efficiently.

Andrew Hagedorn and David Starobinski made some improvements based on Deluge: Rateless Deluge and ACKless Deluge [5]. Their approaches rely on rateless coding to eliminate the need to convey control information of packets retransmission. With this approach, a sensor needs only receiving a sufficient number of distinct, encoded packets to recover code image. ACKless Deluge protocol augments the Rateless Deluge with a packet level forward erasure correction (EFC) mechanism. In [6], the authors proposed ReXOR, which employs XOR encoding in the retransmission phase to reduce the communication cost. Compared with Deluge and Rateless Deluge, ReXOR achieves better network-level performance in both dense and sparse networks.

Current research on reprogramming is almost focus on sender selection mechanism and minimizing message transmission, there are few discussions on optimizing the performance of reprogramming by installing more source nodes in a network. In large scale networks, single-source leads to several problems as follow:

First, the energy consumed is not balanced. Compared to the leaf nodes, the parent nodes may participate more packets transmission and reception. As a result, the energy consumed by the parent nodes is much more than the leaf nodes, the network lifetime would be minimized.

Second, the channel competition is fierce in single-source network, which contributes to much more packet collisions. All the nodes attend to request code image from the single-source node, it may lead to more packet collisions during messages transmission and more need for messages retransmission.

Third, once the connection between the only source node and receivers become terrible because of the unpredictable wireless environment, reprogramming becomes inefficiency, even some nodes are unable to receive completely code image and update their function.

Considering wireless sensor networks are becoming larger and larger, more than one source node is needed to disseminate code image to optimize the performance of reprogramming. We proposed approaches of multi-source reprogramming in large scale wireless sensor networks. Multi-source may minimize interferences while reprogramming, result in a faster time to completion, balance and decrease energy consumed by all nodes in wireless sensor networks.

7.2 Proposed Approaches

In order to solve those problems, we proposed a multi-source scheme. Multiple source nodes can balance energy consumed by all nodes to lengthen network lifetime, and it support pipelining, multiple pipelining without interference can occur simultaneous. In a multi-source network, if one source node is hidden by obstacle, other source nodes can continue transmitting image data, this reduce the impact by obstacles. Multi-source can divide a large scale network into several small networks. It may reduce the time to completion. Therefore, five placement approaches were also discussed: random, maximum, randomly in MDS, maximum in MDS and segmentation. We evaluated those approaches in Sect. 7.3.

7.2.1 System Model

At first, we make few assumptions about the topology of the systems discussed in this article. Like Deluge protocol developed by Hui [3], the nodes are static and network is connected, the code image disseminating speed is constant. All nodes are running the same type of program and it is not possible to selectively transmit program data to a subset of nodes. To update their function successfully, all of the data must be received in its entirety.

The topology of network is $G = (V, E)$, which is a connected graph. The nodes are placed in three ways: straight line, square grid, and random placement. The spacing between nodes is various in square grid placement, and the range of network was fixed in random placement. Each node could be a sender as well as a receiver. The number of nodes, the number of source nodes and the transmission range of nodes are fixed. Each two nodes whose distance is less than the transmission range are neighbors. We specify the code image transmitted is composed of 38 pages, and each page is composed of 47 packets. We adopt deluge as our basic reprogramming protocol.

7.2.2 Placement Approaches of the Source Nodes

In large scale network, the placement approach of source nodes is variety and it influence the performance of reprogramming significantly. Therefore, we developed several methods of multi-source placement as follows.

The first approach is random. Placement of N source nodes is decided by uniform random numbers, it has a high probability that source nodes are placed intensive, receivers may be advertised by more than one source node, which results in greater opportunities for hidden terminal problem, therefore, an efficient sender selection mechanism will be important. This approach does not show apparent advantages, we proposed it simply for comparison.

The second is in a maximum way. This approach need traverse all nodes within the network, and calculate the approximation of hop count from each node to others, the N maximum will be chosen as source nodes. The source nodes are generally placed at the edges of networks, which may promote spatial multiplexing, and it reduces hidden and exposed terminal problems in most instances.

The third is randomly in minimal dominating set (MDS). We should calculate the minimal dominating set of networks and choose N source nodes randomly in it. Minimal dominating set is calculated with the approach proposed in [7]. The nodes in minimal dominating set disseminate data to all nodes with single-hop, we choose the nodes in minimal dominating set as source nodes in order to transmit code image to others with the least hops.

The fourth is the maximum way in MDS. We also calculate the minimal dominating set at first, then we traverse nodes within the minimal dominating set, calculate the hop count from each node to others, the maximum will be chosen as source nodes. We proposed this approach to promote spatial multiplexing under the premise of the least transmit hops.

The fifth is segmentation. We segment the networks into N section, and choose central node of every section as source nodes. In general, each source node transmits code image to its part without interfering with each other.

7.3 Evaluation

Our target platform adopted deluge as a reprogramming protocol, and we used both real deployments of running the code on Telosb-motes (TinyOS 2.1) and simulations of matlab to evaluate our scheme.

Wireless sensor nodes are strictly of energy capacity. We consider total energy consumed by all nodes in the network as a primary metric.

We chose Telosb as our platform. On the Telosb, the components consuming energy include the processor, radio transceiver, nodes and external flash storage (EEPROM). For the purpose of data dissemination, the energy consumed by nodes is inconsiderate. The energy consumed by EEPROM and idle-listening is

Fig. 7.1 Effect of multi-source on time

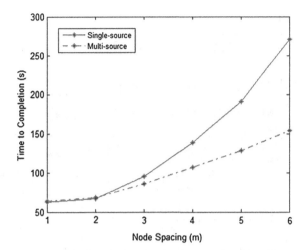

calculated according to literature [8, 9], and the energy model of message transmission is calculated according to literature [10].

Time to completion is another primary metric. Nodes should update their functions quickly once base station sends the command. Furthermore, idle-listening is the major consumer of energy during reprogramming, the longer time to complete results in more energy the nodes spend on the idle-listening state. Obviously, the time to completion influences energy consumption significantly.

Reprogramming requires the program code to be delivered in its entirety. Every byte of the data must be correctly received by all nodes before running. However, wireless communication is unreliable due to possible signal collisions and interference. This makes reliable protocol designs more challenging.

7.3.1 Experiments with Telosb Motes

Our tests were designed to examine the effects of different energy cost and time to completion in multi-source networks and single-source networks. To accurately compare the effects of multi-source and single-source, we fixed the number of receivers (4 nodes). The nodes were placed in a line, one source node was placed at head and the other was the end.

We varied the spacing at 1, 2, 3, 4, 5, 6 m between neighbor nodes to evaluate the performance of multi-source in different configuration of wireless sensor networks. Each experiment with different configuration was executed five times and the results were averaged. Figure 7.1 presents the time to completion, Fig. 7.2a and b presents the messages transmission of 4 receivers, the size of each message is 28 bytes. Because reprogramming run successfully in our experiments, the reliability must be 100 %, so we do not present the reliability.

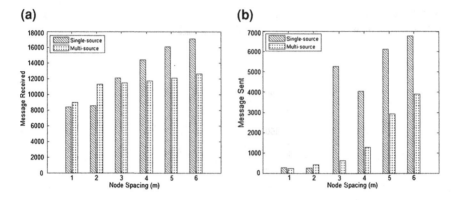

Fig. 7.2 Effect of multi-source on message

In most cases, multi-source performed better in time to completion, and the advantages became more obvious if the spacing was larger. When the spacing was 6 m, the time to completion in multi-source networks was 44.4 % less than the number in single-source networks.

When the node spacing was 1 and 2 m, multi-source need more messages transmission than single-source. The effect of multi-source approaches were most pronounced in sparse network in both time and energy, multi-source showed a large reduction in time to completion and messages transmission. When the node spacing was 6 m, the count of messages received in multi-source networks was approximately 29.4 % less than the number in single-source networks. The messages sent in single-source network showed a dramatic increase with the increase of network size, while it increased gradually in multi-source network. The count of messages received was approximately 42.8 % less than single-source network in sparse networks.

In networks of high density, source nodes can transmit code image directly to nodes, all nodes received code image and updated their function simultaneously, and multi-source nodes did not accelerate process of reprogramming. Furthermore, more source nodes lead to more advertisements, requests and data messages, which resulted in greater opportunity for packet collisions and more messages transmission. Therefore, the performance of single-source actually slightly better than multi-source, single-source required fewer messages sent in dense networks.

In single-source network, one source node transmitted packets to the nearer node at the beginning of reprogramming. After the nearer node received complete pages, it advertised and propagate the newly pages further. The nodes further were idle in most of time. In multi-source network, two source nodes disseminated the code simultaneously without interference, which resulted in a dramatic reduction in time to completion.

In our experiments, we found the nearer node participate more messages transmission and consumed much more energy than the further nodes in single-source network. In multi-source network, the energy consumption is more balanced.

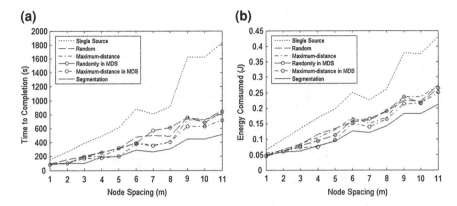

Fig. 7.3 Effect of multi-source in grid network

7.3.2 Simulation Results

We simulated with matlab. In our simulation, we fixed the size of code image, the transmit radius is 8 m. Packet loss rate is base on the model proposed in [11]. We fixed the number of nodes (100) and the number of source nodes (4). We measured energy consumed, time to completion and compared the five approaches.

In our first simulation, we fixed the number of nodes (10×10 grid) and evaluated our approaches by placing the nodes at spacing of 1, 2, 3, 4, 5, 6, 7, 8, 9, 10, 11 m between nodes. Each approach was executed several times, and the results were averaged. Figure 7.3a presents the time to completion, Fig. 7.3b presents the energy consumption.

At all densities, multi-source approaches significantly reduced time to completion and energy consumption in large scale network, and the superiority of multi-source was more prominent in sparse network. The time to completion in multi-source networks was approximately 66.7 % less than the number in single-source networks and the energy consumption was approximately 46.5 % less. We found that the messages transmission count of the nodes near the source were much more than others in single-source networks. The energy consumed by all nodes was more balanced in multi-source networks.

Among the approaches of multi-source, segmentation outperformed other approaches at most densities in term of both energy and time, the time to completion of segmentation was approximately 28.6 % less than the other's and the energy consumption was approximately 16.3 % less. The performance of random was too unstable. The results of each two experiments may be variety. We found the result was terrible while the source nodes were concentrated in the network, the nodes near the source nodes participated much more messages transmission, which may result from greater contention and packet loss. The results of Maximum-distance were stable, but this approach was not efficient. Source nodes were placed at the edges of the networks, it not favored of the transmission of sensor nodes,

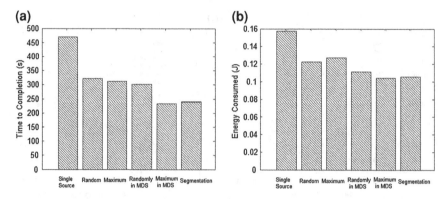

Fig. 7.4 Effect of multi-source in random network

which could transmit data to nodes around it. Random in MDS and Maximum-distance in MDS may be more efficient than segmentation in networks of density 2 and 8, however, segmentation indeed worked better in most cases. This may for the reason that the several parts divided by source nodes were uniform with segmentation in grid networks, which can maximize the function of each source node.

In our second simulation, the nodes placed randomly in a square (30 × 30 m), and we examined each approach several times, the results were averaged. Figure 7.4a presents the time to completion, Fig. 7.4b presents the energy consumption.

The approaches of multi-sources placement had the advantage of both in energy and time. In multi-source networks, maximum in MDS resulted in faster time to completion and less energy consumption. Compared with segmentation, the advantage of maximum in MDS was slight. The time to completion of maximum in MDS was approximately 4.3 % less than the others' and the energy consumption was approximately 1.98 % less. In those experiments, nodes were placed randomly, the network was variety, but the placement of source nodes were changeless with segmentation, it made the performance of segmentation unstable and indeed not as efficient as it worked in grid network. Choosing source nodes in MDS guarantee the hops of transmission and we found that compared with randomly in MDS, maximum in MDS got more stable results.

7.4 Conclusion

In this paper we proposed an efficient multi-source scheme for wireless repro-gramming, it can improve the performance both in energy consumption and time to completion. Moreover, five placement approaches of the source nodes in large scale network were explored. Real-world experiments with telosb motes and simulations with matlab revealed multi-source get a better performance then single source, and through the simulations, we found segmentation got best results in the 10×10 grid network, the performance of Maximum in MDS was best in random placement network.

Acknowledgments This paper is supported by the National Natural Science Foundation of China (60873033), the Natural Science Foundation of Zhejiang Province (R1090569) and Zhejiang Provincial Natural Science Foundation (No. Y1101053) and Information Service Industry Special Fund of Zhejiang Province Economic and Information Commission [(2010) No.360].

References

1. Jeong J, Kim S, Broad A (2003) Network reprogramming. University of California at Berkeley, Berkeley
2. Stathopoulos T, Heidemann J, Estrin D (2003) A remote code update mechanism for wireless sensor networks. Technical report, UCLA, Los Angeles
3. Chlipala A, Hui J, Tolle G (2003) Deluge: data dissemination for network reprogramming at scale. Class Project, Berkeley, University of California, Fall
4. Kulkami S, Wang L (2005) Mnp: Multihop network reprogramming service for sensor networks. In: 25th IEEE international conference on distributed computing systems, pp 7–16
5. Hagedorn A, Starobinski D, Trachtenberg A (2008) Rateless deluge: over-the-air programming of wireless sensor networks using random linear code. In: the seventh international conference on information processing in sensor networks (IPSN'08), pp 457–466 (April)
6. Dong W, Chen C, Liu X, Bu J, Gao Y (2011) A lightweight and density-aware reprogramming protocol for wireless sensor networks. IEEE Trans Mob Comput V10(10), pp 1403–1415 (Oct)
7. Guha S, Khuller S (1996) Approximation algorithms for connected dominating sets. Algorithmica V29(6). (in Chinese)
8. MSP430x161x mixed signal microcontroller. www.ti.com. 2006
9. CC2420 2.4 GHz IEEE 802.15.4/ZigBee-ready RF Transceiver. www.chipcon.com. 2004
10. Zhang Z, Sun Y, Liu Y, Yang T (2009) Energy model in wireless sensor networks. J Tianjin Univ, vol 40(9) (in Chinese)
11. Sun PG, Zhao H, Luo DD, Zhang XY (2007) A communication link evaluation model for wireless sensor networks. J Northeast Univ (Natural Science), vol 28(9) (in Chinese)

Chapter 8
Study on Industrial Cluster Knowledge Emergence Model Based on Asymmetric Influence

Xiaoyong Tian, Huiyuan Jiang, Jin Zha and Linlin Li

Abstract The interaction among various agents of industrial cluster is an important factor which transfers the initial innovation into cluster knowledge. The asymmetry influence functions among agents were given by defining uncertainty of agent's attitude and trust factor. Based on these functions, an Agent simulation model of cluster knowledge emergence was realized. The simulation results show this Agent model is reasonable and effective. The analysis of simulation results lead to conclusions as follows: The uncertainty of agent's attitudes and trust factors are important factors to impact cluster knowledge final emergence's form and speed.

Keywords Industrial cluster · Asymmetric agent influence · Knowledge emergence

8.1 Introduction

Knowledge sharing, as an important internal mechanism in industrial cluster, has become a powerful driving force in the survival and development of cluster. It is an essential factor to the success or failure of industrial cluster. In the cluster, knowledge is scattered in different individuals and organizations, in order to make full use of the role of knowledge to win and maintain the competitive advantage, we must make knowledge transfer in cluster and finally transform into cluster knowledge which can be shared by each agent of cluster.

X. Tian (✉) · H. Jiang · J. Zha · L. Li
School of Transportation, Wuhan University of Technology,
No. 1040 Heping Road, Wuhan 430063, China
e-mail: shinytame@sina.com

W. Lu et al. (eds.), *Proceedings of the 2012 International Conference on Information Technology and Software Engineering*, Lecture Notes in Electrical Engineering 210, DOI: 10.1007/978-3-642-34528-9_8, © Springer-Verlag Berlin Heidelberg 2013

In short, the industrial cluster knowledge is common knowledge that most agents of industrial cluster adopt by sharing and transferring. The initial innovation mastered by a single agent is not industrial cluster knowledge; it can be transformed into cluster knowledge only if it is transferred from the initial agent to another and also be accepted and absorbed. The research indicates that, in the process of industrial cluster knowledge transfer and evolution, the degree of knowledge internalization in most agents will be more easily affected by the agent relationships and its status. It is a typical evolution process of complex system from the mess, disorder, no organization agent knowledge to obvious common industry cluster knowledge. Complex adaptive systems are systems that have a large numbers of components, often called agents, which interact and adapt or learn [1].

Whenever you have a multitude of individuals interacting with one another, there often comes a moment when disorder gives way to order and something new emerges: a pattern, a decision, a structure, or a change in direction [2]. In the modeling process of this complex system, setting up effective agent interactive mode is the key. The industry cluster knowledge sharing model of this paper was realized through the asymmetric influence function among agents which reflect the influence of each other is not symmetrical when they transfer and adopt knowledge in cluster. That means the influence of agent I to agent J is often different from the influence of J to I. And also this paper introduced trust factors among agents, based on which realized the emergence simulation model from agent knowledge to cluster knowledge on the NetLogo platform, and finally do analysis of the simulation results.

8.2 The Model Assumption and Asymmetry Agent Influence Function

8.2.1 Model Assumption

The basis of industrial cluster knowledge sharing model is an attitude of agent to transferred knowledge. There are two representation methods of agent attitude: discrete values and continuous interval. For example, in discrete values, -1 represents "refuse", 0 represents "hesitate", 1 represents "accept", or using other discrete data to show agent's attitude. In the discrete representations, the attitude is not intersect and apart from each other. In the actual social and economic life, the attitudes of most agents in the beginning are not clear and unmistakable before the formation of common industrial cluster knowledge, but many agents transform their attitudes from vague and uncertain to firm one through the public platform or influence among agents. How to get the unclear and changeable agents is more practical significance and applied value.

Therefore, this model use numerical value in continuous interval [0,1] to represent the attitude of agent to initial innovation, it reflects the reality that the attitudes of the majority of agents are not entirely different, where 0 is completely refused, 1 means overall accept and internalize, the values between 0 and 1 represent the various intermediate attitudes.

On the other hand, the agent's insistence to their attitudes are different, it subject to the status of agents in industrial clusters and the mastery of their own situation and market information and so on. This paper study on macro-emergence process of industrial clusters knowledge, so listing all factors is unnecessary. All of these factors are called "uncertainty", with a value in the interval [0, 100 %] to represent the uncertainty of agent's attitude. That is, to agent I, the initial attitude x_i is used a pair of values to represent: x_i (attitude$_i$, uncertainty$_i$), its simple form is $x_i(a_i, u_i)$, and $0 \leq \text{attitude}_i = a_i \leq 1$, $0\% \leq \text{uncertainty}_i = u_i \leq 100\%$.

At the same time, there are cooperation and competition among enterprises in the cluster, which makes trust particularly important in the process of knowledge transfer. This paper introduces the influence of trust, and uses trust factor to measure the degree of mutual trust among agents. Roger and Yushan (2003) insist in their study that trust creates the necessary condition for a high level of knowledge transfer. Enterprises in cluster use strong knowledge spillovers and together-learning to gain more knowledge, and if there are the organizational trust and interpersonal trust, more knowledge transfer will happen [3]. Dhanaraj (2004) confirmed that the trust will positively affect the explicit and implicit knowledge transfer among partners. Knowledge transfer will be hampered or even failed without trust, since the enterprise will lose enthusiasm for knowledge transfer if there is no trust as guarantee [4].

A high degree of trust among agents will make easier to influence each other. Using trust representation mode in the BBK model [5], the trust factor is given a real value in the interval of [0, 100 %]. That is 0 % represents no trust and 100 % means full trust. Usually the degree of mutual trust among agents is not the same and the trust is a cumulative process, trying to determine the degree of trust among all agents is not possible. In this paper, the trust factor is a constant [6], as the general considerations of trust level among agents in industrial clusters. Based on the above assumptions, we define the asymmetric influence functions and establish the interaction criteria on the attitudes of agents.

8.2.2 Asymmetry Agent Influence Function

Suppose the attitude of agent I is x_i (a_i, u_i), and the attitude of the agent J is x_j (a_j, u_j). The trust factor is T.

$$d_{ij} = \min(a_i + u_i, a_j + u_j) - \max(a_i - u_i, a_j - u_j) \tag{8.1}$$

Definition 1 The influence function of the agent I on agent J f (i, j) is:

If $d_{ij} > u_i$, then $f(i,j) = T \times [(d_{ij}/u_i) - 1]$; If $d_{ij} \leq u_i$, then $f(i,j) = 0$ (8.2)

Definition 2 The influence function of the agent J on agent I f (j, i) is:

If $d_{ij} > u_j$, then $f(j,i) = T \times [(d_{ij}/u_j) - 1]$; If $d_{ij} \leq u_j$, then $f(j,i) = 0$ (8.3)

If the agent I and agent J exchange attitudes and influence function is not 0, each attitude has an impact on the other side, so their attitudes adjust as xi'(ai',ui') and xj'(aj',uj'):

$$a_i' = a_i + f(j,i) \times (a_j - a_i); \ u_i' = u_i + f(j,i) \times (u_j - u_i) \qquad (8.4)$$

$$a_j' = a_j + f(i,j) \times (a_i - a_j); \ u_j' = u_j + f(i,j) \times (u_i - u_j) \qquad (8.5)$$

The defined interaction criteria derivates directly two conclusions:

Conclusion 1: In the evolution of industrial cluster knowledge, not only the attitude of agent will interact and change, but also its attitude uncertainty will do.

Conclusion 2: The influence among agents is asymmetrical. If agent I's attitude is more convinced than agent J, the impact of I on J is larger than J on I.

That is, agent's attitude and their tendencies are gradually formed, the more confident agent is easier to affect the less confident agent, which is also consistent with the actual situation.

8.3 Model Simulation and Results Analysis

8.3.1 Model Simulation Algorithm

Using the multi-agent simulation platform NetLogo, Industrial Cluster Knowledge Emergence Model based on asymmetric influence was realized. The model's schematic algorithm as follows:

1. Setting agent's number, agent's attitude ai, agent's uncertainty ui, (i = 1, 2, ..., num), randomly confirming one of agents has initial innovation knowledge and giving it red color, see Fig. 8.1.
2. Randomly selecting an agent J in the neighborhood of agent I, and exchanging their attitudes according to Eqs. (8.4) and (8.5), so their attitudes can update to xi' (ai', ui') and xj' (aj', uj').
3. Make sure that the agent's attitude to innovation knowledge is one of accept, hesitate or refuse, which is represented relatively by red, green and grey.

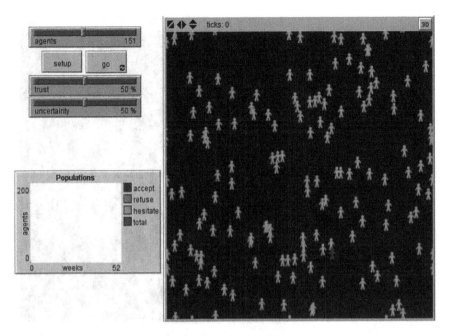

Fig. 8.1 The initial innovation knowledge in industrial cluster

4. Traverse all agents and synchronously update the agent's attitudes to innovation knowledge.
5. All agents move one step randomly in any direction.
6. Repeat 2–5, until emerge the characteristic that the entire industrial cluster owns but a single agent doesn't.

8.3.2 The Emergence of Industrial Cluster Knowledge

In order to focus on the study of emergence mechanism, we make necessary simplification for the formalized description of industrial cluster knowledge during the process of simulation. First, we assume that all the agent's attitudes subject to uniform distribution in [0, 1]. Uniform distribution is a more reasonable assumption in the case there is no priori information. Second, to facilitate the analysis, we assume the same uncertainty which is set to 50 % of all the agent's attitudes and the same trust factor which is set to 50 % among agents whose number is set to 151. As time goes by, the attitudes of agents on innovation knowledge gradually concentrate on the acceptance and refuse and each accounts for approximately 50 %, which is shown in Fig. 8.2. In the case the initial attitude subjects to uniform distribution and the uncertainty of agent on innovation knowledge is same, it indicates the understanding of agent on innovation knowledge is clearer, sideliners who consider whether innovation knowledge adapts to

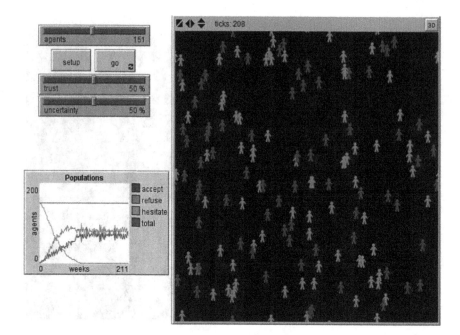

Fig. 8.2 Neutral industrial cluster knowledge emergence

itself are fewer. More and more agents either choose or refuse the innovation knowledge. Industrial cluster knowledge which applies to half of the agents comes into being in the end.

In actual society, the degree that the agent determines its attitude is often different. For example, the agent who has a leadership in the market and knows more about itself and the situation of the market would more firmly insist on its attitude. We gradually change the parameter uncertainty and trust factor, and find that when uncertainty is larger, namely the agent is not sure about the suitability of innovation knowledge, the innovation knowledge is more difficult to form the industrial cluster knowledge that be widely promoted, see Fig. 8.3. When uncertainty is smaller, namely the agent can more confident that the innovation knowledge suit itself, it's very easy to form the industrial cluster knowledge which is acceptable in a wider scope, see Fig. 8.4. The degree of trust among agents will not change the final choice result. But it will impact on the emergence rate of the industrial cluster knowledge. The smaller the trust, the slower is the convergence of the agent's attitudes. That is, the choice of agents will contribute to the forming of the industry cluster knowledge more quickly in a trustful industry cluster (contrast Figs. 8.3 and 8.4).

Through the analysis of the above simulations, we found that the ultimate form of industrial cluster knowledge emergence is determined by the number of the extreme agents in the cluster and the attitudes they hold under the circumstance

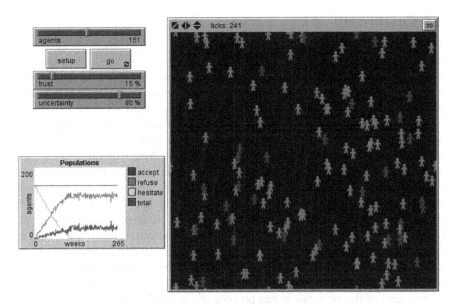

Fig. 8.3 The industry cluster knowledge emergence of initial innovation in small scope

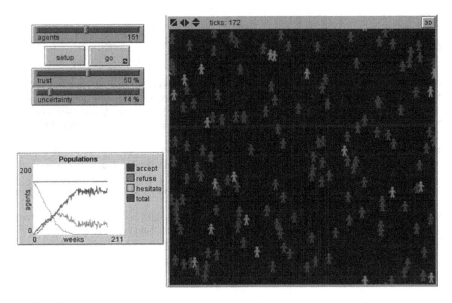

Fig. 8.4 The industry cluster knowledge emergence of initial innovation in wide scope

that the initial attitudes are distributed evenly. The trust factor only affects the convergence speed of industry cluster knowledge emergence and doesn't affect the final selected result of industry cluster knowledge.

8.4 Conclusion

For a specific events or phenomena, the factors which can influence the industrial cluster knowledge emergence are complexly changeable and contain much uncertain factors. However, it implicates some common regular pattern in the process of industrial cluster knowledge emergence from chaos to order.

In this paper, we introduce the asymmetric influence function and trust factor among agents to establish the computer simulation model of the industrial cluster knowledge evolution. The simulation results of the model can describe as follows: with the premise of uniform distribution of the initial attitude, the final selected form of industrial cluster knowledge is determined by the number and attitude of extreme agent and the time period of industrial clusters knowledge emergence can be affected by the trust relationship among agents. From the perspective of control, if we want to promote the initial innovation into the industrial cluster knowledge in a wider scope, in the first, the agents who have a leadership position should be controlled and influenced to accept and apply the initial innovation. The second point is strengthening the trust relationship among agents. At last, the power departments of the government can propose the industrial policy to guide agent's attitude of initial innovation, which also influence the final formation of the industrial cluster knowledge.

References

1. Holland JH (2006) Studying complex adaptive systems. J Syst Sci Complexity 19(1):1–8
2. Miller P (2010) The smart swarm: how understanding flocks, schools, and colonies can make us better at communicating, decision making, and getting things done. Avery, New York
3. Roger JC, Yushan Z (2003) Tacit knowledge transfer and firm innovation capability. J Bus Ind Marketing 18(1):6–21
4. Dhanaraj C, Lyles MA, Steensma HK, Tihanyi L (2004) Managing tacit and explicit knowledge transfer in IJVs: the role of relational embeddedness and the impact on performance. J Int Bus Stud 35:428–442
5. Deane JHB, Hamill DC (1996) Improvement of power supply EMC by chaos. Electron Lett (S0013-5194), 12:1045
6. Hegselmann R, Krause U (2002) Opinion dynamics and bounded confidence models, analysis and simulation. J Artif Soc Soc Simul (S1460-7425), 3(5):162–173

Chapter 9
Study and Optimization of FIR Filters Based on FPGA

Jinlong Wang, Lan Wang and Zhen Yuan

Abstract FIR filter plays a very important role in the digital image and digital signal processing. This paper describes a design and optimization of demoboard based on Xilinx's spartan6 chip FIR filter. First, it needs to select the appropriate FIR filter structure for reducing the waste of FPGA logic resources; then, the FIR filter with multipliers and adders need to be improved and optimized, and FIR filter model is built by using of Simulink of Matlab. Finally, we need to convert FIR model into project file by System Generator, and use the software ISE to edit, synthesize, map and layout for project file. After then, configuration file is generated, and it is downloaded into the FPGA for realizing optimization of FIR filter and hardware design.

Keywords FPGA · Xilinx · Matlab · Simulink · System generator · ISE

9.1 Introduction

Finite Impulse Response (FIR) filter is a basic component in the digital signal processing system, and it has some advantages such as linear-phase, easily hardware implementation and system stability, etc. It has wide application in the field of communications, image processing and pattern recognition, etc. As systems

J. Wang (✉) · L. Wang
Electric Engineer College of Heilongjiang University, Harbin, China
e-mail: wjl5682@126.com

Z. Yuan
Harbin Coslight Electric Automation Co., Ltd, Harbin, China
e-mail: yuanzhenclea@126.com

W. Lu et al. (eds.), *Proceedings of the 2012 International Conference on Information Technology and Software Engineering*, Lecture Notes in Electrical Engineering 210, DOI: 10.1007/978-3-642-34528-9_9, © Springer-Verlag Berlin Heidelberg 2013

increase in broadband, high-speed and real-time signal processing, greater demands are placed on filter processing speed and performance, etc. [1]. The obvious advantage of FPGA is that it can provide parallel for implementing digital signal processing algorithms, can significantly improve data throughput of filter, and can embed a dedicated fast adder, multiplier and input/output devices. Thus it becomes a high-performance digital signal processing ideal device. However, the FIR filter usually calls directly internal adder, multipliers and delay units of FPGA. But the resources of the multiplier are finite and the direct use of the multiplier will cause a great waste of resources [2]. Between this, this paper introduces a optimization method of FIR filter and it has some advantages such as reducing the waste of system resources, increasing system speed and simplifying the system design process.

9.2 Basic Principle, Structure and Algorithm of FIR Filters

9.2.1 The Principle of the FIR Filter

FIR (Finite impulse response filter) is one kind of digital filter, which features that unit impulse response is a limited length series. Thus, the system function is generally written as follows:

$$H(z) = \sum_{n=0}^{N-1} h(n)z^{-n} \tag{9.1}$$

where N is the length of h(n)-the number of taps of FIR filter. In addition, the linear-phase characteristic is one of the advantages of FIR filter, and the unit impulse response are all real numbers and the system function H(z) is stable.

9.2.2 The Structure of FIR Filter

The structure of direct-type of FIR filter is shown in Fig. 9.1. Its output can be expressed as:

$$y(n) = \sum_{i=0}^{N-1} h(i)x(n-1) \tag{9.2}$$

Adder and multiplier are not difficult to realize FIR filter with this structure. Figure 9.1 is a implementation of direct-type of FIR filter. But direct-type of FIR filter is not ideal whether or speed or resource consumption [3].

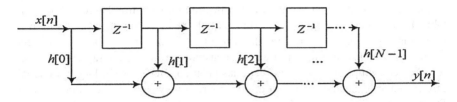

Fig. 9.1 Schematic diagram of direct-type of FIR filter

Because of the coefficient of linear-phase FIR filters possesses symmetry, the structure of Linear symmetric FIR filter has been widely used in projects [4]. The coefficient h(n) of linear-phase FIR filters satisfies the following conditions:

$$h(n) = \begin{cases} h(N-1-n) & 0 \le n \le N-1, \text{Even symmetry} \\ -h(N-1-n) & 0 \le n \le N-1, \text{Odd symmetry} \end{cases} \quad (9.3)$$

where h(n) is real-valued sequence, N denotes the length of h(n). In (9.3), The coefficient h(n) of linear-phase FIR filters possesses even symmetry or odd symmetry. The System function H(z) of linear-phase FIR filters follows [5]:

- When N is even,

$$H(z) = \sum_{n=0}^{N/2-1} h(n) \left[z^{-n} \pm z^{-(N-1-n)} \right] \quad (9.4)$$

- When N is odd,

$$H(z) = \sum_{n=0}^{(N/2-1)-1} h(n) \left[z^{-n} \pm z^{-(N-1-n)} \right] + h\left(\frac{N-1}{2}\right) z^{-(N-1)/2} \quad (9.5)$$

As Eqs. (9.4) and (9.5) are easy to be known, when N is even, H(z) just needs N/2 times multiplication; when N is odd, then simply (N + 1)/2 times multiplication. That is to say, the symmetry of the coefficient of linear-phase FIR filters can halve the times of multiplication, which dramatically reduces resource consumption [6].

9.2.3 The Traditional Algorithm

Distributed Arithmetic has been brought forward to solve this problem of multiplication resources and it is a classical optimization algorithm [7]. This algorithm architecture can effectively convert multiplication into addition based on look-up-table LUT which can quickly get the partial product. But there are still two problems: First, in DA structure, the size of LUT and filter order is exponent relationship. Delay of filter and resource consumption will grow rapidly; Second, resource consumption and speed there has been conflict between larger. In serial DA, resources consume is less but the system runs very slow. The existing parallel DA can solve the speed problem, but it may be achieved at the cost of consumed resources in substantial additional parts. In this paper, the CSD coding algorithm is selected and it will be taken further research to achieve resource-saving and higher speed.

9.3 FIR Filters Design

It needs to be set the appropriate indicators to generate the filter coefficients by use of FDATOOL in Matlab. After then, use of Simulink to build the FIR filter model and simulate it [8]. Finally, we can use System Generator to establish project file which needs to be edit, synthesized, implement and configuration by Software ISE13.4. And the configuration file is downloaded into the FPGA so as to realize automation, modular and hardware. The design process of FIR filter is shown in Fig. 9.2.

In order to further optimize resources and speed, improve resource utilization in FPGA, we use even symmetry of the linear FIR filter as example for re-designing multiply unit and the addition unit.

9.3.1 The Design of Multiplication Units

Multiplication can take advantage of the binary choices and the addition operation to realize multiplication between the filter coefficients and sampling data which can avoid the direct call of the multiplier.

$$x \cdot y = x \cdot \left(\sum_{m=0}^{M-1} s_m \cdot 2^m \right) \qquad (9.6)$$

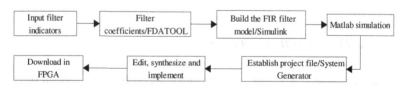

Fig. 9.2 The design process of FIR filter

M is the highest bit of y, x is decimal integer. s_m take 0 or 1 value, when s_m is 0, it isn't involved in computation by shifting multiplier to realize $a \cdot b$, thereby convert Multiplication into add shifting computation. The number of "1" in y decides "add" times. Of crouse, the less number, the better. Use CSD (Canonic Signed Digit) encoding can realize the minimized number. CSD encoding differs from traditional binary encoding, the ranges of its number values are 0, 1 and -1.

-1 usually express $\bar{1}$. This encoding a usual expression with the least nonzero elements and the form is only. The following example shows: take any number $h = 231$. We have assumed to be filter coefficient, filter input is X(n). $231 = (11100111)_2 = (100\,\bar{1}\,0100\,\bar{1})_{CSD}$. Respectively use binary and CSD encoding torealize $231 \cdot X(n)$ in Figs. 9.3, 9.4 and 9.5 as shown.

As you can see from these Figures, binary encoding form needs six adders and six the units of delay time. But CSD encoding just need four adders and four the units of delay time.

To reduce resource consumption, CSD encoding needs to be further optimized. Firstly, coefficient h is separated into several multiplier factors. Secondly, every factor is achieved by CSD encoding. Finally, linear adding structure in Fig. 9.6 is instead of tree. Pay attention to factor values: one is that every factor value should be 2^k or nearly 2^k at the very highest level, k is integer; the other is that the number of factor is minimum. For example, $231 = 33 \times 7$, as shown in Fig. 9.6. Optimized CSD encoding is shown in Fig. 9.7. As you can see from those, it reduced two adds and two units of delay time.

9.3.2 The Design of Adder Units

After CSD encoding, addition operation is needed to realize multiplication finally. The way to realize the structure of addition operation is cumulative and addition tree. If cumulative structure carries on addition, bit wide is needed to be extended by one bit, which increases extra resource consumption; Not only addition tree structure can overcome that shortcoming and save resource consumption, but also realize multichannel parallel operation to improve system speed [9].

Facing to so many additions, the operation priority problem should be concerned in addition tree structure. Adding two different numbers of input, the sign bit of input data is carried on extending or adding zero on low bit. After the number of bits adjusted well, addition will carry out. This process increases resource consumption and logic delay time. Especially when it is high order FIR filter, more computation is needed and consumed extra resource will increase. Therefore, the principle to makesure addition as following: the first step is that the same validate bit and the highest inputs add. In that case, input computation has

Fig. 9.3 Multiplication

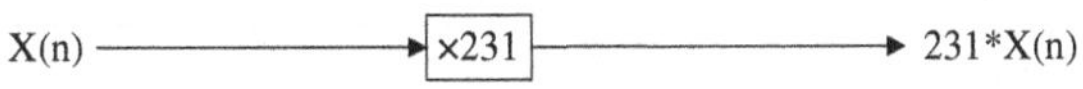

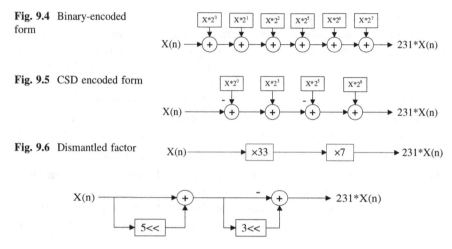

Fig. 9.4 Binary-encoded form

Fig. 9.5 CSD encoded form

Fig. 9.6 Dismantled factor

Fig. 9.7 Optimized CSD encoded form

the first priority; the second step is that input is sorted by the number of validate bit and highest bit closeness. More close, the higher priority. According to this, bit wide of output can be reduced at the very highest level, addition operation can be optimized and logic recourse consumption can be lowered.

9.4 Simulation

Use a 8 order low pass FIR filter as example to simulate the method from this chapter [10].

9.4.1 Model

It needs to be set the appropriate indicators by use of FDATOOL in Matlab [11]. Click on the Design Filter to automatically generate Filter coefficients. As the Table 9.1 shown:

Table 9.1 Filter coefficients

h(n)	Filter coefficients
h(0) = h(7)	0.00875474
h(1) = h(6)	0.04794887
h(2) = h(5)	0.16402439
h(3) = h(4)	0.27927199

Select "Export" under the File menu in FDATOOL and export the coefficient to Workspace. According to actual situation, the quantized and optimized by CSD code coefficient coding as follows:

$$h(0) = h(7) = 2 = 0000\ 0010 \quad h(1) = h(6) = 12 = 0001\ 00\bar{1}0$$
$$h(2) = h(5) = 42 = 00101010 \quad h(3) = h(4) = 72 = 8 \times 9 = (0000\ 1000)$$
$$\times (0000\ 1001)$$

In the Simulink environment, call the Xilinx toolbox module and set up the filter circuit model in Figs. 9.8 and 9.9 as shown.

In Figs. 9.8 and 9.9, the source is the sine wave with 7 and 700 HZ. Mixing for two signals, extract useful 7 HZ low frequency signal from mixed.

As you can see from Fig. 9.10, the first line is 7 HZ signal, the second line is the 700 HZ high frequency signal, the third line is mixing signal from 7 and 700 HZ. The fourth line is the useful signal from the mixed signal and it is clear to see that the extracted low frequency signal is there. It validates the feasibility of the proposed method in this paper.

Click "System Generator" in model to generate ISE project files. In order to achieve the hardware design of FIR filters, edite, synthesize and place and route for project file to generate "bit files" for configuring into the FPGA.

Through PlanAhead, we can view resource consumption of the two models, as shown in Fig. 9.11(a) shows that the resource consumption of direct-type of FIR Filters, (b) presents that the resource consumption FIR filters in this paper.

After synthesizing, RTL block diagram of the FIR filter can be generated, as shown in Fig. 9.12. To make the FIR filter design be automation, simplified, modularity.

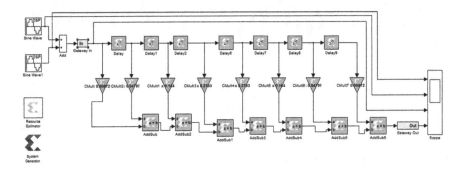

Fig. 9.8 Direct-type of FIR filters

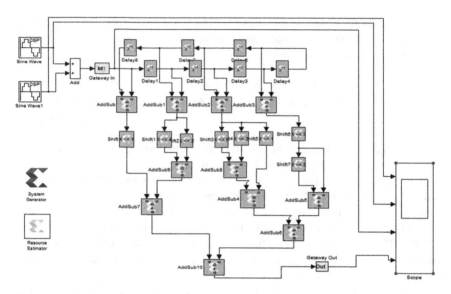

Fig. 9.9 Optimized FIR filters

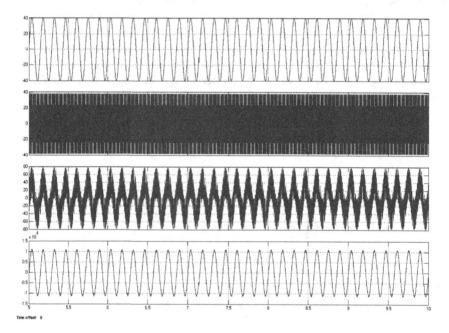

Fig. 9.10 Optimized simulation results

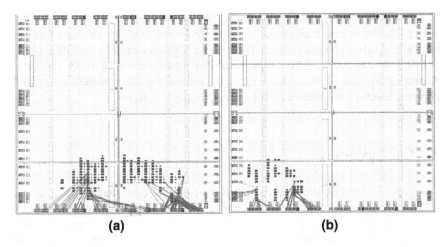

Fig. 9.11   Resource consumption of two types of FIR filters

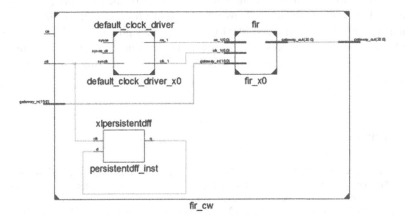

Fig. 9.12   RTL block diagram

Table 9.2   The used space of FPGA chip

FIR	Slices	LUTS	Mults	Muxcys	IOBS
Direct-type of FIR	209	666	8	512	56
FIR methods in this paper	65	242	0	208	38

9.4.2 Simulation Analyses

After simulating 8 order direct-type of FIR Filter and described FIR Filter in this paper, its use of resources is as shown in Table 9.2.

As you can see from Table 9.2, the method described in this paper has greatly reduced their cost resources consumption in FPGA.We don't use multiplier in this method and other resource usage is also relatively much more less than direct-type of FIR Filter. In addition, the proposed method can realize modularity and automation for design of the filter, to greatly reduce the development effort and improve development efficiency for engineering application.

9.5 Conclusions

This paper introduces you to a design method of FIR filter that is low-cost and efficient. Redesign multiplication unit structure and addition unit structure in FIR filters. For some shortcomings of FIR filters design, such as low resource utilization and not ideal operation speed, this method breakthroughs the bottleneck of limited multiplier resources in FPGA; And then using combination of System Generator and ISE realizes automation and modular of the filter design, to avoid several shortcomings for the tedious development process of based on hardware language approach and the complexity of debugging process; Simulation results verify the proposed method that possesses low resource consumption, running speed, high resource utilization and simple operation, while simplifies the process of system design.

References

1. Nekoei F, Kavian YS, Strobel O (2010) Some schemes of realization digital FIR tilters on FPGA for communication applications, 20th international crimean conference on microwave and telecommunication technology (CriMiCo), pp 616–619
2. Mirzaei S, Hosangadi A, Kastner R (2006) FPGA implementation of high speed FIR filters using add and shift method, IEEE
3. Shi D, Yu YJ (2011) Design of linear phase FIR filters with high probability of achieving minimum number of adders. IEEE Circuits Sys Soc 58(1):126–136
4. Benkrid A, Benkrid K (2009) Novel area-efficient FPGA architectures for FIR filtering with symmetric signal extension. IEEE Trans Very Large Scale Integr VLSI Syst 17(5):709–722
5. Jiang X, Bao Y (2010) FIR filter design based on FPGA. In: International conference on computer application and system modeling, pp 621–624
6. Leong PHW (2008) Recent Trends in FPGA architectures and applications, 4th IEEE International Symposium on Electronic Design, Test and Applications, pp 137–141
7. Zhao L, Bi WH, Liu F (2010) Design of digital FIR bandpass filter using distributed algorithm based on FPGA. Electron Meas Technol 30(7):101–104 (In Chinese)
8. Woods R, McAllister J, Lightbody G, Yi Y (2011) FPGA based implementation of signal processing systems. Wiley, Chichester
9. Beyrouthy T, Fesquet L (2011) An event-driven FIR tilter: design and implementation, 22nd IEEE international symposium on rapid system prototyping (RSP), pp 59–65

10. Das J, Wilton SJE (2011) An analytical model relating FPGA architecture parameters to routability. In: FPGA 2011 proceedings of the 19th ACM/SIGDA international symposium on field programmable gate arrays, 27 Apr 2011, pp 181–184
11. Nianxi X (2008) Application of MAT LAB in digital signal processing, 2nd edn. Tsinghua University Press, Beijing, pp 397–400

Chapter 10
ES-VBF: An Energy Saving Routing Protocol

Bo Wei, Yong-mei Luo, Zhigang Jin, Jie Wei and Yishan Su

Abstract Limited Energy is a challenge in Underwater Sensor Network (UWSN). To solve the energy problem in UWSN, this paper puts forward a new energy-aware routing algorithm, called Energy-Saving Vector Based Forwarding Protocol (ES-VBF). The main purpose of the new routing protocol is saving energy. ES-VBF puts both residual energy and localization information into consideration while calculating desirableness factor. By simulation, it shows that the ES-VBF algorithm increases the residue energy, reduces value of mean square error and prolongs the lifetime of network without worsening the packet reception ratio (PRR) apparently.

Keywords Underwater sensor network · VBF · ES-VBF

B. Wei (✉) · Y. Luo
School of Computer Science and Technology, Tianjin University, No.92 Weijin Road, NanKai District, Tianjin, China
e-mail: ChristinaBo2008@gmail.com

Y. Luo
e-mail: luoyongmei@tju.edu.cn

Z. Jin · Y. Su
School of Electronic Information Engineering, Tianjin University, Tianjin, China

J. Wei
School of Computer Science, Beijing University of Posts and Telecommunications, Beijing, China

W. Lu et al. (eds.), *Proceedings of the 2012 International Conference on Information Technology and Software Engineering*, Lecture Notes in Electrical Engineering 210, DOI: 10.1007/978-3-642-34528-9_10, © Springer-Verlag Berlin Heidelberg 2013

10.1 Introduction

The concept of UWSN was first brought up by the AOSN plan, which was developed by American Institute of marine research and MIT. And it has emerged as a powerful technique for lots of applications for underwater environment, such as measurement, monitoring, control and surveillance [1, 2].

Though there are some similarities between underwater sensor network and terrestrial network, UWSN is significantly different from terrestrial network, for example, limited bandwidth and energy, high attenuation and long propagation delay, high error ratio of signal transmission, difficulty of charging and so on [3]. Because underwater environment is very special, underwater sensor network has some characteristics: large scale, self-organization and dynamics. All the factors mentioned above, especially limited energy, would make designing routing protocol for UWSN an enormous challenge.

There are many underwater sensor network routing protocols, including VBF, HH-VBF [4], ACR, FFBR, QELAR [5] and so on, see more in [6]. Vector Based Forwarding Protocol (VBF) [3] which is put forward by Peng Xie, Jun-Hong Cui from University of Connecticut of America could provide strong, expandable, energy saving routing protocol. VBF is based on position information, and decides whether and when to forward received data packets through calculating desirableness factor.

10.2 Energye Saving Routing Protocol ES-VBF

VBF algorithm only takes location information into consideration, i.e. if it has position advantage to transmit data packet without other reference standards. But when there is a node whose location is always the best in routing pipe, it will be selected again and again during routing selection. Thus its energy will be consumed excessively, while other nodes could not transmit data packet even when they still have enough energy due to the position deviation, even though that is very small. All these lead to network energy being consumed unreasonable and uneven node energy consumption. And we know, this defect in algorithm of routing may lead to the result that nodes with position advantage are selected so many times that the energy of these nodes in particular area may be exhausted, causing routing failure and paralysis of network.

So it is necessary to introduce nodes' energy information into VBF protocol. The routing process considers both factors of position and energy consumption comprehensively. Thus it could reduce energy consumption of network, balance the network energy consumption and prolong the lifetime of network. Figure 10.1 shows the routing process of the improved routing protocol ES-VBF. The dotted boxes are added parts compared to VBF.

Fig. 10.1 Routing process of the improved routing protocol ES-VBF

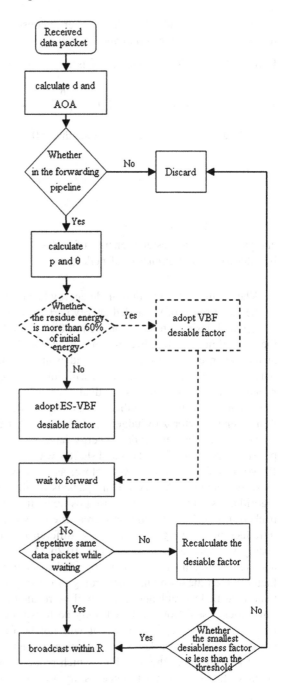

In ES-VBF, we add the value of energy consumption into desirableness factor to decide waiting time. Improved desirableness factor α is got as follows:

Definition if Node residual energy is larger than 60 % of initial energy:

$$\alpha = \left(\frac{p}{w}\right) + \left(\frac{R - d \times \cos\theta}{R}\right) \tag{10.1}$$

if the Node residual energy is smaller than 60 % of initial energy:\

$$\alpha = 0.5 \times \left(1 - \frac{energy}{initialenergy}\right) + \left(\frac{p}{w}\right) + \left(\frac{R - d \times \cos\theta}{R}\right) \tag{10.2}$$

energy   The residual energy of nodes;
initialenergy   Initial energy of nodes;

When using this algorithm, node energy has been consumed to less than 60 %, so the value range of the added energy part is from 0.2 to 0.5. The value range of the original desirableness factor is from 0 to 3. So there is no radical influence on the delay time, which indicates that it is acceptable to balance the network energy at the expense of delay increase. And waiting time $T_{adaptation}$ is inversely proportional to residual energy. Nodes with lots of residual energy will transmit data packet, while ones with less residual energy have to wait. If one node receives same data packets from other nodes during waiting time, it will recalculate desirableness factor α to judge whether to transmit the packet.

Nodes with higher residual energy have smaller desirableness factor, higher priority of forwarding packets and shorter waiting time among neighboring nodes. If two nodes' positions are very close, as is shown in Fig. 10.2a node A and B, these two nodes will forward the packets at almost the same time in the VBF algorithm as a result of their close positions. It is prone to collisions for the two packets at next hop, which also increases energy consumption. However, after the introduction of energy factor, if the difference between two nodes' remaining energy is large, as is shown in Fig. 10.2b (A > B), when calculating the desirableness factor, node A with more residual energy has a smaller desirableness factor, which means higher forwarding priority that waits shorter time to forward than node B. Desirableness factor will be recalculated when node B receives the same packet sent from A. It will only be forwarded when meeting the condition $\min(\alpha_1, \ldots, \alpha_n) < (\alpha_c/2^n)$, if not it will not be forwarded. Generally, the node usually discards the packet when it receives more than two same packets, so as not to cause collision at the next hop, which reduces the times calculating the desirableness factor and avoids unnecessary forwarding energy consumption at the

Fig. 10.2 Analyzing and comparing VBF and ES-VBF

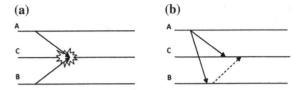

same time. The most important thing is that it reduces the energy consumption of adjacent nodes and avoids nodes forwarding which have less energy but meet the location criterion.

10.3 Simulation Analysis

The simulation uses Aqua-Sim, developed by UCONN, as the simulator. It is An NS-2 based simulator. This paper uses average remaining energy, mean square error of node remaining energy and packet reception ratio as the criteria to analyze the performance of routing protocol.

10.3.1 Energy Analysis

This section analyzes both VBF and the modified routing protocol ES-VBF from nodes in random motion and random distribution. While in the simulation of NS-2, different seeds will generate different results. Then this paper chooses 20 values of seed and makes an average of these obtained results trying to approach the actual results of the network. Parameters of the simulation are shown in Table 10.1:

Table 10.1 Parameter setting

Routing pipe radius (m)	Propagation distance (m)	Scene size (m)
100 Table	100	500 × 500 × 500

10.3.1.1 Static Network

Simulation time: 1000, 2000, 3000, 4000, 5000, 6000, 7000, 8000, 10000 s, initial energy is 150 J, 120 nodes.

As is shown in Figs. 10.3 and 10.4, before 4,000 s, energy consumption of VBF and ES-VBF are close for the reason that there are still more than 60 % of original energy remaining for nodes and the energy factor has not taken effect yet before this time. While after 4,000 s, values of ES-VBF average remaining energy are larger than those of VBF, which indicates that nodes of ES-VBF consume less

Fig. 10.3 Average
remaining energy

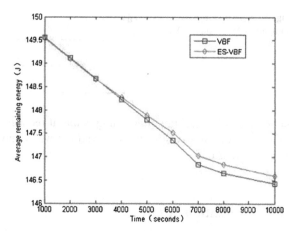

Fig. 10.4 Mean square error
of node remaining energy

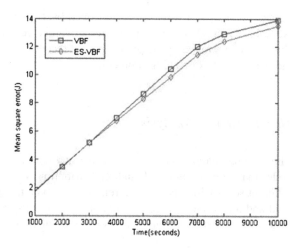

energy and the distribution of energy among nodes is more balanced. What's more, as is shown in Fig. 10.5, the minimum node energy of ES-VBF is bigger than VBF indicating that the use of key nodes is restricted, which prolongs the lifetime of the network.

10.3.1.2 Dynamic Network

Because of the environment's impact, nodes of underwater sensor network will be dynamic, which leads to changes in topology. Therefore, adding movement to the simulation scene to simulate the actual network conditions is better. Considering the actual situation, the node velocity is set from 0.2 to 3 m/s randomly. And other environmental parameters are same with the static simulation.

Fig. 10.5 Minimum node
energy

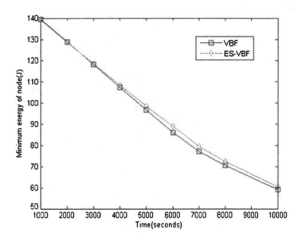

Fig. 10.6 Average
remaining energy

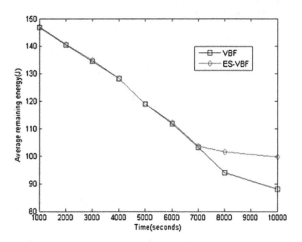

It can be seen from Fig. 10.6, the average remaining energy of nodes in ES-VBF network is slightly larger than the VBF one. While after 6,000 s, the difference becomes larger and shows the sum of energy consumption of nodes in ES-VBF network is smaller than VBF network, which achieves the purpose of saving energy. What's more, as is shown in Fig. 10.7, after 6,000 s, the minimum node energy of VBF network continues falling, however, the trend in ES-VBF network becomes flattened making the survival time of network longer. We got the conclusion that the ES-VBF can reduce the network energy consumption and prolong the network lifetime.

It can be seen from Fig. 10.8, that the energy consumption of nodes is more uniform when using the ES-VBF protocol. This helps to lighten the burden of nodes receiving or forwarding data packets, balancing and saving the network energy consumption, extending the network lifetime.

Fig. 10.7 Minimum node
energy

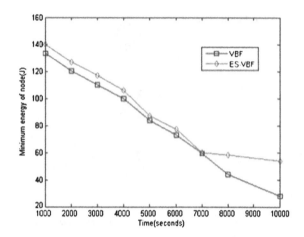

Fig. 10.8 Mean square error
of node remaining energy

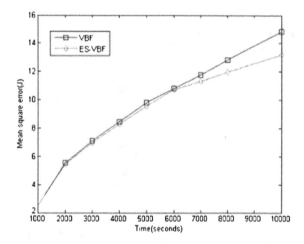

The two curves of the average remaining and mean square error of node remaining energy almost coincide before 6,000 s. It is because that in ES-VBF algorithm, when node energy remained is more than 60 % of the initial it still uses the algorithm of the original VBF protocol. Before 6,000 s the node residual energy does not meet the threshold of the new algorithm, so the curve of ES-VBF agrees with VBF's. But difference between the two curves arises after using the improved algorithm of ES-VBF after 6,000 s.

Relationship of Energy with Data Sending Rate and Number of Nodes

Under the condition of random motion, we change the data sending rate and the number of nodes to analyze the performance of two protocols. The number of

Fig. 10.9 Average
remaining energy

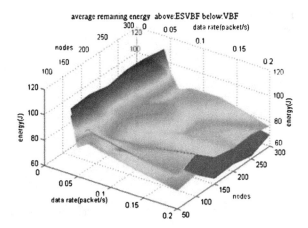

Fig. 10.10 Minimum node
energy

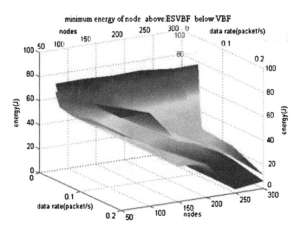

nodes are set to 50, 100, 150, 200, 250, 300. Data sending rate are set to 0.03, 0.04, 0.05, 0.625, 0.083, 0.125, 0.2 packet/s.

It can be seen from Figs. 10.9, 10.10, when changing the packets sending rate and the number of nodes, the average node residual energy and node minimum residual energy of ES-VBF are larger than those of VBF respectively, indicating the validity of the protocol improvements.

10.3.2 Analysis of Packet Reception Ratio

• The figure below is the result of setting node parameter that the initial energy 150 J, 360 nodes, and the simulation time is from 1,000 to 10,000 s:

Packet reception ratio is always an important criterion to measure the network. As is shown in Fig. 10.11, on the one hand, two values of the ratio are basically the

Fig. 10.11 Packet reception ratio

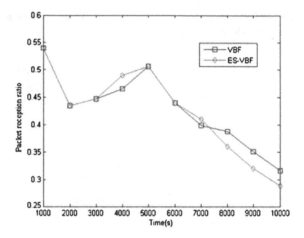

Fig. 10.12 Relationship of packet reception ratio and sending rate

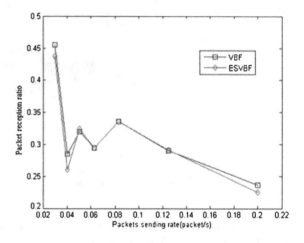

same before 5,000 s. On the other hand, it can also be found that the energy value is not always declining over the time, which is because that nodes' random movement leads to network topology changing.

- The initial energy of node is 120 J, simulation time is 5,000 s. Change the data sending rate and number of nodes to analyze the packet reception ratio of ES-VBF and VBF respectively.

  - Data sending rate: 0.03, 0.04, 0.05, 0.625, 0.083, 0.125, 0.2 packet/s.
  - Number of nodes: 50, 100, 150, 200, 250, 300, 350, 400, 450, 500.

It can be seen in Fig. 10.12, difference of packet reception ratio between VBF and ES-VBF is small, and even during a period two curves almost overlap. From the trend it can be seen that after 0.125 packet/s, if we continue to increase data sending rate value, the reception ratio begins to fall, for that the packet reception ratio is so large that it causes serious packets collisions leading to a large number of packets discarded. So we should appropriately select the data sending rate to obtain high packet reception ratio.

10.4 Conclusion and Prospect

To solve the energy problem in UWSN, this paper puts forward a new energy-aware routing algorithm, called Energy-Saving Vector Based Protocol (ES-VBF). The main purpose of the new routing protocol is saving energy. ES-VBF puts both residual energy and localization information into consideration while calculating desirableness factor, which allows nodes to weigh the benefit for forwarding packets. By simulation, it shows the promising performance in balancing network energy consumption and packet reception ratio.

The proposed ES-VBF routing protocol saves energy efficiently. At the same time, there's a small decline in packet reception ratio, which needs further research aimed at finding a better solution not only reducing energy consumption but also achieving high packet reception ratio.

Acknowledgments This paper is supported partly by NSFC under No. 61162003 and by Qinghai Science Hall under No. 2012—ZR—2989. And thanks to Miss YI Liya for her simulation work.

References

1. Heidemann J, Li Y, Syed A, Wills J, Ye W (2005) Underwater sensor networking: research challenges and potential applications.USC/ISI technical report ISI-TR-2005-603
2. Heidemann J, Stojanovic M, Zorzi M (2012) Underwater sensor networks: applications, advances, and challenges. Philos Trans R Soc A 370:158–175. doi: 10.1098/rsta.2011.0214
3. Xie P, Cui J, Lao L (2006) VBF: vector-based forwarding protocol for underwater sensor networks. Networking 2006. Networking technologies, services, and protocols; performance of computer and communication networks; mobile and wireless communications systems, pp 1216–1221
4. Nicolaou N, See A, Xie P et al (2007) Improving the robustness of location-based routing for underwater sensor networks. In: OCEANS 2007-Europe, 1–6
5. Hu T, Fei Y (2008) QELAR: a q-learning-based energy-efficient and lifetime-aware routing protocol for underwater sensor networks. In 27th IEEE Inter Performance Comput Communi Conference, Austin
6. Ayaz M, Baig I, Abdullah A, Faye I (2011) A survey on routing techniques in underwater wireless sensor networks. J Netw Comput Appl 34:1908–1927

Chapter 11
The Enhancements of UE Mobility State Estimation in the Het-net of the LTE-A System

Haihong Shen, Kai Liu, Dengkun Xiao and Yuan He

Abstract With the introduction of Low Power Node (LPN) into Long Term Evolution-Advanced (LTE-A) system, the network structure of the mobile communication system becomes more complex in the Heterogeneous Network (Het-net). Thus the User Equipment (UE) in the Het-net has new characteristics, which greatly increases the difficulty and complexity of the accurate UE Mobility State Estimation (MSE), and poses a challenge for traditional MSE algorithm. In this paper, the disadvantage of the traditional MSE algorithm has been discussed in the Het-net of LTE-A system. Furthermore, a new enhanced algorithm is introduced based on the traditional MSE algorithm to improve the UE MSE accuracy.

Keywords Het-net · LTE-A · MSE · Handover

Supported By Open Fund (NO.0808) of Key Laboratory of Geo-detection (China University of Geosciences, Beijing), Ministry of Education.

H. Shen (✉) · K. Liu
School of Information Engineering, China University of Geosciences, Beijing, China
e-mail: haihong_shen@163.com

K. Liu
e-mail: kailiu1987@163.com

D. Xiao · Y. He
Huawei Technologies Co. Ltd, Beijing, China

W. Lu et al. (eds.), *Proceedings of the 2012 International Conference on Information Technology and Software Engineering*, Lecture Notes in Electrical Engineering 210, DOI: 10.1007/978-3-642-34528-9_11, © Springer-Verlag Berlin Heidelberg 2013

11.1 Introduction

With the development of mobile communication technology, the velocity of the
User Equipment (UE) that wireless communication systems can support becomes
increasing high. When entering into the long term Evolution (LTE) stage, the
requirements of the speed have reached up to 350 km/h. We know that in order to
enable UE to have a sustainable call when moving, the network must achieve
large-scale coverage. UE will be able to complete several mobility processes,
including cell selection and cell reselection in the IDLE state, and handover in the
RRC-CONNECTED state. In the mobility process UE is configured of some
network parameters. In order to get the optimal mobility performances when UE is
under the different speed in the network, the configuration of these parameters
should be adjusted accordingly.

Therefore, if the network can get the mobile state of UE, it will greatly improve
the mobile performances. In UMTS system, so as to achieve the purpose and
optimize the mobility performances, the mobility state is divided into 2 states. One
is the normal-mobility state and the other one is the high-mobility state. Further in
the LTE system, due to the support to the high speed scenario, the mobility state is
divided into 3 states, including normal- mobility state, medium-mobility state and
high-mobility state.

When developing into the LTE-A system, in order to increase the coverage rate
of high speed data, temporary network coverage and cell edge throughput, the
Heterogeneous Network (Het-net) is introduced with a mix deployment of macro
and low power nodes (LPN), which consist of Pico, H-eNB and Relay.

In heterogeneous network which is shown in Fig. 11.1, Macro cell is regarded
as the initial deployment, and Pico cell, H-eNB and Relay are added for incremental capacity growth, richer user experience and in-building coverage.

In the UMTS system and LTE system, an accurate of UE mobility state estimation (MSE) is easy to achieve through the existing MSE algorithm. However, it
may be not suitable in the Het-net of LTE-A system.

The mobility state estimation specified for 3GPP Release 8 is studied in [1]. The
scheme parameterization has been studied in varying conditions, such as the
number of LPN in macro cell. In this paper, simulation results of the proposed
scheme show that it is an effective solution to evaluate UE mobility state with
other methods in LTE-A system.

The remaining of this paper is organized as follows: Sect. 11.2 introduces
existing MSE algorithm and its disadvantages in Het-net. Section 11.3 describes
the enhanced MSE. Simulation results are provided in Sect. 11.4. Finally, conclusions are drawn in Sect. 11.5.

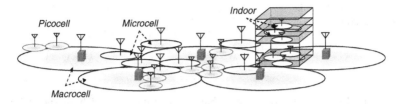

Fig. 11.1 Het-net deployment

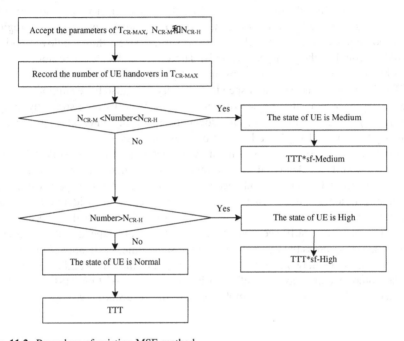

Fig. 11.2 Procedure of existing MSE method

11.2 Existing MSE

In the section, firstly existing MSE algorithm is described, and then its disadvantage in Het-net is elaborated.

Existing Mobility State Evaluation (MSE) algorithm in the UE side has a detailed introduction of the RRC_CONNECTED state and IDLE state in [2]. In this paper, the MSE algorithm is based on the RRC_CONNECTED state.

Traditional MSE has been expatiated in [3], and its core idea has several steps. The procedure is shown in Fig. 11.2.

Firstly, accept the parameters of T_{CR-max}, N_{CR-M} and N_{CR-H} from the network. Secondly, record UE's handover numbers in a specific period of time (T_{CR-max}). Thirdly, compare to a defined threshold in advance (N_{CR-M}, N_{CR-H}) to determine

Fig. 11.3 UE's mobility in different scenarios

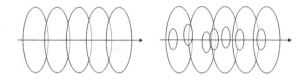

the UE mobility state. Lastly, reduce handover failure rate of high speed UE by scaling down TTT according to different states.

The existing MSE is in fact a method to reduce handover failure rate of high speed UE by scaling down TTT to expedite the handover process. Theoretically, TTT should be scaled to fit the time interval one UE spends in a handover region.

In homogeneous network, all cells have similar sizes, thus handover regions would be roughly the same across cells. The effect of handover region size can then be taken into account to set an initial TTT, and the UE can scale the initial TTT only based on its estimated speed. As history cells also have similar sizes and are all deployed with its unique coverage, UE evaluate its own speed by simply considering the number of history handovers during a time interval. In a word, the existing MSE would work well in homogenous network.

However, cell sizes vary significantly in heterogeneous network (Het-net). While some cells are deployed for coverage extension, many more are overlaid on top of other cells for capacity enhancement. In that scenario, UE speed estimation becomes problematic if only history handover numbers are considered, but history cell types are not taken into account [4, 5]. Comparison with the two charts in Fig. 11.3, we can clearly see that, if the UE at the same speed and in the same direction going across the two networks at the same time, the handover numbers in the heterogeneous network (right) will be far greater than the numbers performed in the homogeneous network (left). So in Het-net, the existing MSE technology has been unable to estimate the mobility state of UE accurately.

11.3 The Proposed Method

The reason that Existing MSE technology can estimate the UE mobility state in homogeneous network precisely is the coverage of each service area is basically the same, so you can plot the number of handovers after a period of time to estimate the mobility state of the UE. However, because the low-power cells are added into the heterogeneous network, we can no longer simply use the number of handovers by the UE after a period of time to estimate the UE mobility state.

In order to deal with this issue, a new method based on conventional MSE is dedicated in [6]. The solution is to count the handovers according to the cell type of the same or different macro coverage and the absolute weights of one or zero. If the target cell and source cell are deployed in the different macro coverage, then the handover is counted and the related weight equals to one. If the target cell and

Fig. 11.4 Procedure of
enhanced MSE

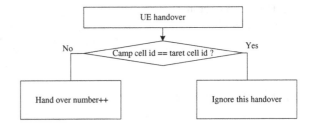

source cell are deployed in the same macro coverage, then the handover is not
counted, i.e., the corresponding weight equals to zero.

Figure 11.4 shows the detail process of the proposed MSE method and the step
is as follows: If the target cell and source cell are deployed in the different macro
coverage, then the handover is counted and the related weight equals to one. If the
target cell and source cell are deployed in the same macro coverage, the handover
is ignored and the corresponding weight equals to zero.

In Het-net scenario, there are several handover situations in Fig. 11.5 of [7],
from the proposed method, we can ignore these situations. Then the mobility state
of UE can be estimated accurately.

11.4 Simulation and Performance

This section provides simulation results for the considered scenarios.

In this section, the detailed simulation assumption is described and simulation
results of the proposed scheme compared with the existing MSE method are
presented. The analysis of simulation results show that the proposed scheme have
a great advantage over the existing MSE.

11.4.1 Simulation Scenario

In this section we simulate the proposed scheme compared with the existing MSE
method with two LPNs and four LPNs per macro cell, and the simulator is
described in detail.

The 19 (sites) * 3 (sectors) topology is used in the simulation. In order for easily
understanding, the network topology is shown as Fig. 1.6. The site indexes are
illustrated, and the bound of sectors is denoted by the dash line.

All the detailed simulation parameters for existing MSE method and proposed
method are shown in the Table 11.1. Other simulation parameters and assumptions
follow the large area system simulation agreements in 3GPP TR 36.839. In this
simulation, inter-site distance (ISD) is 500 m. Carrier frequency is assumed to be
2 GHz according to LTE protocols. There are two carrier bandwidths to be

Fig. 11.5 Several situations
should be ignored

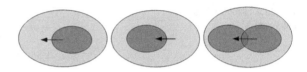

Fig. 11.6 Network topology

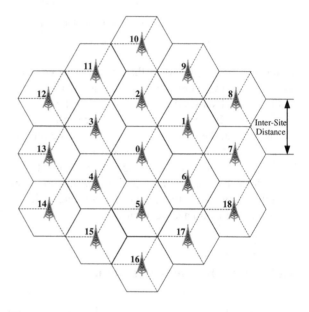

Table 11.1 Simulation assumptions

Items	Description	Item	Description
Cell layout	Hexagonal grid, wrap around	Configuration parameter set	Set 3
Number of sites	19 sites, with 3-sectored antennas at each site	Pico cell placement	Random placed according to 3GPP TR 36.814
ISD	500 m	Pico number per macro	2,4
Channel model	SCM[TR 25.996]	UE speed	30 km/h
sf-High	0.25	sf-Medium	0.5
t-Evaluation	30 s	t-HystNormal	30 s
N_{CR-M}	3	N_{CR-H}	5

considered, which are 3 and 1.4 MHz. The scaling factor for high-mobility state
(sf-High) and medium-mobility state (sf-Medium) is 0.25 and 0.5. The evaluating
duration to enter high-or medium-mobility state (t-Evaluation) and normal-
mobility state (t-HystNormal) is 30 s. The handover number to enter medium-
mobility state (N_{CR-M}) and high-mobility state (N_{CR-H}) is 3 and 5.

Fig. 11.7 Mobility state of
UE with 2 picos per macro

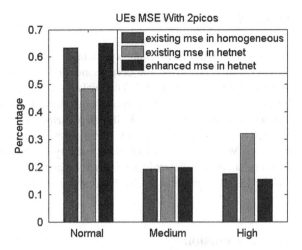

Fig. 11.8 Mobility state of
UE with 4 picos per macro

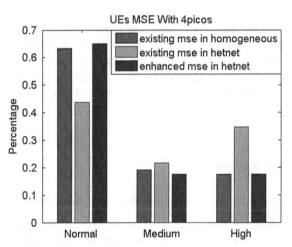

Table 2 Comparison among different method

Solutions	Normal (%)	Medium (%)	High (%)
Existing MSE in homogeneous	63.33	19.12	17.54
Existing MSE in Het-net	48.42	19.65	31.93
Enhanced MSE in Het-net	64.91	19.65	15.44

Table 11.3 Comparison among different method

solutions	Normal (%)	Medium (%)	High (%)
Existing MSE in homogeneous	63.33	19.12	17.54
Existing MSE in Het-net	43.68	21.58	34.74
Enhanced MSE in Het-net	65.07.	17.44	17.49

11.4.2 Simulation Results

The UE distribution of mobility states in different schemes are showed in Figs. 11.7 and 11.8. And the corresponding values of the different solutions are listed in Tables 11.2 and 11.3. The evaluation of high, medium and normal states for the existing MSE in the homogeneous network is the baseline. It can be seen that when using the existing MSE, the distribution of mobility states in Het-net drastically changes and is quite different from the baseline. Furthermore, with the enhanced MSE it is shown that MSE stability can be improved and distribution of mobility states can be maintained as the baseline. Thus the enhanced MSE has advantages of improving the MSE accuracy in Het-net.

11.5 Conclusion

In this paper, the existing MSE algorithm and its disadvantages in Het-net are discussed. Also the principle of proposed method has been shown and the existing MSE of current LTE system are provided as a comparison. Two methods are used at the same time to estimate the mobility state of UE. The simulation results obtained show that the proposed method can improve the accuracy of UE MSE in the Het-net.

References

1. Puttonen J, Kolehmainen N, Henttonen T, Kaikkonen J On idle mode mobility state detection in evolved UTRAN
2. 3GPP TS 36.331 Radio resource control (RRC); protocol specification
3. 3GPP TS 36.304 User equipment (UE) procedures in idle mode
4. R2-114316 Enhancements for UE mobility state estimation. Nokia Siemens Networks, Nokia Corporation
5. R2-120736 Discussion on the scenario for UE mobility state estimation. Fujitsu
6. R2-121250 Further evaluation on enhancements of mobility state estimation in Het-net. Huawei
7. R2-114950 Discussion on the mobility performance enhancement for co-channel Het-net deployment. ZTE

Chapter 12
Efficient Authentication Based on Orthogonality and Timeliness for Network Coding

Ming He, Lin Chen, Hong Wang, Zhenghu Gong and Fan Dai

Abstract Network coding provides a new data transmission paradigm for network protocols. However, the packets-mixture nature makes network coding systems notoriously susceptible to pollution attacks. Previous security solutions will incur high computation and transmission overhead, and they will be worse when facing malicious collusion attacks. In this paper, we propose an efficient authentication scheme, which allows participating nodes to efficiently detect corrupted packets based on orthogonality and timeliness. Our work is the first orthogonality and timeliness based solution to frustrate pollution attacks with arbitrary collusion. The security analysis shows that polluted packets can pass our authentication scheme with a very low probability. We also present simulations of our scheme, and results demonstrate the practicality and efficiency of our scheme.

Keywords Network coding · Security · Pollution attack · Collusion

12.1 Introduction

Network coding was first introduced by Ahlswede et al. [1]. In order to achieve the maximum transmission capacity, network coding allows intermediate nodes to perform coding operations on packets rather than just forward them. With lots of benefits for communication network, network coding has emerged as a promising technique to reduce network congestion [2] and improve the robustness [3].

M. He (✉) · L. Chen · H. Wang · Z. Gong · F. Dai
College of Computer, National University of Defense Technology, Changsha, Hunan, China
e-mail: hemingclear@nudt.edu.cn

W. Lu et al. (eds.), *Proceedings of the 2012 International Conference on Information Technology and Software Engineering*, Lecture Notes in Electrical Engineering 210, DOI: 10.1007/978-3-642-34528-9_12, © Springer-Verlag Berlin Heidelberg 2013

However, because the nature of packet mixture, network coding systems become more susceptible to pollution attacks, where malicious nodes inject corrupted packets into the information flow. So far, a number of schemes have been proposed to address pollution attacks in network coding systems. These schemes can be categorized into two classes.

- *End-to-end schemes* [4, 5]. Polluted messages are detected only at receivers.
- *In-network schemes* [6, 7]. Detection is at all participating nodes.

An end-to-end scheme makes minimal changes to existing network coding algorithms, and only the source and receivers are involved in the detection of polluted packets. However, when adversaries locate their attacks at the bottleneck of network, end-to-end schemes can lead to a worst-case view of adversarial attack. In-network solutions introduce cryptographic approaches by which intermediate nodes verify the integrity of incoming packets. However, there is no practical in-network scheme so far. Previous solutions either provide only a partial solution without enough security ability [4, 5, 7], or result in notable degradation to the available bandwidth [6]. In addition, previous solutions to the problem will be worse when facing arbitrary collusion among malicious nodes.

In this paper, we propose an efficient authentication scheme, which allows participating nodes to efficiently detect corrupted packets based on orthogonality and timeliness of network coding. Our scheme differs from previous schemes in the following primary features.

- *Security against pollution attacks*. It can effectively resist pollution attacks.
- *Low transmission overhead*. Our scheme requires a smaller number of validation packets and it adds no extra bit to source blocks.
- *Low computation overhead*. Our scheme detects corrupted packets based on orthogonality and timeliness to achieve computationally efficient.

The remainder of the chapter is structured as follows. Section 12.2 provides the detail of our scheme. Section 12.3 gives the security analysis of our scheme, while Sect. 12.4 shows simulations of our scheme. Finally, Sect. 12.5 concludes the paper.

12.2 Authentication Based on Orthogonality and Timeliness

We consider a general network coding system consisting of a single source S, multiple receivers $R_1, R_2, \ldots, R_k$, and forwarders for packets. Receivers may also act as forwarders. S divides a sequence of messages into generations, and only messages from the same generation are encoded. Each generation consists of n blocks, and each represented by m elements from the finite field F_q, where q is a prime number. Each block p_i can be viewed as an element in an m dimensional vector space over the field F_q:

$$\vec{p}_i = (p_{i1}, p_{i2}, \ldots, p_{im}), \quad p_{ij} \in F_q, 1 < j < m. \tag{12.1}$$

A generation G consisting of n packets can be viewed as a matrix, with each packet in the generation as a column in the matrix:

$$G = [\vec{p}_1, \vec{p}_2, \ldots, \vec{p}_n]^T. \tag{12.2}$$

The source S forms random linear combinations of native packets $\vec{e} = \vec{c}G = \sum_{i=1}^{n} c_i \vec{p}_i$, where $\vec{c} = (c_1, c_2, \cdots, c_n)$ and $c_i (1 \leq i \leq n)$ is a random element in F_q. The source then forwards packets consisting of $(\vec{c}, \vec{e})$ in the network. We refer to $(\vec{c}, \vec{e})$ as coded packets, and $\vec{c}$ as the global encoding vector. A forwarder node also forms new coded packets by computing linear combinations of linearly independent coded packets it has received and forwards them to the network. When a receiver has obtained n linearly independent coded packets, it can decode them to recover native packets by solving a system of n linear equations. We denote blocks which S is about to disseminate to receivers as X, a $n \times (m + n)$ matrix whose ith row is $\vec{x}_i (1 \leq i \leq n)$.

In the packet distribution of network coding, the linear subspace Π_X spanned by valid packets stays constant. We characterize Π_X with vector $\vec{u}$ randomly chosen from its orthogonal space $\Pi_X^{\perp}$, and participating nodes can check the integrity of a received packet w by verifying whether $w \cdot \vec{u}^T = 0$. This orthogonality principle inspires some security schemes [7, 8]. However, the satisfaction of orthogonality principle does not imply strictly valid of received packets. If malicious nodes collude to obtain orthogonal vectors collected by its neighbors, they can easily find a corrupted vector that satisfies the orthogonality principle.

To solve this problem, we propose an efficient scheme where participating nodes verify packets basing on orthogonality and timeliness of network coding, and the security relies on the asymmetry of time in message authentication.

12.2.1 Parameter Initialization

We note that $X = \{\vec{x}_1, \vec{x}_2, \ldots, \vec{x}_n\}$ span a subspace Π_X. The orthogonal subspace $\Pi_X^{\perp}$ is spanned by orthogonal vectors $U = \{\vec{u}_1, \vec{u}_2, \ldots, \vec{u}_m\}$, and U can be denoted as a $m \times (m + n)$ matrix [7].

- F_q^*: F_q^* is the multiplicative group of field F_q, i.e., $F_q^* = F_q - \{0\}$. F_q^* is a cycle group, and the order of F_q^* is $q - 1$.
- k: k is the generator of F_q^*. i.e., $k \in F_q^*$ and $F_q^* = \{k^j \mid j \in Z\}$.
- Z_q: Z_q is the residue class ring modulo q, i.e., $Z_q = \{0, 1, \cdots, q - 1\}$. The characteristic of F_q is q, so F_q is an extension field of Z_q, denoted as $Z_q \subset F_q$. We can see that $Z_q \subset Z$.

12.2.2 Source Setup

Let G be an active generation, and the source needn't change the mode of sending native data packets. To detect pollution packets, the source periodically computes and disseminates a validation packet U', and we denote each U' as a new version validation packet. The source ensures the authenticity of the validation packet by standard signature scheme [8], e.g., DSA. The detail is as follows.

- The source use Gaussian elimination to find a vector $\vec{u}$ matching $X \cdot \vec{u}^T = 0$. We denote that $\vec{u} = \{u_1, u_2, \ldots, u_{m+n}\}$.
- The source generates a set of nonzero random elements $B = \{b_1, b_2, \ldots, b_{m+n}\}$ in Z_q. The multiplicative inverse of B is denoted as $B^{-1} = \{b_1^{-1}, b_2^{-1}, \ldots, b_{m+n}^{-1}\}$, where $b_i \cdot b_i^{-1} = 1 (1 \leq i \leq m + n)$. Both B and B^{-1} are only kept by the source.
- The source computes validation vector $A = \{a_i = k^{b_i}\}$ $(1 \leq i \leq m + n)$.
- The source computes new orthogonal vector $D = \{d_i = u_i \cdot b_i^{-1}\}$ $(1 \leq i \leq m + n)$.
- The source forms a validation packet $U' = (A, D, c\_time)$, and c_time is the timestamp at the source when U' is created. The source signs U' with standard signature scheme and broadcasts U'.

12.2.3 Packet Verification

Each participating node maintains two packet buffers, verified-buffer and unverified-buffer, which buffer verified and unverified packets, respectively. Each node combines only packets in the verified-buffer to form new packets and forwards such packets. On receiving a new coded packet, a node buffers the packet into unverified-buffer and records the corresponding received time. We denote the data form of both verified-buffer and unverified-buffer as $(buf\_flag, rec\_time, data)$.

Once receiving a validation packet U', a participating node first verifies that U' is signed by the source. If U' is authentic, the node publishes it to its neighbors.

The node uses $U' = (A, D, c\_time)$ to verify those packets in unverified-buffer that were received before U' was created at the source. i.e., $rec\_time \leq c\_time - \Delta$. Δ is the maximum clock difference between the node and the source. The node verifies the integrity of corresponding data block w by checking if it holds the following equation:

$$\prod_{i=1}^{m+n} a_i^{d_i \cdot w_i} = 1 \qquad (12.3)$$

Equation (12.3) holds for any valid w, we have:

$$\prod_{i=1}^{m+n} a_i^{d_i \cdot w_i} = \prod_{i=1}^{m+n} \left(k^{b_i}\right)^{\left(u_i \cdot b_i^{-1}\right) \cdot w_i} = \prod_{i=1}^{m+n} k^{u_i \cdot w_i} = k^{\sum\limits_{i=1}^{m+n} (u_i \cdot w_i)} = k^0 = 1$$

This is simple computation that allows participating nodes to rapidly perform the verification. Valid data blocks are transferred from unverified-buffer to verified-buffer. Data blocks that do not pass the verification are discarded.

Validation packets are not required to be delivered reliably. If a node fails to receive a current version validation packet, it can verify its buffered packets upon the receipt of the next version validation packet. When a receiver receives enough linear independent coded packets that have passed the packet verification, it decodes these coded packets to recover native packets. It verifies native packets using an end-to-end authentication scheme such as traditional digital signature or message authentication code before passing packets to the upper layer protocol. The additional end-to-end authentication is used for addressing the extremely rare occasion when some polluted packet pass our verification at the receiver, which would otherwise cause incorrect packets to be delivered to the upper layer.

12.3 Security Analysis

Recall that validation packets are protected by standard signature scheme, thus the attacker can not inject forged validation packets into the network. The only option left for the attacker is to generate corrupted packets that will pass the verification at honest nodes.

Theorem 1 *Let* $U' = (A, D, c\_time)$ *to be a current version validation packet in generation* G, *and let* w' *to be a polluted packet. The probability that* w' *successfully passes the packet verification of* U' *is at most* $\frac{1}{q^{m+n}}$.

Proof In our scheme, participating nodes use the current version validation packet to verify only packets that are distributed by source node before the creation of those validation packets. Therefore, unlike packets authentication with pure orthogonality principle where the attacker has a chance to find a corrupted vector that satisfies the orthogonality principle, in our scheme, the time asymmetry in packets verification prevents the attacker from computing a suitable polluted packet that will pass the verification algorithm. At best, the attacker can randomly guess the upcoming new version verification value in participating nodes. The number of elements in verification value is $m + n$, so the probability that the attacker successfully guess the upcoming new version verification value is $\frac{1}{q^{m+n}}$, and the probability that a polluted packet successfully passes the packet verification of our scheme is at most $\frac{1}{q^{m+n}}$. $\square$

We should notice that the verification in our scheme will not fail to verify legitimate packets. A packet can be verified using multiple versions of validation packet to further reduce the false negative probability as long as the packet is safe with respect to the newer version of verification packet. Furthermore, the failure of any verification indicates that the packet is polluted. However, our simulation shows that our verification with only one verification value is already sufficient. We should also notice that each verification packet is generated independently with random coefficient. On this premise, it will not help attackers in launching successful attacks that they obtain multiple old version verification packets.

Malicious attackers may attack our scheme itself by preventing participating nodes from receiving verification packets. Because verification packets are flooded among all participating nodes, if the security situation is worse to that attackers are always able to prevent a node from receiving verification packets, attackers can also isolate the node by dropping all data packets as dropping verification packets. The corresponding solutions against verification packets dropping as [9] can be applied to improve the security and it is out of the scope of this paper.

12.4 Simulations

In this section, we study the performance of our scheme. We also compare it with cooperative security scheme introduced by Gkantsidis et al. [7] and null keys scheme introduced by Kehdi et al. [10]. We use the percentage of corrupted nodes as a metric to measure the efficiency of security schemes with network coding. In the simulations, the network consists of 1,000 nodes, and the block size is set to 128. The topology is a directed random graph with one source node. The simulator is round-based, where in each round a node can upload and download blocks. In each round, the malicious node send one polluted message on each of their outgoing links. We randomly choose malicious nodes from the population, and all the results are averaged over several runs.

Figure 12.1 shows that our scheme succeeds in improving the protection against pollution attacks. We should notice that the corruption decreases as we increase the checking probability in cooperative security, but the network performance degrades due to higher computational complexity. A checking probability of 30 % will impose a significant computation overhead in cooperative security. In homomorphic hashing model, a node stops using unsecured blocks when an alert message is received. This is another drawback that decreases the network performance, since non-corrupted blocks can be part of these unsecured blocks. In null keys scheme, we mentioned that increasing the percentage of malicious nodes obviously expands the percentage of corrupted nodes, which shows that null keys scheme become vulnerable when facing multiple attackers. In our scheme, we see that across different percentages of malicious nodes, the percentage of corrupted nodes only increases slightly compared to other solutions.

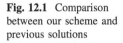

Fig. 12.1 Comparison between our scheme and previous solutions

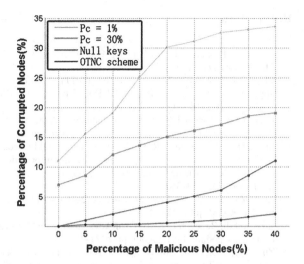

Our scheme can effectively defend against pollution attack with arbitrary collusion among multiple attackers.

12.5 Conclusion

In this paper, we present a computationally efficient security scheme against pollution attacks based on orthogonality and timeliness. Our scheme nicely makes use of the linearity property of random linear network coding, and enables participating nodes to check the integrity of packets without the requirement for a secure channel. Packets authentication is based on the orthogonality principle and the hardness of the discrete logarithm problem, and protection against pollution attacks with collusion is based on the timeliness of orthogonal vectors.

We argue that orthogonality and timeliness of network coding can limit pollution attacks of malicious nodes. Verification packets are signed using standard signature scheme and the impact of signature can be optimized by adjust the time interval of releasing verification packets. Our work is the first orthogonality and timeliness based solution to frustrate pollution attacks with arbitrary collusion and tag pollution. Our simulations show that our scheme effectively limits the pollution and isolates malicious nodes. The stable corruption spread, helps in identifying the locations of the malicious nodes even facing pollution attacks with collusion.

Acknowledgments This work is supported by Program for Changjiang Scholars and Innovative Research Team in University(No.IRT1012), "Network technology" Aid program for Science and Technology Innovative Research Team in Higher Educational Institutions of Hunan Province, Hunan Provincial Natural Science Foundation of China (11JJ7003), "Reconfigurable Network Emulation Testbed for Basic Network Communication" Program for National Basic Research Program of China (973 Program).

References

1. Ahlswede R, Cai N, Li S-Y, Yeung R (2000) Network information flow. IEEE Trans Inf Theory 46:1204–1216
2. Chung K-C, Chen H-C, Liao W (2011) Congestion-aware network-coding-based opportunistic routing in wireless Ad Hoc networks. In: IEEE global telecommunications conference (GLOBECOM 2011), pp 1–5
3. Hnin Yu S, Adachi F (2011) Power efficient adaptive network coding in wireless sensor networks. In: IEEE international conference on communications pp 1–5
4. Silva D, Kschischang FR (2009) Universal weakly secure network coding. In: IEEE information theory workshop on networking and information theory, pp 281–285
5. Koetter R, Kschischang FR (2009) Coding for errors and erasures in random NC. IEEE Trans Inf Theory 54:3579–3591
6. Li Y, Yao H, Chen M, Jaggi S, Rosen A (2010) RIPPLE authentication for network coding. In: INFOCOM, pp 2258–2266
7. Elias Kehdi, Baochun Li (2009) Null keys limiting malicious attacks via null space properties of network coding. In: Infocom, pp 1224–1232
8. Zhao F, Kalker T, Medard M, Han KJ (2007) Signatures for content distribution with network coding. In: IEEE international symposium on information theory (ISIT), pp 556–560
9. Mahmoud ME (2011) An integrated stimulation and punishment mechanism for thwarting packet dropping attack in multihop wireless networks. IEEE Trans Veh Technol 60:3947–3962
10. Gkantsidis C, Rodriguez P (2006) Cooperative security for network coding file distribution. In: INFOCOM, pp 1–13

Chapter 13
Modeling of Foreign Objects Debris Detection Radar on Airport Runway

Dandan Ao, Xuegang Wang and Hong Wang

Abstract Based on the fixed environment of airport runway where foreign objects debris detection radar on airport runway works on, geometric model is established by using method of coordinate system transformation. Meanwhile, approximation of model parameters is used in order to simplify the calculation. This paper analyses echo signal to establish the model of corresponding target signal and uses statistical model to produce clutter. As the main source of clutter is from meadow and runway, the backscatter coefficient of the composite scene is calculated combine with geometric model. Clutter of random sequences and echo is simulated based on MATLAB and the result is given.

Keywords Foreign objects debris (FOD) detection · Millimeter-wave radar · Geometric model · Echo signal

13.1 Introduction

Airport runways foreign object debris (FOD, Foreign Object Debris abbreviation) mean foreign objects that don't belong to airport but may appear in the operation of the airport area and cause loss or harm to the aircraft [1]. Harm brought by FOD not only damage aircraft and take away precious lives, but also accompanied by economic losses. Currently most airports use the approach of artificial visited regularly, which not only time-consuming taking up valuable time of runway use

D. Ao (✉) · X. Wang · H. Wang
University of Electronic Science and Technology of China (UESTC), No.2006 Xiyuan Road, High-tech Western District, Chengdu, Sichuan, China
e-mail: add8825@163.com

W. Lu et al. (eds.), *Proceedings of the 2012 International Conference on Information Technology and Software Engineering*, Lecture Notes in Electrical Engineering 210, DOI: 10.1007/978-3-642-34528-9_13, © Springer-Verlag Berlin Heidelberg 2013

Fig. 13.1 FOD radar
geometry relationship
diagram

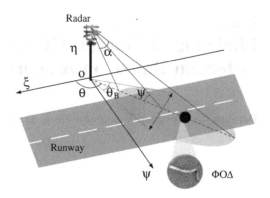

and led to security risks subject to weather and other factors. Therefore, effective and reliable monitoring of FOD is a problem need to be solved urgently.

Since the Concorde Tragedy in 2000, many countries have started to develop FOD detection system. View from several reported FOD monitoring system [2–4], radar monitoring systems for FOD mainly use millimeter-wave radar and commonly use frequency modulated continuous wave (FMCW) radar system.

FOD radar in this paper is fixed radar to scan the land and land clutter is the main factor to affect its detection performance [5]. Therefore, the radar clutter model plays an important role in assessment of FOD radar work performance.

This paper discusses FOD detection radar model. According to the specific work environment of radar, geometric model is established by using coordinate system transformation method. Then echo signal model is discussed, including target echo, clutter and noise, while clutter is focus. The backscatter coefficient of the composite scene is calculated combining with geometric model according to grass-runway composite scene. The paper uses statistical model to produce clutter and discuss its probability density function and power spectral density function.

13.2 Geometric Model

FOD detection radar is land search radar working in airport runway. Its geometric relationship is shown in Fig. 13.1. Taking the shadow of radar on the ground, O, as the origin, the coordinate system is established as shown. In Fig. 13.1, green oval area is land area of radar beam irradiation, h is the height of radar, θ_B is radar beam azimuth width, α is the width of radar beam tilt, Ψ is grazing angle, θ is antenna scan angle. Figure 13.2 is a ground top view. The oval area is formed by two semi-elliptical that have same elliptic short half shaft and different elliptic semi-major axis.

The installation mode of radar is to install several radar sensors along runway. The installation position depends on the advisory bulletin released by the FAA AC 150/5220-24: Sensors are located 50.0 m or more from the runway center line [6].

Fig. 13.2 Ground top view

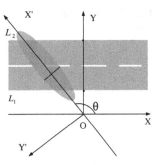

As radar is mounted height limited, thus glancing angle Ψ is very small, such as $\Psi = 2.29°$ when radar is located 50 m from the runway in height of 2 m. Radar only rotated in the horizontal orientation and does not move in the pitch orientation, which means that h, θ_B, α, Ψ would not change, only the antenna scan angle θ varies with time. Therefore, the land area of the radar beam irradiation is fixed.

Coordinate transformation can be used to obtain the elliptic equation and its parameters easily. As shown in Fig. 13.2, when coordinates transformed from XOY into X'OY', origin is unchanged and the straight line connecting the beam center with the origin O will be regarded as the X axis, corresponding to the rotate the XOY coordinate system counterclockwise by angle of θ to obtain X'OY' coordinates. First directly obtain the equation of the ellipse in coordinates X'OY', as following

$$\begin{cases} (x' - x_0)^2/a_1^2 + y'^2/b^2 = 1, & x' \le h/\tan\psi \\ (x' - x_0)^2/a_2^2 + y'^2/b^2 = 1, & x' > h/\tan\psi \end{cases} \tag{13.1}$$

Where $a_1 = h/\tan\psi - h/\tan(\psi + \alpha/2)$, $a_2 = h/\tan(\psi - \alpha/2) - h/\tan\psi$, $b = h\tan(\theta_B/2)/\sin\psi$, $x_0 = h/\tan\psi$ are obtained from geometric relationship.

As radar is installed with height limited and near the runway, α and Ψ would be very small. Thus a_1, a_2, b, x_0, all can be $x_0 = h/\psi$ $a_1 = h/\psi - h/(\psi + \alpha/2)$, $a_1 = h/(\psi - \alpha/2) - h/\psi$, $b = h\tan(\theta_B/2)/\psi$, here α and Ψ use radians unit. As θ_B takes $5°$ and α takes $3°$, parameters error analysis is shown in Fig. 13.1. It can be seen that approximation of parameters is practicable (Fig. 13.3).

Thus, coordinates X'OY' can be directly converted to the coordinate system XOY. Coordinate transformation is shown in Fig. 13.2. Transformation formula is

$$\begin{cases} x' = x\cos\theta + y\sin\theta \\ y' = y\cos\theta - x\sin\theta \end{cases} \tag{13.2}$$

Formula (13.2) and the parameters are substituted into the formula (13.1) to obtain elliptic equations in the XOY coordinates, as shown in formula (13.3).

Fig. 13.3 Parameters error analysis

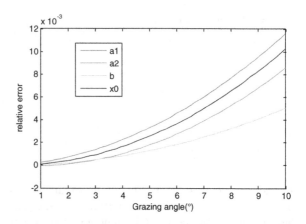

$$\begin{cases} x\cos\theta + y\sin\theta \le h/\psi, \dfrac{(x\cos\theta + y\sin\theta - h/\psi)^2}{[h/\psi - h/(\psi + \alpha/2)]^2} + \dfrac{(y\cos\theta - x\sin\theta)^2}{(h\tan(\theta_B/2)/\psi)^2} = 1 \\[4mm] x\cos\theta + y\sin\theta > h/\psi, \dfrac{(x\cos\theta + y\sin\theta - h/\psi)^2}{[h/(\psi-\alpha/2)-h/\psi]^2} + \dfrac{(y\cos\theta - x\sin\theta)^2}{(h\tan(\theta_B/2)/\psi)^2} = 1 \end{cases}$$

$$(13.3)$$

13.3 Echo Signal Model

The echo signal received by radar includes target echo signal, clutter and noise. Assuming target as point target, noise as an additive white Gaussian noise, then the land clutter echo signal is the focus. Set $S_R(t)$ as the echo signal, then

$$S_R(t) = S_{RV}(t) + S_{RC}(t) + n(t) \tag{13.4}$$

Here $S_{RV}(t)$ is target echo signal, $S_{RC}(t)$ is clutter, n(t) is noise.

13.3.1 Echo Model

LFMCW radar transmit signal in the nth cycle can be presented as

$$S_T(t) = A\cos[2\pi(f_0 t + \mu t^2/2) + \Phi_0], 0 < t < T \tag{13.5}$$

Where A, Φ_0 are respectively the amplitude random phase of the transmission signal, f_0 is carrier frequency, $\mu = $ B/T is frequency modulation slope, B is signal bandwidth, T is scanning cycle.

In detection of FOD, FOD mainly refer to stationary object. Here assuming that target is point target, target echo signal that has distant R from radar is as follows.

$$S_{RV}(t) = KA\cos[2\pi(f_0(t-\tau) + \mu(t-\tau)^2/2) + \Phi_0 + \theta_0] \quad (13.6)$$

Where K is target reflection coefficient, $C = 3 \times 10^8$ m/s is speed of light, $\tau = 2R/C$ is target echo delay, θ_0 is the additional phase shift caused by the object reflector.

Take transmitter signal with the received signal to do mixing, and through low-pass filtering can obtain beat signal, as shown in formula (13.7).

$$S_{IF}(t) = 0.5KA^2\cos[2\pi(f_0\tau + t\mu\tau - \mu\tau^2/2) + \theta_0] \quad (13.7)$$

Thus beat signal frequency can be presented as $f_{IF} = B\tau/T = 2BR/TC$, then $R = f_{IF}TC/2B$. R can be calculated by doing FFT of beat signal to get f_{IF}.

13.3.2 Land Clutter

Land clutter is the main factor affecting FOD detection Performance. The ground reflected clutter signal can be considered as the echo signal of a large number of scattering body superposed on each other within the effective area on the ground of FOD radar irradiation in a certain time.

According to the radar work environment, there are two terrain, grass and runway, in the radar beam irradiation area. In order to more closely simulate clutter, clutter backscatter coefficient need to consider influences from both grass and runway. According to Ref. [7], backscatter coefficient of the composite scene can be expressed as

$$\sigma^0 = (A_{grass}\sigma^0_{grass} + A_{way}\sigma^0_{way})/A \quad (13.8)$$

Here A, A_{grass}, A_{way} express the whole area, and grass and runway area within the beam; σ^0_{grass}, σ^0_{way} express clutter backscatter coefficient of grass and runway.

In order to save the amount of calculation in the simulation, backscatter coefficient, σ^0_{grass}, σ^0_{way}, use empirical model. This paper uses model proposed by Kulemin in Ref. [8] which is presented as

$$\sigma^0 = A_1 + A_2\log\psi/20 + A_3\log f/10 \quad (13.9)$$

Here f is the frequency in GHz, and Ψ is the grazing angle in degrees. The $A_1 - A_3$ coefficients for grass and runway are presented in Table 13.1.

As shown in Fig. 13.4, Elliptical area is the beam irradiation region, L1, L2 is runway edge. Meshing unit is divided within the effective area of radar beam according to radar range resolution ΔR and azimuth resolution $\Delta\theta$. Coordinates of any grid unit A is $A(x_{mn},y_{mn})$.

Table 13.1 The $A_1 - A_3$ coefficients in clutter model

Terrain type	A_1	A_2	A_3
Concrete	−49	32	20
Grass	−(25–30)	10	6

Fig. 13.4 Divide meshing units

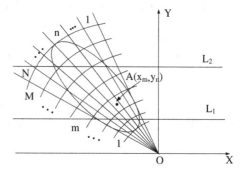

When meet the constant Doppler theory [9], the clutter signal generated on a single grid unit can be presented as

$$S_{RC}(t; m, n) = S_T(t - \tau_{mn})e^{j2\pi f_d(t - \tau_{mn})} \left[\frac{\lambda^2 G(t)^2}{(4\pi)^3 R_{mn}^4} \right]^{1/2} \sqrt{\sigma_{mn}}, \qquad (13.10)$$

$$1 \leq n \leq N, 1 \leq m \leq M$$

Where G(t) is one-way antenna power gain, λ is the radar wavelength, R is the distance from clutter unit to the radar, τ_{mn} is clutter unit two-way delay time, f_d is the Doppler center frequency for clutter unit. σ_{mn} is scattering area of ground clutter within the grid cell, can be presented as

$$\sigma_{mn} = \sigma_{mn}^0 A_{mn}, \quad A_{mn} = \Delta R \Delta \theta \sqrt{x_{mn}^2 + y_{mn}^2} \qquad (13.11)$$

Here m and n indicate geographical location of clutter cell, σ_{mn}^0 is the backward scattering coefficient and A_{mn} is the corresponding ground mesh cell area.

Gaussian distribution to describe the clutter amplitude probability distribution model is used in this paper. The mean and variance of Gaussian distribution is μ and σ^2. The mean of clutter is generally regarded as zero and variance is determined by backscatter coefficient.

As FOD radar is fixed land radar, the clutter power spectrum reflected from various regions of the ground generally is available to be described in the Gaussian function [10]. The power spectral density function can be expressed as

$$S(f) = \exp(-f^2/2\delta_c^2) \qquad (13.12)$$

Here, δ_c is the clutter power spectral broadening impacted by the radar transmit signal and the antenna scans.

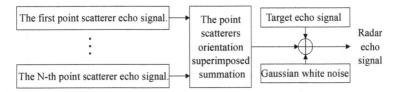

Fig. 13.5 Radar echo simulation flowchart

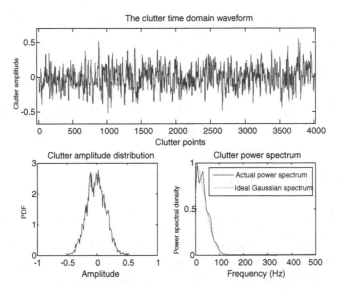

Fig. 13.6 Random sequence consistent with the Gaussian distribution

13.3.3 Noise

In the millimeter wave band, the main source of radar system noise is the internal noise of the radar system which is usually seen as a white Gaussian noise.

By simulating clutter, target signal and noise, we can get the radar echo signal. Figure 13.5 shows a flowchart of the radar echo simulation.

13.4 Simulation

The focus of echo modeling in this paper is clutter. Gaussian white noise sequence is first generated, and then the sequence through a linear filter having a Gaussian spectral Gaussian random sequence. The simulation result is shown in Fig. 13.6. As can be seen, amplitude of generating a random sequence satisfy the Gaussian

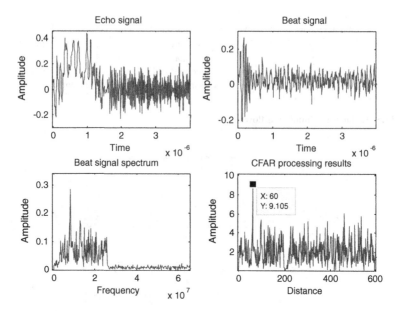

Fig. 13.7 Echo signal and signal processing

distribution, the power spectrum is also consistent with the Gaussian distribution, i.e. the random sequence obtained is correct.

Echo simulation result is shown in Fig. 13.7, including echo signal, the beat signal, the beat signal spectrum diagram and CFAR processing result. A target in distance of 60 m is simulated and can be seen in CFAR processing result.

13.5 Conclusions

FOD radar echo modeling has important guiding significance for study of performance and detection accuracy of radar system, clutter interference. This paper establishes geometric model according to radar work environment, discusses echo signal model, using statistical modeling to model clutter and then do simulation in MATLAB. From simulation results, the clutter sequence obtained is correct.

Acknowledgments This work is partially supported by National Natural Science Foundation funded project (61079006) and the key project of the National Natural Science Foundation of China (61139003)

References

1. Security and Technology Center of Department of Airport CAAC (2009) FOD precaution manual, pp 1–5 (in Chinese)
2. Mazouni K, Kohmura A, Futatsumori S et al (2010) 77 GHz FMCW radar for FODs detection. In: Proceeding of the 7th European radar conference. Paris, France: EuMA, pp 451–454
3. Feil P, Zeitler A, Nguyen TP et al (2010) Foreign object debris detection using a 78 GHz sensor with cosec antenna. In: Proceeding of the 7th European radar conference. Paris, France: EuMA. pp 33–36
4. Zeitler A, Lanteri J, Migliaccio C et al (2010) Folded Reflect arrays with shaped beam pattern for foreign object debris detection on runways. IEEE Trans Antennas Propag 58(9):3065–3068
5. Yonemoto N, Kohmura A, Futatsumori S et al (2009) Broad band RF module of millimeter wave radar network for airport FOD detection system. In: Proceedings of international conference on radar. Bordeaux, France, IEEE
6. AC 150/5220-24, Federal Aviation Administration[S]
7. Chun-xia LI et al (2010) A clutter modeling method in the composite sea-plus-land environment. J Microwaves 26(2):22–26 (in Chinese)
8. Kulemin GP, Tarnavsky EV, Goroshko EA (2004) Land backscattering for millimeter wave radar. In: Proceedings of international conference on modern problems of radio engineering, telecommunications and computer science, Lviv-Slavsko, Ukraine, IEEE, 2004:138–141
9. Mitchell RL (1982) Radar system simulation. CHEN Xun-da. Beijing: Science Press
10. Long MW (1981) The radar scattering properties of land and sea. Science Press, Beijing, pp 26–40

Chapter 14
Min–Max Decoding for Non-Binary LDPC Codes

Leilei Yang, Fei Liu and Haitao Li

Abstract In this paper, we present a partial-parallel decoder architecture based on Min–Max algorithm for quasi-cyclic non-binary LDPC codes. An efficient architecture of the check node elementary processor is designed. The variable node update unit with fully parallel computation is proposed, which has the advantage of low complexity and latency by eliminating forward–backward operation and removing recursive computation among the message vector. Moreover, the FPGA simulation over GF(16) NB-LDPC is given to demonstrate the efficiency of the presented design scheme.

Keywords NB-LDPC · Min–Max · Fully parallel · FPGA

14.1 Introduction

Low density parity check (LDPC) codes were first introduced by Gallager [1]. It was shown that binary LDPC codes, when decoded using the belief-propagation (BP) decoding algorithm, could approach the capacity of the additive white Gaussian noise (AWGN) channel [2]. Non-binary LDPC (NB-LDPC) codes, which are an extension of binary LDPC codes, were first investigated by Davey and MacKay [3]. The nonzero entries in the parity check matrix of a NB-LDPC code are directly replaced by elements in a Galois field. When the code length is small or a higher-order modulation is used in a communication system, NB-LDPC

L. Yang (✉) · F. Liu · H. Li
College of Electronic Information and Control Engineering, Beijing University
of Technology, Beijing, People's of Republic of China
e-mail: leileiyang0319@163.com

W. Lu et al. (eds.), *Proceedings of the 2012 International Conference on Information Technology and Software Engineering*, Lecture Notes in Electrical Engineering 210, DOI: 10.1007/978-3-642-34528-9_14, © Springer-Verlag Berlin Heidelberg 2013

codes perform better than binary LDPC codes. In addition, NB-LDPC codes are more efficient than binary codes for combating mixed types of noise and interferences [4]. Also, it is further shown that NB-LDPC codes have superior performance in the presence of burst errors [5].

Recently, a considerable amount of research effort has already been spent on studying the efficient decoding algorithm for NB-LDPC codes. In [3], an extended sum-product algorithm (SPA), which is originally designed for binary LDPC codes, was proposed for decoding NB-LDPC codes. The computational complexity of the SPA is dominated by $O(q^2)$ sum and product operations for each check node processing, where q is the cardinality of the Galois field. Similar to the min-sum decoding algorithm of binary LDPC codes, the SPA for NB-LDPC codes can also be approximated in the logarithmic domain. The logarithmic SPA does not need complicated multiplications. In [6], the log-SPA and max-log-SPA were presented. Based on max-log-SPA, an extended min-sum (EMS) algorithm, which can reduce the message memory requirement with little performance loss, was proposed in [7]. Only a limited number N_m of reliabilities in the messages is used. The complexity of EMS algorithm is reduced from $O(q)$ to $O(N_m)$.

A more efficient Min–Max decoding algorithm for non-binary LDPC codes is proposed in [8], 'max' instead of 'sum' is computed in the check node processing. As a result, the throughput of the decoder is larger than EMS algorithm. In this chapter, based on the Min–Max algorithm, we present an efficient VLSI implementation architecture of NB-LDPC decoder for the quasi-cyclic (QC) LDPC codes.

The rest of this chapter is organized as follows: In Sect. 14.2, we introduce the simple Min–Max decoding algorithms for NB-LDPC codes. The architecture of the decoder is proposed in Sect. 14.3. The simulations of BER performance and FPGA implementation results are given in Sect. 14.4. Finally, the conclusions are drawn in Sect. 14.5.

14.2 Encoder and Min–Max Decoder for Nb-LDPC

A kind of NB-QC-LDPC coder can be defined by the parity check matrix **H**. As show in Fig. 14.1, the check matrix **H** is divided into 6×24 sub-matrixes whose size is 5×5 Each sub-matrix is either a zero or a shifted identity matrix with non-zero entries replaced by elements of $GF(q)$. In the text of the chapter, the simulation and design are all based on the matrix depicted above. Then, we will discuss the detailed description for Min–Max algorithm for the given NB-QC-LDPC.

We assume that L is the initial message vector computed based on the channel measurements. And let $U_{tp}(V_{tp})$ be the computed variable (check) message vector, $\bar{U}_{tp}(\bar{V}_{tp})$ be the reserved variable (check) message vector with size N_m, $\hat{x}$ be the result of decoding. The detailed Min–Max algorithm is described as follows:

Fig. 14.1 Structure of a (6,24) base matrix

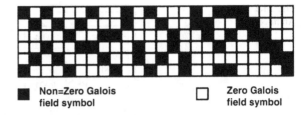

■ Non=Zero Galois field symbol □ Zero Galois field symbol

1. Initialization: Only the N_m largest values of L are copied and signed $\bar{L}$.
2. Variable node update for a degree d_v node:

$$U_{tp}[i_1, \ldots, i_q] = \bar{L}[i_{r1}, \ldots, i_{rN_m}] + \sum_{v=1, v \neq t}^{d_v} \bar{V}_{pv}[i_{s1}, \ldots, i_{sNm}] \tag{14.1}$$

$$(i_1, \ldots i_q) \in GF(q)$$

3. Permutation step (from variable to check nodes):

$$\bar{U}_{pc}[j_1, \ldots, j_{N_m}] = \bar{U}_{vp}[i_1, \ldots, i_{N_m}]$$
$$j(X) = h(X)i(X) \tag{14.2}$$
$$(j_1, \ldots, j_q) \in GF(q)$$

4. In the Min–Max algorithm, the check node update is a sequential implementation of the elementary steps:

$$V_{tp}[j_1, \ldots, j_q] = \min_{j \neq p}(\max_{j \neq p} \bar{U}_{t,j}[j_1, \ldots, j_{N_m}]) \tag{14.3}$$

5. Inverse permutation step (from check to variable nodes):

$$\bar{V}_{pc}[i_1, \ldots, i_{N_m}] = \bar{V}_{cp}[j_1, \ldots, j_{N_m}]$$
$$i(X) = h^-(X)j(X) \tag{14.4}$$

Fig. 14.2 Overall
Architecture

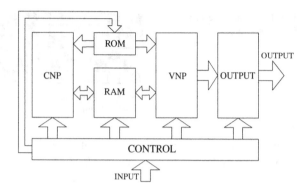

6. Tentatively decoding:

$$\hat{x} = \arg \max_{i \in GF(2^p)} \left\{ \bar{L}[i_1, \ldots, i_{N_m}] + \sum_{v=1}^{dv} \bar{V}_{pv}[i_1, \ldots, i_{N_m}] \right\} \qquad (14.5)$$

14.3 Decoder Architecture and Design

14.3.1 Overall Architecture

The key technology of NB-LDPC coded modulation system is the decoder design
for Min–Max algorithm. The overall architecture of the proposed Min–Max
decoder is shown in Fig. 14.2, which consists of five kinds of module: check node
update module (CNP), variable node update module (VNP), control module,
RAM\ROM and output module. While the number of CNP/VNP used equals to the
row/line weight of check matrix **H**. To reduce the hardware resources and reach
the high throughput, the check nodes update are implemented with a elementary
step, and variable nodes update are implemented with fully parallel architecture.
All message vectors are of 11 bits quantization, with 4-bit for symbol and 7-bit for
LDR. Initial message vector will be written into 24 ROMs, each is of
$5N_m \times 11 = 440$ bits. And each nonzero sub-matrix is distributed a RAM which
stores check\variable node message vector and is of $5N_m \times 11 = 440$ bits. Thus,
we need totally $440 \times 24 + 440 \times 50 = 32{,}560$ bits for variable/check node
message saving.

Fig. 14.3 The recursive structure composed of elementary for 8°

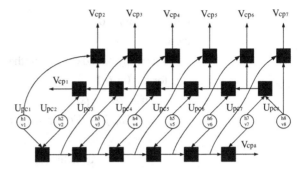

Fig. 14.4 The structure of interpolation sorter

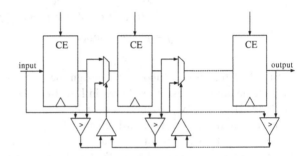

14.3.2 CNP Architecture

To update the check and variable nodes message, forward/backward algorithm was used in [6]. A recursive implementation combined with a forward/backward strategy to minimize the number of elementary steps is shown in Fig. 14.3.

Assuming that an elementary step describing the check (variable) node update has **V** and **I** as input messages and **U** as output message. The vectors **V**, **I** and **U** of size N_m are sorted in decreasing order using interpolation sorter as show in Fig. 14.4. We note also by β_V, β_I and β_U their associated index vectors.

The implementation of check update step will be discussed which is the bottleneck of the algorithm complexity. And we define $S(\alpha_U[i])$ as the set of all N_m^2 possible combinations which satisfy the parity equation $\alpha_U[i] \oplus \alpha_V[j] \oplus \alpha_I[p] = 0$. With these notations, the output message values are obtained:

$$U[i] = \max_{S(a_U[i])} (V[j], I[p]) \quad i \in \{0, \ldots, n_m - 1\} \tag{14.6}$$

We introduce a low computational strategy [9] to skim through the two sorted vectors **V** and **I**, that provides a minimum number of operations to process the N_m sorted values of the output vector **U**. The pseudo-codes of the check node elementary step are given:

Initialization: $i = 0, j = 1;$

loop:

   if $V(i) < I(j)$(boundary goes to the bottom)

      for $k = 0 : j-1$

         if $(a_V(i) + a_I(k)) \notin a_U$

         $\{U, a_U\} \leftarrow \{V(i), a_V(i) + a_I(k)\}$

      $i = i+1$;goto loop

   else(boundary goes to the right)

      for $k = 0 : j-1$

         if $(a_V(k) + a_I(j)) \notin a_U$

         $\{U, a_U\} \leftarrow \{V(j), a_V(k) + a_I(j)\}$

      $j = j+1$;goto loop

Based on the strategy above, we design the efficient architecture of the check node elementary processor as shown in Fig. 14.5. There are two RAMs (RAM.A and RAM.B) of size N_m are used to save the input message vector. The skim Starts from the origin (0,1) of the RAM.A and RAM.B, the minimum of which will be written into interpolation-sorter. Four self-addition and self-subtract are used to generate the read address. The COMP will compare the date from RAM.A and RAM.B and return a control signal to both multiplexer and address generation. Detect unit is used to remove the same symbol and write the different symbol into message vector when the length of vector is less than N_m. For example, if A[i] is smaller than B[j] (while i/j are used to read the data from RAM for compare, ii/jj are used to read the data and add them as the input of sort, and i/j will write into ii/jj after every compare), the decoder will read data from RAM.B by address that is smaller than the jj and read data from RAM.A by address i. And then i will add 1, and j keep unchanging. If A[i] is large than B[j], the decoder will read data from RAM.A by address that is smaller than the ii and read data from RAM.A by address j. And then j will add 1, and i keep unchanging. Then at the beginning of next loop, read from A by address of i, read from B by address of j.

Based on the architecture above, it is necessary to perform about $25 \times 18 \times$ sub-matrixes $= 2,250$ clock cycles to finish check node update each iteration.

14.3.3 VNP Architecture

The variable node update unit with fully parallel computation is presented without forward–backward, which removes recursive computation among the message vector and is of low complexity and latency. In the skim algorithm of VNP, the

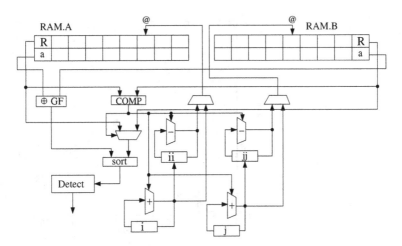

Fig. 14.5 Check node architecture

Fig. 14.6 Variable node architecture

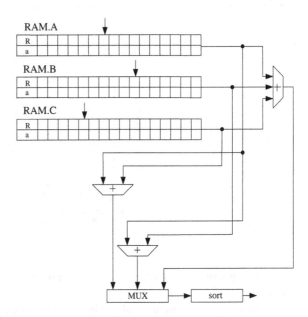

message of the same symbol is added together and the minimal N_m is reserved. So we can simplify the process of skim. In the first step input message **A** (equal to initial message) and **B/C** (from CNP) are written into RAM **A/B/C** by address of symbol $\alpha_A/\alpha_B/\alpha_C$, while the RAM is initialized with the maximal value. In the second step,we use a multiplexer to select the message that the recursive computation needs and the value of the same address will be added. In the third step,

Fig. 14.7 BER for min–max decoer over AWGN

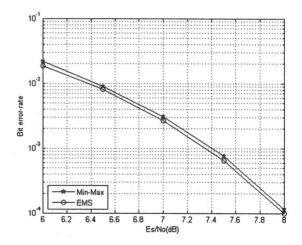

Fig. 14.8 BER for quantization

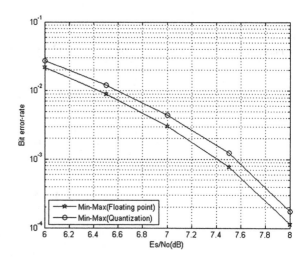

interpolation-sorter is used to copy the N_m minimal values from the result of the second step.

All message vector are normalized by doing a subtraction with the minimal one. Figure 14.6 illustrates the architecture of the variable node processor. Based on the architecture above, it is necessary to perform about 44 × sub-matrixes = 220 clock cycles to finish variable node update each iteration.

Table 14.1 The performance parameters of decoder

Number of Slices	21,561
Number of slice flip flops	22,536
Number of 4 input LUTs	38,244
Number of RAMs	148
Maximum frequency	209 MHz

14.4 Simulation Results

In this section, we evaluate the performance of the proposed decoder design scheme for GF(16) QC-LDPC. The maximum number of iterations in the algorithm has been fixed to 5 and log-density-ratio (LDR) vector is truncated to 8 values. The simulated Bit Error Rate (BER) for 16-QAM modulation and Additive White Gaussian Noise (AWGN) channel is shown in Figs. 14.7, 14.8. In Fig. 14.7, the BER comparison is presented for the Min–Max and EMS algorithm. We can find that the Min–Max performs little degradation of the error performance in the waterfall region, but it can improve the throughput of the decoder. Figure 14.8 gives the simulation results for quantization with 4-bit for symbol and 7-bit for LDR, which performs 0.15 dB degradation of floating-point. In Table 14.1, the synthesis results on Xilinx FPGA are presented. It is shown that the decoder consumes a lot of hardware resources, though we use a lot of multiplexer. In the future of this work, we will explore the efficient way to reduce the implementation complexity.

14.5 Conclusion

In the chapter, we introduce the Min–Max algorithm. Then, the architectural model of a GF(q) LDPC decoder based on the reduced complexity algorithm is proposed. Moreover, we synthesize the generic processor architecture on FPGA. Future work will be devoted to further reduce the hardware resources requirement of the decoder and improve the throughput of the decoder.

The author would like to thank the support of Beijing Natural Science Foundation (Approaching the optimal capacity of wireless networks based on interference alignment) under Grant 4112012, Beijing Municipal Commission of Education project "R&D multimedia trunk wireless access system".

References

1. Gallager RG (1963) Low-density parity-check codes. MIT Press, Cambridge
2. Richardson TJ, Shokrollahi MA, Urbanke RL (2001) Design of capacity-approaching low-density parity check codes. IEEE Trans Inf Theor 47:619–637

3. Davey M, MacKay DJC (1998) Low density parity check codes over GF(q). IEEE Comm Lett 2:165–167
4. Zhou B, Kang J, Song S, Lin S, Abdel-Ghaffar K (2009) Construction of non-binary quasi-cyclic LDPC codes by arrays and array dispersions. IEEE Trans Comm 57(10):1652–1662
5. Song H, Cruz JR (2003) Reduced-complexity decoding of Q-ary LDPC codes for magnetic recording. IEEE Trans Magn 39:1081–1087
6. Wymeersch H, Steendam H, Moeneclaey M (2004) Log-domain decoding of LDPC codes over GF(q). In: The proceedings of IEEE internaional conference on communication, Paris, France, pp 772–776
7. Declercq D, Fossorier M (2007) Decoding algorithms for nonbinary LDPC codes over GF(q). IEEE Trans Comm 55:633–643
8. Valentin Savin and CEA-LETI, Min-Max decoding for non binary LDPC codes. IEEE international conference on ISIT2008, p 960
9. Zhang X, Cai F (2010) Partial-parallel deccoder architecture for quasi-cyclic non-binary LDPC codes. IEEE international conference on communication, Cleveland, USA, p 1506

Chapter 15
A Security Attack Risk Assessment for Web Services Based on Data Schemas and Semantics

Wanwisa Phocharoen and Twittie Senivongse

Abstract Security concerns have been raised by Web services providers and consumers since Web services are vulnerable to various security attacks including counterfeiting, disclosure, tampering, disruption, and breach of information. In particular, Web services can be vulnerable if the schemas of the input data are not strong, giving way to security attacks like command injection and denial of service. This chapter proposes an initial assessment of security attack risks for Web services. The assessment begins with an analysis of the input data schemas that are described in the service WSDL document to determine if they are unconstrained and at risk of command injection and denial of service attacks. Then we determine if such a risk can be mitigated by making use of semantic information that is annotated to the input data elements within the WSDL. If the semantic annotation is stronger than the schema elements themselves, we refer to the case of weak interface design in which a redesign of the service interface with stronger schemas should help reduce attack risks. We also propose a risk assessment model for determining quantitatively the attack risk level of a Web service to guide the provider when considering schema hardening as well as the consumer when selecting between different services.

Keywords Risk assessment · Security attacks · Web services · Ontology

W. Phocharoen · T. Senivongse (✉)
Software Engineering Program, Department of Computer Engineering,
Faculty of Engineering, Chulalongkorn University, 254 Phyathai Road,
Pathumwan, Bangkok 10330, Thailand
e-mail: twittie.s@chula.ac.th

W. Phocharoen
e-mail: wanwisa.p@student.chula.ac.th

W. Lu et al. (eds.), *Proceedings of the 2012 International Conference on Information Technology and Software Engineering*, Lecture Notes in Electrical Engineering 210, DOI: 10.1007/978-3-642-34528-9_15, © Springer-Verlag Berlin Heidelberg 2013

15.1 Introduction

Web services technology is widely adopted by organizations to provide and consume business functionalities through software units over the Internet using established information and communication standards. Among these standards, the technology promotes the use of XML-based service descriptions described in the Web Services Description Language (WSDL) [1]. The WSDL document of a Web service serves as a contract between service providers and consumers, defining the service interface and how to invoke the service. Enclosed in the WSDL is an XML schema which describes the definition of the data elements that are to exchange with the service.

Security concerns have been raised by service providers and consumers as Web services are vulnerable to various security attacks including counterfeiting, disclosure, tampering, disruption, and breach of information. In particular, Web services can be vulnerable if the schemas of the input data are not strong. Even though weak typing of the data elements (e.g. a data element of type string with no value constraints) is defined for ease of use, it can give way to security attacks like command injection and denial of service. This paper proposes an initial assessment of security attack risks for Web services. The assessment is considered initial because it is based on the service WSDL only. It begins with an analysis of the input data schemas within the WSDL document to determine if they are unconstrained and at risk of command injection and denial of service attacks. In a certain case, the risk may be unavoidable if the service really needs to exchange unconstrained data (e.g. a news service which allows an exchange of a string news content of unlimited size). However, the risk is sometimes caused by careless design of the service interface (e.g. an order service which allows a customer ID data to be unconstrained, instead of using an XML schema constraining facet [2] to restrict the data format to certain string values). Therefore we determine if such a risk can be mitigated by making use of semantic information that is annotated to the input data elements within the WSDL. We use SAWSDL [3] which is an annotation mechanism for specifying a mapping between any data schema elements and their corresponding concepts (i.e. terms) within the service domain ontology. That is, if the semantic concepts define more constraints (i.e. are stronger) than the schema elements, we refer to the case of weak interface design in which a redesign of the service interface with stronger schemas should help reduce attack risks. We also propose a risk assessment model for determining quantitatively the attack risk level of a Web service. The assessment result can guide the service provider when considering hardening the data schemas as well as the service consumer when selecting a service with less attack risks among different candidate services.

Section 15.2 of this paper discusses related work and Sect. 15.3 gives a background on WSDL data schemas and semantic annotation and analyzes their relation to command injection and denial of service attacks. Section 15.4 proposes the attack risk assessment model and presents how to apply the model to a service WSDL. The paper concludes in Sect. 15.5 and outlines future work.

15.2 Related Work

A number of research attempts have proposed ways to assess security of Web service provision and some also make security measurable. Based on Web application vulnerabilities and attack patterns as well as vulnerabilities of XML messaging, Yu et al. [4] develop a software vulnerability fault model for Web services in which the vulnerabilities are classified by attack targets and attack types. They also propose a guideline to test and detect vulnerabilities in Web services. Pang and Peng [5] propose a security risk assessment for Web services by analyzing risks to service confidentiality, integrity, and availability. A scanning tool is also used to scan service vulnerabilities, and a survey on security techniques and standards adopted by the service is conducted. Jiang et al. [6] present a Web services security evaluation model which considers the ability of the service to guard against security threats. Other aspects are also considered, including the security measures that the consumer adopts, the provider's reputation in providing secure service, and risk of threats with regard to possibility, criticality, and rate of attacks. Banklongsi and Senivongse [7] presents a security measurement model based on the ability of the Web services to provide countermeasures for various attack types as classified in CAPEC [8]. The approaches taken by all the afore-mentioned research are quite similar in that they consider security aspects that are external to the service (e.g. security attacks) as well as internal (e.g. security provision of the service). Unlike related work, our approach is based on the analysis of WSDL data schemas and their constraints.

Other research work also performs an analysis on WSDL. Hanna and Munro [9] analyze constraints on WSDL input elements to generate test cases for testing service robustness, input manipulation vulnerability, and fault tolerance to invalid input. Similarly, Brinhosa et al. [10] present an implementation of an input validation model to validate Web service input against an input validation specification which is derived from data schemas in a WSDL. Unlike these approaches, the focus of our analysis on a WSDL document is on service security in connection with data schema design flaws.

15.3 Data Schemas, Semantic Annotation, and Attack Types

XML schemas for the data elements and attributes in WSDL may or may not be constrained. The constraints that we consider are cardinality (i.e. the number of possible occurrences of a data element) and facet (i.e. enumerated value, pattern, or length of a data element or attribute) [2]. In addition, different parts of WSDL

Table 15.1 Example of XML data schemas and their counterparts in OWL ontology

Constraint	XML schema	OWL 2 ontology
None	`<xsd:element name = "creditCard" type = "xsd:string"/>`	Declaration(Class(:CreditCard)) EquivalentClasses(:CreditCard DataSomeValuesFrom(:cardType xsd:string))

Enumeration	`<xsd:element name = "creditCard">`
	`<xsd:simpleType >`
	`<xsd:restriction base = "xsd:string">`
	`<xsd:enumeration value = "Amex"/>`
	`<xsd:enumeration value = "MasterCard"/>`
	`<xsd:enumeration value = "Visa"/>...`
		Declaration(Class(:CreditCard))
		EquivalentClasses(:CreditCard DataSomeValuesFrom(:hasCardType :cardType))
		Declaration(DataProperty(:hasCardType))
		DataPropertyRange(:hasCardType :cardType)
		Declaration(Datatype(:cardType))
		DatatypeDefinition(:cardType DataOneOf("Amex" "MasterCard" "Visa"))

Pattern	`<xsd:element name = "customerID">`
	`<xsd:simpleType >`
	`<xsd:restriction base = "xsd:string">`
	`<xsd:pattern value = "[A-Z]{3}\d{10}"> ...`
		Declaration(Class(:CustomerID))
		EquivalentClasses(:CustomerID DataSomeValuesFrom(:id DatatypeRestriction(xsd:string xsd:pattern "[A-Z]{3}\d{10}")))

Length	`<xsd:element name = "orderID">`
	`<xsd:simpleType>`
	`<xsd:restriction base = "xsd:string">`
	`<xsd:maxLength value = "10"/>...`
		Declaration(Class(:OrderID))
		EquivalentClasses(:OrderID DataSomeValuesFrom(:id DatatypeRestriction(xsd:string xsd:maxLength "10"^^xsd:integer)))

Cardinality	`<xsd:element name = "telephone" maxOccurs = "2" minOccurs = "1">`
	`<xsd:simpleType>`
	`<xsd:restriction base = "xsd:string">`
	`<xsd:maxLength value = "14"/>`
	`</xsd:restriction>`
	`</xsd:simpleType>`
	`</xsd:element>`
		Declaration(Class(:Telephone))
		EquivalentClasses(:Telephone ObjectUnionOf(ObjectMaxCardinality(2 :hasNumber) ObjectMinCardinality(1 :hasNumber)))
		SubObjectPropertyOf(:hasNumber owl:topObjectProperty)
		ObjectPropertyRange(:hasNumber DataSomeValuesFrom(:phone DatatypeRestriction(xsd:string xsd:maxLength "14"^^xsd:integer)))

Table 15.2 Attack types and typical severity levels [8]

Attack Type		Severity (score)
Command injection	SQL injection	High (4)
	Blind SQL injection	High (4)
	SQL injection through SOAP parameter tampering	Very high (5)
	Simple script injection	Very high (5)
	XPath injection	High (4)
	XQuery injection	Very high (5)
DoS	Violating implicit assumptions regarding XML content (aka XML denial of service (XDoS))	Very high (5)
	XML ping of death	Medium (3)
	Resource depletion through DTD injection in a SOAP message	Medium (3)
	Resource depletion through allocation	Medium (3)
	Resource depletion through flooding	Medium (3)

elements can be enhanced with semantic information as described in a service domain ontology. Table 15.1 gives an example of XML data constraints and their counterparts in OWL 2 ontology language, using the functional syntax [11].

We address the relation between data schemas and two security attack types, i.e. command injection and denial of service (DoS). They are the most common and very critical attack types [4] and related to weak typing of the exchanged data. Command injection attacks a Web service by embedding special characters or malicious commands in the data sent to the service [12]. This allows the attacker to bypass authentication, gain access to service data, or do further malicious actions. DoS attacks a Web service by overwhelming service processing capability or by installing malicious code [4]. The attacker may cause the service to allocate excessive resources (such as memory) or overwhelm it with oversized payload if there is no restriction on size of the input data. This paper considers six command injection attacks and five DoS attacks, as classified in CAPEC and shown in Table 15.2. We represent CAPEC's five levels of typical severity (Very Low-Very High) in a scale of 1–5 for use in the next section.

WSDL data schemas are vulnerable to these attack types but strong data typing and input validation can mitigate the attacks. We also make use of the semantic information that is annotated to the input data schemas, using the SAWSDL mechanism [3], as a knowledge specification for checking if the schemas are weaker than they should be according to the service domain knowledge. If so, the schemas may put the service at risk but the risks can be mitigated if the service provider can improve the design by hardening the schemas. But if the schemas have no semantic annotation or the annotation is not stronger than the schemas, we say that the risks are unavoidable. Table 15.3 shows a few cases of weak/strong data typing and their relation to attack risks.

Table 15.3 Examples of data schemas and their relation to attack risks

No.	Data schema	Risk of command injection[a]	Risk of DoS[a]	Remark
1	`<xsd:element name = "customerName" type = "xsd:string"/>`	Y (U)	Y (U)	Schema has no restriction on value and size; no annotation
2	`<xsd:element name = "paymentType"> ... < xsd:restriction base = "xsd:string"> <xsd:enumeration value = "Cash"/> <xsd:enumeration value = "Card"/>...`	N	N	Schema defines specific values and cardinality is 1 by default; no annotation
3	`<xsd:element name = "orderID"` *sawsdl:modelReference* `= "http://org1.example.com/ ontologies/CustomerOrderOntology#OrderID" type = "xsd:string"/>`[b]	Y (M)	Y (M)	Schema has no restriction on value and size but annotation is stronger, specifying maxLength; risk can be mitigated by hardening schema

[a] *Y* yes, *N* no, *U* unavoidable, *M* can be mitigated

[b] With an attribute *modelReference* of SAWSDL, the element is annotated by the concept *OrderID* of *CustomerOrderOntology* (shown previously in Table 15.1)

Table 15.4 Security attack risk assessment model

	Unavoidable risk	Risk that can be mitigated
Risk index for command injection[a,b]	$RI_{CI\_U} = \sum_{i=1}^{N_{CI}} S_{CIi} \times \frac{NR_{CI\_U}}{IN}$ (15.1) where NR_{CI_U} is number of input elements that are at risk of command injection and the risk is unavoidable	$RI_{CI\_M} = \sum_{i=1}^{N_{CI}} S_{CIi} \times \frac{NR_{CI\_M}}{IN}$ (15.2) where NR_{CI_M} is number of input elements that are at risk of command injection but the risk can be mitigated
Total risk index for command injection	$RI_{CI} = RI_{CI\_U} + RI_{CI\_M}$ (15.3)	
Risk index for DoS[b,c]	$RI_{DOS\_U} = \sum_{i=1}^{N_{DOS}} S_{DOSi} \times \frac{NR_{DOS\_U}}{IN}$ (15.4) where NR_{DOS_U} is number of input elements that are at risk of DoS and the risk is unavoidable	$RI_{DOS\_M} = \sum_{i=1}^{N_{DOS}} S_{DOSi} \times \frac{NR_{DOS\_M}}{IN}$ (15.5) where NR_{DOS_M} is number of input elements that are at risk of DoS but the risk can be mitigated
Total risk index for DoS	$RI_{DOS} = RI_{DOS\_U} + RI_{DOS\_M}$ (15.6)	
Total risk index	$RI = RI_{CI} + RI_{DOS}$ (15.7)	

[a] N_{CI} is number of command injection attack types (as in Table 15.2) which is 6
S_{CIi} is severity score of command injection attack type i (as in Table 15.2)
[b] IN is number of all input elements in WSDL
[c] N_{DOS} is number of DoS attack types (as in Table 15.2) which is 5
S_{DOSi} is severity score of DoS attack type i (as in 15.2)

Table 15.5 Data schemas and semantics of an order service in relation to attack risks

No.	Input data schema	Semantic annotation	Risk of command injection[a]	Risk of DoS[a]
1	customerName: string	None	Y (U)	Y (U)
2	customerID: string	CustomerID: string, pattern = [A-Z]{3}\\d{10}[b]	Y (M)	Y (M)
3	telephone: string, minLength = 10, maxOccurs = unbounded	Telephone: string, maxLength = 14, ObjectMaxCardinality = 2, ObjectMinCardinality = 1[b]	Y (M)	Y (M)
4	orderID: string	OrderID: string, maxLength = 10[b]	Y (M)	Y (M)
5	itemCode: string, maxOccurs = unbounded[c]	None	Y (U)	Y (U)
6	quantity: integer, maxOccurs = unbounded[c]	None	N	Y (U)
7	paymentType: string, enumeration = Cash, Card	None	N	N
8	creditCard: string	CreditCard: string, DataOneOf = Amex, MasterCard, Visa[b]	Y (M)	Y (M)

[a] *Y* yes, *N* no, *U* unavoidable, *M* can be mitigated
[b] Annotation refers to an ontological concept shown previously in Table 15.1
[c] Occurrence of this data element is considered unbounded as it is nested in another element named orderItem with unbounded cardinality

15.4 Security Attack Risk Assessment Model

The security attack risk assessment model is presented in Table 15.4. It comprises a number of risk indexes, each of which is a multiplication of impact of risks (i.e. severity of attack risks in Table 15.2) and probability of risk occurrence (i.e. the number of vulnerable input data elements out of all input to the service). We define risk indexes for command injection and DoS attacks as well as for unavoidable risks and risks that are caused by careless design and can be mitigated.

As an example, the model is applied to an order Web service. Table 15.5 lists the schemas of eight input data elements in the WSDL, their semantic annotation (if any), and an analysis of whether the schemas are at risk of the two kinds of attacks and whether the risks can be mitigated.

Using the equations in Table 15.4, we calculate the risk indexes for this order service as: $RI_{CI} = 27 \times (2/8) + 27 \times (4/8) = 20.25$, $RI_{DOS} = 17 \times (3/8) + 17 \times (4/8) = 14.88$, and $RI = 20.25 + 14.88 = 35.13$.

15.5 Conclusion

This paper proposes an initial assessment of command injection and DoS attack risks for Web services based on an analysis of WSDL data typing and semantics. Service consumers can initially assess and compare risks involved with the candidate services that are targets of use, while service providers should consider the assessment result and enforce stronger typing as well as proper validation of input. Future work involves development of an automated assessment tool and an experimental study on attack risks in public and corporate services.

References

1. W3C (2007) Web services description language (WSDL) version 2.0 part 0: primer. http://www.w3.org/TR/wsdl20-primer/
2. W3C (2004) XML schema part 2: datatypes second edition. http://www.w3.org/TR/xmlschema-2/
3. W3C (2007) Semantic annotations for WSDL and XML schema. http://www.w3.org/TR/sawsdl/
4. Yu WD, Aravind D, Supthaweesuk P (2006) Software vulnerability analysis for Web services software systems. In: Proceedings of 11th IEEE symposium on computers and communications (ISCC 2006), pp 740–748
5. Pang J, Peng X (2009) Trustworthy web service security risk assessment research. In: Proceedings of international forum on information technology and applications, pp 417–420
6. Jiang L, Chen H, Deng F, Zhong Q (2011) A security evaluation method based on threat classification for Web service. J Softw 6(4):595–603
7. Banklongsi T, Senivongse T (2011) A security measurement model for web services based on provision of attack countermeasures. In: Proceedings of 15th international annual symposium on computational science and engineering (ANSCSE15), pp 593–598
8. Mitre.org (2012) Common attack pattern enumeration and classification (CAPEC) release 1.7.1. http://capec.mitre.org/
9. Hanna S, Munro M (2008) Fault-based web services testing. In: Proceedings of 5th international conference on information technology: new generations, pp 471–476
10. Brinhosa RB, Westphall CM, Westphall CB (2012) Proposal and development of the web services input validation model. In: Proceedings of 2012 IEEE network operations and management symposium (NOMS 2012), pp 643–646
11. W3C (2009) OWL 2 web ontology language primer. http://www.w3.org/TR/owl2-primer/
12. Antunes N, Laranjeiro N, Vieira M, Madeira H (2009) Effective detection of SQL/XPath injection vulnerabilities in web services. In: Proceedings of 2009 IEEE international conference on services computing (SCC 2009), pp 260–267

Chapter 16
Analysis and Modeling of Heterogeneity from Google Cluster Traces

Shuo Zhang and Yaping Liu

Abstract Many Internet-scale services are emerging and developing quickly in recent years, and they may have different performance goals. It is a challenge for a data center to deploy these services and satisfy their performance goals. Servers in a data center are usually heterogeneous, which makes it more sophisticated to efficiently schedule jobs in a cluster. This paper analyzed workload data from publicly available Google cluster traces, and explored two types of heterogeneity from these traces: machine heterogeneity and workload heterogeneity. Based on analysis results, we proposed a heterogeneity model for dynamic capacity provisioning problem in a cluster to deal with these Internet-scale services.

Keywords Trace analysis · Model · Heterogeneity

16.1 Introduction

In recent years, more and more Internet-scale applications are emerging and deployed in mega data centers. On one hand, they may have different performance goals (e.g., low latency, low response time, high throughput, and etc.), and on the other hand, they may have different resource requirements which a data center should satisfy to serve them. For example, CPU-intensive applications (e.g., scientific computing, video encoding) consume more CPU resources, memory-

S. Zhang (✉) · Y. Liu
National University of Defense Technology, Changsha, Hunan 410073, China
e-mail: zhangshuo@nudt.edu.cn

Y. Liu
e-mail: ypliu@nudt.edu.cn

W. Lu et al. (eds.), *Proceedings of the 2012 International Conference on Information Technology and Software Engineering*, Lecture Notes in Electrical Engineering 210, DOI: 10.1007/978-3-642-34528-9_16, © Springer-Verlag Berlin Heidelberg 2013

intensive applications (e.g., web services, multimedia services) consume more memory resources, and other applications may ask for high bandwidth or more disk storage, and etc. So it is a challenge for a data center to run these services and meet their performance goals and requirements.

However, it is common to deploy heterogeneous machines in current data centers, which makes it more difficult to schedule jobs (or tasks) in a cluster. Many data centers are composed of high-performance machines and low-power ones [1] and they become heterogeneous over time due to hardware upgrades and replacement of failed components or systems [2]. And machines in a data center may different in processor (or core) architectures, memory capacity, and I/O capacity [2].

Many researches are trying to improve energy efficiency of data centers to reduce total power consumption. One way to improve energy efficiency is that leveraging machine heterogeneity of data centers, e.g., scheduling applications with different resource requirements to the most suitable machines which will make the cluster's power consumption minimal while meeting the performance goals of applications. It is proved in [3] that a hybrid data center design that mixes low power machines with high performance one scan handle diverse work loads with different service level agreements (SLA) in an energy efficient fashion.

In order to understand and leverage the heterogeneity of a cluster and the diversity of applications, this paper aimed to analyze and explore the heterogeneity from publicly available cluster traces, and propose a heterogeneity model for dynamic capacity provisioning problem in a data center. The remainder of the paper is organized as follows. Section 16.2 presents related work of current research literature. In Sect. 16.3, we present an analysis of publicly available Google cluster-usage traces and propose a heterogeneity model in Sect. 16.4. Finally, we draw our conclusion and further work in Sect. 16.5.

16.2 Related Work

Analysis of cluster trace has become a useful way to understand the feature of the cluster and characteristics of workloads from applications. There is some research on it [4–6], in which they analyzed different cluster trace for different purposes.

Authors of [4] analyzed a large production cluster trace (over 6 h and 15 min) made publicly available by Google, and based on the analysis results, they performed k-means clustering to classify incoming jobs and further do correlation analysis to help capacity planning. However, analysis of Map Reduce logs in Yahoo! cluster (logs for 10 months) resulted in an understanding of characteristics of workloads running in Map Reduce environments [5]. With the goal of better understanding the resource usages in workloads, [6] analyzed a workload trace of 6 compute clusters spanning 5 days. The research above focused on the characteristics of workload, such as task resource demand or usage for CPU, memory or disk, but they didn't consider heterogeneity related to the cluster or workload.

Cluster heterogeneity has been one concern of researchers for many years. Heath et al. [7] proposed a method to distribute the clients' requests to the different cluster nodes in a heterogeneous cluster in order to optimize performance metrics such as energy, throughput and latency. Andrew et al. [8] presented a design of a power-proportional web cluster consisting of heterogeneous machines, which makes use of energy-efficient hardware, and dynamic provisioning and scheduling to handle application workload and minimize power consumption.

In addition to cluster heterogeneity, there is also some research on the heterogeneity of workload [9, 10]. Zhan et al. [9] pointed that heterogeneous workload soften have different resource consumption characteristics and performance goals, and they proposed a resource provisioning solution for dealing with four typical workloads. Koller et al. [10] explored a linear relationship between power consumption and application throughout through experiments using benchmarks and running diverse applications, and proposed a power model to deal with heterogeneous applications.

In contrast to previous work, we consider to explore two classes of heterogeneity from cluster traces: machine heterogeneity and workload heterogeneity.

16.3 Analysis of Heterogeneity in Google Cluster Traces

Google cluster traces we analyzed are available in [11], which record the work of a production cluster at Google for 29 days. And the size of these traces is nearly 39 GB, including more than two thousand files. The detailed analysis of this trace can be found in our previous work [12]. From this analysis, we can obtain the following observations:

1. The traces contain three types of events which record the status and process of the whole cluster: machine event, job event and task event;
2. Machine event records the status (available, off or failed) of machines in the cluster. There are more than 12,000 machines in this cluster, and most of machines are always on during the whole trace period;
3. Job event and task event record the scheduling events in the cluster, which mainly record resource demands and usages. In this cluster, there are 672,003 jobs and 25,462,157 tasks to be recorded over 29 days.

Based on our analysis of Google cluster traces, we found two classes of heterogeneity: machine heterogeneity and workload heterogeneity.

16.3.1 Machine Heterogeneity

From machine event, we can classify all the machines into 10 types based on machine capacity and its platform. The result is shown in Table 16.1.

Table 16.1 Machine heterogeneity in Google cluster traces

Type	Capacity		Platform	Number of machines
	CPU	Memory		
1	0.5	0.25	1	3,864
2	0.5	0.5	1	6,728
3	0.5	0.12	1	54
4	0.5	0.03	1	5
5	0.5	0.75	1	1,003
6	0.5	0.97	1	4
7	0.25	0.25	2	126
8	0.5	0.06	1	1
9	1	1	3	795
10	1	0.5	3	3

Because of some security issue, there is few detailed information about the machine configurations in these traces, and all the information here is normalized between 0 and 1. In Table 16.1, the capacity of machine refers to the amount of CPU and Memory, and the platform means the architecture of the machine which is related to the execute speed and energy consumption, etc.

16.3.2 Workload Heterogeneity

Nowadays many data centers (e.g., Google, Amazon etc.) not only provide Internet applications to serve end users (e.g. user-oriented applications), but also run other applications which are not user oriented (e.g. Map Reduce, cluster management applications, etc.). Different applications may have different workload characteristics, e.g. arrival rate, task duration, resource requirement, and etc. This diversity feature of workload can be found in our analysis.

In the traces analyzed, workload consists of jobs and tasks, and a job is comprised of one or more tasks. And tasks are finally scheduled into machines to execute, so we only discuss workload heterogeneity based on tasks.

According to analysis of task event, we obtain the following observations:

1. There are 4 types of scheduling class, which represents the level of latency-sensitive and affects machine-local policy for resource access;
2. There are 12 types of priority, which means tasks with high priority can get preference for resources over tasks with low priority, and tasks with low priority running in the cluster may be preempted first if the scheduler could not find a machine to run an incoming task with a higher priority.
3. When a task comes into a cluster, it may request for machine resources (e.g., CPU, memory, disk space, etc.). In this paper, we only consider task requests for CPU and memory (we treat task request for CPU and memory together as resource request pairs) because these two resources are scarce and important for

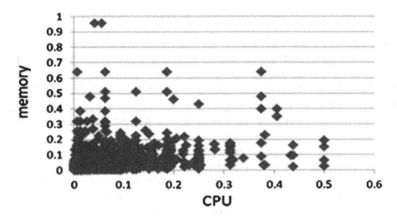

Fig. 16.1 Task statistics of the same resource request pairs

a cluster to schedule tasks. We did statistics to all the same resource request pairs, and got a result (values are normalized between 0 and 1) in Fig. 16.1. There are 10,289 resource request pairs in the trace, and Fig. 16.1 shows that most of the resource requests are small (e.g., most requests for CPU and memory are smaller than 0.2).

4. We can also classify tasks based on task duration, which represents how long a task is running in the cluster once it is scheduled into a machine to execute. We defined a task is a short task if its task duration is no longer than 5 min. The result of task duration analysis is shown in Table 16.2, and shows most of tasks (87 %) are short tasks (percentage refers to the ratio of the number of short tasks with priority over the total number of tasks in these traces multiplied by 100).

16.4 Modeling of Heterogeneity in Google Cluster Traces

In [12], we have proposed a model for dynamic capacity provisioning problem that minimizes the total energy cost while meeting the performance objective in terms of task scheduling delay, but this model is only suitable for homogeneous cluster. So we need to propose a new model for the heterogeneous case. Garg et al. [13] proposed a model for scheduling problem considering a heterogeneous cluster and jobs to achieve minimum worst-case energy consumption overall task type. But their work has an assumption that it need to be known ahead that an upper bound on the total rate at which jobs arrive at the data center, which may not be available in a real cluster.

We formally model the heterogeneity in this section, and our objective is to model a dynamic capacity provisioning problem in a heterogeneous cluster in order to minimal the power consumption of the heterogeneous cluster while minimizing SLA violations in terms of task scheduling delay.

Table 16.2 Analysis of task duration with different priority in cluster traces

Priority	0	1	2	3	4	5	6	7	8	9	10	11
Percentage (%)	17	6.7	4	0.003	54	≈ 0	2.5	≈ 0	0.5	0.2	≈ 0	≈ 0

Table 16.3 Constant variables for modeling the heterogeneity in Google cluster traces

Variable	Meaning
M	The total number of physical machines in a cluster
N	The number of machine types
M_j	The number of physical machine with type $j (1 \leq j \leq N)$
R	The types of resources in one machine
C_j^r	The capacity for resource type r in a machine with type $j (1 \leq r \leq R, 1 \leq j \leq N)$
T	The types of tasks which can be served in this cluster
S_i^r	The request for resource r from a task with type i $(1 \leq i \leq T)$
P^{SLA}	SLA penalty cost for a task if it violates SLA
D	Desired task scheduling delay, as an example of SLA

First, we define some constant variables in Table 16.3.

For simplicity, we assume all the tasks to be served in this cluster have the same desired task scheduling delay D.

Then, we discuss some variables related to time. At time k, the number of machines with type j is denoted as $m_k^j (1 \leq j \leq N)$. We assume that there are n_k tasks coming to this cluster, and we can classify these tasks into different task types using classification method.

Because the execution of tasks in machines with type j is independent from machines with other types, we can divide a cluster to N sub clusters (i.e., a sub cluster represents one machine type), and each type j has a waiting queue, whose length is written as $q_k^j (1 \leq j \leq N)$. And also the usage of resource r at time k in sub cluster j is $G_k^{r,j} (1 \leq r \leq R, 1 \leq j \leq N)$ available from the resource management. So the utilization of each sub cluster j for resource r can be denoted by $U_k^{r,j} (1 \leq r \leq R, 1 \leq j \leq N)$, and its expression is like:

$$U_k^{r,j} = \frac{G_k^{r,j}}{m_k^j C_j^r} \tag{16.1}$$

Like we did in [12], the utilization of the bottleneck resource b in sub cluster j can be calculated as:

$$U_k^{b,j} = \max_{1 \leq r \leq R} \{ U_k^{r,j} \} \tag{16.2}$$

Then the average scheduling delay of each sub cluster j can be expressed as:

$$d_k^j = f\left(U_k^{b,j} \right) = f\left(\max_{1 \leq r \leq R} \left\{ \frac{G_k^{r,j}}{m_k^j C_j^r} \right\} \right) \tag{16.3}$$

Where $f(\cdot)$ is a function that expresses the relationship between the scheduling delay and the utilization of the cluster, and the expression of $f(\cdot)$ can be obtained from experiment.

So the SLA penalty cost for sub cluster j at time k can be expressed as:

$$p_k^j = q_k^j P^{SLA} (d_k^j - D)^+ = q_k^j P^{SLA} \left(f \left(\max_{1 \le r \le R} \left\{ \frac{G_k^{r,j}}{m_k^j C_j^r} \right\} \right) - D \right)^+ \quad (16.4)$$

Here $(x)^+ = \max(x, 0)$.

So the energy consumption of one machine in sub cluster j can be got:

$$e_k^j = E_{idle}^j + \sum_{1 \le r \le R} \alpha^r U_k^{r,j} = E_{idle}^j + \sum_{1 \le r \le R} \alpha^r \frac{G_k^{r,j}}{m_k^j C_j^r} \quad (16.5)$$

E_{idle}^j is a constant, which is the energy consumption when a machine is idle. We assume each machine in the cluster has a power-proportional feature, which indicates energy consumption of a machine is proportional to its utilization.

So the problem can be formulated as an optimization problem as (16.6):

$$\min \sum_{1 \le j \le N} \begin{array}{c} q_k^j P^{SLA} \left(f \left(\max_{1 \le r \le R} \left\{ \frac{G_k^{r,j}}{m_k^j C_j^r} \right\} \right) - D \right)^+ \\ + m_k^j \left(E_{idle}^j + \sum_{1 \le r \le R} \alpha^r \frac{G_k^{r,j}}{m_k^j C_j^r} \right) \end{array} \quad (16.6)$$

subject to:

$$0 \le m_k^j \le M_j (1 \le j \le N) \quad (16.7)$$

The optimal result of (16.6) will be m_k^j (i.e., the number of machines with type j in time k), which means at time k, with this configuration, a cluster can get optimal energy consumption while minimizing SLA violations in terms of task scheduling delay. And according to the value of m_k^j, the cluster can add machines to work if current number of machines is not enough or turn off machines when current number of machines is more than actually needed.

16.5 Conclusion and Future Work

This paper analyzed datasets from Google cluster traces, and explored two types of heterogeneity from these traces. Based on analysis results, we proposed a heterogeneity model for dynamic capacity provisioning problem in a cluster to deploy Internet-scale services.

In future work, we should find an effective method to solve this heterogeneous model. And this model also needs to be improved in future work, such as different type of tasks may have different desired scheduling delay, considering machine has more power states, and etc.

Acknowledgments Special thanks to Qi Zhang, Mohamed Faten Zhani, Prof. Boutaba in University of Waterloo for their kind help. And this work is supported by Program for Changjiang Scholars and Innovative Research Team in University No.IRT1012.

References

1. Ahmad F, Chakradhar ST, Raghunathan A, Vijaykumar TN (2012) Tarazu: optimizing map reduce on heterogeneous clusters. In: Proceedings of ASPLOS 2012, 3–7 Mar 2012, London, UK
2. Nathuji R, Isci C, Gorbatov E (2007) Exploiting platform heterogeneity for power efficient data centers. In: Proceedings of the IEEE international conference on autonomic computing (ICAC), June 2007, Florida, USA
3. Chun B, Iannaccone G, Iannaccone G, Katz R, Lee G, Niccolini L (2009) An energy case for hybrid data centers. In: Proceedings of HotPower 2009, 10 Oct 2009, Big Sky, MT, USA
4. Chen Y, Ganapathi AS, Griffith R, Katz RH (2010) Analysis and lessons from a publicly available google cluster trace. Technical Report UCB/EECS-2010-95, 2010, UC Berkeley, USA
5. Kavulya S, Tan J, Gandhi R, Narasimhan P (2010) An analysis of traces from a production map reduce cluster. In: Proceedings of IEEE/ACM conference on cluster, cloud and grid computing (CCGrid), May 2010, Melbourne, Australia
6. Zhang Q, Hellerstein J, Boutaba R (2011) Characterizing task usage shapes in google compute clusters. In: Proceedings of LADIS, 2–3 Sept 2011, Washington, USA
7. Heath T, Diniz B, Carrera EV, Jr. Meira W, Bianchini R (2005) Energy conservation in heterogeneous server clusters. In: Proceedings of the tenth ACM SIGPLAN symposium on principles and practice of parallel programming, 15–17 June 2005, Chicago, USA
8. Krioukov A, Mohan P, Alspaugh S, Keys L, Culler D, and Katz R (2011) NAPSAC: design and implementation of a power-proportional web cluster, ACM SIGCOMM computer communication review, vol 41(1), pp 102–108
9. Zhan J, Wang L, Li X, Shi W, Weng C, Zhang W, Zang X (2012) Cost-aware cooperative resource provisioning for heterogeneous workloads in data centers accepted by IEEE transactions on computers, May 2012
10. Koller R, Verma A, Neogi A (2010) WattApp: an application aware power meter for shared data centers. In: Proceeding of the 7th international conference on autonomic computing, 07–11 June 2010, Washington, DC, USA
11. Googleclusterdata–google workloads (2011) http://code.google.com/p/googleclusterdata/
12. Zhang Q, Zhani MF, Zhang S, Zhu Q, Boutaba R, Hellerstein JL (2012) Dynamic energy-aware capacity provisioning for cloud computing environments. In: Proceedings of IEEE/ACM international conference on autonomic computing, Sept 2012, California, USA
13. Garg S, Sundaram S, Patel HD (2011) Robust heterogeneous data center design: a principled approach. SIGMETRICS Perf Eval Rev 39(3):28–30

Chapter 17
A String Approach for Updates in Order-Sensitive XML Data

Zunyue Qin, Yong Tang and XiaoBo Wang

Abstract When order-sensitive XML files require insertion updates, the existed labeling schemes can't avoid relabeling or recalculating, leading to higher costs. To solve this problem, this paper proposed a new labeling scheme, Valid ASCII String Labeling Scheme (VASLS). We established a labeling model with separated structure and order information. Then we designed algorithms to realize this scheme and prove its feasibility. The experimental results indicate, this scheme has smaller labeling space, zero sacrifice of query performance and low update costs as a result of no relabeling or recalculations.

Keywords Valid string · Order-sensitive XML document · Update · VAS labeling

17.1 Introduction

With the development of Internet technology, XML data has become the de facto standard of data exchange on the Web. It's of great significance to study XML data query processing. Some labeling schemes such as Regional labeling scheme [1–3]

Z. Qin (✉)
Department of Computer Science, SunYat-sen University, No 108 Ziwu Road,
Zhangjiajie 427000 Hunan, China
e-mail: qzystudy@163.com

Y. Tang
Department of Computer Science, SunYat-sen University,
Guangzhou 510006, China
e-mail: issty@mail.sysu.edu.cn

X. Wang
No 108 Ziwu Road, Zhangjiajie, Hunan 427000, China
e-mail: 635144908@qq.com

W. Lu et al. (eds.), *Proceedings of the 2012 International Conference on Information Technology and Software Engineering*, Lecture Notes in Electrical Engineering 210, DOI: 10.1007/978-3-642-34528-9_17, © Springer-Verlag Berlin Heidelberg 2013

and Prefix labeling scheme [4–7], were proposed to quickly determine the ancestor-descendent and parent–child relationships among nodes. Although these schemes can efficiently process various queries, relabeling and recalculating become necessary when updates come, which sacrifices the performance.

In many applications of XML, like XML publishing systems [1], there's a need to maintain the order of XML files in queries [3]. Therefore, when XML is updated, it's also necessary to maintain the orders of files. A lot of researches have been conducted to keep the document order in XML updating [5, 8–14], however, those approaches for updates enlarge the labeling space, degrade query processing efficiency, or lead to higher cost in recalculating and relabeling. DCPL labeling scheme in document [20] adopts VAS (Valid ASCII String) to completely avoid relabeling and recalculations in insertions, but it leads to the overgrowth of labeling size in frequent deletions and insertions.

This paper focuses on how to process update calculations in insertions without relabeling or recalculation, under zero loss of query performance and smaller labeling space. The contributions of this paper include:

1. A new algorithm, based on VAS, is proposed to slow down the increase of labels in frequent deletions and insertions;
2. We designed algorithms to implement VASLS in which the cost of updating order-sensitive XML documents, with no sacrifice of query performance and smaller labeling space, is lower than that of any other existing scheme;
3. The conducted experimental results show that VASLS can efficiently process order-sensitive queries and updates.

In Dewey scheme [15], a new insertion requires relabeling of the following sibling nodes and descendent nodes so that the order can be maintained. In Regional labeling scheme [2], inserting a new node affects the relabeling of its following and their ancestor nodes, including its own ancestor nodes. To solve this problem, Amgase [8] proposed using float-point for the "start"s and "end"s. However, in frequent insertions, relabeling still cannot be avoided. VLEI [16] and Sedna Scheme [9] adopted a bit sequence code to reduce the cost of updates, but sacrificed the query performance. ORDPATH [13] adopted Even Number Reservation to achieve better updates, but [17] pointed out that the necessary complex calculations for deciding the relationship among nodes degraded the query performance. QED [10] and CDBS [11] could efficiently process dynamic XML data: CDBS could partially avoid relabeling, while QED could completely, at the cost of larger labeling size and inferior query performance. [14] applied Prime number labeling scheme to order-sensitive XML documents, which, with better query performance, recalculates SC values based on the orders of new nodes after new insertions, and costs higher. In 2009, L Xu and others, on the basis of path labeling, proposed DDE Labeling [19], to support update efficiently, but it became more complex to determine the relationship between nodes, and the query performance was degraded. In 2010, Ko H and Lee S proposed IBSL [18] to improve CDBS [11], which could completely avoid relabeling other nodes in updates, and slow the growth of labeling space a little bit with the reuse of the deleted nodes, while the labeling space increased rapidly in accordance with the fan out and depth by applying the Binary String approach to all the path information, thus degraded the query performance.

The rest of the paper is organized as follows: Sect. 17.2 describes the properties of VASLS, and designs algorithms to achieve VAS labeling; Sect. 17.3 presents the results of the conducted experiments; Sect. 17.4 concludes the paper.

17.2 VAS Labeling Scheme

17.2.1 VAS Approach

VAS [20] approach is used to encode ASCII strings. The main feature is that it compares codes based on lexicographical order. With VAS, a new string can be inserted between any two consecutive strings without interrupting the order, relabeling or recalculating the existing VASs.

Definition 1 (*VAS*) String that consists of ASCII from the 33rd to 126th is named as Valid ASCII String (VAS). VAS = {S = s1s2...sm|33 ≤ ASCII(si) ≤ 126}.

Definition 2 (*Lexicographical order* <) Given two VAS strings, S_{left} and S_{right}, S_{left} ∈ VAS and S_{right} ∈ VAS. If S_{left} is exactly the same with S_{right}, they're lexicographically equal. To determine whether S_{left} is lexicographically smaller than S_{right}, i.e., $S_{left} < S_{right}$, the following procedures are performed:

1. For S_{left} and S_{right}, we need to make lexicographical comparisons from the left to the right. If the ASCII of S_{left} is smaller than that of S_{right}, then $S_{left} < S_{right}$, and the comparison is stopped, or
2. If len (S_{left}) < len (S_{right}), S_{left} is the prefix of S_{right}, and the ASCII of the first remaining string of S_{right} is equal to or greater than 34, then $S_{left} < S_{right}$, and the comparison is stopped, or
3. If len (S_{left}) < len (S_{right}), S_{right} is the prefix of S_{left}, and the ASCII of the first remaining string of S_{left} is 33, then $S_{left} < S_{right}$, and the comparison is stopped.

i.e., Given two VAS strings BBCFM and BBCG, BBCFM < BBCG (condition 1), HHVR < HHVR#A (condition 2), and WWER!Y < WWER (condition 3).

17.2.2 VAS Algorithms

With VAS approach, relabeling can be avoided no matter how a new node is inserted, as is shown in Fig. 17.1. Although it generates order-sensitive strings of infinite length, it would lead to the overgrowth of the size of new strings in frequent deletions and insertions. For example, there are two strings AAB and AAC (AAB < AAC). Based on the algorithm in document [20], five strings are inserted consecutively after AAB, and then the sequence (from the smallest to the greatest) would be AAB, AAC!!!!!, AAC!!!!, AAC!!!, AAC!!, AAC!, AAC. When we delete AAC!!!!, AAC!!! and AAC!!, the new sequence would be AAB, AAC!!!!!, AAC!, AAC. Now, a new string is inserted between AAC!!!!! and AAC!, and the new string would be AAC!!!!!# based on the algorithm in [20]. But when the algorithm below is adopted, the new string would be AAC!!, with a small size.

Algorithm: Reuse of Deleted Nodes
Input VAS_{left} and VAS_{right}
Output the new inserted VAS_{new}

1. Begin
2. if (VAS_{left} and VAS_{right} have the same prefix, and the distance between the next strings of them is greater than 1) //Case1
3. VAS_{new} = subString(VAS_{left}, 1, length (commPrefix (VAS_{left}, VAS_{right})+1)) $^{\oplus}$ (char At (VAS_{left}, length (commPrefix (VAS_{left}, VAS_{right})+1))+1);
4. else if (VAS_{right} is the prefix of VAS_{left}, and both of them end up with !) // Case2
5. VAS_{new} = VAS_{right} $^{\oplus}$!;
6. else if (VAS_{left} is the prefix of VAS_{right}, and both of them end up with #)// Case3
7. VAS_{new} = VAS_{left} $^{\oplus}$ #;
8. else if(VAS_{left} is empty but VAS_{right} is not empty) //Case4
9. VAS_{new} = VAS_{right} $^{\oplus}$!;
10. else if(VAS_{left} is not empty but VAS_{right} is empty) //Case5
11. VAS_{new} = VAS_{left} $^{\oplus}$ #;
12. else if (len (VAS_{left}) $\leq$ len (VAS_{right})) //Case6
13. VAS_{new} = VAS_{right} $^{\oplus}$!;
14. else if (len(VAS_{left}) > len(VAS_{right})) //Case7
15. VAS_{new} = VAS_{left} $^{\oplus}$#;
16. else//Case8
17. VAS_{new} = ";
18. return VAS_{new};
19. End

---

i.e. VAS_{left} = AABCDDF, VAS_{right} = AAEK, based on algorithm2 in [20], VAS_{new} = AABCDDF#; by the algorithm in this paper, VAS_{new} = AAC (Case 1); VAS_{left} = AAEM!!!!!, VAS_{right} = AAEM!, by [20], VAS_{new} = AAEM!!!!!#; by the new algorithm, VAS_{new} = AAEM!! (Case2); VAS_{left} = AAEY#, VAS_{right} = AAEY#####, by [20], VAS_{new} = AAEY#####!; by the new algorithm, VAS_{new} = AAEY## (Case3). Apparently, these three cases have shorter new strings.

As is shown in Fig. 17.4, the shadowy nodes are generated by the new algorithm and the dotted ones by the algorithm 2 in [20].

Fig. 17.1 Three insertions

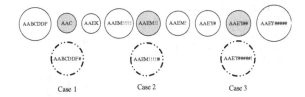

Case 1   Case 2   Case 3

Theorem 1 *The VAS strings generated by the algorithm, Reuse of Deleted Nodes, are lexicographically ordered.*

Proof

(1) Case 1 The common prefix of VAS_{left} and VAS_{right} is commPrefix, and the distance between the next strings of them is greater than 1. Assume that x is the following character of the commPrefix of VAS_{left}, and y is the following character of the commPrefix of VAS_{right}, then there must be a different character, s, between x and y, x<s<y, therefore VAS_{left} <commPrefix $\oplus$ s<VAS_{right}.

(2) Case 2 VAS_{right} is the prefix of VAS_{left}, and both of them end up with !, obviously $VAS_{left} < VAS_{right}{}^{\oplus}!$,and $VAS_{right}{}^{\oplus}!< VAS_{right}$ (condition 3 in definition 2).

(3). Case 3 VAS_{left} is the prefix of VAS_{right}, and both ofthem end up with #, obviously $VAS_{left} < VAS_{left}{}^{\oplus}\#$, and $VAS_{left}{}^{\oplus}\#< VAS_{right}$ (condition 2 indefinition 2).

(4) Cases 4, 5, 6, 7 and 8 have been proved in [20].

17.2.3 VAS Labeling Scheme

Definition 3 For any order-sensitive XML document tree, a labeling model which expresses the labels of nodes' structure information and the relative position of sibling nodes individually is named as Labeling Model with Structure Information and Order Information Separated.

Definition 4 Based on Definition 3, the scheme in which structure information applies Prime Number Labeling, and order information VAS Labeling, is named as VAS Labeling Scheme-Prime (VASLS-Prime Labeling).

Theorem 2 *VASLS-Prime can determine the structure and order relationship among nodes.*

Proof 1. Apparently, in any order-sensitive XML document, for nodes *u* and *v*, VASLS-Prime Labeling can determine the structure relationship, such as ancestor-descendent, parent–child, and so on. 2. In any order-sensitive XML tree, for nodes *u* and *v*, we, with structure information, can get the LCA (Lowest Common

Fig. 17.2 Labeling schemes
based on models of
separation **a** Structure-Prime,
order-BS **b** Structure-Prime,
order-VAS

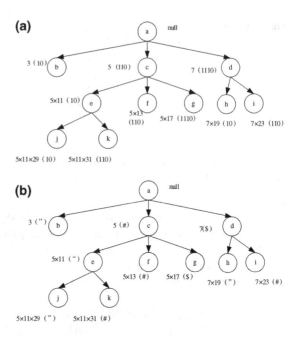

Ancestor) node of the two nodes, **x**. Therefore, with node **x**, we can conclude that
u' and *v*' are both the child nodes of x (if x exists), and the ancestor nodes of *u* and
v (if u and v exist). Based on Definition 4, the order information of *u*' and *v*' adopts
VAS. Based on Theorem 1, *u*' and *v*' are ordered, and the order of *u*' and *v*' can
determine that of *u* and *v*.   □

According to the labeling model with separation of structure information and order
information proposed in Definition 3, the structure information can adopt labeling
schemes like Dewey Labeling and Prime Number Labeling, and the order information can adopt labeling schemes like BS (Binary String) [19] and VAS.

   In Fig. 17.2 a and b, the structure information adopts Prime Number, while the
order information adopts BS Labeling and VAS Labeling. Judging the relationship
between *j* and *g*, we can get their LCA with Prime Number, then, the order
information of *e* is 10 (or" of VAS), and the order information of *g*, 1110(or $ of
VAS), so *e* is before *g*, that is, *j* is before *g*.

Theorem 3 *In the order-sensitive XML documents based on VASLS-Prime
Labeling, inserting nodes into any position needs neither relabeling nor recalculating other nodes.*

Proof Prime Number for the structure information and VAS for the order information. When inserting a child node **v** under node **u**, then the structure information
of new node **v**, structInfo(**v**) = structInfo(**u**) × New prime number. Based on
Algorithm 2, the order information of **v**, orderInfo(**v**) is generated. Thus the
structure information and the order information of **v** need neither relabeling nor
recalculating other nodes.   □

Fig. 17.3 Inserting a new
node n in order-VAS

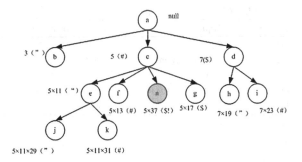

i.e., in Fig. 17.3, inserting new node, **n**, based on Theorem 3, needs neither relabeling nor recalculating other nodes.

Reference [19] points out that with BS Labeling for the order information, the labeling space of order-sensitive XML documents, under bigger fan out, will increase rapidly, therefore we only consider VAS for the order information in the following experimental analysis.

17.3 Performance Analysis

We selected real-world XML documents available as data sets, because they're different in fan out, depth, and total number of nodes. The experiments were conducted on a 2.53 GHz Intel Core 2 Duo CPU with 1 GB of RAM running on Windows XP Professional, and the order-sensitive XML document labeling is stored in a relational database, Oracle 9i. For the sake of experiment needs, we divided the labels into three parts and store them in the table (elementName, structInfo, orderInfo), to get better update performance without query degradation. The data sets are presented in Table 17.1.

17.3.1 Static Performance Research

17.3.1.1 Space Requirements

We compare the space requirements of VAS Labeling with the existed Dewey[15], V-CDBS-Prefix [11], Prime [14], OrdPath [13], IBSL [19] and DDE Labeling [18].

Figure 17.4 shows that the labeling size is determined by the maximum length and depth of labels. As VASLS adds the order information on the basis of the structure information, the length of VASLS is larger than that of Dewey, V-CDBS-Prefix, OrdPath and DDE Labeling, but smaller than that of Prime and IBSL.

Table 17.1 Characteristics of data sets

Dataset	Topic	Total no. of nodes	Max fan out	Max-depth
D1	Reed	10,546	703	4
D2	Sigmondrecord	11,526	67	6
D3	Wsu	74,557	3,924	4
D4	Shakespear plays	179,689	434	6
D5	Nasa	476,646	2,435	8

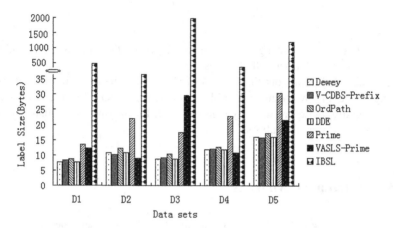

Fig. 17.4 Space for data sets

In data set D3, with bigger maximum fan out, the label of the order information is larger when adopting Algorithm 2. On the whole, the label has a larger size.

17.3.1.2 Query Performance

Given an XPath query, transform it into a corresponding SQL statement according to labeling schemes, and then execute it on the database. i.e., in Fig. 17.2, the SQL statement of VASLS-Prime Labeling Scheme is "select label from nodes N where N.structInfo/5 = 0". In Fig. 17.2, the SQL statement of VASLS-Prime is "select label from nodes N where N.structInfo is prime and c.orderInfo < N.orderInfo".

The query parsing time and query translation time are of the same size for all labeling schemes, therefore we present only the SQL execution time. We used the Shakespeare play data set (D4) for query tests. The test queries are shown in Table 17.2. Fig. 17.5 shows the response time of eight queries. For queries of the structure information (Q1, Q2, Q3, Q4), VASLS-Prime presents the best performance, because the structure information of nodes presented by Prime Number can easily determine the structure relationship between nodes; for queries of the order information (Q5, Q6, Q7), VASLS-Prime is close to other labeling schemes; in

Table 17.2 Test queries on the D4

ID	Queries
Q1	/speaker/ancestor::act
Q2	/play//speech
Q3	/scene/parent::act
Q4	/play//act/person
Q5	/play//act[3]//following::act
Q6	/speech[4]//preceding::line
Q7	/act//following sibling::speech[3]
Q8	/play/act//speech[3]/preceding-sibling::*

Fig. 17.5 Query response time

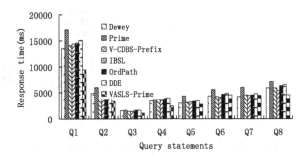

Q8, more tests of structure relationship are needed, thus VASLS-Prime presents better performance.

17.3.2 Dynamic Performance Research

17.3.2.1 Order-Sensitive Leaf Node Updates

The elements in Shakespeare play data set D4 are order-sensitive. Considering the real condition, we studied the update performance of Hamlet in this document. Hamlet node includes a series of order-sensitive act element child nodes, and we inserted a new act in these nodes. When inserting a new node, we counted the number of those nodes that need relabeling. Hamlet has five acts, and we tested the following five cases: inserting an act before act [1], and between act [1] and act [2], act [2] and act [3], act [3] and act [4], act [4] and act [5]. Hamlet node has 6636 child nodes, however, Dewey had to relabel 6605 nodes while Prime had to recalculate the value of SC when inserting an act before act [1]. Therefore, the update performance of them is pretty bad and other labeling schemes, without relabeling and recalculating, share similar update performance, as shown in Fig. 17.6.

Fig. 17.6 Update time for each insertion

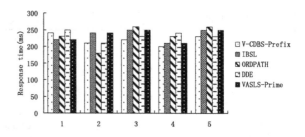

Fig. 17.7 Frequent leaf node updates

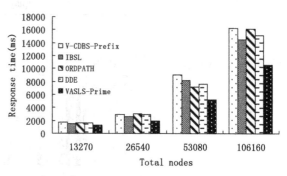

The influence of inserting a subtree in order-sensitive XML file is the same with that of inserting a node. Therefore, we ignore the experiment of inserting a subtree.

17.3.2.2 Frequent Leaf Node Updates

We studied the performance when the nodes were frequently inserted in order-sensitive XML files. Figure 17.7 shows the cases of frequently inserting nodes into different positions in order-sensitive XML files. Hamlet Node has 6,636 child nodes in total. We inserted a node in each pair of consecutive nodes in the 6636 nodes. After repeating the experiment for four times, we got 106,160 nodes in total. Figure 17.7 shows that, VASLS-Prime can considerably reduce the update time. Based on Algorithm 2, VASLS, without recalculating and relabeling, only needs to create new labels to complete the insertions; therefore, VASLS has lower update costs. We did not conduct experiments on the frequent update of Dewey and Prime, mainly because Fig. 17.6 has shown that Dewey and Prime have pretty high costs when inserting a node, and certainly their costs will be higher than any other labeling scheme when inserting a lot of nodes.

17.4 Conclusions

This paper proposed a new labeling scheme based on the separation of structure information and order information, VASLS. When inserting a new node in order-sensitive XML files, VASLS can assign new labels to newly-inserted nodes without revising any previous labels of nodes or recalculating the value of the existing labels. The experiment results have shown that, VASLS doesn't sacrifice query performance and its labeling space is under control. When large scales of insertions occur, VASLS can maintain low update costs. In the future, we will focus on how to reduce the labeling length and process the update of the internal nodes.

Acknowledgments This work is supported by the National Science Foundation of China (60970044, 60736020).

References

1. Agrawal R, Borgida A, Jadadish HV(1989) Efficient management of transitive relationships in large data and knowledge bases. In: ACM SIGMOD, New York, 253–262
2. Li Q and Moon B (2001) Indexing and querying XML data for regular path expressions. In: Prceedings of the 27th VLDB conference 2001, Roma, 361–370
3. Zhang C, Naughton JF, DeWitt DJ, Luo Q, Lohman GM (2001) On supporting Containment queries in relational database management systems. In: ACM SIGMOD, New York, 425–436
4. Abiteboul S, Kaplan H, Milo T (2001) Compact labeling schemes for ancestor queries. In: Proceedings of symposium discrete algorithms (SODA), PA, 547–556
5. Cohen E, Kaplan H, Milo T (2002) Labeling dynamic XML trees. In: Proceedings of symposium principles of database systems (PODS), New York, 271–281
6. McHugh J, Abiteboul S, Goldman R, Quass D, Widom J (1997) Lore: a database management system for semistructured data. In: ACM SIGMOD Record, 2(3):54–66
7. Silberstein A, He H, Yi K, Yang J (2005) Boxes: efficient maintenance of order-based labeling for dynamic XML data. In: Proceedings of the 21st international conference on data engineering (ICDE), Tokyo, Japan, 285–296
8. Amagasa T, Yoshikawa M, Uemura S (2003) QRS: a robust numbering scheme for XML documents. In: Proceedings of the 19st international conference on data engineering (ICDE), Bangalore, India, 705–707
9. Fomichev A, Grinev M, Kuznetsov S (2006) Sedna: a native XML DBMS. In: Proceedings on current trends in theory and practice of computer science (SOFSEM), Berlin Heidelberg, Germany, 272–281
10. O'Neil P, O'Neil E, Pal S, Cseri I, Schaller G (2004) ORDPATHs: insert-friendly XML node labels. In: ACM SIGMOD, Paris, 903–908
11. Li C and Ling TW (2005) QED: a novel quaternary encoding to completely avoid Re-Labeling in XML updates. In: Proceedings of the 14th ACM international conference on Information and knowledge management (CIKM), Bremen, Germany, 501–508
12. Li C, Ling TW, Hu M (2006) Efficient processing of updates in dynamic XML data. In: Proceedings of the 22st international conference on data engineering (ICDE), Atlanta, USA, 13–22
13. Wu X, Lee ML, Hsu W (2004) A prime number labeling scheme for dynamic order-sensitive XML trees. In: Proceedings of the 20st international conference on data engineering (ICDE), Boston, USA, 66–78
14. Ko H, Lee S (2010) A binary string approach for updates. IEEE Trans Knowl Data Eng 22(4):602–608
15. Tatarinov S, Viglas D, Beyer K, Shanmugasundaram J, Shekita E, Zhang C (2002) Storing and querying order-sensitive XML using a relational database system. In: ACM SIGMOD, Wisconsin, p 204–215

16. Xu L, Ling TW, Wu H, Bao Z (2009) DDE: from dewey to a fully dynamic XML labeling scheme. In: ACM SIGMOD, Rhode Island, 719–730
17. Li C, Ling TW, Hu M (2006) Reuse or never reuse the deleted labels in XML query processing based on labeling schemes. In: Proceedings of international conference database systems for advanced applications (DASFAA), LNCS 3882, 659–673
18. Kobayashi K, Liang WX, Kobayashi D, Watanabe A et al (2005) VLEI code: an efficient labeling method for handling XML documents in an RDB. In: Proceedings of the 21st international conference on data engineering (ICDE), Tokyo, Japan, 386–387
19. Li C, Ling TW, Hu M (2008) Efficient updates in dynamic XML data: from binary string to quaternary string. J Very Large Databases 17(3):73–601
20. Qin ZY, Tang Y, Xu HZ (2012) A string approach for dynamic XML document. Appl Mech Mat 220–223:2512–2519

Chapter 18
Fault Diagnosis of Nodes in WSN Based on Particle Swarm Optimization

Chengbo Yu, Rui Li, Qiang He, Lei Yu and Jun Tan

Abstract In the Wireless Sensor Network (WSN), the operation reliability is usually evaluated by processing the measured data of the network nodes. As the problems of the large energy consumption and complex calculation in traditional algorithms, a method for fault diagnosis of nodes in WSN based on particle swarm optimization is proposed in the paper. The range of threshold value is obtained by optimizing the measured data of nodes according to the fast convergence rate and simple rules of characteristics of the PSO. The judgment of the nodes' malfunction is determined by analyzing the relationship between the measured data and the range of threshold value. The experimental results show that the method of fault diagnosis can find the fault nodes promptly and effectively and improve the reliability of WSN greatly.

Keywords Wireless sensor network (WSN) · Fault diagnosis · Particle swarm optimization (PSO)

18.1 Introduction

The research and innovate of WSN nodes' hardware design, computing process, wireless communication, network protocol and energy efficient are put forward constantly in recent years and the demand of the network's reliability and

C. Yu (✉) · R. Li · Q. He · L. Yu · J. Tan
Research Institute of Remote Test and Control, Chongqing University of Technology, Chongqing 400054, People's Republic China
e-mail: yuchengbo@cqut.edu.cn

R. Li
e-mail: 402618917@qq.com

W. Lu et al. (eds.), *Proceedings of the 2012 International Conference on Information Technology and Software Engineering*, Lecture Notes in Electrical Engineering 210, DOI: 10.1007/978-3-642-34528-9_18, © Springer-Verlag Berlin Heidelberg 2013

sustainability is increasing, the fault diagnosis of nodes in WSN plays an important role in understanding the network's state in real time [1]. But the probability of sensor nodes' failure is much higher than other systems due to many inevitably factors and the complex and harsh environment. The fault nodes will reduce the service quality of the whole network and they will also produce or transmit wrong sensor data, these make the surveillance center can't receive the correct detection information and then produce wrong decisions which may cause a heavy loss or even the whole network paralyzed. Therefore, the study for fault diagnosis of nodes in WSN is very necessary.

Aiming at the merits and demerits of some typical algorithms for WSN nodes' fault diagnosis (distributed Bayesian algorithms [2], weighted median fault detection [3] and distributed fault detection [4]), the worldwide scholars study a lot and put forward many other methods. Literature [5] puts forward a solution based on MANNA hierarchical structure topology, this method needs to do centralized diagnosis on nodes through external base stations and the communication cost is large. Literature [6] puts forward a method based on tree structure for WSN fault diagnosis, this method has a high diagnosis precision and good robustness, but the structure is complex. Literature [7] manages the other nodes' operation through clustering nodes, the algorithm has a high precision but it is not available for the network with high failure rate. To solve these problems, this paper proposed a way based on particle swarm optimization to diagnose the fault nodes in WSN. The algorithm is simple, the convergence rate is fast and it can detect the fault nodes effectively.

18.2 Fault Diagnosis Algorithm for WSN Nodes

18.2.1 Algorithm Ideas

Finding the abnormalities timely can ensure the accuracy and reliability of the measurement data through fault diagnosis. The fault diagnosis idea is based on the PSO which has the characteristics of simple calculation, fast convergence velocity and high quality solutions. First, use the PSO to search the measurement data globally and then get a global extremum. Second, confirm the range of threshold value by this global extremum and get the nodes with possible fault. Third, judge these nodes whether are broken or not by the scope of threshold value and Gaussian distribution. Finally, judge which nodes are broken according to outlier data and the sensor nodes' address information.

18.2.2 Particle Swarm Optimization

The particle swarm optimization [8] (PSO) is an evolutionary computation suggested by Dr. Kennedy and Dr. Eberhart and it comes from the imitation of fish and birds' action. The PSO is based on groups and it regards every individual as a particle (point) without volume or quality in D-dimensional searching space, the particles fly at a certain speed and adjust this speed dynamically according to their own or other particles' flying experience. Express the ith particle in groups as $Z_i = (z_{i1}, z_{i2}, \ldots, z_{iD})$ and record the best position (the best fitness) it experienced as $P_i = (p_{i1}, p_{i2}, \ldots, p_{iD})$, called *pbest*. And the best position that the whole particles experienced is called *gbest*. Express the ith particle's speed as $V_i = (v_{i1}, v_{i2}, \ldots, v_{iD})$, and to each iteration, the ith particle's motion in d-dimensional space $(1 \leq d \leq D)$ changes according to follow formulas:

$$v_{id}^{k+1} = \omega v_{id}^k + c_1 r_1 (p_{id} - z_{id}^k) + c_2 r_2 (p_{gd} - z_{id}^k) \tag{18.1}$$

$$z_{id}^{k+1} = z_{id}^k + v_{id}^{k+1} \tag{18.2}$$

In Eqs. (18.1) and (18.2), $i = 1, 2, \ldots, m$, $d = 1, 2, \ldots, D$, k is the iterations, r_1 and r_2 are random numbers in $[0, 1]$, and they are used to keep the groups' diversity. c_1 and c_2 are acceleration constants, they make particles have the ability of self-sum and learn from excellent individuals in the group, then every particle can move to *pbest* and *gbest* faster. The second part in Eq. (18.1) is the cognition part, it represents particles' self-study. The third part is the social part and it represents the sharing and cooperation among particles. Particles in Eq. (18.1) refresh flying speed according to the speed, current position and the distance between their own or others' best experiences, and then they fly to the new position by Eq. (18.2). Parameter ω is the inertia weight, it plays an important role in whether PSO has perfect convergence and makes particles keep inertial motion and have the trend of space extension and the ability of searching new areas. The function of ω is as follows:

$$\omega = \omega_{\max} - \frac{\omega_{\max} - \omega_{\min}}{k_{\max}} \times k \tag{18.3}$$

In Eq. (18.3), $\omega_{\max}$ is the initial weight, $\omega_{\min}$ is the final weight, $k_{\max}$ is the maximum iterations, k is the current iterations [9]. Figure 18.1 is the flow chart of the PSO:

18.2.3 Algorithm Implementation

In the PSO, *pbest* and *gbest* are determined by each particle's fitness. The temperatures which measured by WSN nodes are important indicators that reflecting whether the nodes are broken or not. Temperatures range from 0 to 50 °C in

Fig. 18.1 The flow chart of
the PSO

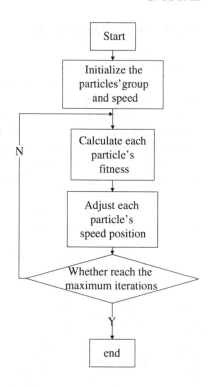

common environment. In the sample space of temperatures, the collected individual samples fluctuate in the vicinity of the value which appears in a large probability. In the process of PSO initialization, give out a poor fitness value which can't be selected as the extremum in iterative computations. The initial fitness value may be replaced by particles in better positions through positions' change in iterative process, but it has no influence on particle swarm's normal evolution [10].

According to the analysis above, this paper uses normalization function to get the ideal fitness function:

$$fitness(t_i) = 51 \times \frac{t_i - t_{min}}{t_{max} - t_{min}} \tag{18.4}$$

In Eq. (18.4), t_i is one sample in the sample space of temperatures, t_{min} is the minimum and t_{max} is the maximum in the sample space of temperatures.

After 100 iterations with PSO, the global extreme value P_g is obtained, and then confirm the scope of temperature threshold value. If temperatures measured are in the scope of threshold value, they are identified as normal nodes, or they are the nodes with possible fault. Figure 18.2 is the flow chart reflects the temperature that the PSO processed, the calculation is as follows:

1. Collect 100 temperatures T_i that the temperature sensor nodes measured as samples, and calculate each particle's fitness. Temperatures of particles with

Fig. 18.2 The flow chart
reflects the temperature that
the PSO processed

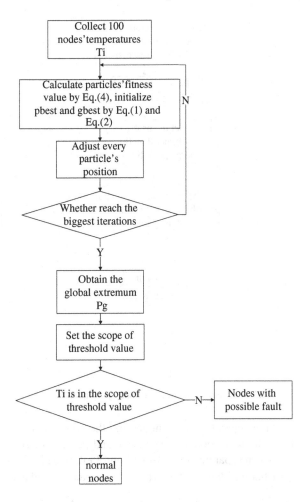

the biggest fitness value are set as the initial *gbest*, and other particles set their
own *pbest* as their initial position.

2. After iteration, choose the particles with the biggest fitness value as global
   extremum *gbest^k* and each particle chooses individual extremum *pbest^k*
   according to its own flying record.
3. Put *pbest^k*, *gbest^k* and z_i in Eqs. (18.1) and (18.2), update each particle's speed
   and position. If the current iterations reach 100 or the minimum error
   requirements, then stop iteration and obtain the global extremum P_g, or switch
   to step (2).
4. Confirm the scope of temperature threshold value through P_g and sensors'
   precision. If T_i is in the scope of threshold value, it's identified as normal nodes,
   or it's the nodes with possible fault.

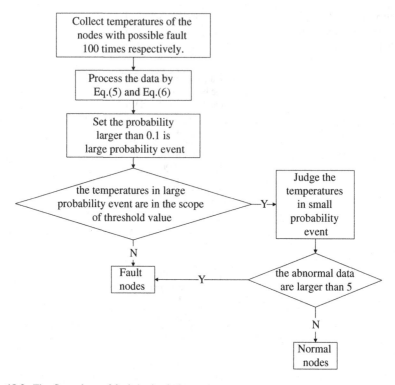

Fig. 18.3 The flow chart of fault nodes judgment

After separate normal nodes and the nodes with possible fault, collect 100 data from the latter nodes respectively and process these data with Gaussian distribution, then separate normal nodes and fault nodes. The temperatures meet Gaussian distribution, there is $X \sim N(\mu, \sigma^2)$, and the formulas are as follows:

$$\mu = \frac{\sum\limits_{i=1}^{k} t_i}{k} \tag{18.5}$$

$$\sigma^2 = \frac{\sum\limits_{i=1}^{k} (t_i - \mu)^2}{(k-1)} \tag{18.6}$$

In Eqs. (18.5) and (18.6), k is represented particles' numbers.

We consider the probability ranges from 0.1 to 1 is large probability event with consideration of sensors' precision. Two situations may appear at present: first, if the temperatures in large probability event aren't in the scope of threshold value, we can judge these nodes as fault nodes directly. Second, if the temperatures in large probability event are in the scope of threshold value, if the abnormal data are larger than 5, we judge them as fault nodes. Otherwise, they are normal nodes. Figure 18.3 shows the flow chart of fault nodes judgment:

But in the experimental process there may have the situation that the data measured are correct but the nodes are broken in practice. Therefore, after the first round the algorithm accomplished, repeat the experimental process, so that it can reduce the wrongful diagnosis phenomenon and improve the accuracy.

18.3 Simulation Test and Result Analysis

This paper uses sensor nodes based on Zigbee wireless communication protocol and collects the temperature data indoors through module JN5139-Z01-M02. Place 100 nodes indoors with sunshine one side and overcast another side. The mean temperature is 23 °C on the day. Analyze the temperature data through the algorithms in the front.

The maximum iterations is set as $k_{max} = 100$, the initial inertia weight is $\omega_{max} = 1.2$, the final inertia weight is $\omega_{min} = 0.9$, the accelerated factors is $c_1 = c_2 = 2$, the temperatures measured from the 100 nodes are expressed by 10-dimensional matrix as follows:

$$
\begin{bmatrix}
21 & 20 & 23 & 23 & 20 & 19 & 20 & 20 & 23 & 20 \\
20 & 23 & 21 & 23 & 27 & 19 & 22 & 20 & 17 & 23 \\
19 & 13 & 23 & 20 & 23 & 22 & 23 & 19 & 23 & 20 \\
26 & 22 & 20 & 23 & 19 & 23 & 20 & 22 & 23 & 19 \\
23 & 20 & 20 & 23 & 20 & 19 & 20 & 23 & 20 & 20 \\
20 & 20 & 23 & 19 & 19 & 42 & 23 & 33 & 20 & 20 \\
23 & 19 & 23 & 20 & 20 & 23 & 20 & 23 & 20 & 23 \\
20 & 19 & 21 & 22 & 22 & 19 & 23 & 45 & 19 & 20 \\
23 & 20 & 23 & 20 & 20 & 23 & 20 & 23 & 20 & 23 \\
20 & 23 & 22 & 23 & 22 & 19 & 23 & 20 & 20 & 19
\end{bmatrix}
$$

Simulate the data through MATLAB and obtain the global extremum value: 21.709. As the sensor nodes' precision is ±5 °C, set the error range as ±2 °C according to the experimental environment and sensor nodes' precision and the threshold range is (19.709, 23.709). Take the nodes addresses with 126F and 195E as examples, the temperatures measured are 17 and 26 °C and they are both out of the threshold range, so collect these two nodes' data 100 times respectively again. After the analysis through Gaussian distribution, we find that to the data measured from the node with address 126F, the data in the large probability event aren't in the threshold range. According to the analysis, this node can be judged as the fault node. And to the data measured from the node with address 195E, all the data in the large probability event are in the threshold range and the abnormal temperature data in the small probability event are 26 and 27°C, the number of the abnormal data is smaller than 5, so we can judge this node as normal node.

In order to improve the diagnosis accuracy, repeat the experimental process and there finds no nodes with misjudgment.

18.4 Conclusion

The experiments indicate that the PSO can be applied to the WSN nodes fault diagnosis effectively and the algorithm is simple, the convergence rate is fast and the diagnosis accuracy is high. But this method is not suitable for the situation that the number of the WSN nodes is large and the failure rate is high. Therefore, the further work is to improve the PSO to expand the scope of application.

Acknowledgments This research is sponsored by Project of Chongqing Technology Research (CSTC, 2011AC2179) and Project of Chongqing Economy and Information Technology Commission (Yu Economy and Information Technology [2010] No. 9) and Project of Chongqing Jiulongpo District Technology Committee (Jiulongpo District Technology Committee [2009] No. 52).

References

1. Ji S, Yuan S, Ma T, Tian W (2010) Method of fault detection for wireless sensor networks. Comput Eng Appl 46(23):95–97, 121 (in Chinese)
2. Bhaskar K, Sitharama L (2004) Distributed Bayesian algorithms for fault tolerant event region detection in wireless sensor networks. IEEE Trans Comput 53(3):241–250
3. Gao J, Xu Y, Li X (2007) Weighted-median based distributed fault detection for wireless sensor networks. J Softw 18(5):1208–1217 (in Chinese)
4. Chen J, Kher S, Arun S (2006) Distributed fault detection of wireless sensor networks. In: Proceedings of the ACM international conference on mobile computing and networking. ACM Press, New York, pp 65–72
5. Ruiz LB, Siqueira IG, Oliveira LB et al (2004) Fault management in event-driven wireless sensor networks. MSWiM '04: Proceedings of the 7th ACM international symposium on modeling, analysis and simulation of wireless and mobile systems
6. Xianghua X, Wanyong C, Jian W, Ritai Y (2008) Distributed fault diagnosis of wireless sensor networks. In: 11th IEEE international conference on communication technology proceedings, pp 148–151 (in Chinese)
7. Asim M, Mokhtar H, Merabti M (2009) A cellular approach to fault detection and recovery in wireless sensor networks, computer science, pp 352–357
8. Huang H (2009) Fuzzy control strategy of the road tunnel ventilation based on particle swarm optimization. Hunan: mechanical vehicle engineering institute of Hunan university (in Chinese)
9. Li M, Wang Y, Nian F (2010) Intelligent information processing and application. Publishing House of Electronics Industry, Beijing 124–127 (in Chinese)
10. Yu C, Zhang Y, Li H (2012) The research of self-calibration location algorithm based on WSN. J Vibr Meas Diagn 32(1):6–10 (in Chinese)

Chapter 19
Beat Analysis Based on Autocorrelation Phase Matrix

Yanzhu Gong, Bing Zhu, Hui Wang and Yutian Wang

Abstract Rhythmic information has been used in the analysis of acoustic music widely, in which the onset extraction and beat tracking technology take the fundamental position. In this paper, we present a novel approach to analysis the beats information in WAV format data which is based on the Autocorrelation Phase Matrix (APM) data structure. This method based on periods clustering combines peak and time to detect beat; then uses dynamical beat tracking technology to specific locations of the notes and beats. The experimental results show that the method used in this paper is able to achieve ideal results in beat analysis.

Keywords Rhythm feature · Autocorrelation phase matrix (APM) · Beat analysis · Dynamic tracking

19.1 Introduction

The analysis of acoustic signals is the basis of the representation and manifestation of music, and the fundament of music cognition, music retrieval, music recovery and so on. Rhythm plays an important part in the musical characteristics. Rhythm generally refers to the regular timing of music events and can be seen as a hierarchical form of notes duration and accents on beats. So this paper focuses on the beat extraction and present the notes layer and the music tempo.

Y. Gong (✉) · B. Zhu · H. Wang · Y. Wang
School of Information Engineering, Communication University of China,
Beijing 100024, China
e-mail: yanzhudxw@cuc.edu.cn

W. Lu et al. (eds.), *Proceedings of the 2012 International Conference on Information Technology and Software Engineering*, Lecture Notes in Electrical Engineering 210, DOI: 10.1007/978-3-642-34528-9_19, © Springer-Verlag Berlin Heidelberg 2013

There have been many attempts to build a computational model to process music signals. In 1998, Scheirer proposed a typical oscillator models [1], it took the signals as inputs to the models and utilized oscillators to highlight the sharp peaks. Cegmil et al. demonstrated a perceptual model [2] based on probabilistic knowledge. He trained the onset sequences using Kalman filter, and finally got the envelope of the tempo in MIDI format data. In 2001, a multi-agent model for MIDI music [3], which was presented by Simon Dixon, got a desired result for the extraction of the beat position. Goto merged his previous models into one [4], he divided the frequency of the signal into seven ranges then detected beat in multi-agent. In 2005, Martin Dostal put forward a model [5] based on genetic algorithm, which took advantage of Artificial Intelligence. Afterwards, Klapuri et al. [6] demonstrated a graphical probabilistic model, Hoffman et al. presented the Codeword Bernoulli Average model [7], which are all the applications of the probabilistic models. While in addition to the above models, there is a typical algorithm widely applied in beat extraction: autocorrelation algorithm. Compared with the above-mentioned models, the advantages of autocorrelation is its good portability and simplicity, thereby the transplantation and application of related ideas have emerged in many methods such as ref [1] and [4].

Autocorrelation first began to be applied in rhythm analysis in the 1990s. Brown [8] weighted the notes by duration, then used autocorrelation information to induct meter. Her research got expected results in spectrum analysis of MIDI format signals. Gouyon and Herrera [9] incorporated autocorrelation into the process of beat detecting. However, Gouyon's focus was on classifying musical chord by beat. And as such, the results only showed the Duple/Triple style the meter is. Toiwiainen and Eerola [10] demonstrated a new autocorrelation-based algorithm (a MIDI-oriented approach) that incorporated event duration and melody to detect beats. In 2005, Eck and Casagrande [11] proposed the concept of APM (Autocorrelation Phase Matrix). By making use of APM data structure and Shannon Entropy to identify metrical structure, he solved the problem that autocorrelation didn't work well for many kinds of music especially for the non-percussive ones in certain. Subsequently in 2006, Eck published another paper based on APM [12] that showed improved prediction results compared with previous attempts. Holzapfel and Stylianou [13] combined the autocorrelation-based approach of Toiviainen with rhythmic similarity and explored the problem of automatic classification in traditional music. In 2011, Ellis et al.[14] overviewed recent music signal processing technologies; the authors specifically addressed the application of autocorrelation algorithm in dynamic tracking.

As mentioned earlier, APM data structure designed by Eck [11, 12] gives a solution to beat induction of non-percussive music signals, compensating the common limitation of autocorrelation-based methods, yet with a lower accuracy of about 65 %. In this paper, we design a procedure that adopts the APM and entropy, but chooses new detection methods to induct beat. After inducting the beats, the signals are tracked with multipath dynamically so that a more desirable outcome could be got.

In Sect. 19.2, the key idea of our algorithm and detailed implementation are introduced.

In Sect. 19.3, our simulation results and discussions are presented.

Section 19.4 is conclusion which is summaries and extensions of our work.

19.2 Algorithm Details

As mentioned earlier, in order to extract beat information, the duration of every notes and accent structure of the notes should be got. Therefore, to obtain the required information, the onset sequence should be got firstly, then the strength level of every notes would be detected, after that we track the music for accurate beat result. At last all the notes and beats events are presented in figure, and the tempo of the music is also presented. Now each step is introduce in detail.

19.2.1 Onset Detection

The first step is pre-processing the music files. Here, wavelet denoising methods are applied to reduce the interference of the signal.

The next is convolving the signal with the first derivation of a Gaussian function in order to obtain the envelope filtered and differentiated of the signal.

After that, threshold is set to search sharp peaks where notes are expected to occur. These results we get are sent to the next stage of beat prediction later.

19.2.2 Beat Prediction

This subsection includes three parts: (1) Calculating envelope, (2) Analyzing beat, and, (3) Tracking. The flow chart shown in Fig. 19.1 illustrates our beat prediction process.

19.2.2.1 Calculating Envelope

Spikes in an autocorrelation align with the periods at which high correlation exists in a signal, hence, many scholars predict beats using the autocorrelation sequence, Whereas, the significant problem of autocorrelation is that it can't extract the beat information of non-percussive music effectively.

In the first part, we utilize the concept of autocorrelation-phase matrix proposed by Eck, and calculate its Shannon entropy, whose formulas are given by Eqs. (19.1) and (19.2) as follows.

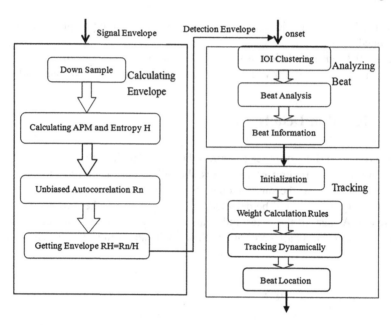

Fig. 19.1 Overview of our beat prediction process

$$A_{\text{unbiased}}(k, \phi) = \sum_{i=0}^{\frac{N}{k}-1} x(ki + \phi)x(k(i + 1) + \phi) \bigg/ \sum_{i=0}^{\frac{N}{k}-1} 1 \qquad (19.1)$$

$$A_{\text{sum}} = \sum A_{\text{unbiased}}(k, :)$$

$$H(k) = -\sum_{i=0}^{N} \frac{A_{\text{unbiased}}(k, \phi)}{A_{\text{sum}}} \log_2 \frac{A_{\text{unbiased}}(k, \phi)}{A_{\text{sum}}} \qquad (19.2)$$

The APM records the beat information in terms of both period and phase. The phase-aligned elements compensate for the defect of traditional autocorrelation-correlated approaches. As entropy indicates the 'disorder' in a system, the distribution of each row of the APM can be obtained using entropy showed by Eq. (19.2).

According to experiments results and corresponding analyses, when the lag k meets the period, the minimum entropy and sharpening autocorrelation value could be got. Thus, our work make use of autocorrelation divided by the entropy as a test envelope (Eq. 19.3) so that the starting time of a beat and the length of it will be maintained.

$$RH = R_n(k)/H(k) \qquad (19.3)$$

19.2.2.2 Analyzing Beat

Beside the method of matching peaks with time into arrays used in [15], we built the method on *IOI* clustering (*IOI* is defined as inter-onset intervals). Furthermore, the array is expanded into 4d to obtain better analysis result

Because music speed commonly appears between 60 and 140 *bpm*, our algorithm is designed within this speed limit. Firstly, the onset sequence is clustered in accordance with 'candidate beat', '2 × candidate beat' and '4 × candidate beat' into three clusters $C1/C2/C3$. After looping through the three clusters, we find all the matches and produce the array as [peak, beat, 2 × beat, 4 × beat] in which the 'peak' is the sum of the corresponding matches' amplitude of envelope *RH*.

Once all the states are covered, the desired elements of beat got the maximum peak, and its corresponding rhythm value can be sent to the following tracking section.

19.2.2.3 Beat Tracking

Till now, the period got from Section 19.2.2 still can't match the beat location exactly, and moreover, the predictions are the correct periods multiplied by $\frac{1}{2}, \frac{1}{3}, 2$ or 3 other than the target results in some situations. This section impact multi-agent beat tracking architecture [16, 17], while more circumstances are considered in tracking to ensure that the results could be more accurate.

The beat locations are determined by first location n and beat values *rhy*. As the path with different n and *rhy* is tracking, scores will be calculated in accordance with the rules depicted in Fig. 19.2. This method considers three states according to the location of the predictions, of which the third state is divided into three parts for different position of predictions with lag *rhy* multiplied by i ($i = 1 : 3$). After that, the highest scoring path is chosen as the best result.

Inner window threshold is set as 0.07 s, and outer window threshold is minimum of *IOI* divided by 1.5.

predictions locates:

$$
\begin{cases}
\text{in inner window:} & w = w + N * 10 \quad N = N + 1 \\
\text{in outer window:} & w = w + 3 \qquad\quad N = 0 \\
\text{out the window:} & \\
& (\text{pre} + \text{rhy} \times i)\,\text{locates}
\begin{cases}
\text{in inner window:} & w = w + N * 5 \\
\text{in outer window:} & w = w + 1 \quad N = 0 \\
\text{out the window:} & w = w - 3 \quad N = 0
\end{cases}
\end{cases}
$$

Fig. 19.2 Rules of score calculation

19.3 Simulation and Results Analysis

19.3.1 Experimental Data

In our work, we test the correctness of the algorithm with 61 monophonic songs including three kinds of musical instruments: piano, violin and flute. All the samples are in WAV format whose speed ranging from 60 to 140 *bpm*.

These files are created by using software Cakewalk, Wingroove and Cooledit. The specific steps are:

1. Get appropriate MIDI files
2. Modify the MIDI files with Cakewalk reserving desired tracks and tones
3. Convert the MIDI files modified to WAV format
4. Select the segmentation needed

19.3.2 Simulation Results

For each piece of the database, the parameters noted in Sect. 19.2 are extracted, and the locations of every note and beat are plotted afterwards. The output of our work is showed with an example named 'Minuetto' whose testing results are depicted in Fig. 19.3. While the horizontal axis represents time, the waveform of 'Minuetto' is plotted along the vertical axis. Furthermore, the onset sequence with magnitude 1 and the beat sequence with magnitude 1.5 are both noted in the figure. It can be seen that our method can detect beat information effectively comparing the results with the score of the signal (Fig. 19.4). The statistical results of the database are listed in Table 19.1.

With regards to this part, the prediction within 2.5 % of the target beat is thought to be correct. According to the experiments our work gets ideal results.

19.3.3 Discussion

As is performed in Sect. 19.3.2, our method works well on the basic meter extraction, and it also has a good ability of robustness.

The statistics in Table 19.1 indicate that the method is more suitable to piano signals than the other two kinds. It's mainly caused by Sect. 19.2.1. For those non-percussive signals, the energy-based onset algorithm can't gain desirable outcome. Although the APM data structure store the beat information of non-percussive music via phase and period, it can't get rid of the errors introduced by the preceding onset detection part.

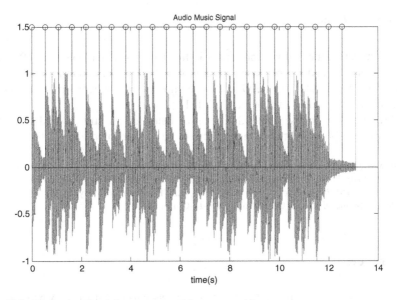

Fig. 19.3 Music signal and the sequence of the notes and beats

Fig. 19.4 Score of 'Minuetto'

Table 19.1 The accuracy of our beat extraction method

Timbre	Accuracy (%)
Piano	90
Flute	71.45
Violin	70

In addition, the APM data structure is calculated from musical segments, once tempo varies, the test results will be affected. Consequently, the method in this paper applies mainly to music in steady speed.

19.4 Conclusion

A beat induction method was presented in this paper which adopts the APM data structure and its entropy proposed by Eck, but chooses our detecting methods to pursuit more accurate results. Overall, the strategy of our work performs competitive with the model of Douglas Eck [11, 12].

Now a more precise method for onset detection is still needed, and the assessment rules for beat tracking could also be further improved.

Acknowledgments This work is supported by the National Natural Science Foundation of China (Grant No. 61071149), the State Administration of Radio Film and Television of China (Grant No. 2011-11) and Program for New Century Excellent Talents of Ministry of Education of the People's Republic of China (Grant No. NCET-10-0749).

References

1. Scheirer ED (1998) Tempo and beat analysis of acoustic musical signals. J Acoust Soc Am 103(1):588–601
2. Cemgil T, Kappen B, Desain P, Honing H (2000). On temp tracking: tempogram representation and kalman filtering. In international computer music conference
3. Dixon S (2001) Automatic extraction of tempo and beat from expressive performances. Int J New Music Res 30(1):39–58
4. Goto M (2001) An audio-based real-time beat tracking system for music with or without drum-sounds. J New Music Res 30(2):159–171
5. Dostal M (2005) Genetic algorithms as a model of musical creativity-on generating. Comput Informatics 22:321–340
6. Klapuri AP, Eronen AJ, Astola JT (2006) Analysis of the meter of acoustic musical signals. IEEE Trans Audio Speech Lang Process 14(1):342–355
7. Hoffman MD, Blei DM, Cook PR (2009). Easy as CBA: a simple probabilistic model for tagging music. In: International conference on digital audio effects
8. Brown JC (1993) Determination of the meter of musical scores by autocorrelation. J Acous Soc Am 94(4):1953–1957
9. Gouyon F, Herrera P (2003) Determination of the meter of musical audio signals: Seeking recurrences in beat segment descriptors. J Acous Soc Am
10. Toiwiainen P, Eerola T (2004) The role of accent periodicities in meter induction: a classification study. In the proceedings of the eighth international conference on music perception and cognition
11. Eck D, Casagrande N (2005) Finding Meter in music using an autocorrelation phase matrix and Shannon entropy. In Proceedings of the 6th international conference on music information retrieval
12. Eck D (2006) Identifying metrical and temporal structure with an autocorrelation phase matrix. Music Perception: An Interdisciplinary Journal 24(2):167–176
13. Holzapfel A, Stylianou Y (2009) Rhythmic similarity in traditional Turkish music. In 10th international society for music information retrieval conference
14. Ellis DPW, Klapuri A, Richard G (2011) Signal Processing for Music Analysis. IEEE J Sel Top Sign Proces 5(6):1088–1109
15. Yang X, Shu P (2009) A kind of music tempo and beat type automatic detection algorithm. Digital Technology and Application (in Chinese)

16. Dixon S, Cambouropoulos E (2000) Beat tracking with musical knowledge. In proceedings of the 14th european conference on artificial intelligence
17. Zeng Q, Ye X, Wu D (2005) Musical rhythm detection based on multi-path search. J Circuits Sys (Chinese) 10(1):119–122

Chapter 20
Improved Serially Concatenated Overlapped Multiplexing System and Performance

Suhua Zhou, Song Gao and Daoben Li

Abstract In order to cut down high complexity for Serially Concatenated Overlapped Code Multiplexing system (SC-OVCDM), this article put forward an improved SC-OVCDM system with constraint length 1. OVCDM code matrix is simplified into multiplexing matrix with one-column; corresponding, its decoding is simplified symbol-level soft demodulation instead of BCJR algorithm, so system complexity is greatly reduced. Therefore, the proposed system structure can realize greater spectral efficiency for low complexity. System performance need joint optimization. Numerical experiments show the effectiveness of the optimal new scheme. When spectral efficiency is 4 bit/s/Hz, it has gain of 0.4 dB in bit error ratio compared with SC-OVCDM. We constructed system with spectral efficiency 5.82 bit/s/Hz, which is approximately 1.9 dB away from Shannon limit at the BER level of 10^{-5}.

Keywords High spectral efficiency · SC-OVCDM · Multiplexing matrix · Symbol-level soft demodulation

20.1 Introduction

High spectral efficiency is one of the most important requirements in future mobile communication system, Overlapped Code Division Multiplexing (OVCDM) [1–3] is a new multiplexing technique, in such scheme, multiple symbol streams are

S. Zhou (✉)
Department of Communication, Century College, Beijing University of Posts and Telecommunications, Beijing City, China
e-mail: 649671211@qq.com

S. Gao · D. Li
School of Information and Telecommunication, Beijing University of Posts and Telecommunications, Beijing, China

W. Lu et al. (eds.), *Proceedings of the 2012 International Conference on Information Technology and Software Engineering*, Lecture Notes in Electrical Engineering 210, DOI: 10.1007/978-3-642-34528-9_20, © Springer-Verlag Berlin Heidelberg 2013

passed through different weighted one-input-one-output convolutional encoders and are summed up together for simultaneous transmission, it can achieve significant coding gain and diversity gain when obtains high spectral efficiency. Since the OVCDM encoder input is modulated symbols such as BPSK [2] or M-ary (M > 2) modulation [4], TCM [5, 6] could be utilized as the error correcting code of uncoded OVCDM for its high bandwidth efficiency. References [7, 8] proposed two kinds of structures between TCM and OVCDM: serial concatenation structure(SC-OVCDM) [7] and Turbo product structure(TPC-OVCDM) [8], this two structures both can obtain significant coding gain compared with uncoded OVCDM system, especially SC-OVCDM. Reference [7] showed that its bit error rate performance is superior to Turbo code plus QAM in LTE with same spectral efficiency, at bit error rate level of 10^{-5}, SC-OVCDM system with short frame length can obtain 0.3 dB coding gain, but good performance is at the cost of higher complexity, which prevented it from practical application in engineering.

In the SC-OVCDM, TCM is employed as outer code, OVCDM code with short constraint length is inner code, a symbol interleaver serially concatenates them, system complexity mainly comes from OVCDM decoding, so in this paper, we improved the structure of inner OVCDM encoder: constraint length is changed into 1, due to reduced constraint length, it is not necessary any more for the receiver to use BCJR algorithm at the symbol level to decode the OVCDM. Instead a simple iterative soft demapper is employed for the iterative detection. Thus the receiver complexity is highly reduced. However, reduction in the constraint length may lead to higher bit error floor, so BCH codes are added in front of simplified SC-OVCDM. Additional, we realized higher spectral efficiency by improving bit rate of TCM encoder in proposed system. We searched optimal structure that spectral efficiency equals to 4 and 5.82 bit/s/Hz. Numerical results show the proposed system is a better application scheme in engineering from aspects of performance and complexity.

20.2 System Model

The block diagram of SC-OVCDM encoder is shown in Fig. 20.1a. Information bits are first encoded by TCM encoder. The output modulated symbol of TCM is then permuted by symbol interleaver before being encoded by OVCDM encoder. Iterative decoding algorithm is applied for decoding SC-OVCDM.

Figure 20.1b illustrates the block diagram of an iterative decoder for SC-OVCDM system. It consists of two Soft-Input Soft-Output (SISO) modules- SISO inner and SISO outer—in a serial concatenation scheme, each one corresponding to a component code. L_u is a sequence of likelihood ratios (LLR) of OVCDM encoder output received from channel. The OVCDM decoder computes a sequence of extrinsic LLRs of the OVCDM encoder input L_{e_i} using BCJR algorithm, and then L_{e_i} is passed through symbol deinterleaver and deinterleaved. The output sequence of deinterleaver is denoted by L_{a_o}. That L_{a_o} enters the TCM decoder for computing the LLRs of both coded symbols and information bits. Extrinsic

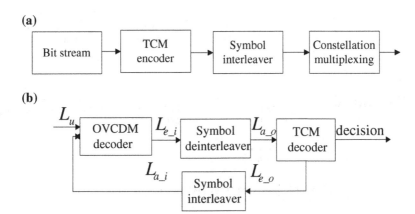

Fig. 20.1 Block diagram of SC-OVCDM system

LLRs of the SISO TCM decoder output L_{e_o} are transformed into L_{a_i} after interleaving and sent to OVCDM decoder. The output LLRs of the information bits of the outer TCM encoder will be used for decision.

The LOG-MAP-based SISO iterative decoding algorithm has high complexity. We know that the decoding complexity is decided by the complexity of the encoder. Therefore, encoding complexity of two encoders and number of decoding iteration numbers jointly decide complexity of the SC-OVCDM. Let decoding iteration numbers is n, then complexity of the SC-OVCDM is $V = (V_{\text{OvCDM}} + V_{\text{TCM}}) \times 2n$, V_{TCM} and V_{OvCDM} represent complexity of the outer TCM code and inner code OVCDM respectively. Where $V_{\text{TCM}} = N_{\text{State}} \times N_{\text{Branch}}$, N_{State} is state number of Trellis and N_{Branch} is branch number. $V_{\text{OvCDM}} = (S_{in})^{N_{\text{reg}}} \circ S_{in}$ is number of input symbol, N_{reg} is number of register. We consider a SC-OVCDM System in [7] and analyses system complexity, outer code TCM, code rate is 2/3, encoder state number is 16, 8PSK signal sets, so $V_{\text{TCM}} = 16 \times 4 = 64$. Inner OVCDM code matrix is 2×2, this mean each convolutional encoder only has one register, there are two registers in inner OVCDM encoder. OVCDM encoder input is 8PSK symbol, so $S_{in} = 8 \times 8 = 64$, $N_{\text{reg}} = 2$, $V_{\text{OvCDM}} = (8 \times 8)^2 = 2^{12}$, therefore total complexity is $V = (V_{\text{OvCDM}} + V_{\text{TCM}}) \times 2n = (2^{12} + 64) \times 2n = (2^{13} + 128)n$, we can see system complexity mainly comes from OVCDM encoder, SC-OVCDM system performance is outstanding in exchange for high-complexity, which is a great impediment for practical applications.

20.3 Improved SC-OVCDM System

Encoder and iterative decoding process of low-complexity SC-OVCDM system are shown in Fig. 20.2. Information bits are first encoded by TCM encoder. The output modulated symbol of TCM such as MPSK modulation, after permuting by

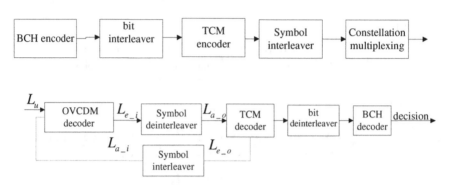

Fig. 20.2 Block diagram of improved SC-OVCDM system

symbol interleave, passes through multiplexing matrix $[a + bi; c + di]$ and is sent to channel. Let $[a + bi; c + di] = [A_1 e^{j\varphi_1}, A_2 e^{j\varphi_2}]$, so this process can be seen as that two parallel TCM output constellation points are transformed by following form: kth $k(k = 1, 2)$ amplitude is enlarged (shortened) as A_k times of original ones, phase is rotated φ_k, then add together. For example, if output constellation of outer encoder is 8PSK modulation, through this inner 2×1 multiplexing matrix, we can form 64 constellations points, efficiency equivalents to the 64 QAM constellation points. The benefits of this treatment is a significant reduction in complexity, while the whole efficiency of the system is not changed, optimal performance can be obtained by adjusting element of multiplexing matrix when outer code and interleaver are fixed. Iterative decoding is similar to that of SC-OVCDM, only difference is that BCJR decoding algorithm of inner OVCDM.

Code is simplified into symbol-level soft demodulation which is main reason to reduce the complexity. In symbol-level soft demodulation Log-likelihood ratio (soft information) of symbol x_l is calculated by the following formula [9, 10]:

$$
\begin{aligned}
LLR(x_l) &= \ln \frac{p(x_l = v_m | s)}{p(x_l = v_n | s)} \\
&= \ln \frac{p(x_l = v_m) \sum\limits_{s_k: x_l = v_m} p(s|s_k) \prod\limits_{i=1, i \neq l}^{m} p(x_i = u_{ik})}{p(x_l = v_n) \sum\limits_{s_k: x_l = v_n} p(s|s_k) \prod\limits_{i=1, i \neq l}^{m} p(x_i = u_{ik})} \\
&= \ln \frac{p(x_l = v_m)}{p(x_l = v_n)} + \ln \frac{\sum\limits_{s_k: x_l = v_m} p(s|s_k) \prod\limits_{i=1, i \neq l}^{m} p(x_i = u_{ik})}{\sum\limits_{s_k: x_l = v_n} p(s|s_k) \prod\limits_{i=1, i \neq l}^{m} p(x_i = u_{ik})}
\end{aligned}
\tag{20.1}
$$

Where, S is received symbol, v_m and v_n is possible input symbol, u_{ik} is value in ith location of transmitted constellation s_k.

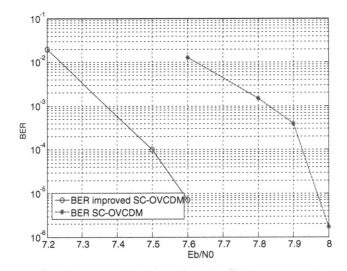

Fig. 20.3 Performance comparison for improved SC-OVCDM and SC-OVCDM

20.4 Simulations and Performance Analysis

The performance of improved SC-OVCDM system depends on the structure of the TCM encoder signal set and multiplexing matrix as well as symbol interleaver, so system parameters must be set for best performance through joint optimization. By far, we have searched optimal structure of improved SC-OVCDM system that spectral efficiency is 4 bit/s/Hz, to show the advantage of this system, computer simulations are performed, and it is compared with the optimal SC-OVCDM system mentioned in [7], detailed simulation parameters are shown in Table 20.1.

$$C_1 = [0.4629 - 0.9017i; 0.5389 + 0.8260i]$$
$$C_2 = \begin{bmatrix} 0.2869 - 0.1606i & -0.1104 + 2.0126i \\ 0.1122 + 1.4716i & -0.2129 - 0.0122i \end{bmatrix}$$

Table 20.1 Parameters improved SC-OVCDM and SC-OVCDM

System	Improved SC-OVCDM	SC-OVCDM
Outer code	TCM: code rate 2/3 8state 8PSK signal set	TCM: code rate 2/3 16state 8PSK signal set
Inner code	C1	C2
Frame length	4,792	4,792
S parameter	22	22
Iteration number	30	15
Channel	AWGN	AWGN
Frame number	10,000	10,000

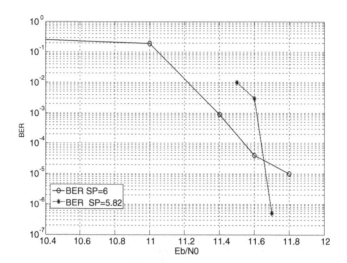

Fig. 20.4 BER performance of improved SC-OVCDM, spectral efficiency is 6 and 5.82 bit/s/Hz
Iteration time is 34, frame length is 13,982

We can see in improved SC-OVCDM system, inner coding matrix C2 is
changed into simple 2×1 complexing matrix C1, at the same time state number
of outer coder TCM is reduced to 8 from 16, which make TCM decoding complexity reduce by half, so whole system complexity has been effectively reduced.

Figure 20.3 illustrates BER performance of the improved SC-OVCDM system
with SC-OVCDM [7] system. Clearly, improved SC-OVCDM has greater advantages, BER begins steep drop from 7 dB and decreases rapidly to 6×10^{-6} till
7.6 dB. In contrast, steep drop threshold of SC-OVCDM system is higher, it just
starts to decline at 7.6 dB, at 8 dB BER can attain 1.7×10^{-6} respectively. We get
following result: when BER is at level of 10^{-5}, improved SC-OVCDM system has
about 0.4 dB gain. Although number of iterations are more than that of SC-OV-
CDM system, for one iteration time is only 1/60 of the latter, so 20 iterations time is
shorter tan that of 15 iterations in SC-OVCDM. However, drop slope of improved
SC-OVCDM is smaller than SC-OVCDM system, so it may exist higher error floor.

In this proposed system, we use a rate 3/4 convolution and 8 state 8PSK signal
set in TCM, spectral efficiency can attain 6 bit/s/Hz, while SC-OvCDM with
spectral efficiency higher than 4 is nearly realized. Multiplexing matrix is C3:

$$C3 = [0.80801 - 0.6144i; 0.98254 - 0.065253i]$$

In order to lower error floor, bit stream after S/P is encoded by BCH codes with
whole code rate is 0.97, so spectral efficiency is reduced to 5.82 bit/s/Hz. These two
system BER performance are shown in Fig. 20.4. We can see that BER both can
attain 10^{-5} level about 11.6 dB, but system of 6 bit/s/Hz clearly has high error floor
from then on, system of 5.82 bit/s/Hz can decrease steeply to 10^{-6} level at 11.7 dB,

Shannon limit is 9.8 dB, so it is approximately 1.9 dB away from Shannon limit at the BER level of 10^{-5}. In short, the proposed system is also quite a good scheme to realize higher than 4 bit/s/Hz for practical application.

20.5 Conclusions

Improved SC-OVCDM system is proposed in this paper, because complexity of SC-OVCDM system mainly comes from inner OVCDM code, this system changes inner OVCDM code into simple 2×1 multiplexing matrix, decoding algorithm is simplified into symbol-level soft demodulation. For its complexity only grows linearly with spectral efficiency, the proposed system structure can realize greater spectral efficiency through joint optimisation, we searched two multiplexing matrixes to obtain better BER performance for different spectral efficiencies, and gave their simulation curves respectively. Results showed improved SC-OVCDM system not only complexity drops significantly but also has greater performance superiority, it is better scheme with practical value.

References

1. Daoben L (2007) A overlapped code division multiplexing method. International Patent, Application No: PCT/CN2007/000536
2. Jiang W, Daoben L (2007) Convolutional multiplexing for multicarrier systems. In: Proceeding of IEEE wireless communications and networking conference, WCNC 2007, 11–15 Mar 2007, Hong Kong, pp 633–638
3. Wei J, Daoben L (2007) Convolutional multi-code multiplexing for OFDM systems. In: Proceeding of IEEE international communications conference, ICC, 2007, 24–28 Jun 2007 Glasgow, pp 4879–4884
4. Xing Y, Yong M et al. (2008) Analysis of overlapped code division multiple access system in Gaussian multiple access channel. In: Proceeding of IEEE international communications conference, ICC 2008, 19–23 May 2008, Beijing, pp 1230–1237
5. Ungerboeck G (1987) Trellis-coded modulation with redundant signal sets, Part I:introduction. IEEE Communications Magazine, 25(2): 5 − 11
6. Ungerboeck G (1987) Trellis-coded modulation with redundant signal sets, Part II: state of the art. IEEE Commun Mag 25(2):12–21
7. Suhua Z, Li F, Daoben L (2009) A novel symbol-interleaved serially concatenated overlapped multiplexing system. In: Proceeding of IEEE international power electronics and intelligent transportation system, PEITS 2009, 19–20 Dec 2009, Shenzhen, China, pp 373–376
8. Suhua Z, Li F, Daoben L (2010) Coded overlapped code division multiplexing system with turbo product structure. J China Univ Posts Telecommun 17(2):25–29
9. ten Brink S, Speidel J, Yan RH (1998) Iterative demapping for QPSK modulation. Electronics Lett 34(15):1459–1460
10. Allpress S, Luschi C, Felix S (2004) Exact and approximated expressions of the log-likelihood ratio for 16-QAM signals, signals. Syst Comput 1:794–798

Chapter 21
Real-Time Face Recognition Based on Sparse Representation

Xiao Chen, Shanshan Qu and Chao Ping

Abstract By coding the input testing image as a sparse linear combination of the training samples via $L1$-norm minimization, sparse representation based classification (SRC) has been successfully employed for face recognition (FR). Particularly, by introducing an identity occlusion dictionary to sparsely code the occluded portions in face images, SRC can lead to robust FR results against occlusion. However, it has not been evaluated in applications with respect to efficiency and image variety. In the proposed system, we have integrated face detection and the sparse presentation for online face recognition, in stead of training and testing face images. The performance of sparse representation has been evaluated in image sequences with the integrated recognition system.

Keywords Face detection · Face recognition · Sparse representation

21.1 Introduction

Human faces are arguably the most extensively studied object in image-based recognition. This is partly due to the remarkable face recognition (FR) capability of the human visual system [1], and partly due to numerous important applications for automatic face recognition technologies. To successfully recognize faces in video, face detector and recognizer need to adapt varying situations. Therefore, technical issues associated with face recognition are representative of object recognition in general.

X. Chen (✉) · S. Qu · C. Ping
Research Institute of Computer Application, China Academy of Engineering Physics,
Mianyang, China
e-mail: derekx.c@gmail.com

W. Lu et al. (eds.), *Proceedings of the 2012 International Conference on Information Technology and Software Engineering*, Lecture Notes in Electrical Engineering 210, DOI: 10.1007/978-3-642-34528-9_21, © Springer-Verlag Berlin Heidelberg 2013

Although facial images have a high dimensionality, they usually lie on a lower dimensional subspace or sub-manifold. Therefore, subspace learning and manifold learning methods have been dominantly and successfully used in appearance based FR [2–9].

Very recently, an interesting work was reported by Wright et al. [10], where the sparse representation (SR) technique is employed for robust FR. In Wright et al.'s pioneer work, the training face images are used as the dictionary to code an input testing image as a sparse linear combination of them via L1-norm minimization. The SR based classification (SRC) of face images is conducted by evaluating which class of training samples could result in the minimum reconstruction error of the input testing image with the sparse coding coefficients. To make the L1-norm sparse coding computationally feasible, in general the dimensionality of the training and testing face images should be reduced. In other words, a set of features could be extracted from the original image for SRC. In the case of FR without occlusion, Wright et al. tested different types of features, including Eigen face, Random face and Fisher face, for SRC, and they claimed that SRC is insensitive to feature types when the feature dimension is large enough.

Although the SRC based FR scheme proposed in [10] is very creative and effective, its capability with holistic features in handling variations of illumination, expression, pose and local deformation is further questioned in [11]. In fact, those variations can be fully demonstrated in video clips. Therefore, the performance of SRC to handle those variations can be actually evaluated in FR in video.

In this paper, we propose to build a FR system by integrating face detector into SRC. By locating potential face regions for recognition, SRC is guided to the right path with a pruned searching space. We will show how the SRC performs in handling FR using surveillance camera. The rest of paper is organized as follows. Section 21.2 briefly reviews SRC and face detectors. Section 21.3 presents the proposed FR system. Section 21.4 conducts experiments and Sect. 21.5 concludes the paper.

21.2 Related Works

In order to implement SRC in real-time applications, we have combined Viola and Jones's face detector with SRC for online face recognition using surveillance camera.

21.2.1 Sparse Representation Based Classification for Face Recognition

The set of the ith training object class is denoted by $A_i = [s_{i,1}, s_{i,2}, \cdots, s_{i,j}]_{m \times n_i}$, where $s_{i,j}, j = 1, 2, \ldots, n_i$ is an m-dimensional vector stretched by the jth sample of the ith class. For a test sample y_0 in this class, it could be well approximated by the

Table 21.1 The SRC algorithm proposed in [10]

1. Normalize the columns of A (case of non-occlusion) or B (case of occlusion)
   to have unit $L2$-norm
2. Solve the $L1$-minimization problem:
$$\hat{a}_1 = \arg\min_{\alpha}\{\|y_0 - A\alpha\|_2^2 + \lambda\|\alpha\|_1\}$$
or
$$\hat{\omega}_1 = \arg\min_{\omega}\{\|y - B\omega\|_2^2 + \lambda\|\omega\|_1\}$$
where $\hat{\omega}_1 = [\hat{\alpha}_1; \hat{\alpha}_e]$ and λ is the positive scalar which balances the reconstructed
   error and coefficients' sparsity
3. Compute the residuals:
$$r_i(y_0) = \left\|y_0 - A\delta_i(\hat{\alpha}_1)\right\|_2, \ i = 1,\ldots,k$$
or
$$r_i(y) = \left\|y - A_e\hat{\alpha}_{e1} - A\delta_i(\hat{\alpha}_1)\right\|_2$$
where $\delta_i(\cdot)$ is the characteristic function which selects the coefficients associated
   with the ith class
4. Output the identity
$$y_0 = \arg\min r_i(y_0) \text{ or } y = \arg\min r_i(y)$$

linear combination of the samples in A_i, i.e. $y_0 = \sum_{j=1}^{n_i} \alpha_{i,j} s_{i,j} = A_i\alpha_i$, where $\alpha_i = [\alpha_{i,1}, \alpha_{i,2}, \cdots, \alpha_{i,n_i}]^T$ is the vector of coefficients. Suppose we have K object classes, and let $A = [A_1, A_2, \cdots, A_K]$ be the concatenation of the n training samples from all the K classes, where $n = n_1 + n_2 + \cdots + n_K$, then y_0 can be written in terms of $y_0 = A\alpha$, where $\alpha = [\alpha_1, \cdots, \alpha_i, \cdots, \alpha_K] = [0, \cdots, 0, \alpha_{i,1}, \alpha_{i,2}, \cdots, \alpha_{i,n_i}, 0, \cdots 0]^T$, which is introduced in [10].

In the case of occlusion or corruption, we can rewrite the test sample y as

$$y = y_0 + e_0 = A\alpha + e_0 = [A, A_e]\begin{bmatrix}\alpha \\ \alpha_e\end{bmatrix} \approx B\omega \tag{21.1}$$

where $B = [A, A_e]_{m \times (n+n_e)}$, the training face image y_0 and the corrupted error e_0 have sparse representations over the training sample dictionary A and A_e, respectively. In [10], the occlusion dictionary A_e was set as an orthogonal matrix, such as identity matrix, Fourier bases, Haar wavelet bases and so on. The SRC algorithm in [10] is summarized in Table 21.1.

21.2.2 Face Detector

Viola and Jones [12] present a new and radically faster approach to face detection based on the AdaBoost algorithm from machine learning. To understand their detector, the concept of boosting needs to be explained first. AdaBoost is a well known algorithm to generate strong learners from weak learners, while providing statistical bounds on the training and generalization error of the algorithm. The weak learners in the Viola and Jones algorithm are based on features of three

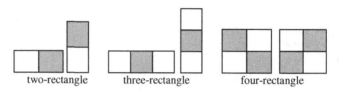

Fig. 21.1 Boxlets for face detection, two-rectangle, three-rectangle, and four-rectangle

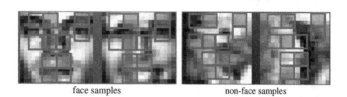

Fig. 21.2 Haar-like features for weak learners training

kinds, which is shown in Fig. 21.1. A two-rectangle feature is the difference between the sum of the values of two adjacent rectangular windows. A three-rectangle feature considers three adjacent rectangles and computes the difference between sum of the pixels in the extreme rectangles and the sum of the pixels in the middle rectangle. A four-rectangle feature considers a 2×2 set of rectangles and computes the difference between sum of pixels in the rectangles that constitute the main and off diagonals. For a 24×24 sub-window there could be more than 180,000 such features.

Base on Adaboost algorithm, the weak learning is applied to image regions to calculate haar like features. In Fig. 21.2, the difference between the sums of pixels within two rectangular regions is computed as features, which are calculated for training images. These features are computed all over a pattern, leading an over-complete representation. Later, weak learners are trained based on different combination of those features and weights for target faces are determined for each weak learner.

The task of the AdaBoost algorithm is to pick a few hundred features and assign weights to each using a set of training images. Face detection is reduced to computing the weighted sum of the chosen rectangle-features and applying a threshold. In particular, for each boxlet j, $f_j(x)$ is computed, where x is the positive or negative example. Each feature is used as a weak classifier. Set threshold θ_j so that most samples are classified correctly:

$$h_j(x, f, \theta) = 1 \; (f_j(x) > \theta_j \text{ or } f_j(x) < \theta_j) \tag{21.2}$$

The obtained weak learners are combined to form a strong learner, as shown in Fig. 21.3. The regions that most likely do not contain faces are filtered out.

With the haar like features and machine learning based on Adaboost algorithm, faces are detected by the cascaded classifier, which is summarized in Fig. 21.4.

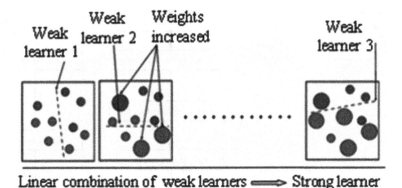

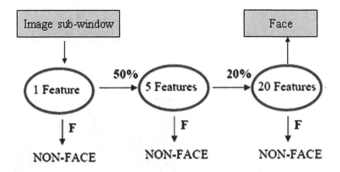

Fig. 21.3 Adaboost weak and strong learners

Fig. 21.4 Face detection by cascaded classifier

A 1-feature classifier achieves 100 % detection rate and about 50 % false positive rate. Using data from previous stage, a 5-feature classifier achieves 100 % detection rate and 40 % false positive rate, 20 % of which is cumulative from previous stage. In the final stage, a 20-feature classifier achieves 100 % detection rate, increasing the false detection by only 2 %.

21.3 Online Face Recognition System

We have integrated Viola/Jones' face detector with SRC for real-time face recognize in surveillance video. The flow chart has been demonstrated in Fig. 21.5. SRC dictionary is formed by training images, which is done off line. When doing online recognition, possible face regions are first detected in the preprocessed image and then these regions will be assigned class by the SRC classifier. In order to stable the recognition, we have introduced a buffer, by which census recognition will be applied to consecutive frames.

Fig. 21.5 Online face
recognition system

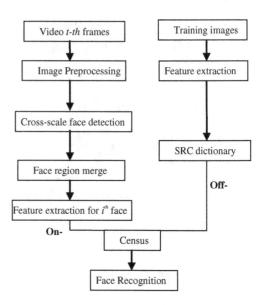

Off-

On-

Fig. 21.6 Mergence of
multiple scale detections

For the input frames, we have preprocessed them to reduce the effects from
noise and illumination. These preprocessing techniques include mean-filter, histogram equalization and image normalization. In original Viola/Jones detector,
multiple detections are generated for the same face region, because faces can be
detected at different scales and at slightly displaced window location. Therefore, a
merge procedure is required to produce single region for a face. In our experiment,
faces detected under multiple scales are merged if there are over 25 % overlapping. An example is shown in Fig. 21.6.

To represent the possible face region for recognition, different features have
been extracted in literature. The dominant approaches choose to construct facial
features adaptively based on the given images, via techniques such as Eigenfaces
[2], Fisherfaces [15], Laplacianfaces [14], and their variants [16, 17] (see Fig. 21.7
for examples). Other methods use fixed filter banks (e.g., downsampling, Fourier,
Gabor, wavelets), which are effective for analyzing stationary signals such as
textures. Reference [13] shows that SRC is insensitive to features type. We will
also test this statement in our online recognition system.

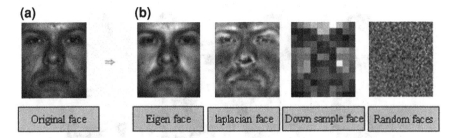

Fig. 21.7 Types of features for face recognition

When doing real-time face recognition, the face regions detected from input video are recognized by searching for the SRC dictionary. However, the recognition could be unstable as processing detected faces, because appearance of the same face varies with illumination, occlusion and view angles. Therefore, the same person could be wrongly classified from time to time. In order to obtain stable recognition, we have set a buffer for the census recognitions on consecutive frames. In particular, SRC results for the consecutive five frames are stored in the buffer and the mostly voted class (or classes) is the recognition result. On the other hand, recognition will only take place when target moving, so that to reduce computational cost.

21.4 Experiments

We have tested our online face recognition system with a stationary webcam mounted on the computer monitor. In order to classify a face detected in a frame with respect to the dictionary, the following basic tasks have to be carried out for each video frame.

Task 1: Face-looking regions are detected using a pre-trained face classifier, the one of which, trained on Haar-like features. Figure 21.8 shows some facial regions detected and the blue dot is the center of detected region.

As can been seen, the face detector responses to face as well as regular patterns. Therefore, the detection should be constrained with motion, since the same face in consecutive frames should not have a large displacement. This motion constraint is detailed in Task 2.

Task 2: Motion information of the target is employed to filter out false faces. In particular, faces should have moved within last several frames. The face detected in current frame should be near the face detected in last frame. Therefore, the coordinates of the center of detected region in frame is recorded. The detection is positive if the centers of consecutive frames are within certain value. In our experiments, this value is determined as 10 with respect to a frame rate of 20 fps. Motion constraint is illustrated in Fig. 21.9, where the red dot the previous face center.

Fig. 21.8 Real-time face detection with webcam

Task 3: Census recognition is applied by using the buffering process. For five consecutive frames, SRC is applied for detected faces. The classification information is accumulated and the mostly voted class is calculated. The census recognition is shown in Fig. 21.9, where the assigned class is determined.

With the motion constrain and census recognition, the false detection rate are dramatically reduced and our system can achieve real-time implementation.

In the experiment, we have tested on the webcam data, when 10 subjects have passed by the webcam from time to time. As they walking by, facial variations including illumination change, expressions, and facial disguises are involved. For the sake of efficiency, frames take the dimension of 165 × 120, and all converted to grayscale. We have followed [13] to select four feature space dimensions: 30, 54, 130, and 540, which correspond to the down-sample rations 1/24, 1/18, 1/12, and 1/6, respectively. The recognition rate in our system is shown in Table 21.1. By a PC with 2 core 2.4 GHz CPU, we can achieve accurate and 20 fps high speed face recognition. The detector works well when faces are with moderate expression and partially occlusions.

Results in Table 21.2 is in line with the conclusion of [13], the choice of features becomes less important than the number of features used with sparsity

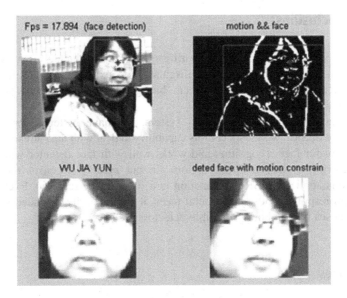

Fig. 21.9 Different stages of real-time face recognition

Table 21.2 Recognition rate with respect to different feature dimensions

Feature dimension	30	56	120	504
Recognition rate (%)				
Eigen	92.2	94	95.16	96.34
Laplacian	91.4	92.75	93.31	95.73
Random	89.8	91.27	95.5	97
Downsample	91	92.3	93.26	96.1

properly harnessed. Our system responses to frontal-like faces, namely rotation angles are not bigger than ±30. However, it performances well on different expressions. SRC works well for detected faces and it can achieve good recognition rate without any post-processing of the face image. It is noted that performance of this system is also subject to distance between a person and webcam, as well as his moving velocity. Performance of SRC dramatically reduced when the distance reach certain value for each feature dimension.

SRC is efficient for face recognition, as it associates face regions with classes stored in SRC dictionary. However, it does not have a rejection process when the input is never seen, since it always assigns a class to the input. In situations where non-seen faces appear from time to time (say surveillance camera), SRC itself cannot serve as trigging a security system. In our system we have added a buffer (census recognition) to cope with this problem. On the other hand, a SRC performance depends on the training images, due to its mathematic base. It requires the size of the training database.

21.5 Conclusions

We evaluated the proposed system on different conditions, including variations of illumination, expression as well as disguise. The experimental results clearly demonstrated that the proposed system has good recognition rates and little computational cost.

With our online face recognition system, we have better understand the performance of SRC with respect to its capability to deal with data streams. It is an efficient tool for face recognition and works well with face detector without any post-processing of detected face regions. In our system, a census recognition procedure is applied for SRC to function in a surveillance scenario. Even though our system responses mostly to frontal faces, it can be easily extended to handle multiple poses by combining multiple detectors.

References

1. Sinha P, Balas B, Ostrovsky Y, Russell R (2006) Face recognition by humans: nineteen results all computer vision researchers should know about. Proc IEEE 94(11):1948–1962
2. Turk M, Pentland A (1991) Eigenfaces for recognition. J Cogn Neurosci 3:71–86
3. Belhumeur PN, Hespanha JP, Kriengman DJ (1997) Eigenfaces vs. fisherfaces: recognition using class specific linear projection. IEEE Trans Pattern Anal Mach Intell 19:711–720
4. Yang J, Yang JY (2003) Why can LDA be performed in PCA transformed space. Pattern Recogn 36:563–566
5. Tenenbaum JB, deSilva V, Langford JC (2000) A global geometric framework for nonlinear dimensionality reduction. Science 290:2319–2323
6. Roweis ST, Saul LK (2000) Nonlinear dimensionality reduction by locally linear embedding. Science 290:2323–2325
7. He X, Yan S, Hu Y, Niyogi P, Zhang HJ (2005) Face recognition using laplacianfaces. IEEE Trans Pattern Anal Mach Intell 27:328–340
8. Chen HT, Chang HW, Liu TL (2005) Local discriminant embedding and its variants. In IEEE computer society conference on computer vision and pattern recognition
9. Yang J, Zhang D, Yang JY, Niu B (2007) Globally maximizing, locally minimizing: unsupervised discriminant projection with applications to face and palm biometrics. IEEE Trans Pattern Anal Mach Intell 29:650–664
10. Wright J, Yang A, Ganesh A, Sastry S, Ma Y (2009) Robust face recognition via sparse representation. IEEE Trans Pattern Anal Mach Intell 31:210–227
11. Yang M, Zhang L (2010) Gabor Feature based sparse representation for face recognition with Gabor occlusion dictionary.In European conference on computer vision
12. Viola P, Jones M (2001) Rapid object detection using a boosted cascade of simple features. In IEEE computer society conference on computer vision and pattern recognition
13. Yang AY, John W (2007) Feature selection in face recognition: a sparse representation perspective. Technical report, UC, Berkeley
14. He X, Yan S, Hu Y, Niyogi P, Zhang H (2005) Face recognition using Laplacianfaces. IEEE Trans Pattern Anal Mach Intell 27(3):328–340
15. Belhumeur P, Hespanda J, Kriegman D (1997) Eigenfaces vs. Fisherfaces: recognition using class specific linear projection. IEEE Trans Pattern Anal Mach Intell 19(7):711–720

16. Kim J, Choi J, Yi J, Turk M (2005) Effective representation using ICA for face recognition robust to local distortion and partial occlusion. IEEE Trans Pattern Anal Mach Intell 27(12): 1977–1981
17. Li S, Hou X, Zhang H, Cheng Q (2001) Learning spatially localized, parts-based representation.In IEEE international conference on computer vision and pattern recognition

Chapter 22
Impact Analysis of Reducing Inbound Information Rate on the RDSS System Inbound Capacity

Jingyuan Li, Long Huang, Weihua Mou and Feixue Wang

Abstract User terminal transmitting power is an important indicator of the RDSS system design. Reducing the transmitting power can lower the transmitting signal exposure opportunities, but also reduce the size of the user terminal. However, reducing the user terminal transmitting power requires to reduce the counterpart inbound information rate, causing the growth of an inbound signal duration. Which increasing the probability of signal overlap within inbound signal frame duration, increased inbound signal multiple access interference. While the growth of an inbound signal duration causes higher demands of data processing capabilities in unit time for the RDSS central station. In a Poisson distribution hypothesis of the inbound signal, this paper analyses the impact of multiple access interference on the RDSS inbound capacity, and obtained the relationship between inbound capacity and the number of signal processing terminals. Simulation results show that as the information rate reduced to 1/8, the aggravation of the multiple access interference caused by the lower information rate will not lead to the RDSS inbound capacity reduction, on the contrary inbound capacity increased by 9 %, the cost is the user terminal numbers of the central station increase by 6.8 times.

Keywords Radio determination satellite service (RDSS) · Multiple access interference · Inbound capacity · Short burst

J. Li (✉) · L. Huang · W. Mou · F. Wang
Satellite Navigation Research and Development Center, School of Electronic Science and Engineering, National University of Defense Technology, Changsha 410073, China
e-mail: jingyuanlee.ljy@gmail.com

W. Lu et al. (eds.), *Proceedings of the 2012 International Conference on Information Technology and Software Engineering*, Lecture Notes in Electrical Engineering 210, DOI: 10.1007/978-3-642-34528-9_22, © Springer-Verlag Berlin Heidelberg 2013

22.1 Introduction

Radio determination satellite service (RDSS) has the capabilities of fast positioning, high-accuracy timing and short-message communication, which is the indispensable component of the future Compass global navigation satellite system [1, 2]. The main characteristic of the new generation RDSS are large capacity, security and quickly link and reduced terminal size [3]. Lower the transmitting power can reduce the size of the user terminal, while reduce the exposure opportunities of user transmitting signals, increase the difficulty of enemy detect and improve the security of link. The minimum transmitting power of the user terminal is determined by the G/T of the satellite inbound link and the demodulation threshold carrier-to-noise power ratio of the inbound signal. Therefore, the transmitting power of the user terminal can be reduce more by increase the G/T of satellite or reduce the carrier-to-noise power ratio of the signal demodulation threshold. When inbound signal demodulator bit error rate is invariable, reducing inbound information rate is the only way to reduce the carrier-to-noise power ratio of the signal demodulation threshold.

RDSS inbound signal is a short burst signal, when the information frame length is determined, the lower the information rate, the longer the duration of information frame. When multiuser transmitting signal is random inbound, as the information frame duration of the inbound signal growing longer, it will cause two problems. First, the probability of signal overlap within one information frame duration growing bigger, it will aggravate the inbound signal multiple access interference and reduce the received equivalent carrier-to-noise power ratio, which limits the RDSS inbound signal capacity. Second, as the processing time occupied by an inbound signal frame of signal processing terminals increased, it need more signal processing terminals when the inbound capacity unchanged.

Therefore, the impact of multiple access interference on the RDSS inbound capacity, and the relationship between the inbound information rate and the system capacity, as well as the relationship between the inbound information rate and the number of signal processing terminals of central station with the multiple access interference, are the research emphasis of this paper.

22.2 Section Determination of the Number of Simultaneous Inbound Users

As the character of short burst of the RDSS inbound signal, it become relatively difficulty to determinate the number of simultaneous inbound users. Refer to the analysis of immediately rejected queuing system with limited users for communication network, assume a Poisson distribution of inbound signal, the probability of having k times callings within the time interval of Δt is [4]:

$$P_k(\Delta t) = \frac{(\lambda \Delta t)^k}{k!} e^{-\lambda t} \tag{22.1}$$

where λ is the average number of call in the unit time, which is the calling rate. The corresponding accumulation probability equation can expressed as:

$$P_K(\Delta t) = e^{-\lambda \Delta t} \sum_{i=0}^{K} \frac{(\lambda \Delta t)^i}{i!} \tag{22.2}$$

Figure 22.1 is the accumulation probability curve of inbound times within interval of the entire 25 ms when the calling rate is different and the frame length is 25 ms, it can compute the inbound times when the accumulation probability reach 99.9 %, which is the approximation of the simultaneous inbound user numbers within the duration of an inbound signal frame.

According to [3], the RDSS inbound signal frame duration range at the existing inbound information rate is from 25 to 250 ms. When the minimum, maximum and average frame length is 25, 250 and 140 ms respectively, the calling rate is 1,000, 2,000, 3,000 times/s respectively. Table 22.1 show the inbound times when the accumulation probability reach 99.9 %, actually the frame length is among the two cases that show in the table.

It can calculate the relationship between the RDSS inbounds capacity and the number of simultaneous inbound users for certain frames duration by the above method. Each inbound signal is not simultaneous inbound at a certain frame duration, Table 22.1 show the number of simultaneous inbound users when the number of signal overlap is maximum at a certain frame duration, that is the serious situation of multiple access interference. Next, we will analyze the impact of multiple access interference on the inbound signal equivalent carrier-to-noise power ratio.

Fig. 22.1 Cumulative probability distribution curve of inbound times within 25 ms for different calling rate

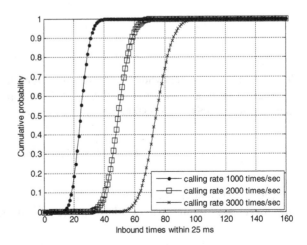

Table 22.1 Inbound times of 99.9 % cumulative probability for different calling rate and different frame length

Calling rate (times/sec)	Inbound times of 99.9 % cumulative probability		
	Frame length 25 ms	Frame length 140 ms	Frame length 250 ms
1,000	42	178	300
2,000	73	333	571
3,000	103	485	836

22.3 Analysis of Inbound Signal Multiple Access Interference

Assume the number of simultaneous inbound user is K, the receive signal can expressed as:

$$r(t) = e^{j\theta_1} s_1(t - \tau_1) + w(t) \tag{22.3}$$

$$s_k(t - \tau_k) = \sqrt{2P_k} d_k a_k(t - \tau_k) e^{jw_c t} \quad k = 1, 2, \ldots, K \tag{22.4}$$

$$w(t) = \sum_{K=2}^{K} e^{j\theta_k} s_k(t - \tau_k) + n(t) \tag{22.5}$$

where $w(t)$ is the interference term of receive signal, it is sum of multiple access interference and white noise.

Due to the longer spreading codes and the plenty of simultaneous inbound signal, the multiple access interference of inbound signal can be approximated as Gaussian noise [5, 6], and the equation of equivalent carrier-to-noise power ratio can expressed as [7, 8]:

$$\left(\frac{C_s}{N_0} \right)_{eff} = \left(\frac{C_s}{N_0} \right) \left[\eta^{-1} + \frac{C_I}{N_0} \frac{\kappa_{si}}{\eta^2} \right]^{-1} \tag{22.6}$$

The fraction of signal power that remains after band limiting filter can defined as:

$$\eta = \int_{-b}^{b} G_s(f) df \tag{22.7}$$

b is the double-side bandwidth of the receiver front-end filter.

And the spectral separation coefficient (SSC) for the punctual channel equivalent carrier-to-noise ratio estimation can be defined as:

$$\kappa_{si} = \int_{-b}^{b} G_i(f) G_s(f) df \tag{22.8}$$

where $G_s(f)$ and $G_i(f)$ are defined as the power spectral density normalized to unit power over infinite bandwidth of the signal and multiple access interference respectively.

$$G_i(f) = G_s(f) = T_c \sin c^2(\pi f T_c) \qquad (22.9)$$

where the signal carrier power is $C_s = P_1$, the multiple access interference power is $C_I = \sum_{k=2}^{K} P_k$.

According to Eq. (22.6) and the number of simultaneous inbound user that obtained by Table 22.1, it can obtain Multiple access interference on the impact of RDSS inbound capacity. Assume the RDSS inbound information rate is R_b, we will analyze the change of system inbound capacity when information rate is R_b, $R_b/2$, $R_b/4$, $R_b/8$ respectively.

22.4 Inbound Capacity of Different Information Rate

Due to the uncertainty of inbound signal frame duration, we will analyse the case of the average frame length is 140 ms that calculate by the inbound information rate, and other situations are similar to this. The demodulator bit SNR threshold can be determined after the inbound information demodulator bit error rate determined. Assume the inbound information rate is different, the demodulation threshold carrier-to-noise power ratio can be expressed as $(C/N_0)_{limit} = R_b(E_b/N_0)_{limit}$ The $(C/N_0)_{limit}$ decrease 3 dB when the information rate R_b decrease one half. The RDSS inbound information rate R_b corresponding demodulation threshold carrier-to-noise power ratio is 42.5 dBHz.

When the information rate reduced to $R_b/2$, $R_b/4$, and $R_b/8$, the corresponding average frame durations are increased to 280, 560, 1120 ms respectively. When the information rate reduced to $R_b/2$, $R_b/4$, $R_b/8$, the corresponding simultaneous inbound user numbers are shown in Table 22.1 when the average frame inbound at different call rate. According to equivalent carrier-to-noise power ratio Eq. (22.6), and the information rate is R_b, $R_b/2$, $R_b/4$, $R_b/8$ that obtain by Tables 22.1 and 22.2 it can obtain the relationship between multiple access interference and the inbound capacity by the maximum number of simultaneous inbound user of average frame.

The existing information rate R_b corresponds to the demodulation threshold carrier-to-noise power ratio 42.5 dBHz and the average frame is 140 ms. Figure 22.2 shows the multiple access interference equivalent carrier-to-noise power ratio with different inbound calling rates. The result indicates that as the increasing of the inbound calling rate, the number of simultaneous inbound users within the duration

Table 22.2 Inbound times of 99.9 % cumulative probability for different calling rate and different frame length

Calling rate (times/sec)	Inbound times of 99.9 % cumulative probability		
	$R_b/2$ (280 ms)	$R_b/4$ (560 ms)	$R_b/8$ (1,120 ms)
1,000	333	635	1,225
2,000	635	1,225	2,388
3,000	931	1,808	3,541

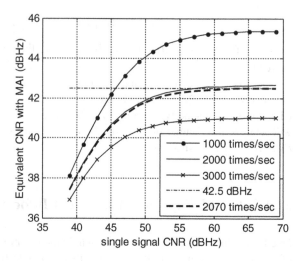

Fig. 22.2 Inbound signal equivalent CNR curve for different calling rate with information rate R_b, average frame length 140 ms

of an inbound signal frame is also increasing, which will lead to decrease the multiple access interference equivalent carrier-to-noise power ratio. The multiple access interference equivalent carrier-to-noise power ratio exists a maximum; it cannot increase the multiple access interference equivalent carrier-to-noise power ratio by increasing the single signal carrier-to-noise power ratio (for example, increasing the inbound signal transmitting power and the transponder gain). Therefore, the multiple access interference lead to the system exist a maximum inbound calling rate after determining the demodulation carrier-to-noise power ratio threshold. The bold dashed line shown in Fig. 22.2 denotes the maximum inbound calling rate caused by the multiple access interference in the carrier-to-noise power ratio threshold, which the value is 2,070 times/sec.

Figures 22.3, 22.4 and 22.5 respectively, indicate the equivalent carrier-to-noise power ratio with multiple access interference for different inbound calling rates. It is noteworthy that the corresponding equivalent carrier-to-noise power ratio reduces 3 dB when the information rate reduces by half, but the average fame increases twice compared to the original. When the information rate reduced to $R_b/2$, $R_b/4$ and $R_b/8$, the demodulation threshold carrier-to-noise power ratio is 39.5, 36.5 and 33.5 dBHz respectively. The average frame is 280, 560, 1120 ms respectively.

The curve trends of Figs. 22.3, 22.4 and 22.5 are consistent with Fig. 22.2. The results in Fig. 22.2 show that when the information rate reduce to $R_b/2$, $R_b/4$, $R_b/8$, the maximum inbound calling rate with multiple access interference is 2,190, 2,245, 2,296 times/sec respectively. It is clear that as the information rate decreases, the inbound capacity gradually increased and the range of growth is small. As the information rate reduced to 1/8, the maximum inbound capacity caused by multiple access interference changed from 2,070 to 2,296 times/sec and the inbound capacity increased by 9 %.

Fig. 22.3 Inbound signal equivalent CNR curve for different calling rate with information rate $R_b/2$, average frame duration 280 ms

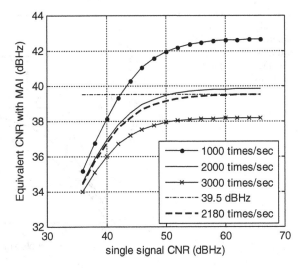

Fig. 22.4 Inbound signal equivalent CNR curve for different calling rate with information rate $R_b/4$, average frame duration 560 ms

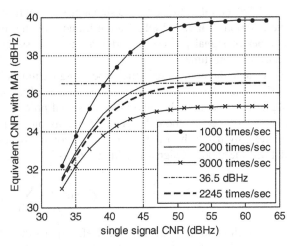

22.5 The Relationship Between the Inbound Capacity and the Number of Signal Processing Terminals

Based on the equivalent carrier-to-noise power ratio threshold after reduced inbound information rate, which determinate the maximum inbound calling rate with multiple access interference. This result is obtained in premise of only ensure the quality of inbound signal, it do not consider the date processing ability of RDSS central station. The inbound information duration will be long when reduce inbound information rate, while the date processing time of handling a frame date will be long, which causing higher demands of data processing capabilities for RDSS central station.

Fig. 22.5 Inbound signal
equivalent CNR curve for
different calling rate with
information rate $R_b/8$,
average frame duration
1120 ms

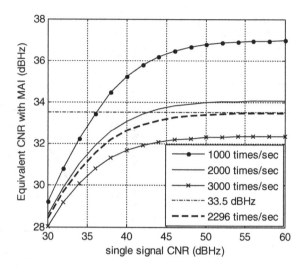

RDSS central station inbound signal processing model is shown in Fig. 22.6. As the duration of inbound information frame growing longer, it need increasing the number of signal processing terminals to process the same number of inbound signals in unit time.

Assume the number of signal processing terminals is m, the total calling rate is λ, the total call capacity can be defined as $\alpha = \lambda/\mu$, $1/\mu$ expressed as each inbound signal that occupied the time mean of signal processing terminals, it equal to the average frame of inbound signal. In a Poisson distribution of incoming signal context, by using the Erlang formula of communication network theory, we can defined the inbound signal blocking rate as [3]:

$$P_c(\alpha, m) = (\alpha^m/m!) \bigg/ \left(\sum_{i=0}^{m} \alpha^i/i! \right) \tag{22.10}$$

According to Eq. (22.10), the Ref. [3] assume the blocking rate is 0.001. Figure 22.7 is the curve of calling rate versus the number of signal processing terminals, when the blocking rate is 0.001. It can be seen that with the increase of the signal processing terminal numbers, the blocking rate showed linear growth trend. The lower the inbound information calling rate, the more signal processing terminals are required in the same calling rate.

Fig. 22.6 Inbound signal
processing model of RDSS
central station

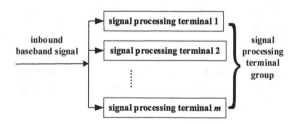

Fig. 22.7 Curves of calling rate versus signal processing terminals for different information rate with call blocking rate 0.001

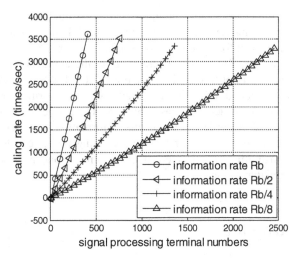

Table 22.3 The relationship between signal processing terminals and calling rate for different information rate with call blocking rate 0.001

Number of signal processing terminals	Calling rate			
	R_b	$R_b/2$	$R_b/4$	$R_b/8$
128	1,002	501	251	126
237	2,000	1,000	500	250
450	——	2,003	1,002	501
856	——	——	2,002	1,001
1,597	——	——	——	2,000

It can obtain by Table 22.3 The relationship between signal processing terminals and calling rate for different information rate with call blocking rate 0.001, as the information rate reduced by 1/2, the corresponding calling rate also reduced by 1/2 when the number of signal processing terminals is fixed. When the inbound information rate is R_b, $R_b/2$, $R_b/4$, $R_b/8$ respectively and calling rate is 2,000 times/sec, the requirement of the signal processing terminals number is 237, 450, 856, and 1,597 respectively. It is clear that the number of signal processing terminals increased by 6.8 times when the information rate reduced to $R_b/8$.

22.6 Conclusion

The inbound signal information rate determines the inbound signal equivalent carrier-to-noise power ratio threshold. This threshold determines the maximum number of simultaneous inbound users with the multiple access interference, and determines the RDSS inbound capacity. The size and power consumption of the

RDSS central station equipment are unlimited, by increasing the number of signal processing terminals, it can meet the inbound capacity requirements when the inbound information rate is reduced. Under the premise of unlimited signal processing ability of the central station, the aggravation of multiple access interference will not lead to the reduction of the RDSS inbound capacity, on the contrary the inbound capacity increased. As the information rate reduced to 1/8, the inbound capacity increased by 9 %. The impact of multiple access interference is general accordance when the information rate reduced by 1/2 every time. Reducing the user terminal transmitting power can lower the probability of the transmitting signal detected by the enemy and is beneficial to the user terminal miniaturization.

References

1. Tan SS (2008) Development and thought of compass navigation satellite system. J Astronaut 29:391–396 (in Chinese)
2. Briskman RD (1990) Radio determination satellite service. Proc IEEE 78:106–1096
3. Tan SS (2010) The engineering of satellite navigation and positioning. National Defense Industry Press, Bejing (in Chinese)
4. He XS (2010) Random process and queuing theory. Hunan University Press, Changsha (in Chinese)
5. James Caffery J, Stüber GL (2000) Effects of multiple-access interference on the noncoherent delay lock loop. IEEE Trans Commun 48:19–2109
6. Li JY (2006) Study of the multiuser detection and signal to interference ratio estimation in satellite CDMA systems. National University of Defense Technology, Changsha (in Chinese)
7. Betz JW, Kolodziejski KR (2009) Generalized theory of code tracking with an early-late discriminator part I: lower bound and coherent processing. IEEE Trans Aerosp Electron Syst 45:50–1538
8. Betz JW, Kolodziejski KR (2009) Generalized theory of code tracking with an early-late discriminator part II: noncoherent processing and numerical results. IEEE Trans Aerosp Electron Syst 45:64–1551

Chapter 23
SNR Estimation Algorithm for Single Signal with Strong Multiple Access Interference

Jingyuan Li, Yuanling Li, Shaojie Ni and Guangfu Sun

Abstract Effective power control was required to improve system capacity and time synchronization accuracy of the CDMA earth station time synchronization networks. It usually uses the equivalent carrier-to-noise power ratio measurement for the CDMA communication system power control. As the receiver equivalent carrier-to-noise power ratio measurement was coupled of the multiple access interference, it cannot accurately estimate the strength of the earth station transmitting power. This paper analyzes the statistical characteristics of both the code tracking loop punctual channel measurements and the early-late phase discriminator channel measurements. By using the early-late phase discriminator output value, increased the multiple access interference noise observations. While using both the punctual channel equivalent carrier-to-noise power ratio and the early-late phase discriminator channels equivalent carrier-to-noise power ratio, we derive the estimate of single signal carrier-to-noise power ratio. The estimates accurately reflect the strength of earth station transmitting power, simulation results show that the algorithm has good estimation performance, both the estimated bias and the estimated standard deviation of the algorithm are less than 0.5 dB, providing high-precision observations for the earth station power control.

Keywords Multiple access interference · Signal to noise ratio estimation · Non-coherent delay lock loop · Power control

J. Li (✉) · Y. Li · S. Ni · G. Sun
Satellite Navigation Research and Development Center,
School of Electronic Science and Engineering, National University of Defense Technology,
Changsha 410073, China
e-mail: jingyuanlee.ljy@gmail.com

W. Lu et al. (eds.), *Proceedings of the 2012 International Conference on Information Technology and Software Engineering*, Lecture Notes in Electrical Engineering 210, DOI: 10.1007/978-3-642-34528-9_23, © Springer-Verlag Berlin Heidelberg 2013

23.1 Introduction

The CDMA earth station time synchronization network using GEO satellite transponder is widely used in the Compass satellite navigation [1, 2], the Chinese Area Positioning System (CAPS) [3] and Two-Way Satellite Time and Frequency Transfer system(TWSTFT) [4]. The time synchronization network shown in Fig. 23.1 often has a star network topology. Each earth station transmits a CDMA spectrum signal to the GEO satellite, so the GEO satellite transponder downlink signal includes the entire K earth stations spread spectrum signal. When the number of Earth station is considerable large and the power of transmitted signal is strong, the multiple access interference of received signal will be much more serious.

Due to the instability of the transmitting power of the power amplifier, the fluctuation of the space link attenuation and significant attenuation of signal transmission caused by harsh weather conditions such as rainstorm and so on, power imbalance of the central station received signals from other earth stations happens frequently. In order to improve the capacity of system and the accuracy of time synchronization measurement, effective power control of the earth stations transmit signal is necessary.

The methods of power control for DS-CDMA communication systems mainly include centralized power control [5] and the distributed power control [6, 7]. The feedback of these power control algorithms was the received signal-to-noise-plus-interference ratio (SNIR).

The central station receivers usually only use the punctual channel accumulation used to estimate the received signal SNIR [8, 9]. When the earth station transmitting power is very strong and imbalance, the SNIR estimates of each earth stations estimated by the central station are coupled with strong multi-access interference. It cannot obtain the accurate estimation of the relative strength of earth station transmitting power by the value of these SNIR estimates. Therefore, it led to the inaccurate power control of the central station to other earth stations.

Fig. 23.1 Earth station time synchronization network using TWSTFT

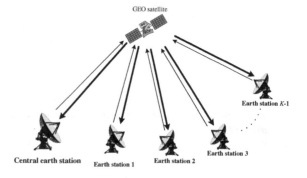

The central station needs to obtain a more accurate measurement for effective power control. When there is only one received signal, the single signal SNR estimate is the ideal power control measurement. Therefore, how to estimate the single-channel carrier-to-noise power ratio of the received signal with strong multiple access interference is the goal of this paper. By analyzing the statistical properties of the code tracking loop punctual channel measurements, as well as the early and late channel measurements, we can use the early-late loop code phase discrimination outputs to increase the noise observation equation of the multiple access interference. While taking advantage of the equivalent carrier-to-noise power ratio of the punctual channels, early-late loop code phase discrimination channel, we can work out the estimate value of single signal carrier-to-noise power ratio.

23.2 Signal Model

Assume the number of users that into the station at the same time is K, the received signal can be expressed as:

$$r(t) = e^{j\theta_1} s_1(t - \tau_1) + w(t) \tag{23.1}$$

$$s_k(t - \tau_k) = \sqrt{2P_k} d_k a_k(t - \tau_k) e^{jw_c t}, \; k = 1, 2, \ldots, K \tag{23.2}$$

$$w(t) = \sum_{K=2}^{K} e^{j\theta_k} s_k(t - \tau_k) + n(t) \tag{23.3}$$

where $w(t)$ is the received signal interference, it is sum of multiple access interference and white noise.

The principle of non-coherent delay-locked loop for signal receiving shows in Fig. 23.2, the accumulation of the early and late channel are used to generate the code phase discriminator error for pseudocode phase tracking. The punctual channel accumulation is used for signal carrier phase tracking, data demodulation, and the signal and noise power ratio SNIR estimation.

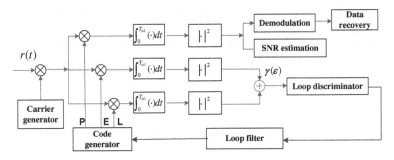

Fig. 23.2 Non-coherent delay-locked loop

The accumulated value of punctual channel can be expressed as:

$$\lambda_n(\Delta\tau_1) = \left| \frac{1}{T}\int_0^T e^{j\theta_1} s_1(t-\tau_1)a(t-\hat{\tau}_1)e^{-jw_ct} + \frac{1}{T}\int_0^T w(t)a(t-\hat{\tau}_1)e^{-jw_ct}\right|^2 \quad (23.4)$$

Where $\Delta\tau_1 = \hat{\tau}_1 - \tau_1$ is the code phase tracking error of user one.

The punctual channel output SINR is defined as:

$$\rho_n(\Delta\tau_1) = |E\{\lambda_n(\Delta\tau_1)\}|^2 / Var\{\lambda_n(\Delta\tau_1)\} \quad (23.5)$$

where $E\{\lambda_n(\Delta\tau_1)\}$ and $Var\{\lambda_n(\Delta\tau_1)\}$, respectively, can be expressed as [10]:

$$E\{\lambda_n(\Delta\tau_1)\} = \left| C_s \int_{-b}^{b} G_s(f)e^{j2\pi f \Delta\tau_1} df\right|^2 + \frac{C_s}{T}\int_{-b}^{b} G_w(f)G_s(f)df \quad (23.6)$$

$$Var\{\lambda_n(\Delta\tau_1)\} = \left[\frac{C_s}{T}\int_{-b}^{b} G_w(f)G_s(f)df\right]\left[\left| C_s \int_{-b}^{b} G_s(f)e^{j2\pi f \Delta\tau_1} df\right|^2 + \frac{C_s}{T}\int_{-b}^{b} G_w(f)G_s(f)df\right]$$

$$(23.7)$$

Here, the punctual channel output SNIR achieves a maximum at $\Delta\tau_1 = 0$, so that Eq. (23.5) becomes:

$$\rho_n = T\frac{C_s}{N_0}\left|\int_{-b}^{b} G_s(f)df\right|^2 \bigg/ \left(\int_{-b}^{b} G_s(f)df + \frac{C_I}{N_0}\int_{-b}^{b} G_i(f)G_s(f)df\right) + 1 \quad (23.8)$$

where b is the one-side bandwidth of the receiver front-end filter, $G_s(f)$ and $G_i(f)$ are defined as the power spectral density normalized to unit power over infinite bandwidth of the signal and multiple access interference respectively.

$$G_i(f) = G_s(f) = T_c \sin c^2(\pi f T_c) \quad (23.9)$$

where the signal carrier power $C_s = P_1$, the multiple access interference power $C_I = \sum_{k=2}^{K} P_k$.

The fraction of signal power that remains after bandlimiting can be defined as:

$$\eta = \int_{-b}^{b} G_s(f)df \quad (23.10)$$

and the spectral separation coefficient (SSC) for the punctual channel equivalent carrier-to-noise ratio estimation can be defined as:

$$\kappa_{si} = \int_{-b}^{b} G_i(f)G_s(f)df \tag{23.11}$$

Substituting (23.10) and (23.11) into the expression (23.8), and omit the last one :

$$\rho_n = \left(T \frac{C_s}{N_0} \eta^2 \right) \Big/ \left(\eta + \frac{C_I}{N_0} \kappa_{si} \right) \tag{23.12}$$

The equivalent carrier-to-noise ratio of the corresponding punctual channel:

$$\left(\frac{C_s}{N_0} \right)_{eff} nbsp; = \alpha_1 = \frac{\rho_n}{T} = \frac{C_s}{N_0} \left[\frac{1}{\eta} + \frac{C_I}{N_0} \frac{\kappa_{si}}{\eta^2} \right]^{-1} \tag{23.13}$$

23.3 SNR Estimation Algorithm for Single Signal

It can be seen that when the users of the system are numerous and each user signal power is strong, we cannot get the single signal carrier-to-noise power ratio C_s/N_0 of the signal we desire to track from ρ_n. To solve this problem, we need an equation that includes C_s/N_0 and independent with ρ_n. It shows from the derivation of ρ_n, that the statistical properties of the early and late channel accumulations are consistent with the punctual channel accumulations. The SNIR estimates of early and late channel are related with the punctual channel SNIR estimates, thus it cannot use those SNIR estimates to solve the C_s/N_0.

By analyzing the statistical properties of the difference $\gamma(\varepsilon)$ between the early and late channel accumulations, we can finally solve the C_s/N_0. The variance of $\gamma(\varepsilon)$ is obtained by the previous results [11, 12].

$$Var\{\gamma(\varepsilon)\} = \frac{8C_s^3}{T} \left(\int_{-b}^{b} G_s(f) \cos(\pi f D)df \right)^2 \int_{-b}^{b} G_s(f)G_w(f) \sin^2(\pi f D)df \left[1 + \frac{\int_{-b}^{b} G_w(f)G_s(f) \cos^2(\pi f D)df}{TC_s \left(\int_{-b}^{b} G_s(f) \cos(\pi f D) \right)} \right] \tag{23.14}$$

Using the equation above and the punctual channel coherent accumulations (23.6), the equivalent signal to noise ratio of the early-late loop code phase discrimination output can be defined as:

$$\rho_{NCDLL}(\Delta\tau_1, D) = |E\{\lambda_n(\Delta\tau_1)\}|^2 \Big/ Var\{\gamma(\varepsilon)\} \tag{23.15}$$

When the code phase tracking error $\Delta\tau_1 = 0$, and in the limit as the early-late spacing D becomes vanishingly small, we get:

$$\rho_{\text{NCDLL}} = \frac{\frac{T}{8C_s}\left[C_s\left(\int_{-b}^{b}G_s(f)df\right)^2 + \frac{1}{T}\int_{-b}^{b}G_w(f)G_s(f)df\right]^2}{\left(\int_{-b}^{b}G_s(f)df\right)^2\int_{-b}^{b}f^2G_w(f)G_s(f)df}\left[1 + \frac{\int_{-b}^{b}G_w(f)G_s(f)df}{TC_s\left(\int_{-b}^{b}G_s(f)df\right)^2}\right] \tag{23.16}$$

the spectral separation coefficient (SSC) for the NCDLL equivalent carrier-to-noise ratio estimation can be defined as:

$$\chi_{si} = \int_{-b}^{b} f^2 G_i(f)G_s(f)df \tag{23.17}$$

and β is the bandlimited rms bandwidth of the modulation with unit power, defined by

$$\beta^2 = \frac{1}{\eta}\int_{-b}^{b} f^2 G_s(f)df \tag{23.18}$$

Ignore the noise term $\frac{1}{T}\int_{-b}^{b}G_w(f)G_s(f)df << 1$, we get:

$$\rho_{\text{NCDLL}} = \left(\frac{T}{8}\frac{C_s}{N_0}\eta^2\right)\left(1 + \frac{1}{\rho_n}\right)\bigg/\left(\eta\beta^2 + \frac{C_I}{N_0}\chi_{si}\right) \tag{23.19}$$

According to Eq. (23.19) and the statistical properties of the difference $\gamma(\varepsilon)$ between the early and late channel accumulations for coherent delay-locked loop, we can also get the coherent delay-locked loop equivalent carrier-to-noise ratio by the same method above:

$$\rho_{\text{CDLL}} = \left(\frac{T}{8}\frac{C_s}{N_0}\eta^2\right)\bigg/\left(\eta\beta^2 + \frac{C_I}{N_0}\chi_{si}\right) \tag{23.20}$$

Equations (23.19) and (23.20) reveal that NCDLL error variance is the product of CDLL variance and a "squaring loss" that is greater than unity. According to Eq. (23.13), we can define the CDLL equivalent carrier-to-noise ratio as:

$$\left(\frac{C_s}{N_0}\right)_{\text{CDLL}} = \alpha_2 = \frac{8\rho_{\text{CDLL}}\beta^2}{T} = \frac{C_s}{N_0}\left[\frac{1}{\eta} + \frac{C_I}{N_0}\frac{\chi_{si}}{\eta^2\beta^2}\right]^{-1} \tag{23.21}$$

and the NCDLL equivalent carrier-to-noise ratio is:

$$\left(\frac{C_s}{N_0}\right)_{\text{NCDLL}} = \alpha_2 \left[1 + \frac{1}{\rho_n}\right] \tag{23.22}$$

Combined solution of Eqs. (23.13) and (23.21), we get the single signal carrier-to-noise power ratio:

$$\frac{C_s}{N_0} = \alpha_1 \left(\frac{1}{\eta} + \left[1 - \frac{\alpha_1}{\alpha_2}\right] \frac{\kappa_{si}}{\eta} \left[\frac{\alpha_1 \kappa_{si}}{\alpha_2} - \frac{\chi_{si}}{\beta^2}\right]^{-1}\right) \tag{23.23}$$

23.4 Simulation Results

According to Eq. (23.23), we get the formula of single signal carrier-to-noise power ratio with strong multiple access interference. We can estimate the carrier-to-noise power ratio of a single signal with multiple access interference using the second- and fourth-order moments estimator of literature [13], which has the minimum estimated standard deviation of four moments of second-order moment estimation. Table 23.1 shows the symbol definitions and parameters in the simulation:

The actual carrier-to-noise ratio of single signal is ρ_0, the simulation performance evaluation criteria are given below:

Single estimation:

$$\text{MEAN}\{\hat{\rho}\}(i), \quad i = 1, 2 \ldots L \tag{23.24}$$

Estimation mean:

$$\text{MEAN}\{\hat{\rho}\} = E\{\hat{\rho}\} = \frac{1}{L} \sum_{i=1}^{L} \hat{\rho}(i) \tag{23.25}$$

Estimation bias:

Table 23.1 Simulation parameter values

Parameter name	Parameter value
PN code rate R_c	5.115 Mcps
Symbols rate	100 kbps
CDMA users	30
Front-end bandwidth	7 MHz
Early-late spacing	0.5 chip
Coherent integration time	0.01 ms
Coherent integration numbers for a single estimation	100
Estimation times	1,000

$$\Delta\{\hat{\rho}\} = \text{MEAN}\{\hat{\rho}\} - \rho_0 \tag{23.26}$$

Estimation standard deviation:

$$\text{STD}\{\hat{\rho}\} = \sqrt{\frac{1}{L-1}\sum_{i=1}^{L}(\hat{\rho}(i) - \text{MEAN}\{\hat{\rho}\})^2} \tag{23.27}$$

Each estimate uses 100 relatively cumulative value in simulation, namely 1 ms data. The 1000 estimations averaged within 1 s, we get the carrier-to-noise power ratio estimation. If the power change of each earth station is relatively stable at the practical system, we can use much longer data for the corresponding average.

The estimation mean is the carrier-to-noise power ratio estimate used by a practical system. The estimation bias is the size of the deviation between the carrier-to-noise estimate and the true value. Estimation standard deviation is the deviation between the estimates. It reflects the stability of the estimation method.

Figure 23.3 shows four curves, two are the single signal carrier-to-noise estimates and punctual channel SINR estimates by use Eq. (23.23), the other two are the equivalent carrier-to-noise estimates of the coherent delay-locked loop and non-coherent delay-locked loop by definition of Eqs. (23.21) and (23.22). The results show that as the single signal carrier-to-noise power ratio increased from 45 to 60 dBHz, the bias of the punctual channel SINR estimates, CDLL equivalent carrier-to-noise ratio estimates, NCDLL equivalent carrier-to-noise ratio estimates to the true value of the single signal carrier-to-noise power increase gradually. Therefore, the SINR estimates cannot be a true reflection of the transmitting power of each earth station.

The single signal carrier-to-noise power ratio estimation method proposed in this paper can reflect the true change of the single signal carrier-to-noise power ratio. Figure 23.4 shows the comparison between the single signal carrier-to-noise ratio estimation bias and all kinds of SNIR estimation bias. It can be seen from the figure, when the single signal carrier-to-noise power ratio increased from 45 to 60 dBHz,

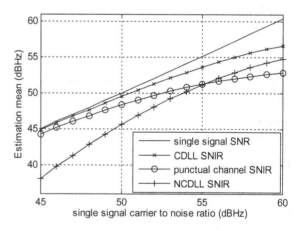

Fig. 23.3 Single signal carrier-to-noise power ratio estimation value and all kinds of SNIR estimation values

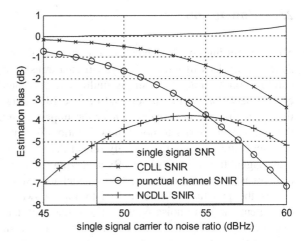

Fig. 23.4 Single signal carrier-to-noise power ratio estimation bias and all kinds of SNIR estimation bias

the single signal carrier-to-noise power ratio estimation method to get the results with the true value of deviation is smaller, the maximum deviation of less than 0.5 dB. And the results of other types of SNIR estimate of method have deviation compare to the true value of the single signal carrier-to-noise power ratio.

Figure 23.5 shows the comparison of estimating standard deviation between the single signal carrier-to-noise ratio estimates and various types of SNIR estimates. The results show that the estimated standard deviation of single-channel signal carrier is relatively large when the actual signal carrier-to-noise power is relatively large. The single-channel signal estimated standard deviation close to 0.5 dB when the actual carrier-to-noise power ratio is 60 dBHz. Various SNIR estimated standard deviation is relatively small.

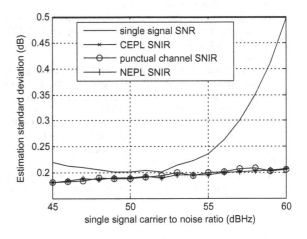

Fig. 23.5 Single signal carrier-to-noise power ratio is estimated standard deviation of all kinds of SNIR estimated standard deviation comparison

23.5 Conclusion

The transmitting power imbalance of earth stations between the time synchronization networks will reduce the system data transmission capacity and time synchronization accuracy with strong multiple access interference. Therefore, power control between central station and other earth stations is very important in the time synchronization network. Under the impact of multiple access interference, the equivalent carrier-to-noise power ratio of central station estimated from received signal does not accurately reflect the relative strength of each earth station's signal transmitting power. By analyzing the statistical properties of punctual channel as well as the early and late channel accumulations, we get two equations, one is the punctual channel carrier-to-noise power ratio, and the other is the equivalent carrier-to-noise power ratio of the early-late loop code phase discriminations. By solving these two equations, we get the single signal carrier-to-noise power ratio C_s/N_0, which is the carrier-to-noise power ratio when the multi-access interference does not exist. The estimation deviation and the estimation standard deviation are both less than 0.5 dB in this method, which accurately reflect the relative strength of the transmitting power of each earth station. It provide more accurate measurements for power control of central station, which can effectively improve the accuracy of time synchronization network power control between earth stations.

References

1. TAN Shusen (2008) Development and thought of compass navigation satellite system. J Astronaut 29:391–396(in Chinese)
2. TAN Shusen (2010) The engineering of satellite navigation and positioning. National Defense Industry Press, Bejing (in Chinese)
3. Yuanfa JI, Xiyan S (2009) Analysis on the positioning precision of CAPS. Sci China, Ser G: Phys, Mech Astron 52:328–332(in Chinese)
4. Jiang Z, Petit G (2009) TWSTFT data treatment for UTC time transfer. In: 41st annual precise time and time interval (PTTI) meeting, pp 409–420
5. Yang C-Y, Chen B-S (2010) Robust power control of CDMA cellular radio systems with time-varying delays. Sig Process 90:363–372
6. Gross TJ, Abrão T, Jeszensky PJE (2011) Distributed power control algorithm for multiple access systems based on Verhulst model AEU. Int J Electron Commun 65:72–361
7. Sung CW, Leung K–K (2005) A generalized framework for distributed power control in wireless networks. IEEE Trans Inf Theor 51:35–2625
8. Ren G, Chang Y, Zhang H, Zhang H (2003) A novel SIR estimation technique in CDMA systems. Acta Electronica Sinica 31:1461–1464(in Chinese)
9. Xu H, Zheng H (2005) A new blind SNR estimation algorithm for BPSK signals. J China Inst Commun 26:123–126(in Chinese)
10. Betz JW (2000) Effect of narrowband interference on GPS code tracking accuracy. In: National technical meeting of the institute of navigation, pp 16–27

11. Betz JW, Kolodziejski KR (2009) Generalized theory of code tracking with an early-late discriminator part I: lower bound and coherent processing. IEEE Trans Aerosp Electron Syst 45:1538–1550
12. Betz JW, Kolodziejski KR (2009) Generalized theory of code tracking with an early-late discriminator part II: noncoherent processing and numerical results. IEEE Trans Aerosp Electron Syst 45:1551–1564
13. Jingyuan LI (2006) Study of the multiuser detection and signal to interference ratio estimation in satellite CDMA systems. National University of Defense Technology, Changsha (in Chinese)

Chapter 24
Use of Technology to Increase Functional Autonomy and Facilitate Communication for People with Disabilities: Design of an Emerging "Dis(ease)ability" Model

Raffaella Conversano and Gaetano Manzulli

Abstract Disability is not a choice, but a trauma that no one ever wants to live. Effectively connecting with peers and successfully participating in the processes of learning is an 'indispensable condition for not being discriminated against. While research continues to generate data to accurately diagnose and improve the inclusion of those with disabilities, such efforts sometimes contribute to an incorrect focus on the "disability" as a result of disease or disorder of the person. Although well-intentioned, unintentional discrimination may be the result. To counter such erroneous behavior, undergirded with field-proven educational theory, we began a complex search for solutions, built with the help of various technology skills that place focus on an array of options for differing needs, thus maintaining the dignity of each learner.

Keywords Smart technologies · Language communication · New media · APP · Disabilities

24.1 The State of the Art

Multiple fault lines that mark distances and differences between social groups, physical and mental health conditions and impairments in bodily and neurological functions and structures are one of the main reasons of inequality and

R. Conversano (✉)
Università La Sapienza, Roma, Italy
e-mail: raffaellaconversano@tin.it

G. Manzulli
Istituto Pacinotti, 74121 Taranto, Italy
e-mail: gaetano.manzulli@pacinottitaranto.it

W. Lu et al. (eds.), *Proceedings of the 2012 International Conference on Information Technology and Software Engineering*, Lecture Notes in Electrical Engineering 210, DOI: 10.1007/978-3-642-34528-9_24, © Springer-Verlag Berlin Heidelberg 2013

discrimination, which are often associated with forms of "stigma" gravely injurious to the dignity of the human person. New technologies are helping learners overcome many limitations and encourage the expression of learners' potential in many dimensions, with the infinite panoply of possibilities they offer. Slowly, they help improve inter-functional communication and learning. The importance to fully understand the possible technological applications, especially when used for the information needs of people with limited autonomy was the inspiration of our operational work. This approach led to the pragmatic implementation of our theory, making extraordinary implications for us, yet completely unpredictable. To claim that its applicability, from our point of view, could undermine what has been theorized and realized up to now, for and on behalf of the use of new technologies, and to say the least restrictive, because we wanted to shake the foundations of what was stated by theorists and empiricists. Therefore, the preeminent value we have chosen to deepen and transmit is an understanding of how each individual labeled "disabled," as distinguished from "normal," regardless of gravity the of the diseases that afflict them, are hidden from the great potential of ideas and solutions for the social life of everyone.

24.2 Our Theory: The Dis(ease)ability

In everyday reality school problems emerge simultaneously with difficulty in learning. The movement toward inclusion and integration. Is unfortunately floundering in seeking solutions. Especially needed are paths of mediation between teachers' knowledge and skills of experts in the field, who are almost always seen as the only real resources. In this field, the urgent need to meet the challenges that learning difficulties and mental disability generally reveal, convinced advocates that educating through what is said, but more by what you do, puts in place the need to develop an increased awareness and shared reflection (between teachers and "experts") on the issues. According to a reinterpretation, employing a Holistic point of view, different ways and styles of learning are embraced as a resource for addressing the clinical pathology. Such a perspective of diversity should be the guideline observed and kept as a reference point to calibrate and adjust the method and teaching, thus encouraging the process of both teaching and learning. The idea is 'simple: instead of restricting the scope of autonomy of the person, partly because of one's clinical and/or perceived pathological status (i.e., disease), raise awareness of the potential of technology to create ease-ability. Indeed, although the appearance of disability involves both the subject and its surrounding world, the construction of technological devices enables personal autonomy, while triggering universes of relationships, feelings, and hopes, since the device has the strong power to overshadow the impairment, thus bringing the learner into the world of the learning community. Disregarding or not observing this basic thought, trivializes the clinical diagnosis as a barrier to the social and individual reinforcing the common stereotype of disability. In fact, the term we coined, *Dis(ease)ability*, redefines the real obstacles that children, students and people with disabilities generally live and/or encounter. Such difficulties include not just the deficit but anxieties, fears, adjustments to their environment, interactions, and reactions of those around them. Despite their disadvantages, they are still able to reach varying degrees of personal autonomy. This term provides a clearer focus on the learner's difficulties (i.e., dis-ease), thus affording an opportunity to "see" the problem and possible solutions to more autonomy in their entirety, in order to create a space, physical or virtual, where experience and professionalism can find the right expression with respect to roles and responsibilities. The strong point of reference with regard to this new term—*Dis(ease)ability* is to act on false interpretations that may have been perceived of such learners' behaviour. Working with a person with a disability, however, serious the diagnosis, should not preclude respect and the developmental potential to lead a normal life: in other words, because of the deficit, mental physical or both, one should not be singled out as "one who is not able to understand or live" according to common standards of normalcy. The positive implication of this term and 'given by the *synergy*—that is, the optimal integration of several elements intended to achieve a common goal-to obtain an overall result more satisfactory than that which would be obtained separately: open the *Dis(ease)ability* technologies, designed as a resource for everyone including cultural differences and/or social, must be understood as an element of wealth in human ethics. Our starting point and inspiration 'was not so much to consider that people with disabilities and/or otherwise disadvantaged people can learn different habilitation like everyone else; that's obvious. We have reinterpreted the relationship that the school environment must have with the world of media and communication, in order to use new and creative technologies to promote efficient and effective learning of knowledge, mastery of approaches, monitoring and evaluation of impact of training, promoting specific skills in the areas of labor discipline. Our intent was not to teach the use of media. The new generations do not need them, let alone people with difficulties, who have been the first to appreciate them. Our emphasis was focused on how to understand and capture their experiences and how our educational project could become the new teaching. Media Education, as we reinterpreted it, refers to a new educational and technological reality, a true resource in the teaching process that has allowed us to create models and styles of teaching/learning that are not only innovative, but based on processes of collaborative communication and bi-directional, as they have been outlined in recent years with distance education. These concepts were reiterated in the interview broadcast that Dr. Conversano issued within the program "Different from Whom?" Aired on Rai Radio 1 on 20/11/2010, whose podcast can be listened to via the link published in the collection of websites (see references).

24.3 Our Doubts: Promoting Autonomy or Complicating the Reality?

"According to some authoritative texts of technical aeronautics, the bumblebee cannot fly because of the shape and weight of their bodies, in relation to the wing surface. But the bumblebee does not know and therefore continues to flow." (I. Sikorsky). Even in the discussion of these lines of Igor Sikorsky the key to the application of dis(ease)ability (the important belief that has guided our steps), we identity, capacity, behaviours and environments, inverted with respect to limiting beliefs. Reversing the route allowed us to achieve common goals for all, with full consideration of success. The pervasive presence of disability or rather, the concrete contexts of *Dis(ease)ability* in everyday social life, poses problems in individuals of various sizes: the goal was to design a functional technological application in order to facilitate access to social communication, while acquiring, a personalized approach necessary to compensate for what the disability is missing in the individual, as this would increase his gap of psychological distress to others, but we wanted to reinforce that phase between what the individual as a whole was able to do or might have done and what he still might be able to achieve, despite the disabling situation, with tutorial support technologies; thus moving the attention from what the individual can do alone (area of individual competence) to what could be done if supported by any help. So, even when technology is essentially designed for the field of electronics or computer, one could change any element to move matters from a state of passivity and dependence to one of independence and wellness (area of proximal development). In teaching what the person is already capable of doing one risks discouragement or devaluation of self-generating ideas. Strengthening one's healthy part becomes essential. Unfortunately, various types of clinical and pathological disabling put into place a mode of being wherein the "disabled" had to communicate with us in a unique and standardized manner, creating ongoing conflicts, depression, anxieties and frustrations with closures and escapes by the so-called "sick" and defeated by so-called "normal." A re-reading reverse would simplify everything. The application of our theory—the dis(ease)ability embraces a dynamics approach with regard to the differences not only in general, but detaching in fact empiricist lines of "psycho-technology" theorized by de Kerckhove aims, through pragmatic approaches instrumentally, to make concrete and operational the Vygotskijan theory of the "zone of proximal development." In fact, interacting on the connection between language and mental organization, in its application being able to edit and/or speak at a deep level of an individual's psyche, we have acted strategically in the solution of needs not always met, because they are based exclusively on complex operation, thanks to the integrated skills of clinical, educational and technological domain. Conceived in its universal approach, this theory is well connected to life not only with the disabled but its plural facets. See for example the stranger with the difficulty of understanding the linguistic codes of communication; if he encounters a context that does not support the idea and recognition of these codes, a condition of living large with dis(ease)ability, it contributes to a further state of estrangement, exploitability and social distance. The same goes for those who have suffered disabling injuries that lead them to be excluded so temporarily or permanently from the context in which one belongs. Our goal 'was to act on those structurally critical factors that normally affect the efficient approach to integrating, implementing solving assumptions for technological assistance to foster communication, experience it and create new bases for its transferability to all operational contexts of everyday social life. Moreover, its focus is part of the heuristics. The heuristic process relies on intuition and temporary circumstances, in order to generate new knowledge as a method or approach to problem solving. In fact, the dis(ease)ability theory indicates, as this heuristic, the avenues and possible strategies to be activated to make progressive, developmental and applied intercommunication' with disabilities' or approaches to life contexts.

24.4 Our Facts: Testing and the Project

This project is a natural evolution of two prior works: "The enchanted maze" presented last year to the 9th Annual International Conference on Communication and Mass Media, and the project "The Virtual Media as a tool for development and integration of pupils with disabilities," conducted with a network of schools from Taranto. The project currently discussed focused upon 50 students with intellectual disabilities, but also those with specific disorders and learning disabilities (ASD, dyslexia, dysgraphia, dyscalculia), often due to a maturational delay, with a low investment and low motivational experiential background. This educational project is strongly oriented to the adoption of specific technologies for teaching and the use of technologically advanced software platforms for the integration of children with various intellectual disabilities. The knowledge that different and innovative ways of support can facilitate training and educational opportunities, thus enabling focused and accessible recovery, led the teachers–authors of this initiative to design an educational system with technology intended to pursue the following objectives:

- Finding solutions and innovative development of new technologies applied to teaching Information and Communication;
- Support for pupils with temporary or permanent disabilities by making accessibility and usability of information compliant with W3C standards; and
- Development of autonomy in learning and experiential and motivational involvement of the disabled person. The logic followed in the project was to develop a tool to foster the development of autonomy in learning of students with disabilities who, in addition to learning problems, have difficulties in the application of knowledge and in carrying out tasks. Or use of instruments such that they are not able to act on their own. The project aims to strengthen the capacity of abstraction and logical thinking—reinforcing basic skills and operational real capabilities of the recipients, through the use of innovative

technology that decreases the difficulties the disabled pupil experiences, thus reducing the gap with the class group. These students, in fact, often present a low level of self-esteem and psycho-emotional and behavioral immaturity and for this reason, in any educational activity, they are usually restrained by fear of failure, as argued by the engineer Mr. Manzulli in his radio interview released within the program "Service Area" of October 2, 2011 aired on Rai Radio 1 in Italy. The podcast can be heard via the link provided in the site links. Each disabled person is "foreigner" in the entire world because they often live in their isolation also among their own countrymen. Therefore, the issues addressed in this article are aimed at both Italian and foreign people with disabilities, who share such concerns. The design concept, therefore, is based on the premise of making contents (that are often difficult to understand and redress when using traditional approaches to teaching) most easily assimilated by the disabled population. Moreover, understanding the didactic message is strongly affected by attentional liability, and so the student needs constant reminders to focus attention. Often, the teacher provides such a function. Technology, used inno- vatively and effectively, can provide such a function and thus reduces the gap between the disabled and the world around him, eliminating the isolation and encouraging his integration. It should be noted, furthermore such students are often low performing in tasks of working memory, so it is important to provide a tool that gives the opportunity for self-control, resulting in the possibility of error correction in real time, given the immediacy with which feedback, positive or negative, is presented. Thus they are able to independently manipulate and transform information. This tool would help the user become a more indepen- dent, active, constructive learner. From the point of view of technological innovation, the objective is to promote learning through the use of a mode of communication adopted by the latest models of mobile phones such as smart phones, because, when used in communication activities and educational applications, smart phones allow the creation of a learning environment whose characteristics are the immediacy of understanding, ease of use and the adoption of mechanisms, which they are motivated to use. In fact, this generation of students already uses the Internet and the new frontiers of communication, as a natural space for learning and play; so that they have been called *digital natives* [1] since they are growing inside of a world already digital. The involvement of the student with a disability in the use of such tools would allow the ability to break down the "digital divide" often due to the not always practical use of technology; that is, what is referred to as isolation technology of which one is often the victim. Applying language and actual tools close to the current way of communicating in everyday life may help overcome any personal isolation such a student might feel in the classroom. The now-essential use of new commu- nication technologies can, therefore, truly innovate learning environments by rethinking the paradigms that underlie the educational process: if the traditional paradigm was based on the transfer of knowledge from teacher to learner, in the new reading it relies on the mechanisms of constructivist knowledge. In fact, "virtual" learning environments promote new learning systems based on relativistic models of knowledge where, through cooperation and communication based on the use of images, sounds and electronic messaging, students find creative ways to improve their knowledge, using the community as a meeting place for the production, distribution and management of knowledge but also as a space for socializing (social learning) and virtual dimension of confrontation and dialogue. The first experience made with the introduction of mobile phone use at school was conducted in a multiethnic class of variable composition that saw the presence of several students with disabilities. After conducting a technical investigation of cell types and how they had worked, a calendar was created of "slang" common to all. In a very simple and natural way, we realized that students had become more independent (typing is easier than writing) and did not lose any opportunity to engage in this new teaching methodology in order to use their phones. Moreover, in a subsequent theatrical experience in school, students texted the various adjustments to the beats of the script, procured by the Bluetooth various mp3 sound files and created the backstage photos by participating actively in the creation of the show. Translated into educational terms, this teaching method provides, in addition to a particular attention to the relational and affective aspect (with the activation of mechanisms to facilitate group dynamics), the exposure to various subjects and planned routes in a form simplified to facilitate both learning content and increasing autonomy. Teachers can now choose the content and set minimum targets, with specific interventions coordinated not only by the support teacher, but also by use of various methods mainly for tutoring, teaching multimedia, orientating activities that are of an interactive and multidisciplinary nature. Accordingly, they may realize special orientation activities for children with disabilities. It is in this context that the school's new mission can be realized: to direct its pedagogical activity as educational action. In reference to our current research project, after success in the use of the techniques mentioned above, attention has been given to some limitations that the current technology of the smart phone shows when users are disabled. For example, for visually impaired users the graphic richness of a smart phone can be a daunting exercise because the display appears to be small compared to that of a computer. Those who are suffering from impaired mobility of the limbs cannot use their hands properly, so correctly using the property of a capacitive screen smart phone (sensitivity to small current of fingertips) is challenging. Facing these obvious difficulties, the project idea initially saw the realization of a software platform that, through a dedicated operating system, could reconstruct on a computer or a tablet PC, the exact re production of a smart phone and its unique operating characteristics. Thus, using the tools made for disabilities currently available as accessories for the PC (for each type of disability there is a specific aid, a tool able to facilitate the movement or improve the vision of the screen) everyone can use functions such as navigating with the touch while on the screen everything is exactly as in the phone screen: all functions, including the telephone can be switched from the PC, through an appropriate interface with SIM cards. The interface with the computer is not an amazing thing. There are many programs that today allow one to drive the most common mobile phones with one's computer, even Mac OS X lets one write SMS messages directly from the keyboard with many Bluetooth phones. For a disabled person, however, the simplicity and flexibility of a smart phone interface directly on the computer would enable them to use their mobile phones with display and control systems they already routinely use, without suffering the limitations of the available solutions today. Moreover, the evolution of the project, with the advent of the Tablet PC, consisted of thinking about an application that would allow disabled children to improve communication with others and with their teacher using the screen of I-Pad and its touch screen functionality. The interface to touch, and devices like the I-Pad have been created for the intuitive ability of a user of 2 years. In essence, the application uses a kind of "slang" of images that helps the disabled to express themselves through some buttons, making the device a real communication tool, allowing them to interact with the class. The project includes an enforcement phase of testing in a properly equipped classroom, but soon the application will be available in all stores. The application has been made with the development environment Xcode 4.2 for Leopard and Lion operating systems, using the Software Development Kit available from Apple on devices with IOS5 (I-Pad, I-Phone and Mac).

24.5 Conclusion

Special education does not walk on roads other than the "normal," and the study of disability and that of the person, even if they belong to different areas of knowledge and action, require professionally valid input, but based on choice or need. Developing concrete skills should be contextualized by giving precedence to appropriate action in relation to the emerging needs of individuals in their existential dimension, pedagogical-didactic and clinical education. Finding the best solutions to help build these skills is a challenging and difficult task, which raises many questions and problems to be addressed because each student has specific characteristics and difficulties that make one unique and different from others. So the best teachers are able to cut their work around the specific characteristics of each learner, where such action cannot be exhausted in sterile technological performance. Innovation is not just about managing to have good insights but new ideas to improve the old to understand and anticipate even if the path is still all uphill, complicated, full of disappointments, true and false hopes that we, with perseverance, we have decided to dedicate our lives, daily supported by the words that still vibrate to the graduates of the late Steve Jobs at Stanford University in 2005: "... If you cannot find what suits you, keep looking, do not stop ...", [in order to] provide better quality of life for our less fortunate friends or, to use McLuhan's aphorism: "... If you do not like our idea .... no matter .... We have many others ...!"

Reference

1. Prensky M (2001) Digital game-based learning

Useful Links

2. RAI RADIO 1: "Area di servizio"—Interview Gaetano Manzulli http://www.rai.it/dl/radio1/
   2010/popup.html?t=Area%20di%20Servizio%
3. RAI RADIO 1: "Diversi da chi?"—Interview Raffaella Conversano www.radio.rai.it/radio1/
   diversidachi/view.cfm?Q_EV_ID=321903
4. HANDIMATICA 2010—Migrants and disabled Seminar: Technology mediation and medi-
   ators http://www.youtube.com/watch?v=DCU0A74cwfo

Chapter 25
A Rectification Method for Removing Rolling Shutter Distortion

Gang Liu, Yufen Sun and Xianqiao Chen

Abstract We propose a rolling shutter video rectification method that can deal with both camera translation and rotation for videos obtained from unknown sources. As the exact distortions caused by rolling shutter are too complex, this method aims to remove the majority of the distortions. A 2D rotation model is used to approximately represent the motion of each frame. The parameters of this model are solved by minimizing the measurement constraints on point correspondences. To relax the restriction on the form of frame motion, the frame motion computation is performed sequentially. Experiments show that our method is comparable to the 3D models that require camera calibration.

Keywords CMOS image sensor · Rolling shutter distortion · Image rectification · Post-processing technique

25.1 Introduction

Nowadays, CMOS sensors are widely used in video cameras because of their low prices and the ability to integrate with image processing circuits. A CMOS camera commonly adopts rolling shutter to read and reset the rows of the sensor array sequentially, making the capture time of each row slightly after the capture time of the previous row. When camera or object moves during image capture, the scene imaged by each row presents relative motion, resulting in geometric distortion

G. Liu (✉) · Y. Sun · X. Chen
Intelligent Transportation System Research Center, Wuhan University of Technology, 430063 Wuhan, China
e-mail: liu_gang@whut.edu.cn

W. Lu et al. (eds.), *Proceedings of the 2012 International Conference on Information Technology and Software Engineering*, Lecture Notes in Electrical Engineering 210, DOI: 10.1007/978-3-642-34528-9_25, © Springer-Verlag Berlin Heidelberg 2013

235

called rolling shutter distortion. These distortions degrade the quality of images and making objects in videos appear dynamic non-rigid deformation.

Removing rolling shutter distortion needs to compute the relative displacements between rows in a frame. This computational task is more difficult than video stabilization in which only the relative motions between frames are computed. Recently the problem of removing rolling shutter distortions from videos has been studied by many researchers [1–8]. These researches focused on the distortions caused only by camera motion. 2D image processing methods were proposed to process the distortions induced by smoothly varying translation parallel to the image plane [2, 3, 6, 7]. The more complex distortions caused by camera rotation was studied using 3D rotation models [4], with a requirement of camera calibration.

In this paper, we propose a rectification method that can handle both camera translation and camera rotation, without requiring any information about the camera. Since the general image transformation is very complex, we do not intent to estimate the motion of each pixel exactly, but only to compute approximate frame motions that can be used to remove the dominating distortion caused by rolling shutter. A 2D rotation model is used to represent the transformation between two successive frames. The parameters of the model are solved by minimizing the measurement constraints on point correspondences. Based on the rotation model, the distortion caused by rolling shutter can be revealed. Experiments show that our method is comparable to the 3D rotation model [4] that beat other 2D methods in most situations.

We introduce the related work in Sect. 25.2. Section 25.3 gives our 2D rotation model and analyzes why the model works for our task. Our rectification method is described in section four. Section five analyzes the experimental results. At last, section six concludes the paper.

25.2 Related Work

Mechanical methods have been designed to avoid rolling shutter distortion by using a mechanical global shutter or adopting a mechanical image stabilization system. However, the mechanical global shutter will reduce the fill factor of image sensors, and the mechanical image stabilization system is not equipped in most consumer cameras. Post-processing methods are thus required to rectify the distorted videos.

The most important thing in rectifying rolling shutter distortions is to find the relative pixel displacement during the image exposure interval. This displacement is equal to the temporal integral of the image motion between the time instants at which the pixels are imaged. If a camera undergoes a constant translation parallel to the image plane, the constant motion velocity of the image can be directly computed using the global motion between two successive frames [2, 3, 7]. Liang et al. proposed a method that can deal with smoothly varying translation parallel to the image plane [6]. The high-resolution motion velocity of frames is computed using global motion estimation and parametric curve fitting. These methods suppose the points on the same row have the same motion, and cannot be used to rectify distortion induced by camera rotation.

Forssén and Ringaby are the first to compute 3D camera motion for rolling shutter rectification [4, 8]. As the motion of image is the result of camera motion through perspective projection, their method has a clear physical meaning. They used spherical linear interpolation to interpolate the camera rotation from the knots represented by rotation parameters, and then solved the parameters using inter-frame correspondences. This method can effectively recover the rolling shutter images distorted by camera rotation. But it needs camera calibration, which is not convenient in some situations. To facilitate controlled evaluation of different algorithms, Forssén and Ringaby generated synthetic rolling shutter sequences that can be downloaded from [9].

When a camera is attached to a moving vehicle, the camera may vibrate at a higher frequency than the frame rate of the camera. Baker et al. studied this problem using translational model and affine model [1]. The camera jitter was solved by optimizing the measurement constraints of the correspondences sub-sampled from the optical flow fields.

Cho and Hong used affine matrix to represent the transformation between two distorted frames [2]. Their derivation was based on uniform frame translation.

Experiments showed that the video stabilization methods proposed by Liu et al. could reduce rolling shutter distortions [10, 11]. Though the theoretical explanation for why this happened was not clear.

25.3 An Image Motion Model

We first analyze the relationship between rolling shutter distortions and image motion. Then a 2D rotation model is introduced.

25.3.1 Rolling Shutter Distortions

To rectify a distorted rolling shutter frame, the middle row of the frame is normally chosen as the reference. As show in Fig. 25.1, the aim of the rectification is to obtain a frame as if it is imaged by a global shutter camera at the time instant when the middle row of the frame is imaged. We define the time interval between the sampling instants of two successive rows as the unit time. The displacement of a pixel p_i during the time interval between the instant r_i it is imaged and the reference instant r_m is:

$$d_i = \int_{r_i}^{r_m} v(t)dt, \tag{25.1}$$

where $v(t)$ represents the motion velocity of the pixel. If the inter-frame delay of the camera is zero, the time interval $r_m - r_i$ is at most half of the frame period.

Fig. 25.1 The image
rectification model

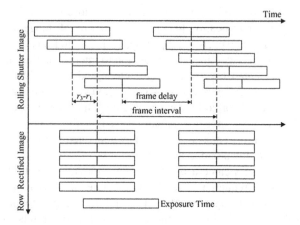

A nonzero value of inter-frame delay will decrease this rate further. So if the
displacement of pixel p_i during the time interval between frames is small, the
distortion d_i will be smaller, at most half of the former. Thus we can only consider
large displacements of pixels to remove obvious distortions.

25.3.2 A 2D Rotation Model

For two successive frames captured by a global shutter camera, the major motion
of each pixel is rotation around a point in the image plane or translation along a
direction. If a frame rotates around a point $\mathbf{c}$ in the image plane through an angle θ,
pixel $\mathbf{x}$ in this frame will move to a new position $\mathbf{x}'$, which is:

$$\mathbf{x}' = f(\mathbf{x}, \theta, c_x, c_y) = \begin{bmatrix} \cos\theta & -\sin\theta \\ \sin\theta & \cos\theta \end{bmatrix} \left(\begin{bmatrix} x \\ y \end{bmatrix} - \begin{bmatrix} c_x \\ c_y \end{bmatrix} \right) + \begin{bmatrix} c_x \\ c_y \end{bmatrix}. \qquad (25.2)$$

This equation models in-plane rotation. If the motion is translation, the pixels
can be thought as rotating around a far away point located on the direction that is
perpendicular to the motion direction. Then Eq. (25.2) can be used to approximately model translation if we choose an appropriate rotation center.

25.4 Image Rectification

Suppose we have point correspondences $\mathbf{x}$ and $\mathbf{x}'$ in two successive frames, then:

$$\mathbf{x}' = \mathbf{x} + \int_r^{N+r'} v(t, x)\, dt \qquad (25.3)$$

where r and r' are the row number of pixel $\mathbf{x}$ and $\mathbf{x}'$, N is the number of rows in a frame. In Eq. (25.3), we suppose the inter-frame delay is zero. If we have a set of data correspondences, the measurement constraint of the correspondences is:

$$E = \sum_{k=1}^{N} d(\mathbf{x}_k - \mathbf{o}_k)^2 + d(\mathbf{x}'_k - \mathbf{o}'_k)^2 \tag{25.4}$$

where $\mathbf{x}_k$ and $\mathbf{x}'_k$ are point correspondence, $\mathbf{o}_k$ and $\mathbf{o}'_k$ are relocated points. In Eq. (25.4), $\mathbf{o}'_k$ is computed by $\mathbf{o}_k$ and $v(t)$ using Eq. (25.3), $\mathbf{o}_k$ and $v(t)$ are parameters needed to be determined by minimizing the measurement constraint. If we don't add any constraint on $v(t)$, the parameters are hard to be solved even when we have enough correspondences.

25.4.1 Motion Computation

To compute the frame motion for image rectification, we first use a strong assumption to initialize the computation, and then use a weak assumption to finish the computation.

25.4.1.1 Initialization

If a frame uniformly rotates around a fixed center (c_x, c_y) during the capture interval of two successive frames, the rotation velocity v_r is:

$$v_r = \theta/N, \tag{25.5}$$

where θ represents the rotation angle of the frame in one frame interval. Then Eq. (25.2) can be rewritten as:

$$\mathbf{x}' = f(\mathbf{x}, (N + r' - r)v_r, c_x, c_y). \tag{25.6}$$

Using Eq. (25.6), θ, c_x and c_y can be easily solved by minimizing the measurement constraint. We use the sparse Levenberg-Marquardt algorithm given by Hartley and Zisserman [12] to solve this optimization problem.

Though the assumption of uniform rotation in two frame intervals seems strong, it can be satisfied by some frames in a video captured by a camera undergoing smoothly varying motion. To find such frames, we compute the measurement constraint for each pair of successive frames. The two frames giving the minimal value of E are considered to satisfy the assumption. They are chosen as the initial frames with their motion solved.

25.4.1.2 Sequential Computation

Now we can relax our assumption by assuming a frame uniformly rotates around a fixed center in only one frame interval. Based on the motion of the initial frames, we compute the motion of other frames in two directions, which is illustrated in Fig. 25.2.

For the frames behind the initial frame k, we first transform the frame k to the time instant when the last row was imaged:

$$\mathbf{x}^{new} = f(\mathbf{x}, ((N-r)/N)\theta(k), c_x(k), c_y(k)) \qquad (25.7)$$

Then the distances between the points in correspondences are only caused by the motion of frame $k+1$. The relationship between the corresponding points $\mathbf{x}^{new}$ and $\mathbf{x}'$ is:

$$\mathbf{x}' = f(\mathbf{x}^{new}, (r'/N)\theta(k+1), c_x(k+1), c_y(k+1)) \qquad (25.8)$$

Now we can get $\theta(k+1)$, $c_x(k+1)$ and $cy(k+1)$ by minimizing the equation (25.4).

To find the rotation of frames captured before frame k, we first rotate frame k to the instant before the first row was imaged:

$$\mathbf{x}^{new} = f(\mathbf{x}, (-r'/N)\theta(k), c_x(k), c_y(k)) \qquad (25.9)$$

The relationship between $\mathbf{x}^{new}$ and the corresponding point $\mathbf{x}'$ in frame $k-1$ is:

$$\mathbf{x}' = f(\mathbf{x}^{new}, (-(N-r')/N)\theta(k-1), c_x(k-1), c_y(k-1)) \qquad (25.10)$$

This process is going on until every frame motion is solved.

25.4.1.3 Finial Solution

When we compute the motion of frame $k+1$ or $k-1$ based on the motion of frame k, the error of the latter may decrease the quality of the computation. To handle this problem, for each frame we compare the solution computed in the initialization step with the solution acquired in the sequential computation step. The one with smaller value of E is chosen as the finial solution.

Fig. 25.2 Sequential computation of frame motion. The frame k is the initial frame with known motion

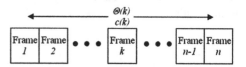

25.4.2 Frame Rectification

To remove the rolling shutter distortion in frame k, the corrected position of pixel $\mathbf{x}$ is:

$$\mathbf{x}^{cor} = f(\mathbf{x}, ((N/2 - r)/N)\theta(k), c_x(k), c_y(k)) \qquad (25.11)$$

Because $\mathbf{x}^{cor}$ may not fall on the image sampling grid, we perform cubic interpolation to acquire the recovered image. The function griddata in Matlab is used to do this work.

25.5 Experimental Analysis

In our experiments, point correspondences are computed by the optical flow algorithm proposed by Sun et al. [13]. The flow fields over the whole frame are sampled uniformly to obtain the point correspondences. Thus the acquired correspondences cover the whole frame.

25.5.1 Accuracy Comparison

We use the synthetic dataset provided by Forssén and Ringaby [9] to compare our method with their 3D models [4]. These 3D models beat previous 2D rolling shutter rectification methods on this dataset.

We use the thresholded Euclidean colour distance to compute the rectification accuracy as in [4]. The threshold is also set to 0.3. Figure 25.3 shows the results on

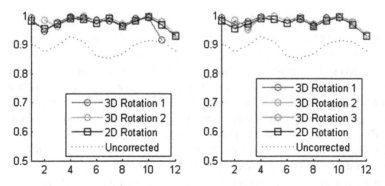

Fig. 25.3 Sequence rot1_B0. *Left* the 3D rotation model with 2-frame reconstruction window, our 2D rotation model, and uncorrected frames. *Right* the 3D rotation model with 3-frame reconstruction window, our 2D rotation model, and uncorrected frames

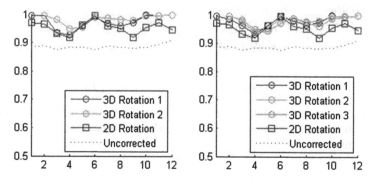

Fig. 25.4 Sequence rot2_B40. *Left* the 3D rotation model with 2-frame reconstruction window, our 2D rotation model, and uncorrected frames. *Right* the 3D rotation model with 3-frame reconstruction window, our 2D rotation model, and uncorrected frames

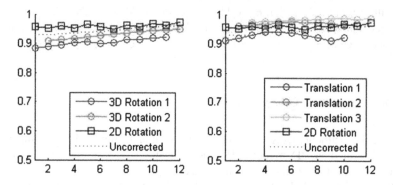

Fig. 25.5 Sequence trans1_B40. *Left* the 3D rotation model with 2-frame reconstruction window, our 2D rotation model, and uncorrected frames. *Right* the 3D translation model with 3-frame reconstruction windows, our 2D rotation model, and uncorrected frames

sequence rot1_B0 in the dataset. In this sequence, the camera rotates in a spiral fashion. Our method gets nearly the same result as the 3D rotation model. Figure 25.4 gives the results on sequence rot2_B40. This sequence simulates a complex camera rotation in 3D space. The performance of our 2D rotation model is a little worse than the 3D rotation model. In Fig. 25.5, we compare the 3D rotation model and the 3D translation model with our 2D model on sequence trans_B40, in which the camera makes a pure translation. Our model is better than the 3D rotation model, and is comparable to the 3D translation model. In Fig. 25.6, we compare the 3D rotation model and the 3D full model with our 2D model on sequence trans_rot1_B40. This sequence is generated by adding a translation to the sequence rot2_B40. The accuracy of our 2D model is similar to that of the 3D rotation model, and is much better than the 3D full model.

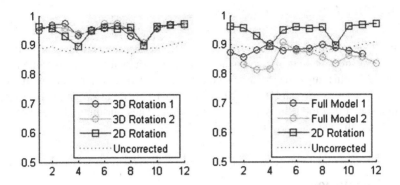

Fig. 25.6 Sequence trans_rot1_B40. *Left* the 3D rotation model with 2-frame reconstruction window, our 2D rotation model, and uncorrected frames. *Right* the 3D full model with 2-frame reconstruction window, our 2D rotation model, and uncorrected frames

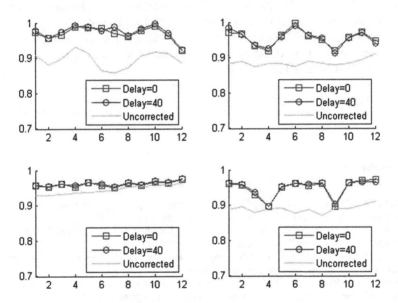

Fig. 25.7 Accuracy comparison with inter-frame delay = 0 vs. inter-frame delay = 40row. *Top left* Comparison on sequence rot1_B40. *Top right* Comparison on sequence rot2_B40. *Bottom left* Comparison on sequence trans1_B40. *Bottom right* Comparison on sequence trans_rot1_B40

From above figures we can see that our 2D model works well for both rotation and translation. Even for complex camera rotation, it is comparable with the 3D method that requires camera calibration.

25.5.2 The Inter-Frame Delay

In our 2D rotation model, we simply assume the inter-frame delay is zero. In this section, we show that our method is not sensitive to the value of inter-frame delay. Figure 25.7 illustrates the accuracy comparison between the results with inter-frame delay set to zero and the results with inter-frame delay set to the true value. It can be seen that the results are nearly the same. Thus our method does not need to compute the value of the inter-frame delay.

25.6 Conclusion

In this paper, we propose a 2D rectification method for removing rolling shutter distortion caused by camera motion. This method approximately models the frame motion as 2D rotation. The parameters of the model are solved by minimizing the measurement constraint on point correspondences. Experiments show the method can effectively remove distortions caused by camera rotation and translation. Without requiring any information of cameras, the method can be used to process videos obtained from unknown sources.

Acknowledgments This work was supported by the National Natural Science Foundation of China (51179146) and the Fundamental Research Funds for the Central Universities (2012-IV-041).

References

1. Baker S, Bennett E, Kang S, Szeliski R (2010) Removing rolling shutter wobble. In: IEEE conference on computer vision and pattern recognition, pp 2392–2399
2. Cho WH, Hong KS (2007) Affine motion based CMOS distortion analysis and CMOS digital image stabilization. IEEE Trans Consum Electron 54:833–841
3. Chun JB, Jung H, Kyung CM (2008) Suppressing rolling-shutter distortion of CMOS image sensors by motion vector detection. IEEE Trans Consum Electron 54:1479–1487
4. Forssén PE, Ringaby E (2010) Rectifying rolling shutter video from hand-held devices. In: IEEE conference on computer vision and pattern recognition, pp 507–514
5. Geyer C, Meingast M, Sastry S (2005) Geometric models for rolling-shutter cameras. In: International workshop on omni vision
6. Liang CK, Chang LW, Chen H (2008) Analysis and compensation of rolling shutter effect. IEEE TIP 17:1323–1330
7. Nicklin SP, Fisher RD, Middleton RH (2007) Rolling shutter image compensation. RoboCup 2006: Robot Soccer World Cup X 4434: 402–409
8. Ringaby E, Forssén PE (2012) Efficient video rectification and stabilisation for cell-phones. IJCV 96:335–352
9. Ringaby E Rolling shutter dataset with ground truth. (http://www.cvl.isy.liu.se/research/rs-dataset)
10. Liu F, Gleicher M, Jin H, Agarwala A (2009) Content-preserving warps for 3d video stabilization. In: ACM SIGGRAPH, p 44

11. Liu F, Gleicher M, Wang J, Jin H, Agarwala A (2011) Subspace video stabilization. ACM Trans Graph 30:1–10
12. Hartley R, Zisserman A (2011) Multiple view geometry in computer vision. Cambridge University Press, Cambridge
13. Sun D, Roth S, Black MJ (2010) Secrets of optical flow estimation and their principles. In: IEEE conference on computer vision and pattern recognition, pp 2432–2439

Chapter 26
Design and Implementation
of Electrolyzer Simulation System

Xingquan Cai and Kun Zang

Abstract In this paper, we present one method to develop and design an inexpensive electrolyzer simulation system, which can be applied to the management of the electrolyzer production. Firstly, we explore the framework of the electrolyzer simulation system. Our system is divided into several modules, including resource loading, scene rendering, scene interaction, collision detection and data visualization module. All these modules are designed particularly. Finally, we present the results and the results show that our method is feasible and valid. Our electrolyzer simulation system has been used in practical projects to manage the aluminum production and aluminum production education.

Keywords Electrolyzer · Resource loading · Scene rendering · Data visualization

26.1 Introduction

The fused-salt electrolysis with cryolite and alumina is the most commonly used in production technologies of aluminium electrolysis [1]. Aluminium is produced in the electrolysis cell with the raw material of alumina and electrolyte of molten cryolite. It's above the temperature of 950 °C. Since the environment of aluminium electrolysis is dangerous, manual collection of aluminium-related data is challengeable. It's good for the management and supervision in the production of

X. Cai (✉) · K. Zang
College of Information Engineering, North China University of Technology, No.5
Jinyuanzhuang Road, Shijingshan, Beijing 100144, China
e-mail: xingquancai@126.com

W. Lu et al. (eds.), *Proceedings of the 2012 International Conference on Information Technology and Software Engineering*, Lecture Notes in Electrical Engineering 210, DOI: 10.1007/978-3-642-34528-9_26, © Springer-Verlag Berlin Heidelberg 2013

aluminium with the use of smart components such as sensor, and the technology of virtual reality. And then display the result in the monitor in 3D ways.

So in this paper, we develop an electrolyzer simulation system to manage the aluminium production. In order to implement electrolyzer simulation system, we firstly explore aluminium electrolytic and relevant work of virtual reality. Then we provide the design methods of electrolyzer simulation system, including resource loading, scene rendering, scene interaction, collision detection, and data visualization module. Finally, we present the experimental data and draw the conclusion.

26.2 Related Work

With the development of computer technology, automatic control and aluminum electrolytic technology, computers have been used in the intelligent control and management of the production of the aluminum electrolytic. There are several research on aluminum production visualization [2–10]. In 2009, Shenyang Institute designed SY-CKJ350A intelligent slot control system, which controls the electrolyzer production by using the network information [2]. In 2011, JiaXin wu [3] used the software ANSYS to establish the thermal structure model of aluminum reduction cell anode and cathode. It studies the numerical simulation of thermal field and stress field. This can help adjust electrolyzer lining structural optimization design and parameters of the electrolysis process, etc. In 2012, Jianhong Li [4] adopt equidistant voltage drop method in working condition to make an industrial test on electrolyzer anode guide current, by 300 and 400 KA series combination of current cathodes.

These systems just manage the process of aluminum production in two dimensions and do not achieve real-time rendering and supervision in three dimensions. So in this paper, we make the full effort to research and design a system which can display and manage aluminum production process more easily and realistically.

26.3 Design of Electrolyzer Simulation System

26.3.1 System Framework

Visualization of three-dimensional electrolyzer system uses virtual reality technology to achieve the 3D scene rendering, interaction and aluminum data visualization. The system is divided into five modules, that is, resource loading module, scene rendering module, scene interaction module, collision detection module, data visualization module. The resource loading module manages the three-dimensional model, including the plant, electrolyzers, carbon rod, electrolyte, etc. The scene

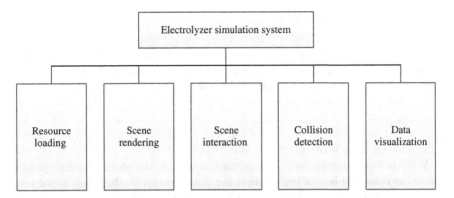

Fig. 26.1 Framework of our electrolyzer simulation system

rendering module organizes and renders the scene. The scene interaction module completes the interaction of the mouse, keyboard, and other interactive devices. The collision detection module completes the collision between the role and the scene and mouse picking. The data visualization module completes data shows in chart form, which facilitates the operator to observe and analyze the data in the process of electrolysis. Figure 26.1 shows the framework of electrolyzer simulation system.

26.3.2 Resource Loading

Resource loading module is mainly responsible for the loading and management of the three-dimensional model resource files. These resource files include three-dimensional model, the material, the Overlay scripts, fonts, Shader procedures and so on. A unified resource management method is necessary for the reason of the low speed capacity of reading and writing from the hard disk.

The models used in the system include factory, plant, electrolyzers, carbon rod, electrolyters and so on. The resource files needed to load include the model materials of factory, plant, electrolyzers, carbon rod, electrolyte, the overlay script fonts, Shader program, etc. In general, the model files include mesh files, the corresponding material files, and some texture mapping files (png, jpg, tga, etc.). The material files stores several material and each material contains the information of refractive index, reflectivity and others. The mesh files contain a number of submeshes and each submesh has a corresponding material. The system has designed two special management classes, that is, ResourceManager and ResourceGroupManager, which achieve effective management of resources and real-time operating system.

26.3.3 Scene Rendering

The scene rendering module is mainly responsible for management and rendering of the scene. The effect of scene rendering has a direct impact on the realism of the system. Because of the big number of resources in the scene, if importing all models into the system one-time, the efficiency of loading resources and real-time rendering will be reduced, resulting in a scene stuck and running slowly of the system and so on. Therefore, render the scene with the methods of loose octree scene management strategy and levels of detail strategy.

What is loose octree scene management strategy is shown below. Firstly, calculate the bounding box of the whole scene. And then divide the whole scene into many small nodes with iterative segmentation. The number of polygons in each node is roughly the same and there will be no node generated in the polygon occupied space. And lastly go on with cut operations, depending on the resources in the cone. This system designs the OctreeSceneManagement class to complete the management of the scene.

Levels of detail strategy improved the efficiency of the real-time rendering scene by simplifying the surface detail successively and reducing the complexity of the geometry of the models under the condition of not affecting the effect of rendering the scene. The system, defining by the mesh interface LOD level, according to the definition of the LOD level and the distance between the camera and models, automatically select the appropriate LOD level model.

26.3.4 Scene Interaction

The scene interaction module is mainly responsible for interaction among mouse, keyboard, track ball and other interaction devices, completing scene interactive roaming and data display. It provides users with better sense of immersion and realism. Interaction devices can be used to interact with the system, roaming the scene models such as factory, factory building, cell, cell internal facilities and so on freely. In order to realize the interaction of the scene better, this system designs OIS (Object Oriented Input System), which is responsible for the interactive message processing between interactive devices and system.

26.3.5 Collision Detection

The collision detection module completes the collision between the role and the scene and mouse picking. The collision between the role and the scene is the collision between the camera and three dimensional models, such as the plant, electrolytic cells, carbon rod and so on. What mouse picking is that selecting objects in the scene by interactive devices and displaying the corresponding data.

Implementing the collision between the role and scene, the system uses the method of bounding box detection. What bounding box detection is that constructing each object model's bounding box in the scene by the way of descripting a complex object model with the simple bounding box approximately which is slightly larger than the model in volume and simpler in geometric characteristics [5]. And then whether the bounding box of the role interacts with the bounding box of the model determines the collision.

The system uses the coordinate conversion comparison method and ray intersection method to realize mouse picking. What the coordinate conversion comparison method is that firstly getting the point's position which is to be crawled real-time, and then figuring out the three dimensional space of the position, and lastly comparing the position with the model's position. If the comparison result is within the range of error allowed, the object model is crawled; if not, the object model is not crawled. This method is simple but less efficient. What the ray intersection method is that firstly launching a ray from the camera's position to the goal position, and then calculating the number of triangles which are intersected with the ray, and lastly figuring out the object which the triangle belongs to.

26.3.6 Data Visualization

The data visualization module completes data shows in chart form, which facilitates the operator to observe and analyze the data in the process of electrolysis. This module can not only display key data real-time changes in the process, but also generate the analysis chart, such as line charts, graphs, bar charts, etc., to facilitate the observation and analysis of data. To realize the module, the system has designed ManualObject class to generate graphics information and convert the information to mesh and display the mesh in the monitor.

26.4 Result

In order to verify the feasibility and effectiveness of the design of electrolyzer simulation system, the system has been achieved. The system is under the windows platform using visual studio 2010 development environment, based on OGRE rendering engine. We also used Oracle database to manage the data of aluminum production. The hardware environment for the system is CPU Intel Pentium 2.8 GHz, graphics card NVIDIA GeForce 6800 GS, memory 2.0 GB and operating system Windows 7.

The system has a complete implementation, and has been tested and run. After the system starts up, the menu navigation interface will be shown. The effect is

Fig. 26.2 Our system is smooth running

shown in Fig. 26.2a. The system menu navigation provides four options, "Roaming", "Configuration Tool", "Help" and "Exit". Click on "Exit" button, and the system will be shut down. Click on "Help" button, and there will be a window pop up, which describes how to use the system. Users can be familiar with every part of the system and mode of operation through the texts in the window. Click on "configuration tool" button, and the system will pop up a configuration window, which is used to configure the system and display options, such as full-screen, vertical distribution, resolution of the system and so on. Click on "Roaming" button, and the system will hide the navigation menu and then enter the loading and display module to display the scene.

The system will render the scene according to the default camera position, direction of observation and model space position. The effect is shown in Fig. 26.2b. The figure shows the terrain, high mountains, trees, flowers and aluminum electrolysis production workshop model. It is a more realistic simulation of a realistic scene. After loading the scene, the system provides users interaction to change the camera's position and viewing direction. For example, you can use the up and down buttons to move the camera around the location; you can use the left and right buttons to rotate the camera for changing the spatial location; you can use the left mouse button to change the camera's viewing direction; you can use the right mouse button to change the vertical position of the camera in the scene. These interactive operations are convenient to roam the scene.

In order to be able to select specific models, the system achieves mouse picking function. The effect is shown in Fig. 26.2c. The figure shows how to pick up a flower model which is next to the production plant. Move the mouse and click the left mouse button to pick up the models. If succeeded in picking, the model will show its bounding box; if not, there will be no change in the scene.

Cameras will crash with the model during the move in the scene if entering the collision space of the model. The collision effect is that the camera will stay in the space before the collision and do not encroach on the internal model. Figure 26.2d shows the collision between the camera and the tree.

Move the camera and enter the internal plant. There are many cell model put in order. The effect is shown in Fig. 26.2e. In this scene, the mouse can pick up any of the cell model.

Select an electrolyzer model through the mouse and press number 1 button in the keyboard to switch to the internal electrolyzer structure. The effect is shown in Fig. 26.2f. In this figure, we can clearly see the internal electrolyzer structure. The system provides interactive operation to move the camera through the up and down button on the keyboard.

26.5 Conclusions and Future Work

In order to implement electrolyzer simulation system, we firstly explore aluminum electrolytic and relevant work of virtual reality. Then we provide the design methods of electrolyzer simulation system, including resource loading, scene rendering, scene interaction, collision detection, and data visualization module. Finally, we present the experimental data and analysis. The experimental result shows that our method is feasible and effective. Our system is running stably, and has been used in management and education of aluminum production.

Our future work is focus on rendering the scene more efficiently, visualization the aluminum production data in different way, etc.

Acknowledgments This work was supported by National Natural Science Foundation of China (No. 51075423), PHR(IHLB) Grant (PHR20100509, PHR201008202), and Funding Project of Beijing Municipal Education Committee (No. KM201010009002).

References

1. Liu Y, Li J (2008) Modern aluminum electrolysis. Metallurgical Industry Press, Beijing
2. Li J (2012) Aluminum electrolytic computer control technology. J China Nonferrous Met 66–67
3. Jia X, Li J, Tuo P, Li H (2011) A numerical simulation to thermal field and stress field of aluminum cell cathode and anode. J Met Mater Metall Eng 15–18
4. Li J, Tu G, Qi X, Mao J, Lu D, Feng N (2012) Real-time monitoring and analysis on fluctuation state of liquid aluminium in 300 and 400 kA aluminium reduction cell. J Light Met 30–34
5. Wang X, Wang M, Li C (2010) Research of collision detection algorithms based on AABB. J Comput Eng Sci 59–61
6. Chen R, Chen Z, Di Y, Zheng X, Chen X (2009) Treatment of landfill leachates by electrodialysis in three-chamber cell divided by two membranes. Chin J Appl Chem 26:1336–1340
7. Men C, Ding X (2010) Optimized control of reduction cell based on data mining. J Chongqing Univ Technol 89–93
8. Ding J, Zhang J, Ji F, Zhang H (2011) Design and production practice of curved cathode 300 kA Cell. J Kunming Univ Sci Technol 15–20
9. Wu Y, Li S, Lang J, Liu Z (2011) Numerical simulation of electric field in sodium electrolytic cell. J Nonferrous Met 26–29
10. Jia R, Zhang H, Liu Z, Wu Y, Wang L (2012) 3D simulation of magnetic field on 40 kA sodium electrolysis cell. J Nonferrous Met 21–23

Chapter 27
Counter Based Energy Saving Scheme for Sleep Mode in IEEE802.16e

Leilei Yang and Haitao Li

Abstract The sleep mode is utilized to reduce the energy consumption of the mobile station and the resource occupation of the air interface of serving base station in IEEE802.16e network. In order to improve the efficiency of energy saving, a practical sleep mode based on the counter is proposed. User's work mode can be determined according to the user's active level and the frequent that user switch between wake mode and sleep mode can be reduced in the proposed scheme. Simulations have been undertaken to verify the proposed scheme and analyze the effect of the threshold of counters on mean energy consumption and mean response time. The simulation results show that, the proposed scheme compared with the standard method can save at least 30 % of energy.

Keywords IEEE802.16e · Sleep mode · Energy consumption · Response time

27.1 Introduction

Energy consumption is mainly decided by the amount of connection time when the mobile station actually switches on its baseband and RF circuitry. In order to reduce the connection time, IEEE802.16e standard defines sleep mode which is a state that MS conducts pre-negotiated absence periods from the Serving Base and turns off baseband and RF circuitry temporarily.

L. Yang (✉) · H. Li
College of Electronic Information and Control Engineering, Beijing University of
Technology, Beijing city, China
e-mail: leileiyang0319@163.com

W. Lu et al. (eds.), *Proceedings of the 2012 International Conference on Information Technology and Software Engineering*, Lecture Notes in Electrical Engineering 210, DOI: 10.1007/978-3-642-34528-9_27, © Springer-Verlag Berlin Heidelberg 2013

Quality of Service (QoS) definition is one advantage of IEEE 802.16e reflected in its power saving schemes. In sleep mode, three types of power classes are defined to fit diverse characteristics of different services. Power saving class type I is for traffic connections of Non-Real-Time Varying Rate (NRT-VR) and Best Effort (BE) service types. Type II is for traffic connections of Real-Time Varying Rate (RT-VR) and Unsolicited Grant Service (UGS) service types. Type III is for management operations and multicast traffic connections [1].

Based on the traffic connections which serviced by class type I, we draw the conclusion that class type I could afford some transmission delay. And besides MS can receive data frames instead of the message *MOB_TRF_IND* in listening interval. According to the two characteristics, sleep mode based on the counter is proposed.

The remainder of this chapter is organized as follows: the conventional sleep mode is introduced in Sect. 27.2. The proposed sleep mode scheme is proposed in Sect. 27.3. In Sect. 27.4, the Markov's analysis model is utilized to explore the performance of the proposed scheme. The simulation results are presented in Sects. 27.5 and 27.6 is the conclusion.

27.2 Sleep Mode in IEEE802.16e

In this chapter we mainly focus on power saving of type I for the IEEE 802.16e network. The definition of type I is illustrated in Fig. 27.1. When MS without data transmission during a long interval, it sends request message *MOB_SLP_REQ* to BS for requesting to switch to sleep mode. Upon receiving the approval response message *MOB_SLP_RSP* from BS, the MS enters into the sleep mode. The response message includes four common parameters: start frame number for first sleep window, minimum sleep window size (t_{min}), maximum sleep window size (t_{max}) and the listening window size (L). Once MS received the response message, sleep mode is activated.

The whole sleep mode is made up of one or more sleep cycles, and each sleep cycle (SC) consists of a sleep window and a listening window. When the sleep

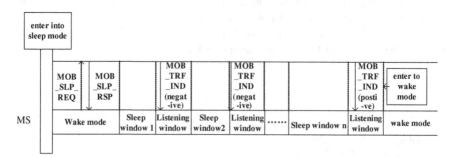

Fig. 27.1 Sleep mode in IEEE802.16e

interval is time out, MS turns into listening window waiting for traffic indication message *MOB_TRF_IND*. If new frames arrive during the sleep interval, the BS buffers the packets destined to the MS and send positive *MOB_TRF_IND* message in the listening window to notify that the MS should retreat from sleep mode and get back to wake mode. If not, BS will send negative *MOB_TRF_IND* message to notify that the MS could continue next sleep cycle. The listening interval L is always set to a fixed value. Whereas the sleep interval t_n of different sleep cycles changes from t_{min} to t_{max} differently. The standard adopts exponential increase algorithm, in which the sleep interval t_n is calculated as Eq. (27.1). The initial sleep interval is set to t_{min}. After the first sleep cycle, each next sleep interval is double length of previous one until t_{max}. Then the next sleep interval is keeping t_{max} until the MS wakes up [2].

$$t_n = \begin{cases} 2^{n-1}t_{min}, & 2^{n-1}t_{min} < t_{max} \\ t_{max}, & 2^{n-1}t_{min} \geq t_{max} \end{cases} \qquad (27.1)$$

27.3 Proposed Scheme

MS can transmit data frames when listening window comes [3]. Based on this characteristic, we utilize listening interval to transmit the data frames instead of the message *MOB_TRF_IND*. If there are few service streams to be transmitted by MS, it is not necessary to switch the work mode of MS from sleep mode to wake mode. On the other hand, if there are many data frames at the transmitter, BS can transmit data frames through more than one sleep cycles continuously. Based on the above observation, we can draw the conclusion that the number of sleep cycles used to transmit data frames can be suggested the MS whether or not switch to wake mode. We can set a couple of counters *CMS* and *CBS* beside MS and BS respectively, which are used to record the number of sleep cycles for transmitting data frames. If the BS has any buffered packets destined to MS during the listening window, counters *CMS* and *CBS* will add synchronously until reach the threshold m. It means that the MS has received data from the BS for m sleep cycles consecutively. It indicates that the MS should switch to wake mode in order to avoid delaying data transmission. If the BS has no data transmit to MS during the listening window, counter *CBS* and *CMS* are reset. The operation of the proposed scheme is shown in Fig. 27.2.

27.4 Markov's Analysis Mode

In order to study the relationship among the threshold of counters, mean energy consumption function $E(energy)$ and mean response time function $E(t)$, we established Markov's analysis model according to the action of MS, as Fig. 27.3.

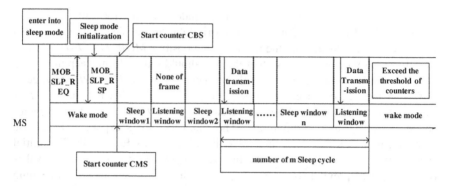

Fig. 27.2 Proposed sleep mode

The action of MS can be denoted by a two-dimensional function $\{s(t), m(t)\}$. $s(t)$ is the state of MS at the t moment [4]. If $S(t) = $ 'S', it means that MS is in the sleep mode, and if $S(t) = $ 'W', it means that MS is in the wake mode. $m(t)$ is a stochastic process to denote the sleep cycle of the MS. The nth sleep cycle can be expressed by $m(t) = n \ (0 \le n \le \max)$

Assuming that the rate of data frames arrival to MS follows Poisson process [5]. The probability function of receiving k frames within the interval t can be written as:

$$P\{N(s+t) - N(s) = k\} = \frac{(\lambda t)^k e^{-\lambda t}}{k!}, \quad k = 0, 1, \cdots \qquad (27.2)$$

where λ (frames per second) indicates the rate of data frame.

If there is no frames arrival, the probability function is derived by:

$$P\{N(s+t) - N(s) = 0\} = e^{-\lambda t} \qquad (27.3)$$

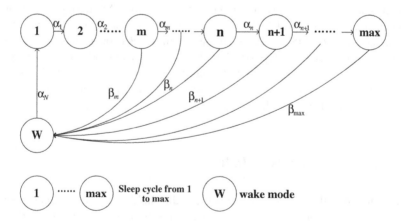

Fig. 27.3 Markov's analysis model

Otherwise, at the case of at least one frame arrival, the probability function is

$$P\{N(s+t) - N(s) \neq 0\} = 1 - e^{-\lambda t} \tag{27.4}$$

Each sleep cycle (SC) consists of a sleep interval and a listening interval. Then the length of SC can be written as:

$$SC_n = t_n + L \tag{27.5}$$

The variable β_n in Fig. 27.3 is used to denote the probability that MS changes from nth sleep cycle to wake mode. β_n is derived by,

$$\beta_n = \begin{cases} 0, & (n < m) \\ \prod_{i=n-m+1}^{n} (1 - e^{-\lambda(t_i+L)}) & (m \leq n \leq max - 1) \\ 1, & (n = max) \end{cases} \tag{27.6}$$

According to Eq. (27.6), we consider the following three cases. (1) If $n < m$, MS won't switch to the wake mode. (2) If $m \leq n \leq max - 1$, the number of frames arriving continuously exceed the counter's threshold m. Then MS will switch to the wake mode. (3) If $n = max$, MS only can switch to wake mode.

The variable α_n in Fig. 27.3 is used to denote the probability that MS changes from n-th sleep cycle to $(n + 1)$th sleep cycle. Then we consider the following two cases (1) If $1 \leq n < m$, MS only can switch to next sleep cycle. (2) If $m \leq n \leq max - 1$, MS can switch to next sleep cycle or wake mode. Based on the above two cases and β_n obtained from Eq. (27.6), α_n can be derived by,

$$\alpha_n = \begin{cases} 1, & (1 \leq n < m) \\ 1 - \beta_n, & (m \leq n \leq max - 1) \end{cases} \tag{27.7}$$

The probability that MS switches from wake mode to sleep mode is denoted as α_N. Since MS and BS exchange control message for entering into sleep mode, α_N can be assumed as 1.

Therefore, the steady state probabilities can be calculated by the following Eq. (27.8).

$$\begin{cases} \pi_N + \sum_{i=1}^{max} \pi_{S,i} = 1 \\ \pi_{S,n} = \alpha_{n-1} \pi_{n-1}, & (1 < n \leq max) \\ \pi_{S,1} = \alpha_N \pi_N \\ \pi_N = \sum_{i=m}^{max} \beta_i \pi_i \end{cases} \tag{27.8}$$

where, $\pi_{i,j} = \lim_{t \to \infty} P\{s(t) = i, m(t) = j\}(i \in \{W, S\}, 0 \leq j \leq max)$ are the steady state probabilities for the sleep mode at cycle j. π_N is the steady state probability in wake mode.

By solving Eq. (27.8), we obtain closed-form expressions for the steady state probability $\pi_{S,n}$ and π_N.

$$\pi_N = \left\{ (m+1) + \sum_{j=1}^{max-m} \prod_{r=1}^{j} \left[1 - \prod_{i=r}^{m+r-1} \left(1 - e^{-\lambda t_i} \right) \right] \right\}^{-1} \quad (27.9)$$

$$\pi_{S,n} = \left\{ \prod_{r=1}^{n-m} \left[1 - \prod_{i=r}^{m+r-1} \left(1 - e^{-\lambda t_i} \right) \right] \right\} \pi_N \quad (m < n \leq max) \quad (27.10)$$

$$\pi_{S,n} = \pi_N \quad (n \leq m) \quad (27.11)$$

The values of λ, m and max are known for the realistic WiMAX network, thus, the steady state probabilities can be calculated out. Considering the stochastic of Poisson distribution, the arrival time of data frame in the same sleep cycle can be assumed to follow uniform distribution. If sleep cycle is SC_n, statistically packet latency between packet arrival time and packet transmission time equals $SC_n/2$ [6]. The mean response time $E(t)$ is written by:

$$\begin{aligned}
E(t) &= \sum_{i=1}^{max} \frac{SC_i}{2} \times \pi_{S,i} \\
&= \sum_{i=1}^{max} \frac{t_n + L}{2} \times \pi_{S,i}
\end{aligned} \quad (27.12)$$

E_S and E_L are defined to donate the energy consumption per second in the sleep interval and the listening interval respectively. The mean energy consumption. $E(energy)$ in the sleep mode is derived by [7]:

$$E(energy) = \sum_{i=1}^{max} \left(\sum_{k=1}^{i} \left(\frac{L}{SC_k} \times E_L + \frac{t_k}{SC_k} \times E_S \right) \right) \times \pi_{S,i} + \pi_N \times E_L \quad (27.13)$$

Equations (27.12) and (27.13) show the relation among function $E(energy)$, mean response time function $E(t)$ and m. In order to prove more clearly, the simulation is undertaken shown in the next section.

27.5 Simulation Results

The simulation parameters are listed as follows: $t_{min} = 10\,ms$, $t_{max} = 160\,ms$, $L = 5\,ms$, $E_S = 100\,mW$, $E_L = 1\,W$, and λ(frame/s) increases from 0 to 80. The simulation results of proposed scheme compared with exponential increase algorithm and linear increase algorithm are shown in Figs. 27.4 and 27.5 respectively.

Fig. 27.4 Energy
consumption comparison of
different algorithms

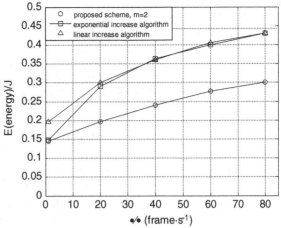

Fig. 27.5 Response time
comparison of different
algorithms

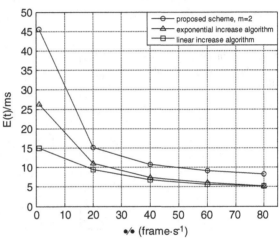

With the increase of λ, the results indicate that the mean energy consumption of proposed scheme decreases nearby 30 % compared with others, whereas the mean response time increases by 50 %.

Obviously, the listening interval is used to transmit the data frame instead of the message MOB_TRF_IND, which makes the probability that MS switches to longer sleep cycle higher than the others. The proposed scheme tends to a longer and longer SC window, resulting in longer packet latency and lower energy consumption.

Although the results are not very perfect, power saving class type I is for traffic connections of Non-Real-Time Varying Rate (NRT-VR) and Best Effort (BE) service types. Therefore some transmission delay is permitted. Within the endurable response time, the proposed scheme can achieve the energy-saving goal.

In order to evaluate the influence of the threshold m, we set m to 2, 3 and 4 respectively. The simulation results of the mean response time and mean energy consumption are shown in Figs. 27.6 and 27.7 respectively.

Fig. 27.6 Response time
comparison of different m

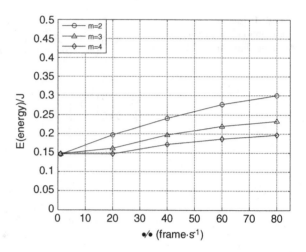

Fig. 27.7 Energy
consumption comparison of
different m

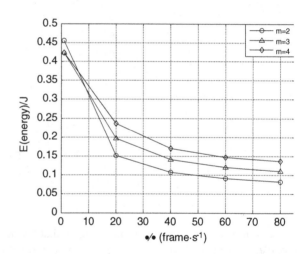

The simulation results show that, with the increase of m, the energy consumption becomes lower and lower. Compared with the exponential algorithm, when m equals 3, the energy consumption reduces by 45 %; when m equals 4, the energy consumption reduces by 53 %. However, the mean response time become longer and longer, so m can not be set too large. Though saving class type I can afford some transmission delay, we still need to set m as larger as possible to balance power saving and response time.

27.6 Conclusion

The proposed scheme can avoid MS switching between wake mode and sleep mode and minimize signaling overhead. Then the relationship among mean energy consumption, mean response time and the threshold of counters is derived by means of Markov's analysis model. After that, we compare the performance of the proposed scheme with the exponential increase algorithm and linear increase algorithm. Simulation results indicate that, with the increase of m, mean energy consumption becomes lower and lower, but the response time becomes longer than the others. It limits the capability of power saving. However, the proposed scheme can easily balance response time and energy consumption by adjusting m in the affordable response time.

The author would like to thank the support of Beijing Natural Science Foundation (Approaching the optimal capacity of wireless networks based on interference alignment) under Grant 4112012, Beijing Municipal Commission of Education project "R&D multimedia trunk wireless access system".

References

1. IEEE 802.16-2009 IEEE Standard for Local and metropolitan area networks. May 2009
2. Li L, Jin-sheng L, Pei-lin H (2007) Research and improvement of sleep mode in IEEE 802.16e. Chin J Comput 30(1):146–152
3. En-jie L, Jie ZH, Wei-li R (2011) A counter-driven adaptive sleep mode scheme for 802.16e networks. IEEE VTC2011-Spring, Budapest, Hungary, 15–18 May 2011
4. Jian-bin X, Yan-feng ZH, Zhan-ting Y (2009) Adaptive data rate algorithm in IEEE 802.16e sleep mode operation. J Jilin Univ (02):241–246
5. Xiao Y (2005) Energy saving mechanism in the IEEE 802.16e wireless MAN. IEEE Commun Lett 9(7):595–597
6. Shao-fei L, Jian-xin W, Yue-juan K (2010) An efficient energy saving mechanism for sleep mode in IEEE 802.16e networks. In: Proceedings of the 6th international wireless communications and mobile computing conference, 28 June–2 July 2010, Caen, France
7. Cho S, Youngil K (2007) Improving power savings by using adaptive initial-sleep window in IEEE 802.16e. In: Proceedings of IEEE VTC2007-Spring, 22–25 April 2007, pp 1321–1325

Chapter 28
An Object Encapsulation Approach for P2P Resource Sharing Platform

Shufang Zhang, Jun Han and Fei Jiang

Abstract This paper designed and built a distributed hash table based computing resource sharing platform named OE-PRSP. By employing P2P services, OE-PRSP allowed users to submit jobs to be run in the system and to run jobs submitted by other users on any resources available over the Internet, essentially allowing a group of users to form an Ad hoc set of shared resources. The experimental results obtained via simulations show that the system can reliably execute scientific applications on a widely distributed set of resources with good load balancing and low matchmaking cost and that OE-PRSP has good efficiency, load balancing and scalability.

Keywords Structured peer-to-peer · Decentralized system · Resource sharing; scalability · Volunteer computing · Cycle stealing

S. Zhang (✉) · F. Jiang
School of Mechanical and Electronic Engineering, Suzhou University,
49 Middle Bianhe Road, Suzhou City, Anhui Province, China
e-mail: zhshf_sztc@126.com

F. Jiang
e-mail: feiwuhan@126.com

J. Han
School of Information Science and Technology, University of Science
and Technology of China, Hefei, China
e-mail: hanjun_sztc@126.coml

W. Lu et al. (eds.), *Proceedings of the 2012 International Conference on Information Technology and Software Engineering*, Lecture Notes in Electrical Engineering 210, DOI: 10.1007/978-3-642-34528-9_28, © Springer-Verlag Berlin Heidelberg 2013

28.1 Introduction

Among peer-to-peer resource-sharing platforms, centralized unstructured peer-to-peer network Napster can not resist the attack from malicious nodes, so it is lake of strong robustness; distributed unstructured peer-to-peer network Gnutella or kaZaA have solved the single point of failure problem, but the expansion of the system is rather poor because of message flooding problem. Subsequent structured distributed Harsh table DHC solved the aforesaid problem [1–3].

In this paper we propose an Object Encapsulation P2P Resource Sharing Platform (OE-PRSP), which is a system established on the application of DHC. OE-PRSP is based on Pastry, and change from the node ID mapping to the object encapsulation based mapping; the difference between OE-PRSP and other related projects is that OE-PRSP is mainly used in the field of parallel and distributed computing, instead of data/file sharing, live video/on-demand field. Current peer-to-peer computing resources sharing projects always require the independence between main task and subtasks, so they mainly support the primary masterslave style of parallel computing [4–9]. OE-PRSP is a distributed object programming model, so it allows the direct or indirect communication between entities by remote method invocation, which makes OE-PRSP also support distributed computing with data-dependent problems, and makes the scope of its application becomes sufficiently widespread. In this paper, we will use the principle of object moving and physical location adjacent to make OE-PRSP gathering and sharing idle computing resources on the Internet more efficiently [10].

Then we test the performances of OE-PRSP by simulations, and the results show that OE-PRSP has good efficiency, load balancing and scalability.

28.2 The Architecture of OE-PRSP

28.2.1 Platform Overview

OE-PRSP is a non-centralized peer-to-peer computing resource sharing platform. It consists of a large-scale voluntary free node on the Internet, it also provides a distributed object encapsulation based programming model which can perform a parallel distributed computing by the inner nodes communication mechanism.

The difference between OE-PRSP other existing computing resources sharing platform is that it can support communication between parallel sub-task operations, and the architecture is different from the voluntary computing platform using C/S model, has the characteristics of self-organization, non-concentration, scalability and large-scale. By using the direct communication between the peer entities, OE-PRSP can solve more parallel computing problems, such as the distributed data-dependent traveling salesman problem and the distributed genetic algorithm. So we can fully excavate and make full use of the free resources on Internet.

28.2.2 *Pastry Network*

OE-PRSP platform is based on distributed hash table Pastry network, and the Pastry network consists of k-bit network nodes which has the only NodeID voluntarily. NodeID distribution is of non-centralized manner and need to across the address space to achieve load balancing.

Network message can be addressed to the address of space of the NodeID, when passing messages the route model of Pastry is O (LogN) jump, the node has the location informations of both physical location and virtual location. Physical location indicates in physical network the minimum delay to reach the nearest location. If good physical location data is lacked, the system will assume the two nodes close to the IP address are the physical locations.

28.2.3 *Addressing Objects and Programming Model*

OE-PRSP can create objects which are located on Pastry network that consists of voluntary machines. The communication between these objects are achieved by remote method invocation RMI (remote method calls) and realized by the RMI technology of Java and .NET technology of Microsoft. Even so, when we use OE-PRSP we need not to specify a specific server but to choose a machine from voluntary machines. In spite of the joining and leaving of nodes are very frequent, in OE-PRSP the voluntary machine that can control given remote object can be changed over time.

When a object is created, it will be assigned an object ID which is similar with the nodes in Pastry, then a construction message will be sent to the object ID and inform the nearest node that accept the message, then the node will create a remote object instance. At this node the information need to be informed is just the object ID. If another node joins the network, it will be assigned a node ID (also relatively close to the object), and then the object will be transferred to the new node. OE-PRSP can complete the transfer automatically; this is similar with the log and checkpoints which can improve the reliability of system.

28.3 Improvement of Communication Mode

OE-PRSP has some improvement than in Pastry in communication mode. In addition to use object encapsulates, OE-PRSP improve the system's communication mode by object group and the physical proximity principle.

28.3.1 Allocation of Object Identifier

Object ID does not require a centralized generation. A single customer or node need to activate the collection of relative remote objects frequently. And the remote object needs not random communication, so it's a good manner to use the adjacent neighbor nodes. By using this non-random way, we can make sure that the objects can be assigned in the same physical host even if the numbers of objects are far more than that of entities. Since there is frequent joining and exiting of nodes, so the mapping and location information of objects will not be affected. So this proximity principle of object ID is one-dimensional and the communication mode between objects is folded into a one-dimensional space by developers. The programming interface API which is to assign object ID inform system to launch a unified object distribution function. Start Layout method requires a parameter N (N is the number of objects in objects group), in this way system will cover all the address space correctly. After this call, the next N objects will be activated automatically, and the object ID will be created according to this mode. Code is as follows:

```
Island [] islands = new Island [numIslands];
OE-PRSP Channel.Current.StartLayout(numIslands);
For (int i = 0; i < numIslands; i++)
Islands[i] = new Island (i).
```

Then the system will use different random offset address in order to avoid the conflict between the original object ID and the current object ID. In Sect. 28.4.1 the experimental result indicates the communication efficiency and load balance performance by using this method to assign object ID.

28.3.2 Object Groups

As previously mentioned by adding the object communication opportunities in the same machine we can improve the efficiency of the platform, but we can not make sure at the same machine all the time. In this section we use the method of object groups to make the objects in the same object group communicate each other frequently and ensure they are assigned to the same node. We can realize object groups by adding extra bits to the object ID and lengthen length of node ID, converting the object ID to a decimal, not a integer. The objects with the same integer part will be mapped to the same node.

If a group of objects were assigned to the same integer part, this group of objects will be mapped to a single machine, even if the network has a large number of other free nodes. The actual physical nodes can also change over time. But the objects are grouped together, improving the parallelism. Another optimization approach is to encapsulate these object groups to a single remote object container, the message can be transmitted foreward through the object. The advantage of this method is that each object in the group maintains a private addressable remote client.

28.3.3 Distance Proximity Principle

By adding the communicate opportunities of two objects in the same machine we can improve the efficiency of the platform, but some objects may be positioned to different machines. If they can not locate on the same machine, the physical proximity principle in this section ensures that they are in physically similar positions. To accomplish this function OE-PRSP continue to allocate object ID according to the method in Sect. 28.3.1, but we change the node ID so that the nodes which match well will be very close physically. If two objects communicate frequently, they will be allocated well-matched object ID, and these two objects may be controlled at the same node or close nodes, meaning close physical locations.

By using IP address when we describe node ID we can ensure well-matched node ID are close physically. The shortcoming of this method is that sometimes node ID will not distribute to the entire address space. To achieve this technique, OE-PRSP uses initialized node ID and current node ID. Initialized node ID is generated by the IP address. The current node ID is used by the node. When normal communication massage is routed by the ID of current node, the joining message will be routed to the node that is initialized nearest.

28.4 Experiments and Performance Analysis

28.4.1 Construct the Simulator

We construct a Pastry network with simulator and simulate a set of objects in one-dimensional nearest neighbor configuration space of periodic communications. OE-PRSP can improve the efficiency and performance of Pastry network, and the improvement can be analyzed by the following two performance metrics. The first indicator is the communication efficiency, which indicates the ratio of sending messages between objects when network hops are using message. It indicates the number of hops every message needs to cross network. If the ratio is relatively large, it means some messages can be transferred without network. The simulator doesn't consider the physical communication price between different nodes, because Pastry network has considered the physical position price of nodes when they are selecting by network hops.

As mentioned earlier OE-PRSP has been trying to reduce the number of hops which network require. So the second character is load balance rate, which shows the balance of performance of objects when getting through the network. It is calculated by the standard deviation statistics of the number of objects on each node. If the load balancing ratio is relatively small, it indicates that each node has about the same load.

28.4.2 The Results by Using Object ID

The first test shows the performance of Pastry network which is improved by OE-PRSP platform using object package and object ID. Object ID can be a good way to improve the efficiency of network communications, especially in the condition when object number is more than node number, load balancing rate can be improved obviously and along with the increase of nodes, the number of network is increasing, the rate of improvement of load balancing effect is reducing.

Figures 28.1 and 28.2 show the improvement by object ID. Dotted line represents the result which is not optimized; the solid line shows the optimal results. Each line expressed an independent test result when different numbers of Voluntary machines add in OE-PRSP platform.

The results of this experiment using distance proximity. The second experiment tested the proximity principle of Pastry network performance improvements. As we have described, the principles can be a good method to improve load balancing. The improvement will get weak and the communication efficiency will slightly decreases with the increase of nodes. This improvement is directly related to the rate of load balancing, the network can use the smallest number of hops to route messages to the destination node.

Figures 28.3 and 28.4 show the experimental results, which dotted line represents the result without using the proximity principle, the solid line represents the result when use proximity principle. Each line is that a different number of voluntary drive to join OE-PRSP platform independent test results. Each line expressed an independent test result when different numbers of voluntary machines add in OE-PRSP platform.

Tests run a total of three times, respectively, using normal strategy, enhanced object ID strategies and physical proximity principle. Test using a different number of objects, the length of the different processing steps executed repeatedly. In the same situation tests are run several times and we use the average of the results. Test also set a different message size, message size base 1 KB, although this setting does not affect the results much sense. We use two processing length, the first is about 0.12 s to complete each object, the second for each object is about 0.39 s to complete. Figure 28.3 shows the experimental results running short process length. Figure 28.4 shows the experimental results running long process length.

Fig. 28.1 Communication efficiency improvement by object ID package

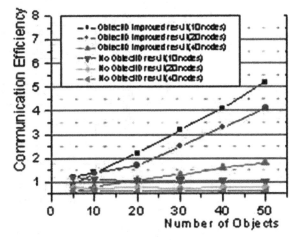

Fig. 28.2 Load balance rate improvement by object ID package

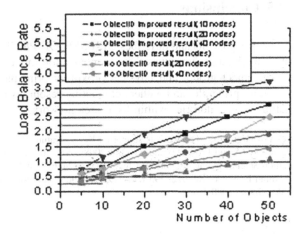

Fig. 28.3 The experimental results running short process length

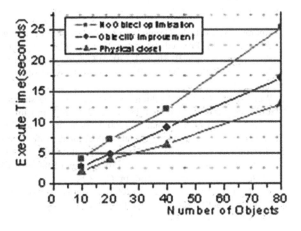

Fig. 28.4 The experimental results running long process length

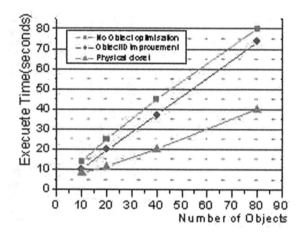

28.5 Conclusions

This paper presents and describes OE-PRSP. OE-PRSP is based on structured peer on network share, improves the communication mode of system by encapsulated object, object group, and physical proximity principle, and the use of physical proximity principle, and gathers computing resources better in the same research institution or enterprise within local area network.

Acknowledgments The work is supported by the Nature Science Research Project of Suzhou University (Anhui Province) under Grant (2008yzk08, 2009yzk15, and 2011cxy02), the National Natural Science Foundation of P.R.China (61174124, 61074033), the Natural Science Foundation of Anhui Province (KJ2012Z393,KJ2012Z397) and Technology Planning Project of Suzhou city under Grant (2008-22).

References

1. Andrade N, Costa L, Germogmio G (2005). Peer-to-peer gridcomputing with the our grid community. In: Proceedings of the 23rd Brazilian symposium on computer networks
2. Boyapati C, Liskov B, Shrira L (2003) Ownership types for object encapsulation. In: POPL 2003: the 30th ACM SIGPLAN-SIGACT symposium on principles of programming languages
3. Wei W, Shi J (2011) Building computing resource sharing platform using structured P2P approach. Appl Res Comput 28(1):223–225, 229 (in Chinese)
4. Luther A, Buyya R, Ranjan R (2005) Alchemic: a NET-based enterprise grid computing system. In: Proceedings of the 6th international conference on internet computing
5. Angelino C, Canonic M, Guazzone M, et al (2008). Peer-to-peer desktop grids in the real world: the share grid project. In: Proceedings of the 8th IEEE international symposium on cluster computing and the grid, pp 609–614
6. Fox G, Pallickara S, Rao Xi (2005) Towards enabling peer-to-peer grids. J Concurrency Comput: Pract Experience 17:1109–1131

7. Andradea N, Brasileiroa F, Cirnea W, Mowbrayb M (2007) Automatic grid assembly by promoting collaboration in peer-to-peer grids. J Parallel Distrib Comput 67(8):957–966
8. Morrison JP, Coghlan B, Shearer A, Foley S, Power D, Perrott R (2006) WEBCOM-G: a candidate middleware for grid-Ireland. Int J High Perform Comput Appl 20(3):409–422
9. Madsen J, Venkat J (2001) Is distributed computing commercially viable. Internet World 7(7):14
10. Zhou J, Mo R, Wang M (2010) SDDG: Semantic desktop data grid. In: 3rd international conference on Information sciences and interaction sciences (ICIS), 2010, pp 240–245

Chapter 29
A Method of Optical Ultra-Wide-Band Pulse Generation Based on XPM in a Dispersion Shifted Fiber

Shilei Bai, Xin Chen, Xinqiao Chen and Xiaoxue Yang

Abstract A method is proposed to generate UWB monocycle pulse using cross-phase modulation (XPM) in a dispersion shifted fiber (DSF). In this method, the low-power continuous wave probe light is phase modulated by the high-power pump light because of the XPM in a dispersion shifted fiber, then the following fiber Bragg grating (FBG) serves as a frequency discriminator, which convert the PM to IM. Finally, a UWB monocycle signal is achieved. Two UWB monocycle pulses are obtained by Optisystem7.0. The impacts of the input signal pulse width, the reflectivity of FBG and using different frequency discriminators on the generated UWB monocycle pulse are numerically simulated and studied. The results show that this scheme has good tolerance to the input signal pulse width. Optical Gaussian band-pass filters, wavelength division multiplexer and FBG can be used as the frequency discriminators. FBG has more advantages due to the flexibility to adjust the waveform of the generated UWB pulse signal by changing the reflectivity of FBG.

Keywords Ultra-wideband-signal · Dispersion shifted fiber · Cross-phase modulation · Fiber bragg grating

S. Bai · X. Chen (✉)
College of Information Engineering, Communication University of China, No.1, Dingfuzhuang East street, Chaoyang District, Beijing, China
e-mail: nuowei066@163.com

S. Bai
e-mail: bsl@cuc.edu.cn

X. Chen · X. Yang
No.1, Dingfuzhuang East street, Chaoyang District, Beijing, China
e-mail: chenxinqiao9999@163.com

X. Yang
e-mail: yang_xiaoxue_cuc@sina.com

W. Lu et al. (eds.), *Proceedings of the 2012 International Conference on Information Technology and Software Engineering*, Lecture Notes in Electrical Engineering 210, DOI: 10.1007/978-3-642-34528-9_29, © Springer-Verlag Berlin Heidelberg 2013

29.1 Introduction

Ultra-wideband impulse radio has been receiving considerable interest of late as a highly promising solution for its applications in short-range, high-capacity wireless communication systems and sensor networks because of advantages such as high data rate, low power consumption, and immunity to multipath fading. The Federal Communications Commission (FCC) of the U.S. has approved the unlicensed use of the UWB from 3.1 to 10.6 GHz with a power density lower than −41.3 dBm/MHz [FCC, part 15] [1–3]. In order to adapt to long distance transmission, a new type of UWB transmission system which use low loss optical fiber is proposed (UWB-over-fiber). The generation of UWB signal is the key technology in UWB-over-fiber system.

Recently the generation of UWB signal in the optical domain has attracted great interest by offering the advantages of low-loss and long-distance transmission of optical fiber. Many approaches to generating UWB signals in the optical domain have been reported [4–11]. An optical UWB pulse generator based on the optical phase modulation (PM) to intensity modulation (IM) conversion by use of an electrooptic phase modulator (EOPM) and FBG serving as an optical frequency discriminator was proposed and experimentally demonstrated [4]. Yao propose a scheme to generate UWB monocycle signals based on XPM of a semiconductor optical amplifier (SOA) [5]. Wang proposed UWB signals based on cross-gain modulation (XGM) of the SOA [6]. Dong proposed and demonstrated a simple scheme to generate UWB monocycle pulses based on the gain saturation effect of the SOA [7]. A method to generate UWB pulses based on chirp-to-intensity conversion using a distributed feedback (DFB) laser whose driving current is modulated by the electrical data signal was proposed [8]. Lin proposed UWB monocycle signals generated by using a gain-switched Fabry–Perot laser diode and a microwave differentiator [9]. An approach to generate UWB pulses based on an SOA and a electro-absorption modulator (EAM) [10]. A novel to generate UWB pulses by using a polarization modulator (PolM) and an FBG is proposed and experimentally demonstrated [11].

In this paper we propose a scheme to generate UWB monocycle pulse based on XPM of the DSF. According to the method, the phase modulation is achieved by simply using XPM of the DSF, without using electrooptic phase modulator. It significantly reduces the complexity of the system. By locating the cross-phase modulated probe at opposite linear slopes of the FBG reflection spectrum, UWB pulses with inverted polarity were obtained. This feature makes the implementation of the pulse polarity modulation (PPM) scheme possible in the optical domain, by simply switching the wavelength of the probe. The FBG, serving as a frequency discriminator, also serves as an optical bandpass filter to remove the residual pump and the amplified spontaneous emission (ASE) noise from the EDFA. It reduces the interference and improves stability of the system.

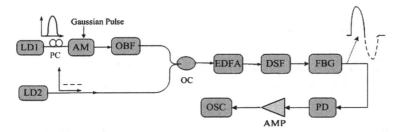

Fig. 29.1 Principle of UWB monocycle pulse generation based on XPM of the DSF (*LD1* laser diode1, *LD2* laser diode2, *PC* polarization controller, *AM* amplitude modulation, *OBF* optical bandpass filter, *OC* optical coupler, *EDFA* erbium doped fiber amplifier, *DSF* dispersion shifted fiber, *FBG* fiber Bragg grating, *PD* photo diode, *AMP* amplifier, *OSC* oscilloscope)

29.2 Operating Principle

The principle of UWB monocycle pulse generation based on XPM of the DSF is presented in Fig. 29.1. The pump light which emitted by LD1 propagates through PC is modulated by electrical Gaussian pulse in AM. To control the pulse width, a optical bandpass filter is incorporated after the AM. The pump light and a continuous-wave (CW) probe which transmitted by LD2 are applied to the EDFA to adjust the optical power via a 3 dB coupler. The two lights are then injected into a DSF that serves as the nonlinear medium, to achieve optical XPM. The phase modulated probe is then converted to an intensity-modulated pulse at an FBG. When the cross-phase modulated probe is located at the right linear slope of the FBG reflection spectrum, equivalent to first-order derivative of Gaussian pulse, the phase-to-intensity conversion is achieved, then the positive polarity UWB monocycle pulse can be obtained. If the cross-phase modulated probe is located at the left linear slope of the FBG reflection spectrum, the negative polarity UWB monocycle pulse will be got. This feature makes the implementation of the pulse polarity modulation (PPM) scheme possible in the optical domain. The FBG also serves as an optical bandpass filter to remove the residual pump and the amplified spontaneous emission (ASE) noise from the EDFA. It reduces the interference and improves stability of the system.

29.3 Simulation Design

The Simulation of UWB monocycle pulse generation based on XPM of the DSF is presented in Fig. 29.2. Simulation parameters set as follows: LD1 generate CW beams at 1,561.5 nm. Pseudo-random sequence is defined as "0000010000000000" (one "1"per 16 bits) at 13.5Gbit/s, the equivalent repetition rate is 0.84 GHz. The input signal pulse width is 0.50 bit. The central wavelength of the OBF is tuned at 1,561.5 nm, and its 3 dB bandwidth is 0.32 nm. The DSF has a chromatic dispersion

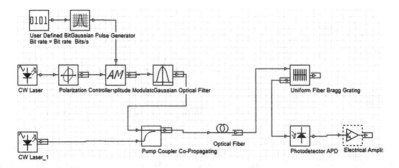

Fig. 29.2 Simulation of UWB monocycle pulse generation based on XPM of the DSF (*BPG* bit sequence pulse generator, *GPG* Gaussian pulse generator, *LD1* laser diode1, *LD2* laser diode2, *PC* polarization controller, *AM* amplitude modulation, *OBF* optical bandpass filter, *OC* optical coupler, *EDFA* erbium doped fiber amplifier, *DSF* dispersion shifted fiber, *FBG* fiber Bragg grating, *PD* photo diode, *AMP* amplifier)

of 5.6 ps/nm/km at 1,556 nm, a nonlinear refractive index of, and an effective mode-field area (MFA) of. The central wavelength of the FBG is tuned at 1,556.6 nm, its 3 dB bandwidth is 0.35 nm and the reflectivity of the FBG is 66 %.

29.4 The Main Factors Affect the Generated UWB Signal

29.4.1 The Influence of the Signal Polarity

There are two opposite polarity in UWB monocycle pulses. The polarity of the UWB signal depends on the transfer function of the frequency discriminator and the wavelength of the probe. When the cross-phase modulated probe is located at the right linear slope of the FBG reflection spectrum, an ideal positive polarity UWB monocycle pulse can be got. The waveform and electrical spectra are presented in Fig. 29.3. The upper FWHM of the generated positive polarity UWB monocycle pulse is equal to 40 ps, lower FWHM is equal to 48 ps, the central frequency is 7 GHz, the fractional bandwidth is 143 %, Close to the FCC's UWB template waveform and spectrum specifications. If the cross-phase modulated probe is located at the left linear slope of the FBG reflection spectrum, a negative polarity UWB monocycle pulse can be obtained. The waveform and electrical spectra are presented in Fig. 29.4. The upper FWHM of the generated negative polarity UWB monocycle pulse is equal to 49 ps, lower FWHM is equal to 43 ps, the central frequency is 6.95 GHz, the fractional bandwidth is 145 %, Close to the FCC's UWB template waveform and spectrum specifications.

From the Fig. 29.3, 29.4, we can find that if the cross-phase modulated probe at opposite linear slopes of the FBG reflection spectrum, the UWB monocycle pulses with inverted polarity will be obtained. This feature makes the implementation of

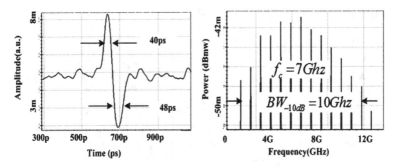

Fig. 29.3 Waveform and spectra of the UWB positive pulse

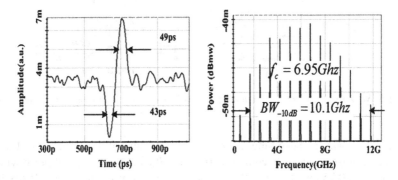

Fig. 29.4 Waveform and spectra of the UWB negative pulse

the pulse polarity modulation scheme possible in the optical domain, by simply switching the wavelength of the probe. As shown in Figs. 29.3 and 29.4, the waveforms of the generated UWB monocycle pulses show a little asymmetry. For the UWB monocycle pulses with inverted polarity, a slight asymmetry is observed, which is due mainly to the self phase modulation (SPM) and cross-gain modulation (XGM) in the DSF, which leads to the pulse broadening and distortion.

29.4.2 The Influence of the Input Pulse Width

The input signal width is an important parameter to affect the waveform of the generated UWB monocycle pulse signal. Keeping the other parameters fixed, the input signal pulse width is set at 0.50, 0.75 and 1.00 bit respectively. As shown in Fig. 29.5a, c and e, when the input signal pulse width is increasing, the generated UWB monocycle pulse signal upper/lower pulse width is increasing, the waveform distortion of the UWB signal is more serious. As shown in Fig. 29.5b, d and f, when the input signal pulse width is increasing, high frequency components of the

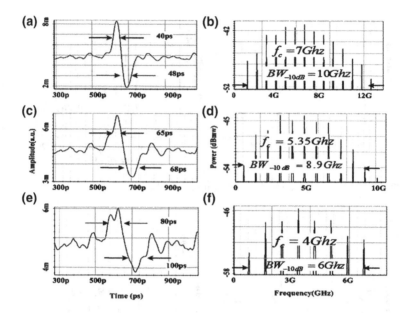

Fig. 29.5 Waveforms and RF spectrum for generated monocycle pulse in cases of different input signal width

RF signal power is becoming lower, the bandwidth of the generated UWB monocycle pulse signal is becoming narrower. When the input signal pulse width is set to 0.50, 0.75 and 1.00 bit, the center frequencies of the relative UWB pulse signals are 7, 5.35 and 4 Ghz, and the relative bandwidths are 143, 166 and 150 %, they all meet the criterion of FCC.

The results show that this scheme has good tolerance to the input signal pulse width. However, when the pulse width increases to a certain level, the waveform distortion of the UWB signal becomes serious. This is because as the pulse width increases, the XGM and SPF effect in DSF will further increase, the waveform distortion of the UWB signal will become more serious. Therefore, when the input pulse width is set at 0.50 bit, an ideal UWB monocycle pulse signal can be got.

29.4.3 The Influence of the Reflectivity of FBG

The FBG is serving as a frequency discriminator, the reflectivity of the FBG is an important parameter to affect the waveform of the generated UWB monocycle pulse signal. Keeping the other parameters fixed, the reflectivity of the FBG is set at 30, 50, 66 and 80 % respectively. As shown in Fig. 29.6a, b, c and d, the generated UWB monocycle pulse signal upper/lower FWHMs are 33 and 43 ps, 37 and 46 ps, 40 and 48 ps, 36 and 51 ps respectively. As shown in Fig. 29.6, when the FBG reflectivity increases, the upper amplitude of the generated UWB

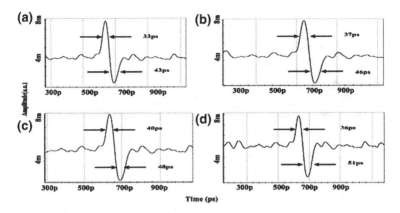

Fig. 29.6 Waveforms for generated monocycle pulse in cases of different FBG reflectivity

monocycle pulse signal will decrease, the lower amplitude will increase; the upper FWHM of the generated UWB monocycle pulse signal will first increase to a certain level and then decrease, the lower FWHM will increase.

The results show that when the reflectivity of the FBG is set at 66 %, the lower amplitude is close to upper amplitude, the difference between upper amplitude and lower amplitude is smaller, the generated UWB monocycle pulse signal shows good symmetry. Therefore, when the reflectivity of the FBG is set at 66 %, an ideal UWB monocycle pulse signal can be got.

29.4.4 The Influence of Using Different Frequency Discriminators

In theory, any filtering device can be used as the frequency discriminator, but using different frequency discriminators will influence the performance of the generated UWB monocycle pulse signal. In this scheme, we adopt the commonly used commercial optical filters as frequency discriminators. Keeping the other parameters fixed, optical Gaussian band-pass filter, DWDM and FBG are used as the frequency discriminators respectively. The central wavelength of the DWDM channel which is used to frequency discriminate is tuned at 1,556.6 nm and 3 dB DWDM channel bandwidth is 0.35 nm. The central wavelength of the FBG and optical Gaussian band-pass filter are also tuned at 1,556.6 nm and their 3 dB bandwidths are 0.35 nm. In the simulation, optical Gaussian band-pass filter, DWDM and FBG are used as the frequency discriminators to frequency discriminate respectively. As shown in Fig. 29.7a, b and c, the generated UWB monocycle pulse signal upper/lower FWHMs are 39 and 42 ps, 38 and 40 ps, 40 and 48 ps respectively. We can find that when optical Gaussian band-pass filter and DWDM are used as the frequency discriminators, the generated UWB

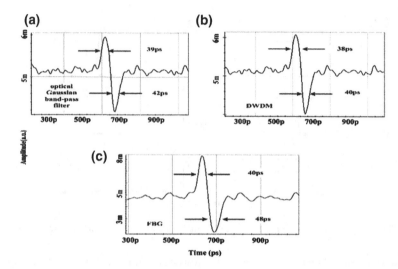

Fig. 29.7 Waveforms for generated monocycle pulse in cases of different frequency discriminators

monocycle pulse signals are almost the same; when FBG is used, the waveform of the generated UWB monocycle pulse shows a little different, for the waveform of the generated UWB pulse signal can be adjusted by changing the reflectivity of FBG.

The results show that optical Gaussian band-pass filter, DWDM and FBG can be used as the frequency discriminators, achieving the conversion from PM to IM. FBG has more advantages due to the flexibility to adjust the waveform of the generated UWB pulse signal by changing the reflectivity of FBG.

29.5 Conclusion

A method of optical ultra-wide-band pulse generation based on XPM in a dispersion shifted fiber is proposed. According to the method, the phase modulation is achieved by simply using XPM of the DSF, without using electrooptic phase modulator. It significantly reduces the complexity of the system. By locating the cross-phase modulated probe at opposite linear slopes of the FBG reflection spectrum, UWB pulses with inverted polarity were obtained. This feature makes the implementation of the pulse polarity modulation (PPM) scheme possible in the optical domain, by simply switching the wavelength of the probe. The FBG, serving as a frequency discriminator, also serves as an optical bandpass filter to remove the residual pump and the amplified spontaneous emission (ASE) noise from the EDFA. It reduces the interference and improves stability of the system. By using the software of Opti-system, the impacts of the input signal pulse width, the reflectivity of FBG and using different frequency discriminators on the generated UWB monocycle pulse are numerically simulated and studied. The results show that this scheme has good tolerance to the input signal pulse width. Optical Gaussian band-pass filters, wavelength division multiplexer and FBG can be used as the frequency discriminators. FBG has more advantages due to the flexibility to adjust the waveform of the generated UWB pulse signal by changing the reflectivity of FBG. The XGM and SPF effect in DSF can influence the simulation results, and how to minimize the influence and waveform distortion is the focus of our future research.

References

1. Porcine D, Research P, Hirt W (2003) Ultra-wideband radio technology: potential and challenges ahead. Commun Mag 41(7):66–74
2. Aiello GR, Rogerson GD (2003) Ultra-wideband wireless systems. Microwave Mag 4(2):36–47
3. Yang L, Giannakis GB (2004) Ultra-wideband communications: an idea whose time has come. Signal Process Mag 21(6):26–54
4. Zeng F, Yao JP (2006) Ultrawideband impulse radio signal generation using a high-speed electrooptic phase modulator and a fiber-Bragg-grating-based frequency discriminator. Technol Lett 18(19):2062–2064
5. Zeng F, Wang Q, Yao J (2007) All-optical UWB impulse generation based on cross-phase modulation and frequency discrimination. Opt Lett 43(2):121–122
6. Wang Q, Zeng F (2006) Optical ultrawideband monocycle pulse generation based on cross-gain modulation in a semiconductor optical amplifier. Opt Lett 31(21):3083–3085 November
7. Dong J (2007) All-optical ultrawideband monocycle pulse generation utilizing gain saturation of a dark return-to-zero signal in a semiconductor optical amplifier. Opt Lett 32(15):2158–2160
8. Torres-Company V, Prince K, Monroy IT (2008) Fiber transmission and generation of ultrawideband pulses by direct current modulation of semiconductor lasers and chirp-to-intensity conversion. Opt Lett 33(3):222–224
9. Lin WP, Chen JY (2005) Implementation of a new ultrawide-band impulse system. Photon Technol Lett 17(11):2418–2420
10. Wu B, Wu J, Xu K et al (2010) Photonic ultra-wideband monocycle pulses generation using semiconductor amplifier and electro-absorber in parallel. Chin Opt Lett 8(9):902–905
11. Wang Q, Yao J (2007) An electrically switchable optical ultrawideband pulse generator. J Lightwave Technol 25(11):3626–3633

Chapter 30
Extraction of Acoustic Parameters for Objective Assessment of Nasal Resonance Disorder

Ning Li

Abstract The assessment methods of Nasal Resonance Disorder has mainly contained the subjective and objective ones at present. In this paper, the acoustic parameters for objective assessment were extracted. This paper focused the computing methods with the technique of the processing of the voice signal, by which the sounds with oral-nasal separate recorded were analyzed. It contained pre-processing of voice signals, detecting of the voiced and unvoiced, Linear Prediction (LP) Analysis and computing of formants and other parameters separately. Finally, the nasal formants were applied in an experiment between normal children and ones with nasal resonance disorder. The results show that there is significant difference of the first nasal formant between normal children and ones with hyponasality or hypernasality. That is, the acoustic parameters are effective for objective assessment of Nasal Resonance Disorder.

Keywords Nasal resonance disorder · Linear prediction analysis · Formant

30.1 Introduction

Nasal resonance disorder is one type of speech disorder in Speech Pathology and Audiology. The assessment methods of nasal resonance disorder contain the objective ones and the subjective ones. The objective assessment methods are to provide objective acoustic parameters as data support by objective measurement

N. Li (✉)
Department of Speech and Hearing Rehabilitation, Key Laboratory of Speech and Hearing Sciences Ministry of Education, East China Normal University, Shanghai, China
e-mail: b03214419@163.com

W. Lu et al. (eds.), *Proceedings of the 2012 International Conference on Information Technology and Software Engineering*, Lecture Notes in Electrical Engineering 210, DOI: 10.1007/978-3-642-34528-9_30, © Springer-Verlag Berlin Heidelberg 2013

technology. At present, the objective assessment methods, the acoustic parameter of which is nasalance or nasal flow rate, have been combined with the subjective ones for assessing nasal resonance disorder. However, nasalance, as the ratio of nasal energy and nasal energy plus oral energy, is one of time-parameters. Because resonance belongs to the acoustic characteristic of frequency domains, frequency-parameters are more reasonable for the assessment of resonance function. The formants, which are one kind of frequency-parameters, are the most important parameters for assessing resonance function. The formants describe the transfer characteristic of the vocal tract. A series of pulse signals produced by acoustic sound source get the modulation of the vocal tract, thus some harmonic components of Source spectrum are strengthened, and others are reduced. A series of resonance frequency, the formants, then generate [1]. However, the oral formants are mainly studied currently both at home and abroad, and the few studies have paid attention to the nasal formants. This paper focused the computing methods with the technique of the processing of the voice signal, by which the sounds with oral-nasal separate recorded were analyzed. Then formants and their energy of nasal and oral signals were picked up separately, which provided data for further experimental research.Finally,the first nasal formant (NF_1), one of the extracted parameters, is tested by an experiment as an example.

30.2 Principles and Analysis

30.2.1 Analysis Flow

The extraction process of acoustic parameters for the objective assessment of nasal resonance disorder includes the following four parts—the acquisition and pre-processing of the oral-nasal separate voice signals, Linear Prediction Analysis of nasal and oral signals, the identification for type of nasal and oral signals, and the extraction of characteristic parameters. The analysis flow is shown as Fig. 30.1.

30.2.2 Acquisition and Pre-processing of Voice Signals

The acquisition of the oral-nasal separate voice signals is standardized recording of the speech signals. This step is accomplished by the hardware. Because the analyzed speech signals are oral-nasal separate, the general microphones cannot meet the requirements. We use 'special helmet pickup' (Tiger Electronics Co., Ltd.) to acquire the oral-nasal separate speech signals. It contains the helmet and the clapboard, and the former can fix the relative position between speakers and clapboards, then the latter can ensure both oral signals and nasal signals are independent, no interfering with each other(as shown Fig. 30.2).

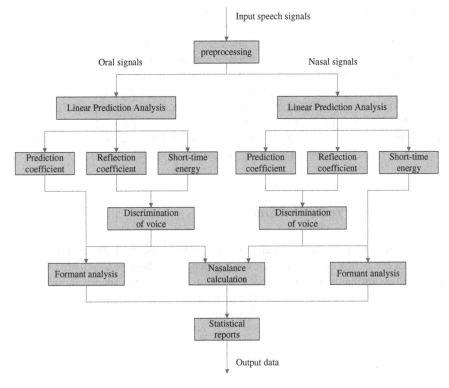

Fig. 30.1 The analysis flow

Fig. 30.2 Special helmet pickup

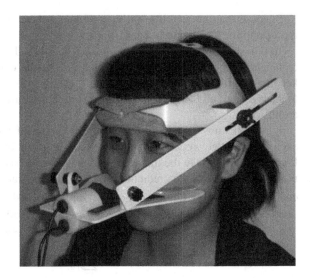

The pre-processing of speech signals includes pre-filtering and amplification to the oral-nasal separate speech signals, and A/D conversion. This step is also achieved by the hardware. We use 'DLA01' (Two-channel pre-filter amplifiers, Tiger Electronics Co., Ltd.) for pre-filtering and amplification, in order to prevent power frequency interference and aliasing, and implement audio pre-amplification. The cut-off frequency of 'DLA01' is 60 Hz (Low frequency) and 5 kHz (High frequency), the stop-band attenuation is 55 dB, and the band pass flatness is ±0.2 dB. Moreover, the magnification is set to an optional parameter, divided into four levels of 5, 10, 15 and 20 dB. The A/D conversion is achieved by 'Creative Sound Blaster 5.1'. Finally the sample frequency of speech signals is 11 kHz.

30.2.3 Linear Prediction Analysis

The role of Linear Prediction Analysis is calculating the prediction error energy and linear prediction coefficients of every speech frame, both of which are the base of extracting formants. Before Linear Prediction Analysis, the speech signals need to be processed to adapt this algorithm, such as elimination of DC offset (1), normalization of signals (2), and window framing.

$$X(i) = X(i) - \frac{\sum_{i=1}^{n} X(i)}{n} \tag{30.1}$$

$$X(i) = 32000 \times \frac{X(i)}{MAX(X(i))} \tag{30.2}$$

Suppose that the length of the speech signal should be n, and the amplitude of the i-sampling point should be $X(i)$.

In this study, the window function for Linear Prediction Analysis is set to the rectangular window, the length of which is 256, the overlap of which is 128 for keeping continuity of analysis. So the speech signal is divided into several frames to analyze, and all the following analysis procedures are processed in the frame as a unit.

It is well known that the mathematical model of the speech signal includes excitation model, vocal tract model, and radiation model [2]. Linking Linear Prediction Analysis to the mathematical model of the speech signal, the vocal tract model is regarded as the prediction filter, and the excitation is prediction error. The coefficients of the prediction filter, $\widehat{A}(z)$, are estimated by minimizing the average squared prediction error, $E[\widehat{e}^2(n)]$, where $\widehat{e}(n)$ is difference between the real speech signal s(n) and the predictive output $\widehat{s}(n)$. This step is shown as Fig. 30.3.

$$\widehat{A}(z) = \sum_{i=1}^{p} \widehat{a}(i)z^{-i} \tag{30.3}$$

$$\widehat{s}(n) = \sum_{i=1}^{p} \widehat{a}(i)s(n-i) \tag{30.4}$$

$$\hat{e}(n) = s(n) - \hat{s}(n) =$$
$$s(n) - \sum_{i=1}^{p} \hat{a}(i)s(n-i) \tag{30.5}$$

where $\hat{a}(i)$s are coefficients of the prediction filter, and p is the order of Linear Prediction Analysis. It is defined as 12 in this study, for it is related to the length of vocal tract.

The coefficients of the prediction filter, $\hat{A}(z)$, can be solved by the autocorrelation method, as well as the prediction error, $\hat{e}(n)$. the correlation function is formulated as

$$\Phi_{ss}(i,k) = E[s(n-i)s(n-k)] \quad 0 \le i \le p, 0 \le k \le p \tag{30.6}$$

$$
\begin{bmatrix}
\Phi_{ss}(1,1) & \Phi_{ss}(1,2) & \Phi_{ss}(1,3) & \cdots & \Phi_{ss}(1,p) \\
\Phi_{ss}(2,1) & \Phi_{ss}(2,2) & \Phi_{ss}(2,3) & \cdots & \Phi_{ss}(2,p) \\
\Phi_{ss}(3,1) & \Phi_{ss}(3,2) & \Phi_{ss}(3,3) & \cdots & \Phi_{ss}(3,p) \\
\cdot & \cdot & \cdot & & \cdot \\
\cdot & \cdot & \cdot & & \cdot \\
\cdot & \cdot & \cdot & & \cdot \\
\Phi_{ss}(p,1) & \Phi_{ss}(p,2) & \Phi_{ss}(p,3) & \cdots & \Phi_{ss}(p,p)
\end{bmatrix}
\times
\begin{bmatrix}
\hat{a}(1) \\
\hat{a}(2) \\
\hat{a}(3) \\
\cdot \\
\cdot \\
\cdot \\
\hat{a}(p)
\end{bmatrix}
=
\begin{bmatrix}
\Phi_{ss}(1,0) \\
\Phi_{ss}(2,0) \\
\Phi_{ss}(3,0) \\
\cdot \\
\cdot \\
\cdot \\
\Phi_{ss}(p,0)
\end{bmatrix}
$$
$$\tag{30.7}$$

$$\overrightarrow{\Phi}_{ss} \times \overrightarrow{a} = \overrightarrow{\phi}_{ss} \tag{30.8}$$

Equation (30.7) is the Yule-Walker equation, of which Eq. (30.8) is the matrix–vector representation. Where $\overrightarrow{\Phi}_{ss}$ is named as the correlation matrix.

30.2.4 Identification for Type of Speech Signals

The identification for type of speech signals is the voiced and unvoiced classification. The speech signals are classified into voiced and unvoiced according to the vibration characteristic of the vocal folds when phonating. The voiced and unvoiced classification must be done before the formant extraction since the formants are the feature parameters of the voiced speech [3, 4]. In this study, we use the prediction error energy which is calculated by Linear Prediction Analysis and

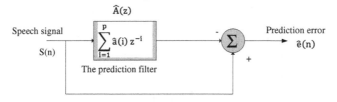

Fig. 30.3 The process of linear prediction analysis

the first reflection coefficient to classify the speech signals as voiced, because there is significant difference between the prediction error for the voiced and the unvoiced speech. The first reflection coefficient is formulated as

$$r_1 = \frac{R_{ss}(1)}{R_{ss}(0)} \tag{30.9}$$

where r_1 is the first reflection coefficient, the autocorrelation functions of the numerator and denominator are respectively as

$$R_{ss}(0) = \frac{1}{N} \sum_{n=1}^{N} s^2(n) \tag{30.10}$$

$$R_{ss}(1) = \frac{1}{N} \sum_{n=1}^{N} s(n)s(n+1) \tag{30.11}$$

where N is the number of samples in one analysis frame, and s (n) is the speech signal.

The definite classification rules are that this frame is voiced if the first reflection coefficient is greater than 0.2, and the prediction error energy is greater than twice the threshold; or if the first reflection coefficient is greater than 0.3, and the prediction error energy is greater than the threshold, and the previous frame is also voiced. If the above conditions are not valid, then this frame is unvoiced.

In this step, '1' and '0' represents the voiced and unvoiced respectively. Then a number string is obtained such as '0011110', and each number is associated with one frame.

30.2.5 Extraction of Characteristic Parameters

30.2.5.1 Formant Extraction

The formant includes frequency and bandwidth parameters [5]. The method of obtaining them is calculating and obtaining the roots of the prediction filter, $\hat{A}(z)$. Since the coefficients, $\hat{a}(i)$, has been obtained, and the order of Linear Prediction Analysis is 12, there are 12 roots, some of which belong to the vocal tract model and others belong to the excitation model or radiation model. The relationships between the roots and the possible formants are defined as follows,

$$2\pi T F_i = \theta_i \tag{30.12}$$

$$e^{-B_i \pi T} = r_i \tag{30.13}$$

So,

$$F_i = \frac{\theta_i}{2\pi T} \tag{30.14}$$

$$B_i = \frac{-\ln r_i}{\pi T} \qquad (30.15)$$

where T is the sampling period, which is the inverse of the sampling frequency; F_i is the possible formant frequency; B_i is the possible formant bandwidth; θ_i is the angle of the root in radian; and r_i is the magnitude of the root [6].

For the accuracy and continuity of formant extraction, some necessary check need to process. For example, stability check, formant frequency and bandwidth restriction, spurious check and formant deletion, allocation, or interpolation.Finally, the nasal and oral formants' values are acquired and plotted in a figure as Figs. 30.4 and 30.5. Figure 30.4 shows the formants of/ma/that a normal adult pronounces, and Fig. 30.5 shows that of/ba/.The significant low frequency resonance phenomenon occurs when a normal adult is pronouncing/m/but not/b/.

30.2.5.2 Formant Energy Concentration Rate Analysis

The linear prediction spectrum can be revealed by LP analysis. The n-formant energy is the area between the n-formant bandwidth locating in the n-formant on the spectrum. The total energy of speech signals is the total area of the spectrum. The rate of formant energy concentration is the percentage of the formant energy and the total energy of speech signals [7, 8]. It is formulated as

$$C_{f_n} = 100\% \times A_n / A_{total}(n = 1, 2, 3) \qquad (30.16)$$

Where
A_n is the n-formant energy
A_{total} is the total energy.

30.3 Application Experiment

These parameters can be applied to assess nasal resonance disorder. The first nasal formant (NF_1), one of the parameters, is tested by an experiment as an example.

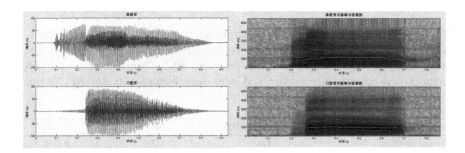

Fig. 30.4 The formants of/ma/that a normal adult pronounces

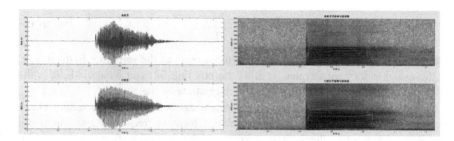

Fig. 30.5 The formants of /ba/ that a normal adult pronounces

30.3.1 Objects and Methods

In this experiment, the objects are normal children and ones with nasal resonance disorder, both of which were 3–5 years old. And the nasal resonance disorder contained hyponasality and hypernasality. The number of normal children is 30 of each age, and 15 of each gender among them. The number of children with nasal resonance disorder is a total of 30, among which 18 children are diagnosed with the hypernasality, and 12 children are diagnosed with the hyponasality. All objects' native language is Mandarin.

The Chinese syllables,/ma/and/ba/are selected as the recording materials, because/ma/, that contain nasal ingredient, apply to the hyponasality; reversely, /ba/apply to the hypernasality. We select syllables instead of separate nasal or plosive because the separate consonant cannot turn into a word in Chinese language. All objects must record the two syllables.

The syllables are recorded standardly by 'special helmet pickup' (Tiger Electronics Co., Ltd.), and the voice signals are acquired at 11 kHz, 16 bits signal resolution. The acoustic parameters are extracted on the basis of the second section. Then the first nasal formant (NF$_1$) are imported to the Statistical Package for Social Sciences (SPSS Inc, ChicagoIII, Illinois) for statistical analysis using two independent sample T test. Because gender and age did not affect the results in previous experiments we did, we don't consider them in this experiment.

30.3.2 Results and Discussion

NF$_1$ values of/ma/are compared between normalchildren and ones with hyponasality, and values of/ba/are compared between normal and hypernasality. The results are respectivelyshown in Tables 30.1 and 30.2. There are significant differences between normal children and ones with hyponasality for NF$_1$ of/ma/

Table 30.1 The T test results of nf$_1$ of/ma/

Groups	Mean (Hz)	SD.(Hz)	P
Normal children	471	169	P = 0.032
Children with hyponasality	970	105	

Table 30.2 The T test results of nf_1 of/ba/

Groups	Mean (Hz)	SD.(Hz)	P
Normal children	940	272	P = 0.000
Children with hypernasality	425	74	

(P < 0.005), and the latter's mean is much larger than the former's. There are also significant differences between normal children and ones with hypernasality for NF_1 of/ba/(P < 0.001), andthe latter's mean is much smaller than the former's.

When the children with hyponasality pronounce nasals, the nasals could contain less nasal ingredient or be made non-nasals. However, the NF_1 values of nasal ingredient are much smaller than non-nasal '/a/', so the NF_1 values of/ma/that the children with hyponasality pronounce are much larger. Reversely, the non-nasals could contain nasal ingredient or be made nasals when the children with hypernasality pronounce non-nasals. So the NF_1 values of/ba/that the children with hypernasality pronounce are much smaller as a result of nasal ingredient influence.

30.4 Conclusion

In this study, the formants and other frequency-parameters are extracted by analyzing the sounds with oral-nasal separate recorded, and the computing methods are studied on the basis of the technique of the processing of the voice signal. Finally, it can provide data for further experimental research, and contribute parameters to the objective assessment of nasal resonance disorder such as NF_1. The most important is to obtain the parameters of the nasal speech, which can emphasize the acoustic characteristics of nasal speech, and reflect nasal resonance function more directly.

References

1. Jiang SC, Gu R (2005) Speech and language nosography. Science Press, Beijing (in Chinese)
2. Li Z (2003) Speech signal processing. Machinery Industry Press, Beijing (in Chinese)
3. Childers DG (1990) Speech processing and synthesis for assessing vocal disorders. IEEE Eng Med Biol Mag 9(1):69–71
4. Atal BS, Rabiner LR (1976) A pattern recognition approach to voiced-unvoiced-silence classification with application to speech recognition. IEEE Trans Acoust, Speech Sig Process 24(3):201–212
5. Ladefoged P (2003) Phonetic data analysis. Blackwell publishing, Oxford
6. Fant G (1983) The vocal tract in your pocket calculator, STL-QPSR, vol 2(3). Royal Institute of Technology, Stockholm, Sweden
7. Hong-ping Z, Ma L, Luoyi (2003) The correlation between spectrum manifestations of nasalized vowels and listener judgments of hypernasality. China J Oral Maxillofacial Surg 1(1):56–58 (in Chinese)
8. Li JF, Liu JH (1997) Study of testing Velopharyngeal function via the third formant frequency of chinese vowel. Western China J Oral Sci 15(4):325–327 (in Chinese)

Chapter 31
Congestion Distance Based BitTorrent-Like P2P Traffic Optimization

Qian He, Yanlei Shang, Yong Wang and Guimin Huang

Abstract A large number of P2P traffic brings great impact on the bearer network, so how to utilize network bandwidth efficiently has become a common trade problem of Internet Service Providers (ISP) and terminal users. A P2P traffic optimization model is derived from the aims of ISPs and users. Minimizing the congestion distance is chosen as the main objective which not only can represent the requirements of two sides, but also can be distributed implemented easily. According the optimization model, Bittorrent-like P2P system is improved. The distance aware peer selection and secondary sorting choking/unchoking algorithms are proposed. The experiments results show that this optimization can reduce inter autonomous system (AS) traffic effectively and respond to network congestion automatically.

Keywords P2P · Traffic optimization · Congestion distance · Bittorrent

31.1 Introduction

Peer-to-Peer (P2P) technology develops very fast recently, and it is widely used in content sharing, instant messaging and so on. The Internet traffic distribution results, given by Ipoque in 2008/2009, showed P2P systems contribute 42.5–70 %

Q. He (✉) · Y. Wang · G. Huang
Key Laboratory of Cognitive Radio and Information Processing, Ministry of Education,
Guilin University of Electronic Technology, 541004 Guilin, China
e-mail: treeqian@gmail.com

Q. He · Y. Shang
State Key Laboratory of Networking and Switching Technology, Beijing University of Posts
and Telecommunications, 100876 Beijing, China

W. Lu et al. (eds.), *Proceedings of the 2012 International Conference on Information Technology and Software Engineering*, Lecture Notes in Electrical Engineering 210, DOI: 10.1007/978-3-642-34528-9_31, © Springer-Verlag Berlin Heidelberg 2013

to the overall network traffic [1]. While utilizing P2P to accelerate the content transmitting, it also impacts network bandwidth very much and has put Internet Service Providers (ISP) in a dilemma [2, 3]. Among them, BitTorrent(BT) is arguably the biggest constituent of P2P traffic, which contributes 30–80 % to the whole P2P traffic [1]. How to use network bandwidth reasonable while giving full play to the P2P performance has become a big problem concerned by the ISPs and ordinary users together.

BT is one of the typical and most popular P2P content sharing systems, and it can also support video on demand and live multimedia streaming after modified. All these P2P systems mainly based on BT protocol are BT-like systems in this paper. In the normal BT, the peers fetch neighbors from the tracker randomly and choose communication objective peers with "tit for tat". It doesn't consider the network layer topology. The random selection algorithm of P2P application is one of the main reasons causing large P2P traffic on backbone, so the network topology optimization are becoming a hot research field [3, 4]. After analyzing the actual BT traffic logs, Karagiannis et al. [2] think the probability of requesting the same file in the same ISP is 30–70 %. The Verizon network field testing also found that the average P2P connections after 5.5 metro jumping, which does not affect the performance of reduced to 0.89 jump [3].

There have been several proposals on locality-aware P2P solutions. Reference [5] takes advantage of the cosine similarity of CDN redirection to guide partner selection, while Ref. [6] leverage a distance dataset extracted from publicly available information. Meanwhile, Aggarwal et al. incorporate Oracle [7] and last-hop bandwidth of each peer to judge the link quality. Furthermore, Refs. [8, 9] both designed and implemented real locality-aware BT clients. These researches may give us the foundation to optimize BT-liked P2P traffic deeply. However, there is little work on the P2P traffic optimization description and simplifying, especially for BT-like P2P applications as a whole.

In order to optimize the BT-like P2P traffic, a general optimization model and improving BT algorithms are proposed in this paper. The optimization model starts from the requirements of ISPs and normal P2P users, in which the fast download speed, inter-AS traffic and backbone congestion are discussed in the objectives. And then, the special congestion distance is concluded for distributed implementation. Based on minimizing congestion distance, BT is as the example to be simulated and analyzed for optimizing BT-like P2P traffic optimization.

31.2 P2P Traffic Optimization Model

For the P2P users, they normally concentrate on fast download speed. On the other hand, ISPs hope to be reward from P2P, but not to get such bad affection as high traffic impression, low transfer performance, congestion and so on. The fast download speed and the more P2P users mean there may be more P2P traffic flow on the backbone, so the requirements of users and ISPs are often conflicted. The backbone network links are used to connect each ISP which topology can be described by a graph $G(V,L)$, where V is the backbone nodes set and L is the key Links connecting ASes set. Because the network topology, the bandwidth between ASes, and the upload/download speed of each peer are relatively fixed, the P2P traffic optimization model is a multi objectives optimization with constraints, which can be defined as follows:

$$
\begin{aligned}
&\min P_m, P_c, I_c \\
&s.t. \forall i, f_i + \tau_i \leq b_i \\
&\qquad u_i \leq pu_i \\
&\qquad v_i \leq pd_i
\end{aligned}
\tag{31.1}
$$

Where, f_i, τ_i and b_i are the P2P traffic, background traffic and bandwidth of link i; u_i and v_i are the upload and download rate of peer i; pu_i and pd_i are the max upload and download rate of peer i; P_m is the mean download rate of all the peers, $P_m = \left(\sum_i d_i\right)/n$; P_c is the variance of download rate of all peers which can represents the fair of peers, $P_c = \left(\sum_i (d_i - P_m)^2\right)/n$. The P_m and P_c represents the users' requirements, and the bigger P_m and the litter P_c are better. The congestion reflection index (I_c) is used to represent a function of traffic and congestion.

31.2.1 Congestion Distance Based Optimization

Because this rate information is changing constantly, the Eq. 31.1 may be weak for practice in a big P2P network. It should be decoupled and simplified deeply to satisfy the requirements of distributed implementation.

Proposition 1 *When the fixed-size data is distributed, the more key links are passed through, the inter-AS traffic is bigger.*

Proof Assuming peer A wants to distribute x M bytes data to peer B, the inter-AS traffic f_i equals to the value that is gotten from $d * h$, where $h = |G| + 1$, $|G|$ is the number of AS passed through. Therefore, we can get: $f_i = x \cdot (|G| + 1) = x \cdot |V|$. Therefore, the inter-AS traffic f_i is proportional to the number of key links $|V|$; the more key links passed through, the inter-domain traffic is greater.

Based on Proposition 1, we know that the aim of inter-AS traffic can be convert into the function of the number of key links, $|V|$ is relatively stable after the route is determined and it can be tested by trace-route tools or sensors set on the key links [9]. Integrating the affection of congestion, the congestion distance (d_c) is the summary of d and the accumulation of all the key links' congestion mapping value:

$$
d_c = |V| + \delta \sum_{i \in L} c_d^i.
\tag{31.2}
$$

Where c_d^i is the congestion level of link i. If no congestion, $c_d^i = 0$; with the more serious congestion c_d^i become 1,2,3, and so on. δ is a congestion adaption gene to adapt the congestion affection.

Proposition 2 *The congestion distance d_c can be compute by this equation*:

$$d_c = d + \delta \cdot \sum_i c_d^i \cdot n_i \qquad (31.3)$$

Proof $\sum_{i \in L} c_d^i$ equals the summary c_d^i of all the backbone, it can be the summary c_d^i of multi congestion levels. That is to say, we can compute the congestion separately, and sum them together finally. From the Eq. 31.2, we can get this: $\sum_{i \in L} c_d^i = \sum_i c_d^i \cdot n_i$, and then $d_c = d + \delta \cdot \sum_i c_d^i \cdot n_i$.

31.3 BT Traffic Optimization Scheme

BT is a P2P file-sharing application designed to distribute large files to a large of users efficiently. BT consists of seed, peer and tracker. During the neighbor selection process, the peer learns from the tracker the knowledge of other peers in the same group of peers interested in the same file and a peer list is returned by the tracker. Upon receiving the peer list, a peer connects with the majority of them. It then sends to its neighbors the bitfield messages. If a neighbor has needed pieces, it informs the neighbor with an interested message. The peer choking/unchoking is the decision process made by a peer about which of its "interested" neighbor it should send data to. In the next, assuming the congestion distance is already obtained from other traffic optimization architecture, how to use the congestion distance to improve the performance of BT is discussed.

31.3.1 Minimizing Congestion Distance Peer Selection

During the neighbor selection process, the peer learns from the tracker the knowledge of other peers in the same swarm, which is the group of peers interested in the same file. The tracker returns a peer list to the inquiring peer, and if the population of the swarm exceeds this number, the selection is entirely random. In this paper, we replace the random peer selection with the selection that minimizes d_c. The optimization of peer selection consists of tracker and peer, and they are similar in our procedure. The tracker selection procedure is given in Fig. 31.1. After peer queries the tracker, the tracker not only returns the peer list in the aforementioned fashion, but also attaches d_c in an add-on field.

Fig. 31.1 The flow of peer requesting neighbors

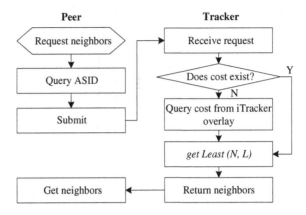

31.3.2 Secondary Sorting Choking/Unchoking Algorithm

The secondary sorting choking/unchoking algorithm is modified from "*Tit-for-Tat*". The new method not only considers the peers' contribution data (*CD*), but also considers d_c. Firstly, the neighbors are sorted by *CD*, a rank of each peer (r_u) can be gotten. Then, the final rank (r) can be obtained:

$$r = r_u + \beta \times d_c \tag{31.4}$$

Where, β is a gene to adapt the affection of congestion distance. The bigger β, the more important is d_c. We set β equals to the number of unchoking peers (default 4) in order to proof the communication costs are preferred. Only the d_c of peers is equaled, the Tit-for-Tat mechanism works. Finally, we keep the original BitTorent optimistic unchoking and boycott to neglect policy intact, for the purpose to help bootstrap brand new peers and avoid the peers with higher costs will choke forever. The improved choking/unchoking algorithm is as follows:

Input: the local candidate peers *PL*

Output: the unchoking peers *unchokeL*

(1) *if* |*PL*| < *ucNum, ucNum*: the number of unchoking Peers
(2) *return PL*;
(3) *else {*
(4) *Sort PL by d_u, and refresh PL.urank.*
(5) Integrate d_c and refresh the secondary rank *PL.rank* using the Eq. 31.4.
(6) Select the *ucNum* Peers from PL with least *PL.rank*.
(7) *return unchokeL*;
(8) *}*

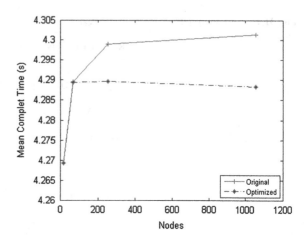

Fig. 31.2 The mean
complete time t_m

31.4 Experiments Analysis

The performances of Optimized BT are evaluated in the General Peer-to-Peer
Simulator (GPS) [10]. All the experiments are run on a computer with Inter Core 2
Q9400 2.66 GHz CPU and 4 GB Memory where the windows 7 operation and 1.5
JDK are installed.

31.4.1 Inter-AS Traffic

Assuming there are no congestions in all the links, we monitor the indexes
including the mean t_m and variance t_c of each peers' complete time and I_f
according the Eq. 31.1. Referring to the Ref. [10], two-level Transit-Stub topologies are generated using GT-ITM. There are four different topologies with 16,
64, 252 and 1054 peers, A transit domain means a AS, and its number is 4, 8, 18
and 34. The bandwith between AS is 1000 Mbps, bandwidth between transit and
stub is 100 Mbps and there are no connections within stubs. BT Peers are randomly attached to non-transit nodes. All the documents are 500 MB in size. In the
simulation, t_c of the optimized is similar to the original. Figure 31.2 gives the
mean complete time which shows that the optimized BT has better download
speed for peers. Figure 31.3a gives the total inter-AS traffic and Fig. 31.3b shows
the decreased ratio (R) of inter-AS traffic about optimized BT. The optimized
BT has litter inter-AS traffic than the original and R become larger with the
more peers.

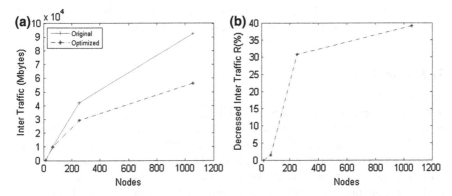

Fig. 31.3 Inter-AS traffic analysis. **a** The inter AS traffic: I_f. **b** The decreased ratio of inter-AS traffic: R

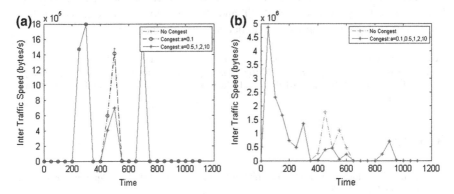

Fig. 31.4 The P2P traffic flow rate on congestion links. **a** Link1_7. **b** Link1_18

31.4.2 Congestion Response

The moderate congestion events are added to Link1_7 and Link 1_18 from the simulation time 300.0–600.0. In the simulation, the peer download rate can be reduced very much during the congestion time in the optimized BT system. From Fig. 31.4a and b, the mean rate γ of Link 1_7 can be reduced by 46.6 %, and γ of Link 1_7 can be reduced by 71.2 %. On the other hand, we can find that $\delta = 0.5$ or 1.0 decreased more than $\delta = 0.1$. The bigger δ means it is more sensitive to the congestion so the bigger δ can bring more flow is avoided. However, if δ reach to the big value, it will not be changed any more. When $\delta = 0.5, 1, 2$ in the Fig. 31.4a, they have the same results on link1_7. From Fig. 31.4b, we can find that the traffic on link1_18 always is same under various δ. This is because how δ works is based on whether the congestion distance will affect the selection algorithms and choking/unchoking algorithm or not. Since the decreased traffic will alleviate congestion and BT chooses other links may fast peers' download speed, it is useful for backbone to respond to congestion automatically.

31.5 Conclusions

In this paper, a congestion distance based BT-like P2P traffic optimization scheme is proposed to make full of P2P application and reduce the bad impacts of P2P traffic. Importing the aims of ISP and P2P users, a general P2P traffic optimization model is obtained and the congestion distance becomes the main optimization object which is suitable to use for distributed peers. The distance aware peer selection and secondary sorting choking/unchoking algorithms are proposed for improving BT. In the large scaled network, the optimized BT has remarkable advantages comparing to the original and the inter AS traffic can be reduced remarkably. In the congestion simulation, the optimized BT can respond network congestion automatically. In the future, we will try to modify real BT and deploy it to the planetLab, and then analyze its performance under the real network environment.

Acknowledgments This work was partly supported by the National Natural Science Foundation of China (61201250, 61063038, and 61163058), Important National Science & Technology Specific Project (2012ZX03006001-005) and Guangxi Natural Science Foundation of China (2012GXNSFBA053174).

References

1. Schulze H, Mochalski K (2009) Internet study 2008/2009. https://portal.ipoque.com/downloads/index/get/id/265/
2. Karagiannis T, Rodriguez P, Papagiannaki K (2005) Should internet service providers fear peer-assisted content distribution? In: Internet measurement conference, Berkeley, CA, pp 1–14
3. Yang P, Xu L (2011) On tradeoffs between cross-ISP P2P traffic and P2P streaming performance. Comput Netw 55(18):4033–4042
4. Xie H, Yang YR, Krishnamurthy A et al (2008) P4p: provider portal for applications. In: Proceedings of the ACM SIGCOMM 2008. Seattle, pp 351–362
5. Choffnes DR, Bustamante FE (2008) Taming the torrent: a practical approach to reducing cross-ISP traffic in peer-to-peer systems. In: Proceedings of ACM SIGCOMM, vol 38(4), pp 363–374
6. Hsu CH, Hefeeda M (2008) ISP-friendly peer matching without ISP collaboration. In: Proceedings of the ACM CoNEXT conference
7. Aggarwal V, Feldmann A, Scheideler C (2007) Can ISPs and P2P users cooperate for improved performance? ACM SIGCOMM Comput Commun Rev 37:29–40
8. Lin M, Lui J, Chiu D (2010) An ISP-friendly file distribution protocol: analysis, design and implementation. IEEE Trans Parallel Distrib Syst 21(9):1317–1329
9. Liu B, Cui Y, Lu Y et al (2009) Locality-awareness in BitTorrent-like P2P applications. IEEE Trans Multimedia 11:361–371
10. Yang W, Abu-Ghazaleh N (2005) GPS: a general peer-to-peer simulator and its use for modeling BitTorrent. In: Proceedings of IEEE computer society's annual international symposium on modeling, analysis, and simulation of computer and telecommunications systems. Atlanta, pp 425–432

Chapter 32
Unified Continuous Collision Detection Framework for Deformable Objects

Xiaoyang Zhu, Shuxin Qin, Yiping Yang and Yongshi Jiang

Abstract Various types of advanced continuous collision detection algorithms have significantly improved the quality of physically-based animation and robot navigation planning. However, these continuous collision detection methods are not implemented in a unified software framework, which brings along lots of disadvantages such as breaking connection of different modalities, lack of module reuse and inconvenience to method comparison. This paper discusses continuous collision detection process from the viewport of data flow and implements an extensible Unified Collision-Detection Software Framework (UCSF). The goal of this framework is to make the development of collision detection for deformable objects much easier, and implement a set of popular collision detection algorithms, so that it is convenient for researchers to compare against. The overall design and certain key technologies are introduced in detail. Presented experiment examples and practical applications commendably demonstrate the validity of this framework.

Keywords Continuous collision detection · Bounding volume hierarchies · Deformable objects · Framework · Self-collision · GPU

X. Zhu (✉) · S. Qin · Y. Yang
Integrated Information System Research Center, Institute of Automation,
CAS, Beijing, China
e-mail: xiaoyang.zhu@ia.ac.cn

X. Zhu · S. Qin · Y. Yang · Y. Jiang
CASIA—Beijing CAS X-Vision Digital Technology Co., Ltd. Joint Lab of Information Visualization, #95 East Zhongguancun Road, Haidian, Beijing, China

W. Lu et al. (eds.), *Proceedings of the 2012 International Conference on Information Technology and Software Engineering*, Lecture Notes in Electrical Engineering 210, DOI: 10.1007/978-3-642-34528-9_32, © Springer-Verlag Berlin Heidelberg 2013

32.1 Introduction

Collision detection between two objects or within an object is a fundamental technique that widely used in virtual reality applications, physically-based animation and robot navigation planning. Collision detection methods can be classified into two categories: discrete collision detection (DCD) and continuous collision detection (CCD). The key difference of two approaches is that DCD considers collisions only at discrete time instances, while CCD model the object motion between successive time instances as a piecewise linear path and check collisions along the path. While DCD can be performed quite efficiently, it does not prevent two objects passing through each other and may miss collision during this time period. On the other hand, CCD does not miss any interference between two time steps, but typically requires much more computation cost compared to DCD and have not been widely used in interactive applications.

Many approaches have been presented to accelerate the performance of CCD by designing algorithms on rigid and deformable objects [1–6], but there are many problems which still remain unresolved or can be improved. One serious problem is that there is no unified software platform covering a wide spectrum of CCD methods. Lack of unified framework brings along lots of disadvantages, such as breaking relationship between different modalities, duplicate efforts during code development and inconvenience to method comparison. Moreover, classic methods are coded again and again by different researchers just for comparison against their own methods. Aforementioned requirements make consolidating existing CCD algorithms to a unified software framework necessary.

This paper analyses CCD process from the standpoint of work flow and implements a unified collision detection software framework (UCDF), which provides an abstract software solution for collision detection. In Sect. 32.2, we will first introduce the continuous collision detection problem. Section 32.3 will give the practical implementation details of UCSF. In Sect. 32.4, we present benchmarks, while the last section is about final conclusions as well as an outlook onto the work ahead.

32.2 Continuous Collision Detection

In continuous collision detection, collision events are detected for objects in close proximity between two consecutive discrete time steps. For interactive rendering and animation, objects are most often discretized into triangular meshes. A mesh M consists of a set of triangle faces F, a set of edges E, and a set of vertices V. We denote M as (F, E, V) [7]. Each triangle has a set of features: the triangle itself, its three vertices and its three edges. In order to identify intersecting features at the first time-of contact (ToC) during a time interval between two discrete time steps, CCD methods model continuous motions of features by linearly interpolating positions of primitives between two discrete time steps.

There are two types of feature pairs that are handled in CCD: vertex-face (VF) and edge–edge (EE). For a pair of triangles, there are six VF pairs and nine EE pairs. The problem of many efficient culling algorithms has been designed for CCD between rigid and deformable models. Their main purpose are to reduce the number of the elementary tests, since the exact VF and EE tests need to solve cubic equations.

32.3 Software Framework Implementation

This section is divided into four parts to introduce basic structure of UCSF in detail. Section 32.3.1 introduces overall design of this framework. Sections 32.3.2, 32.3.3 and 32.3.4 are about object level, triangle level and feature level culling methods respectively.

32.3.1 Software Framework Overview

UCSF is actually an abstract encapsulation of necessary continuous collision detection modules, which include model module, I/O interface module, algorithm module, etc. Figure 32.1 shows main modules of UCSF. At a high level, our UCSF method consists of two parts: CPU-based culling and GPU-based acceleration calculation.

Our UCSF method first performs broad-phase processing by using bounding volume hierarchies (BVH) or spatial partitioning techniques to find potentially colliding object pairs. Then we employ normal cone and star contour based culling methods to skip "uninteresting" zones and concentrate only on triangle parts which are likely to collide. Traditionally, after collecting the triangle pairs that are potentially intersecting, the feature pairs of these triangles are directly sent for

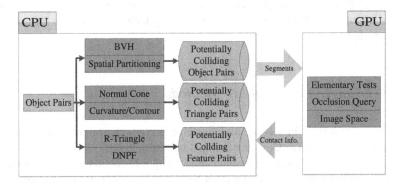

Fig. 32.1 Pipeline of continuous collision detection for deformable objects

computation of collision information. Many feature pairs, however, end up being processed repeatedly because may be shared by more than one triangle pairs. So we use deforming non-penetration filters (DNPF) [4], orphan sets [4], and representative-triangles (R-Triangles) [1] method to remove redundant feature pairs and reduce the number of elementary tests. When collecting enough potentially intersecting feature pairs, we send them to GPUs and perform occlusion query and elementary tests constructed from the triangles using GPUs. We only send colliding primitives and their contact information (e.g. a contact point and normal) at the first ToC to CPUs after finishing the tests.

32.3.2 Object Level Culling Methods

Pairwise processing is expensive, and the idea of broad-phase processing is to restrict pairwise tests to objects near enough that they could possible intersect. Generally, there are two types of object level culling method, i.e. spatial partitioning and bounding volume hierarchies.

Spatial partitioning techniques provide broad-phase processing by dividing space into regions and testing if objects overlap the same region of space. Because objects can only intersect if they overlap the same region of space, the number of pairwise tests is drastically reduced. There exist three types of spatial partitioning methods: grids, trees, and spatial sorting.

Bounding volume hierarchies (BVH) have been widely used as high-level culling methods. Bounding volume hierarchies have been widely used as object-level culling methods. Some widely used bounding volumes include spheres [8], axis-aligned bounding boxes (AABBs) [9], oriented bounding boxes (OBBs) [10], discretely oriented polytopes (k-DOPs) [11], etc. For a big number of objects, we could build a well-organized hierarchy of these objects optimized by the bounding box testing, with optionally dynamical updating of this hierarchy whereas the objects are moving or are added between each frame. As the data of each object is rather independent, a new object can easily be inserted at any time in the scene.

Our UCSF method first performs broad-phase processing by using bounding volume hierarchies (BVH) or spatial partitioning techniques to find potentially colliding object pairs.

32.3.3 Triangle Level Culling Methods

In case of clothing simulation and skeleton-driven character animation, the mechanical behaviour is mainly determined by multiple self-collisions [12]. Take advantage of the surface curvature and connectivity information, normal cone tests are used to cull mesh regions that cannot have self-collisions during discrete self-collision tests [13, 12]. Several methods aim at improving the culling efficiency of self-collision detection, including continuous normal cone tests (CNC), star-contour tests and subspace min-norm certification tests.

In order to use the normal cone for CCD, [4] computes a normal cone that bounds the normals of the deforming triangles in the entire interval. Let $\mathbf{a_0}$, $\mathbf{b_0}$, $\mathbf{c_0}$ and $\mathbf{a_1}$, $\mathbf{b_1}$, $\mathbf{c_1}$ to be positions of the vertices of the triangles at time frame 0 and 1, respectively, as shown in Fig. 32.2a. Assuming the vertices of the triangle are under linearly interpolating motion, then normal $\mathbf{n_t}$ at time t can be given by the equation:

$$\mathbf{n_t} = \mathbf{n}_{_0} \cdot B_0^2(t) + (\mathbf{n}_{_0} + \mathbf{n}_1 - \delta)/2 \cdot B_1^2(t) + \mathbf{n}_1 \cdot B_2^2(t), \qquad (32.1)$$

where $\mathbf{n}_0 = (\mathbf{b}_0 - \mathbf{a}_0) \times (\mathbf{c}_0 - \mathbf{a}_0)$, $\mathbf{n}_1 = (\mathbf{b}_1 - \mathbf{a}_1) \times (\mathbf{c}_1 - \mathbf{a}_1)$, $\delta = (\vec{\mathbf{v}}_\mathbf{b} - \vec{\mathbf{v}}_\mathbf{a}) \times (\vec{\mathbf{v}}_\mathbf{c} - \vec{\mathbf{v}}_\mathbf{a})$ and $B_i^2(t)$ is the ith basis function of the Bernstein polynomials of degree 2. We can use these vectors to construct a CNC for each triangle in the interval, as shown in Fig. 32.2b. Then, the CNCs will be merged as described in [13] by traversing the hierarchy in a bottom-up manner while the BVH is refitting. Thus, this algorithm skips efficiently non interesting parts of the mesh and concentrates only on parts which are likely to collide.

32.3.4 Feature Level Culling Methods

Traditionally, for meshed surfaces, after collecting the triangle pairs that are potentially interacting, the feature pairs of these triangles are directly sent for the computation of collision information. Many feature pairs end up being processed repeatedly because may be shared by more than one triangle.

We follow the idea presented in [1, 14]. Initially, we keep a triangle list which contains all triangles of the object. We check each triangle in turn. If all of its

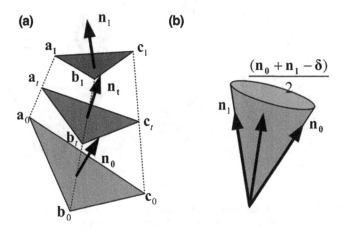

Fig. 32.2 Continuous normal range of a deforming triangle

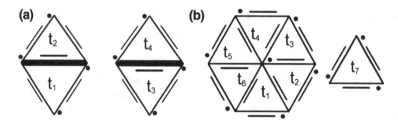

Fig. 32.3 **a** shows a feature assignment for an EE contact. **b** shows the same for a VF contact

edges are not assigned yet, we assign all the edges to it. At the same time, we also assign its unassigned vertices. After that, the triangle is then removed from the list. Then we check the remaining triangles in the same way. We improve the ratio of the number of edges to the number of assigned triangles. At run time, if the bounding volumes of two triangles overlap but they are not assigned any vertices or edges, we ignore such triangle pair. By employing this feature assignment scheme, we would decrease the total number of potentially colliding triangle pairs. So, the speedup factor in the elementary test processing would be improved, as shown in Figs. 32.3, 32.4.

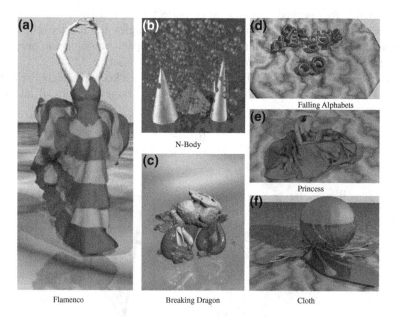

Fig. 32.4 Benchmarks: all the benchmarks have multiple simulation steps. We perform continuous collision detections and compute the first time-of-contact

Table 32.1 Average query time of continuous collision detection

Benchmarks	Query time (ms)	Benchmarks	Query time (ms)
Flamenco	106.4	Falling alphabets	4.3
N-Body	51.7	Princess	16.5
Breaking dragon	3865.5	Cloth	129.1

32.4 Experiment Benchmarks

We can testify this framework by performing some experiments of CCD. The following benchmarks show UCSF's great power of comparing different methods. These experiments are performed on a Windows-PC, with an Intel i5 2.66 GHz quad-core CPU, a NVIDIA GT540 GPU card and 4 GB main memory. The computation was all carried out in double precision floating point for solving the cubic equations. For the update and traversal of bounding volume hierarchies, we implemented them based on Intel SSE instruction set.

We had implemented most of the method presented in this paper and in order to test the performance of different algorithm, we used six different benchmarks, arising from different simulations with different characteristics. The average continuous collision detection time of our UCSF for the six benchmarks is shown in Table 32.1.

- **Flamenco**: An animated sequence of a flamenco dance with 26K vertices, 75K edges and 50K triangles, consisting of 350 frames.
- **N-Body**: In the dynamic scene, there are 300 balls and 5 cones interacting with each other. This scene has 18K vertices and 34K triangles.
- **Breaking Dragon**: In this simulation, a bunny model is dropped on top of the dragon model and the dragon model breaks into many pieces. This model has 193K vertices and 253K triangles.
- **Falling Alphabets**: Multiple characters intersect with a bowl. This model has 3K vertices and 5K triangles.
- **Princess**: A dancer with a flowing skirt. This model has 60K vertices and 40K triangles.
- **Cloth**: We drop a cloth on top of a rotating ball. This model has 46K vertices and 92K triangles and the simulation results in a high number of self-collision.

32.5 Conclusion

In this paper, we have proposed a continuous collision detection framework for deformable objects. Our method has considered both inter-collision detection and self-collision detection. The experimental results above demonstrate that this framework is well suited to implement and accelerate computationally intensive CCD. Moreover, three sub-frameworks and lots of advanced schemes are presented in detail, which will be meaningful for other engineers.

Further extension to other operating systems will be a meaningful work. In addition, integration of more modalities and more practical algorithms to UCSF will be a long-term task.

References

1. Curtis S, Tamstorf R et al (2008) Fast collision detection for deformable models using representative-triangles. In: Proceedings of the 2008 symposium on interactive 3D graphics and games, Redwood City, California, ACM, pp 61–69
2. Hutter M, Fuhrmann A (2007) Optimized continuous collision detection for deformable triangle meshes. Proceedings of WSCG'07, pp 25–32
3. Kopta D, Ize T et al (2012) Fast, effective BVH updates for animated scenes. In: Proceedings of the ACM SIGGRAPH symposium on interactive 3D graphics and games. Costa Mesa, California, ACM, pp 197–204
4. Min T, Curtis S et al (2009) ICCD: interactive continuous collision detection between deformable models using connectivity-based culling. IEEE Trans Vis Comput Graph 15(4):544–557
5. Tang C, Li S et al (2011) Fast continuous collision detection using parallel filter in subspace. In: Symposium on interactive 3D graphics and games. San Francisco, California, ACM, pp 71–80
6. Tang M, Manocha D et al (2011) Collision-streams: fast GPU-based collision detection for deformable models. In: Symposium on interactive 3D graphics and games. San Francisco, California, ACM, pp 63–70
7. Baciu G, Wong WS-K (2002) Hardware-assisted self-collision for deformable surfaces. In: Proceedings of the ACM symposium on virtual reality software and technology, Hong Kong, China, ACM, pp 129–136
8. Bradshaw G, O'Sullivan C (2004) Adaptive medial-axis approximation for sphere-tree construction. ACM Trans Graph 23(1):1–26
9. Bergen GVD (1998) Efficient collision detection of complex deformable models using AABB trees. J Graph Tools 2(4):1–13
10. Gottschalk S, Lin MC et al (1996) OBBTree: a hierarchical structure for rapid interference detection. In: Proceedings of the 23rd annual conference on computer graphics and interactive techniques, ACM, pp 171–180
11. Klosowski JT, Held M et al (1998) Efficient collision detection using bounding volume hierarchies of k-DOPs. IEEE Trans Vis Comput Graph 4(1):21–36
12. Volino P, Thalmann NM (1994) Efficient self-collision detection on smoothly discretized surface animations using geometrical shape regularity. Comput Graph Forum 13(3):155–166
13. Provot X (1997) Collision and self-collision handling in cloth model dedicated to design garments
14. Wong WS-K, Baciu G (2006) A randomized marking scheme for continuous collision detection in simulation of deformable surfaces. In: Proceedings of the 2006 ACM international conference on virtual reality continuum and its applications. Hong Kong, China, ACM, pp 181–188

Chapter 33
Application of Data-Based Parameter Optimizing Approaches in IC Manufacturing System Simulation Model

Rong Su, Fei Qiao, Yumin Ma, Keke Lu and Liang Zhao

Abstract When building an Integrated circuit (IC) manufacturing system simulation model, the practices normally simplify some complex features to ensure build ability and run ability. However, this brings the consequence of model imprecision; parameters disable to represent actual situation, etc. This paper proposes parameter optimizing approaches in IC manufacturing system simulation model based on the offline weighted average method and Kolmogorov–Smirnov (K–S) test method. They are on the basis of the processing time's feature, for the parameter reflects the overall character of the sample's former performance and at the same time, the system has a great amount of historical data. The approaches are to optimize the whole model by optimizing the significant processing time parameter. It turns out that both the weighted average method and the K–S test based method can improve the precision of model.

Keywords Data-based · Model parameterization method · Weighted average method · K–S test

R. Su (✉) · F. Qiao · Y. Ma
CIMS Research Center, Tongji University, Shanghai Cao'an road No. 4800 Tongji University Jia ding campus Electronic and Information building 660, Shanghai, China
e-mail: surong522@yahoo.com.cn

K. Lu · L. Zhao
Advanced Semiconductor Manufacturing Co.,Ltd., Shanghai, China
e-mail: ke_ke_lu@asmc.com.cn

L. Zhao
e-mail: liang_zhao@asmc.com.cn

W. Lu et al. (eds.), *Proceedings of the 2012 International Conference on Information Technology and Software Engineering*, Lecture Notes in Electrical Engineering 210, DOI: 10.1007/978-3-642-34528-9_33, © Springer-Verlag Berlin Heidelberg 2013

33.1 Introduction

The features of complicated processing flow, re-entrant, batch processing equipment and single wafer processing equipment coexistence have made the Integrated circuit (IC) manufacturing system the third sort of manufacturing system after job shop manufacturing system and flow shop manufacturing system. IC manufacturing system has also been recognized as one of the most complex production systems in the world [1]. Those features bring in plenty of assumptions and constraints when building a simulation model. Meanwhile, most current IC manufacturing system simulation model parameters do not make full use of real-time data, simulation data and historical data which related to the production process and store in production management system [2]. Those facts disable the model's ability of changing with the actual situation dynamically. The rigid simulation model cannot make a rapid and effective response when actual productive conditions changes. Under such condition, the dispatch given out by the simulation model is low in precision and cannot guide practical production.

Traditional complex manufacturing system simulation model optimization method mainly contains operation research method, Petri Net, heuristic method, artificial intelligence method, etc. Although these methods have made a lot of achievements [3–6], the integrated circuit manufacturing system's unique feature of high complexity and high degree of coupling makes traditional optimization methods difficult to achieve the ideal effect. Under this situation, more and more scholars set their sights on another optimization way: data-based model optimization method. Some scholars at home and abroad have obtained certain progress by using data-based model optimization method such as characteristic analysis, principal component analysis, statistics methods, neural network, rough set, support vector machine and fuzzy set mining historical data, real-time data and simulation data which is related to the production process and store in a large manufacturing production management system [7–9].

Based on data-based model optimization theory, this paper proposes several offline solutions which can optimize the simulation model through optimizing the critical parameter by using the historical data. It turns out that the solutions that this article mentioned can relieve the rigidity of model parameters and improve the accuracy of dispatching plan.

33.2 Problem Statement

We make the following statements to facilitate our discussions.

(1) There is a successfully built IC manufacturing system simulation model.
(2) There are one or several parameters which obviously influence the output of simulation model but unable to reflect the real production line situation.

(3) The parameters that we mentioned above have the feature that reflects the overall character of the sample former performance.
(4) There are abundant historical data and real-time data which relate to the parameters above are stored.

33.3 Solution Introductions

• Normal Average Solution

For the given N data, namely λ_i, $i \in [1, N]$, the normal average solution can be expressed as

$$\bar{\lambda} = \frac{1}{N} \sum_{i=1}^{N} \lambda_i \tag{33.1}$$

In this paper, the value calculated by normal average solution is the default value when the model is built. We put this solution here as a reference standard to check the other two solutions' performance.

• Weighted Average Solution

For the given N data, namely λ_i, $i \in [1, N]$, the weighted average solution can be expressed as

$$\bar{\lambda} = \sum_{i=1}^{N} f_i(t) * \lambda_i \tag{33.2}$$

Among which, $f_i(t)$ represents weighted value, and it is the function of the time. The advantage of the solution is that $f_i(t)$ and can be chose differently according to different requirements. The following are the optimizing steps:

Step1. We do some data pre-processing work before the analysis starts, aiming at wiping off the singular points in the data sequence.
Step2. This step is to determine the value of the weight. Here we give out the weight value by experience from the production line.
Step3. Calculate the value of every parameter and replace them into the model. Run the simulation model again to get the final dispatch.

• Kolmogorov–Smirnov Test Based Solution

In Statistics, the Kolmogorov–Smirnov (K–S) test is based on the empirical distribution function (ECDF) which is used to decide if a sample comes from a population with a specific distribution. The Glivenko theorem points out that the empirical distribution is the consistent estimate of the theoretical distribution, so

Kolmogorov dug deeper in this area and came out with a distribution goodness-of-fit Kolmogorov–Smirnov test based on this theorem. It is used to describe two independent statistical samples' similarities. Assume that:

$$X_1, \cdots, X_m \overset{iid}{\sim} F(x), \ Y_1, \cdots Y_m \overset{iid}{\sim} G(x) \tag{33.3}$$

And the whole sample is independent, $F(x)$ and $G(x)$ are continuous distribution function, the test issue that we are interested is that:

$$H_0 : \ F(x) \equiv G(x) \ \leftrightarrow H_1 : F(x) \neq G(x) \tag{33.4}$$

According to the Glivenko theorem, it is practicable to use empirical distribution function to approximate theoretical distribution function.

$$D = \max_{i,j}\{|F_m(X_{(j)}) - G_n(Y_{(j)})|\} \tag{33.5}$$

To verify the above assumption, $F(x)$ and $G(x)$ represent X simple and Y simple's empirical distribution respectively, $X_{(j)}$ and $Y_{(j)}$ represent X simple and Y simple's order statistics respectively

$$\text{Prob}(D) = Q_{ks}(\lambda) = 2\sum_{j=1}^{\infty}(-1)^{j-1}e^{-2j^2\lambda^2} \tag{33.6}$$

Among which,

$$\lambda = \left[\sqrt{N_e} + 0.12 + \frac{0.11}{\sqrt{N_e}}\right]D \tag{33.7}$$

$N_e = \frac{mn}{m+n}$, Apparently, if two independent sample are similar, when Statistics distance $D \rightarrow 0$, $p \rightarrow 1$, and vice versa. K–S test can be used to answer whether data are from a normal distribution, a log-normal distribution, an exponential distribution, a logistic distribution or a Weibull distribution.

The processing can be expressed as following steps:

Step1. We do some data pre-processing work as the weight average solution does for the same reason.

Step2. Normalization the data. Using the following formula

$$y = (x - \text{MinValue})/(\text{MaxValue} - \text{MinValue}) \tag{33.8}$$

Step3. Then we test whether the data sequence fit the standard normal distribution. If the data comes from a standard normal distribution, we treat arithmetic average value of the simple as representative value of the overall size of the features; else if the data does not come from a standard normal distribution, then test if the data comes from a standard log-normal distribution, if so, we treat geometrical average value of the simple as representative value of the overall size of the features; else if the data neither comes from a standard normal distribution nor a standard log-normal distribution, we treat median of the simple as representative value of the overall size of the features.

Step4. We replace the value obtained by former step to simulation model and run the model again to get the dispatch so that we can see the performance of the solution.

For all the pre-processing work this paper mentioned is done by Visual Studio 2008 together with SQL Server 2008. The realization is simple with the help of open source component offered by Microsoft. Also some of the analyzing work is done by Visual Studio together with Matlab, like K–S test verification, for Matlab has some very useful tools, we can just invoke the functions can use them. We design and realize the functions we need in a M file and then deploy it to a dynamic link library, so we can operate it in the Visual Studio to finish the batch work. The usage of the software provides quite a lot convenience. This solution has a significant means for the companies that want to improve their output with none or few inputs.

33.4 Actual Production Line Verification

33.4.1 Research Object Introduction

Shanghai advance Semiconductor Co., LTD has already built an integrated circuit simulation model to simulate the practical production line process. According to the simulation, the model will finally give out a dispatch which is planning to replace the artificial dispatching to guide the practical production line. We evaluate the precision of the model by calculating the coincidence between simulation dispatch and the practical production line's artificial dispatch. By analysing it we know that recipe's processing time has significant influence on precision of the model's dispatch, at the same time, the parameter of the recipe's processing time are given arithmetic average value directly when the simulation model being built, so the parameter failed to represent the general characteristics of reaction production line accurately, for the reasons above we chose processing time as the research object of this section.

33.4.2 Performance Index Introduction

The performance indexes that this paper used are proposed by the company according to actual production line's needs, aiming at evaluating the precision and the reliability of the model's come out by calculating the coincidence between simulation dispatch and the artificial dispatch at the same initial conditions and the same circumstance. The performance indexes can be measured by the proportion of the count which is gained by simulation in the count practical production line under the same simulation conditions. The performance indexes that this paper mentioned are labeled as precision.

33.4.3 *Implementation and Result Analysis*

The company provides the data needed for analyzing in format of CSV file which are called local files. Then the developer will load them into database, and the data from the database will provide all data the simulation model needs. The process of the optimizing can be described as following: firstly, we loaded the data which related with processing time to database, 710227 records in total. Secondly, we chose the repeated recipe and left 1873 different recipes' information and 589572 records in total, after that did some pre-processing of the data aiming at eliminating noise; In the end, we calculated the parameter processing time by normal average solution, weighted average solution and K–S test based solution, and updated the database, reload the optimized data to model and ran the simulation model again to gain the different dispatch for evaluating the performance. The develop platform that we used here is Visual Studio 2008; the programming language is C#; we also used Matlab to complete the data analysis part, and the database is SQL Server 2008.

The Figs. 33.1, 33.2 indicate the performance of three solutions. Figure 33.1 shows the precision of 5 in. production by the means of normal average solution, weighted average solution and K–S test based solution under the circumstance of

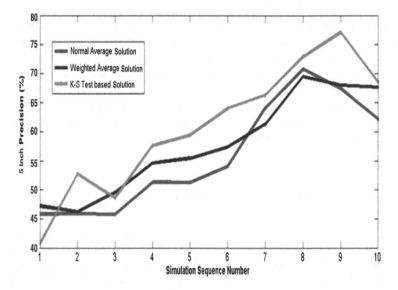

Fig. 33.1 5 in. precision figure

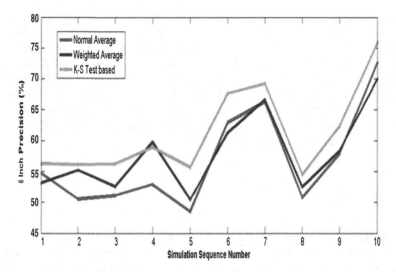

Fig. 33.2 6 in. precision figure

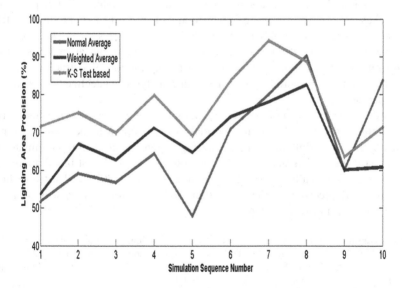

Fig. 33.3 Lighting area precision figure

10 groups of different initial conditions, while Fig. 33.2 shows the 6 in. production's precision under the same condition of the 5 in.'s. From the figures we can easily find out that weighted average solution and K–S test based solution have better performance than the original normal average solution, and K–S test based solution's performance is the best. Meanwhile the weighted solution's performance is unstable, but the entire precision still has a better trend than the normal average solution.

In this paper, we also took the precision of lighting machining area into consideration, because this area contains more than 35 % output in dispatch. Figure 33.3 shows under the above-mentioned initial condition, the parameter processing time using the different ways to calculate and the related precision performance. And the figure points out that the K–S test based solution has the best performance by a big step of 21.22 % and then the weighted average solution.

To sum up, the K–S test based solution and the weighted solution offer a better precision performance of the simulation model's output. And the K–S based solution has a more obvious improvement.

33.5 Conclusion

This paper proposed several data-based approaches for IC simulation model parameter optimizing. The solutions are based on parameter's feature, utilize abundant of historical data and simulation data, and handle data with different ways, namely normal average solution, weighted average solution and K–S test based solution. Finally we obtained better parameter values. We applied the approaches in Shanghai advance Semiconductor Co., LTD's production line simulation model, and it turned out that the approaches this paper proposed can improve the precision of the model. Another advantage of our approaches is that we operate the data offline, so we avoid modifying the model's structure. This is very significant for the company because offline analysis can decline the cost of the development.

However, the precision of the simulation model are increasing at the cost of the time. In this paper case, the original normal average approach needs about 10 min to handle the data, while the weighted average approach needs 5 more minutes to operate, still the performance is not so obvious. As for K–S test based approach, the performance is really obvious but at a time cost of 25–30 min. Therefore, how to improve the precision of the simulation model at a low time cost is what we should focus on in the next step.

Acknowledgments This paper is supported by China National Natural Science Foundation (61034004, 61273046), Science and Technology Commission of Shanghai (10DZ1120100, 11ZR1440400), Program for New Century Excellent Talents in University (NCET-07-0622) and Shanghai Leading Academic Discipline Project (B004).

References

1. Chenjiang X, Fei Q, Li L (2005) The research and implementation of a scheduling and simulation system for semiconductor production line. Syst Simul Technol 1(3):168–172
2. Qidi W, Fei Q, Li L, Ying W (2009) Data-based scheduling for complex manufacturing processes. Acta AutomationSinica 35(6):807–813 (in Chinese)

3. Tang LX, Huang L (2007) Optimal and near-optimal algorithms to rolling batch scheduling for seamless steel tube production. Int J Prod Econ 105(2):357–371
4. Yuxi S, Naiqi W (2011) Cycle time analysis for wafer revisiting process in scheduling of single-arm cluster tools. In: Proceedings of IEEE international conference on robotics and automation. Shanghai, China, pp 5499–5504
5. Qidi W, Fei Q, Li L, Zuntong W, Bin S (2011) A fuzzy petri net-based reasoning method for rescheduling. Transact Inst Meas Control 33:435–455
6. Yoon HJ, Shen WM (2008) A multiagent-based decision-making system for semiconductor wafer fabrication with hard temporal constraints. IEEE Trans Semicond Manuf 21(1):83–91
7. Liu M, Shao MW, Zhang WX, Wu C (2007) Reduction method for concept lattices based on rough set theory and its applications. Comput Math Appl 53(9):1390–1410
8. Weckman GR, Ganduri CV, Koonce DA (2008) A neural network job-shop scheduler. J Intell Manuf 19(2):191–201
9. Lee KK (2008) Fuzzy rule generation for adaptive scheduling in a dynamic manufacturing environment. Appl Soft Comput 8(4):1295–1304

Chapter 34
An Electrical Connector Locating Method Based on Computer Vision

Haozheng Yan and Yifei Feng

Abstract Electrical Connector plugging in–out operation by robot is critical content of spacecraft on-orbit self-maintenance. A new electrical connector locating method is proposed based on computer vision. This method recognizes an circular connector by itself combined shape characteristics. The binocular camera measures the center and the normal vector of connector combing with the recognition results. Finally, the camera measures the connector gap position and determines the position and attitude of the connector. Experiment proves the measure result is accurate and the method is effective.

Keywords Connector · Image recognition · Locating

34.1 Introduction

With the development of spacecraft, the demand for itself maintenance on-orbit is growing strongly. Electrical Connector plugging in–out operation by robot is involved during module replacement procedure. For the purpose, the computer recognizes and locates the connector of the module, then controls the manipulator completing the plugging in–out operation.

H. Yan (✉)
Image Processing Center of Beihang University, Beijing, China
e-mail: yhz541@163.com

Y. Feng
School of Mechanical Engineering, University of Science
and Technology Beijing, Beijing, China
e-mail: yifei.feng@hotmail.com

W. Lu et al. (eds.), *Proceedings of the 2012 International Conference on Information*   321
Technology and Software Engineering, Lecture Notes in Electrical Engineering 210,
DOI: 10.1007/978-3-642-34528-9_34, © Springer-Verlag Berlin Heidelberg 2013

ETS-VII satellite mission launched in Japan in 1997 includes the task of power connector plugging in–out using the on-orbit manipulator [1]. In this mission, a camera is mounted on the end effector of the manipulator. It uses two kinds of cooperative markers to measure the camera's position and attitude relative to markers. Because the target object's position and attitude relative to markers are known, the computer can solve the target object's position and attitude in the camera coordinate. Then the manipulator controller generates a motion path to capture the object. However, as long as the target object moves an unknown displacement, this vision system can't work properly.

A new locating method of connector is proposed. The method only uses the shape characteristics of the connector without markers. It is a kind of non-cooperative vision locating method.

34.2 The Connector Recognition

An connector with the characteristics of rectangle and circle is illustrated in Fig. 34.1. Obviously, the connector can be recognized by the combined shape characteristics.

The metal around the connector is highly reflective. The connector is mounted on the flatbed which reflects few light in order to recognize the connector conveniently.

34.2.1 Image Preprocessing

The captured raw image has obvious distortion, so the raw image need to be rectified. Then we use the Gaussian method to remove the noise to the rectified image. The preprocessing result of the Fig. 34.1 is illustrated in the Fig. 34.2.

Fig. 34.1 Raw image

Fig. 34.2 Preprocessing
image

34.2.2 Image Segmentation Based on Adaptive Binary

Adaptive binary method transforms a grayscale image to a binary image according
to the formulae:

$$dst(x, y) = \begin{cases} 255 & 38; if \; src(x, y) > T(x, y) \\ 0 & 38; otherwise \end{cases}$$

where $T(x, y)$ is the threshold. The threshold is defined as following:

$$T(x, y) = means(x, y) + C$$

where mean(x,y) is the average of pixel (x,y) neighbourhood, C is constant. The
result is illustrated in the Fig. 34.3.

34.2.3 Rough Image Recognition

Firstly, label every connected region in the binary image. Then, find the external
contour and all the corresponding internal contours of every labelled connected
region [2]. Then, determine whether it is a connector according to the shape
characteristics of the external and internal contours.

Calculate the area of the external contour and note it as area1. Then, calculate
the area of the minimum externally connected rectangle of the contour and note it
as area2. ρ represents rectangle degree:

Fig. 34.3 Adaptive binary

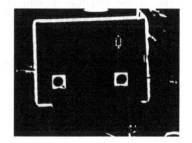

$$\rho = \frac{\text{area}1}{\text{area}2}$$

When ρ is larger than the given threshold value, the external contour is determined as a rectangle. Because the projection of a 3D circle is a ellipse in the camera's imaging plane, determine whether the internal contour is a ellipse. Get a fitted ellipse based on the internal contour with the least-squares method [3]. Note the error of the fitted ellipse as E and its definition is given as follow [4]:

$$E = \frac{1}{N \times B} \sum_{k=1}^{N} \left\| P_k - P_k^* \right\| \tag{34.1}$$

E	is the error of the fitted ellipse
N	is the number of points on the contour
P_k	is the kth point on the internal contour
P_k^*	is the point on the fitted ellipse contour and has the shortest distance to P_k
$\left\| P_k - P_k^* \right\|$	is the distance between P_k and P_k^*
B	is the semi-minor axis of the fitted ellipse

If E is smaller than given threshold, this internal contour is determined as an ellipse and its connected region is a connector.

34.2.4 Precise Image Recognition

The rough recognition of the connector is accomplished through analysis of the combined shaped characteristics. However the recognition result here is on the base of self-adaptive binaryzation which is unable to genera precise contour. In order to get a precise one, apply the region growing method to the problem [5]. First calculate the center of the contour which has been determined as an ellipse. Then choose this center as the growing point to carry on region growing and when the growth ends extract the external contours of the region. Fit the external contour into an ellipse using the least-squares method and get five parameters of this ellipse. These five parameters are the center- (x_c, y_c), the length of its semi-major axis—A, the length of its semi-minor axis—B as well as the vectorial angle from x-axis to the minor axis of the ellipse. With these parameters, the equation of the ellipse can be written as follow:

$$\frac{(dx \sin \alpha - dy \cos \alpha)^2}{A^2} + \frac{(dx \cos \alpha + dy \sin \alpha)^2}{B^2} = 1 \tag{34.2}$$

where $dx = x - x_c$; $dy = y - y_c$; (x, y) is the point on the ellipse.

34.3 The Connector Locating by the Binocular Camera

The position and attitude of the connector is determined by its center, normal vector and the rotation angle. In this article, we use the binocular camera to measure these information [6].

34.3.1 Calculate the Center and Normal Vector

The projection ellipse of the connector hae been detected using former method. Calculate the connector center in the camera coordinate by the ellipse center's image position in both left and right images and the known parameters of binocular camera.

In order to determine the attitude, calculate the normal vector of the connector end face. Find four points A,B,C,D in the connector circle adaptively and calculate these four points' positions in the camera coordinate. Note the normal vector as $\vec{n}$. It can be solved as: $\vec{n} = \overrightarrow{BC} \times \overrightarrow{AD}$.

The four points' solving process is illustrated in Fig. 34.4. Solve the highest and lowest points of the fitted ellipse in the left image. Note both points' ordinates relative to ellipse center as maxdy and mindy. Then solve the corresponding four points' image coordinates whose ordinates relative to ellipse center are mindy/2 and maxdy/2 respectively. When the distance between the camera and connector changes, the size of the fitted ellipse changes. So adaptively finding four points by this method can improve the accuracy of the normal vector.

The two cameras of binocular camera in this article is distributed horizontally. The same points' ordinates in the left and right images are equal. Substitute four points' ordinates into the right ellipse equation to solve the corresponding four points'(a', b', c', d') coordinates in the right image. The ellipse equation is given as formula 34.2.

Finally, solve the four points' (A,B,C,D) 3D position in the camera coordinate by the four pairs of corresponding points.

Fig. 34.4 Four pairs of points

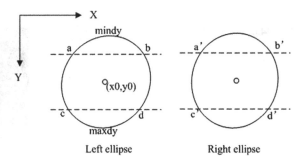

Left ellipse   Right ellipse

Fig. 34.5 Gap

34.3.2 Calculate the Rotation Angle

In this paper, the connector's center and normal vector is presented in the right camera coordinate. In order to determine the attitude of the connector, detect the connector's gap rotation angle around the normal vector. It is illustrated in the Fig. 34.5.

However, the gap is so small that only when the camera moves near it, the gap can be detected. Mount the camera on the end axis of the manipulator and control the right camera's optical axis and the normal vector to be coincident. Control the camera to move along its right optical axis until it is away from the connector about 30 cm.

Because the right optical axis coincide the normal vector, the connector's face is parallel to the image plane and the projection of the connector's internal circle is almost a circle. It is illustrated in the Fig. 34.5.

Transform the gray image into binary image. Extract the internal contour of the connector's region and fit a circle with the contour. Calculate the nearest distance of every point in the contour away from the fitted circle. We consider the point which has biggest distance away from the fitted circle as the gap position. The result is illustrated in the Fig. 34.6.

In the Fig. 34.6, the gap is pointed by the thin line. We define the angle from X-axis to the thin line as the rotation angle. In order to present connector's attitude conveniently, we define the connector coordinate as the origin is the connector center, the Z axis direction is same with normal vector, the X axis direction is from the center to the gap. So the right camera coordinate and the connector coordinate is linked by the rotation angle.

Fig. 34.6 Detect gap

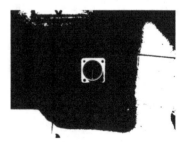

34.4 Experiment

34.4.1 The Locating Accuracy

Let binocular camera measure the connector's center and normal vector at the distance about 100, 70, 50 cm respectively. We control the camera to move along a square line at each distance. The square line is illustrated in Fig. 34.7. 9 labelled points are included in the square line. Let the camera measure the connector at each point.

In order to analyze the measure error, we paste a template of 3 X markers near the connector as in Fig. 34.8. Because the position between the template and the connector is known, we can get the connector's position and attitude indirectly by measuring the template's position and attitude. We define center measure error as the distance between the direct and indirect measure centers, the normal vector measure error as the angle between the direct and indirect measure vectors. The measure error is illustrated in Figs. 34.9 and 34.10.

We can measure the rotation angle of connector and calculate the rotation matrix from connector coordinate to the camera coordinate when we control the camera move along the normal vector. We note the rotation matrix as R1. We can also get the rotation matrix indirectly by measure the template and note it as R2. The R1 should be theoretically equal to R2. We define the rotation matrix error as E2:

$$E2 = \|R1 - R2\|$$

where $\|\ldots\|$ presents the matrix Frobenius norm.

We control the camera rotate along its right optical axis, then measure the rotation matrix R1 and R2. The error E2 is not larger then 0.12.

34.4.2 Analysis

Compare the locating error in the Figs. 34.9 and 34.10. It is obvious that the error in the distance about 1 m is bigger than that in the distance 0.7 and 0.5 m. When the camera is away from connector 1 m, the projection of the connector is very small, the contour is not accurate, so the locating error becomes bigger.

Fig. 34.7 Square line

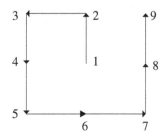

Fig. 34.8 Paste X markers

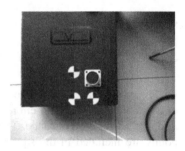

Fig. 34.9 Center error

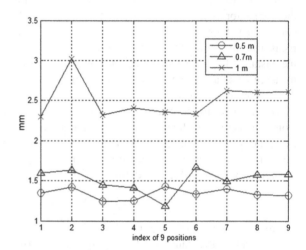

Fig. 34.10 Vector degree error

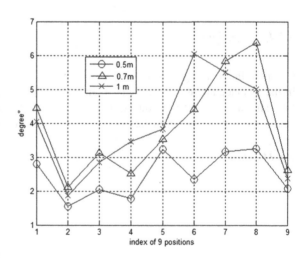

We assume that the fitted ellipse center is the projection of the connector circle center. In reality, they are same only when the image plane is parallel to the connector face. In general, they has a small difference which can cause measure error [7]. We will solve this error in the future research.

The measured normal vector always has a small offset degree to the true one due to noise. When the camera moves along the measured normal vector near the connector, the image plane is not absolutely parallel to the connector face, the rotation angle of the gap has a measure error. So we use the rotation matrix error which is caused by the rotation angle error and the normal vector error to evaluate the accuracy.

34.5 Conclusion

To plug in–out the connector by the manipulator automatically, we proposed a non-cooperative vision locating method. This method recognizes the connector only by its combined shape characteristics without markers. Then, calculate the center and normal vector with the recognition result. In order to solve the attitude of the connector, we detect the connector gap to measure the rotation angle. Experiment proves the method is an effective and high-precision locating solution.

Acknowledgments This research is supported by the National High Technology Research and Development Program of China (863 Program) with the reference number 2011AA7045040.

References

1. J Y, X C Spacecraft on-orbit service technology. China aerospace press, Beijing, pp 65–68 (in Chinese)
2. Gray B, Adrian K Learning OpenCV. O'Reilly, Massachusetts, pp 260–264
3. WAFRB (1995) A buyer's guide to conic fitting. In: British machine vision conference, Birmingham
4. Nguyen TM (2009) A real-time ellipse detection based on edge grouping. IEEE transactions on systems, man, and cybernetics. Texas
5. Gonzalez RC, Woods RE (2008) Digital image processing. Publishing House of Electronics Industry, Beijing, pp 613–615
6. Songde M, Zhengyou Z (2003) Computer vision. Science Press, Beijing, pp 72–75(in Chinese)
7. Guangjun Z (2005) Machine vision. Science Press, Beijing, pp 56–63 (in Chinese)

Chapter 35
Analyses of Security Issues with RFID Application Under U-Healthcare System

Jung Tae Kim

Abstract Medical industries have advanced as IT (Information technology) is fused with other technology such mobility devices and wireless communication. RFID (Radio frequency identification) offers a great potential in identifying and tracking patients, objects, and assets. The technology has been applied in hospital management and is valuable for quickly retrieving patient information and monitoring patient locations in the hospital. The use of a mobile device in healthcare system under wireless network environment gives an opportunity to offer better services and eliminates many manual operations in checking and processing for patients and staffs such as doctors and nurses. The applications of pervasive healthcare services for wireless and mobile networks are needed many requirements such as secure information exchange, reliable remote control, confidential data storage, effective mobility management, rapid emergency response, and continuous monitoring of a patient's medical records. Several vulnerabilities and security requirements related to mobile device should be considered to implement mobile services in the hospital environment. Secure authentication and protocol with mobile agent for applying ubiquitous sensor network under healthcare system surroundings is proposed and analysed in this paper.

Keywords Mobile agent · RFID · Privacy and healthcare system

J. T. Kim (✉)
Department of Electronic Engineering, Mokwon Uiversity, 800, Doan-dong,
Seo-Ku, Daejeon 302-729, Korea
e-mail: jtkim3050@mokwon.ac.kr

W. Lu et al. (eds.), *Proceedings of the 2012 International Conference on Information Technology and Software Engineering*, Lecture Notes in Electrical Engineering 210, DOI: 10.1007/978-3-642-34528-9_35, © Springer-Verlag Berlin Heidelberg 2013

35.1 Introduction

Today's society is rapidly moving toward to the elder people's society. The rapid changes of modern styles give a new chance to build new market and industry. Ubiquitous technologies based on mobile devices and sensor nodes are able to manage healthcare information. As information technologies are developed rapidly, E-healthcare system has been realized. Recently the trend of healthcare system has moved to U-healthcare system because of smart equipment and devices with low computing power. Radio frequency identification (RFID) is an automated data capture technology that uses low-power radio waves to communicate between readers and tags [1]. RFID technology is also adapted to U-healthcare system to reduce manual handling error, monitor patient's medical information, process efficiency and track patient's location. To apply this system, new issues are also induced such as security, authentication and safety [2]. Advanced technologies and existing medical technologies should be combined properly to meet requirements for service efficiency, accuracy and clinical significance. This fusion technology is called ubiquitous healthcare system, and it is highlighted that U-healthcare system makes better and rich quality of life as it improves constraints and medical treatments by joining a living space and a medical treatment together. Any vulnerability of security may cause a leakage of the personal information. Therefore, security protocols should be prepared prior to the implementation of U-health. The owner of personal information will also suffer from hackers and malicious attackers [3]. For this reason, the Department of Health and Human Services (HHS) in the United States has recently issued "an interim final rule regulating when and how patients must be notified if their healthcare information has been exposed in a security breach by hospitals, physician offices and other healthcare organizations". Security issues could be occurred with sharing information among interconnected hospitals. Secure access of electronic healthcare records (EHR) which may be distributed across healthcare units should be considered. Healthcare networks based on electronic or mobile devices are established by connecting general clinic center, hospitals and national/private medical centers. However, health information which is stored in a healthcare center is usually accessible only to authorized staff that of center in traditional healthcare systems [4]. For improvement of medical service quality in hospitals and enhancement of safety control for patients, integrating RFID technologies into medical industries has progressively become a trend on these days. To effectively incorporate RFID techniques with the existing HIS (Hospital Information Systems) is gradually developed [5].

The remainder of this paper organized as follows. Section 35.1 is the introduction. Section 35.2 provides related works of application of RFID for fusion technologies. Section 35.3 presents the proposed mechanism, reviews analysis of protocol and discusses the various security and privacy issues including the associated attack. Section 35.4 provides the conclusions.

35.2 Related Works

Information related with healthcare service is very confidential and sensitive in the real world. The healthcare systems are divided into several part system in hospitals. Therefore, divided security mechanism is required for mutual security. In general, we should consider wireless communication between user's tag and mobile agent and wire connection between mobile device and database. In general, firewall and traffic analysis can be used for the protection of sensible information from non-authorized attacks over Internet protocol. It can be occurred authentication and security problem in this situation [6]. Healthcare staffs increasingly require medical data delivered in real time to support their decision making process. The adoption of mobile devices allows this process to activate concurrently. Olla surveyed that healthcare applications based on mobile agent can be utilized in the future with fusion technology because of advanced technologies such as nanotechnology, device miniature, device convergence, high-speed mobile networks, and advanced medical sensors [7]. Wei Tounsi proposed secure communication for home healthcare system based on RFID sensor network especially to evaluate key exchange protocols with resource-constrained devices [8]. Lee proposed privacy management for medical service application using mobile phone collaborated with RFID Reader. This paper aims to provide privacy protection services by adopting a privacy protection system in the mobile RFID service network [9]. Jelena Misic introduced enforcing patient privacy in healthcare WSNs (Wireless sensor network) using ECC (Elliptic curve cryptosystem) Implemented on 802.15.4 beacon enabled clusters. He also analyzed ECC and hardware platform for healthcare wireless sensor networks. Elliptic curve cryptography for key distribution is used to decrease energy consumption compared to the better known RSA algorithm [10]. Lhotska et al. described security recommendations for implementation in distributed healthcare systems. In paper, he deals with privacy preserving and security in distributed platform for medical domains [11]. Chiu proposed the exchange of medical images through a MIEP (Medical Image Exchange Platform), especially replacing ad-hoc and manual exchange practices and also show how contemporary technologies of Web services and water-marking can help archive images with layered implementation architecture [12]. There have been many studies with regards to secure health informatics until now. One of them, "Open and Trusted Health Information Systems/Health Informatics Access Control (OTHIS/HIAC)" research was carried out by Liu et al. [13]. A viable solution to provide appropriate level of secure access control for the protection of private health data is proposed in this work. The mobile technology based on wireless communication gives an improvement of the overall quality of healthcare information delivery. But privacy and security matters will be incurred in spite of benefits of the technology. As security issues are of crucial importance for the healthcare technology, authentication and protocols for establishing secure healthcare systems should be accompanied and considered. Any breaching of confidentiality and integrity of the information may be induced privacy and vulnerability for both patients and staffs in hospital. A VPN (Virtual Private Network) is a private network application that utilizes the public Internet as a wide area network backbone and is used by SSL (Secure socket layer) protocol. This technology is called ubiquitous healthcare system and can communicate effectively and securely information in wireless and wire mediums between medical organizations, insurance companies, government agencies and users. However, this remarkable technology can be useless without consideration of security issues. If security and privacy problems occur, personal information and medical records will be exposed in danger and captured by unauthorized person [14].

35.3 Proposed Protocols

At the initial model of U-health system, M-health system is designed as an enhancement of E-health system supported by wireless EMR (Electronic Medical Record) access. Figure 35.1 depicts a concept of the network topology for the U-health system. It represents a brief network topology for a virtual hospital. This network will be modified and extended based on the requirement of the security issues and protocol requirements [15]. To overcome the additional vulnerabilities, wireless security architecture should be designed with essential requirement for wireless EMR access. Especially, patient privacy concerns are very important in HIS (Hospital information system) environment. The information must be secured permanently, either when transmitted or stored in databases.

At the end of network topology, mobile device and wireless access point are utilized to implement a visual interaction between the end device and the database servers through network. Authentication server is located in the middle of the data transmission for access control in authentication process. To identify security vulnerabilities and threats, security solutions and compact protocols design should be suggested to mitigate all kinds of risks as follows [16].

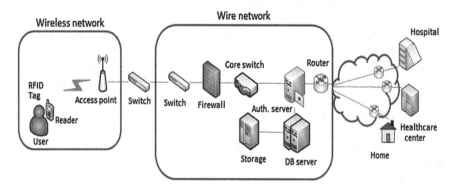

Fig. 35.1 Example of topology of ubiquitous healthcare system

- Establishment of security policy
- Analysis of a variety of security technologies
- Implementation of security primitives in the network
- Implementation of security features in the application

Table 35.1 shows structure of secure layer and its characteristics.

To mitigate aforementioned problems, we proposed a security mechanism that protects EMR and suitable to the ubiquitous medical service. The mechanism consists of numerous security measures such as authentication and cryptographic algorithm. According to structure of security level, the measures can be categorized into three security layers: authentication based on network, authentication based on application, and database protection. Figure 35.2 depicted authentication procedures of EAP-TLS in a wireless communication network.

The WEP (Wired Equivalent Privacy) protocol was designed to provide the same level of privacy as a wired network. Due to low complexity of security concerns over the WEP standard, many researchers continue to debate whether WEP alone is sufficient for HIPAA (Health Insurance Portability and Accountability Act) transmission security or not. Consequently, the healthcare delivery organization should use a combination of WEP and other security protocols for wireless networks. In order to avoid potential security breaches such as authentication and privacy protection, the existing Healthcare agent architecture should be considered some special security requirements. The HIPPA privacy and security regulations are two crucial provisions in the protection of healthcare privacy. Privacy regulations create a principle to assure that patients have more control over their health information and limits on the usage and disclosure of health information. The security regulations give a guideline the provision which is implemented to guard data integrity, confidentiality, and availability. Undoubtedly, the cryptographic mechanisms are well defined to provide suitable solutions. Wei-Bin proposed a cryptographic key management solution for HIPAA Privacy and security regulations [17]. Generally speaking, basic healthcare architecture is included four configurations and connected with proper protocols. When sensor nodes connect to the Internet from the front of the sensor nodes, there are the issues of network security. These are the problems that the construction of

Table 35.1 Structure of secure layer for service

Mobile device	Network	Database
RFID, NFC security issues	Two way authentication	Private data protection
Privacy, light weight protocol	Digital certification, IPSec., encryption engine	EMR data security
Limited hardware resources and memory	SSL channel	Authentication mechanism
Low memory and power consumption	Challenge response protocol	Encryption engine.

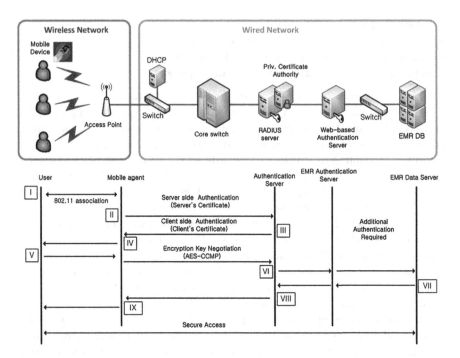

Fig. 35.2 Procedures of EAP-TLS with WPA2 in a wireless communication network

U-healthcare network must be considered [18]. To understand the authentication scenario based on networks, following user authentication process is presented in Fig. 35.2 and these steps are as follows.

Step I: When a user tries to access the mobile device, a pseudo random code is requested for verifying the user. After verification, the user must prove their identity for accessing mobile device.

Step II: In the mobile device, User identification (ID and PW) is separately encoded with hash function. In this mechanism, the hashed ID and PW are separately encrypted again. At this moment, the default key is used. Next, the encrypted user information is forwarded to the web-based authentication server. The encrypted user identification is encrypted one time more in the network layer before sending.

Step III: The web-based authentication server decrypts C(id) first and checks the session connection for the user. Sessions for each user can be distinguished by H(id). To understand the initial scenario for authentication, we assume that the session for the user is not created yet. After session checking fails, the server decrypts C(pw) according to the identification and authentication module. The server authenticates by matching the encrypted hash value with user information in the database of the server. The user's information in the database is also encoded by hash functions to match the ID and PW. If authentication is successfully processed, the server replies with a message R (authen_succ) "Authentication Success" with an encrypted random number for a new secret key. At the encryption process, the new secret key(n) is encrypted with the (k + 1). The value of the new secret key is always increased by random number from the original value. The mobile device, in turn, expects the key with the increased value. This process protects unauthorized attack.

Step IV: The mobile device can be authenticated and decrypts a new secret key by an increased value of k which forms (k + 1). The new secret key is stored instead of the used secret key.

Step V: When the user attempts to send a query to the database, the query and ID are separately encrypted. H(id) is needed for checking the session of this user. Then, the mobile device sends the encrypted H(id) and the encrypted query to the web-based authentication server.

Step VI: The web-based authentication server decrypts C(id) first and checks the session. On succeeding the session confirmation, the server automatically decrypts and forwards the query to the database. This feature reduces the workload on the communication channel. While the query is being sent, H(id) is concurrently transmitted. If the session is expired, the server replies the message (R(sess_exp)) "Session expired" to the mobile agent and requires a user identification again.

Step VII: The database firstly checks the requestor's security level by H(id). The requestor's role and security level must be appropriated to access the resources requested. If authentication is confirmed, database searches and replies to the request for sufficient response. The H(id) is sent with the response.

Step VIII: The server should determine who requested the response. It can be achieved by H(id) included in the response. After then, the server encrypts a response with new secret key (n + 1) and sends the status of the query R(q_succ), and the encrypted result of query.

Step IX: The mobile checks the status of query first. If the query is successfully processed, the status is R(q_succ). Finally, the mobile device decrypts the response by the expected secret key which has been increased (n + 1) and displays the result to the user. The authentication is successfully connected.

35.3.1 Authentication Based on Network

As a first step of security, well-organized authentication processes based on network should be prepared against various threats especially for wireless networks. In recent years, hospitals are also introduced wireless communication systems. However, not many hospitals are aware of the security issues because their working process is mainly focused on emergency than security. This may result in a security problem such as information leakage. Therefore, we will firstly review a suitable wireless security mechanism for the hospital. To begin with, the organizational characteristics of the hospital should be analyzed before selecting the wireless mechanism. With mobile devices in HIS, the staff will be able to access EMR systems to view or update patient's medical records. The staffs are usually permanent users rather than temporary users. It means that they are suitable to use a digital certificate rather than using an ID and Password, which are more vulnerable to be compromised for network access. To utilize digital certificate, the EAP (Extensible Authentication Protocol) can be selected and then it became the wireless user authentication protocol in IEEE 802.1x. EAP is also an encapsulation protocol compatible with various links and authentication methods [19]. Table 35.2 describes the EAP authentication methods. As shown in Table 35.2, there are three methods which offer mutual security levels. To select the proper method, compatibility with the hospital environment should be analyzed. EAP-MD5 should not be employed because it has a weak security feature whereas the rest of them provide the mutual security mechanism. To apply EAP-TLS for a wireless network, implementation of WLAN standard is required. Wi-Fi Protected Access2 (WPA2) is one of the standards in which EAP-TLS can be implemented. Also, WPA2 employs AES-CCMP to overcome vulnerabilities of other wireless communication standards [18, 20].

35.3.2 Authentication Based on Application

User must be authenticated by web-based authentication process for accessing to EMR despite user connects to the network through network authentication. The security guidance is republished by HIPAA in Unites State of America. Report advises to two-factor authentication. These are the usual authentications based on challenge-response handshake and session key agreement during the authentication process. Secure communication with a session key can communicate with confidentiality [21]. In the mechanism we proposed, data transmission between mobile device and web-based authentication server is implemented by SSL channel for security. SSL channel automatically generates and exchanges keys. In general, symmetric cryptography implemented by OpenSSL is chosen for cryptographic primitive because public key scheme decrypts at speeds more than the maximum 100 times slower than symmetric key scheme. Symmetric key ciphers have a problem which is termed as an incipient key distribution issue, but it is solvable. The solution of incipient key distribution issue is related to the process of new staff activation in HIS. By supplying a mobile device for new staff, administrator must set a randomly generated value for a secret key up to the mobile device count and the web-based authentication server. AES 128-bits version was chosen for encryption algorithm due to its decryption performance at speed. The 128-bit AES (Advanced Encryption Standard) performs better than the 192-bit and 256-bit version of DES or of AES [22] (Table 35.3).

Table 35.2 Structure of secure layer for service

Item	Client authentication of server	Server authentication of client	Dynamic WEP	DOS attack	Allocation of server resource	
					Before authentication protocol	Among authentication protocol
MD5	Password	No	Impossible	0	User's password	Non-conditional allocation if requested
EAP-TLS	Certificate	Certificate	Possible	0	None	Non-conditional allocation if requested
EAP-TTLS	Certificate	Password	Possible	0	User's password	Non-conditional allocation if requested

Table 35.3 Comparison of U-healthcare network and IPV6 [18]

Demand of U-healthcare network	Feature of IPv6
A large number of biomedical sensors	Large IP space
Sensor's IP automatically	Stateless auto-configuration
Real-time transmission	QoS functions
Sensors hand off and roaming	Mobile IPv6
Bio-medical information on internet	IPSec, AH, ESP
IPv6 and IPv4 translation	Dual stack, tunneling, translators

35.3.3 Benefits, Barriers and Attacks on RFID

Deploying RFID technology in the healthcare industry for promoting patient's data and records in hospital is a complex issue since it involves technological, economic, social, and managerial factors. Table 35.4 summarizes the major barriers, benefits and attacks from collected literature [23].

The Health Information System (HIS) based on the wireless network infrastructure is generally adopted nowadays. Immediate access to the Electronic Medical Record (EMR) database which is offered with the mobile device may reduce the processing time to access and update EMR database. With U-healthcare system, patient monitoring service in real time should be analyzed. Small sensors are attached on a patient's body to transmit biomedical signals to the remote hospitals. However, E-health system may have some limitation and issues to be solved for the patient's safety. Even though many works for applying U-healthcare system have been done for many years, security threats and issues with RFID technology are still unsolved. RFID has little memory resources and static 96-bit identifier in current ultra-cheap tags. As IC fabrication process is developed, hundreds of bits with tag can be used in the future. Little computational power is needed in tag and several thousand gates (mostly for basic functionality) can be designed in application until now. No real cryptographic functions are possible in current technology with limited sources. We should consider price pressure and restricted resources to implement cryptography engine [24].

Table 35.4 Benefits, barriers and attacks of RFID applications in healthcare system [24]

Increased safety or reduced medical errors	Interference	Denial of service
Real-time data access	Ineffectiveness	Physical attack
Time saving	Standardization	Tag cloning attack
Cost saving	Cost	Replay attacks, spoofing attack
Improved medical process	Privacy and legal issues	Side channel attack
Other benefits: improve resource utilization	Other barriers: lack of organizational support, security	Tag tracking

35.4 Conclusions

The use of the mobile device in the hospital environment offers an opportunity to deliver better services for patients and staffs. Mobile technology offers advantages to improve patient's data and reduce costs for processing time. We analyzed the performance of security related to issues by means of information security and privacy. Neither a symmetric nor an asymmetric cryptographic deployment is necessarily with lightweight algorithm in user's device. In future work, we will develop test bed with embedded RFID system in ubiquitous healthcare system to estimate performance and related problems.

Acknowledgments This research was supported by Basic Science Research Program through the National Research Foundation of Korea (NRF) funded by the Ministry of Education, Science and Technology (grant number: 2012-0007896)

References

1. Park Y-J et al (2012) On the accuracy of RFID tag estimation functions. J Inf Commun Convergence Eng 10(1):33–39
2. Xiao Y et al (2006) Security and privacy in RFID and applications in telemedicine. IEEE Communications Magazine, pp 64–72
3. Lim S (2010) Security issues on wireless body area network for remote healthcare monitoring. In: IEEE international conference on sensor networks, ubiquitous, and trustworthy computing, pp 327–332
4. Huang YM, Hsieh MY, Chao HC, Hung SH, Park JH (2009) Pervasive, secure access to a hierarchical sensor-based healthcare monitoring architecture in wireless heterogeneous networks. IEEE J Sel Areas Commun 27(4):400–408
5. Yao W, Chu C-H, Li Z (2010) The use of RFID in healthcare: benefits and barriers. In: IEEE international conference on RFID technology and applications, pp 128–1342
6. Kwock DY et al (2011) Secure authentication procedure with mobile device for ubiquitous health environments. In: International conference of Korea institute of maritime information and communication society (ICKIMICS2011), vol 4(1)
7. Olla P (2007) Mobile health technology of the future: creation of an M-health taxonomy based on proximity. Int J Healthc Technol Manag 370–387
8. Tounsi W et al (2010) Securing the communications of home health care systems based on RFID sensor networks. In: 8th annual communication networks and services research conference, pp 284–291
9. Lee B, Kim H (2007) Privacy management for medical service application using mobile phone collaborated with RFID reader. In: Third international IEEE conference on signal-image technologies and internet-based system. pp 1053–1057
10. Misic J (2010) Enforcing patient privacy in healthcare WSNs using ECC implemented on 802.15.4 beacon enabled clusters. In: Sixth annual IEEE international conference on pervasive computing and communications, pp 669–686
11. Lhotska L et al (2008) Security recommendations for implementation in distributed healthcare systems. ICCST 2008, pp 76–83
12. Chiu DKW et al (2007) Protecting the exchange of medical images in healthcare process integration with web services. In: Proceedings of the 40th Hawaii international conference on system sciences, pp 1–10

13. Liu V, Franco L, Caelli W, May L, Sahama T (2009) Open and trusted information system/ health informatics access control (OTHIS/HIAC), Australasian Society for Classical Studies (ASCS), pp 99–108

14. Kwon S-H, Park D-W (2012) Hacking and security of encrypted access points in wireless network. J Inf Commun Convergence Eng) 10(2):156–161

15. Su C-J, Chen B-J (2010) Ubiquitous community care using sensor network and mobile agent technology. In: Symposia and workshops on ubiquitous, autonomic and trusted computing, pp 99–104

16. Acharya D (2009) Security in pervasive health care networks: current R&D and future challenges. In: Eleventh international conference on mobile data management, pp 305–306

17. Lee W-B (2008) A cryptographic key management solution for HIPAA privacy/security regulations. IEEE Transaction on Information technology in Biomedicine 12(1):34–44

18. Hung C-C, Huang S-Y (2010) On the study of a ubiquitous healthcare network with security and QoS. IET :139–144

19. Savola RM (2009) Current and emerging security, trust, dependability and privacy challenges in mobile telecommunications. In: Second international conference on dependability, pp 7–12

20. Jo H-K, Lee H-J (2007) A relay-transmission of the RFID tag ID over the wireless and TCP/ IP with a security agent, KES-AMSTA 2007. LNAI 4496:918–927

21. Li Z, Shen H, Alsaify B (2008) Integrating RFID with wireless sensor networks for inhabitant, environment and health monitoring. In: 14th IEEE international conference on parallel and distributed systems, pp 639–646

22. Yao W, Chu C-H, Li Z (2010) The use of RFID in healthcare: benefits and barriers. In: IEEE international conference on RFID-technology and applications, pp 128–134

23. Chien H-Y (2008) Varying pseudonyms-based RFID authentication protocols with DOS attacks resistance. In: IEEE Asia-Pacific services computing conference, pp 507–615

24. Hsu C, Levermore DM, Carothers C, Babin G (2007) Enterprise collaboration: on-demand information exchange using enterprise databases, wireless sensor networks, and RFID systems. IEEE Trans Syst Man Cybern Part A: Syst Hum 37(4):519–532

Chapter 36
Performance of Low-Density Lattice Codes in COFDM Systems

Shuanlei Hu and Lianxiang Zhu

Abstract Orthogonal Frequency Division Multiplexing (OFDM) is suitable for high-bit-rate data transmission in multipath environments. Many error-correcting codes have been applied in COFDM to reduce the high bit error rate (BER). Recently, low-density lattice codes (LDLC) attracted much attention. LDLC was proposed by Sommer in 2007, and the performance is very close to Shannon limit with iterative decoding. We applied LDLC in COFDM system and the simulation results showed that the LDLC is effective to improve the BER of the system in multipath environments.

Keywords Low-density lattice codes · COFDM · Iterative decoding

36.1 Introduction

In the future, high quality communications are attractive to people, high-bit-rate transmission is required. Orthogonal Frequency Division Multiplexing technology divides the bandwidth into numbers of narrowband sub channels, in which the parallel signals are transmitted in a lower rate. In that way, it turns the high-bit-rate signals into low ones [1]. OFDM is a very suitable technique for high-bit-rate data transmission in multipath environments. In other words, the OFDM technique can reduce the number of the inter symbol interference (ISI) which emerges in the multipath environments, but the subcarriers may lose completely when the amplitudes become too small [2]. Hence, the overall BER turn out to be very high

S. Hu (✉) · L. Zhu
Chongqing University of Posts and Telecommunications, Chongqing, China
e-mail: hushuanlei@163.com

W. Lu et al. (eds.), *Proceedings of the 2012 International Conference on Information Technology and Software Engineering*, Lecture Notes in Electrical Engineering 210, DOI: 10.1007/978-3-642-34528-9_36, © Springer-Verlag Berlin Heidelberg 2013

even most of the subcarriers are transmitted without ISI. So error correcting is essential to avoid the high BER which was caused by the weak subcarriers. And many codes have been applied in the OFDM system, such as convolution codes, Turbo codes, R-S codes, LDPC [3], and so on.

Recently, LDLC has attracted much attention especially in the field of coding theory. LDLC was proposed by Sommer in 2007 and the performance is very close to Shannon limit with iterative decoding. LDLC is a kind of lattice which was defined by a sparse inverse generator matrix in which the nonzero elements only take a small part of the total number. Sommer proposed this lattice construction, described its iterative decoder which is similar with decoder in LDPC, and also gave extensive convergence analysis. The results showed that LDLC decoder has a linear complexity [4].

The paper is organized as follow, the basic definitions of LDLC are described in Sect. 36.2, the LDLC iterative decoding algorithm is described in Sect. 36.3, the structure of the LDLC-COFDM is given in Sect. 36.4 and Sect. 36.5 describes the simulation results.

36.2 Low Density Lattice Codes

36.2.1 Basic Definitions of LDLC

36.2.1.1 Lattice Codes and Parity Check Matrix for Lattice Codes

An n dimensional lattice in R^m is defined as the set of all linear combinations of a given basis of n linearly independent vectors in R^m with integer coefficients. The matrix G, whose columns are the basis vectors, is called a generator matrix of the lattice. A lattice code of dimension n is defined by a lattice G in R^m and a shaping region B, where the code words are all the lattice points that lie within the shaping region.

An n dimensional lattice code is defined by its $n \times n$ lattice generator matrix. Every codeword is formed by $x = G \cdot b$, where b is a vector of integers. Therefore, $G^{-1} \cdot x$ is a vector of integers for every codeword. We defined the parity check matrix for the lattice code by $H \triangleq G^{-1}$. And each check matrix has a unique factor graph that corresponding to [5, 6].

36.2.1.2 Low Density Lattice Codes

Definition 1 (*LDLC*): An n dimensional LDLC is an n-dimensional lattice code with a non-singular lattice generating matrix G satisfying $|\det(G)| = 1$, for which the parity check matrix H is sparse. The ith row degree r_i is defined as the number

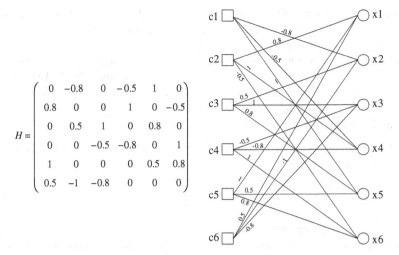

$$H = \begin{pmatrix} 0 & -0.8 & 0 & -0.5 & 1 & 0 \\ 0.8 & 0 & 0 & 1 & 0 & -0.5 \\ 0 & 0.5 & 1 & 0 & 0.8 & 0 \\ 0 & 0 & -0.5 & -0.8 & 0 & 1 \\ 1 & 0 & 0 & 0 & 0.5 & 0.8 \\ 0.5 & -1 & -0.8 & 0 & 0 & 0 \end{pmatrix}$$

Fig. 36.1 The check matrix and factor graph

of nonzero elements in ith row of H, and the ith column degree c_i is defined as the number of nonzero elements in ith column of H.

Definition 2 (*magic square LDLC*): An n dimensional regular LDLC with degree d is called magic square LDLC if every row and column of the parity check matrix H has the same d nonzero values except for a possible change of order and random signs.

36.2.1.3 Factor Graph of LDLC

The factor graph of an LDLC is defined similarly to LDPC: it is a graph with variable nodes at one side and check nodes at the other side. Each variable node corresponds to a single element of the codeword $x = G \cdot b$. Each check node corresponds to a check equation. An edge connects check node c_i and variable node x_j if and only if $H_{ij} \neq 0$.

For example, Fig. 36.1 shows a parity check matrix of a magic square LDLC with dimension n = 6, degree d = 3 and the nonzero values are [1.0, 0.8, 0.5].

36.2.2 Iterative Decoding Algorithm

Assume that the codeword $x = G \cdot b$ was transmitted, where b is a vector of integers. We observe the noisy codeword $y = x + n$, where n is a vector of i.i.d Gaussian noise samples with common variance σ^2, and we need to estimate the integer valued vector b. Our decoder will not estimate directly the integer vector b. Instead, it will estimate the probability density function of the codeword vector. And the generate sequence is $[h_1, h_2, \ldots h_{d-1}, h_d]$.

The iterative algorithm is most conveniently represented by using a message passing scheme over the bipartite graph of the code, similarly to LDPC. The messages sent by the check nodes are periodic extensions of PDFs. The messages sent by the variable nodes are single PDFs.

The decoding algorithm is as follows:

1. Step1—initialization: each variable node x_k sends message to all the check nodes that connect to it.

$$f_0^k(x) = \frac{1}{\sqrt{2\Pi\sigma^2}} e^{-\frac{(y_k - x)^2}{2\sigma^2}} \quad (36.1)$$

2. Step2—message update:
   Check node message update: each check node c_l sends to all the variable nodes that it connects to, and that updating includes three basic steps: convolution, stretching and periodic extension.

$$\tilde{P}_j(x) = f_1\left(\frac{x}{h_1}\right) \otimes \ldots \otimes f_{j-1}\left(\frac{x}{h_{j-1}}\right) \ldots \otimes f_{j+1}\left(\frac{x}{h_{j+1}}\right) \ldots \otimes f_d\left(\frac{x}{h_d}\right) \quad (36.2)$$

$$p_j(x) = \tilde{P}_j(-h_j x) \quad (36.3)$$

$$Q_j(x) = \sum_{i=-\infty}^{\infty} p_j\left(x + \frac{i}{h_j}\right) \quad (36.4)$$

Variable node message update: each variable node x_k sends to all the check nodes that it connects to, and that updating includes two basic steps:

$$\tilde{f}_j(x) = e^{-\frac{(y_k - x)^2}{2\sigma^2}} \prod_{l=1,\, l\neq j} Q_l(x) \quad (36.5)$$

$$f_j(x) = f_j(x) = \frac{\tilde{f}_j(x)}{\int_{-\infty}^{\infty} \tilde{f}_j(x)dx} \quad (36.6)$$

3. Step3—final decision: after iteration step as we explain before, it comes to the last two steps: product and normalization.

$$\tilde{f}_j^{final}(x) = e^{-\frac{(y_k - x)^2}{2\sigma^2}} \prod_{l=1} Q_L(x) \quad (36.7)$$

$$\hat{b}_j = \arg\max_x \tilde{f}_{final}^{(j)}(x) \quad (36.8)$$

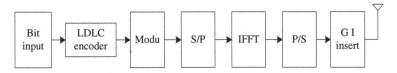

Fig. 36.2 Transmitter model

36.3 LDLC-COFDM System Model

Figures 36.2 and 36.3 show the model of the LDLC-COFDM system. At the transmitter, information bits are encoded at the LDLC encoder and modulated at the modulator to get waveform. After the serial-to-parallel conversion, the OFDM sub-channel modulation is implemented by using an inverse fast Fourier transform (IFFT) and assigned to some OFDM symbols for the purpose of compensating two dimensional errors in the OFDM system. On a frequency-selective fading channel the guard interval is inserted for the purpose of eliminating the ISI caused in the noisy channel.

At the receiver, the guard interval is removed after passing a frequency-selective fading channel. After the serial-to-parallel conversion, the OFDM sub channel demodulation is implemented by using a fast Fourier transform (FFT). The received OFDM symbols generated by the FFT are demodulated at the demodulator. The demodulated bits are decoded with each LDLC encoded block and data bits are restored [7].

36.4 Simulation Result

We present the results of our computer simulation. Table 36.1 shows the simulation parameters. The multipath condition used in our simulation is an additive white Gaussian noise (AWGN) channel. Note that we use a check matrix whose dimension and degree are (100, 3) and (1,000, 5). And we set the maximum number of iterations to 100.

Figure 36.4 shows the distribution of the average number of iterations at which the decoder stops at different Eb/N0 for the LDLC-COFDM system on an AWGN channel. We can see that as Eb/N0 become larger, the average number of iterations becomes smaller to convergence, but the rate of convergence becomes slower. We can also see that, as for LDLC with different dimensions, the number of iterations to get stop shows a lot difference. Generally speaking, LDLC with large dimension need less number of iterations to get convergence. When Eb/N0 is larger than 0.5, the maximum number of iterations is no more than 100 (so we choose the maximum number of iteration as 100 to make sure that it get the convergence results at last). Note that when the Eb/N0 equals 0.5, it takes about 95 iterations to get the convergence result, we may image that if Eb/N0 becomes

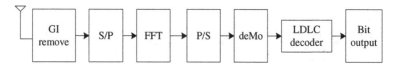

Fig. 36.3 Receiver model

Table 36.1 Simulation parameters

Code length	500 and 1,000
Number of subcarrier	64
Bandwidth	40 MHZ
Guard interval	0.25 μs
Delay interval	0.125 μs
Channel model	AWGN

Fig. 36.4 Average iterative number

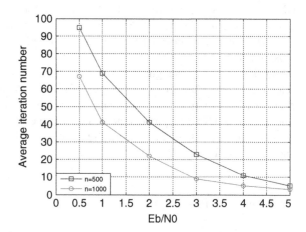

even smaller the result after 100 iterations may be not a convergence anymore. It also shows that when Eb/N0 is large than 3.0 the number of iteration is less than 10(n = 1,000), that is because the number of errors become small, it need not so many numbers of iterations any more.

Figure 36.5 shows the BER of the LDLC-COFDM on an AWGN channel for various numbers of iterations in the decoding algorithm. We can see that as the number of iterations increases, the BER of the LDLC-COFDM is improved. And the BER converges if the LDLC-COFDM converges at 100 iterations. This is because the errors that cannot be corrected with 100 iterations would not be corrected even if we increase the number of iterations more. As for LDLC with different dimensions, the large one need less number of iteration to converge compared with the small one.

Figure 36.6 shows the BER performances of the LDLC-COFDM in an AWGN channel compared with the BER performances of the LDPC-COFDM in the same AWGN channel. In order to have a clear comparison with the performance LDPC,

Fig. 36.5 BER at different
number of iteration

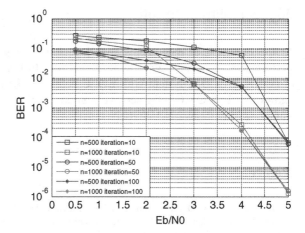

Fig. 36.6 System BER
compared with LDPC

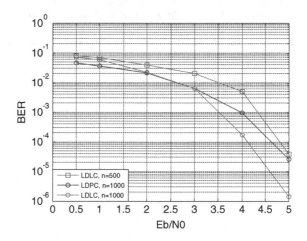

we simulate LDPC in the COFDM system in AWGN channels. We can see that
compared with LDPC, LDLC with n = 500 have a worse performance as
Eb/N0 < 5.0 and the maximum distance between the performance of LDPC and
LDLC is about 1.0 db, but as Eb/N0 increase to 5.0 their BER gap turn out to be
much more closer, about 0.2 db. As for LDLC with n = 1,000, the result turn out
the similar trend, and they have a very close BER as Eb/N0 is about 2.5 db, but
when Eb/N0 > 2.5 LDLC has a better performance and the distance trends to be
larger, we can preestimate that the gap between LDPC and LDLC can become
even more larger. It also can be seen that LDLC with large dimensions
(n = 1,000) has an obviously better performance than the small one (n = 500), so
we can image that if the dimension of LDLC become even larger, the performance
will be much better.

36.5 Simulation Result

LDLC is a kind of new error correcting code whit great potential; it has good performance, especially whit large dimension. In this paper, we propose a new structure of OFDM system with LDLC acting as the channel coding scheme. It shows that LDLC-COFDM achieve a good error rate performance with a small number of iterations on AWGN channel.

References

1. Proakis JG (2001) Digital communications. McGraw-Hill, New York
2. Weinstein SB, Ebert PM (1971) Data transmission by frequency-division multiplexing using the discrete Fourier transform. IEEE Trans Commun Technol 5:628–634
3. Gallager RG (1962) Low density parity check codes. IEEE Trans Inf Theory 8(1):21–28
4. Sommer N, Feder M, Shalvi O (2007) Low density lattice codes. IEEE Trans Inf Theory 54:1561–1585
5. Conway JH, Sloane NJA (1999) Sphere packings, lattices and groups. Springer, New York
6. Erez U, Zamir R (2004) Achieving 1/2 log (1+ SNR) on the AWGN channel with lattice encoding and decoding. IEEE Trans Inf Theory 50:2293–2314
7. Peng F (2004) Coding and equalization techniques for OFDM fading channels. Master's thesis. The University of Arizona, Tucson

Chapter 37
A Secure RFID Communication Protocol Based on Simplified DES

Zhao Wang, Wei Xin, Zhigang Xu and Zhong Chen

Abstract As a non-contact auto-identifying technology, RFID has played an important role in our daily life. However, the tradeoff between limited computing resource and high security requirement has become its bottleneck for wider use and application. Current mechanisms cannot give an effective solution to use limited computing resource to achieve a high goal of security. This paper proposes a novel secure protocol of communication between tags and readers. The protocol, which is based on EPC global UHF protocol and simplified DES, can lower the size of tag circuit and enhance the security of RFID reading and writing process.

Keywords RFID security · EPC · Improved DES

37.1 Introduction

Radio frequency identification (RFID) [1] is a kind of technology based on transmitting information by radio frequency signal to identify and control objects. It has been widely used in many aspects of industry and daily life, such as the identification of goods and certifications, supply chain management, library management and access system. A RFID system [2–4] is usually made up of tags, readers, and back-end database which is usually installed in a controller (PC) [5]. Controller and reader communicate with each other by a secure channel. The reader and the tag communicate with each other by radio frequency signal

Z. Wang (✉) · W. Xin · Z. Xu · Z. Chen
Key Laboratory of High Confidence Software Technologies (PKU), MoE, School of Electronics Engineering and Computer Science, Institute of Software, Peking University, No. 5 Yiheyuan Road, Haidian District, Beijing 100871, People's Republic of China
e-mail: wangzhao@infosec.pku.edu.cn

W. Lu et al. (eds.), *Proceedings of the 2012 International Conference on Information Technology and Software Engineering*, Lecture Notes in Electrical Engineering 210, DOI: 10.1007/978-3-642-34528-9_37, © Springer-Verlag Berlin Heidelberg 2013

(insecure channel). With the development of RFID and its wide application, the problem of privacy and security is getting more and more serious day by day. As an open wireless environment, all secret attackers can obtain the information stored in the tags, which makes the leakage of sensitive information such as money, drugs, and even locations of a user. In other words, current RFID system cannot effectively protect the data in tag. If the information stored in tags is stolen or modified, the loss that it brings may be immeasurable. Considering that the communication between tags and readers are open and the channel is insecure, therefore, there are many kinds of attacks with respect to them, such as forging fake RFID tags and readers, sniffing, physical attack, replay attack and so on. Therefore, when adopting RFID, the potential security threats need to be analyzed carefully and some security measures need to be taken, one of which is to design RFID secure communication protocol and it only allows the authenticated object to communicate with each other. This paper proposes a novel secure protocol of communication between tags and readers. The protocol, which is based on EPC global UHF protocol and improved DES, can lower the size of tag circuit and enhance the security of RFID reading and writing process.

To the best of our knowledge, there is not any study, raising a communication protocol similar with the protocol in this paper, on enhance the security of communication in RFID. The main contributions of this paper are as follows:

1. We propose a novel secure RFID communication protocol, and we are the first to introduce EPC global UHF protocol and DES algorithm into the design of RFID communication protocol.
2. We analyze the security and performance of the protocol in this paper and present its effectiveness and performance against some attacks.

Outline: The rest of this paper is organized as follows. We present the related work in Sects. 37.2 and 37.3 describes the design rationale of our protocol. Then, the evaluation is discussed in Sect. 37.4. Finally, we present conclusion and future work in Sect. 37.5.

37.2 Related Work

37.2.1 RFID System

A RFID system [2–4] is usually made up of tags, readers, and back-end database, as shown in Fig. 37.1. The reader is the device of identifying electronic tags. The power of tags is provided by the electromagnetic waves that the reader sends so that the tags can send their data back and get results stored in the database. The database is also in charge of the management of all the tags and readers according to the communication.

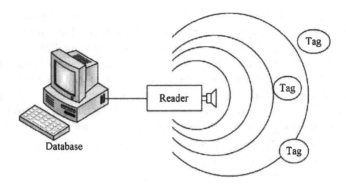

Fig. 37.1 Structure of RFID

The frequency of current tags is always 13.56 MHz; however, the tags working in ultra high frequency (UHF) 915 MHz have a longer transmission distance. According to the source of the power, the tags can be divided into two types: active tags and passive tags. Active tags have battery inside, while passive tags' power comes from readers. Therefore, 915 MHz passive tags have lower cost and longer life and distance, therefore have better application future.

37.2.2 EPC Global Protocol

Currently there is no unified RFID standard; most RFID systems adopt the standard of ISO 18,000 series [6] standard and UID standard in Japan. In EPC global Class-1 Generation-2 protocol [7], readers manage tags by three basic operations: Select, Inventory, and Access. Tags have seven states according to readers' operations: Ready, Arbitrate, Reply, Acknowledge, Open, Secured and Kill, as shown in Fig. 37.2.

When the tag is powered on, it is in the state of Ready [8]. When it receives the request of reader, it chooses a unique tag to access according to the anti-collision

Fig. 37.2 Communication between readers and tags in EPC global class-1 generation-2 protocol

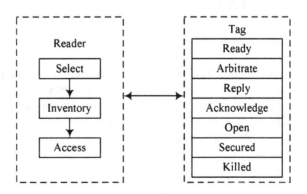

algorithm and steps into the state of Arbitrate. At that time if the reader sends a legal request again, the tag will return a random number (RN), and gets into the state of Reply. The reader will send a request which contains RN*. The tag will compare RN with RN*, if they are equal then the tag will transmit the information of PC and EPC that it stores to the reader in RFID system, and it will goes into the state of Arbitrate. The reader can continue to send request to tag so as to make it be in the state of open, and the reader can read and write the tag with the command of Read and Write. If the owner of the reader has access password, he can make the tag into the state of Secured, or he can make the tag into the state of Killed according to the command of kill.

37.3 RFID Security Protocol

37.3.1 Design Goal

According to RFID system structure and protocol analysis, the design goals of the communication protocol in this paper are as follows:

1. Adopt the authentication mechanism between tags and readers to prevent illegal readers from obtaining tag information or tampering tag data;
2. Avoid the plain text transmission during the communication process, because of low cost and power and limited resource of RFID tags, proper encryption algorithm needs to be considered to achieve the above goals;
3. Apart from the communication between readers and tags, the back-end database management has also played an important role in the security of RFID system. Attacks with respect to this part will cause the leakage of data and in-formation related to large quantities of tags and secret keys, which may lead to immeasurable loss. Therefore, the database management should be enhanced.

37.3.2 Protocol Design

37.3.2.1 Data Format

In the communication between readers and tags, our RFID communication protocol adopts the mutual authentication mechanism. The authentication from readers to tags can prevent illegal readers from attacking tags to obtain information and can prevent unauthorized readers from tracking tags. The cipher in tags and digital signature technology can prevent secret information in tags from being obtained illegally. The data formats in back-end database and tags are shown in Fig. 37.3.

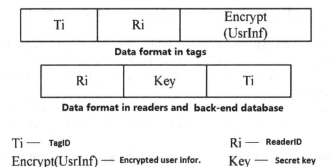

Data format in tags

Ri	Key	Ti

Data format in readers and back-end database

Ti — **TagID**   Ri — **ReaderID**

Encrypt(UsrInf) — **Encrypted user infor.**   Key — **Secret key**

Fig. 37.3 Data format in RFID system

37.3.2.2 Authentication Process

First the tag does authentication to the reader, the process is as follows (shown in Fig. 37.4):

1. The reader sends a request;
2. When the tag receives the request signal, it sends a random number back and does encryption Encrypt (RN, Ri) by its storing legal readers ID(Ri) and the random number;
3. When the reader receives the random number, it does the same encryption as in the tag to get Encrypt*(RN, Ri) based on the reader's ID, and sends the encryption result to the tag. The tag will compare it with Encrypt (RN, Ri) obtained in step 2, if they are the same then the reader is legal and the tag goes to the next operation, otherwise the tag goes to the initial state.

The process of authentication from the reader to the tag is as follows:

1. After the reader passing the authentication, the tag sends request to the reader;

Fig. 37.4 The mutual authentication process between tags and readers

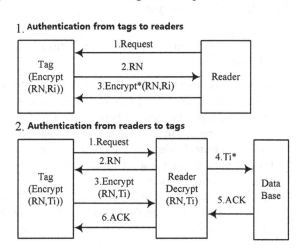

2. When the reader receives the request signal, it sends a random number RN back to the tag;
3. The tag does encryption by its receiving random number and its ID(Ti), and sends the result Encrypt (RN, Ti) back to the reader;
4. The reader decrypts the cipher with RN and gets Ti*, and transmits it to the back-end database;
5. The back-end database checks whether the tag's ID exists quickly. If the ID exists, then the tag is a legal tag and the database returns an acknowledgement (ACK);
6. After the reader's regarding the tag as a legal tag, the reader can access, read or write the tag. The whole authentication process has finished.

37.3.3 Security Enhancement by Simplified DES

During the above authentication process, Encrypt is a encrypted function which can be implemented by several good encryption algorithm such as ECC, RSA and AES. However, these algorithms have high costs and don't fit in the design of low-cost RFID tag circuit. The design goal of DES is to be implemented easily. It has high processing speed and good security. The core of DES is a ring function. The plain text firstly is permutated, and then is operated by 16 round ring functions. The last step is an inverse permutation to get the final result. During each round, the keys are different.

In traditional DES, S box has taken large area and high cost. Therefore, this paper improves S box and uses single S box to replace 8 traditional ones. The extended 48b is divided into 8 blocks and uses a multiple selector to reduce the circuit scale. Lower size will cause the reducing of cost.

37.4 Security Analysis

The protocol proposed in the above part of this paper can prevent several attacks to enhance the security.

1. Prevent eavesdropper from eavesdropping. Normally the communication between database and readers are based on wires which can be regarded as secure channel. The channel between tags and readers may become the object of eavesdropping. In the above design, there is no plain text transmission in unsecure channel. Therefore, the eavesdropper cannot get useful information.
2. Prevent unauthorized readers from reading or writing tags, and prevent forged tags information from being read.
3. Prevent attackers from making replay attacks. Because of the random number that has been sent during the beginning of sending requests, even if the attacker can get the data of transmission, he cannot launch effective attacks.

Use DES based encryption. The attacker cannot find useful information from transmitted cipher. The disadvantage of this protocol is that the length of secret keys is a little small.

37.5 Conclusion

As a non-contact auto-identifying technology, RFID has played an important role in our daily life. However, the tradeoff between limited computing resource and high security requirement has become its bottleneck for wider use and application. Current mechanisms cannot give an effective solution to use limited computing resource to achieve a high goal of security. This paper proposes a novel secure protocol of communication between tags and readers. The protocol, which is based on EPC global UHF protocol and improved DES, can lower the size of tag circuit and enhance the security of RFID reading and writing process.

Acknowledgments Zhao Wang is the corresponding author of this paper. We would like to thank the anonymous reviewers for their helpful suggestions. This work was supported in part by the NSFC under grant No. 61170263 and PKU-PY2011-007.

References

1. Mao G, Fidan B, Anderson BDO (2007) Wireless sensor network localization techniques. Comput Netw 51(10):2529–2553
2. Kim YH, Jo HK (2009) Patient information display system in hospital using RFID. cIn: Proceedings of the 2009 international conference on hybrid information technology, ser. ICHIT'09, New York pp 397–400
3. Miller MJ, Ferrin DM, Flynn T, Ashby M, White KP Jr, Mauer MG (2006) Using RFID technologies to capture simulation data in a hospital emergency department. In: Proceedings of the 38th conference on winter simulation, ser. WSC'06, Winter simulation conference pp 1365–1370
4. Wu F, Kuo F, Liu L-W (2005) The application of RFID on drug safety of inpatient nursing healthcare. In: Proceedings of the 7th international conference on electronic commerce, ser. ICEC'05, New York pp 85–92
5. Bailey DV, Boneh D, Goh E-J, Juels A (2007) Covert channels in privacy-preserving identification systems. In: Proceedings of the 14th ACM conference on computer and communications security, ser. CCS'07, New York pp 297–306
6. Lee J EL-Khatib K (2009) A privacy-enabled architecture for an RFID based location monitoring system. In: Proceedings of the 7th ACM international symposium on mobility management and wireless access, ser. MobiWAC'09, New York pp 128–131
7. Saad MK, Ahamed SV (2007) Vulnerabilities of RFID systems in infant abduction protection and patient wander prevention. SIGCSE Bull 39:160–165
8. Sumi M, Soujeri EA, Rajan R, Harikrishnan I (2009) Design of a Zigbee-based RFID network for industry applications. In: Proceedings of the 2nd international conference on security of information and networks, ser. SIN'09, New York pp 111–116

Chapter 38
Novel Discrimination Algorithm for Deceptive Jamming in Polarimetric Radar

Can Huang, Zhuming Chen and Rui Duan

Abstract A discrimination algorithm for extracting targets in the deceptive jamming environment is proposed in this paper. While the Sinclair Scattering Matrix (SSM) of radar target correlates with the operating frequency, the SSM modulated on the decoy in the jammer is conventionally stored in memory and is independent with frequency. Based on the full polarization receiving and transmitting radar system, this frequency-dependent polarization character of potential radar target is employed in the paper to discriminate expected targets from false ones. Simulation results indicate the validity of the discrimination algorithm.

Keywords Polarization · Discrimination · Sinclair scattering matrix · Frequency diversity

38.1 Introduction

Electronic warfare is a war without smoke of gunpowder which happens all the time, no matter in pre-war time, in war time or in peace time. All kinds of ECM which pose a great threat to radar to work properly are put into use. Because of this, means to permit radar to operate successfully in an ECM environment, which is known as ECCM, is essential.

The most fundamental way to against digital deceptive jamming is to find out the difference between decoy signals and the radar target echo signal, and then enhance the target echo or suppress the decoy signals. Polarization, together with

C. Huang (✉) · Z. Chen · R. Duan
School of Electronic Engineering, UESTC, Chengdu 611731 Sichuan, China
e-mail: 1905860726@qq.com

W. Lu et al. (eds.), *Proceedings of the 2012 International Conference on Information Technology and Software Engineering*, Lecture Notes in Electrical Engineering 210, DOI: 10.1007/978-3-642-34528-9_38, © Springer-Verlag Berlin Heidelberg 2013

the time, frequency, spatial, and bearing descriptors of radar signals, completes the information which can be obtained from target returns [1]. The difference between deception jamming signals with the target echo signal may appear in one of domains mentioned above or some of the domains. With the development of DRFM and DDS, it's difficult to distinguish target echo signal from deception jamming signal in time domain, frequency domain and spatial domain. However, the difference in polarization domain is obvious. Some discrimination algorithm for single polarization deceptive jamming is already exist [2, 3]. However, the discrimination algorithm for full polarization deceptive jamming is rare.

The purpose of the paper is to propose a new method for anti jamming with polarimetric processing techniques for multiband waveform target returns. In the paper, the model of digital deceptive jamming and the deceit effect are analyzed firstly; then the multiband waveform signal model and signal processing method is given; the result is proved by simulation finally.

38.2 Full Polarization Deceptive Jamming Model

The polarization information of radar target can be expressed by Sinclair Scattering Matrix (SSM) [4]. It is a complex matrix determined by space position, the carrier frequency of radar signal, target's shape, size, material, etc. The traditional SSM is a 2-dimension complex square matrix. The SSM of real target changes with the carrier frequency of radar signal [5]. So the SSM is expressed as:

$$S(\omega) = \begin{bmatrix} S_{HH}(\omega) & S_{HV}(\omega) \\ S_{VH}(\omega) & S_{VV}(\omega) \end{bmatrix} \tag{38.1}$$

where ω is the carrier frequency of radar signal. From the physical sense, S_{HV} is the horizon-polarization component of back scattering which is produced by verticality-polarization radar transmit wave. The other three parameters have the similar meaning.

The traditional jamming is single polarization jamming. The SSM of single polarization jamming is singular. But the SSM of real target is a nonsingular matrix which is satisfies reciprocity. Beside the difference on SSM, the cross-polarization component are different between real target and single polarization jamming [2, 3]. So it is relatively easy to identify single polarization jamming with polarization information.

Full polarization deceptive jamming (FPDJ) is more effective to deceive radar relative to single polarization jamming. With the help of full polarization process technology, the SSM of decoy produced by FPDJ satisfies reciprocity and non-singularity. So the traditional discrimination algorithm can't distinguish FPDJ form radar target. The principium of FPDJ is analyzed as follows.

The FPDJ receive signals by two-orthogonal-channel polarization antenna. FPDJ measure the amplitude ratio and phase difference of the two signals which is received by the antenna to gain the polarization information of the radar transmit signal. After front end processing of radar signal, choose the signal whose signal and noise ratio is large and send it to digital signal processor. In order to get a deceptive jamming which is similar to real target echo in time domain, frequency domain and doppler domain, baseband signal is delay, amplitude and phase modulated by digital signal processor. The baseband signal is up convert and divided into two-way by power divider. According to polarization scattering characteristics of deceptive target and the measured polarization of the radar signal, the polarization state of transmit signal is determined. Then, the two-way signals are amplitude and phase modulated to finish the polarization domain modulation. Finally, the two-way signals pass through a power amplifier and transmit by two-orthogonal-channel polarization antenna.

Assume the polarization vector of the signal $S_r(t)$ received by FPDJ is h_{rm}. The Sinclair scattering matrix of simulated target is S_J. Then the polarization vector of the signal transmitted by FPDJ is:

$$h_{tm} = S_J \cdot h_{rm} \tag{38.2}$$

The signal $S_r(t)$ received by FPDJ change to $S_t(t)$ after it is modulated by digital signal processor. Then the waveform of the signal transmitted by FPDJ can be expressed as:

$$E_t(t) = S_t(t)h_{tm} \tag{38.3}$$

The waveform of the signal transmitted by FPDJ is quite similar to real target echo. So the traditional discrimination algorithm can't be used to distinguish FPDJ form radar target. Novel effective discrimination algorithm is needed.

38.3 Full Polarization Discrimination Algorithm

FPDJ can produce false targets deceptive jamming which is quite similar to real target while the signal transmitted by radar is narrow-band signal. When the signal transmitted by radar is multiband signal, the false target deceptive jamming produced by FPDJ has a little difference with real target. Because the SSM of false targets produce by FPDJ is constant while the SSM of real target is changed with the carrier frequency of radar signal. The greater the frequency interval between different carrier frequencies is, the changes of the scattering matrix of real target are more obvious. Detail discrimination algorithm is analyzed as follows.

The transmitted signal is multiband signal and the signal on different carrier frequency process individually. Filter is applied to separate the signal on different carrier frequency. Simultaneous scattering matrix measurement is applied to get the SSM of target [6, 7]. Firstly, polarization vector tensor product receiving mode is used to receive signals. Two-orthogonal-channel polarization antenna is used to receive signals and calculate the polarization vector h_s of the received signal. Then tensor product process method is used to process h_s with receiving polarization h_r. The mathematical process is expressed as:

$$R = h_s \otimes h_r^T \tag{38.4}$$

where $\otimes$ is the tensor product process method. The 2×2 dimension matrix R is called polarization vector tensor product matrix.

Assume the SSM of radar target is $S(\omega)$. Several carrier frequencies whose frequency interval is large enough were chosen. Radar transmits signals with $2M$ different polarization states at each carrier frequency. The polarization vector of radar transmits signals is $h_{tm} = [\cos \theta_m, \sin \theta_m \cdot e^{j\varphi_m}]^T$, $(m = 1, 2, \ldots, 2M)$. The polarization vector of receiving polarization is $h_{rm} = [\cos \gamma_m, \sin \gamma_m \cdot e^{j\phi_m}]^T$, $(m = 1, 2, \ldots, 2M)$. Polarization vector tensor product receiving mode is used to receive target echo. For one of carrier frequencies, the polarization vector tensor product matrix measured is:

$$
\begin{aligned}
\hat{R}_m &= \left(E_{Sm} + \begin{bmatrix} nH_m \\ nV_m \end{bmatrix} \right) \cdot h_{rm}^T = \left(AS(\omega)h_{tm} + \begin{bmatrix} nH_m \\ nV_m \end{bmatrix} \right) \cdot h_{rm}^T \\
&= AS(\omega) \begin{bmatrix} \cos \theta_m \cos \gamma_m & \cos \theta_m \sin \gamma_m e^{j\phi_m} \\ e^{j\varphi_m} \sin \theta_m \cos \gamma_m & e^{j\varphi_m} \sin \theta_m \sin \gamma_m e^{j\phi_m} \end{bmatrix} + \begin{bmatrix} nH_m \\ nV_m \end{bmatrix} \cdot h_{rm}^T
\end{aligned} \tag{38.5}
$$

where $E_{Sm} = AS(\omega)h_{tm}$ is the polarization vector of target echo. A is the magnitude of signal. $[nH_m \quad nV_m]^T$ are the vectors of observation noise. The vectors of observation noise are independent of each other and followed $N(0, \sigma^2 I_{2\times2})$ Gaussian distribution. σ^2 is the variance of noise.

If the polarization vector of receiving polarization h_{rm} is matched with the polarization vector of radar transmits signal h_{tm}. Then $h_{rm} = h_{tm}^*$. So Eq. (38.5) changes as follow:

$$
\begin{aligned}
\hat{R}_m &= AS(\omega)h_{tm}h_{tm}^T + \begin{bmatrix} nH_m \\ nV_m \end{bmatrix} \cdot h_{tm}^T \\
&= AS(\omega) \begin{bmatrix} \cos^2 \theta_m & \cos \theta_m \sin \theta_m e^{-j\varphi_m} \\ e^{j\varphi_m} \sin \theta_m \cos \theta_m & \sin^2 \theta_m e \end{bmatrix} + \begin{bmatrix} nH_m \\ nV_m \end{bmatrix} \cdot h_{tm}^T
\end{aligned} \tag{38.6}
$$

The $2M$ different polarization states h_{tm} consist of M pairs of orthogonally polarization states: $h_{tA}^{(n)}$ and $h_{tB}^{(n)}$ $(n = 1, 2, \ldots, 2M)$

So $\left(h_{tA}^{(n)} \right)^H h_{tB}^{(n)} = 0$ and $h_{tA}^{(n)} \left(h_{tA}^{(n)} \right)^H + h_{tB}^{(n)} \left(h_{tB}^{(n)} \right)^H = I_{2\times2}$

Then the polarization vector tensor product matrix extracted from the echo of orthogonally polarization state transmit signals are added together, i.e.,

$$\hat{R}_S^{(n)} = \hat{R}_n + \hat{R}_{n+m} = AS + \begin{bmatrix} nH_{An} \\ nV_{An} \end{bmatrix} \cdot \left(h_{tA}^{(n)} \right)^H + \begin{bmatrix} nH_{Bn} \\ nV_{Bn} \end{bmatrix} \cdot \left(h_{tB}^{(n)} \right)^H \tag{38.7}$$

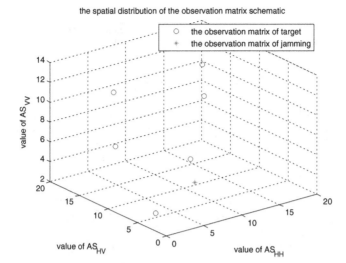

Fig. 38.1 The spatial distribution of the observation matrix schematic in the ideal situation

$\begin{bmatrix} nH_{An} & nV_{An} \end{bmatrix}^T$ and $\begin{bmatrix} nH_{Bn} & nV_{Bn} \end{bmatrix}^T$ are independent of each other and follow $N(0, \sigma^2 I_{2\times2})$ Gaussian distribution. According to Eq. (38.7), observation matrix $\hat{R}_S^{(n)}$ is the unbiased estimation of the SSM of radar target.

For a multiband transmit signal, the observation matrix $\hat{R}_S^{(n)}$ of FPDJ is converged on one point in the four-dimensional complex space while the observation matrix $\hat{R}_S^{(n)}$ of real target is irregular distributed in the four-dimensional complex space. In this way, real target echo can be distinguished from FPDJ.

38.4 Simulations and Results

In order to testify the discrimination algorithm, simulations are performed. Six groups of transmit signals with different carrier frequency is adopted in simulations. The carrier frequencies of the signals are 1, 2, 3, 4, 5 and 6 GHz. Four pairs of orthogonally polarization state of transmit signals is used in every carrier frequency in order to obtain the observation matrix.

The simulated SSM of FPDJ is set as [0.7 0.3; 0.3 0.7]. For convenience, the SSM of real target is set as six groups of random matrix and the magnitude of signal is set as eight. In order to display the result clearer, three-dimensional space is used to replace the four-dimensional complex space. The values of three (of four) numbers in the observation matrix make up three-dimensional space.

The ideal situation is considered first. Assume the observation noise does not exist (Fig. 38.1).

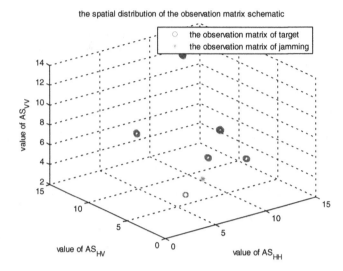

Fig. 38.2 The spatial distribution of the observation matrix schematic when the SNR is 40 dB

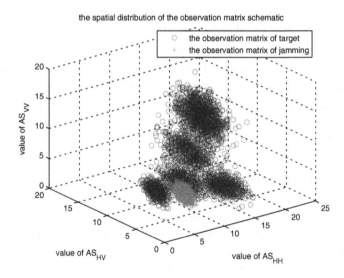

Fig. 38.3 The spatial distribution of the observation matrix schematic when the SNR is 10 dB

The observation matrix of target is irregular distribute and the observation matrix of jamming is converged on one point in the figure. Thus, decoy can be easily distinguished from the target.

Then, the situation that the observation noise exists should be considered. Assume the signal-to-noise ratio is 40 and 10 dB.

Figures 38.2 and 38.3 illustrates the effect of the observation noise. The observation noise can affect the result of discrimination algorithm. The result is worse when the signal-to-noise ratio is low.

38.5 Conclusions

For full polarization deceptive jamming, a new discrimination algorithm is proposed in this paper. Computer simulation is used to verify the validity of the discrimination algorithm. The theory of FPDJ is analyzed. According to the difference of the SSM of real target and the SSM of the simulated target produced by FPDJ, the FPDJ can be distinguished from target echo. The performance of discrimination algorithm is determined by the signal-to-noise ratio of the receiving signal. So, the way to improve the signal-to-noise ratio is the next step of work.

Acknowledgments This work was supported by the National Defense Advanced Research Foundation of China (9140A07030211DZ0211).

References

1. Giuli D (1986) Polarization diversity in radars. Proc IEEE 74(2):245–269
2. Neri F (2001) Anti-monopulse jamming techniques. Microw Optoelectron Conf 2:45–50
3. Dai H, Wang X, Li Y (2011) Novel discrimination method of digital deceptive jamming in mono-pulse radar. J Syst Eng Electron 22(6):910–916
4. Poelman AJ (1981) Virtual polarization adaptation: a method of increasing the detection capabilities of a radar system through polarization-vector processing. IEEE Proc Radar Signal Process 128(5):261–270
5. Zhang Z, Li Y et al (2005) Statistical characteristics and processing of instantaneous polarization. National Defence Industry Press, Beijing (in Chinese)
6. Giuli D, Facheris L, Fossi M (1990) Simultaneous scattering matrix measurement through signal coding. In: Proceedings of IEEE 1990 international radar conference. Arlington, VA, pp 258–262
Giuli D, Fossi M (1993) Radar target scattering matrix measurement through orthogonal signals. IEE Proc-F 233–24

Chapter 39
ZigBee Improving Multicast Routing Algorithm

Leqiang Bai, Jialin Wang and Shihong Zhang

Abstract In order to improve ZigBee network forward efficiency, this paper presents ZigBee improving multicast routing algorithm (ZIMR). On the basis of Z-Stack multicast algorithm, ZIMR algorithm combines with the ZigBee distributed addressing assignment scheme, optimizing the multicast forwarding node selection and making a node send the information to more destination nodes. Simulation results prove that under the same circumstances the existing algorithm for multicast forwarding consumes more forwarding nodes, wastes the energy of nodes and affects the network performance; ZIMR algorithm reduces the number of the forwarding nodes, saves the energy of the whole network and improves the stability of the network. ZIMR algorithm is superior to the existing ZigBee multicast algorithm.

Keywords ZigBee · Multicast · Routing · Forwarding node

39.1 Introduction

ZigBee is a personal area network protocol based on the IEEE 802.15.4 standard [1]. The standard was approved in 2003 and defined the two lower layers: the physical (PHY) layer and the medium access control (MAC) layer. The ZigBee Alliance provides the network (NWK) layer and the framework for the application layer. The application layer framework is comprised of the application support sub-layer

L. Bai (✉) · J. Wang · S. Zhang
Information and Control Engineering Faculty, Shenyang Jianzhu University, Shenyang 110168, China
e-mail: baileqiang@sjzu.edu.cn

W. Lu et al. (eds.), *Proceedings of the 2012 International Conference on Information Technology and Software Engineering*, Lecture Notes in Electrical Engineering 210, DOI: 10.1007/978-3-642-34528-9_39, © Springer-Verlag Berlin Heidelberg 2013

(APS), the ZigBee device objects (ZDO) and the manufacturer-defined application objects. The ZigBee device is divided into the ZigBee Coordinator (ZC), ZigBee Router (ZR) and ZigBee End Device (ZED) [2]. ZigBee Coordinator controls the network and communicates with the other devices.

In tree and mesh topologies, ZigBee Coordinator must be responsible to initiate a network and install some parameters of the network which can also be expanded by the routers. The routing nodes play a crucial role in the whole network communication. Energy is the reliable guarantee of the wireless sensor network, minimizing energy consumed by sensing and communication to extend the network lifetime is an important design objective [3]. Improving the ZigBee transmission method is an effective way to improve ZigBee energy utilization.

Many multicast routing protocols have been proposed, such as ODMRP [4], ODMRP is a mesh-based, rather than a conventional tree-based, multicast scheme and uses a forwarding group concept and it is well suited for ad hoc wireless networks with mobile hosts where bandwidth is limited, topology changes frequently, and power is constrained. DODMRP [5] and ODMRP-LR [6] improve the ODMRP. The CAPM [7] generalizes the notion of core-based trees introduced for internet multicasting into multicast meshes that have much richer connectivity than trees. But these protocols are not suitable for directly applying in ZigBee network. Z-Cast [8] routing algorithms has been proposed for ZigBee network. However, the energy optimization is not taken into consideration to improve the use of ZigBee network.

This paper presents a ZigBee improving multicast routing algorithm (ZIMR). On the basis of Z-Stack multicast algorithm, ZIMR algorithm considers the range of the node of sending information to select the forwarding path and different multicast groups using different routing algorithms, saving the energy of the whole network.

39.2 Algorithm Design and Description

39.2.1 Distributed Address Allocation

ZigBee applies the distributed addressing assignment scheme to assign the nodes' addresses which are unique in the network. Every device knows its 1-hop neighbors and stores neighbor information. Given values for the maximum number of children a parent may have, nwkMaxChildren (C_m), the maximum number of routers a parent may have as children, nwkMaxRouters (R_m) and the maximum depth in the network, nwkMaxDepth (L_m) [2]. If knows a router's address, its children nodes' addresses are confirmed by the node address block Cskip(d), the formula to calculate Cskip(d) has been given in (39.1).

Fig. 39.1 Tree forwarding node algorithm

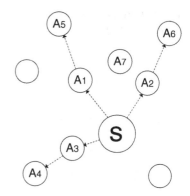

$$Cskip(d) = \begin{cases} 1 + C_m(L_m - d - 1), & if \quad R_m = 1 \\ \frac{1 + Cm - Rm - Cm*Rm^{Lm-d-1}}{1-Rm}, & otherwise \end{cases} \quad (39.1)$$

A parent device that has a Cskip(d) value greater than 0 shall accept child devices and assign addresses to them differently. If a node's level is d and its address is A_{parent}, then the address of its nth children router node is known, as given by the Eq. (39.2):

$$A_n = A_{parent} + Cskip(d) \times R_m + n \quad (39.2)$$

Where $1 \leq n \leq (C_m - R_m)$ and A_{parent} represents the address of the parent. In this way, ZigBee Coordinator for each ZigBee device is assigned a fixed ID.

39.2.2 ZigBee Improving Multicast Routing Algorithm

Different algorithms on the impact of the forwarding nodes are different. As shown in Fig. 39.1, S is the source node and A_1, A_2, A_3, A_4, A_5, A_6 are the destination nodes. S cannot send the information directly to A_4, A_5, A_6, requiring forwarding nodes. A_1, A_2 send the information to A_5, A_6. As shown in Fig. 39.2, S sends the information to A_7. A_7 sends the information to A_5, A_6. ZIMR algorithm saves the number of forwarding nodes.

In order to achieve ZIMR algorithm, it requires defining three parameters, including the forwarding priority judgment (P), state of receiving information (R) and state of transmitting (T). The meanings of P, R, and T parameters are listed in Table 39.1.

The ZIMR algorithm is two parts which are looking for a forwarding node algorithm and sending information algorithm. We randomly select a node as a source. The source node sends the information to the destination nodes of its 1-hop neighbors. These nodes' R is set to 1. The source's T is set to 1. Then ZIMR

Fig. 39.2 ZigBee improved
multicast routing algorithm

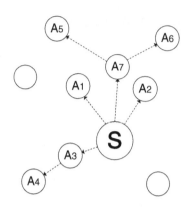

Table 39.1 The meanings of P, R and T parameters

Parameter	Meaning
P	The greater P the higher the priority, 0 is the lowest priority, -1 means waiting for judgment
R	1 means that it has accepted the information, 2 means that it will not accept the information, 0 means waiting for judgment
T	1 means that it has transmitted the information, 0 means that it has not transmitted the information

algorithm looks for a forwarding node. Looking for a forwarding node algorithm is described as follows.

All nodes of the source's 1-hop neighbors judge the priority of forwarding based on ID. To judge whether a node has judged the priority of forwarding, P is -1 means waiting for judgment. If P is not -1, the node need not to judge. Ignore the node and judge from next node. Record of the number of destination nodes' R is 0 which are the node's 1-hop neighbor, the more the number of the nodes, the more priority P. If the number is 0, the node's P is set to 0 and the R is set to 2. The node does not receive information. If two nodes' priorities are same, judge by ID. The smaller the ID is, the greater the priority is. When the source node's all 1-hop neighbors' P is not -1, selection of the node whose P is max as forwarding node. Then the forwarding node' P is set to 0. Other nodes' P is set to -1 to prepare for the next round to select forwarding node. Then the forwarding node sends information to its 1-hop neighbor destination node. Sending information algorithm is described as follows.

Forwarding node receives multicast information from the source. Forwarding node's R is set to 1. Forwarding node sends multicast information to its 1-hop neighbors destination nodes whose R is 0. Then these destination nodes' R is set to 1. Forwarding node's T is set to 1. Forwarding node's information sending is

Fig. 39.3 ZigBee improving
multicast routing algorithm

> ZigBee improving multicast routing algorithm:
> The source node sends the information to the
> destination nodes of its 1-hop neighbors;
> The destination nodes' R= 1;
> The source's T= 1;
> If source's 1-hop neighbor nodes' P = -1;
>   The node as potential forwarding node, P=0;
>   If potential forwarding node destination neighbor's R=0;
>   Potential forwarding node's P=P+1;
>   End if
> End if
> If P=0
>   R=2;
> End if
> Choose max's P accept information as forwarding
> node
> Other nodes' P= -1
> R= 1;
> P= 0;
> Forward massage to forwarding node's 1-hop
> neighbors destination nodes whose R= 0;
> Destination nodes' R= 1;
> Forwarding node's T= 1;

completed. Then all nodes of the source's 1-hop neighbors judge the priority of forwarding again based on ID. Until all source's 1-hop neighbor nodes' P are 0.

The source node's work is finished. According to ID, ZIMR algorithm looks for a new node whose T is 0 and R is 1 as a new source of information node. Until all nodes' R are not 0, the whole network' information sending is completed. ZIMR algorithm is shown in Fig. 39.3.

39.3 Simulation Results and Analysis

This paper designs a set of simulation models to simulate the operation of ZIMR algorithm on the basis of MATLAB platform. The simulation scene is generated within a 100 × 100 m square area in which the number of the ZigBee nodes varies from 50 to 200. Destination nodes randomly generated by a certain percentage of the nodes. Randomly selected node can make the results more convincing. The maximum transmission range of each node is 25 m. The maximum number of the children nodes Cm and the network depth Lm are 4 and 4. This paper assumes the nodes are in the ideal case. Physical layer and data link layer are without serious packet loss. As shown in Figs. 39.4 and 39.5 the hollow points are ZigBee ordinary nodes. Cross nodes represent the destination nodes. Solid points represent

Fig. 39.4 100 ordinary
nodes 20 destination nodes
topology of Z-Cast

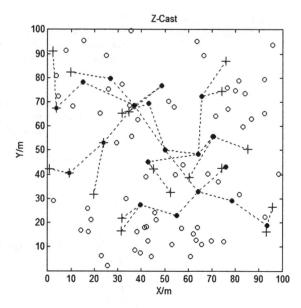

Fig. 39.5 100 ordinary
nodes 20 destination nodes
topology of ZIMR algorithm

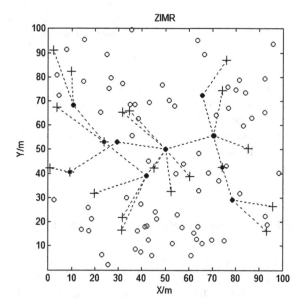

the forwarding nodes. Black dashed line indicates the sending of information
between nodes. This paper selects important representative parameters to assess
the performance of ZIMR algorithm compared with the other algorithm. The
parameter is the number of forwarding nodes which can measure the communication energy efficiency. As shown in Fig. 39.4, the information is radioactively
transferred to the surrounding, but the sending information back to the source's

Fig. 39.6 Different number of ZigBee nodes which 10 % are destination nodes of two algorithms

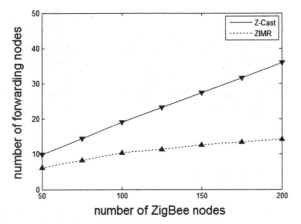

Fig. 39.7 100 ZigBee nodes which destination nodes are from 10 to 50 % of two algorithms

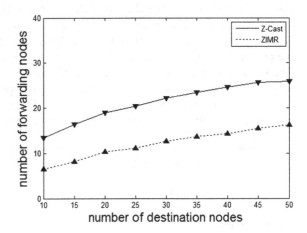

range. Nodes and energy are wasted, affecting network's reliability. As shown in Fig. 39.5, the ZIMR algorithm node judges the gain and loss before sending information.

This paper chooses 7 size scenes whose number of the nodes is from 50 to 200 to simulate each algorithm. Each case randomly selects 10–50 % as the destination nodes. Two algorithms are respectively simulated 100 times to calculate the average values of the parameter. Figure 39.6 presents the simulation result of a 7 size scenes which 10 % are destination nodes of two algorithms. Figure 39.7 presents the simulation results of 100 ZigBee nodes which destination nodes are from 10 to 50 % of two algorithms.

As shown in Figs. 39.6 and 39.7, the solid line indicates the number of forwarding nodes of Z-Cast algorithm. Dashed line indicates the number of forwarding nodes of the ZIMR algorithm. As shown in Fig. 39.6, in the case of the same size the number of forwarding nodes of Z-Cast algorithm is the higher, and it continues to rise with the increase of the ZigBee nodes number. In the same circumstances, forwarding node increases slowly of ZIMR algorithm. In a small number of ZigBee nodes, the numbers of two algorithms' forwarding nodes are almost same, but with the increasing of the numbers of ZigBee nodes, the number of forwarding nodes of ZIMR algorithm is less than Z-Cast algorithm. As shown in Fig. 39.7, we get the conclusion that in the case the number of ZigBee nodes is 100, the increase of the number destination nodes has the same impact on the two algorithms, while ZIMR algorithm saves half of the forwarding nodes than Z-Cast algorithm.

39.4 Conclusion

This paper proposes ZigBee improving multicast routing algorithm. ZIMR algorithm combines with the traditional wireless sensor multicast algorithm and makes full use of ZigBee distributed address characteristic. Based on the judging information which source node sends to the neighbor nodes and feedback information which source node receives from the neighbor nodes, ZIMR algorithm can make the priority judgment and then find the best forwarding node to send information. The simulation results prove that, compared with Z-Cast algorithm, ZIMR algorithm can reduce ZigBee multicast routing and improve the multicast routing forwarding efficiency and saves more energy. The conclusion drawn from this work can be useful for the information send of a large scale ZigBee network.

Acknowledgments This research was financially supported by the Open Research Fund from the State Key Laboratory of Rolling and Automation, Northeastern University, Grant No.: 2011002.

References

1. IEEE Standard 802 (2003) Part 15.4
2. Alliance ZB (2008) ZigBee-2007 specification
3. Wang L, Xiao Y (2006) A survey of energy-efficient scheduling mechanisms in sensor networks. Mob Netw Appl 11(5):723–740
4. Lee SJ, Su W, Gerla M (2002) On-demand multicast routing protocol in multihop wireless mobile networks. Mob Netw Appl 7(6):441–453
5. Tian K, Zhang B, Mouftah H, Zhao Z, Ma J (2009) Destination-driven on-demand multicast routing protocol for wireless ad hoc networks. In: Proceedings 2009 IEEE international conference on communications, pp 4875–4879
6. Naderan-Tahan M, Darehshoorzadeh A, Dehghan M (2009) ODMRP-LR: ODMRP with link failure detection and local recovery mechanism. In: Proceedings of the 2009 8th IEEE/ACIS international conference on computer and information science, pp 818–823
7. Madruga EL, Garcia-Luna-Aceves JJ (2001) Scalable multicasting: the core-assisted mesh protocol. Mobile Networks Appl 6(2):151–165
8. Gaddour O, Koubâa A, Cheikhrouhou O, Abid M (2010) Z-Cast: a multicast routing mechanism in ZigBee cluster-tree wireless sensor networks. In: Proceedings-2010 IEEE 30th international conference on distributed computing systems workshops, pp 171–179

Chapter 40
A Real-Time and Reliable Transmission Scheme in Wireless Sensor Networks for Sandstorm Monitoring

Jinze Du, Yubo Deng, Yi Yang and Lian Li

Abstract Sandstorm is a ubiquitous disastrous weather phenomenon. Many domestic and foreign scholars have paid great attention to forecasting and preventing sandstorm. In this paper, we presented a hierarchical and clustering network topology. In order to meet the complex communication medium in sand environment conditions, we adopted the method of node self-regulating it's power to ensure reliable communication. We analyzed and calculated the overall reliability of the network, finding that the cluster head node is the key factor to dominate the overall reliability. Ensuring the cluster head node can reliably transfer data to the base station, we proposed a real-time and reliable communication scheme. Compared to other routing algorithms, the communication mechanism has the ability to adapt to extreme environments and ensures real-time data transmission and a higher reliability of the entire network in the sandstorm environment.

Keywords Sandstorm · Wireless sensor networks · Timeliness · Reliability

40.1 Introduction

Sandstorm is a common natural phenomenon, related to the special geographical and meteorological conditions. Many domestic and foreign scholars have paid great attention to forecasting and controlling sandstorm. Several sandstorm

J. Du (✉) · Y. Deng · Y. Yang · L. Li
Lanzhou University, Room 516, Feiyun Building, No 222 Tianshui South Road,
Chengguan Zone, Lanzhou City, China
e-mail: dujz05@163.com

W. Lu et al. (eds.), *Proceedings of the 2012 International Conference on Information Technology and Software Engineering*, Lecture Notes in Electrical Engineering 210, DOI: 10.1007/978-3-642-34528-9_40, © Springer-Verlag Berlin Heidelberg 2013

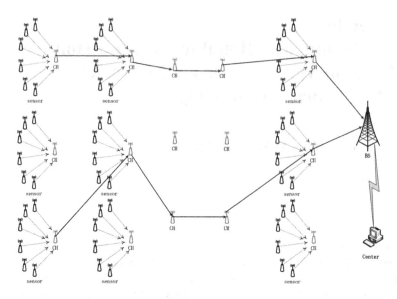

Fig. 40.1 Network topology model

forecasting systems have been deployed to serve different regions and countries [1]. In order to make up for the defects of traditional monitoring system, we use wireless sensor networks to monitor and predict the sandstorm, as well as tracking its movement tendency timely. In this paper, our goal is to design a real-time and reliable communication mechanism in wireless sensor networks; the communication mechanism is mainly to achieve two objectives, real-time data transmission and higher transmission reliability. In order to ensure real-time transmission and higher reliability, we adopted the following policy: Hierarchical and clustering network topology, Reliable communication strategy, Queue-based routing protocol, Data aggregation method, Overall reliability analysis.

40.2 Network Topology and Reliability Analysis

In this section, we presented a WSN model and we analyzed and calculated the overall reliability of the network.

40.2.1 Network Topology Model

We constructed a hierarchical network topology, as is illustrated in Fig. 40.1. The entire network consists of three levels. Data acquisition nodes, also called data sensor nodes are deployed at the bottom of the network. Cluster head node is in the

Fig. 40.2 The performance
of R_{BS} with R_{CH}

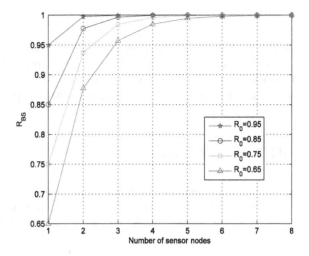

Fig. 40.3 The performance
of R_{BS} with R_0

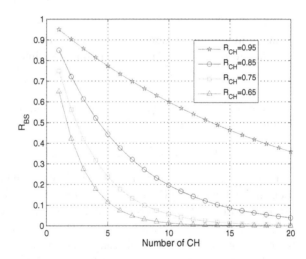

Fig. 40.4 System
architecture design

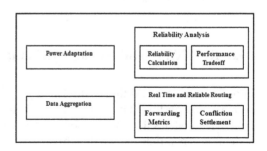

middle layer of the network, the highest level of network is the base station. A cluster consists of a bunch of sensor nodes and a cluster head node. Data sensor nodes, located at identical cluster, are proximately dotted into a ring-like shape. CH node is located approximately at ring's centre. The bottom sensor nodes only transmit data to their CH node. CH nodes forward data to the BS, by way of multi-hop. The BS continues to transmit data to the data processing centre.

40.2.2 Overall Reliability Analysis

According to the network topology, at least a node can transfer data to the base station within its cluster, the data of this cluster is considered to be transmitted reliably. Therefore, we can calculate the reliability of a cluster's data transmission:

$$R_{cluster} = 1 - \prod_{i=1}^{n} (1 - R_i) \tag{40.1}$$

In our model, we assumed that the probability of all collection nodes are equal, $R_i = R_0$. So above formula is further simplified into:

$$R_{cluster} = 1 - (1 - R_0)^n \tag{40.2}$$

The probability of the data inside a cluster successfully delivered to the base station is calculated as:

$$R_{BS} = (R_{CH})^k \times R_{cluster} = (R_{CH})^k \times [1 - (1 - R_0)^n] \tag{40.3}$$

In Fig. 40.2, we assumed that the probability of the bottom nodes successfully transmitting data to the CH node tends to 1, i.e. $1 - (1 - R_0)^n \to 1$. In Fig. 40.3, we assumed the data transformed successfully between the CH and the base station, i.e. $(R_{CH})^k \to 1$. Comparing to the CH, the impact of the bottom nodes to the reliability of the entire network can be neglected. Therefore, the real-time and reliable communication scheme mainly set the focus on improving the reliability of transmission of the CH node.

40.3 Real-Time and Reliable Scheme

In order to achieve real-time and reliability of the network, we add the following design based on the above network model. In Fig. 40.4, RTRS consists of four functional modules that include power adaptation, data aggregation, performance tradeoff and real-time and reliable routing.

40.3.1 Pattern Switching

Based on the Wang's research and analysis on the communication mediums in the sandstorm environment, we further analysed the switch process between the four patterns [2]. Through the calculation, using the same transmit power in the AA pattern, the transmission distance is about $R_{AA} > 100$ m, in the SS pattern, the transmission distance is about $R_{SS} < 10$ m, in the AS pattern and SA pattern, the transmission distance is about 30 m $< R_{AS} = R_{AS} < 70$ m. In this transmission scheme we regarded $R_{AA} = 100$ m, $R_{SS} = 8$ m, $R_{AS} = R_{AS} = 50$ m, we can take AA mode as the standard, and calculate the power values respectively, need to be adjusted in the transmission process.

40.3.2 Routing Metrics

In the process of the selection of the next hop node, the next hop node must be in the middle area of the source node and the base station. To achieve this objective, we only need to ensure that:

$$d_{iB} > d_{jB} \tag{40.4}$$

$$d_{iB} = \sqrt{(x_i - x_B)^2 + (y_i - y_B)^2} \tag{40.5}$$

$$d_{jB} = \sqrt{(x_j - x_B)^2 + (y_j - y_B)^2} \tag{40.6}$$

where d_{iB}, d_{jB} are distance, which node i and node j to the base station, respectively [3].

After selecting the next hop node, we will consider link quality, transmission path and the idle state as the main factors for the calculation of the routing metric. Forwarding node selection will follow the routing metric formula:

$$FS = \max\left(0.4 \times LQ + 0.5 \times \frac{R}{R_m} + 0.1 \times \frac{C_{idle}}{C_{full}}\right) \tag{40.7}$$

In the process of data transmission, link quality is directly related to the data transmission rate and energy utilization [4]. In the paper [5], the author deduced the link quality formula, based on the link layer model.

$$LQ = \left[1 - \left(\frac{8}{15}\right)\left(\frac{1}{16}\right)\sum_{j=1}^{16}(-1)^j\binom{16}{j}\exp\left(20\gamma(d)(\frac{1}{j} - 1)\right)\right]^{176} \tag{40.8}$$

In the process of routing metric, link quality, transmission path and the idle state will be considered comprehensively. Therefore, the selected path is not

Fig. 40.5 Schematic
diagram of the routing metric

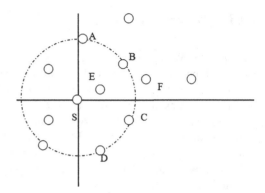

necessarily the shortest path. As shown in Fig. 40.5, when selecting the forwarding node, source node s first calculate the FS values of all nodes within the transmission radius, and then adopt the one with the maximum FS value as the forwarding node. In the figure, nodes A, B, C, D, E, F all within the perception radius of the source node s. Nodes A, B, C, D have the equal distance value from the source node, which is closer to the source node than the node F. But the source node may choose the node F as the forwarding node as long as the node F has the maximum FS value.

40.3.3 Tradeoff Strategy

In any data transmission mechanism, it is impossible to achieve a high reliability and a wonderful timeliness simultaneously. We considered the interaction between the two performances in our mechanism and we can make a compromise with the two kinds of performance according to our needs in practical applications [6].

When we need to ensure a higher real-time performance, the delay of the whole network must be lower than a predetermined value. In this case, we defined the delay of the whole network not greater than $t_{\max}$. The delay CH nodes send data to the base station is t_B, and there is limit for two variables.

$$t_B < t_{\max} \tag{40.9}$$

After obtaining the distance from source i to the base station and the distance from i to the node j, the total number of hops from source i to the base station can be calculated [7].

$$N_{hop} = \left\lceil \frac{d_{iB}}{d_{ij}} \right\rceil \tag{40.10}$$

The data transmit delay from node i to the node j:

$$t_{ij} = t_q + t_{tran} + t_{proc} + t_{prop} \tag{40.11}$$

where t_q is queue time before the data transmission. t_{tran} is the transmission time from the node's receiver to the node's data processing centre. t_{proc} is the node's internal data processing delay. t_{prop} is the transmission time from the sender of the node i to the receiver of the node j. This delay can be calculated as [8]:

$$t_{proc} = \frac{d_{ij}}{C} \quad C = 3 \times 10^8 \text{ m/s} \tag{40.12}$$

The total number of hops and the total time to meet the following relationship:

$$N_{hop} \times t_{ij} = t_B < t_{max} \tag{40.13}$$

$$\left[\frac{d_{iB}}{d_{ij}} \right] < t_{max} \tag{40.14}$$

The distance of one hop and the total delay meet the following relationship:

$$\left[\frac{d_{iB}}{d_{ij}} \right] \times \left(t_q + t_{tran} + t_{proc} + \frac{d_{ij}}{C} \right) < t_{max} \tag{40.15}$$

Solve the above inequality by computer:

$$d_{ij} > d_{min} \tag{40.16}$$

In order to ensure the delay constraints, we just need to make the following relationship established:

$$d_{ij} = R > d_{min} \tag{40.17}$$

The restricted routing metric inequality is as follows:

$$\begin{cases} FS = \max\left(\alpha_1 \times LQ + \alpha_2 \times \frac{R}{R_m} + \alpha_3 \times \frac{C_{idle}}{C_{full}} \right) \\ d_{min} < d_{ij} < R_m \\ d_{iB} > d_{jB} \end{cases} \tag{40.18}$$

This routing metric restrictions can guarantee a higher real-time performance, make the network satisfy the requirement of the minimum delay, but reduce the reliability of the network. In our routing metric, per-hop transmission distance cannot be less than the predetermined minimum distance value. Therefore, the original routing metric formula will be altered. The value of d_{ij} was limited. When selecting the next-hop node to forward the data, the source node S will calculate the distance from the nodes within its perception radius to itself. If the forwarding node doesn't satisfy the formula $d_{min} < d_{ij} < R_m$, would be discarded. Shown in Fig. 40.6, node E and node D may be selected as the preferred forwarding node, but both of them do not meet distance constraints, so they will not be enrolled for the next routing metric calculation. Similarly, when selecting the second hop node, the node F, node G and node H also be abandoned. At the last, there are three possible paths. They are $S \to A \to M \to BS$, $S \to B \to N \to BS$, $S \to Q \to BS$

Fig. 40.6 Schematic
diagram of the restricted
routing metric

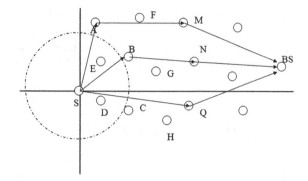

Fig. 40.7 Packet delivery
ratio VS source rate

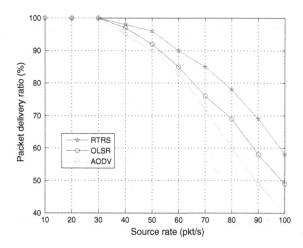

respectively. One of them would be chosen as the ultimate path for data
transmission.

40.4 Performance Evaluation

In the simulation experiments, we compared RTRS with Ad Hoc on Demand
Distance Vector (AODV) and Optimized Link State Routing (OLSR) in the reliability and timeliness performance meticulously.

The reliability of the communication mechanism is measured by the packet
delivery ratio [9]. In Fig. 40.7, we have compared the changes of the packet
delivery ratio of the RTRS, AODV and OLSR in the case of a varied source rate.
The RTRS maintained a higher packet delivery success rate by mitigating the
impact of a varied source rate. In Fig. 40.8, with the increase of the link interruption rate, the packet delivery ratio of the AODV and OLSR decreased rapidly
sharply. Conversely, in our RTRS, the algorithm can overcome link interruption.

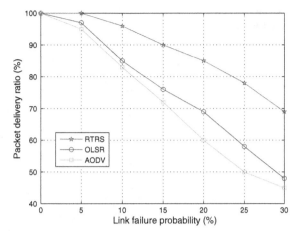

Fig. 40.8 Packet delivery ratio VS link failure probability

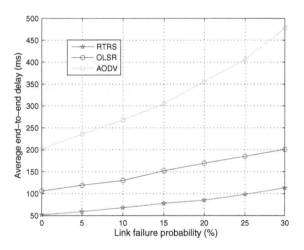

Fig. 40.9 Packet delivery ratio VS number of CH

The impact of the number of nodes in the network to the packet delivery ratio is shown in Fig. 40.9. The increase of node's number will improve the reliability of the entire network. The more nodes, the source node will have more choices to select in the process of the forwarding nodes selecting. Therefore, the packet delivery ratio will be improved. In the case of fewer nodes, the RTRS algorithm has superiority over AODV and OLSR, but this superiority is getting smaller and smaller with the increase of the number of nodes.

The average end-to-end delay of data transmission was calculated to compare the timeliness performance of the three algorithms [10]. As shown in the Fig. 40.10, the average end-to-end delays of the three algorithms present an increasing tendency with the increase of the source rate. RTRS has a distinct advantage over AODV and OLSR. In Fig. 40.11, the changes of the time delay of the three algorithms with the impact of the link interruption are illustrated. Among

Fig. 40.10 Average end-to-end delay VS source rate

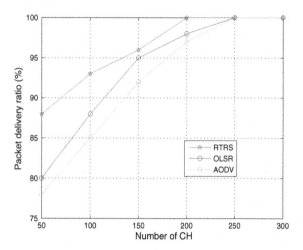

Fig. 40.11 Average end-to-end delay VS link failure probability

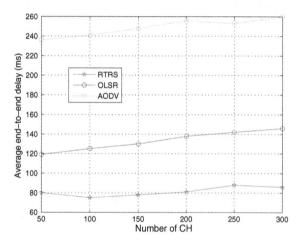

the three algorithms, the RTRS routing algorithm has the ability to overcome the communication link interrupt. AODV and OLSR are very sensitive to the link interruption. In Fig. 40.12, the changes of the time delay of the three algorithms with the impact of the number of the CH nodes are illustrated. The results show that the number of the CH nodes has a little influence on the time delay. The time delay of AODV and OLSR is just showing a slow increasing trend with the number of nodes increasing. In the RTRS, sometimes the time delay drop with the number of CH nodes increasing. When the number of nodes increases from 250 to 300, the time delay drops from 88 to 86 ms.

Fig. 40.12 Average end-to-end delay VS number of CH

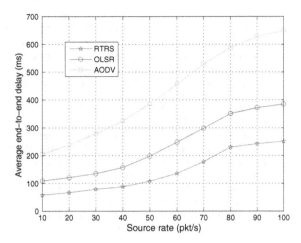

40.5 Conclusion

In this paper, we first constructed a hierarchical topology model of the wireless sensor networks for sandstorm monitoring. In order to meet the complex communication medium in sand environment conditions, we adopted the method of node self-regulating it's power to ensure reliable communication. In order to ensure the cluster head node can reliably transfer data to the base station, we proposed a real-time and reliable communication scheme. In the mechanism, we adopt the method of routing metric to allow the source node select the optimal next hop node in order to ensure the data is real-time and reliable transmitted. Solving the contradiction between the timeliness and reliability, we proposed a compromise strategy. Our simulations on the RTRS demonstrated that RTRS had a low transmission delay and a higher packet delivery ratio in the extreme communication environment, comparing to the AODV and OLSR. The RTRS can tolerate complex and volatile communication medium, but also overcome the unfavorable factors brought about by the node damage.

In the future work, we will consider using the mobile nodes in the sandstorm environment. Adopting the mobile node will create additional communication opportunities between nodes. Another concept, opportunity communication will be applied to early warning and monitoring of the sandstorms.

References

1. Wang P et al (2011) On network connectivity of wireless sensor networks for sandstorm monitoring. Comput Netw 55:1150–1157
2. Al-Karaki J, Kamal A (2008) Efficient virtual-backbone routing in mobile ad hoc networks. Comput Netw 52(2):327–350

3. Gungor VC, Akan OB, Akyildiz IF (2008) A real-time and reliable transport (RT) 2 protocol for wireless sensor and actor networks. IEEE/ACM Trans Netw 16(2):359–370

4. Wang Y, Vuran MC, Goddard S (2009) Cross-layer analysis of the end-to-end delay distribution in wireless sensor networks. In: Proceedings of IEEE real-time systems symposium (RTSS'09), Washington, pp 138–147

5. Mahmood H, Comaniciu C (2009) Interference aware cooperative routing for wireless ad hoc networks. Ad Hoc Netw 7(1):248–263

6. Mohanoor A, Radhakrishnan S, Sarangan V (2009) Online energy aware routing in wireless networks. Ad Hoc Netw 7(5):918–931

7. Rangarajan H, Garcia-Luna-Aceves J (2007) Efficient use of route requests for loop-free on-demand routing in ad hoc networks. Comput Netw 51(6):1515–1529

8. Naserian M, Tepe K (2009) Game theoretic approach in routing protocol for wireless ad hoc networks. Ad Hoc Netw 7(3):569–578

9. Souihli O, Frikha M, Hamouda M (2009) Load-balancing in MANET shortest-path routing protocols. Ad Hoc Netw 7(2):431–442

10. Ivascu G, Pierre S, Quintero A (2009) QoS routing with traffic distribution in mobile ad hoc networks. Comput Commun 32(2):305–316

Chapter 41
4D-Trajectory Conflict Resolution Using Cooperative Coevolution

Jing Su, Xuejun Zhang and Xiangmin Guan

Abstract Conflict resolution becomes a worldwide urgent problem to guarantee the airspace safety. The existing approaches are mostly short-term or middle-term which obtain solutions by local adjustment. 4D-Trajectory conflict resolution (4DTCR), as a long-term method, can give better solutions to all flights in a global view. 4DTCR involved with China air route network and thousands of flight plans is a large and complex problem which is hard to be solved by classical approaches. In this paper, the cooperative coevolution (CC) algorithm with random grouping strategy is presented for its advantage in dealing with large and complex problem. Moreover, a fast Genetic Algorithm (GA) is designed for each subcomponent optimization which is effective and efficient to obtain optimal solution. Experimental studies are conducted to compare it to the genetic algorithm in previous approach and CC algorithm with classic grouping strategy. The results show that our algorithm has a better performance.

Keywords 4D-Trjactory · Conflict resolution · Cooperative coevolution · Evolutionary computation

J. Su (✉) · X. Zhang · X. Guan
School of Electronics and Information Engineering, Beihang University,
Beijing 100191, People's Republic of China
e-mail: sujing@ee.buaa.edu.cn

J. Su · X. Zhang · X. Guan
National Key Laboratory of CNS/ATM, Beijing 100191, People's Republic of China

W. Lu et al. (eds.), *Proceedings of the 2012 International Conference on Information Technology and Software Engineering*, Lecture Notes in Electrical Engineering 210, DOI: 10.1007/978-3-642-34528-9_41, © Springer-Verlag Berlin Heidelberg 2013

41.1 Introduction

In recent years, the aviation industry develops at an astonishing speed. With the increase of transport volumes, the density of flights in airspace jumps to a dangerous level, where the minimum safety intervals between flights are difficult to be kept. Consequently, the probability of flight conflict increases, and conflict resolution to guarantee the flight safety becomes a worldwide urgent problem. Many methods have been proposed to solve this problem.

The flight conflict resolution here can be viewed as a high dimensional optimization problem. Most related researches are short-term or middle-term optimizations which are implemented by adjusting the local route, instant velocity, or instant heading angle of particular flights [1–6]. Researches of long-term optimization, which can provide the global optimal solutions to all of the flights, are rare. However, local adjustment can eliminate present conflict, yet it would probably bring new conflicts from other flights. As the 4D-Trajectory[1] (4DT) concept proposed by FAA in the NextGen [7], the time of flight arriving at the particular nodes in the network can be handled accurately. That makes it easy to eliminate all of the conflicts by modulating the 'time' of the flights and achievable to give global optimal solutions.

The problem in this paper is the 4DT conflict resolution (4DTCR) by optimization and scheduling the flights. 4DTCR is a difficult problem to deal with. Firstly, it is a large-scale combinatorial optimization problem [8] which involves numerous variables and large solution space brought by the national flights. Secondly, the objective function is discrete and nonlinear. So it is hard to be handled by classical approaches.

There are only two existing 4DTCR approaches [8, 9]. One is based on GA which is widely used in numerical and combinatorial optimization problems; the other is based on Constraint Programming (CP) with a branching scheme. However, neither of them could solve the large-scale combinatorial optimization problem effectively. As the scale of the problem increases, the computation complexity dramatically jumps and it quickly becomes computationally intractable [1]. So the two approaches use a sliding forecast time window to reduce the dimensions of variables. Nonetheless, they just give the solutions of the flights in the sliding window at one time rather than all of the flights in a global view. Thus, a global optimization is expected.

Cooperative coevolution [10, 11] is a general framework using a divide-and-conquer strategy [12]. In CC, the variables to be optimized are decomposed into different subcomponents first. The subcomponents evolve mostly separately using an Evolutionary Algorithm (EA) with the only cooperation in fitness evaluation. CC has been realized to be an efficient tool for large and complex problems. Therefore, a CC framework with random grouping strategy as the decomposition

---

[1] 4D-Trajectory can be described as (x, y, z, t), where t is the time flight arrives at position (x, y, z).

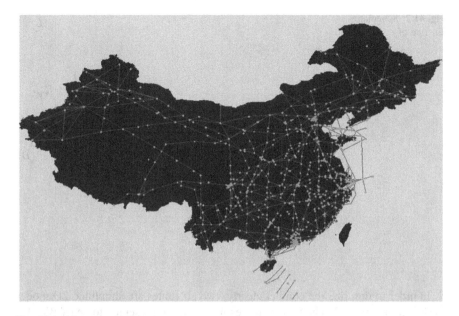

Fig. 41.1 Air routes of China

method and a fast GA as the subcomponent optimization (GACC-R) is proposed in this paper.

The rest of this paper is organized as follows: Sect. 41.2 gives a model about problem formulation. Section 41.3 presents the details of our approach. Section 41.4 is the results of our computational studies and the analysis about them. Section 41.5 gets some conclusions about this paper with some remarks and future researches.

41.2 Modeling the Problem

Concerning China's civil aviation data, there is a complex air route network with more than 5,000 air routes and 1,000 waypoints, as shown in Fig. 41.1. Besides, there are more than 7,000 flight plans everyday, which result in thousands of possible conflicts.

In this paper, it is supposed there are n flights $(F_1, F_2, F_3, \ldots, F_n)$ flying along the specific air routes according to the flight plans. Air routes are known to be composed of air route segments, and these segments are combined by waypoints. Here it is supposed that there are only two flight levels on one air route segment, and the flights whose directions are opposite will fly on the two levels respectively. So there will not be a conflict between two flights on the opposite directions.

Fig. 41.2 Schematic
diagram of cross routes

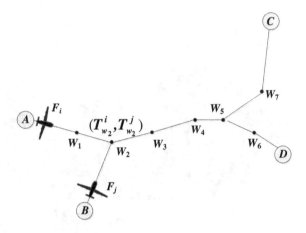

In order to describe the conflict clearly, two air routes are simplified 2D model as shown in Fig. 41.2. The two air routes A_i and A_j are presented to be $(A, W_1, W_2, W_3, W_4, W_5, W_6, D)$ and $(B, W_2, W_3, W_4, W_5, W_7, C)$. A, B, C, D are four airports as well as the starts or terminals of the two air routes, and $W_1, W_2, W_3, W_4, W_5, W_6, W_7$ are the waypoints. As shown in the figure, there is an intersection (W_2, W_3, W_4, W_5) between A_i and A_j. It is supposed that two flights F_i and F_j are flying along A_i and A_j respectively with the same and constant velocities. So the conflict can only occur at the cross point W_2. That means there may be a conflict only when

$$|T_{w_2}^i - T_{w_2}^j| < \tau \tag{41.1}$$

where T_{w2}^i and T_{w2}^j are the time F_i and F_j arrived at W_2, and τ is the minimum safe time interval.

The conflicts are eliminated by adjusting the flight delays in this paper. The flight delay set D is defined by $D = \{\delta_i, \forall i \in [1, n]\}$, where δ_i presents the departure delay of the ith flight. For each δ, it must satisfy a constraint of $\delta \in [0, \delta_{max}]$, where δ_{max} is the allowable maximum delay.

With consideration of reducing cost for airlines, the objective in this paper is to eliminate conflicts and reduce flight delays. So the objective function is defined by

$$\text{Max } F = \frac{1 - \frac{1}{n}\sum_{i=1}^{n}\left(\frac{\delta_i}{\delta_{Max}}\right)}{1 + NC} \tag{41.2}$$

where NC is the total conflict number of the flights. The objective function shows that fewer conflicts and less flights average delay mean better performance.

41.3 Optimization Framework

As a large-scale combinatorial problem with discrete and nonlinear objective function and tight interaction among variables, 4DTCR is hard to handle by classical approaches. So a 4DTCR approach based on CC with the random grouping strategy is proposed, which can be summarized by three steps elaborated as follows.

41.3.1 Problem Decomposition

A critical step in CC framework is problem decomposition [14]. A good problem decomposition method should put interrelated variables into the same subcomponent, and keep the interdependencies among variables from different subcomponents minimal. According to the existing algorithms [11, 13, 15, 16], there are two classic grouping strategies, i.e. the one-dimensional based and splitting-in-half, yet they can't handle nonseparable problem efficiently. Fortunately, the random grouping strategy has been proved as an efficient solution to nonseparable problem since it can effectively increase the chance of putting interaction variables in the same subcomponent [12].

41.3.2 Subcomponent Optimization

Subcomponent optimization is another significant step of CC. It affects the quality of solutions and the speed of finding solutions. Here, a fast GA is applied by each subgroup.

1. Selection

   The classic tournament selection is adopted in this paper.

2. Crossover

   The crossover operator specially designed for 4DTCR has a great advantage to find solution without conflict quickly. It is introduced from [8, 17] with some improvements.

   The local fitness of each flight in the subcomponent is defined by:

$$F_i = \frac{1 - \frac{\delta_i}{\delta_{\max}}}{1 + NC_i} \tag{41.3}$$

where NC_i is the number of remaining conflicts involving flight i.

For the kth group of arbitrary individual A and B, each of local fitness F_{Ai}^k and F_{Bi}^k are compared accordingly. If $F_{Ai}^k > F_{Bi}^k$, the children inherit from parent A, that is A_i. If $F_{Bi}^k > F_{Ai}^k$, the children inherit from B_i. If $F_{Ai}^k = F_{Bi}^k$, the children are got by combination of parents with a random decimal α. The crossover is operated with a probability p_c.

3. Mutation

The mutation operator in [17] is used for reference. As higher fitness means better performance in this paper, the variable whose local fitness is lower than the mutation parameter ε will mutate.

41.3.3 Subcomponents Coevolution

Since interdependencies may exist between subcomponents, coevolution is essential in capturing such interdependencies during optimization. The cooperation between subcomponents here only occurs during fitness evaluation. The subcomponents evolve serially [11, 13].

41.4 Experimental Studies

41.4.1 Experimental Setup

In this paper, the flights in the busiest 4 h of a day are chosen. During the 4 h, there are about 2004 flights which are distributed in about 5,485 air routes among the whole country according to the flight plans. The velocities of all flights are set to be 450 km/h, and the minimum safe time interval $\tau = 60$s. δ_{max} is set to be 90 min, and the value interval of δ is 0.25 min. The number of subcomponent $m = 2$. Number of generations in each cycle for a subgroup was set to be once only. For each algorithm, the results were collected and analyzed on the basis of 15 independent runs. The parameters applied in all experiments are presented in Table 41.1.

41.4.2 Comparing GACC-R to Other Methods

The first experiment aims to evaluate the performance of GACC-R by comparing it with existing algorithm. Since CP is known as an efficient heuristic approach to solve small-scale problem, and evolutionary algorithms perform better in the large-scale problem, an improved GA in Ref. [8] is chosen to be the comparing experiment.

Table 41.1 Parameters of the experiments

Parameters	Description	Value
Popsize	Population size	80
Maxgen	Max generation	800
p_c	Crossover probability	0.35
p_m	Mutation probability	0.4
ε	Mutation parameter	0.5

Table 41.2 Results of the comparison of GACC-R, GACC-H and GA

Items	GA	GACC-H	GACC-R
Best fit	0.00304	0.908648	0.939368
Average fit	0.002545	0.493906	0.712317
Variance of fit	1.8419e-007	0.1006	0.0736
Mean delay (min)	15.8877	8.599967	5.527625
Number of conf	329	1.667	0.556

The second experiment is designed to evaluate the efficacy of the random grouping strategy by contrasting other grouping strategies. As mentioned in Sect. 41.3.1, there are two comparable grouping strategies, i.e. one-dimensional based strategy and splitting-in-half strategy. To 4DTCR, one-dimensional based strategy is obviously inapplicable because of the numerous variables to evolve serially. So the CC framework with splitting-in-half grouping strategy and the fast GA (GACC-H) is designed to be the contrast object.

The results calculated based on 15 independent runs were counted statistically with their best fitness[2] value (*Best fit*), average fitness value (*Average fit*), variance value of fitness (*Variance of fit*), the mean delay of 2004 flights (*Mean delay*) and average number of conflicts (*Num of conf*) for each method, which are shown in Table 41.2. The evaluation processes of the methods are presented in Fig. 41.3.

According to the fitness function mentioned in Sect. 41.2, larger fitness value in the range of [0, 1] means better performance. As shown in Table 41.2, GACC-R obvi-ously outperformed than GA and GACC-H in all the evaluation items except for the variance of fitness. GACC-R has a mean delay less than 6 min with only 0.556 conflicts. The best fitness value has reached to 0.939368, which means no conflict and little delay. The little variance of fitness, 0.0736, shows the great robustness. The results manifest that GACC-R has great capability to eliminate the conflicts, which is important in real applications.

To be specific, though GA is the most robust, the optimization effects of CC based algorithms are much better than GA on other performances. The numbers of

---

[2] The 'fitness' mentioned in the evaluation items refers to the fitness value of the final generation of each run.

Fig. 41.3 The evaluation process of GACC-R, GACC-H and GA

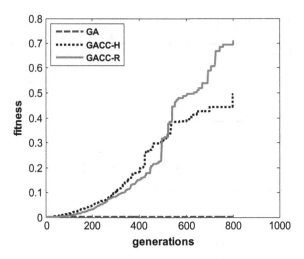

conflicts show the sharp difference between GA and CC based algorithms. These can fully indicate that the decomposition of CC framework has great efficiency to solve large and complex problem. When the solution space was huge, GA has a very low efficiency in finding solutions and is easy to fall into local optimum.

Compared each evaluation item, GACC-R performs better than GACC-H. It is because 4DTCR is a large-scale combination optimization problem, and the variables have tight interactions. The random grouping strategy has been proved to improve the quality of resolutions by increasing the chance to put interaction variables in the same subcomponent.

41.5 Conclusions and Future Work

Since the existing approaches cannot provide satisfying solutions to the large-scale and complicated problem, this paper proposes a GACC-R algorithm for 4DTCR. The key points of GACC-R include the random grouping strategy and the fast GA in subcomponent evolution. A theoretical analysis was given to illustrate how such strategy helps capturing the interdependencies of variables and how the fast GA improves the solutions. Then the results of experiments confirmed our analysis that GACC-R is a very effective and efficient approach in handling large-scale problems.

Although GACC-R performs quite well in our problem, the optimization speed still can be enhanced. As random grouping strategy is the current effective way in blind grouping, there may be a better grouping method if the interdependencies of variables can be known. So finding the interaction among the variables and then make a suitable grouping method to our problem are the future work.

Acknowledgments The authors would like to acknowledge the National High Technology Research and Development Program of China (Grant No. 2011AA110101), the National Basic Research Program of China (Grant No. 2011CB707000) and Specialized Research Fund for the Doctoral Program of Higher Education (Grant No. 20101102110005) for supporting this work.

References

1. Archibald JK, Hill JC, Jepsen NA, Stirling WC, Frost RL (2008) A satisficing approach to aircraft conflict resolution. IEEE Trans Syst Man Cybern C Appl Rev, 38(4):510–521
2. Mao ZH, Dugail D, Feron E (2007) Space partition for conflict resolution of intersecting flows of mobile agents. IEEE Trans Intell Transp Syst 8(3):512–527
3. Wollkind S, Valasek J, Ioerger T (2004) Automated conflict resolution for air traffic management using cooperative multi-agent negotiation. AIAA guidance, navigation, control conference exhibition, providence, RI, paper AIAA-2004-4992
4. Hwang I, Tomlin C (2002) Protocol-based conflict resolution for finite information horizon. In: Proceedings of American control conference, vol 1. Anchorage, AK, 748–753
5. Mondoloni S, Conway S (2001) An airborne conflict resolution approach using a genetic algorithm. Nat Aeronautics Space Admin Washington Tech. Rep. NASA-AIAA-2001-4054
6. Vivona R, Karr D, Roscoe D (2006) Pattern based genetic algorithm for airborne conflict resolution. AIAA guidance, navigation, control conference exhibition, keystone, CO, paper AIAA 2006-6060
7. Concept of operations for the next generation air transportation system, joint planning and development office, FAA (2007)
8. Durand N, Allignol C, Barnier N (2010) A ground holding model for aircraft deconfliction. Digital avionics systems conference (DASC), 2010 IEEE/AIAA 29th, 2.D.3-1–2.D.3-10
9. Durand N, Allignol C (2009) 4D-Trajectory deconflicton through departure time adjustment. In: Proceedings of ATM'09,8th USA/Europe R&D seminar on air traffic management, Napa
10. Potter M (1997) The design and analysis of a computational model of cooperative coevolution, Ph.D. Thesis, George Mason University
11. Liu Y, Yao X, Zhao Q, Higuchi T (2001) Scaling up fast evolutionary programming with cooperative coevolution. In: Proceedings of 2001 congress on evolutionary computation, Seoul, pp 1101–1108
12. Yang Z, Tang K, Yao X (2008) Large scale evolutionary optimization using cooperative co-evolution. Inf Sci 178(15-1):2985–2999
13. Potter M, Jong KD (1994) A cooperative coevolutionary approach to function optimization. In: Proceedings of 3rd conference on parallel problem solving from nature, vol 2, pp 249–257
14. Potter M, Jong KD (2000) Cooperative coevolution: an architecture for evolving coadapted subcomponents. Evol Comput 8(1):1–29
15. Shi Y, Teng H, Li Z (2005) Cooperative co-evolutionary differential evolution for function optimization. In: Proceedings of 1st international conference on natural computation, pp 1080–1088
16. Sofge D, Jong KD, Schultz A (2002) A blended population approach to cooperative coevolution for decomposition of complex problems. In: Proceedings on 2002 congress on evolutionary computation, vol 1, pp 413–418
17. Durand N, Alliot JM, Noailles J (1996) Automatic aircraft conflict resolution using genetic algorithms. In: Proceedings of the symposium on applied computing, Philadelphia ACM

Chapter 42
A Method of Road Recognition Based on LIDAR Echo Signal

Qinzhen Zhang and Weidong Deng

Abstract Road recognition technology is a difficult area of unmanned vehicle for autonomous navigation. Combining with the practical application of driving environment, this paper propose a method of road recognition based on laser infrared radar (LIDAR) of the unmanned vehicle. Firstly a clustering analysis with the radar echo intensity data is processed. Then the road is divided into region of interest (ROI) and non-ROI combining with the radar distance information. Finally the road information is managed by the method of straights fitting curve to accomplish the road recognition in real time. The tests on running vehicle show that the proposed method of the road recognition can work efficiently in real traffic environment.

Keywords LIDAR · Echo intensity · Road recognition · Navigation of unmanned vehicles

42.1 Introduction

Road recognition, which is a major research content of unmanned vehicles for autonomous navigation, is a difficult and heat field in the study of unmanned vehicles environment perception. Generally road recognition is a main factor

Q. Zhang (✉)
College of Information Engineering, Capital Normal University, Beijing 100048, China
e-mail: zhangqinzhen1986@163.com

W. Deng (✉)
Beijing Aerospace Guanghua Electronics Technologies Limited Corporation, Beijing 100854, China
e-mail: dengweidong2010@sina.com

W. Lu et al. (eds.), *Proceedings of the 2012 International Conference on Information Technology and Software Engineering*, Lecture Notes in Electrical Engineering 210, DOI: 10.1007/978-3-642-34528-9_42, © Springer-Verlag Berlin Heidelberg 2013

reflecting the level of intelligent vehicles. Recently many domestic and foreign scholars have made a lot of effects and developed different techniques of road recognition based on the computer vision. Those techniques are usually on the basis of different land line styles (continuous or discontinuous styles), various road models (2D, 3D, lines and curves) and different segmentation technology (Hough Transform, template match or neural network). Conclusively, the techniques of road recognition mainly include the recognition methods based on road models, the methods based on road feature and the method based on color images [1–5].

At present, there are few researches based on sensor in the study of road recognition. A method checking road border based on LIDAR was proposed by Yu in [6]. According to the certain border height between road region and non-region, the method take advantage of border line and collect border data depth by Kalman filter algorithm to detect the road border. The method has advantage in instantaneity since it adopts LIDAR as the sensor. However, because it is on the basis of altitude difference between road and off road, the method is difficult to detect road when there is no or little altitude difference.

Based on mobile robot localization of LIDAR, the approach proposed by Wang [7] collects the LIDAR data to build local map, and then achieve finally make a global map. The foundation of the approach proposed in [8] is car tracing by LIDAR for road recognition. It's worth noting that, the above mentioned methods make use of LIDAR distance data for road and vehicle recognition based on cluster. Due to the limited information got, all the above methods have disadvantage in reliability and narrow application scope.

It's come to our attention that different materials possess different echo reflectance. The LIDAR is capable of collecting the distance information and echo intensity information. Meanwhile, echo intensity information, especially reflectance information, reflects the nature of target material. As an active sensor, LIDAR is hardly suffered the effect of the outside.

This paper introduces LD-LRS3100 two-dimensional LIDAR with information of echo intensity and distance for road recognition. The works in the paper are focusing on two aspects. Firstly the road echo intensity experiment characteristic analysis is processed. Secondly the paper proposes a road recognition method based on the laser radar echo intensity.

42.2 Working Principle and Characteristics of LD-LRS3100

The basic working principle of LD-LRS3100 is to measure the time intervals from the launch of a laser beam to the receiving reflected light. The formula for Laser Ranging is: $s = v \times t/2$, (Where v is the speed of light, t is the laser time of flight). Its features are as follows:

- Based on time of flight measurement principle, Non-contact measurement.
- Two-dimensional scanning, 300° scanning.
- Three protected areas and an arbitrary collection of graphics can be set.
- With filtering to prevent interference from objects such as rain, snow, etc.
- High resolution, the angle resolution is up to 0.125.
- LD-LRS3100 scanning angle: Max: 300°, and the scanning angle can be adjusted in setting Parameters.
- Adjustable angular resolution: 0.125/0.25/0.5/1.0°.
- Response Time: 100–200 ms,range (Max/10 % reflectivity): 250/80 m.
- Resolution/System Error: 3.9 mm/typ.±25 mm;Scanning Frequency: 5–10 Hz.

The returned data include the distance value and echo intensity. The five radars are fixed in the first half of the car, then the intelligent vehicle driving environment are made to form 360° detection range. The com servers are applied to receives the laser radar data, and the terminal control platform are developed using VC ++6.0 to deal with the laser radar data.

42.3 Characteristics of Road Echo Intensity

To achieve road recognition of unmanned vehicle, road echo intensity characteristics in different environments (such as in full sunlight, half sunlight, full shadow) should be studied firstly. In addition, because the unmanned vehicle should finish road recognition when moving, the different characteristics of road echo intensity with different speed need be analyzed further.

The experimental analysis in this section provides instructions for the road recognition approach based on echo signal. The figures testing LIDAR echo intensity under the different environment and speed are presented as following.

Figures 42.1 and 42.2 show that the LIDAR echo intensity are almost not affected by the influence of illumination; even the road is not illuminated at night, the LIDAR echo intensity is still valid. Therefore, in this respect, the approach is superior to the methods based on the visual. In addition, when there are distinct differences between road and non-road material, the corresponding echo intensity are markedly different. This property provides a basis for the approach using echo intensity to realize road identification.

The wave of LIDAR echo intensity in Fig. 42.4 show more intense than that in Fig. 42.3, the reason is that the vehicle vibrate when vehicle move. However, these effects can be reduced or even eliminated by road recognition algorithm in the next section.

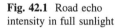

Fig. 42.1 Road echo intensity in full sunlight

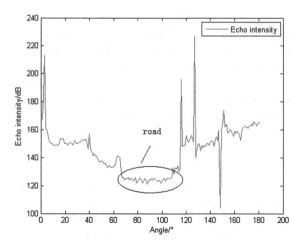

Fig. 42.2 Road echo intensity in full shadow

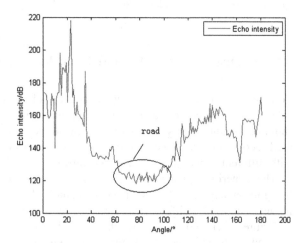

42.4 Road Recognition Algorithm Based on the Echo Intensity

From the above series of experiments, it obtains that the echo intensities of the same material at different distances (the angle of incidence of the laser) are different. The corresponding echo intensity is greater when the incident angle is greater. Different materials have different echo intensity and can basically reflect the surface reflectivity of materials. When the LIDAR is fixed, it has a certain angle of incidence. From the experiments for various road conditions, it concludes that the road echo intensity is in a small range of fluctuation under the same traffic. Meanwhile, the corresponding echo intensity is significantly different when the roads and non-road materials have considerable discrepancies.

Fig. 42.3 Echo intensity
under speed 0 km/h

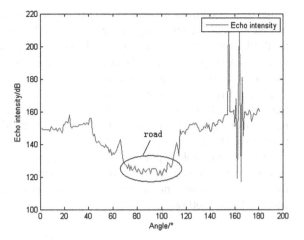

Fig. 42.4 Echo intensity
under speed 60 km/h

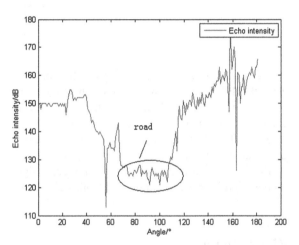

Consequently combined with the path distance information of LIDAR, this article proposes a road recognition algorithm based on the LIDAR echo intensity. The algorithm mainly includes the following procedures: LIDAR echo intensity data clustering, zoning, segment split and line segment notation.

42.4.1 LIDAR Echo Intensity Data Clustering

The previous test of LIDAR echo strength show that the same kinds of material objects have the analogous echo intensity. Based on this property, the paper proposes a clustering method using the means of step by step. According to the actual road conditions, LD-LRS3100 LIDAR works by scanning from right to left, and the LIDAR scanning angle is selected as 180°, the angular resolution is chosen as 1°, the range resolution is set as 1 cm. In other words, at intervals of 1°, it returns a distance value and the corresponding echo intensity. Consequently there are 181 data of the distance and the echo intensity.

In the LIDAR coordinate system, the distance and echo intensity are all returned based on the polar coordinates (i, ρ_i) where ρ_i is the data of the echo intensity obtained from the LIDAR for the ith scanned point; and i is the scanned point with 180 points.

Firstly a threshold value δ is fixed (its specific value can be given in experiment). Then the first echo intensity data is taken as the calculating basis. Secondly the difference of the mean value of this temporary cluster and the next data is counted by the following;

$$
\Delta = \left| \rho_{i+1} - \frac{\sum_{k=j}^{i} \rho_k}{i - j} \right| \tag{42.1}
$$

where Δ is the difference, j is the first echo intensity data of the temporary cluster. Then the difference and the given threshold value δ are compared. If the difference is greater than the threshold value, the echo intensity data is put into the different categories. If it is in the range of the threshold, the data is placed into the same category. Thirdly the mean value of this temporary cluster is calculated, and then the mean and difference of the next adjacent point are compared.

Repeat this process until 181 points are judged, and then the classification number and its corresponding data points are recorded.

42.4.2 Zoning

To ignore the non-interest area and improve data processing speed, regional division is disposed to divide into region of interest (ROI) and non-ROI using the clustering results of previous step. Considered as noise points, the clusters containing one or two data are excluded firstly. Since the installation angle and height of the LIDAR are certain, the distance value the road surface returns is within a certain scope when the road slope angle is very small.

Through angle correction, the distance information from the LIDAR can be roughly classified the scope of the road. Then the ROI are carved out by searching the echo intensity data in this scope.

42.4.3 Segment Split

The ROI and non-ROI are divided by the zoning procedure. Then one or multiple segments are used to represent the ROI cluster $B_i(i = 1, 2...N)$. To realize the linear segmentation, the steps are presented as following:

1. In the cluster B_i, these distance data are processed to fit a straight line using least squares firstly. Then the different distances from every point to this straight line are calculated. Assume that the straight line is given by the following;

$$Ax_i + By_i + C = 0 \tag{42.2}$$

The different distances can be calculated as follows;

$$d = \frac{|Ax_0 + By_0 + C|}{\sqrt{A^2 + B^2}} \tag{42.3}$$

where (x_0, y_0) is the coordinate of the point. Then the point with the longest distance is picked up. If the longest distance is bigger than the threshold σ which can be one-tenth of the length of the linear, the zone is divided into two parts.

2. Repeat step 1 until M linear regions $L_i(i = 1, 2...M)$ are divided. Each element of linear regions can be expressed as a straight line. Then the linear region and the echo intensity of priori objects are matched.

42.4.4 Line Segment Notation

By the previous steps, the road can approximately be distinguished. Finally through mapping the corresponding angle to the distance information from the LIDAR, a straight line can be processed by the least squares fitting based on these distance data. Then the road can be represented as follows

$$y = a_0 + ax \tag{42.4}$$

For precision measurement of the obtained N set of data (x_i, y_i), i = 1,2..., N, where x_i value is considered to be accurate. Then using least squares method to fit the observed data straight line, the linear estimation of parameter can be obtained by the following;

$$\sum_{i=1}^{N} [y_i - (a_0 + ax_i)]^2 |_{a=\hat{a}} \tag{42.5}$$

That requests the observation y_i deviation of sum of squares to minimum.

$$\frac{\partial}{\partial a_0} \sum_{i=1}^{N} [y_i - (a_0 + ax_i)]^2|_{a=\hat{a}} = -2 \sum_{i=1}^{N} (y_i - \hat{a}_0 - \hat{a}x_i) = 0 \qquad (42.6)$$

$$\frac{\partial}{\partial a_1} \sum_{i=1}^{N} [y_i - (a_0 + ax_i)]^2|_{a=\hat{a}} = -2 \sum_{i=1}^{N} (y_i - \hat{a}_0 - \hat{a}x_i) = 0 \qquad (42.7)$$

After finished computation, the normal equations are given as follows;

$$\hat{a}_0 = \frac{(\sum x_i^2)(\sum y_i) - (\sum x_i)(\sum x_i y_i)}{N(\sum x_i^2) - (\sum x_i)^2} \qquad (42.8)$$

$$\hat{a} = \frac{N(\sum x_i y_i) - (\sum x_i)(\sum y_i)}{N(\sum x_i^2) - (\sum x_i)^2} \qquad (42.9)$$

The straight line is the actual road.

42.5 Experimental Results and Analysis

Take the road condition in Fig. 42.2 as an experimental scene to illustrate the proposed approach. Figure 42.5 shows a regional division based on the distance information of the LIDAR, the red part of which is a ROI. Figure 42.6 shows the corresponding result of the road recognition algorithm based on the echo intensity.

From the experimental result, it can be seen that the radar echo intensity data is segmented to 13 categories by the clustering analysis. There are three categories representing the interest area. Using the segment splitting, one category is picked

Fig. 42.5 Zone division

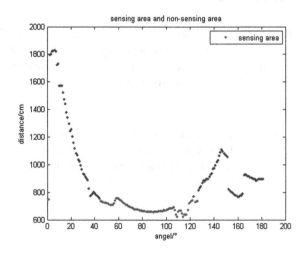

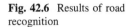

Fig. 42.6 Results of road recognition

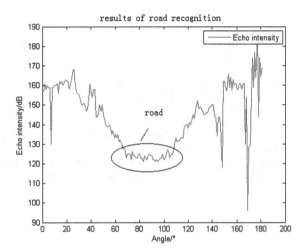

up by comparing the width of actual road to be recognized as the road. The LIDAR angles of two edges of the road are 65 and 97° respectively. According to the distance information from the LIDAR, the road width calculated is approximately 4 m. This distance is consistent with the width of the actual road.

42.6 Conclusions

In this paper, road echo intensity characteristics in different environments are presented firstly. Then combined with its distance information, a road recognition method based on the echo intensity is proposed to achieve road recognition of unmanned vehicle. The experiments are conducted on an actual vehicle. The results indicate that the method is feasible and can effectively identify the traffic road.

Moreover, as an initiative sensor that is almost free from impacts of the external environment, LIDAR possesses a clear advantage comparing with the visual. In addition, since the proposed method comprehensively uses the echo intensity information and distance information, its robustness and instantaneity have all been improved.

References

1. Yu T, Wang R, Gu B (2006) Survey on thr vision-based recognition methods of intelligent vehicle road boundaries and lane markings. J Highway Transp Res Dev 23(1):39–147 (in Chinese)
2. Wu Z, Sun H (2009) Lane mark identification algorithm based on fast line detection. Comput Technol Dev 19(5):48–53 (in Chinese)

3. Bertozzi M, Broggi AG (1998) A parallel real-time stereo vision system for generic obstacle and lane detection. IEEE Trans Image Process 7(1):62–81
4. Li D, Chai Y, Yin H (2008) A method of road recognition based on color image edge detection. Comput Eng Appl 33(28):177–183 (in Chinese)
5. Jiang GY, Choi TY, Hong SK et al (2000) Lane and obstacle detection based on fast inverse perspective mapping algorithm. In: IEEE international conference systems, man and cybernetics. Nashville, USA, pp 2969–2974
6. Yu C (2008) Road curbs detection based on laser radar. Chin J Electron Devices 31(3):756–759 (in Chinese)
7. Wang Y, Li Y, Tang X (2009) Localization and mapping using laser scanner for mobile robot. Robot Technol 25(5):227–229 (in Chinese)
8. Gan Z, Wang C, Yang M (2009) A method for vehicle tracking and recognition based on scanning laser radar. J Shanghai Jiaotong University 43(6):923–926 (in Chinese)

Chapter 43
Peer Segment Cache Replacement Based on Supply–Demand for Hybrid P2P VoD Services

Huang Guimin, Li Xuanfeng, Zhou Ya, Liu Pingshan and Dai Yunping

Abstract Peer-to-Peer (P2P) networks have been widely used in Video-on-Demand (VoD) services in recent years. Peers in a P2P VoD system usually have limited cache spaces. When the cache space of a peer is full, the peer needs to choose which segments should be replaced. In this paper, we propose a cache replacement strategy based on supply and demand balance theory of economics in a hybrid P2P VoD architecture, which is called SD Cache. Peers in this architecture are divided into super nodes and normal nodes. Super nodes manage the supplies and demands of segments from peers and utilize a distributed averaging algorithm to compute the segments' demand degree. Peers cache and replace segment according to the segment's demand degree, which can ensure the supplies for a segment to be proportional to the number of received demands. Through simulations, we show that our proposed algorithm outperforms other algorithms (e.g. RANDOM, FIFO, LRU, LFU) in terms of reducing server load, improving the hit rate, and reducing startup delay.

Keywords Peer-to-peer · Video-on-demand · Cache replacement · Difference between supply and demand · Urgency degree

This work was supported by the National Natural Science Foundation of China (No. 61063038), the Foundation of Key Laboratory of Guangxi Trusted Software (No. kx201114), and the Foundation of Key Laboratory of Guangxi Wireless Broadband Communication and Signal Processing (No. 11108).

H. Guimin (✉) · L. Xuanfeng (✉) · Z. Ya · L. Pingshan · D. Yunping
Research Center on Data Science and Social Computing, Guilin University of Electronic Technology, Guilin, China
e-mail: gmhuang@guet.edu.cn

L. Xuanfeng
e-mail: lixuanfeng06@163.com

W. Lu et al. (eds.), *Proceedings of the 2012 International Conference on Information Technology and Software Engineering*, Lecture Notes in Electrical Engineering 210, DOI: 10.1007/978-3-642-34528-9_43, © Springer-Verlag Berlin Heidelberg 2013

43.1 Introduction

Recently, the Peer-to-Peer (P2P) technology has been widely applied to streaming media Video-on-Demand (VoD) services. The P2P VoD architectures can be broadly classified into three categories: tree-based, mesh-based and hybrid systems. Many tree-based overlays are proposed, such as P2Cast [1] and OStream [2]. They build an application layer multicast tree to provide basic stream and search for appropriate patching streams for later coming peers from the root peer. But this one-to-one data delivery model is not efficient enough in the heterogeneous network. In mesh-based systems, peers establish and terminate peer relationships dynamically, such as pcVoD [3] and PROMISE [4], which can improve bandwidth utilization by fetching data from multiple data suppliers. However, random seeking data to support Video Cassette Recorder (VCR) operations in pure mesh-based model may lead to an unacceptable latency, and always cause download imbalance. Hybrid architecture is proposed in [5], which can support a large number of clients at a low overall system cost. Peers in this overlays are organized in a network aware fashion, in which nearby peers are grouped into a cluster. Powerful peers help to route the request and search for media segments. In this paper, we combine P2P techniques with the client/server architecture to build our hybrid P2P VoD system.

Among all above mentioned P2P architectures, each peer is required to contribute a small amount of storage to cache the watched media segments which construct a virtual P2P based network storage to ensure the continuity of playback, and reducing the load of media servers. Taking the PPLive [6] as an example, each peer dedicates about 1 GB cache space to cache previously watched segments and supply those to its neighbors. Therefore, peers can obtain some media segments directly from other peers instead of downloading from servers. But the capacity of media files is large generally. While a peer can only contribute limited cache space, therefore, each peer needs to selectively cache some watched segments. When the peer's cache space is full, it will discard some old segments to make room for new segments. However, discarding an old segment will decrease the replica of that segment in P2P network which may be required by others, so those peers will be influenced accordingly. How do the peers decide to cache which segments on which media and discard which existed segments? In order to reduce the impact of discarding old segments, we propose a segment cache replacement strategy, called SDCache, which is based on supply and demand balance theory of economics. Under this strategy, the demand degree of a segment is determined by evaluating its supply–demand condition and urgency degree. A segment with higher demand degree is required more by other peers. According to the demand degree of segments, peers can decide which segments to be replaced when the cache space is full.

The rest of this paper is organized as follows. In Sect. 43.2, we state the related work. In Sect. 43.3, we present the hybrid P2P VoD topology. In Sect. 43.4, we propose the segment caching replacement algorithm based on supply–demand model. In Sect. 43.5, we evaluate the performance of our proposed algorithm. Finally, we conclude this paper in Sect. 43.6.

43.2 Related Work

Mostly used traditional cache replacement algorithms are the FIFO, LRU, LFU and the improved algorithm of LRU-K [7]. Unlike the traditional caching replacement algorithms, Hefeeda in [5] proposes a cache scheme in a hybrid system. This scheme focuses on peers' bandwidth and cache capacity, but seldom concerns about data's history accessing feature. A structure P2P media system PROP is proposed in [8], which focuses on media or segment with middle popularity. But it does not consider the impact of capacity of storage on usability. Yiu in [9] proposes a distributed segment storage scheme based on segment popularity in VMesh. Peers firstly prefetch those segments with higher supply deficit to be cached. The supply and demand estimates of each segment are computed and updated in a distributed approach. But it takes no account of peers' upload bandwidths. Ying in [3] proposes a cache strategy based on the demand ratio of stream. It requires multiple replicas of a segment to ensure a demand peer with enough download speed from several providing peers. But this approach only takes streams with high priority into consideration, and also not for the merely popular movies or VCR operations frequently occurred. Wu in [10] proposes a replication optimal strategy to keep the number of replicas proportional to the deficit bandwidth, but it needs all peers to report their cached bitmaps to the tracker server. This centralized approach is not preferred in P2P systems because a large amount of bitmap traffic would congest the network near by the tracker server. Cheng in [11] implements Single Video Caching (SVC) strategy and Multiple Video Caching (MVC) strategy respectively in GridCast network. The experimental results show that replicas have great impact on load balance.

In this paper, a supply–demand model-based distributed peer segment cache replacement strategy is proposed to reduce the startup delay, promote the hit rate and reduce server load in a super node based hybrid P2P VoD architecture. According to the historical performance and abilities of peers, the P2P networks can be abstracted to be a super node hierarchical structure to manage both the segment demands from peers and the supplies of segments cached by peers. Peers cache and replace segments by the priority rank of the segments' demand degrees, which can ensure the actually available replicas for a segment to be proportional to the number of received requirement.

43.3 System Architecture and Client Model

43.3.1 System Architecture

A tracker server is introduced in our system which can well adjust the performance of the system and afford more reliable services for peers. Besides, the system adopts a set of media servers which hold all media files and provide media data for peers when a peer can't request data from other peers. Nodes in the system are divided into super nodes and normal nodes. The super nodes which have longer online time and better bandwidth than normal nodes are deployed in layer-2, accordingly, the layer-3 overlay consist of normal nodes. All nodes login in tracker server which allocates a unique ID for the new joining nodes and records their login information, in the meantime, the new node should establish a connection with the nearest super node. Those normal nodes connected with the same super node are organized into an application group, called a cluster. Each super node maintains an index of the media available in its network cluster and their locations, and also builds connection with tracker server and some other several neighbored super nodes.

When a streaming session starts, the request node runs a protocol involving three phases: requesting, streaming and caching. In phase I, the request node first sends a lookup request to its super node. If the required segment exists in this cluster, the super node will return a list of candidate supplying nodes, otherwise, the super node will send the request to its neighbors. If the above process all failed, the request node will send the request to media servers. In phase II, after the required segment successfully found, the request node will build a connection with the supplying node owning best performance to acquire the segment. In phase III, the cache space is divided into two parts, the playback buffer and the disk storage. The received segment will firstly cached in playback buffer. When the segment has been watched, then the request node will decide to whether to cache or discard according to the SDCache algorithm (Sect. 43.4.3).

43.3.2 Client Model

Some client models are established according to peers' features and people's habits.

- Super Node's Select Tactic

After observing [12], we can see that there exists stable nodes in P2P systems. We set C_s as the cache size, T_n as the online time remained during each session, B_n as the average upload bandwidth, N_n as the average number of connections, and W_i as the comprehensive performance of peer i, which can be calculated by formula

(43.1). Parameters, $\alpha, \beta, \gamma, \lambda$, are weight factors. The peers which have better comprehensive performance will be selected as super nodes.

$$W_i = \alpha C_s + \beta T_p + \gamma B_p + \lambda N_n, (\alpha, \beta, \gamma, \lambda \subseteq (0,1), \alpha + \beta + \gamma + \lambda = 1) \quad (43.1)$$

- Probability Prediction of Peer Online

Peers in VoD systems are highly autonomous and may join or leave at will. A peer online prediction model is constructed to predict the availability of this peer. The peer's history and current online status is used as the value of predictions for looking ahead periods. We set a_k as the length of time between joining and leaving of a peer which logins in the system at the kth time slot, and $a_{(k-i)}$ denotes the length of online time at the $(k-i)th$ time slot. Then we can use the historical information to forecast the length of online time at nth time slot by: $a_n = \sum_{i=1}^{k} \lambda_i a_{n-i}$, λ_i is a set of constants. We denote the probability of this peer is still staying in the system at the next moment as $Ponline$, and set t as the duration of the nth time slot after the peer joining, then the $Ponline$ can be calculated by:

$$Ponline = \begin{cases} \dfrac{a_n - t}{a_n}, (a_n > t) \\ \dfrac{t - a_n}{t}, (a_n < t) \end{cases} \quad (43.2)$$

43.4 Segment Cache Replacement Strategy

43.4.1 Definition

In this paper, a cache replacement strategy based on supply and demand balance theory of economics is designed. According to this theory, when the demand and supply of a segment is in equilibrium, the performance of the whole system can be optimized. In order to maximally utilize the peers' limited cache space, the number of segments' supply should be proportional to its demand rates. However, peers always have hierarchy, some peers may have limited upload bandwidth, so it may cause some segments can not be scheduled even though it is stored by peers. Therefore the segment's supplies should be amended by its actually available replicas. Besides, those segments after the watching point will have high scheduling urgency degrees. So the segments' demands also should be amended by their urgency degrees. Assume that the media servers have V distinct videos, and video j has the length of L_j. Each segment size is l, which may not suit for the last segment. Then the total number of segments belonging to movie j, M_j, is given as

$M_j = \left\lceil L_j / l \right\rceil$. The total number of segments of all movies is $M = \sum_{j=1}^{V} M_j$. The popularity of segment follows a Zipf distribution [13]. If all the segments are sorted in a descending order, the popularity of segment k, p_k, can be calculated by:

$$p_k = \frac{1/k^\alpha}{\sum_{i=1}^{M} 1/i^\alpha} \tag{43.3}$$

First, we set some definitions as follows:

Segment's demand: segment's demand is defined as the ratio that the times of this segment requested is relative to the total times of all segments requested over a period of time. We set $F_{j,k}(t)$ as the statistics value of the segment k belonging to movie j request at the time t, so the demand of segment k at the time t, $R_{j,k}(t)$, can be calculated by:

$$R_{j,k}(t) = \frac{F_{j,k}(t)}{\sum_{j=1}^{V} \sum_{k=1}^{M_j} F_{j,k}(t)}. \tag{43.4}$$

Segment's supply: Similarly, the ratio that the number of this segment cached is relative to the total replicas stored by all peers of the same cluster over a period of time, is called as segment's supply. We set $I_{j,k}(p)$ as the cache status of segment k belonging to movie j stored by peer p. If the segment k is stored by peer p, the $I_{j,k}(p)$ will be set to 1, if not, it will be put to 0. We denote $Q_{j,k}(i)$ as the number of replicas of segment k belonging to movie j stored by normal nodes within the cluster i, which is recorded by super node i, so $Q_{j,k}(i) = \sum_{p=0}^{N_i} I_{j,k}(p)$. As peers have hierarchy, different peers have different service abilities, some peers even only download from others but never share resources with others. In order to ensure the availability of resources, the $Q_{j,k}(i)$ can be improved by: $Q_{j,k}(i) = \sum_{p=0}^{N_i} \Phi_p \times I_{j,k}(p)$.

Here $\Phi_p = Ponline_p \times W_p$. The Φ_p is the probability of the practical availability of peer p, and W_p is the comprehensive performance of peer p, which is calculated by formula (43.1), $Ponline_p$ is the probability of ongoing online of peer p, which is calculated by formula (43.2). The supply of segment k belonging to movie j in cluster i, $S_{j,k}(i)$, can be calculated by:

$$S_{j,k}(i) = \frac{Q_{j,k}(i)}{\sum_{i=1}^{N_c} \sum_{j=1}^{V} \sum_{k=1}^{M_j} Q_{j,k}(i)}. \tag{43.5}$$

Segment's urgency degree: most people are used to watching movies sequentially, so those segments after the watching point will have more probability to be requested. We set the urgency window size is m, then the urgency degree of next m segments belonging to movie j, $E_{j,(k+m)}$, can be calculated by:

$$E_{j(k+l)} = P_{seqplay}(l)R_{j(k+l)}, (l = 1, 2, \ldots, m) \tag{43.6}$$

Here $P_{seqplay}(l) = \prod_{l=1}^{m} p_{k+l}$. The $P_{seqplay}(l)$ is the probability of segment sequence watched by nodes and p_k is the popularity of segment k by formula (43.3).

Segment's demand degree: segments not only have different demands and supplies, but also have different urgency degrees. We define the demand degree of segment k belonging to movie j, $D_{j,k}$, in formula (43.7), which reflects the caching utilities of segments.

$$D_{j,k} = S_{jk} - R_{jk} + E_{j,k} \tag{43.7}$$

The status of segments' supply, demand and demand degree, all can be described by two-dimensional arrays. Take the segments' demand as an example, it can be described as: $R(t) = \{ (R_{1,1}, R_{1,2} \cdots R_{1,M_1}), \cdots (R_{j,1}, R_{j,2} \cdots R_{j,M_j}), (R_{V,1}, R_{V,2} \cdots R_{V,M_V}) \}$. Each super node stores it and updates it when the super peer receives a request from other peers. Each normal node stores a segments' supply array S and demand degree array D, and updates it periodically.

43.4.2 Statistics for Segment's Supply and Demand

The cluster-based segment dispersion strategy can dynamically adjusts the available supplies according to the peers' demands. Let $D(i)$ be the demand degree of cluster i. The super nodes periodically compute the global demand degree with its neighbors by the distributed consistency algorithm. The process is as follows:

1. each normal node periodically reports the segments' cache status $I_{j,k}$ to its super node;
2. each super node calculates the segments' demand status array R according to formula (43.4) and counts supply status array S within its cluster according to formula (43.5) and computes the demand degree of segments D according to formula (43.7) within the last time slot;
3. super node i sends $D(i)$ to one of neighbors' j;
4. if super node j receives $D(i)$, it updates $D(j)$ to $D(i) + \theta[D(i) - D(j)]$, $\theta \in (0, 1)$. Then super node j returns $\theta[D(i) - D(j)]$ to super node i;
5. if super node i receives $\theta[D(i) - D(j)]$, the super node i update $D(i)$ to $D(i) + \theta[D(i) - D(j)]$.

43.4.3 Supply–Demand Replacement Strategy

Nodes cache or replace segments according to these segments' demand degree. The steps of SDCache dynamic caching replacement algorithm is to be run by peers, see below:

Step1: A new joining node firstly gets the segments' demand degree D from its super node in the same cluster;

Step2: When a new segment i is received, the node will check whether it is within the scope of playback window, if it is in the scope of playback window and the playback buffer is not full, the segment i will be cached in playback buffer, then the node updates the segments' supply array S and exits the whole process. While if the playback buffer is full, the node will replace the earliest watched segment, assuming segment k here, go to next step. If the new segment j is out of the scope of playback window, go to next step;

Step3: When the temp segment j or k accesses the disk cache replacement module, the node first check the disk storage whether it is full or not. If not, the temp segment will be stored in disk storage, then the node updates the segments' supply array S and exits the whole process. Otherwise, go to step 4;

Step4: If this node received request of segment scheduling from other peers in the last time slot, it will calculate all the cached segments' urgency degree array E according to the formula (43.6), otherwise put the E to 0. and then calculate D by formula (43.7), then go to next step;

Step5: Sort the segments' demand degree D from high to low, go to next step;

Step6: If the demand degree of this temp segment is lower than any others which stored in the disk storage, the node will replace the segment which has highest demand degree value with this temp segment, then this node updates S.

43.5 Simulation Experiment and Results Analysis

Simulation experiments are conducted to evaluate the basic characteristics of our cache replacement strategy, which are implemented by PeerSim [14]. In our simulation, peers are connected on the basis of a hybrid P2P VoD model, which are divided into super nodes and normal nodes. We assume a P2P VoD service with 3,000 peers and 5,000 videos, the number of peers in each cluster is 100, the number of request sent to super node by a peer per time slot is 60, the urgency window size is 3. In the initialization process, the comprehensive performance of each node is initialized randomly, and each video has a random length from 1G to 2G. The media's demands and segments' demands are followed by Zipf law with an experience parameter $\alpha = 0.271$. In the simulations, we focus on the

Fig. 43.1 Cache hit ratio
with different cache size

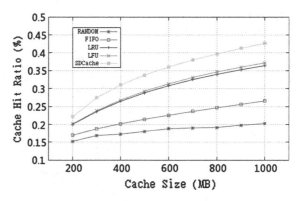

Fig. 43.2 Startup delay
versus the number of segment
demand

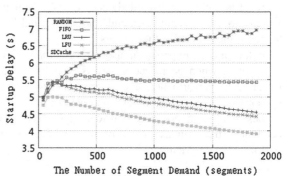

performance advantage of the SDCache comparing with FIFO, RANDOM, LRU
and LFU.

Figure 43.1 contrasts the cache hit ratio of SDCache with several other different
algorithms. For each strategy, with the cache size increasing, the cache hit ratio
also raise. Figure 43.1 shows the SDCache has the highest cache hit ratio, the
second is LFU. Because the SDCache takes many factors into full consideration,
such as the segment popularity, the performance features of peers and neighbors'
cache situations. Along with the growth of cache size, the peers can cache more
watched segments in their cache space. When the cache space is full, peers select
the segment with lowest demand degree to replace them. Therefore, the demand
degree of segments in cache space will become larger than before, and these
segments can be visited more by others. However, the FIFO, LRU and LFU
algorithms only consider the popularity of segments, and ignore cooperative
caching with neighbors, and RANDOM even ignores the popularity of segments.

Figure 43.2 shows the comparison of different strategies' startup delay at the
different demands of segment when the peer cache size is 1G. When the number of
segment demand is small, the segment is rarely cached by neighbors. And it is
mainly downloaded from media servers, so its startup delay jumps up. However,
the segments are cached randomly in RANDOM algorithm even their demand is
small, so the startup delay is shorter than others but SDCache. In SDCache, the

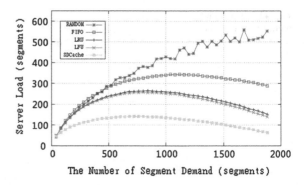

Fig. 43.3 Server load versus the number of segment demand

segment is cached or discarded decided by its demand degree. With the gradually increasing of segments' demand, the startup delay decrease in SDCache and it decrease more notably than others.

In the end, the experiment also contrasts the server load at the different demand of segment when peers cache size is 1G. Figure 43.3 shows that SDCache has better cache performance, and in the same condition, it can efficiently reduce the server load better than other approaches. Because in SDCache strategy, peer caches a segment not only considering the segments' demand, but also taking account of the replicas cached by its neighbors. So the pressure of media server will not change greatly in SDCache.

43.6 Conclusions and Future Work

In this paper, we propose a novel distributed segment cache replacement strategy called SDCache, which is based on supply and demand balance theory of economics under a hybrid P2P VoD architecture to improve the overall P2P VoD system performance. The SDCache not only considers the segment's supply and demand, but also takes account of segment's urgency degree and the practical availability, which realizes cooperative cache among peers. Super nodes utilizes a distributed averaging algorithm to compute the segments' demand degree. All peers first cache those segments with higher demand degrees, which make the valid supplies of every media segment cached in the whole system and their segments' demands in direct proportion. In experiments of contrast with other cache algorithms, SDCache is much better in enhancing the cache hit ratio, shortening the startup delay and decreasing the download pressure of media servers. Beyond this work, there are still some further problems to be studied. Firstly, we should evaluate our proposed algorithm in realistic situations. Secondly, the segment cache replacement strategy should be integrated with segment scheduling strategy to achieve better performance. We will treat them as our future research work.

References

1. Guo Y, Suh K, Kurose J (2003) P2Cast: peer-to-peer patching scheme for VoD service. WWW 2003, Budapest, Hungary, pp 301–309
2. Cui Y, Li B, Nahrstedt K (2004) Ostream: a synchronous streaming multicast in application layer overlay networks. IEEE JSAC, 91–106
3. Ying LH, Basu A (2005) pcVOD: internet peer-to-peer video-on-demand with storage caching on peers. In: The eleventh international conference on distributed multimedia systems DMS'05, Canada
4. Hefeeda M, Habib A, Boyan B, Xu D (2003) PROMISE: peer-to-peer media streaming using collect cast. Technical report, CS-TR 03-016, Purdue University. Extended version
5. Hefeeda MM, Bhargava BK, Yau DKY (2004) A hybrid architecture for cost-effective on demand media streaming. Comput Netw 44(3)
6. Huang Y, Fu TZ, Chiu D-M, Lui JCS, C. Huang. (2008) Challenges, design and analysis of a large-scale P2P-VoD system. ACM SIGCOMM
7. O'Niel E, O'Niel P, Weikum G (1993) The LRU-k page replace-ment algorithm for database disk buffering. ACM SIGMOD Conference, pp 297–306
8. Guo L, Chen S, Zhang X (2006) Design and evaluation of a scalable and reliable P2P assisted proxy for on demand streaming media delivery. IEEE Trans Knowl Data Eng 18(5):669–682
9. Yiu WP, Jin X, Chan SH (2007) VMesh: distributed segment storage for peer-to-peer interactive video streaming. IEEE J Sel Areas Commun 25(9):1717–1731
10. Wu W, Lui JCS (2012) Exploring the optimal replication strategy in P2P-VoD systems: characterization and evaluation. INFOCOM'11
11. Bin C, Stein L, Hai J, Zheng Z (2008) Towards cinematic internet video-on-demand. EuroSys, 109–122
12. Feng W, Jiangchuan L, Yongqiang X (2008) Stable peers: existence, importance, and application in peer-to-peer live video streaming. IEEE INFOCOM
13. Breslau L, Cao P, Fan L, Phillips G, Shenker S (1999) Web caching and zipf-like distributions. Evid Implications, pp 126–134
14. Montresor. A, Jelasity M (2009) PeerSim: a scalable P2P simulator. In: The 2009 IEEE international conference on peer-to-peer computing, pp 99–100

Chapter 44
Methods for Source Number Estimation in Underdetermined Blind Sources Separation

Hui Guo, Yong-qing Fu, Yan Liu and Jian-qiang Wang

Abstract Estimate the number of source signals is a necessary prerequisite for underdetermined blind sources separation (UBSS). The accuracy of sources number estimation has influence to the correctness of the sources separation. For this, a new algorithm—Hough-Windowed is proposed based on the assumption that the source signals are sparse. First, the algorithm constructs straight line equations of the observed signals based on Hough transformation. In order to obtain cluster areas, histogram is cumulated by windowed in transform domain. Then estimate the maximum of every cluster area. The number of different maximum is the number of source signals. Simulation results show the validity and expansibility of the algorithm. At the same time, compared with Potential function, the algorithm reflects the better noise immunity and the lower sparse sensitivity.

Keywords Underdetermined blind sources separation · Estimate the number of source signals · Sparse representation · Hough transformation · Histogram · Windowed method

H. Guo (✉) · Y. Fu
College of Information and Communication Engineering, Harbin Engineering University, Harbin 150001, China
e-mail: chinamengh823@126.com

Y. Fu
e-mail: yqingfu@yahoo.com

Y. Liu · J. Wang
EMC Laboratory, Beijing University of Aeronautics and Astronautics, Beijing 100191, China
e-mail: liuyan0202@sohu.com

J. Wang
e-mail: 370430239one@gmail.com

W. Lu et al. (eds.), *Proceedings of the 2012 International Conference on Information Technology and Software Engineering*, Lecture Notes in Electrical Engineering 210, DOI: 10.1007/978-3-642-34528-9_44, © Springer-Verlag Berlin Heidelberg 2013

44.1 Introduction

Unknown sources can be recovered only from observations without prior knowledge of mixing channel and source signals by using blind sources separation (BSS). Therefore, the research of it has made great contributions up to now [1–3]. But these research should know the number of source signals firstly. Estimate the number of the source signals is a necessary prerequisite for BSS. The accuracy of the source signals' number estimations has influence to the correctness of the sources separation. As a case of BSS, underdetermined blind sources separation (UBSS, the number of observation signals is less than that of source signals) is no exception. Because of practical and challenging, UBSS is currently receiving increased interests. The mathematics model of UBSS is

$$\mathbf{X}(t) = \mathbf{A}\mathbf{S}(t) + \mathbf{N}(t), \quad t = 1, \ldots, T \tag{44.1}$$

where $\mathbf{X}(t) = [x_1(t), x_2(t), \cdots, x_m(t)]^T$ is the vector of observed signals. $\mathbf{A} = [a_1, a_2, \cdots, a_n] \in \mathbf{R}^{m \times n}$ is the mixture matrix with $\mathbf{a}_i = [a_{1i}, a_{2i}, \cdots, a_{mi}]^T, i = 1 \ldots, n$. $\mathbf{S}(t) = [s_1(t), s_2(t), \ldots, s_n(t)]^T$, is the vector of source signals. $\mathbf{N}(t) = [n_1(t), n_2(t), \cdots, n_m(t)]^T$ is noise. Generally, we suppose noise doesn't exist. In this paper considering m is less than n, namely UBSS.

Because of lack of information, it is difficult to estimate the number of source signals. At present, the clustering algorithms [4, 5] are general methods for UBSS sources' number estimation based on sparse representation. In this paper, sparse representation of signals still be adopted. The difference is not directly cluster the observed signals, avoiding global maximum.

44.2 Sparse Representation of Underdetermined Blind Sources Separation

In order to solve the problem that estimating the number of source signals, suppose the source signals are sparse. Sparse signal is the one whose most samples are almost zeros, while a few samples are far different from zero. If some source signals aren't sparse in time-domain, we can make them sparse through some transformation, such as, short-time fourier transform (STFT) [6], Gabor transformation [7].

Here, we suppose that the source signal is nonzero and the other source signals are zero or are near to zero in time. Thus, Eq. (44.1) can be written as:

$$\begin{bmatrix} x_1(t) \\ x_2(t) \\ \vdots \\ x_m(t) \end{bmatrix} = \begin{bmatrix} a_{1i} \\ a_{2i} \\ \vdots \\ a_{mi} \end{bmatrix} s_i(t) \Rightarrow \frac{x_1(t)}{a_{1i}} = \cdots = \frac{x_m(t)}{a_{mi}} = s_i(t) \qquad (44.2)$$

From the above equation, a conclusion can be drawn, the sampling time in which the source signal $s_i(t)$ is playing a leading role can determine a straight line through origin [8]. That is to say, the number of lines is equal to the number of the number of source signals. Therefore, the key is how to get the lines' number by clustering $x(t)$ in all time. For this, a new algorithm called Hough-Windowed is proposed. The following are introduced in detail.

44.3 Hough-Windowed

From the above section, observed signals $x(t)$ should be clustered in the whole sampling time. Our Hough-Windowed algorithm is divided into two steps. First, construct straight line equations of the observed signals based on Hough transformation then obtain cluster areas by windowed method. All maximum values' number of cluster areas is the number of source signals.

44.3.1 Hough Transformation

In image processing, Hough transformation is used for image matching. It makes a given shape in the original image centered on certain position in transform domain, forming a peak point. Without loss of generality, it also can detect the straight line. Thus, This method can be introduced to estimation of source signals.

First, the straight line equations of the observed signals were constructed [9], written as:

$$\begin{bmatrix} x_m(t) \\ x_{m-1}(t) \\ \vdots \\ x_3(t) \\ x_2(t) \\ x_1(t) \end{bmatrix} = D \begin{bmatrix} \cos \varphi_{m-1}^t \\ \sin \varphi_{m-1}^t \cos \varphi_{m-2}^t \\ \vdots \\ \sin \varphi_{m-1}^t \sin \varphi_{m-2}^t \cdots \sin \varphi_3^t \cos \varphi_2^t \\ \sin \varphi_{m-1}^t \sin \varphi_{m-2}^t \cdots \sin \varphi_3^t \sin \varphi_2^t \cos \varphi_1^t \\ \sin \varphi_{m-1}^t \sin \varphi_{m-2}^t \cdots \sin \varphi_3^t \sin \varphi_2^t \sin \varphi_1^t \end{bmatrix} \qquad (44.3)$$

where $D = \sqrt{\sum_{i=1}^{n} x_i^2(t)}$ is the distance form observed point to origin. φ_{n-1}^t is the angle between x_n axis and straight line which is constructed by observed point and origin at the t moment. $\varphi_i^t(i = n - 2, \ldots, 1)$ is the angle between x_{i+1} axis and projection line which is from the line formed by observed point and origin to $i + 1$ dimension space. $\Phi(t) = (\varphi_1^t, \varphi_2^t, \ldots, \varphi_{n-1}^t) \in [0, \pi)^{n-1}$ represent the variables in the transform space.

In order to get the distribution of variables, we let $k = 1, \ldots, \pi/h$ be the number of subareas which belong to histogram. h is quantization step of Hough transformation. Import the basis function

$$\phi_{(k_1, \cdots, k_{n-1})}(\mathbf{h}, \mathbf{u}) = \begin{cases} 1 \; , \; [u_i/h] = k_i - 1 \\ 0 \; , \; elsewhere \end{cases} \tag{44.4}$$

where $[.]$ represent the greatest integer less than. $i = 1, \ldots, n - 1$, contains all of the variable dimensions. Therefore, the number of variables which fall into the subarea of histogram is

$$N_k = \sum_{t=1}^{T} \phi_{(k_1, \cdots, k_{n-1})}(\mathbf{h}, \Phi(t)) \tag{44.5}$$

44.3.2 Windowed Method

When the above steps are completed, get the distributional number of variables by using histogram in the transform domain. It is very easy to fall into the global maximum value, causing the error estimation, if directly estimate the peak of histogram. In order to solve the problem, the windowed method was proposed. The purpose of windowed is smoothing the abnormal value and getting cluster areas. Generally, i dimensional observed signals using $i - 1$ dimensional window. Whether the histogram is high-dimensional or low-dimensional, all the windows are collectively referred to as the hypercube window.

Recalculate classification number of variables after windowed method, that is

$$N_{j+1} = \sum_{k=1+jd}^{L+jd} w_k N_k \tag{44.6}$$

where $j = 0, \cdots, [\pi/hd] - [L/d] + 1$ is classification number of variables after recalculation. Let L and d be the hypercube window length and moving step. $w_k = N_k / \sum_{k=1}^{\pi/h} N_k$ is weight, obtained by the ratio of variable number in every subarea to the total variables number.

From the above algorithm, the number of source signals is gotten, that is the number of cluster area.

44.4 Experiments and Results

44.4.1 Pretreatment

In order to prepare the mix, the following steps were followed:

- If source signals aren't sparse in time-domain, adopt STFT with Hanning windows to make them sparse in frequency-domain.
- The l_2 norm representation of the observed signals can be obtained by reducing the effect of noise.
- In order to simplify the calculation, the data were treated as follow:

$$\hat{\mathbf{x}}(t) = sign(x_1(t)) \times \mathbf{x}(t), \ t = 1, \cdots, T \tag{44.7}$$

where $sign(\cdot)$ express the sign of.

44.4.2 Experiment on Sound Sources with Different SNR

In order to show the anti noise performance of the algorithm in this chapter, we take $m = 2$, $n = 4$ ($2m4s$), namely there are two observed signals and four source signals, when SNR $= 40$dB and SNR $= 5$ dB. The source signals is voice signals (see Fig. 44.1a), and the mixing matrix is generated by MATLAB function randn() as follows:

$$A = \begin{bmatrix} -0.5087 & 0.1774 & 0.9977 & -0.8510 \\ -0.8609 & 0.9841 & -0.0681 & -0.5252 \end{bmatrix}.$$

When SNR $= 40$ dB, the scatter plot mixture signals in time-domain is shown the Fig. 44.1b. After STFT transformation, the scatter plot in frequency-domain just as Fig. 44.1c. Figure 44.1d shows the histogram of variables. In order to reflect the excellent anti-noise performance of algorithm, compared the proposed algorithm and the potential function method. The results of comparison are shown in other subgraph of Fig. 44.1. In the simulations, the parameters are selected as shown in figure legend.

From the above simulation results, it is obvious that the cluster effect of the proposed algorithm is better than which of the potential function method. Although the potential function method for cluster effect is also good when

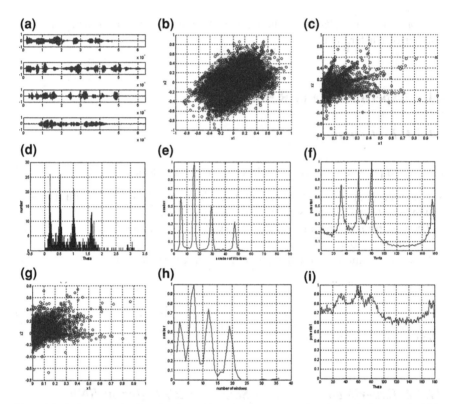

Fig. 44.1 Estimation of voice signals number (2m4s) **a** Four voice signals **b** Scatter plot X in time domain (SNR = 40 dB) ,**c** Scatter plot Xin frequency domain (SNR = 40 dB), **d** Histogram of Φ in transform domain **e** Hough-windowed (SNR = 40 dB, *Th* = 0.1) **f** Potential function (SNR = 40 dB, *N* = 540, *λ* = 45) **g** Scatter plot X in frequency (SNR = 5 dB) **h** Hough-windowed (SNR = 5 dB, *Th* = 0.3) **i** Potential function domain (SNR = 5 dB, *N* = 540,*λ* = 45)

SNR = 40 dB. Under the condition of SNR = 5 dB, Hough-Windowed algorithm still can be seen that there are four clear clustering area. The simulation experiments show that the proposed algorithm has a good anti-noise performance.

As a reference, the setting of the different parameters are shown in Table 44.1. From this table, we can see that the lower signal-to-noise ratio, the greater the parameter. This can be seen the principle of parameter selection. The Hough-Windowed algorithm still can estimate the number of signal sources correctly when SNR is lower than 3 dB. The data shows the excellent anti-noise performance of this algorithm once again.

Table 44.1 Parameter selection and estimation results (2m4s)

Quantization step	Sparsity measure	Signal to noise ratio (dB)	Window width	Moving step	Threshold	Local peak	Sources number
$\pi/540$	0.9	40	6	3	0.1	0.5959;1.0000;0.4384;0.4521	4
		30				0.9118;1.0000;0.5588;0.5196	4
		20	10	8	0.2	0.7556;1.0000;0.64444;0.7778	4
		15				0.7879;1.0000;0.7273;0.5758	4
		10				0.8947;1.0000;0.7368;0.7368	4
		5	20	15	0.3	0.5556;1.0000;0.6111;0.6111	4
		3				0.5625;1.0000;0.4375;0.5625	4
		0				0.5000;1.0000;0.6667;0.7500;0.5000	5

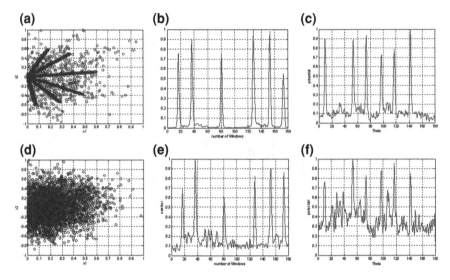

Fig. 44.2 Estimation of simulated signals number (2m6s), **a** Scatter plot X in frequency domain ($Q = 0.9$) **b** Hough-windowed ($Q = 0.9$, $Th = 0.3$) **c** Potential function ($Q = 0.9$, $N = 540$, $\lambda = 45$) **d** Scatter plot X in frequency domain ($Q = 0.6$) **e** Hough-windowed ($Q = 0.6$, $Th = 0.35$) **f** Potential function ($Q = 0.6$, $N = 540$, $\lambda = 45$)

44.4.3 Experiment on Random Sources with Different Sparsity

In this experiment, we first construct sparse signal. Construction method is shown as follow:

$$\mathbf{S} = rand(n, T) * \text{Max}(0, sign(rand(n, T) - Q)) \tag{44.8}$$

where $0 \le Q \le 1$, the larger the Q value, the higher the sparsity. The mixing matrix is randomly taken as:

$$\mathbf{A} = \begin{bmatrix} -0.2889 & 0.7865 & -0.4583 & -0.6627 & -0.9861 & -0.1186 \\ -0.9574 & -0.6176 & 0.8888 & -0.8865 & -0.1660 & 0.9929 \end{bmatrix}.$$

Figure 44.2a and d show the scatter plot mixture signals in frequency-domain when Q is 0.9 and 0.6. Similarly, compared the proposed algorithm and the potential function method.

When $Q = 0.9$, the sparsity of signals is excellent. Under this condition, a good cluster result is obtained by using Hough-Windowed algorithm, but there are some interference in potential function method. This situation is more apparent in the condition of $Q = 0.6$. The data in Table 44.2 can be further described that the proposed algorithm reflects the lower sparse sensitivity. This means that Hough-Windowed algorithm relax sparsity of source signals.

Table 44.2 Parameter selection and estimation results (2m6s)

Quantization STEP	Signal to noise ratio (dB)	Sparsity measure	Window width	Moving step	Threshold	Local peak	Sources number
$\pi/540$	40	0.9	15	10	0.3	0.7250;0.9000;0.7250;1.0000;0.7750;0.8000	6
		0.8				0.8085;0.8723;0.8085;1.0000;0.8511;0.7872	6
		0.75	25	20	0.35	0.7907;1.0000;0.7209;0.9767;1.0000;0.8140	6
		0.7				0.6327;0.7755;0.7143;1.0000;0.6327;0.6327	6
		0.65				0.8286;0.9429;0.8571;0.8571;1.0000;0.7714	6
		0.6				0.8710;1.0000;0.8065;0.8387;0.8065;0.9032	6
		0.55	30	25	0.4	0.9259;0.8519;0.9259;0.8148;0.9630;1.0000	6
		0.5				0.8846;0.6923;0.7692;1.0000;0.8462	5

44.5 Conclusions

This chapter gives a novel algorithm to estimate the number of source signals for underdetermined blind sources separation. Because the algorithm adopts Hough transformation and windowed method, so the feasible and outstanding performance of the Hough-Windowed algorithm is expressed from the simulation results. In addition, it is easy to expand the algorithm to high dimensional underdetermined blind sources separation problems.

Acknowledgments The authors would like to thank for the projects supported by National Natural Science Foundation of China (Grant No. 61172038 and 60831001). Thanks should also be given to the EMC Laboratory of Beijing University of Aeronautics and Astronautics, which provides a good environment and convenient conditions for this research.

References

1. Hyvarinen A, Oja E (2000) Independent component analysis: algorithms and applications. Neural Networks 13(4–5):411–430
2. Hyvarinen A (1999) Fast and robust fixed-point algorithms for independent component analysis. IEEE Trans Neural Networks 10(3):626–634
3. He ZS, Xie SL, Fu YL (2006) Sparse representation and blind source separation of ill-posed mixtures. Sci China Ser F-Inf Sci 49(5):639–652
4. O'Grad YPD, Pearlmutter BA (2004) Hard-lost: modified k-means for oriented lines. Proceedings of the Irish signals and systems conference, pp 247–252
5. Bofill P, Zibulevsky M (2001) Underdetermined blind source separation using sparse representations. Signal Process 81(11):2353–2362
6. Bofill P, Zibulevsky M (2000) Blind separation of more sources than mixtures using sparsity of their short-time fourier transform. Second international workshop on independent component analysis and blind signal separation. Espoo, pp 87–92
7. Cordero E, Nicola F, Rodino L (2009) Sparsity of Gabor representation of Schrodinger propagators. Appl Comput Harmonic Anal 26(3):357–370
8. Shindo H, Hirai Y (2002) Blind source separation by a geometrical method. Proceedings of the international joint conference on neural networks. Honolulu, pp 1108–1114
9. Fu N, Peng X, (2008) K-hough underdetermined blind mixing model recovery algorithm. J Electron Meas Instrum. 22(5):63–67. (in Chinese)

Chapter 45
Frequency Stability Measurement Method Including Anomalies Detection

Hang Gong, Xiangwei Zhu, Feixue Wang and Fan Chen

Abstract The measurement of frequency stability will significantly deviate from the true value if anomalies occur. This paper proposes a frequency stability measurement method including anomalies detection. Firstly the impact of anomalies such as phase outliers, phase jumps, frequency outliers and frequency jumps on frequency stability measurements is analyzed. Then an anomalies detection method using kurtosis of third difference sequence as test statistic is proposed, followed by an anomalies identification method according to the characteristic of anomalies. Based on this, anomalies elimination method on the difference sequence is discussed, and Allan variance calculation method based on the adjacent second difference sequence of clock offset is derived. Finally, the integrated frequency stability measurement solution is given, and its effectiveness is verified by experiment. The methods in this paper can be used for long-term continuous online frequency stability measurement system.

Keywords Frequency stability measurement · Anomalies detection · Anomalies elimination · Kurtosis detection · Allan deviation · Dual-mixer time difference · Third difference

45.1 Introduction

Frequency stability of local oscillator is very important for communication systems. The abnormal measurement results often occur due to environment interference, power supply outage, equipment malfunctions, test cables moved and

H. Gong (✉) · X. Zhu · F. Wang · F. Chen
Satellite Navigation R&D Center, School of Electronic Science and Engineering, National University of Defense Technology, Changsha, Hunan 410073, China
e-mail: gonghang@nudt.edu.cn

W. Lu et al. (eds.), *Proceedings of the 2012 International Conference on Information Technology and Software Engineering*, Lecture Notes in Electrical Engineering 210, DOI: 10.1007/978-3-642-34528-9_45, © Springer-Verlag Berlin Heidelberg 2013

other factors when measuring frequency stability of frequency sources. It is inevitably arise for applications requiring long-term continuous monitoring such as long-term stability measurement of atomic clocks of ground stations or satellites, but a greater price will always be paid to repeat the measurement. Present universal frequency stability measurement devices such as TSC5125 and PN9000 generally do not have the ability of anomalies detection and elimination, and postprocessing method cannot satisfy continuous online measurement requirement. Thus a method of frequency stability measurement which can eliminate anomalies effects and can be used for online measurement is urgently desired.

In this paper, the impact of a variety of anomalies on the frequency stability measurement is analyzed firstly, then anomalies detection, identification and elimination methods are proposed, and an integrated online measurement solution is presented, finally the effectiveness of the methods are verified by experiment.

45.2 Impacts of Anomalies on Frequency Stability Measurement

Generally, the clock offset between frequency device under test (DUT) and the reference is firstly measured by time difference measurement method such as dual-mixer time difference measurement, and then the frequency stability based on the clock offset, or equally phase, can be calculated [1]. Sometimes, anomalies of the clock offset occur when the environment abnormalities present. The usual anomalies include: phase outliers, phase jumps, frequency outliers, frequency jumps and frequency drifts. These anomalies may occur alone or sequentially. The first four kinds of anomalies should be eliminated when calculating frequency stability, but the frequency drifts should be retained in the calculation as they are the inherent characteristic of the clock. This paper only discusses anomalies detection and elimination method of phase outliers, phase jumps, frequency outliers and frequency jumps.

Firstly we will analyze how anomalies impact on frequency stability measurement. Among the four kinds of anomalies, frequency outliers and phase jumps have the same effect on clock difference, so they are treated as same anomalies in this paper. The following discussion is based on actual normal 10 MHz clock offset measurement of a cesium atomic clock with an active hydrogen maser as reference. The measurement duration is 12 h, and anomalies are added at the time of 4 and 8 h, in which the amplitude of phase jumps is ± 5 ns, and frequency jumps are set to $\pm 1 \times 10^{-12}$. Figures 45.1, 45.2, 45.3, 45.4, 45.5, 45.6 are clock offset plots and the corresponding frequency stability measurement results of the above clock with anomalies.

Figure 45.4 indicates that phase outliers have impact on short-term stability measurement. Frequency jumps affect the long-term stability measurements as shown in Fig. 45.6. Figure 45.5 shows that short-term and long-term stability will

Fig. 45.1 Clock with phase outliers

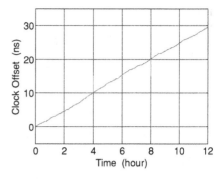

Fig. 45.2 Clock with phase jumps or frequency outliers

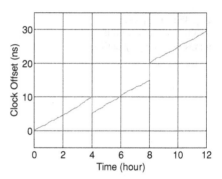

Fig. 45.3 Clock with frequency jumps

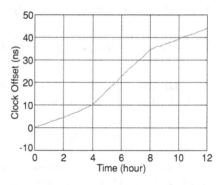

both be affected by phase jumps and frequency outliers. Thus the frequency stability measurement will significantly deviate from the true value if clock observations contain all of these four kinds of anomalies. Therefore, we need to eliminate the impact of these anomalies on frequency stability measurement when they actually exist.

Fig. 45.4 Frequency
stability of clock with phase
outliers

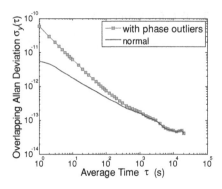

Fig. 45.5 Frequency
stability of clock with phase
jumps or frequency outliers

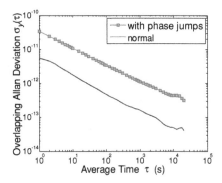

45.3 Measurement Method Including Anomalies Detection

45.3.1 Anomalies Detection Method

As mentioned above, the direct observation of frequency stability measurement is
clock offset of DUT relative to the reference clock:

$$x(n) = x_0 + y_0(n - n_0) + \frac{1}{2}d_0(n - n_0)^2 + \varepsilon_x(n) \tag{45.1}$$

The first three variables are deterministic component of DUT clock, in which x_0
is the phase offset, y_0 is the frequency offset, d_0 is linear frequency drift, and
$\varepsilon_x(n)$ is the random component of DUT, i.e. the phase noise.

There are two requirements for anomalies detection in frequency stability
measurement as follows:

- The first three components of (45.1) are usually presented in DUT clock. Due to
  the presence of frequency offset and frequency drift, the test statistic needs to
  contain the modelling and estimation of these parameters if it is constructed
  directly based on clock observation [2]. Therefore, a test statistic which is free
  from the frequency offset and frequency drift is desired;

Fig. 45.6 Frequency
stability of clock with
frequency jumps

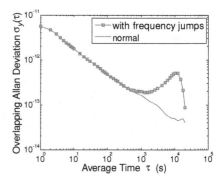

Table 45.1 The third differences of 4 kinds of anomalies

Anomaly type	Clock offset	1st difference	2nd difference	3rd difference
Phase outlier				$s_1(n)$
Phase jump/frequency outlier				$s_2(n)$
Frequency jump				$s_3(n)$

- On the other hand, actual anomalies may contain all the four above types, which have different impact on clock offset. The detection algorithm will be very complex if we construct test statistic for each anomaly respectively [3, 4]. As a result, a test statistic which is adapted for all the four kinds of anomalies is required.

Based on the above requirements, to eliminate the impact of frequency offset and frequency drift, the third difference between adjacent points of clock offset sequence with the minimum sampling interval of τ_0 are taken firstly. Equation. (45.2) shows that, the deterministic component including the frequency offset and frequency drift is eliminated except the random component after third differential. On the other hand, as be shown in Table 1, all the four kinds of anomalies are converted to outliers form after third differential, which can be handled by the same method just as detecting outliers. So we take the third difference of clock offset sequence as a new observation to construct test statistic.

$$z(n) = \Delta^3 x(n) = \varepsilon_x(n+3) - 3\varepsilon_x(n+2) + 3\varepsilon_x(n+1) - \varepsilon_x(n) \qquad (45.2)$$

In the case of a sufficient number of observations, the random component sequence $\{\varepsilon_x(n)\}$ of normal clock offset approximates to a Gaussian noise, and is dominated by white phase modulation noise or white frequency modulation noise at sample interval of 1 s [5], then by Eq. (45.2), the distribution of third difference sequence $\{z(n)\}$ is also distributed as Gaussian approximatively. The distribution

Fig. 45.7 Distribution of the
third difference of normal
clock offset

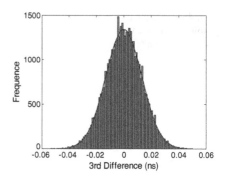

Fig. 45.8 Normal
probability plot of third
difference

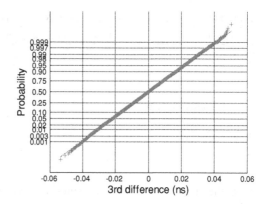

of the third difference of the normal cesium clock offset in Fig. 45.1 is shown in
Fig. 45.7, in which the fitting probability density function agrees well with
Gaussian distribution and performs as a straight line in normal probability paper of
Fig. 45.8, and the correlation coefficient with the normal distribution is close to 1.
All the above facts indicate that the assumption that $\{z(n)\}$ approximates to
Gaussian distribution is tenable.

Based on the above assumption, the anomalies detection problem of this paper
is converted to the problem of detecting abnormal signal $s_1(n)$, $s_2(n)$ and $s_3(n)$ of
Table 45.1 in noise sequence with approximate normal distribution. The latter is a
multiple hypothesis testing problem with unknown noise variance, unknown signal
amplitude and unknown arrival time. So the test statistic derived by generalized
likelihood ratio test contains the estimation of these parameters, which lead to
complex implementation [6], and the detection performance will not be much
better than single point detection because the signal length is very short.

If the abnormal signal is treated as consecutive outliers, and the signal characteristic is used for anomalies type identification, according to the approximate
Gaussian distribution characteristic of $\{z(n)\}$, kurtosis detection method can be
used in which the test statistic is the kurtosis of the sequence as follows[7]:

$$T(n) = \frac{n \sum_{i=1}^{n} (z_i - \bar{z})^4}{\left[\sum_{i=1}^{n} (z_i - \bar{z})^2 \right]^2} > \gamma, \quad \gamma = B(\alpha) \tag{45.3}$$

When $T(n)$ exceeds the threshold, the farthest point $z(k)$ from the mean $\bar{z}$ is judged as a outlier, where γ is the detection threshold determined by α, and $B(\alpha)$ is derived from the kurtosis of standard Gaussian distribution.

Since $\{z(n)\}$ is only approximate normal distribution, so it also has many stragglers except the actual outliers. Latter analysis will indicate that eliminating outliers will introduce error in frequency stability calculation. Therefore the false alarm should be reduced as low as possible, i.e. the probability of judging stragglers to outliers should be reduced. Table 45.1 show that the four kinds of anomalies appearing in the third difference are all consecutive outliers, which can be used for identification outlier types as follows:

- When four consecutive outliers are detected, a phase outlier occurs at the first outlier point;
- When three consecutive outliers are detected, a phase jump or frequency outlier occurs at the first outlier point;
- When two consecutive outliers are detected, a frequency jump occurs at the first outlier point;
- When a single isolated outlier is detected, a straggler occurs.

With identification of the anomalies types, the false alarm probability of this anomalies detection method is α^2, and α is called the detection probability in this paper.

The above method can be implemented easily, but detection performance declines a bit because only third difference distribution of sequence is used, and signal parameters are not estimated.

45.3.2 Anomalies Elimination Method

Anomalies should be eliminated when they are detected. If the elimination is taken on the clock offset sequence, the outliers should be estimated accurately, and the frequency offset and frequency drift effects need to be compensated, also the estimation of the frequency offset and frequency drift need to achieve sufficient accuracy to prevent anomalies elimination greatly impact on the frequency stability calculation. For the purpose of easily implemented, the values of frequency offset, frequency drift and anomalies are not estimated. On the other hand, in order to adapt to four different kinds of anomalies, they should be eliminated in the difference sequence, with the method of setting the outliers to 0 directly.

The frequency stability of frequency sources is usually characterized by Allan deviation, overlapped Allan deviation and modified Allan deviation. Overlapped

Allan variance is commonly used in frequency stability measurement [8], which is defined as follows:

$$\sigma_y^2(\tau) = \frac{1}{2(N - 2m)\tau^2} \sum_{k=1}^{N-2m} (x_{k+2m} + 2x_{k+m} + x_k)^2 \tag{45.4}$$

where $\tau = m\,\tau_0$, and τ_0 is the minimum sampling interval.

Equation (45.4) shows that the overlapped Allan variance is the statistics of the second difference of clock offset. So anomalies should be eliminated in the second difference sequence. But the elimination must be taken on the second difference sequence $\{d(n)\} = \{x(n+2) - 2x(n+1) + x(n)\}$ with sampling interval of τ_0. Therefore the second difference sequence with sampling interval of $m\,\tau_0$ need to be calculated from $\{d(n)\}$ to get the overlapped Allan variance. Equation (45.5) shows the calculation method. The frequency stability measurement values can be obtained by Eqs. (45.4) and (45.5) after elimination of outliers.

$$\underbrace{x_{k+2m} - 2x_{k+m} + x_k}_{\delta_k} = \underbrace{[1\ 2\ \cdots m-1\ m\ m-1\ \cdots 2\ 1]}_{M} \cdot D^T \tag{45.5}$$

where $D = [d_{k+2m-2}\ d_{k+2m-3} \cdots d_{k+m}\ d_{k+m-1}\ d_{k+m-2} \cdots d_{k+1}\ d_k]\cdot$.

Frequency stability calculation errors introduced by the above elimination method will be analysed as follows.

Let $\{\delta(n)\} = \{x(n+2m) - 2x(n+m) + x(n)\}.d_k$ is corrected to 0 when anomaly x_k is detected, supposing that the correcting error is e_k, then errors of $\{\delta(n)\}$ introduced by this correction can be deduce from Eq. (45.5) as following:

$$\varepsilon_\delta(n) = \begin{cases} [n - (k - 2m + 2)]e_k & n \in [k - 2m + 2, k - m + 1] \\ m - [n - (k - m + 1)]e_k & n \in (k - m + 1, k] \\ 0 & else \end{cases} \tag{45.6}$$

$\varepsilon_\delta(n)$ is a trigonometric function with maximum value of me_k occurring at $n = k - m + 1$, thus great errors will arise when the average time $m\,\tau_0$ is very large, and then measurement results of long-term frequency stability will be corrupted.

e_k could be reduced by one or two orders of magnitude if d_k is estimated, but it still will be amplified when m is large enough, so the stability measurement will not be improved significantly. In this case, the impact of errors can be eliminated by directly removing $\{\delta(n), n \in [k - 2m + 2, k]\}$ which are corrupted by e_k, but it will reduced by $2m-1$ points of $\{\delta(n)\}$ observations for every anomaly x_k, thus will cause confidence reduction of frequency stability measurement with larger average time $m\,\tau_0$ [8].

A feasible method to mitigate the above shortcomings is that we can eliminate anomalies on the first difference sequence $\{y(n)\} = \{x(n+1) - x(n)\}$, and derive second difference sequence $\{d'(n)\}$ from it, then calculate frequency stability by Eqs. (45.4) and (45.5). Supposing that the error of correcting y_k to 0 is e_k when

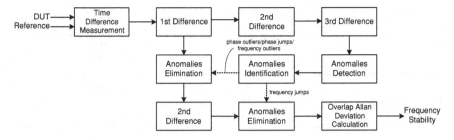

Fig. 45.9 Integrated measurement method of frequency stability

anomaly x_k is detected, then errors of $\{d'(n)\}$ introduced by this correction can be deduced as follows:

$$\varepsilon_d(n) = \begin{cases} e_k & n = k \\ -e_k & n = k+1 \\ 0 & else \end{cases} \tag{45.7}$$

There are two adjacent error components in the $\{d'(n)\}$ with equal amplitude and opposite sign, then $\{\delta'(n)\}$ deriving form $\{d'(n)\}$ by Eq. (45.5) will has constant error of e_k or $-e_k$ for every m when $n[k-2m+2,k]$. It will not affect the stability calculation as the value of e_k is the same order of magnitude as δ_k.

Anomalies such as phase outliers, phase jumps and frequency outliers can be eliminated on the first difference sequence $\{y(n)\}$ using the above method. But frequency jumps cannot be taken that way, because $\{y(n)\}$ still contains jumps, so they still need to be removed in $\{d(n)\}$. In despite of this, confidence reduction caused by this method can be accepted because frequency jumps will not occur frequently in common measurement.

45.3.3 Integrated Measurement Method of Frequency Stability

Based on the above analysis, an integrated frequency stability measurement method including anomalies detection is given in Fig. 45.9. Firstly, the clock offset between DUT and reference is observed though time difference measurement method such as dual-mixer time difference measurement and third difference is derived from clock offset. Then anomalies can be detected and identified based on the third difference sequence. According to identification results, anomalies of phase outliers, phase jumps and frequency outliers are eliminated on the first difference sequence, and frequency jumps are removed from the second difference sequence. Finally, the frequency stability of DUT can be calculated using Eqs. (45.4) and (45.5) based on the second difference sequence with anomalies elimination.

Table 45.2 Anomalies added in experiment

Anomaly type	Anomalies time(h)	Anomalies amplitude
Phase outlier	2	+0.5 ns
	4	−0.5 ns
Phase jump/frequency outlier	6	+0.5 ns
	8	−0.5 ns
Frequency jump	10	$+1 \times 10^{-10}$
	11	-1×10^{-10}

Fig. 45.10 Clock with anomalies

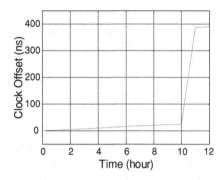

45.4 Verification Experiment

In this section we will verify the above method by the following experiment. Firstly we will measure the 10 MHz clock offset of a cesium atomic clock experiencing no anomalies though dual-mixer time difference measuring device with an active hydrogen maser as reference, and the observation duration is 12 h. Then anomalies of phase outliers, phase jumps, frequency outliers and frequency jumps as shown in Table 45.2 are added in the observation. Finally we use the above method to detect, identify and eliminate anomalies, and then calculate the frequency stability.

Figure 45.10 shows clock offset with these anomalies. Frequency stability derived from Fig. 45.10 significantly deviates from normal clock as shown in Fig. 45.13. Figure 45.11 shows that four kinds of anomalies are converted to outliers form after taking third difference, which can be used for anomalies detection by the method in this paper.

Figure 45.12 shows the results of anomalies detection and identification. The detection probability is set to 0.01, so the false alarm probability is 1×10^{-4}. From Fig. 45.12 we can see that the kurtosis statistic greatly exceeded the threshold when outliers occurred, and all the 6 added anomalies have been detected correctly. Some stragglers are also detected because the third difference is not a standard Gaussian sequence. But the stragglers did not cause false alarm because anomalies identification algorithm worked. Anomaly value of 2, 3 and 4 in Fig. 45.12 represents that anomalies are frequency jumps, phase jumps/frequency

Fig. 45.11 Third difference
sequence of clock

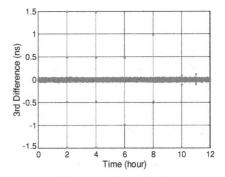

Fig. 45.12 Anomalies
detection and identification
result

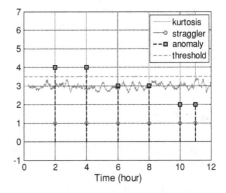

Fig. 45.13 Frequency
stability after anomalies
eliminated

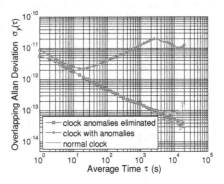

outliers, and phase outliers, respectively. The types of the 6 anomalies have been correctly identified by the proposed method as shown in Fig. 45.12.

Figure 45.13 shows the frequency stability measurement result after anomalies elimination, which agrees well with normal clock, but the confidence of the measurement with larger average time $m \, \tau_0$ was reduced a bit just because some observations were removed when frequency jumps occurred. All the results above have indicated the effectiveness of the method in this paper.

45.5 Conclusion

Anomalies in frequency stability measurement greatly increased the measurement errors. This paper analyzed the impact of four anomalies on frequency stability measurement firstly, then proposed an anomalies detection and identification method with kurtosis of the third difference of clock as test statistic. According to this, an anomalies elimination method based on the first or second difference sequence was presented, and an Allan deviation calculation method based on the adjacent second difference sequence of clock was derived. Finally, the above methods were verified in frequency stability measurement experiment with all the four kinds of anomalies. These methods are meaningful for designing long-term continuous online frequency stability measurement system.

As a certain number of observations were removed when handling frequency jumps anomalies, the confidence of large average time decreased than normal measurement. Anomalies elimination method with less error will be further studied in the future.

References

1. Allan DW, Daams H (1975) Picosecond time difference measurement system. Proceedings of the 29th annual frequency control symposium, Army signal engineering laboratories, New Jersey, pp 404–411
2. Galleani L, Tavella P (2012) Detection of atomic clock frequency jumps with the kalman filter. IEEE Trans UFFC 59:504–509
3. Nunzi E, Galleani L, Tavella P, Carbone P (2007) Detection of anomalies in the behavior of atomic clocks. IEEE Trans Instrum Meas 56:523–528
4. Riley WJ (2008) Algorithms for frequency jump detection. Metrologia 45:154–161
5. Allan DW (1987) Time and frequency (time domain) characterization, estimation and prediction of precision clocks and oscillators. IEEE Trans UFFC 34:647–654
6. Kay SM (1998) Fundamentals of statistical signal processing; detection Theory, prentice-hall, New Jersey, pp 253–260
7. Chinese national standard gb/t 4883–2008. statistical interpretation of data—detection and treatment of outliers in the normal sample (in Chinese)
8. Riley WJ (2008) Handbook of frequency stability analysis, US Government Printing Office, Washington, pp 6–42

Chapter 46
Mobile Peer Management in Wireless P2P Resource Sharing System

Hui Zhou and Jie Yang

Abstract The common characteristics of peer-to-peer (P2P) overlay networks and wireless multi-hop network, such as self-organization, decentralization, hop-by-hop message transmission mode and high degree of dynamicity, lead to research of operating wired P2P applications on wireless multi-hop networks. A peer location management mechanism was proposed to address the effect of peer mobility on P2P resource sharing system in a two-tier Wireless Mesh Network (WMN). A manager router which assigned to each mobile peer takes the responsibility of peer location, reduces deletion and republishing operations for indexes of shared resources due to the movements of peers. The simulation results show peer location management mechanism is effective in decreasing message overhead and increasing accurate ratio of searches in P2P resources sharing system with high churn rates.

Keywords Wireless mesh networks · Peer to Peer networks · File sharing · Mobile peer · Location management

46.1 Introduction

According to the feature of WMN, a P2P resource sharing system has been proposed in a two-tier WMN [1]. The lower tier is composed of mobile mesh clients, which provide resources to be shared in the P2P system; the upper tier is composed

H. Zhou (✉) · J. Yang
Key Laboratory of Broadband Wireless Communications and Sensor Networks,
Wuhan University of Technology, 122 Luoshi Road, Wuhan 430070, China
e-mail: zhouhuiwhut@163.com

W. Lu et al. (eds.), *Proceedings of the 2012 International Conference on Information Technology and Software Engineering*, Lecture Notes in Electrical Engineering 210, DOI: 10.1007/978-3-642-34528-9_46, © Springer-Verlag Berlin Heidelberg 2013

of stationary mesh routers, which implement a DHT used to locate resources within the network.

In the P2P resource sharing system based on a two-tier WMN, each router maintains a local information table of the resources shared by all the clients which register to it. The joining in/moving/leaving behaviors of a client and adding/removing shared resources are represented by index changes of resources catalog in the router. In this paper, we focus on mobile peers in the lower tier. A peer location management technique is proposed to reduce the adverse effects of network churn on performance of wireless P2P resource sharing system.

46.2 Related Works

For realizing wireless P2P resource sharing, some researchers improved the existing approaches in wired networks to work efficiently on wireless networks by taking the features of wireless environment and mobile peers into account, for example, a modification of Gnutella in [2] and an extension of Pastry in [3]. Among all approaches for wired networks, most attention has been paid to enabling an improved Chord algorithm work efficiently in wireless networks.

Reference [4] proposed an improved Chord algorithm which took advantage of routing protocol from network layer in MANETs. When the lookup message invoked by a peer is sent to a remote peer multiple steps away in logical ring, the routing of messages in network layer can help it diverge from its path at some intermediate if the routing information finds a shorter way to its final destination. The authors in [5] proposed a new routing modified scheme MA-Chord which bases on a unicast scheme with DHT. It combines Ad hoc On-Demand Distance Vector (AODV) Ad hoc routing and Chord overlay routing at the network layer to provide an efficient primitive for key-based routing in MANETs. Random land-marking is used for MA-Chord to assign cluster-based address to peers. Cluster heads dedicate in managing the network to reduce overhead. Some researchers focus on more reasonable Chord ring topology to alleviate the mismatch problem between overlay topology and physical topology. C-Chord, a P2P overlay network strategy based on mobile cellular networks, is formed master Chord ring by base stations and formed slave Chord ring by the mobile terminal nodes in every cellular region. C-Chord model uses the station as the landmark node and cellular region as the landmark district by landmark cluster method, which not only implements nodes that their physical location are close are proximate in logical location, but also solves single node failure in landmark cluster method [6]. PLH Chord is a physical location-based hierarchical Chord algorithm. The base station in wireless network was added to Chord as the physical location of the boundary of punctuation. The ID of a peer consists of two parts. The former part which obtained from a nearby base station represents the peer location information, while the latter part is calculated by hash function based on its uniform identifier. PLH Chord not only improves the

Fig. 46.1 Architecture of P2P resource sharing system for a two-tier WMN

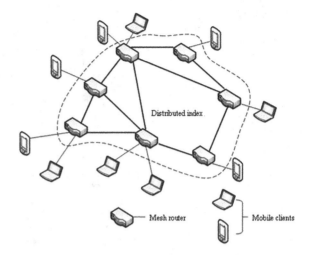

Distributed index

Mesh router — Mobile clients

efficiency of resource queries but also solves the problem that the physical location of nodes and logical position does not match [7].

46.3 System Structure

The architecture of P2P resource sharing system for a two-tier WMN is shown in Fig. 46.1. The lower tier is composed of mobile mesh clients, which provide resources to be shared in the P2P system; the upper tier is composed of stationary mesh routers, which implement a DHT used to locate resources within the network.

We assume that all mesh routers are distributed in a square deployment region and routers are aware of their position in the region. According to Chord, a router is allocated an integer ID in $[0, 2^m-1]$ by hashing. Then routers arrange in a ring according to the ascending order of their IDs. Resource keys are also mapped into the same ID space. Each key resides on the router with the smallest ID larger than the key. The overlay network topology of P2P file sharing system is shown in Fig. 46.2. The capital letters A–G and lowercase letters a–g express mesh routers in the upper tier and mobile clients in the lower tier respectively. In the paper hereinafter, we use the term peer to refer exclusively to a mobile client in the lower tier. When a peer intends to join in the system, it connects to a mesh router within its communication range and sends the catalog of the resources that it wants to share with others to the router. This router is called as the catalog router (CR) of this peer. Then the CR maps the key of each resource into the ID space, informs the router with the smallest ID larger than the key to be responsible for the location information of the resource one by one. The responsible router is called as resource management router (RMR). Each router is not only the CR of all peers that register with it, but also the FMR of all resources which keys resides on the part of responsibility in overlay ring. When a peer wants a specific resource, it informs its

Fig. 46.2 Overlay network
topology of P2P resource
sharing system

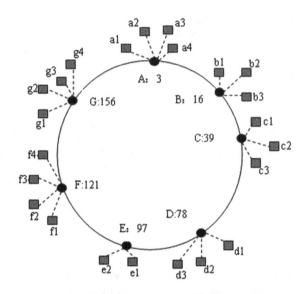

CR, the CR checks its resource catalog to find whether the resource can be took from a local registered peer. If yes, the CR responds to the originator peer. Otherwise the CR sends a lookup message for the resource key. Middle routers forward the message according to their local finger tables until the message arrives at the FMR of the resource. Location information of the resource is obtained by the originator peer.

46.4 Peer Location Management

The two-tier P2P resource sharing system uses mesh routers to construct a structured P2P overlay network, search shared resources in place of peers. Mesh routers have minimal mobility and no constraint on power consumption, so the overlay topology is fairly stable. However, as the shared resources providers, peers in the lower tier, such as laptop, PDA, mobile phone, have limited energy, unreliable link, and high mobility. When a peer moves or switches off, the information of resources shared by this peer, which managed by CR and RMRs, are altered accordingly: peer's CR modifies the resources catalog information and all RMRs modify the resources location information. When a peer moves out of the range of its CR, the CR holds that the peer leaves the system, so it deletes the catalog of resources shared by the peer, and informs all RMRs to delete the location information of resource which it is responsible for. After a time, the peer moves into the range of another router, it registers to the router and join in the system again. The router, as the new CR of the peer, saves the catalog of resources, and informs all RMRs to save the new location information of resource which it is responsible for. The movement of this peer results in a series of modifications in SpiralChord that

Fig. 46.3 ID distribution of peers in the overlay network topology

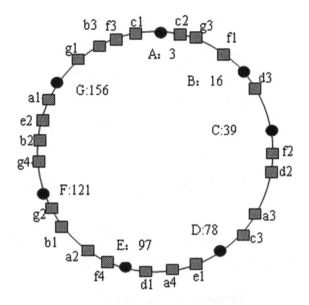

bring about the increase of message overhead. If there are more shared resources and more mobile peers in the system, amount of message is larger which finally causes the performance degradation of Chord. To solve this problem, we assign each peer a management router (PMR) that is responsible for the location information of a peer.

A peer's MAC address is mapped into an integer in $[0, 2^m-1]$, call peer ID. The router in Chord with the smallest ID larger than the peer ID is the PMR of this peer. The ID distribution of all peers is shown in Fig. 46.3.

The router A has four registered peers a1−a4. According to peer IDs, the PMR of a1 is router G, the PMR of a2 is router F. Each router is the CR of all peers that register with it, the RMR of all resources which keys resides on the part of responsibility in overlay ring, the PMR of all peers which IDs resides on the part of responsibility in overlay ring. Expect for finger table, each router maintains three other tables: resource information table, resource location table, and peer location table. The three tables of router B is shown in Fig. 46.4.

After being assigned a PMR, the location of a peer is managed by its PMR. The location information of this peer saved by its PMR includes the keys of resources shared by this peer, the CR of this peer. A status identification of a resource is used to represent whether the resource can be downloaded at this moment in resource information table and peer location table. When the identification is 1, it means that the corresponding resource can be downloaded. When the identification is 0, it means that the corresponding resource can not be downloaded at this moment.

In resource location table of RMR of a shard resource, location information of a resource is the PMR of the resource providing peer, not the CR of the resource providing peer. The PMR of a peer is engaged in related resource sharing operation.

Resource information table

peer	Resource Key	Status
b1	30	1
	45	1
	12	0
b2	76	1
	89	0
b3	114	0

Resource location table

Resource (key)	Router (ID)
12	121
5	156
7	3
9	97

Peer location table

Peer (ID)	Location information		
	Resource key	Status	Router (ID)
5	88	1	39
	32	1	
8	43	0	156
	14	0	
	25	0	
12	65	1	121

Fig. 46.4 Three local index tables in router B

46.4.1 Resource Publication

The PMR publishes shared resources to Chord in place of CR, steps as follows:

1. After joining in the system, a peer informs its CR to saves catalog of the resources that it wants to share with others, and sends a message containing its MAC address to CR.
2. CR updates its resource information table by adding the keys of resources which shared by the peer. The status identifications of these resources are set to 1.
3. CR calculates the peer's ID by hashing the MAC address of the peer, launches a lookup for this ID in SpiralChord to find out the PMR of the peer, and sends the information of resources shared by the peer to PMR.
4. PMR updates its peer location table by adding the CR ID of the peer, the keys of resources which shared by the peer. The status identifications of these resources are set to 1. Then PMR informs to all RMRs of resources to save the location information of resource which it is responsible for.
5. Each RMR updates its resource location table by adding PMR's ID.

We give an example. When a new joined peer a1 wants to publish a resource with key is 5, its CR, router A updates its resource information table with an entry (peer a1, key 5, status 1). Then router A finds that the PMR of a1 is G according to the MAC address of a1. G updates its peer location table, the entry is (peer ID 140, key 5, status 1, router ID 3). Then G finds that the RMR of the resource is B according to the key of the resource. B updates its resource location table by adding an entry is (key 5, router ID 156).

46.4.2 Resource Search

CR searches a resource in Chord in place of a peer, steps as follows:

1. A peer informs its CR of the keyword of the desired resource.
2. CR calculates the resource's key by hashing the keyword of the resource, checks its resource catalog to find whether the resource can be took from a local registered peer. If yes, the CR responds to the peer. Otherwise the CR launches a lookup message for the resource key in SpiralChord to find out the RMR of the resource.
3. RMR receives the message, checks the resource location table to get the information of PMR of the peer that provides the resource.
4. PMR receives the message, checks the peer location table to get the information of CR of the peer that provides the resource.

46.4.3 Resource Deletion

The causes of a resource being deleted from the P2P resource sharing system may be movements/failure of the providing peer or purposely canceling the shared resource by the providing peer.

1. When a peer does not want to share some resources with others, it notifies its CR. The CR set the status identifications of these resources to 0 in its information table. When a peer moves or switch off, the CR receives the leaving message from the peer, or does not receive the OK-message from the peer which used to maintain contact for a long time, The CR set the status identification of all resources shared by the peer to 0 in its information table.
2. If the resource are re-shared by the peer, or the peer which left or switch off rejoins the system by register to the same CR within the specified time T, the CR changes the status identifications of the resources to 1. If the duration of the 0 status of a resource exceeds T, the CR deletes the entry related to the resource in resource information table.
3. Every time the status identifications of resources are changed, the CR notifies the PMR of the peer which provides the resources.
4. The PMR of the peer changes the status identifications of resources in peer location table according to the message from the CR. If the duration of the 0 status of a resource exceeds T, the PMR deletes the entry related to the resource in peer location table.
5. If the PMR deletes an entry related to a resource in peer location table, it sends a deleting message to the RMR of the resource. The RMR of the resource deletes the entry related to the resource in resource location table.

46.4.4 Peer Movements

When a peer moves, it leaves out of the range of its CR. After a time, it moves into the range of another router, it registers to the router and join in the system again, steps as follows:

1. Old CR changes the status identification of all resources shared by the peer to 0 in resource information table, and informs the PMR of the peer.
2. PMR sets the location information of the peer as unknown, the status identification of all resources shared by the peer as 0 in peer location table.
3. New CR saves the catalog of its resources, and informs the PMR of the peer.
4. PMR get the message from the new CR within T, sets the location information of the peer as the new CR, the status identification of all resources shared by the peer as 1 in peer location table.

When peer a1 moves, it leaves out of range of A, A changes the status identification of all resources shared by a1–0 in resource information table, and informs the PMR of a1, router G to update peer location table. The update entry is (peer ID 140, key 5, status 0, router ID unknown). After a time, a1 move into the range of D, D updates its resource information table by adding a new entry (peer a1, key 5, status 1), and informs the PMR of a1, router G to update peer location table again, the update entry is (peer ID 140, key 5, status 1, router ID 78).

46.4.5 Resource Download

When a search for a certain resource succeeds, the process of download begins. While downloading the resource, the process may be interrupted because of resource deletion.

1. The requesting peer finds the download of a resource stops, it contacts the PMR of the providing peer and wants to get a notification when the status identification of the resource changes to 1.
2. The status identification of the resource in peer location table changes from 0 to 1 within T, the PMR sends a notification containing the CR of the providing peer to the requesting peer.
3. PMR finds the duration of the 0 status of this resource exceeds T, its deletes the entry related to the resource in peer location table. Then it sends a notification to the requesting peer.
4. If the case of 2 occurs, the requesting peer continues to download the resource by contacting with the CR of the providing peer.
5. If the case of 3 occurs, the requesting peer re-initiatives a search message for the resource to find whether there is another providing peer and complete the download.

Fig. 46.5 Message overhead
in static network

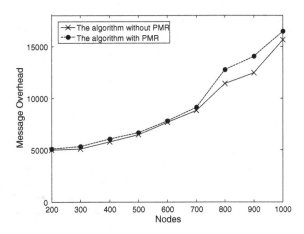

46.5 Simulations

We simulated a two-tier P2P resource sharing system in wireless mesh network
using OverSim. The algorithm without PMR and algorithm with PMR is compared
to evaluate the effects of peer location management on improving the performance
of wireless P2P resource sharing system. We assume there are 50 routers in the
upper tier, and the number of peers in the lower tier varies from 200 to 1,000. Each
peer has 10 resources shard with others. We set the simulation time interval to
600 s machine time, consider two kinds of network configuration, a static network
configuration in which all peers are static, and a dynamic network configuration in
which the probability of a peer moving is 80 %.

Figure 46.5 shows the total number of messages exchanged by peers during the
simulation time interval in the static network. After assigning a PMR to each peer,
a resource publishing message is sent to CR, to PMR which then forwards to all
RMRs, instead of directly be sent to RMRs from CR in the algorithm without
PMR. A resource searching message is sent to RMR, to the PMR, finally to CR,
instead of be sent to RMR, CR in the algorithm without PMR. More messages are
exchanged in the algorithm with PMR, but the number increase of messages is
limited. Figure 46.6 shows the total number of messages exchanged by peers
during the simulation time interval in the dynamic network, the algorithm with
PMR is more effective in reducing message overhead, the message overhead can
be less 1/3 than the algorithm without PMR in the best situations. In the algorithm
without PMR, a mobile peer moves to the communication area of a new CR from
the communication area of the old CR, the old CR sends 10 messages to notify the
RMRs of the 10 shared resources of the mobile peer to delete the invalid location
information of corresponding, resource. After a time, the new CR sends 10 messages to notify the RMRs of the 10 shared resources of the mobile peer to save the
new location information of corresponding, resource. In the algorithm without
PMR, when the same thing happens, the old CR only sends one message to notify

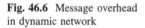

Fig. 46.6 Message overhead
in dynamic network

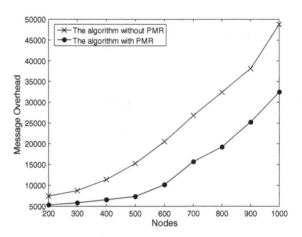

the PMR of the mobile peer to delete the invalid location information of the mobile
peer. After a time, the new CR sends one message to notify the PMR of the mobile
peer to save the new location information of the mobile peer. PMR of a mobile
peer can play an active role in decreasing the message overhead caused by the
movements of peers in the upper tier.

When a resource search message finds the RMR of the resource, the search
process succeeds. The resource message originator then finds the resource provider
according to the location information in the RMR. In the algorithm with PMR, the
originator contacts the PMR of the provider, then the CR of the provider according
to the location information of the resource in the RMR. In the algorithm without
PMR, the originator directly contacts the CR of the provider according to the
location information of the resource in the RMR. If the originator finds that the
resource information table of the CR contains the information of the wanted
resource and the status identification of the resource is 1, it means that the search
process gets the accurate location of the provider. Figure 46.7 shows the average
accurate ratios of the two algorithms are both 100 % in static network no matter
how many peers there are. When all peers are stationary, the location information
of the resources in the RMRs are not invariable, originators can get accurate
location of the provider. Figure 46.8 shows the average accurate ratios of the two
algorithms in dynamic network. The average accurate ratio of the algorithm with
PMR is higher than that of the algorithm with PMR. Because of the movements of
peers, originators get the obsolete location information of resources from the
RMRs before the RMRs are notified to update the location information of
resources in the algorithm without PMR, so the average accurate ratio is low. In
the algorithm with PMR, the location information of resources in the RMRs are the
PMRs of the resource providers, the PMR of a mobile peer dose not changed with
the movements of the peer. Originator can always get the latest location of the
provider from the PMR of the provider, and the average accurate ratio of the
algorithm with PMR is still 100 % in dynamic network.

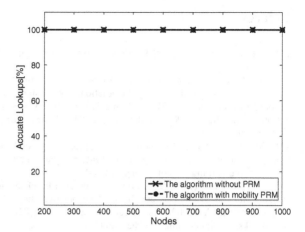

Fig. 46.7 Average accurate ratio in static network

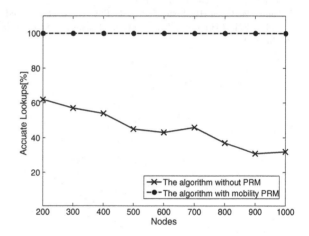

Fig. 46.8 Average accurate ratio in dynamic network

46.6 Conclusions

Mobility of peers, mass data, and limited transmission ability of wireless network become the major difficulties in enabling P2P resource sharing in wireless environment. The paper realizes a P2P resource sharing system in a two-tier WMN network. In this paper, we focus on the effect of mobility of peers in lower tier on performance of resource sharing system. A PMR assignment method is proposed to be responsible for the location management of mobile peers. The simulation results show that the algorithm with PMR is more effective in reducing message overhead and alleviating the decrease of accurate lookup ration caused by movements of mobile peers than the algorithm without PMR.

References

1. Ian A, Wang XD, Wang WL (2005) Wireless mesh network a survey. Comput Netw 47: 445–487
2. Conti M, Gregori E, Turi G (2005) A cross-layer optimization of gnutella for mobile ad hoc networks. In: Proceedings of the international symposium on mobile ad hoc networking and computing (MobiHoc), Urbana-Champaign IL, 2005:343–354
3. Zahn T, Schiller J (2005) MADPasry: A DHT substrate for practicably sized MANETs. In: Proceedings of ASWN
4. Qi M, Hong J (2007) MA-Chord: a new approach for mobile ad hoc network with DHT based unicast scheme. In: Proceedings of 2007 International conference on wireless communications, networking and mobile computing, WiCOM
5. Liu CL, Wang CY, Wei HY (2008) Mobile chord: enhancing P2P application performance over vehicular ad hoc network. IEEE globecom workshops, GLOBECOM, United States
6. Zou DY, Gan Y (2009) The research of a structural P2P model based on mobile cellular networks. In: Proceedings of the 2009 international conference on networking and digital society, China, 64–67 May 2009
7. Tang CW, Chen HD, Shao YQ (2010) Physical location-based hierarchical chord algorithm in mobile P2P. J Huazhong Univ sci technol (natural science edition), 38:21–24, August 2010 (in Chinese)

Chapter 47
One ARGIS Prototype Based on Handheld Mobile Terminal

Lei Liu, Xiang Ren, Min Sun and Na Dong

Abstract In recent years, handheld mobile terminals have developed higher capability of calculation and information processing. In this paper we propose an Augment Reality Geographical Information System (ARGIS) by combining AR technology with simplified Three-Dimensional Geographical Information System (3DGIS). Real-time video streams are augmented with kinds of invisible geo-information, using data from 3DGIS and data captured by built-in sensors. Thus the ARGIS can provide outdoor users with application services such as 3D-Navigation, information retrieval and information communication among users and the server etc. We designed a corresponding prototype system for outdoor navigation in the Peking University campus, to manifest the feasibility and vast potential for future development of the system.

Keywords Augment reality · Handheld mobile terminal · 3DGIS · ARGIS

L. Liu (✉) · X. Ren · M. Sun · N. Dong
The Institute of RS & GIS, Peking University, 100871 Beijing, China
e-mail: spring5689@pku.edu.cn

X. Ren
e-mail: xelmirage@pku.edu.cn

M. Sun
e-mail: sunmin@pku.edu.cn

L. Liu
The 66240 Troop, Beijing Military Region, 100042 Beijing, China

W. Lu et al. (eds.), *Proceedings of the 2012 International Conference on Information Technology and Software Engineering*, Lecture Notes in Electrical Engineering 210, DOI: 10.1007/978-3-642-34528-9_47, © Springer-Verlag Berlin Heidelberg 2013

47.1 Introduction

Most handheld mobile terminals such as smart-phone and Tablet PC now provide Map navigation. However, the practical applications are confined to 2D-geo-information such as birds-eye view of the environment, which has poor performance in interaction and environmental relevance. During the last decade, outdoor AR technology has aroused extensive concern among scholars. Meanwhile many related systems are designed and used in urban planning, navigation and industrial support, such as Touring Machine by Columbia University in 1999 [1], RobiVision by Zillich in 2000, UbiCom in 2002 [2], Vesp'R by Gerhard Schall in 2008 [3], SiteLens in 2009 [4]. With the booming development of mobile terminals and increasing demand for accuracy in pose tracking and the augmented effect, many scholars turn to explore the Mobile Augmented Reality technology. But AR-researches in combination with 3DGIS are limit, most of which focus on navigation application.

In the paper we combine 3DGIS with AR technology. On the one hand, users are allowed to observe the real-world objects, which can simplify the visualization processing, on the other hand, 3DGIS can present invisible information or virtual objects, which include our Interests such as the annotations of buildings and simple wireframe models etc. In fact, AR system can augment any information that is available in 3DGIS.

The advantages of ARGIS research are as follows: (1) It can provide more intuitional experience for users than the 2D map. The augmented geo-information in the real-time video stream is not available in the real scene such as historical information. (2) Based on 3DGIS, it can provide the function of spatial analysis and improve the users spatial cognition. (3) Through the information exchange between mobile terminals and the server, it can share the real-time information for cooperative work and provide rich visual information which is important to emergency decision-making. (4) Because the mobile platform has advantages in portability, low energy consumption and widespread application, it has broad prospect in public applications and commercialization.

47.2 Related Work

Early outdoor AR research is mainly focused on real-time registration in 3D, which is based on combination of multi-sensors technology and vision tracking technology. The combination aimed to superimpose the augmented virtual objects upon the real scene accurately. The research includes those mentioned in [4–7] and so on. The registration error comes from the sensor data errors, which are inevitable: (1) GPS measurement error; (2) IMU measurement error, especially time draft error of gyro; (3) Errors from camera calibration and GPS/IMU mount and et al. Therefore, coarse registration is achieved by only using sensor data.

Meanwhile, vision tracking technology can provide stable registration result for indoors augmented reality. However, for outdoors environment, it only works on ideal cases. Despite that many scholars have proposed a lot of methods to tackle the outdoor registration by vision tracking combined with GPS/IMU sensors [8–10], the problem remains a challenge. The reasons mainly include: (1) There are many complex environments that vision tracking become inefficacious, such as: trees, monochrome walls, dark objects or far views et al; (2) Vision tracking are only suitable for continuous and relatively slow view changes, which cause limitation to user actions; (3) Cumbersome computing is needed for vision tracking, which needs much power consumption, and limits outdoor application of AR. Up to now, with the improvement of mobile hardware performance, the research in outdoor AR system has become a hot topic in mobile AR technology.

Early research in handheld mobile AR was focused on desktop systems: (1) In 2005, a tennis game based on ARToolKit, which was designed by Enrysson, was embedded in a SYMBIAN mobile phone [11]. (2) In 2007, the WikEye system was designed by Rohs, which registered the virtual objects through tracking map grid and provided simple spatial analysis such as distance measurement [12]. (3) In 2008, the Studierstube Tracker was designed by Wagner, a marker library supporting many different types of markers on mobile phones [13]. (4) In 2008, Wagner built an outdoor augmented reality system for mobile phones that matched phone-camera images against a large database of location-tagged images using a robust image retrieval algorithm, which was based on the SURF [14].

In recent years, the improvement of mobile hardware promotes the development of outdoor AR system, which has been used in many fields such as underground infrastructure visualization, urban planning and street view [3, 15, 16].

To our experience, the Augmented Reality system is still lack of using large 3DGIS data. Many vision systems mentioned above are based on 2D geo-information, which is poor in augmented information processing of interest objects. Some of them operate CAD models in office-like environment, which are lack of geo-referenced and related information [17, 18]. 3DGIS can describe the physical space as the VR space. 3D geo-information support is essential in outdoor AR system used for large-scale scenes, which can improve users spatial cognition. The research on using geo-information in outdoor AR system was carried out earlier. In 1999 Thombas developed the ARQuake system [19], one of the first systems that allows users to play AR games outdoor, using Campus Map for guiding. In 2002, Siyka Zlatanova developed Ubicom system [2], which achieved accurate registration by matching lines extracted from video stream and lines retrieved from the 3DGIS. In 2004, Gerhard Reitmayr and Dieter Schmalstieg demonstrated the usage of AR for collaborative navigation and information browsing tasks in an urban environment [5]. In the system Pocket D-GPS and InertiaCube2 orientation sensor were used for helmet tracking and 3D registration. Meanwhile, they developed a 3-tier architecture data server, based on XML, to release the shared data between clients and to communicate with remote server. In 2007, Sun used tracking artificial markers in 3DGIS to improve accurate registration. But in those researches mentioned above, problems still remain: (1) The function of 3DGIS is poorly integrated into AR system. (2) Taking portability and power

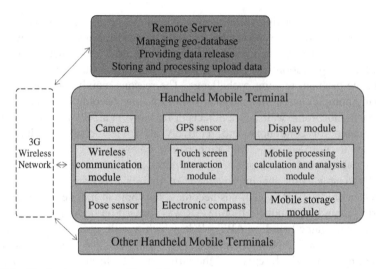

Fig. 47.1 The hardware composition of the prototype

consumption into account, equipments such as laptops and sensors are too complex and overweight for long-playing outdoor application.

This paper proposes an ARGIS prototype embedded in a mobile terminal that enables users to inquire 3DGIS information and process space analysis. Vision tracking technology only works on ideal cases, unsuitable for outdoor application. Meanwhile the registration errors from GPS to IMU measurement are inevitable. Therefore, we designed a reliable interactive method to realize the fine registration. Moreover, we achieved a client–server architecture based on 3G wireless network, to realize the communication in teamwork and transmit outdoor information from mobile terminals to the server.

47.3 Construction Method of ARGIS

Figure 47.1 shows the structure of outdoor ARGIS based on handheld mobile terminals. The handheld terminal used in our prototype system should have functions of GPS positioning, pose measurement, E-compass, Wireless data communication and powerful computation performance, such as IPad etc. In consideration of outdoor navigation, core database of geo-information is provided by the remote server, while the data used frequently by clients is stored in the mobile terminal. We designed the structure of data distribution on reasons such as: (1) The current terminals provides large storage capacity, the related data for clients in a fixed city can be stored and processed in the terminal. (2) The sever can release the updated data to clients for team-cooperation and information-sharing.

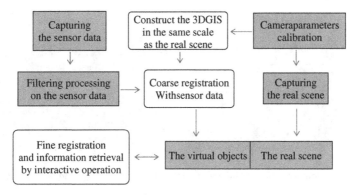

Fig. 47.2   The data processing of the prototype

47.4 Augmented Method of Geo-Information

In order to provide the function of navigation and information browsing in an urban environment, we focus on overlaying 3D-building models on the real scene. The AR processing includes three aspects: (1) 3D modeling is to express information of both the real world objects and any other virtual objects; (2) The visualization of geographical annotations is to display the object properties, which we are interested in; (3) The stabilization of registration aims to improve the matching effect. All the above mentioned are discussed below. Figure 47.2 illustrates the steps of the data processing.

47.4.1 3D Modeling

In order to get vivid visual effect, we commonly use 3D-software (such as AutoCAD, 3DMAX, Sketch up et al.) to process artificial-interactive modeling. In ARGIS, wireframe models are sufficient to assist users in identifying the 3D objects and cognizing the relationship between the objects, as well as to process spatial analysis such as distance measurement et al. There is no need of detailed modeling. In this paper, we adopted a similar method as Google Earth. Terminals can download the 3DGIS data from remote server, which are described in XML [5].

47.4.2 Annotation Visualization Algorithm

In this paper, we mainly discuss the building annotation. In order to display the building information and improve the matching effect between the augmented

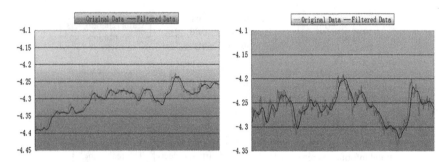

Fig. 47.3 The contrast tables in different sample rates, N = 10 (*left*) and N = 25 (*right*)

objects and the real scene, the prototype system displays the annotation inside the projection range of the wire-frame model on the screen.

Before displaying the annotation, we judge every visible building in the view to calculate the coordinate of the building annotation. Assuming that there are N visible vertices of a building in the view, the coordinates for [Xi, Yi, Zi], the corresponding annotation located in the coordinate of [Xm, Ym, Zm, 1], which is the centroid of the building viewing area:

$$X_m = \frac{1}{n}\sum_{i=1}^{n}X_i \quad Y_m = \frac{1}{n}\sum_{i=1}^{n}Y_i \quad Z_m = \frac{1}{n}\sum_{i=1}^{n}Z_i \qquad (47.1)$$

Assuming that the projective transformation matrix of the viewpoint is P, camera transformation matrix is K, the coordinate of the annotation is M, we can get the screen coordinate of annotation through the formula:

$$[x, y, 1]^T = KPM \qquad (47.2)$$

Because the system only shows the nearest annotation to the viewpoint, there is no problem of note overlapping. Meanwhile the system can adjust the long annotation to the proper position.

47.4.3 Stabilization of Registration

In order to overcome the registration errors from GPS to IMU measurement of the handheld mobile terminal, we must process the sensors' data in specific methods to achieve a better registration. On the one hand, we process the pre-rectification to offset the magnetic declination of the compass sensor; on the other hand, we use median filter to process the data from the pose sensor. In Fig. 47.3, we compare the sampling effect in different sample rates.

The greater the sampling amount is, the better smooth effect we will get. But that will increase the signal delay. So we adopted the sample rate at N = 25.

Fig. 47.4   The screenshot of the terminal (*left*) and the server(*right*)

47.4.4 Interactive Registration

Most handheld mobile terminals come with the interactive function, which allows users to touch the screen to achieve. We set up the message response between macro commands and button-touch events, which includes: the network connection, the screenshots of specific AR scene, data download and upload et al.

In the handheld mobile terminal, the sensor errors reduce the measurement accuracy specially in position and orientation. Although we adopt filter algorithm to enhance the stability of sensor signals, the registration error still exists. In order to reduce the errors, we propose the interactive registration method. After analysis in the sources of errors, we find that the main factors are GPS horizontal measurement error and IMU measurement error. In those errors mentioned above, the drift error of gyro will accumulate with the working time, which is due to the low-end gyro in the handheld mobile terminal. To overcome that error, we must adopt the interactive registration method.

The result of registration errors is mainly the horizontal offset. The offset comes from translation error and rotation error. Because the component in the view direction impacts little on the translation, we only adjust the horizontal position and the horizontal orientation to reduce the error.

47.5 Experiment and Discussion

We developed an ARGIS prototype system based on the ASUS TF101, which equipped a series of sensor such as a build-in camera, a BCM47511 GPS module and a MPU-3050 IMU module etc. The system of TF101 is Android4.0. After calibration and actual measurement, the accuracy of GPS ranges between 5 and 10 m and the accuracy of IMU is ± 5 °C.

We used 3G wireless network to construct the communication of the CS frame. The server provides the information-sharing service. In the experimental process,

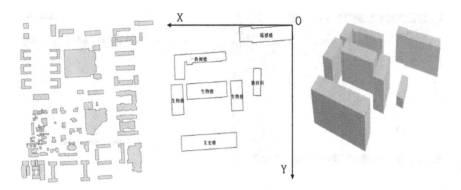

Fig. 47.5 The constructing process of 3D models

we uploaded the screenshot to the server, then the updated information was displayed to clients automatically. Figure 47.4 shows the experiment effect.

We digitalized the topographic map of Peking University to construct the 3DGIS, which applied sample wire-frame models of buildings. The constructing process is shown in Fig. 47.5. On the one hand, a few building models are stored in mobile terminals to be displayed rapidly, on the other hand, mobile terminals can request for data and upload the necessary data to the server.

The augmented information consists mainly of annotation and wire-frame models. As shown in Fig. 47.6, we provided interactive correction in 3D-registration to reduce the registration error. Meanwhile, we made a comparison in day and night navigation to manifest the feasibility of the system in Fig. 47.7, which is convenient to guide the user to the target at night.

Sum up the experimental results above, the prototype system achieves mostly the anticipated goal. The performance of the system can still be improved in several aspects.

(1) The positioning accuracy and the signal delay of the sensors

The performance of the GPS chip in AUSU TF101 is poor. Its positioning accuracy is between 5 and 10 m. The blocking of satellites by high buildings prohibits the use of the system, which will lead signal delay. All mentioned above will cause errors in the ARGIS registration. In the future, we can use the external equipment of high-precision GPS receiver and D-GPS technology to improve the positioning accuracy. We can predict the navigating direction in GIS for accessorial positioning. Meanwhile, we can use the hybrid location technology to overcome the signal delay.

(2) The registration accuracy

The IMU in AUSU TF101 is cheap, so the registration error is inevitable. In addition, other factors could bring the registration error such as: the map data error, the camera parameters measurement error, the ignored topographic relief and the ignored rotation error etc. In order to improve the registration accuracy, the

Fig. 47.6 The registration effect of non-interactive correction (*left*) and interactive correction (*right*)

Fig. 47.7 The comparison between day-navigation (*left*) and night-navigation (*right*)

external equipment of the high-precision IMU, such as MTI, is necessary. Meanwhile, the high-precise DEM data can be used to reduce the error from the ignored topographic relief.

(3) Network communication

At present, the prototype system only builds the communication between server and clients. In the future, we will design the network communication among distribute clients to meet the needs of different applications.

47.6 Conclusion

In the paper, we have designed an ARGIS prototype based on AUSU TF101, which could overlay 3DGIS information on the real-time video stream captured by the terminal's camera. The system can be used for outdoor navigation and geo-information browsing.

This paper has completed works as following. Firstly we have designed an ARGIS prototype on Android-based mobile terminals, and solved the key problems such as: the sensor data application and adjustment, the 3D registration and overlaid display, communication based on 3G wireless network etc. Secondly we combined 3DGIS and AR technology to provide outdoor navigation and information browsing. The virtual camera is set at the same angle of view as users', which reduces the requirements for users' geo-knowledge, so the system has the advantage of universalness and operability. Finally we realized communication among terminals and server through 3G network, which will contribute to team-cooperation and information-sharing.

Acknowledgments We thank NSFC and MOST of China for their financial support (40971240 and No. 2007AA12Z234) to this work.

References

1. Feiner S, MacIntyre B, Höllerer T, Webster A (1997) A touring machine: prototyping 3D mobileaugmented reality systems for exploring the urban environment. In: Proceedings of first IEEE international symposium on wearable computers (ISWC '97), Cambridge, MA, pp 74–81
2. Zlatanova S, Heuvel F (2002) 3D GIS for outdoor AR applications. In: Proceedings of the third international symposium on mobile multimedia systems and applications, Delft, The Netherlands, pp 117–124
3. Schall G, Mendez E, Kruijff E, Veas E, Junghanns S, Reitinger B, Schmalstieg D (2009) Handheld augmented reality for underground infrastructure visualization. Pers Ubiquitous Comput 13:281–291
4. White S, Feiner S (2009) SiteLens: situated visualization techniques for urban site visits. In Proceedings: ACM CHI, Boston, MA, pp 1117–1120
5. Reitmayr G, Schmalstieg D (2004) Collaborative augmented reality for outdoor navigation. In: Proceedings of the symposium on location based services and TeleCartography
6. Green SA, Billinghurst M, Chen X, Chase GJ (2008) Human-robot collaboration: a literature review and augmented reality approach in design. Int J Adv Rob Syst 5(1):1–18
7. Dahne P, Karigiannis JN (2002) Archeoguide: system architecture of a mobile outdoor augmented reality system, in mixed and augmented reality ISMAR 2002. In: Proceedings international symposium on, pp 263–264
8. Schall G, Wagner D, Reitmayr G, Taichmann E, Wieser M, Schmalstieg D, Hofmann-Wellenhof B (2009) Global pose estimation using multi-sensor fusion for outdoor augmented reality in mixed and augmented reality ISMAR 2009. 8th IEEE international symposium on, pp 153–162
9. Hedley NR, Billinghurst M, Postner L, May R, Kato H (2002) Explorations in the use of augmented reality for geographic visualization, Presence: teleoper Virtual Environ, vol 11, pp 119–133
10. Uchiyama H, Saito H, Servières M, Moreau G (2008) AR representation system for 3D GIS based on camera pose estimation using distribution of intersections. In: Proceedings ICAT, pp 218–225
11. Henrysson A, Billinghurst M, Ollila M (2005) Face to face collaborative AR on mobile phones, in mixed and augmented reality. In: Proceedings 4th IEEE and ACM international symposium on, pp 80–89
12. Hecht B, Rohs M, Schöning J, Krüger A (2007) Wikeye—using magic lenses to explore georeferenced wikipedia content, in pervasive. Workshop on pervasive mobile interaction devices (PERMID)

13. Wagner D, Langlotz T, Schmalstieg D (2008) Robust and unobtrusive marker tracking on mobile phones, in mixed and augmented reality ISMAR. 7th IEEE/ACM international symposium on, pp 121–124
14. Takacs G, Chandrasekhar V, Gelfand N, Xiong Y, Chen W-C, Bismpigiannis T, Grzeszczuk R, Pulli K, Girod B (2008) Outdoors augmented reality on mobile phone using loxel-based visual feature organization. In: Proceedings of the 1st ACM international conference on multimedia information retrieval. Vancouver, British Columbia, Canada
15. Allen M, Regenbrecht H, Abbott M (2011) Smart-phone augmented reality for public participation in urban planning, in OzCHI. In: 11 Proceedings of the 23rd australian computer-human interaction conference
16. Tokusho Y, Feiner S (2009) Prototyping an outdoor mobile augmented reality street view application.pdf
17. Höllerer T et al (1999) Exploring MARS: developing indoor and outdoor user interfaces to a mobile augmented reality system, computers and graphics, vol 23, 6 Dec, pp 779–785
18. Piekarski W, Thomas B (2009) ARQuake: the outdoor augmented reality gaming system, communications of the ACM [4] White S, Feiner S SiteLens. In: Proceedings situated visualization techniques for urban site visits. ACM CHI. MA, Boston, pp 1117–1120
19. Zlatanova S (2002) Augmented reality technology. GIS R, vol 17, p 75

Chapter 48
CTFPi: A New Method for Packet Filtering of Firewall

Cuixia Ni and Guang Jin

Abstract The firewall is often seen as the first line of defence in ensuring network security of an organization. However, with the rapid expansion of network, the species and numbers of network attacks continue to increase. And network traffic is also growing markedly. Traditional network can no longer meet the requirements of preventing attacks in high-speed network. Therefore, in order to improve the performance, this paper proposes a new packet filtering method. It is CTFPi, the combination of traditional firewall and Pi (Path identifier) for Packet Filtering. The principle of this scheme is to map the source IP addresses and destination IP addresses to Pi and then use Pi to replace them. Experiments show that our method not only can adapt to the high-speed network requirements, but also be better to prevent attacks, especially with forged packets.

Keywords Firewall · Pi · Rule optimization · Packet filtering

48.1 Introduction

In recent years, with the development of computer networks, the processing and transmission of network information are no longer limited by geography or time. Networking has become an irresistible trend. The Internet has entered in all aspects

C. Ni (✉) · G. Jin
Faculty of Information Science and Engineering, Ningbo University,
No 818 Feng Hua Road, Ningbo, China
e-mail: nicuixianihao@163.com

G. Jin
e-mail: jinguang@nbu.edu.cn

W. Lu et al. (eds.), *Proceedings of the 2012 International Conference on Information Technology and Software Engineering*, Lecture Notes in Electrical Engineering 210, DOI: 10.1007/978-3-642-34528-9_48, © Springer-Verlag Berlin Heidelberg 2013

and areas of social life, which has a huge impact on people's lives and increasingly become the focus of people's attention. With the continuous development and wide application of computer and network technologies, network security issues are also appearing and attract the attention of all trades and professions.

According to [1], over the past decade, network security vulnerabilities have increased from 300 to over 6,000, which have caused a large number of security incidents and seriously economic and social impacts. Moreover, according to [2], DoS [3] attack continues to grow in size and complexity. The maximum instantaneous attack traffic reaches up to 49 Gbps and the speed increment is over 104 %.

The above data show that the network security threats exist everywhere and develop toward diversification and intelligent. Therefore, in order to defend against attacks, many security techniques like IDS, firewalls, antivirus software, etc., appeared continuously. For every enterprise, it wants to protect its internal network from external attackers and ensure that confidential information transmitted in internal network is safe. So, the firewall technique is used widely to ensure the security of an internal network. Firewall technology, known as the first line of network security [4], plays an important role in protecting internal network. The firewall rule table is often a specific reflect of corporate security policies and is an important basis for the firewall to filter packets. Therefore, the configuration of the firewall policies directly affects the efficiency of firewall. The optimization of filtering policy configuration is critically important to provide high performance packet filtering particularly for high speed network security [5]. Based on previous studies, this paper proposes a new firewall optimization method, CTFPi, for packets filtering. Experiments show that our approach greatly improves the time cost and acceptance ratio of legitimate packets. What's more, CTFPi shows a perfect result even with much forged packets. In Sect. 48.2, some existing packet filtering schemes are described briefly. Then we show the specific scheme of CTFPi and have discussed the key problem in Sect. 48.3. In Sect. 48.4, in order to verify the effectiveness of CTFPi, we conduct several experiments and analyze the results. Then we make some conclusions in the end.

48.2 Related Works

We will introduce some important packet filtering schemes and Packet marking schemes for both of them are involved in this paper. The filtering rule of firewall is very important in packets filtering. Figure 48.1 shows some of the typical packet filtering scheme. Packet Marking schemes are used to resist attacks and some of the famous schemes are shown in Fig. 48.2. We will describe them in detail later.

Ternary CAMs—TCAM (Ternary Content Addressable Memory) is a typical hardware-based packets filtering [6] and popular used in high-speed routers. It supports parallel seeking and the speed is faster. However, due to hardware-based characteristic, the method has some defects.

Fig. 48.1 Packet filtering
schemes

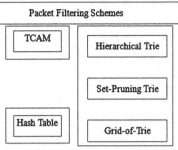

Fig. 48.2 Packet marking
schemes

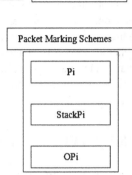

Hash Table—Hash table is a special data structure which plays an important role in packets filtering schemes. A representative one-dimensional packet filtering scheme was proposed in [7]. It builds a hash table for each different prefix of packets. The firewall stores labels in hash table and use Binary-Search algorithm to seek the table. It is different from the general filtering algorithms. The time complexity is rather low. A scalable prefix method was presented in [8]. It reduces the size of the hash table in order to improve the algorithm's time complexity. Based on the previous researches, a multi-dimensional packet filtering method was proposed in [9]. Its efficiency is higher comparing with the one-dimensional packet filtering method.

Trie—The main methods include Hierarchical Trie, Set-Pruning Trie (Tsuchiya P (1997) A algorithm for table entries with non-contiguous wildcarding, unpublished report, Bellcore) and Grid-of-Trie [10]. Hierarchical Trie builds a Trie for each dimension according to left "0" and right "1". This method uses the relationship between the prefix, but ignores the inclusion relationship between the prefixes. So we may traverse the entire Trie in the process of matching packets. In order to reduce trace back, Set-Prunning copies the relevant nodes of the second Trie under the first Trie nodes. However, it needs more storage space. Grid-of-Trie takes full advantage of null pointers of the first dimension and uses the null pointers point to the second dimension rather than copying the nodes of the second Trie. Thus it greatly improves the search efficiency.

Packet Marking Schemes—In order to defense DDos attack, Yaar [11] proposes the new packet marking method. It is a per-packet deterministic mechanism. The main idea of Pi is to carry out the last n (n = 1 or 2) bits of MD5 value of the routers' IP address and then insert them into the packets' ID field. This allows victim to take a proactive role in defending against a DDos attack by using the Pi marks. But this scheme has security risk. Then [12] presents StackPi to improve this drawback. This method supposes that the last several routers do not mark. However, it also has security vulnerabilities with an n = 2 bit. So Jin Guang [13] puts forward OPi, short for optimal path identification, to overcome the shortcoming. This method takes full use of ID field and contains more effective path information. Therefore, OPi scheme can be better to withstand attacks compared with former packet marking schemes.

From above analysis, the study of packet filtering and Pi mechanism are independent. Considering the functionality of firewall, our goal is to enhance the ability of a firewall to resist attacks and speed up filtering. Therefore, we propose a new filtering scheme that is CTFPi, the combing of traditional firewall and Pi. The method not only can reduce the dimension of the firewall's filtering field and improve the filtering efficiency, but also enhance the ability of the firewall to resist attacks.

48.3 CTFPi Scheme

48.3.1 Specific Scheme

Figure 48.3 shows the specific scheme. The first step is to generate the two fields, source IP address and destination IP address, of firewall rules and Pi according to the given data information. And each firewall rule should map to the relevant Pi.

Then, the completely firewall rule table can be composed from the generated rules. Next, we use the generated Pi table to replace the source IP address and destination IP address information in the firewall rule table. In that way, we can get new firewall rule table and it can be used to filter packets for firewall. When new packet arrival, it should carry the Pi information carried out by the routers along the path and then compare with the new firewall rule table. The firewall can decide to accept or drop the packet according to the predefined policies.

Fig. 48.3 The specific scheme

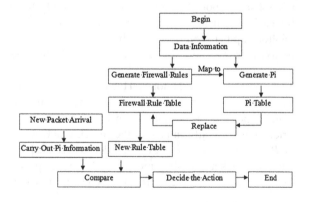

48.3.2 *Filtering Rules and Pi*

In order to make experiments more convincing, we must map the source and destination IP addresses of firewall rules to Pi and use Pi to replace them. However, the source and destination IP address are not completely relevant to the Pi. So how to solve the mapping problem becomes a critical issue.

Firstly, we must observe the characteristics of the selected data. Each data is a complete path composed by a number of the IP address of routers and the first IP addresses are the same. The main functionality of firewall is to protect LAN security. Then, we can consider the issue from the macro view. Namely, we can regard the beginning IP address as the destination IP address and the last IP address as the source IP address LANs respectively. In this way, the source IP address and destination IP address of the firewall policies have been solved. Furthermore, we can generate Pi as path information. Therefore, we can generate firewall policies and Pi information for each given path at the same time. And in that way we can solve the key problem raised above.

48.4 Simulation Experiments and Analysis

48.4.1 *Data Set*

In order to evaluate CTFPi, we select authoritative data from CAIDA [14] to conduct our simulations. The Skitter's apa20030505 data set is used as an example and its structure is showed as follows.

   C   203.181.248.27   12.0.1.28   1052382443   179.371   18   203.181.248.1
203.181.249.118 202.239.171.65 203.181.97.125 203.181.96.97 203.181.100.10
203.181.104.74 205.171.4.205 205.171.205.29 205.171.14.161 205.171.1.166
12.122.11.217 12.122.10.5 12.122.10.49 12.122.11.2 12.127.0.21

   I   203.181.248.27   12.0.134.10   1052378331   188.234   20   203.181.248.1
203.181.249.118 202.239.171.65 203.181.97.142 203.181.96.229 210.132.91.102
203.181.104.66 205.171.4.205 205.171.205.29 205.171.14.161 205.171.1.166
12.122.11.217 12.122.10.5 12.122.10.49 12.122.11.6 12.123.0.129

   N   203.181.248.27   12.0.192.2   1052368827   0.000   0   203.181.248.1
203.181.249.118 202.239.171.65 203.181.97.125 203.181.96.97 203.181.100.2
203.181.104.74 205.171.4.205 205.171.205.29 205.171.14.161 205.171.1.166
12.123.13.66 12.122.10.5 12.122.10.57 12.122.11.2 12.123.0.121

C represents the path is complete and can be used in experiments. I represents the path is incomplete and imply the lack of the router address. N represents the IP path not responded.

Fig. 48.4 The handle
process of data set

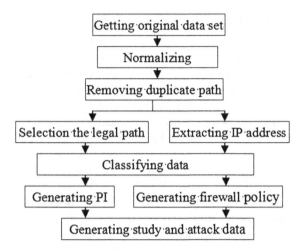

48.4.2 The Handle Process of the Data Set

Figure 48.4 shows the handling process of the data set. Firstly, the data set should be normalized. Then, duplicate paths should be removed. After that, the data set becomes available and the distribution of hop count is shown in Fig. 48.5. Next, we should separate the legitimate and malicious paths and generate Pi and firewall rules. At last, the study and attack data needed in the filter stage is generated.

48.4.3 Experiment Scene

The experiment is divided into two cases. The first case includes two scenes and there are no forged packets. In the first scene, we deploy 50 % legitimate and 50 % malicious paths respectively. In the second scene, the deployment ratio of legitimate paths and attack paths is 25 and 75 % respectively. We compare the drop ratio for malicious packets in the first scene and acceptance ratio for legitimate packets in the second scene. We set two sets of thresholds (0, 0.25, 0.5, 0.75 and 0,

Fig. 48.5 Hop count
distribution

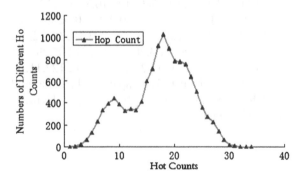

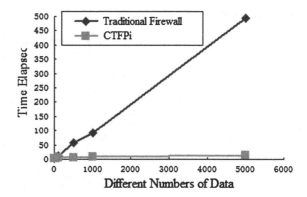

Fig. 48.6 Time comparison

0.2, 0.4, 0.6, 0.8) in the two scenes. Then we select 10,50,100,500,1000 and 5000 data to compare the filtrating time.

In the second case, there are forged malicious packets and this case also includes two scenes. In the first scene, we deploy 20 % forged malicious packets and set two sets of thresholds (0, 0.25, 0.5, 0.75 and 0, 0.2, 0.4, 0.6, 0.8) to compare the drop ratio of malicious packets. We deploy 1/3 forged malicious packets in another scene and test the acceptance ratio of legitimate packets under one set of threshold (0, 0.25, 0.5, 0.75).

48.4.4 Experimental Results

There are two phases in the simulation experiments: learning stage and attacking stage. In the learning stage, each user sends 1 legitimate packet and each attacker sends 3 malicious packets. The victim will define the action (drop or accept) of the filtering rules based on the behaviours. We select different number of data to test the time needed in the filtering process. The results are shown in Fig. 48.6.

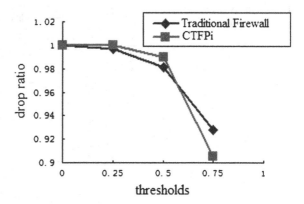

Fig. 48.7 Drop ratio comparison (I)

Fig. 48.8 Drop ratio
comparison (II)

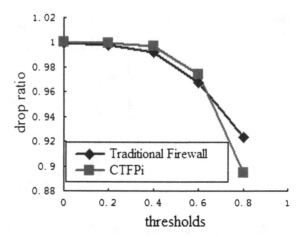

Fig. 48.9 Acceptance ratio
comparison (I)

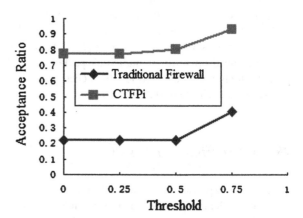

Fig. 48.10 Acceptance ratio
comparison (II)

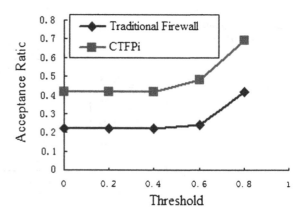

Fig. 48.11 Drop ratio
comparison (I)

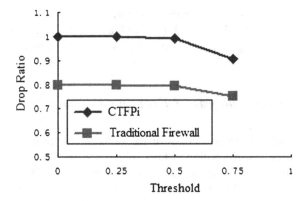

Fig. 48.12 Drop ratio
comparison (II)

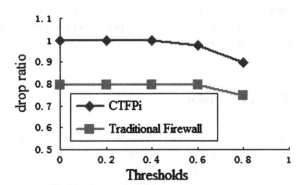

Fig. 48.13 Acceptance ratio
comparison

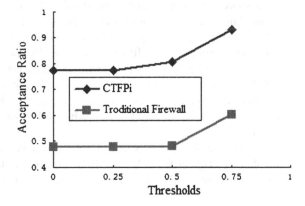

In the first scene of the first case, Figs. 48.7 and 48.8 show the comparison
results of drop ratio according to different thresholds. In the second scene we
tested the acceptance ratio and the results are shown in Figs. 48.9 and 48.10.

We also compare the drop ratio in the first scene of the second case and the results are shown in Figs. 48.11 and 48.12. Figure 48.13 shows the acceptance ratio comparing with the results in the second scene of the second case.

48.4.5 Analysis of Experimental Results

The experiments compare the different performance in CTFPi with traditional firewalls. According to Figs. 48.6, 48.9 and 48.10, CTFPi has clear advantage on time and acceptance ratio of legitimate packets. The drop ratio shown in Figs. 48.7 and 48.8 of malicious packets is not particularly obvious. The drop ratio and the acceptance ratio have greatly improved and Figs. 48.11, 48.12 and 48.13 demonstrate the results. The experiments are entirely reasonable because of the effects of Pi are not always better comparing with other filtering policies but only when the forged packets exist.

48.5 Conclusions

This paper proposes the combination of traditional firewall and Pi and the application of Pi in firewall filtering. A lot of experiments are conducted to test the performances of CTFPi. Compared with traditional firewall, experiments show that our scheme has the following advantages:

- CTFPi reduces the dimension of the field of firewall rules. The two-dimension field, source IP address and destination IP address, can be replaced by a decimal value. Thus the dimension of firewall rules can be decreased.
- CTFPi speeds up filtering. CTFPi only need to compare a decimal number instead of comparing two IP addresses. The filtering speed will be significantly faster and can be better adapt to the high-speed network requirements.
- CTFPi is stronger to withstand attacks. It is not safe to take IP address as filtering rules since attackers can forge IP address. Our scheme can be better to prevent attacks for it uses Pi carried by packet instead of IP address to filter packet.
- CTFPi reduces the storage space of firewall rule table. We can see that the same number of original firewall rules is much greater than the new firewall rules from the point of file size. This is a considerable improvement for limited resources.

From above analysis, we can get the conclusion that CTFPi is better to adapt to high-speed network requirements and can be more effectively to protect LAN from being attacked by external malicious attackers. In the future, we will further improve our scheme as much as possible.

Acknowledgments This research was supported in part by Major Projects of National Science and Technology (2011ZX03002-004-02), Zhejiang Provincial Technology Innovation Team (2010R50009), Natural Science Foundation of Zhejiang Province (LY12F02013), Ningbo Natural Science Foundation (2012A610014), Ningbo Municipal Technology Innovation Team (2011B81002), The Graduate Teaching Innovation on Ningbo University (2011004).

References

1. (2010) CERT. http://www.cert.org/stats/
2. (2009) Arbor networks worldwide infrastructure security report http://www. Arbornetworks. co-m/en/research.html
3. Distributed Denial of Service (DDoS), Attacks/tools http://staff.washington.edu/dittrich/misc/ddos/
4. Arshad M, Nessa S, Khan L, Al-Shaer E et al (2010) Analysis of firewall policy rules using traffic mining techniques. Protoc Technol IJIPT 5:3–22
5. Hamed H, Al-Shaer E (2006) On autonomic optimization of firewall policy configuration. J High Speed Netw Spec Issue Secur Policy Manag 13:209–227
6. Mckeown N (2001) Algorithms for packet classification. IEEE Network 15:24–32
7. Waldvogel M, Varghese G, Turner J (2001) Scalable high speed prefix matching. Int ACM Trans Comput Syst 19:440–482
8. Kim K, Shni S (2003) IP lookup by binary search on length. In: IEEE international symposium on computer and communication
9. Lu H, Sahni S (2007) O(logW) multidimensional packet classification. IEEE/ACM Trans Networking 15:462–472
10. Srinivasan V, Varghese G, Suri S (1998) Fast and scalable layer for switching. In: Proceedings of ACM SIGCOMM
11. Yaar A, Perrig A, Song D (2003) Pi: a path identification mechanism to defend against DDoS attacks. In: Proceedings of IEEE symposium on security and privacy, pp 93–97
12. Yaar A, Perrig A, Song D (2006) StackPi: new packet marking and filtering mechanisms for DDoS and IP spoofing defense. IEEE J Sel Areas Commun 24:1853–1863
13. Guang J, Jianggang Y, Yuan L et al (2008) Optimal path identification to defend against DDoS attacks. J Commun 29(9):46–53 (in Chinese with English abstract)
14. http://www.caida.org

Chapter 49
Semantic Web Service Automatic Composition Based on Discrete Event Calculus

Kai Nie, Houxiang Wang and Jiao He

Abstract A semantic Web service automatic composition method based on discrete Event Calculus is proposed aiming at the issues of service AI planning composition such as large number of services and the confine of sequence composition process. Firstly, the extension of EC to DEC is present. Then the eight basic semantic Web service composition processes and their IOPE are modeled based on the actions, fluent and axioms of DEC. The service composition process is divided into two steps, abstract service planning and instance execution. And the service automatic composition framework is introduced. Also the abduction DEC planning method and semantic matching method of instance execution are given. The comparison indicate the superiority of this method: it solves the confine of sequence composition process of classic AI planning composition method with Event Calculus' presentation of compound action, concurrent action, continuous action, knowledge of the agent and the predicate number of the DEC is much smaller than EC which speed the service discovering and composition.

Keywords Semantic web service · Automatic service composition · Intelligent planning · Discrete event calculus · Abduction

49.1 Introduction

Emerging semantic Web services standards as OWL-S [1] enrich Web service standards like WSDL and BPEL4WS with rich semantic annotations to facilitate flexible dynamic web services discovery, invocation and and composition. The semantic

K. Nie (✉) · H. Wang · J. He
College of Electronic Engineering, Naval University of Engineering,
No 717, Jiefang Dadao Road 430033 Wuhan, People's Republic of China
e-mail: 1999104133@163.com

W. Lu et al. (eds.), *Proceedings of the 2012 International Conference on Information Technology and Software Engineering*, Lecture Notes in Electrical Engineering 210, DOI: 10.1007/978-3-642-34528-9_49, © Springer-Verlag Berlin Heidelberg 2013

description of function interface and behaviour of Web service can be provided by OWL-S, but it does not contain automated reasoning mechanism and does not support automatic service discovery and composition of semantic Web services. The key technology of automatic service composition is using semantic match and automated reasoning mechanism. The automatic semantic Web service composition methods contain method based on workflow, method based on intelligent planning, method based on formal description and so on [2]. The intelligent planning method shortening as AI is very popular and mature in Web service composition and can be implemented in some tools such as JSHOP and Prolog. The AI composition methods contains Hierarchy Task Network (HTN) [3], Situation Calculus [4], PDDL [5], dynamic, description logic [6], theorem prover [7] and so on. But they have two shortcomings: large number of services and the confine of sequence composition process.

Event Calculus is one of the convenient techniques for the automated composition of semantic web services. The Event Calculus (EC) is a temporal formalism designed to model and reason about scenarios described as a set of events whose occurrences have the effect of starting or terminating the validity of determined properties of the world. The one that will be used in this paper is the one defined by Shanahan [8]. The Event Calculus is based on first-order predicate calculus, and is capable of representing actions with indirect effects, actions with non-deterministic effects, compound actions, concurrent actions, and continuous change. Agarwal [9] presents a classification of existing solutions into four categories (approaches): interleaved, monolithic, staged and template-based composition and execution. Okutan [10] proposes the application of the abductive Event Calculus to the web service composition and execution problem about the interleaved and template-based type. Ozorhan [11] presents a monolithic approach to automated web service composition and execution problem based on Event Calculus. But they all have large numbers of predicates and the compose speed of them is slow. The discrete Event Calculus (DEC) has been proven to be logically equivalent with EC, when the domain of time points is limited to integers [12]. The DEC can eliminate the triply quantified axioms that lead to an explosion of the number of predicates.

In this paper, the Abductive Planning of the discrete Event Calculus is used to show that when atomic services are available. The service composition process is divided into two steps, abstract service planning and instance execution. And the service automatic composition framework is introduced and semantic matching method of instance execution is given.

49.2 Discrete Event Calculus

49.2.1 Event Calculus

The Event Calculus is based on first-order predicate calculus, and is capable of representing a variety of phenomena, including actions with indirect effects, actions with non-deterministic effects, compound actions, concurrent actions, and

Table 49.1 Event calculus predicates in classical logic

Predicate	Meaning
Happens (E, T)	Event E occurs at time T
Initially$_P$ (F)	Fluent F holds from time 0
Initially$_N$ (F)	Fluent F does not hold from time 0
Holds At (F, T)	Fluent F holds at time T
Initiates (E, F, T)	Event E initiates fluent F at time T
Terminates (E, F, T)	Event E terminates fluent F at time T
Clipped (F, T_0, T_1)	Fluent F is terminated some time in the interval $[T_0\ T_1]$
Declipped (F, T_0, T_1)	Fluent F is initiated some time in the interval $[T_0\ T_1]$
Releases (E, F, T)	Fluent F is not constrained by inertia after event E at time T

continuous change. The basic ontology of the Event Calculus comprises actions or events (or rather action or event types), fluents and time points. Table 49.1 introduces the language elements of the Event Calculus.

The axioms of the Event Calculus relating the various predicates together are as follows:

$$holdsAt(F, T) \leftarrow initially_P(F) \land \neg clipped(F, 0, T) \tag{49.1}$$

$$holdsAt(F, T) \leftarrow happens(E, T_0) \land initiates(E, F, T_0) \land T_0 < T \\ \land \neg clipped(F, T_0, T) \tag{49.2}$$

$$clipped(F, T_0, T_1) \leftrightarrow \exists E, T\, happens(E, T) \ \land T_0 \leq T \land T \leq T_1 \\ \land terminates(E, T, F) \tag{49.3}$$

$$declipped(F, T_0, T_1) \leftrightarrow \exists E, T\, happens(E, T) \ \land T_0 \leq T \land T \leq T_1 \\ \land initiates(E, F, T) \tag{49.4}$$

49.2.2 DEC

The axioms of DEC utilize a subset of the EC elements (Table 49.1), that is happens, holds At, initiates and terminates. The axioms that determine when a fluent holds, are defined as follows:

$$holdsAt(F, T + 1) \leftarrow happens(E, T) \land initiates(E, F, T) \tag{49.5}$$

$$holdsAt(F, T + 1) \leftarrow holdsAt(F, T) \land \neg \exists E\, happens(E, T) \land terminates(E, F, T) \tag{49.6}$$

Compared to EC, DEC axioms are defined over successive timepoints. Additionally, the DEC axioms are quantified over a single time point variable. Therefore the number of predicates is substantially smaller than EC. Axioms (49.6), however, contain the existentially quantified variable E. Each of these axioms will be transformed into $2^{|\varepsilon|}$ clauses. Moreover, each clause will contain a large number of disjunctions. To overcome the creation of $2^{|\varepsilon|}$ clauses, we can employ the technique of subformula renaming, as it is used in [12]. According to this technique, the subformula $happens(E, T) \wedge initiates(E, F, T)$ in (49.5), is replaced by a utility predicate that applies over the same variables, e.g. $startAt$ (E, F, T). A corresponding utility formula, i.e. $startAt(E, F, T) \leftrightarrow happens$ $(E, T) \wedge initiates(E, F, T)$, is then added to the knowledge base.

In order to eliminate the existential quantification and reduce further the number of variables, we adopt a similar representation as in [12], where the arguments of initiation and termination predicates are only defined in terms of fluent and time points—represented by the predicates initiated At and terminated At respectively. As a result, the domain-independent axioms of DEC presented above are universally quantified over fluents and time points. The axioms that determine when a fluent holds are thus defined as follows:

$$holdsAt(F, T + 1) \leftarrow initiatedAt(F, T) \tag{49.7}$$

$$holdsAt(F, T + 1) \leftarrow holdsAt(F, T) \wedge \neg terminatedAt(F, T) \tag{49.8}$$

49.3 Service Automatic Composition Method Based on DEC

49.3.1 The Basic Service Processes Translation to DEC

In OWL-S the composite processes are composed of sub-processes and specify constraints on the ordering and conditional execution of the sub-processes. The minimal set of control constructs according to includes Sequence, Split, Split +Join, Any-Order, Choice, If–Then-Else, Repeat-While and Repeat-Until. These constructs are translated automatically into compound events in our framework.

Translations of the sequence control construct:

$$axiom(happens(pSequenceExample(Inputs, Outputs), T_0, T_m),$$
$$[happens(pProcess_1(Inputs_1, Outputs_1), T_1, T_2), happens(pProcess_2(Inputs_2, Outputs_2), T_3, T_4), \ldots,$$
$$happens(pProcess_n(Inputs_n, Outputs_n), T_x, T_y), before(T_0, T_1), before(T_1, T_3), \ldots, before(T_x, T_m)])$$
$$Input_i \subseteq Inputs \cup \bigcup_{j=1}^{i-1} Outputs_j \, for \, i = 1, \ldots, n \, Outputs \subseteq \bigcup_{i=1}^{n} Outputs_i$$

Translations of the Split control construct:

$$axiom(happens(pSplitExample(Inputs, Outputs), T_0, T_m),$$
$$[happens(pProcess_1(Inputs_1, Outputs_1), T_1, T_2), happens(pProcess_2(Inputs_2, Outputs_2), T_3, T_4), \ldots,$$
$$happens(pProcess_n(Inputs_n, Outputs_n), T_x, T_y), before(T_0, T_1), before(T_0, T_3), \ldots, before(T_0, T_x)])$$

$$Inputs_i \subseteq Inputs \text{ for } i = 1, 2, \ldots, n \quad Outputs \subseteq \bigcup_{j=1}^{n} Outputs_j$$

Translations of the Split–Join control construct:

$$axiom(happens(pSplitJoinExample(Inputs, Outputs), T_0, T_m),$$
$$[happens(pProcess_1(Inputs_1, Outputs_1), T_1, T_2), happens(pProcess_2(Inputs_2, Outputs_2), T_3, T_4), \ldots,$$
$$happens(pProcess_n(Inputs_n, Outputs_n), T_x, T_y), before(T_0, T_1), before(T_0, T_3), \ldots, before(T_0, T_x),$$
$$before(T_2, T_m), before(T_4, T_m), \ldots, before(T_y, T_m)])$$

Translations of the If–Then–Else control construct:

$$axiom(happens(pIfThenElseExample(Inputs, OutPuts), T_0, T_m),$$
$$[happens(jpl\_pIfCondition(Inputs_1), T_1, T_2), happens(pThenCase(Inputs, Outputs), T_3, T_4),$$
$$before(T_0, T_1), before(T_2, T_3), before(T_4, T_m)])$$

$$axiom(happens(pIfThenElseExample(Inputs, OutPuts), T_0, T_m),$$
$$[happens(jpl\_pElseCondition(Inputs_1), T_1, T_2), happens(pElseCase(Inputs, Outputs), T_3, T_4),$$
$$before(T_0, T_1), before(T_2, T_3), before(T_4, T_m)])$$

Translation of the Repeat-While control construct:

$$axiom(happens(pRepeatWhileExample(Inputs, OutPuts), T_0, T_m),$$
$$[happens(jpl\_pLoopCondition(Inputs_1), T_1, T_2), happens(pWhilecase(Inputs, Outputs), T_1, T_3),$$
$$before(T_0, T_1), before(T_1, T_2), before(T_1, T_3), before(T_3, T_m)])$$

49.3.2 Web Service Automatic Composition Framework

The Web service automatic composition framework is in Fig. 49.1. The framework is divided into two steps: abstract service planning and instance execution. The steps of the workflow are: convert OWL-S descriptions and user inputs and outputs to the discrete Event Calculus axioms; execute the abductive theorem prover to generate plans; convert the generated plans to graphs; convert the selected graphs to OWL-S service files; execute the selected OWL-S service composition.

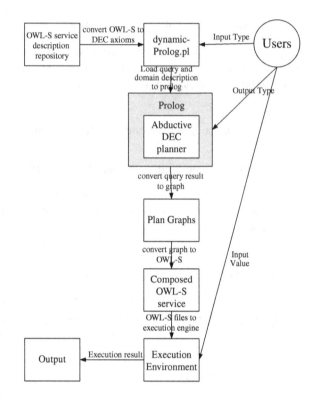

Fig. 49.1 The service automatic composition framework based on DEC

49.3.3 Abstract Service Planning

49.3.3.1 Translation of IOPE to DEC Axioms

The input and output of Web service translation to DEC are as follows:

$$axiom(initiatedAt(web\_service\_event(I_1, I_2, \ldots, I_k, O_1, O_2, \ldots, O_j), known(out\_parameter\_name_i, O_i), T),$$
$$[holdsAt(known(in\_parameter\_name_1, I_1), T), \ldots, holdsAt(known(in\_parameter\_name_k, I_k), T),$$
$$holdsAt(precondition_1, T), \ldots, holdsAt(precondition_p, T)])$$

The precondition of Web service translation to DEC is as follows:

$$axiom(happens(pPreconditionExample(Input_1, Input_2), T_1, T_N),$$
$$[jpl\_pPrecondition(Input_1, Input_2), atom\_number(Input_1, Arg_1),$$
$$atom\_number(Input_2, Arg_2), Arg_1 < Arg_2, [Other\ Events\ and\ Temporal\ Orderings\ ]\ldots])$$

The effect of Web service translation to DEC is as follows:

$$axiom(initiatedAt(web\_service\_event(I_1, I_2, \ldots, I_k, O_1, O_2, \ldots, O_j), effect\_e_i, T),$$
$$[holdsAt(known(in\_parameter\_name_1, I_1), T), \ldots, holdsAt(known(in\_parameter\_name_k, I_k), T),$$
$$holdsAt(precondition_1, T), \ldots, holdsAt(precondition_p, T)])$$

Here *effect_e_i* is the Prolog translation of the effect name taken from the *has effect* section of the service profile.

49.3.3.2 Abductive DEC Planning Algorithm

Abduction is used over the Event Calculus axioms to obtain partially ordered sets of events. Abduction is handled by a second order Abductive Theorem Prover (ATP) in [13]. The ATP tries to solve the goal list by proving the elements one by one. During the resolution, abducible predicates are stored in a residue to keep the record of the narrative. The narrative is a sequence of time-stamped events. In the definition of the predicate abduce, GL denotes the goal list, RL represents the residue list, NL represents the residue of negated literals, G is the axiom head and AL is the axiom body.

$$AH \leftarrow AB_1 \wedge AB_2 \wedge \ldots \wedge AB_N \quad axiom(AH, [AB_1, AB_2, \ldots, AB_N])$$

$$abduct(GL, RL) \leftarrow abduct(GL, \Box, RL, \Box) \quad abduct(\Box, RL, RL, NL)$$

$$abduct([holdsAt(F, T)|GL], CurrRL, RL, NL) \leftarrow axiom(initially(F), AL),$$
$$irresolvable(clipped(0, F, T), CurrRL, NL), append(AL, GL, NewGL),$$
$$abduct(NewGL, CurrRL, RL, [clipped(0, F, T)|NL])$$

$$abduct([holdsAt(neg(F), T)|GL], R1, R3, N1, N4) \leftarrow axiom(initially(neg(F)), AL),$$
$$irresolvable(declipped(0, F, T), CurrRL, NL), append(AL, GL, NewGL),$$
$$abduct(NewGL, CurrRL, RL, [declipped(0, F, T)|NL])$$

49.3.4 Generating Output Graphs from the Results of the Planner

The Service-Oriented C^4ISR system on the naval fields is taken as an example for generating output graphs from the results of the planner. Fig. 49.2 is the generating graph. There eight services are as follows: the surface target detection service W_1, the underwater target detection service W_2, the information process service W_3, the fire control compute service W_4, the missile launch service W_5, the torpedo launch service W_6, the efficiency evaluate service W_7, the repeat attack service W_8, *St* and *En* are the start and end services. The W_1 and W_2 are concurrent, then the W_3 is in sequence. The W_5 and W_6 are choice relation. The W_8 is loop with W_4, W_5, W_6 and W_7, and k is a small integer. The abstract service of the process is corresponding to an instance service set, and the QoS of the instance services are different according their equipments such as execute time, cost, and accuracy.

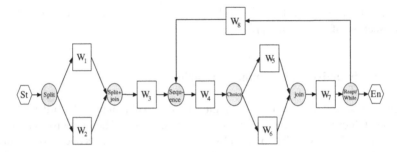

Fig. 49.2 Graph model of service-oriented C[4]ISR system on the naval fields

49.3.5 The Instance Service Matching Method

The instance service matching process is divided into two steps: the local semantic matching and the global QoS-ware matching. The local semantic matching is not only considering the input and output matching, but also the local semantic matching. The global QoS-ware matching is translated into a multi-objective services composition optimization with QoS constraints. The multi-objective estimation of distribution algorithm based on Independent Component Analysis is utilized to produce a set of optimal Pareto services composition with constraint principle by means of optimizing various objective functions simultaneously.

49.4 Conclusion

A semantic Web service automatic composition method based on discrete Event Calculus is proposed aiming at the issues of service AI planning composition method such as large number of services, the confine of sequence composition process. The eight basic semantic Web service composition processes and their IOPE are modeled based on the actions, fluent and axioms of DEC. The service composition process is divided into two steps, abstract service planning and instance execution. And the service automatic composition framework is introduced. Also the abduction DEC planning algorithm and semantic matching method of instance execution are given. The comparison indicate the superiority of this method: it solves the confine of sequence composition process of classic AI planning composition method with DEC' presentation of compound action, concurrent action, continuous action, knowledge of the agent, the predicate number of the DEC is much smaller than EC which speed the service discovering and composition.

References

1. Martin D, Burstein M, McDermott D et al (2007) Bringing semantics to web services with OWL-S. World Wide Web J 10(3):243–277
2. Charif Y, Sabouret N (2006) An overview of semantic web services composition approaches. Electron Notes Theoret Comput Sci 33–41
3. Sirin E, Parsia B, Wu D et al (2004) HTN planning for Web service composition using SHOP2. J Web Sem 1:377–396
4. Mcilraith S, Son T (2002) Adapting go log for composition of semantic web services. In: Proceedings of the eighth international conference on principles of knowledge representation and reasoning, France
5. Klusch M, Gerber A, Schmidt M (2005) Semantic web service composition planning with OWLS-XPlan. In: Proceedings of the 1st international AAAI fall Symposium on agents and the semantic web. Arlington, pp 117–120
6. Wan Changlin, Han Xu, Niu Wenjia et al (2010) Dynamic description logic based web service composition and QoS model. Acta Electronica Sinica 38(8):1923–1928 (in Chinese)
7. Rao JH, Kungas P, Matskin M (2006) Composition of semantic web services using linear logic theorem proving. Inf Syst 4–5:340–360
8. Shanahan M (1999) The event calculus explained. In: Artificial intelligence today: recent trends and developments. Springer, Berlin 409–430
9. Agarwal V, Chafle G, Mittal S et al (2008) Understanding approaches for web service composition and execution. In: Proceedings of the first Bangalore annual compute conference. ACM, New York, pp 1–8
10. Okutan C, Cicekli NK (2010) A monolithic approach to automated composition of semantic web services with the event calculus. Knowl-Based Syst 23:440–454
11. Ozorhan EK, Kuban EK, Cicekli NK (2010) Automated composition of web services with the abductive event calculus. Inf Sci 180:3589–3613
12. Mueller ET (2008) Event calculus. In: Handbook of knowledge representation, vol 3. Elsevier, Amsterdam, pp 671–708
13. Shanahan MP (2000) An abductive event calculus planner. J Logic Prog 44:207–240

Chapter 50
Research and Application of Dual Drive Synchronous Control Algorithm

Herong Jin

Abstract According to the real time rapidity characteristic of generalized predictive control, dual drive control synchronous subsystem based on generalized predictive is designed. Based on coordinated control principle, a fuzzy controller is added between the channels of two drive subsystems. Aiming to minimize the corresponding state difference of two subsystems, control scheme with synchronous error non-equal common feedback fuzzy correction is adopted, which keep the output of two synchronous subsystems consistent, improve the response speed of synchronous error to system, and reduce the synchronous error of two drive subsystems effectively. The simulation results show that synchronous control strategy presented in this paper improve the response speed of synchronous error to system, reduce the synchronous error of two motors effectively, and provide advantageous guarantee for the high precision of large rotary dual drive synchronous systems.

Keywords Dual drive synchronous control · Coordinated control · Common-mode feedback · Fuzzy controller

50.1 Introduction

Compared to the mechanical synchronous drive's shortcomings such as complex structure, inconvenience of installation and maintenance, big noise in working and poor flexibility, the intelligent synchronous control has an extensive application

H. Jin (✉)
Hebei Provincial Key Laboratory of Parallel Robot and Mechatronic System,
Key Laboratory of Advanced Forging & Stamping Technology and Science,
Yanshan University, Ministry of Education of China, 438, Hebei Avenue,
Qinhuangdao 066004, People's Republic of China
e-mail: ysujhr@ysu.edu.cn

W. Lu et al. (eds.), *Proceedings of the 2012 International Conference on Information Technology and Software Engineering*, Lecture Notes in Electrical Engineering 210, DOI: 10.1007/978-3-642-34528-9_50, © Springer-Verlag Berlin Heidelberg 2013

prospect because of its advantages including simple structure, convenient adjustment and noiseless action, etc.

With the continuous development and improvement of the control theory, the
researches about synchronization of two systems are paid extensive attention to all
fields [1–4]. The existing results show that for two transmission systems that have
the identical structure, they can be synchronized by adding to the proper feedback
signals between the two subsystems. As the basis of the dual-drive system synchronization, the performance and efficiency of control methods is crucial for the
systems to achieve synchronization [5, 6].

50.2 Synchronization Strategy of the Dual-Motor Drive Control

To improve the overall performance of the dumper, the rotors driving uses the
scheme that reducer connect to two motors, but there is a problem that how to
ensure two motors driving synchronously. If the system uses a master–slave mode,
the load disturbances in the running process of two motors are different. If we
select a reference model and make two motors change following the reference
model, the dynamic speed processes of two motors are still inconsistent and the
synchronization error is large, so it can't meet the requirements of the control. In
this paper, we use two generalized predictive sub-controllers and the synchronization error being produced from the control project, at the same time, we can
adjust two sub-controllers respectively so that it can achieve the purpose of synchronization. Two controllers can not only inhibit the disturbances itself, but also
have the ability to coordinate. The synchronization strategy of the coordinated
control in this paper is shown in Fig. 50.1.

Controller 1 and controller 2 control their own drive subsystems respectively,
theoretically, the input signals of two subsystems are the same, and therefore, it
belongs to the equivalent synchronization control. If the system parameters always
remain stable, the outputs of these two channels will be synchronized basically.

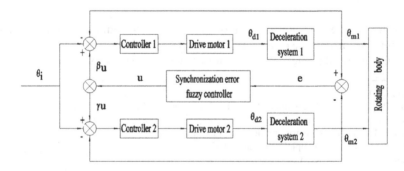

Fig. 50.1 Principle block diagram of synchronization drive system

But during the work process of system, there are some factors such as machining accuracy, environmental impact, nonlinear, time-varying of electrical and mechanical system parameters and external load disturbance, so the outputs of two synchronization subsystems will have deviations and it will produce synchronization error e. We can use the control scheme that fuzzy correction of non-attainment and common feedback in synchronization error to keep two subsystems accuracy synchronous, so that it can achieve the target that the difference between the corresponding states of two subsystems is minimum. The synchronization error can produce outputs through the fuzzy controller, one of outputs βu is treated as the additional signal acting on the synchronization subsystem 1, and the other γu is treated as the additional signal acting on the synchronization subsystem 2 (including $0 \leq |\beta|, |\gamma| \leq 1$, and $|\beta| + |\gamma| = 1$, $\beta \cdot \gamma \leq 0$, if the dynamic characteristics of subsystem 1 are higher than β, take a large value, otherwise, take a small value), two additional signals lower the high output, increase the low output, finally it makes θ_{d1}, the output of synchronization subsystem 1 and θ_{d2}, the output of synchronization subsystem 2 keep consistent. Through the effect of the fuzzy controller, it can eliminate synchronization errors which two channels produce in the equivalent way because the system characteristics are different, then the coordination control between two subsystems will be achieved and the purpose of synchronization control will come true. Compared to decoupling control, this method is a synchronization control method based on coordinated control.

50.3 The Design of Synchronization Fuzzy Controller Based on T-S Model

We can use a method that synchronization error fuzzy correction for the synchronization accuracy of two motor-driven and use a control scheme that fuzzy correction of non-attainment and common feedback synchronization error in the whole control system. The goal of synchronization error fuzzy correction control is to make the difference of the corresponding states of two motors minimum and it can achieve compensation through non-attainment and common feedback in the synchronization error.

50.3.1 Fuzzy Correction Control Scheme of Synchronization Error

The fuzzy control is a computer control method based on fuzzy set theory, fuzzy linguistic variables and fuzzy logic. From the point of controller intelligence, it belongs to the field of intelligent control and now it has been an important and effective form of intelligent control currently, especially the integration of new disciplines such as fuzzy control, neural network, genetic algorithms and chaos theory, is showing its great potential. The typical applications are cement kiln control, spacecraft flight control, robot control, vehicle speed control, elevator control, current and reactor control.

The fuzzy T-S model is a nonlinear model which is proposed by Takagi and Sugeno in 1985. It's easy to express the dynamic characteristics of complex systems. For the non-linear dynamic systems in different regions, we can use the T-S fuzzy model to establish local linear model, then connect each local linear model with a fuzzy membership function. With the overlay each other between multiple rules, we can get a non-linear mapping. After obtaining the whole fuzzy nonlinear model, we can conduct the control design and analysis for the complex systems. The rule i of the model which is based on description language rules can be written as

$$R^i : \text{If } x_1 \text{ is } A_1^i, \ x_2 \text{ is } A_2^i, \ldots, x_m \text{ is } A_m^i,$$

$$\text{Then } y^i = p_0^i + p_1^i x_1 + p_2^i x_2 + \ldots + p_m^i x_m (i = 1, 2, \ldots, n)$$

Where, R^i denotes section i of the fuzzy rules; $A_j^i (j = 1, 2, \ldots, m)$ is fuzzy subset, the parameters in its membership function are premise parameters; x_j is the section j of input variable; m is the number of input variables; p_j^i is the conclusion parameter; y^i is the output of section i of rules. If the input fuzzy vectors are given $(x_1^0, x_2^0, \cdots\cdots, x_m^0)$, the output $\hat{y}$ can be obtained through weighted combination of the output of each rule y_i

$$\hat{y} = \sum_{i=1}^{n} G_i y_i \bigg/ \sum_{i=1}^{n} G_i \tag{50.1}$$

Where, n is the number of the fuzzy rules, G_i denotes the true value of R_i, which is the section i of rules and can be expressed as the following equation

$$G_i = \prod_{j=1}^{m} A_j^i \left(x_j^0 \right) \tag{50.2}$$

By the T-S fuzzy rule expression, we can know the inference algorithm of T-S fuzzy model is basically same as the inference algorithm of other fuzzy models like Mamdani in the process of fuzzy input and fuzzy logic operations and the main difference is the form of output membership functions. The fuzzy set is only used in the premise of the T-S model and the conclusion is the linear combination of the input true values. It makes fuzzy reasoning simplified greatly.

According to the control rules and fuzzy sets membership grade, the fuzzy controller uses a lookup table method translate the fuzzy control rules into a lookup table, then use the lookup table to get the control amount. The control uses two-dimensional fuzzy controller, there are two inputs, let e be a speed error, $\dot{e}$ is

Table 50.1 Fuzzy control table

E EC	PB	PM	PS	PZ	NZ	NS	NM	NB
PB	$g_1(e)$	$g_1(e)$	$g_2(e)$	$g_3(e)$	0	0	0	0
PM	$g_1(e)$	$g_1(e)$	$g_2(e)$	$g_3(e)$	0	0	0	$g_3(e)$
PS	$g_1(e)$	$g_2(e)$	$g_2(e)$	0	0	0	$g_3(e)$	$g_2(e)$
ZE	$g_1(e)$	$g_3(e)$	$g_3(e)$	0	0	0	$g_3(e)$	$g_1(e)$
NS	$g_2(e)$	$g_3(e)$	0	0	0	$g_3(e)$	$g_2(e)$	$g_1(e)$
NM	$g_3(e)$	0	0	0	$g_3(e)$	$g_2(e)$	$g_1(e)$	$g_1(e)$
NB	0	0	0	0	$g_3(e)$	$g_2(e)$	$g_1(e)$	$g_1(e)$

the rate of change of speed error, the output variables are the current signals inputting into the inverter.

Formulation of control rules is the key of fuzzy controller design, if we use T-S fuzzy model, the fuzzy rules can be set as follows

$$R_i : \text{ If e is A and } \dot{e} \text{ is B, THEN } u_f = g(e)$$

where, A and B is the fuzzy subset, g (e) is a continuous function of e, R_i is the section i of control rules, the If statement is the premise of control rules, the then statement is the result of control rules. The fuzzy control rule is obtained through comprehensive induction according to the operating experience and in-depth analysis of the control system. The fuzzy control rules are shown in Table 50.1.

In the table, NB = Negative Big, NM = Negative Middle, NS = Negative Small, NZ = Negative Zero, ZE = Zero, PZ = Positive Zero, PS = Positive Small, PM = Positive Middle, PB = Positive Big, $g_1(e)$, $g_2(e)$, $g_3(e)$ are taken as $K_0 e$, $0.6 K_0 e$, $0.3 K_0 e$, according to the T-S fuzzy reasoning algorithm, we can obtain the fuzzy query table. By selecting a suitable $g_i(e)$, we can know the output of fuzzy controller

$$u_f = K_u e \cdot f(e, \bar{e}) \tag{50.3}$$

In the equation, K_u is fuzzy controller's output variable scale factor, $f(e, \dot{e})$ is function related with e and $\dot{e}$, $0 \le f(e, \dot{e}) \le 1$.

From the Eq. (50.3), it's clear that the controller output is a piecewise linear function of system error, it can be regarded as a proportional controller with a variable scale factor essentially. The variable scale factor is a nonlinear function about the error and its change rate. So the controller has certain ability to adaptive and it will have a better control performance in theory. To make the fuzzy controller has both control and learning capabilities, we add an adaptive link to it. The link includes three units which are performance testing, calibration of the control amount and control rules correction. To improve the control accuracy and make the system be a fuzzy control system of no difference, we can add an integral separation part to it. The integral separation part only works when the error is very small, its purpose is to eliminate systematic errors, when the error is large, the link doesn't work to cancel the integral role and avoid integral saturation phenomenon.

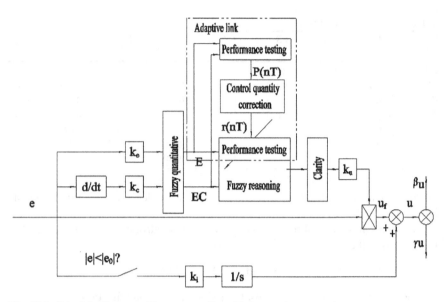

Fig. 50.2 Principle diagram of fuzzy controller with integration separating segment

Therefore, the control structure block diagram of fuzzy controller is shown in Fig. 50.2, k_e and k_c are error and quantitative factors of error change rate, k_i is the integral coefficient.

50.3.2 Stability Analysis of Synchronization System

We can set up respectively that the closed-loop transfer function of system 1 and system 2 are

$$\begin{cases} G_1(s) = \frac{Z_1(s)}{N_1(s)} \\ G_2(s) = \frac{Z_2(s)}{N_2(s)} \end{cases} \tag{50.4}$$

In which, $N_1(s)$ and $N_2(s)$ are respectively closed-loop characteristic style of the subsystem channel 1 and subsystem channel 2.

$$\begin{cases} \theta_{m1}(s) = G_1(s)(\theta_i(s) + \beta \cdot f(e, \dot{e}) K_u e(s)) \\ \theta_{m2}(s) = G_2(s)(\theta_i(s) + \gamma \cdot f(e, \dot{e}) K_u e(s)) \end{cases} \tag{50.5}$$

In which, β is the eliminating error distribution coefficient of subsystem channel 1; γ is the eliminating error distribution coefficient of subsystem channel 2. β and γ must satisfy the following conditions

$$\begin{cases} 0 \leq |\beta|, |\gamma| \leq 1 \\ |\beta| + |\gamma| = 1 \\ \beta \cdot \gamma \leq 0 \\ \beta < 0 \quad (\theta_{m1} > \theta_{m2}) \\ \beta > 0 \quad (\theta_{m1} < \theta_{m2}) \end{cases} \tag{50.6}$$

$$e(s) = \theta_{m1}(s) - \theta_{m2}(s) = \frac{G_1(s) - G_2(s)}{1 - \beta \cdot f(e, \dot{e})K_u G_1(s) + \gamma \cdot f(e, \dot{e})K_u G_2(s)} \theta_i(s) \tag{50.7}$$

Error transfer function is

$$G_e(s) = \frac{e(s)}{\theta_i(s)} = \frac{Z_1(s)N_2(s) - Z_2(s)N_1(s)}{N_1(s)N_2(s) - \beta \cdot f(e, \dot{e})K_u Z_1(s)N_2(s) + \gamma \cdot f(e, \dot{e})K_u Z_2(s)N_1(s)} \tag{50.8}$$

Characteristic equation of the system is

$$N_1(s)N_2(s) - \beta \cdot f(e, \dot{e})K_u Z_1(s)N_2(s) + \gamma \cdot f(e, \dot{e})K_u Z_2(s)N_1(s) = 0 \tag{50.9}$$

From (50.5) and (50.8)

$$\begin{cases} \theta_{m1}(s) = Z_1(s) \cdot \dfrac{(N_2(s) + (\gamma - \beta) \cdot f(e, \dot{e})K_u Z_2(s))\theta_i(s)}{N_1(s)N_2(s) - \beta \cdot f(e, \dot{e})K_u Z_1(s)N_2(s) + \gamma \cdot f(e, \dot{e})K_u Z_2(s)N_1(s)} \\ \theta_{m2}(s) = Z_2(s) \cdot \dfrac{(N_1(s) + (\gamma - \beta) \cdot f(e, \dot{e})K_u Z_1(s))\theta_i(s)}{N_1(s)N_2(s) - \beta \cdot f(e, \dot{e})K_u Z_1(s)N_2(s) + \gamma \cdot f(e, \dot{e})K_u Z_2(s)N_1(s)} \end{cases} \tag{50.10}$$

From the above equation, the error signal allocation method respectively change the pole-zero of subsystem 1 and system 2, thus changing the transient corresponding to make the outputs of system 1 and system 2 be the same, then eliminate the error between the systems. Because $f(e, \dot{e})$ is a nonlinear function about e and $\dot{e}$, that the pole-zeros of subsystem 1 and subsystem 2 are changing in dynamic process.

50.4 Simulation of Synchronous System

To validate that whether the proposed synchronization principles based on coordination control can meet the requirements of dual-drive system of large-scale rotation, we need to do simulation experiment for swivel dumper synchronous drive system.

In two states of control synchronization and no synchronization correction, the 30 Hz sinusoidal signal is given. The synchronization errors of two channels are large in Fig. 50.3. In contrast, the synchronization errors of control synchronous drive system reduce greatly in Fig. 50.4. Simulation results show that the proposed

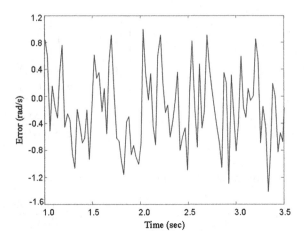

Fig. 50.3 Error curve of synchronization drive

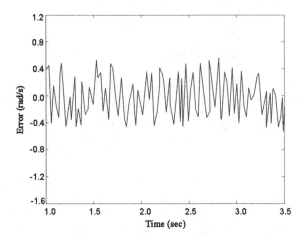

Fig. 50.4 Error curve of synchronization system without control drive system under control

control synchronization scheme based on fuzzy correction of non-attainment and common feedback is feasible in theory. It has a high synchronous precision and suitable for large-scale rotary dual-drive synchronous drive system entirely.

50.5 Conclusion

Simulation results show that the proposed control synchronization scheme in this paper is completely effective for simultaneous dual-motor drive system. It improves the system response speed of the synchronization error and reduces the synchronous error of two motors effectively. At the same time, it provides a favorable guarantee for the high precision of large rotary two-drive synchronous system and develops new ideas for other control synchronization system.

Acknowledgments This paper is supported by Hebei province Science Technology Research and Development Program (No. 12215604D)

References

1. Peng JP, Ding EJ, Ding M (1996) Synchronizing hyperchaos with a scalar transmitted signal. Phys Rev Lett 76(2):904–907
2. Kocarev L, Parlit U (1996) Generalized synchronization, predictable, and equivalence of unidirectionally coupled dynamical systems. Phys Rev Lett 76(3):1816–1819
3. Pecora LM, Carroll TL (1990) Synchronization of chaotic systems. Phys Rev Lett 64(2):821–824
4. Viera M d S, Lichtenberg AJ, Lieberman MA (1992) Synchronization of regular and chaotic systems. Phys Rev A 46(12):7359–7362
5. Cuomo KM, Oppenheim AV (1993) Circuit implementation of synchronized chaos with applications to communications. Phys Rev Lett 7:65–68
6. Morgül ō (1998) On the synchronization of logistic maps. Phys Rev A 247(10):391–396
7. Jackson EA, Grosu I (1995) An open-plus-closed-loop(opcl) control of complex dynamical systems. Phsica D 85(1):1–9

Chapter 51
Traffic Flow Analysis and Prediction Based on GPS Data of Floating Cars

Xiangyu Zhou, Wenjun Wang and Long Yu

Abstract Due to the limited amount of sensor infrastructure and its distribution restrictions, it is a challenge for the traffic related groups to estimate traffic conditions. The GPS log files of floating cars have fundamentally improved the quantity and quality of wide-range traffic data collection. To convert this data into useful traffic information, new traffic models and data processing algorithms must be developed. To conduct a comprehensive and accurate traffic flow analysis, this paper proposes a traffic flow analysis and prediction process based on GPS data of floating cars. For the key links involved in the process, new models and algorithms are implemented. Based on the actual data of a city, we validate the effectiveness of the design process, and provide a reliable solution for urban traffic flow analysis.

Keywords Floating car · GPS trajectory · Traffic flow · Analysis and prediction

51.1 Introduction

The obtaining of real-time traffic information has become an important part of the Intelligent Transportation Systems (ITS). With the key technologies development of ITS, it's possible to offer real-time, dynamic and predicted traffic information for travellers by information collection, processing and analysis. Floating car

X. Zhou (✉) · L. Yu
School of Computer Software, Tianjin University, No. 92, Weijin Road,
Nankai, Tianjin 300072, People's Republic of China
e-mail: zhouxiangyu1987@hotmail.com

W. Wang
School Computer Science and Technology, Tianjin University, No. 92, Weijin Road,
Nankai, Tianjin 300072, People's Republic of China

W. Lu et al. (eds.), *Proceedings of the 2012 International Conference on Information Technology and Software Engineering*, Lecture Notes in Electrical Engineering 210, DOI: 10.1007/978-3-642-34528-9_51, © Springer-Verlag Berlin Heidelberg 2013

information collection is a new traffic flow information collection technology developed with ITS. To ensure the operation safety and scheduling efficiency of the floating cars, GPS devices are equipped. These devices send the status information of the vehicles to the management centre as certain frequency. This information records into log files, which contain the trajectories of the vehicles. Through vehicle trajectory analysis, not only improve the scheduling efficiency of the taxi dispatch system [1], but also provide road traffic conditions for the relevant departments, give them the decision support of traffic control.

In the remain of this paper, related works are summarized in Sect. 51.2 and data model is introduced in Sect. 51.3. We propose traffic flow analysis and prediction process based on GPS data of floating cars in Sect. 51.4. Section 51.5 is the summary.

51.2 Related Works

The feasibility of road traffic flow monitoring and prediction based on the car GPS devices has been validated through a series of experiments. In 2008, Saurabh and his partner carried out the "Mobile century" field experiments [2]. The experiment confirms that GPS-enabled cell phones can realistically be used as traffic sensors to detect the traffic flow, while preserving individuals' privacy. In 2009, Calabrese used the taxi GPS data in Shenzhen to do the city-wide traffic modeling and traffic flow analysis [3] and forecast the traffic conditions about the city's daily movements [4]. Fabritiis' paper [5] proposed two algorithms, pattern matching and Artificial Neural Network (ANN), which used for short-term (15–30 min) road travel time prediction. Daniel used the experiment data from "Mobile Century", put forward an Ensemble Kalman Filtering (EnKF) approach to highway traffic estimation using GPS enabled mobile devices [6].

The floating car GPS data processing is the main work in traffic flow analysis. Map matching is one of the key technologies [7]. We can mine individual life pattern based on location history data [8], so as the taxi travel pattern. During the data processing, the data can be used to mine the floating car's frequent pattern and build the trajectory model. It is a great improvement for data processing.

51.3 Data Model

The source data involved in traffic flow analysis and prediction process consists of two parts: the floating car GPS log files and road network. The GPS log file is a summary of all the floating car location information sent in accordance with the time sequence. Its major fields include device terminal id, car state, position information, message record time, car velocity and direction. Road network is from maps. It consists of three granularities: road section, road, road network.

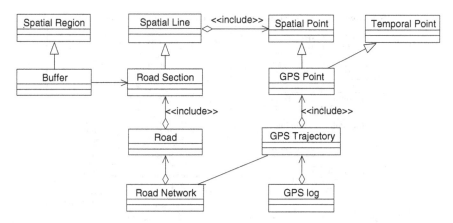

Fig. 51.1   The structure of data model

51.3.1   The Schema of Data Model

Figure 51.1 shows the structure of the data model in the experiment.

Basic parts:

Spatial Point: consists of two floating-point values; denote the longitude and latitude value of the spatial point.

Temporal Point: to describe a moment of time.

Spatial Line: The connection between the start and end points. Spatial line has direction; its direction is from a starting point to an end point.

Spatial Region: a range of continuous and not separated spatial units in the geographic space.

Road data parts:

Road Section: The smallest granularity of the road data set, which is segmented by the intersection of the roads. It is the basic observation unit to calculate the road traffic state.

Road: The basic unit of the roads, which usually have a common name and consist of a continuous segment of the road.

Road network: refers to a certain network of connected road.

Buffer: The affected areas or services areas of a geospatial objects. In this article it specifically refers to the sections around the roads with a certain distance.

GPS data parts:

GPS points: the basic unit which composites the floating vehicle GPS data, record the location of the vehicle running state, time, instantaneous speed, direction and other basic information.

GPS trajectory: the GPS points from the same terminal in accordance with the sort of time series.

GPS log file: a summary of all the floating car location information in time sequence.

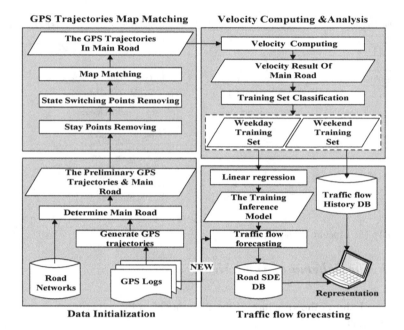

Fig. 51.2 Traffic flow analysis and prediction process

51.4 Traffic Flow Analysis and Prediction Process

Figure 51.2 shows the overall process of the traffic flow analysis and prediction by using the GPS trajectory data of floating cars. It can be divided into the following four parts: data initialization, GPS trajectories map matching, velocity computing and analysis and traffic flow prediction.

51.4.1 Data Initialization

The log files are collected from all of the terminals in floating cars. It sorted in accordance with the time series, if not do data processing accordance with the terminal number, trajectories can't be isolated from the data. The first step of data initialization is to generate GPS trajectories.

We selected one month taxi GPS log files as the data set, which is from more than 5,000 taxis in one city. According to the results, there were 3,800–4,000 floating car GPS trajectories. Some trajectories is too short to analysis, therefore the trajectories with more than 200 record points is selected. Its number ranges from 3,400 to 3,500, 86–89 % of the total number.

Table 51.1 The pseudocode of stay point exclusion

StayPoint_ Exclusion (P, distThresh, timeThreh)
Input: GPS Log file P, distance threshold distThreh and time threshold timeThreh。
Output: the point set which has exclusion the stay points, PSet={P}
1. i=0, pointNum = |P|; // the total point number of the log file
2. while i <pointNum do,
3.   j:=i+1;
4.   while j <pointNum do,
5.   dist=Distance(p_i, p_j); // Calculate the distance between two points
6.   if dist>distThresh then
7. $\Delta T = p_j.Time - p_i.Time$;
8.   if ΔT>timeThreh then
9.   PSet.delete({p_k | i<=k<=j});
10.   i:=j; break;
11.   j:=j+1;
12 return PSet.

51.4.2 GPS Trajectories Map Matching

For floating car daily operation characteristics, their GPS trajectories' map matching degree is better than these from cell phones or other private cars, and it has certain regularity. But at the same time, the floating car also has some special behavior patterns, such as parking to wait for the guest. It makes the data has particularity. In this experiment, we make full use of advantages, and avoid the negative effects in the analysis.

51.4.2.1 Stay Points Exclusion

The definition of stay point is from [8]. A stay point stands for a geographic region where a user stayed over a certain time interval. Every stay point in taxi GPS trajectory has its special semantic. It has two situations of stay points. One likes people enter a building and lose satellite signal over a time interval until coming back outdoors. The other situation is when a user wanders around within a certain geospatial range for a period.

In this experiment, we adjust the stay points detection algorithm from [8] based on the characteristics of the experimental data (the pseudocode of the algorithm can see Table 51.1). The stop points from the GPS trajectory of floating car represent the cars' behavior patterns, such as taxi driver waits for guests. The GPS points included in stay points always have a low-velocity state, as shown in Fig 51.3b. The purpose to use stay point exclusion is to exclude the impact from these special behavior patterns, when doing the traffic flow analysis. In our experiment, the time threshold value is set in 5 min and the spatial threshold value is set in 100 m. The velocity threshold is set below 5 km/h. By this algorithm for

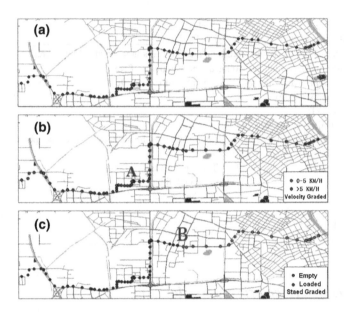

Fig. 51.3 The map matching of GPS trajectories

data processing; the interference data (such as the points in position A from Fig. 51.3b, which are clearly showed in the circle in Fig. 51.4a) can be excluded.

51.4.2.2 State Switching Points Exclusion

Floating Car GPS trajectory can often reflect the characteristics that some ordinary vehicles do not have. The floating car equipped with GPS terminal can record the state of the car; Fig. 51.3c shows the trajectory and state record of one taxi. When some state switching situation, some experimental interference data will be record. For example, when a taxi state changes from empty to loaded, it may record several zero-velocity GPS points; these points can't reflect the proper traffic flow at that time. Based on the velocity value of the points before and after the state switching points at the time sequence, to determine whether these points should be excluded or not. The basic process is as below:

(1) According to the time sequence, traverse the GPS points in one trajectory and find the State switching points p_i, p_j $(i < j)$;
(2) If p_i. $Velocity = 0$, exclude p_i from the data set and search forward, exclude the continuous points with a 0-value velocity;
(3) If p_j. $Velocity = 0$, exclude p_j from the data set and search backward, exclude the continuous points with a 0-value velocity;
(4) Traverse every trajectory as the rule, and get the results data set $\{p\}$.

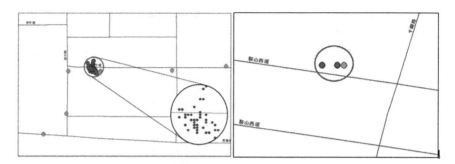

Fig. 51.4 The stay points and state switching points **a** stay points, **b** state switching points

By this algorithm for data processing, the circle in Fig. 51.4b points can be excluded as the interference data.

51.4.2.3 Map Matching

Map matching technology is one of the key technologies of floating car data processing. Only if a vehicle has been determined in which road, the data can be used to the traffic flow analysis. The purpose of map matching is to ensure the accuracy of the clustering data set and to improve the calibration accuracy. In order to improve the efficiency of map matching, map-matching algorithm based on the buffer and direction angle has been designed. The basic idea of the map-matching algorithm based on the buffer and direction angle are as follows:

(1) The floating car GPS data which has been pre-processed is regarded as the initial data source of the map matching;
(2) For the set of road sections $N = \{Rs_1, Rs_2 \ldots Rs_n\}$, every Rs_i ($1 <= i <= n$) set the buffer width equal to Rs_i. *Width*;
(3) Using the buffer tool in Arc Map Toolbox, based on the buffer width of every road section, to generate the buffer Rs_i. *Buffer* for Rs_i;
(4) If point fall out of all the road section buffer, it will be excluded from the initial data source;
(5) The remaining data then is regarded as the data set to do map matching, $MP = \{p_1, p_2 \ldots p_n\}$, point number $PN = |MP|$.
(6) For $p_j \in MP$ ($1 <= j <= PN$), count the buffer number p_j. *BufferNo* it located;
(7) If p_j. *BufferNo* $= 1$, then the p_j located on the only road section; else if p_j. *BufferNo* > 1, compare the angle between the instantaneous velocity's direction of the point and the central axis' direction of every road section the point located in, and choose the corresponding road section of the minimum angle as the point located road section.
(8) To every GPS point in data set, do as 7 until the end.

Fig. 51.5 The velocity
estimation on one weekday

51.4.3 Velocity Computing and Analysis

51.4.3.1 The Velocity Estimation of Road Sections

Based on the map matching result, the average velocity of the road section can be estimated. The vehicle density of one road section in one period can intuitively reflect the traffic condition. However, the traffic density is difficult to obtain. So we use the average velocity to reflect the traffic condition. When floating cars account for more than a certain proportion of the total volume of traffic on the road section [6], the speed estimation can be better. The formula (51.1) below is used to calculate the average velocity:

$$V_{L,T} = \sum_{i=1}^{m} \frac{\sum_{j=1}^{P_i} V_{i,j}}{P_i} \tag{51.1}$$

where $V_{i,j}$ represents the velocity of jth GPS point from floating car i, P_i represents the GPS point number of floating car i on road section L. $V_{L,T}$ represents the average velocity on road section L, during period T.

Figures 51.5 and 51.6 are respectively shows one road's velocity estimation result on one weekday and weekend.

The author analyzed the historical GPS data of taxis and observed that the traffic pattern of weekday is different from that in weekend [9]. According to the experimental results; we analyzed two sorts of charts.

Based on the chart analysis, we can find that every road shows different traffic flow characteristics on weekdays and weekends. For example, as figures show above, traffic flow has obvious trough mainly around 08:00, 12:00 and 18:00, when the velocity is less than 10 km/h. While the traffic flow has no obvious crests and troughs at daytime, the average velocity is almost all less than 10 km/h.

According to the conclusion, we classify the calculation results to weekday data set and weekend data set, as the training set of SVM model.

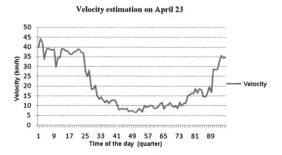

Fig. 51.6 The velocity estimation on one weekend

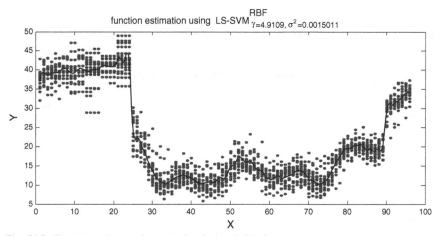

Fig. 51.7 The regression result on weekend via LS-SVM

51.4.4 Traffic Flow Prediction

Based on the calculation and classification results in Sect. 51.4.3, we used least squares support vector machine (LS-SVM) regression model for short-term traffic flow prediction. Urban road traffic system is a system with time-varying, nonlinearity, high dimension and randomicity. LS-SVM uses quadratic loss function, and transfers the optimization solution into linear equation solution, not the quadratic programming solution. Constraint condition also has become equality constraints rather than inequality constraints. When the data quantity is huge, it has the advantages of less computing resource, solving faster and convergence speed faster. Therefore it's better to fit the traffic system time series data, which have nonlinear and non-stationary characteristics. When using the LS-SVM to forecast traffic flow, first, according to the historical data to train support vector machine, get the relationship between input and output; when given the corresponding input,

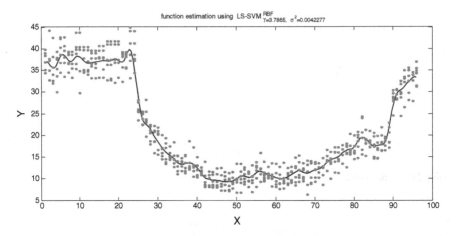

Fig. 51.8 The regression result on weekend via LS-SVM

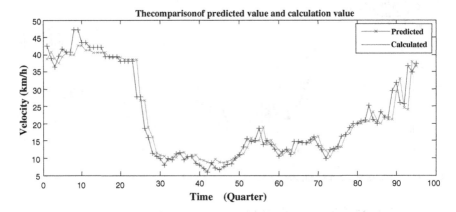

Fig. 51.9 The comparison of predicted value and calculation value

it can get predicted results. We use the LS-SVM lab toolbox developed by SuyKens J.A.K to doing the calculation work.

Based on the calculating results in 51.4.3, we trained the SVM of weekday and weekend respectively, Figures 51.7 and 51.8 show the results.

According to the training result of the LS-SVM regression prediction model, we chose one weekday's data as experiment data, used the data in one quarter to predict the next quarter result. The prediction results and the actual calculation value are as Fig. 51.9, prediction results is close to the true value.

51.5 Conclusion

To road traffic condition estimation and traffic flow prediction problem, we focus on the traffic flow analysis and prediction method. By using GPS log files from taxi operators in one city, we took the operational process of data processing, data analysis, traffic flow estimation and prediction as basic line, combined with trajectory model building by frequent pattern mining, map matching, ITS traffic flow model prediction method based on knowledge discovery, etc., build up the general framework of traffic flow analysis and prediction business process. The aim is accurately and fast on traffic flow condition estimation and prediction, to takes it as the decision support of transportation planning, public transportation operation management, travel planning, transportation induction and traffic control.

The trajectory model building based on the frequent pattern mining is introduced into the data processing and road matching. By analyzing the frequent patterns from taxi history trajectory, which can reflect the behavior patterns of the taxis, we aim to optimize the trajectory data. For the traditional map matching method, it is a good supplement. To a certain extent, it can optimize traffic flow estimation results and reduce noise data to estimate effects.

In the short-term traffic flow prediction process, the application of short-time traffic flow LS-SVM regression forecast method helps to better deal with traffic flow. With the establishment of LS-SVM regression prediction model method, we predict the future traffic flow changes by using the current data as input.

Next we will focus on the time complexity optimization of data analysis process and prediction model training; we will devote ourselves to the storage mode optimization for intermediate results. By make full use of the Hadoop framework, we will realize the distributed processing of huge amounts of data. The forecast result will be closely integrated with business applications. Through the application platform, our aim is to realize traffic management decision support.

References

1. Santani D, Balan RK, Woodard CJ (2008) Spatio-temporal efficiency in a taxi dispatch system. Singapore Management University, Singapore
2. Amin S, Andrews S, Arnold J (2008) Mobile century-using GPS mobile phones as traffic sensors: a field experiment. In: World congress on ITS, pp 16–20
3. Calabrese F, Reades J, Ratti C (2010) Eigenplaces: segmenting space through digital signatures. IEEE Pervasive Comput 9(1):78–84
4. Calabrese F, Colonna M, Lovisolo P (2007) Real-time urban monitoring using cellular phones: a case study in Rome. IEEE Trans intelligent transportation systems 12(1):141–151
5. Fabritiis CD, Ragona R, Valenti G (2008) Traffic estimation and prediction based on real time floating car data. In: Proceedings of the 11th international IEEE conference on intelligent transportation systems. pp 197–203
6. Work D B, Tossavainen O, Blandin S (2008) An ensemble Kalman filtering approach to highway traffic estimation using GPS enabled mobile devices. In: Proceedings of the 47th IEEE conference on decision and control. pp 5062–5068

7. Brakatsoulas S, Pfoser D, Salas R (2005) On map-matching vehicle tracking data. In: Very large data bases (VLDB). pp 853–864
8. Ye Y, Zheng Y, Chen YK, Xie X (2009) Mining individual life pattern based on location history. In: Proceedings of the international conference on mobile data management 2009 (MDM 2009). pp 1–10
9. Yuan J, Zheng Y, Zhang CY (2010) T-drive: driving directions based on taxi trajectories. In: Proceedings of the 18th SIGSPATIAL International conference on advances in geographic information systems. pp 99–108

Chapter 52
Customers' E-Satisfaction for Online Retailers: A Case in China

Ruimei Wang, Tianzhen Wu and Weihua Wang

Abstract Online retailers are facing increasing competitions, thus it is very important for them to attract and retain online customers. However, many factors can affect customer satisfaction in online retailing (E-satisfaction) environment. This chapter investigates the factors and their impacts on customer E-satisfaction in online retail (B–C, C–C) in China. Based on the literature review and an exploratory study with 50 online customers, we identified a number of key factors that potentially impact on customer's E-satisfaction and proposed a conceptual model for E-satisfaction. This model was used to examine how the factors in B–C and C–C e-retailing context influence E-satisfaction. A large scale survey was conducted with 387 online customers. The analytical results demonstrated that customers' perceived time value and customers' experience have positive influence on E-satisfaction. Findings of this research have made valuable contributions to both E-satisfaction research and e-commerce success practice.

Keywords E-Satisfaction · Online retail · E-quality

52.1 Introduction

According to the China Internet Network Information Centre (CNNIC), Internet retail sales for 2009 were 250 billion RMB, or 68 % higher than 2008 sales of 148 billion RMB in China. This rapid growth of e-retailing reflects the compelling

R. Wang · T. Wu · W. Wang (✉)
College of Economics & Management, China Agricultural University,
No.17. Qinghua Donglu, Haidian District, Beijing 100083,
People's Republic of China
e-mail: cauweihua@163.com

W. Lu et al. (eds.), *Proceedings of the 2012 International Conference on Information Technology and Software Engineering*, Lecture Notes in Electrical Engineering 210, DOI: 10.1007/978-3-642-34528-9_52, © Springer-Verlag Berlin Heidelberg 2013

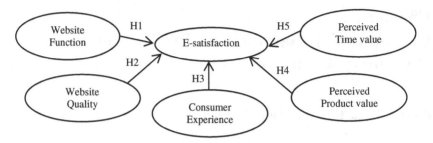

Fig. 52.1 The proposed model of the E-satisfaction in online retailing

advantages over conventional brick-and-mortar stores, including greater flexibility, enhanced market outreach, lower cost structures, faster transactions, broader product lines, greater convenience, and better customization.

However, e-retailing also faces its own challenges. Internet makes the world smaller, competing businesses in the world of electronic commerce are only a few mouse clicks away. As a result, consumers are able to compare and contrast competing products and services with minimal expenditure of personal time or effort. E-tailors need satisfied customers who can bring benefits, but it is very difficult to maintain the satisfaction of the customers, to use the same online retailers repeatedly. Therefore, it is very important for online retailers to attract and retain more online users.

Customer satisfaction can be defined as customers' evaluations of a product or service with regard to their needs and expectations [1]. Scholars are quite interested in the conceptualization and measurement of service quality as it leads to customer satisfaction. Service quality is defined as the outcome measure of effective service delivery, and it occurs when customers receive service that exceeds their expectations [2]. Oliver claimed that quality is the antecedent of satisfaction [3]. Zeithaml and Bitner stated that service quality assessment focuses primarily on dimensions of service, and perceived quality is not only a component of customer satisfaction but is also influenced by product quality, price, customer factors, and situational factors [4]. In the virtual environment, Anderson and Srinivasan defined E-satisfaction as "the contentment of the customer with respect to his or her prior purchasing experience with a given commerce firm" [5].

52.2 Research Model and Hypotheses

Our proposed model presents E-satisfaction as influenced by relationship among perceived product value, perceived time value, website function, website quality, customer experience. We will present our hypotheses through explaining the relationships among E-satisfaction, perceived product value, perceived time value, website function, website quality, and customer experience. The proposed model is outlined in Fig. 52.1.

52.2.1 E-Satisfaction

Customer's satisfaction is a general antecedent of loyalty [4, 6, 7]. Satisfaction has been defined as the perception of pleasurable fulfilment in the customers' transaction experiences [8]. We conceptualize E-satisfaction as a cumulative construct, which is "based on the sum of satisfaction on each purchase and consumption experience with a good or service over time" [5]. Satisfied customers tend to have higher usage of service [9], possess stronger repurchase intention, and are often eager to recommend the product or service to their acquaintances [10] than those who are not satisfied. Dissatisfied customers are more likely to search for alternative information and switch to another retailer, and is also more resistant to developing a close relationship with the retailer [11]. Satisfaction with e-retailers, like satisfaction with traditional retailers, is not derived solely from the customer's satisfaction with the product purchased. The dominant characteristics of e-store satisfaction are customers' experience, site design (include website function and website quality [12–14]) [15]. Thus we propose: H1 Website function has a positive effect on E-satisfaction; H2 Website quality has a positive effect on E-satisfaction; H3 Customers' experience has a positive effect on E-satisfaction.

52.2.2 Perceived Product Value and Perceived Time Value

The value concept is multi-faceted and complex. First, the term 'value' is extremely abstract and polysemous in nature: it has different meanings not only for consumers [16] but also among researchers [17] and even for practitioners [18]. Perceived value in brick-and-mortar contexts has received much attention in recent years [19–21]. Indeed, Cronin et al. [22] argue that the study of perceived value has dominated research in the services literature. While a number of conceptual models of value have been put forward [23], value is most typically presented as acquisition, transaction, in use, or redemption-based [24]. Both for marketing practitioners and researchers, the construct of perceived value have been identified as one of the more important measures [22, 25, 26]. From a managerial point of view, it is linked to marketing strategies such as market segmentation [27]. Among the latest attempts, Sweeney and Soutar [23] perceived value scale (so-called PERVAL scale) ought to be mentioned: based on Sheth et al. [26] who's work identifies four dimensions: emotional value, social value, and two types of functional value (price/value for money and performance/quality).According to the literature and survey, we propose: H4 perceived product value has a positive influence on E-satisfaction; H5 perceived Time value has a positive influence on E-satisfaction.

52.3 Methodology

52.3.1 Survey

Our research subjects were adults in China who had at least 1 year's online shopping experience. In order to determinate the factors which can impact the customer E-satisfaction in online retailing. We survey 50 customers who had an online purchase experience. Before data collection, we trained 2 postgraduate student interviewers by explaining the purpose and content of the survey. Two trained interviewers first asked if the interviewees had purchased products from online retailer, and only those who had online purchase experience were asked to participate in the interview with a 'familiar e-commerce website' in mind. The survey took approximately 20 min, and then the interviewees were asked to return the questionnaire after completing it.

The results of survey shows that the factors which impact customer E-satisfaction in online retailing in China has not special factors different from the literature, factors brand, commitment, advertisement, friends recommend and shopping habit had very low frequency in the survey although some reviewers thought them important, so we will not consider these factors in this chapter.

52.3.2 Data Collection

The questionnaire initially included 25 items, for each item, Five-point Likert Scales were utilized and sere anchored for 1 "not very important" and for 5 "very important". Table 52.1 presents the constructs, variables and survey items used in the questionnaire.

The sample was the people who have online purchase experience at least twice. We send them a questionnaire by QQ and email randomly, if they have more than twice online purchase experience and had a website in their mind when they want to buying something, we ask them to answer the questionnaire and return it.

3,000 questionnaires were initially distributed by QQ (An MSN type software) and email. A total of 432 questionnaires were collected, of which, 387 were valid, resulting in a valid rate of 89.3 %.

Table 52.1 Research constructs and related survey items

Variables	Survey items
Perceived product value	The products I bought from this website are excellent value
	The products I bought from this website are excellent value for money
Perceived time value	This website saves me more time
	This website is very convenient to use
Website function	This website has navigation function
	This website has a search tool that enables me to locate products
	This website has E-service when I am in trouble buying products
	This website has e-community that customers can share about the website/products online with other customers of the website
	This website enables me to assess the products/services or see other customers assessment for products/services of the website
Website quality	The website responds very fast
	The website design is concision
	I feel that this is a very engaging website
Consumer experience	This website always contacts me
	This website makes purchase recommendations that match my needs
	This website often recommends me some related and interesting products
Satisfaction	Shopping at this website is fun
	I am very satisfied when shopping on this website

Table 52.2 CFA factor loadings of the original scale (N = 387)

	Perceived product value (PPV)	Perceived time value (PTV)	Website function (WF)	Website quality (WQ)	Customer's experience (CE)
Product quality (PPV1)	0.588				
Product price (PPV2)	0.554				
Save my time (PTV1)		0.754			
Convenience (PTV1)		0.663			
Interaction (WF1)			0.458		
E-service (WF2)			0.458		
E-community (WF3)			0.491		
Fast speed (WQ1)				0.689	
Concision (WQ2)				0.716	
Navigation (WQ3)				0.651	
Good customer relationship (CE1)					0.786
Recommendation is my need (CE2)					0.643
Recommendation (CE3)					0.746

Model fit χ^2 = 610.190, df = 312, χ^2/df = 1.96, GFI = 0.87, RMSEA = 0.06, AGFI = 0.84, NFI = 0.74, CFI = 0.85. All factor loadings are significant at p = 0.001 level

Table 52.3 Standardized parameter estimates of the hypothesized paths

Hypothesis	Causal path	Parameter	Path coefficient	P	Result
H1	PPV→E-satisfaction	$\lambda 1$	0.099	0.258	Rejected
H2	PTV→E-satisfaction	$\lambda 2$	0.231	0.002	Supported
H3	WF→E-satisfaction	$\lambda 3$	−0.006	0.881	Rejected
H4	WQ→E-satisfaction	$\lambda 4$	0.111	0.072	Rejected
H5	CE→E-satisfaction	$\lambda 5$	0.373	***	Supported

52.4 Data Analysis and Results

52.4.1 Measurement Model

We first developed the measurement model by conducting confirmatory factor analysis (CFA). The SEM was then estimated for hypotheses testing. The models were assessed by the maximum likelihood estimation method using AMOS 16. To evaluate the fit of the models, a Chi-square with degrees of freedom, normed fit index (NFI), GFI, adjusted goodness of fit index (AGFI), comparative fit index (CFI), and RMSEA were employed. A good fit is normally deemed to exit when NFI, GFI and CFI were all greater than 0.9, AGFI was greater than 0.8, and RMSEA was less than 0.08. As show in Table 52.2, the model fit indices demonstrate the model as having moderate fit to the data.

52.4.2 Hypothesis Testing

Table 52.2 shows the various fit indices calculated for the model, a comparison of all fit indices with their corresponding recommended values indicated a good model fit ($\chi 2/df = 1.96$, NFI = 0.74, GFI = 0.87, AGFI = 0.84, CFI = 0.85, RMSEA = 0.06), while NFI value of 0.74 was at a marginal acceptance level. Table 52.3 shows the structural model estimates, where the estimate parameters were standardized path coefficients, and all path coefficients, except for the path of perceived product value to E-satisfaction, website function to E-satisfaction and website quality to E-satisfaction, were significant at the p = 0.005 level.

H1, H3, H4 posited that PPV, WF, WQ has positive influence on E-satisfaction, but the results in Table 52.3 does not provided support for these hypotheses ($\lambda 1 = 0.099$, $\lambda 2 = 0.231$, $\lambda 4 = 0.111$). H2, H4 proposed that PTV, CE has positive influence on E-satisfaction, and the results in Table 52.3 provided support for this hypothesis ($\lambda 5 = 0.373$, $\lambda 2 = 0.231$).

52.5 Discussion and Conclusions

Result of hypothesis testing suggested that perceived time value and Customer's experience have significant positive influences on E-satisfaction. The first and most important factor is the customers' experience ($\lambda 5 = 0.373$) of the online retailers. Perceived time value ($\lambda 2 = 0.231$) is the second important factor to E-satisfaction in online retailing.

This study suggests that online retailers should consider focusing more on customers' experience on their way to improve customers' satisfaction. On one hand, online retailers must continuously work at customer information analysis, especially product information, to guarantee the website recommendation match customers' needs. On the other hand, good customer relationship must be considered to sustain customers' E-satisfaction. Thus, this study suggests that strategies must pay more attention to the after-sold services, particularly with regards to the fast answering for refunding or changing, friendly when refunding or changing, handling of complaint fast and delivery fast, these may need the online retailers to train their employees or to negotiate with their vendor for after-sold services. Although it is not easy, they must do, unless they will lost their customers again.

This study suggests that online retailers should consider the customers' perceived time value customers may consider the perceived benefit of continuing a purchase. The strategies for retaining high internet experience customers should be based on attempts to increase benefit by making more convenience to buy, remembering the customers' hobbies and making them deem they are the unique person to the e-retailer, saving customers' more time in the process of purchase.

The findings suggest that website function and website quality are not positively related to customers' E-satisfaction. Hence, there is no need to enhance the website features to attract and retain customers.

The survey results revealed that about 70 % of respondents expressed their E-satisfaction to Taobao (a C–C online retailer). This finding will suggest that e-marketers, especially the B–C e-marketers, need to explore further why Chinese customers like Taobao more than other online retailers.

Acknowledgments This work was partially supported by Key Discipline of Shanghai Education Commission funded, Shanghai Ocean University. P. R. China (No.B-5308-11-0303).

References

1. Oliver RL (1980) A cognitive model of the antecedents and consequences of satisfaction decisions. J Mark Res 17(4):460–469
2. Parasuraman A, Zeithaml VA, Berry LL (1988) SERVQUAL: a multiple-item scale for measuring consumer perceptions of service equality. J Retail 64(1):12–40
3. Oliver RL (1993) A conceptual model of service quality and service satisfaction: compatible goals, different concepts. Adv Serv Mark Manag 2:65–85
4. Zeithaml V, Berry L, Parasuraman A (1996) The behavioral consequences of service quality. J Mark 60:31–46
5. Anderson RE, Srinivasan SS (2003) E-satisfaction and e-loyalty: a contingency framework. Psychol Mark 20(2):123–138
6. Balabanis G, Reynolds N, Simintiras A (2006) Bases of e-store loyalty: perceived switching barriers and satisfaction. J Bus Res 59(2):214–224
7. Bloemer J, Ruyter K (1998) On the relationship between store image, store satisfaction and store loyalty. Eur J Mark 32(5/6):499–513
8. Oliver RL (1997) Satisfaction: a behavioral perspective on the consumer. The McGraw-Hill Companies Inc, New York
9. Ram S, Jung H (1991) "Forced" adoption of innovations in organizations: consequences and implications. J Prod Innov Manage 8(2):117–126
10. Zeithaml VA, Parasuraman A, Malhotra A (2002) Service quality delivery through web sites: a critical review of extant knowledge. J Acad Mark Sci 30(4):362–375
11. Srinivasan SS, Anderson R, Ponnavolu K (2002) Customer loyalty in e-commerce: an exploration of its antecedents and consequences. J Retail 78(1):41–50
12. Chang HH, Chen SW (2008) The impact of customer interface quality, satisfaction and switching costs on e-loyalty: internet experience as a moderator. Comput Hum Behav 24(6):2927–2944
13. Chang HH, Chen SW (2009) Consumer perception of interface quality, security, and loyalty in electronic commerce. Inf Manag 46(7):411–417

14. Cyr D, Hassanein K, Head M, Ivanov A (2009) Perceived interactivity leading to e-loyalty: development of a model for cognitive-affective user responses. Int J Hum Comput Stud 67(10):850–869
15. Szymanski DM, Hise RT (2000) E-satisfaction: an initial examination. J Retail 76(3):309–322
16. Zeithaml VA (1998) Consumer perceptions of price, quality, and value: a means-end model and synthesis of evidence. J Mark 52:2–22
17. Lai AW (1995) Consumer values, products benefits and customer value: a consumption behavior approach. Adv Consum Res 22:381–388
18. Woodruff RB (1997) The next source for competitive advantage. J Acad Mark Sci 25(2): 139–153
19. Parasuraman A, Grewal D (2000) The impact of technology on the quality-value-loyalty chain: a research agenda. J Acad Mark Sci 28(1):168–174
20. Sirohi N, McLaughlin EW, Wittink DR (1998) A model of consumer perceptions and store loyalty intentions for a supermarket retailer. J Retail 74(2):223–245
21. Sweeney JC, Soutar GN, Johnson LW (1999) The role of perceived risk in the quality-value relationship: a study in a retail environment. J Retail 75(1):77–105
22. CroninJr JJ, Brady MK, Hult GTM (2000) Assessing the effects of quality, value, and customer satisfaction on consumer behavioral intentions in service environments. J Retail 76(2):193–218
23. Sweeney JC, Soutar GN (2001) Consumer perceived value: the development of a multiple item scale. J Retail 77(2):203–220
24. Koo DM (2006) The fundamental reasons of e-consumers' loyalty to an online store. Electron Commer Res Appl 5(2):117–130
25. Holbrook MB (1999) Consumer value: a framework for analysis and research. Rutledge press, London
26. Sheth JN, Newman BI, Gross BL (1991) Why we buy what we buy: a theory of consumption values. J Bus Res 22(2):159–170
27. Tellis GJ, Gaeth GJ (1990) Best value, price-seeking, and price aversion: the impact of information and learning on consumer choices. J Mark 54(2):34–45

Chapter 53
Improvement of Azimuth Super Resolution of Radar via Generalized Inverse Filtering

Chenghua Gu, Xuegang Wang, Wang Hong and Xuelian Yu

Abstract Radar angular resolution is constrained by its beam width. This paper concentrates on obtaining better angular resolutions than the beam width of real aperture radars by a signal processing method that is generalized inverse filtering in time domain. Deconvolution in the time domain with cyclic convolution can reduce computational complexity and is easier to be implemented by programing. In the same time, this approach can overcome the defect of the no spectral inverse and ill-posed problem in the frequency domain and can clearly distinguish the two or more targets within the same beam in the same range unit.

Keywords Deconvolution · Super resolution · Generalized inverse · Filtering · Cyclic convolution

53.1 Introdution

Radar angular resolution refers to the ability that can distinguish the two or more than two goals within the same beam in the same range unit [1]. Generally radar angular resolution is radar antenna 3 dB beam width and is limited by the radar

C. Gu (✉) · X. Wang · W. Hong · X. Yu
University of Electronic Science and Technology of China, NO 2006 Xiyuan AVE,
High-Tech Zone, Chengdu, Sichuan 611731, China
e-mail: guchenghua521@163.com

X. Wang
e-mail: xgwang@ee.uestc.edu.cn

W. Hong
e-mail: whcd@live.cn

W. Lu et al. (eds.), *Proceedings of the 2012 International Conference on Information Technology and Software Engineering*, Lecture Notes in Electrical Engineering 210, DOI: 10.1007/978-3-642-34528-9_53, © Springer-Verlag Berlin Heidelberg 2013

aperture. It is difficult to resolve the two or more targets within the same beam width independently, then just a mixed target. Enlarging antenna aperture and increasing emission frequency is the approaches to improve angular resolution, but it is difficult to implement by the limited of technology, work requirements and environ-mental factors in practice [2]. Radar azimuth super-resolution is an approach that can resolve two or more targets within the same beam in the same range unit with the same hardware condition. It is a method of signal processing to realize bearing super resolution technology [3]. The paper introduces the theory of generalized inverse filtering [4] proposed by doctor Kong Fannian. Cyclic convolution is used instead of linear convolution, which can reduce the operation and is easier to be implemented.

53.2 Generalized Inverse Filtering

Generalized inverse filtering accomplishes signal processing in the time domain, avoiding the problem of no spectral inverse and ill-posed problem existed in the frequency domain. Any element $h(n)$, N is the cycle period, if $n > N$, then $h(n)$ can be replaced by $h(n - N)$, the operation with cyclic convolution can be expressed as:

First: Make the input signal change the symbol by every interval to generate a new function. $h^*(n) = (-1)^n h(n)$, $h^*(n)$ convolutes $h(n)$ can get a new convolution kernel:

$$h_1(n) = \sum_{m=0}^{N} h^*(m)h(n - m) \quad y_1(n) = h^*(n) \otimes y(n)$$

Second: Repeat the process of the first step, then

$$h_1^*(n) = (-1)^{n/2} h_1(n), \quad y_2(n) = h_1^*(n) \otimes y_1(n)$$

Recursion: Assume $N = 2^I$, to a pair of input and output it always can find the function: $h_i^*(n) = (-1)^{n/2^i} h_i(n)$

then it gets a new pair output of convolution kernel:

$$h_{i+1}(n) = h_i^*(n) \otimes h_i(n), \quad y_{i+1}(n) = h_i^*(n) \otimes y_i(n)$$

After $I = \log_2 N$ time convolution operation later, the last convolution kernel is the Δ function h_I, and the output y_I is as the follow:

$$h_I = h \otimes h^* \otimes h_1^* \otimes \cdots \cdots \otimes h_{I-1}^*(n), \quad y_I = y \otimes h^* \otimes h_1^* \otimes \cdots \cdots \otimes h_{I-1}^*(n)$$

According to the definition of impulse response, the output is $y_I(n) = Ax(n)$, A is a constant.

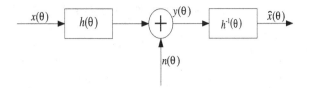

Fig. 53.1 Inverse filtering estimation system

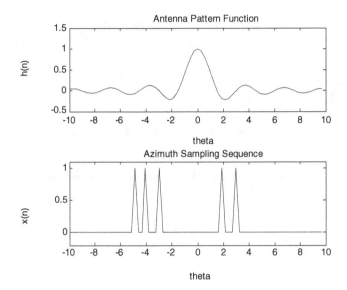

Fig. 53.2 Antenna pattern function, azimuth sampling sequence $[-5 \ -4 \ -3 \ 2 \ 2.5]$

53.3 Deconvolution with Generalized Inverse Filter

Radar to the fixed targets field azimuth scanning can get azimuth echo sequence $y(\theta)$. Antenna pattern is the classical sinc function: $h^{-1}(\theta)$ is the inverse system of $h(\theta)$, the solution of the inverse system can be realized in time domain, also can be realized in the frequency domain (Fig. 53.1). But it may exist the problem of no spectral inverse and ill-posed problem, generalized inverse filtering in time domain can avoid these problems [5]. Using the generalized inverse filtering can get the result such as the follow:

$$h_i^*(\theta) = (-1)^{n/2^i} h_i(\theta), \ h_{i+1}(\theta) = h_i^*(\theta) \otimes h_i(\theta)$$
$$h_I = h \otimes h^* \otimes h_1^* \otimes \cdots \cdots \otimes h_{I-1}^*(\theta) = A\delta(\theta),$$
$$y_I = y \otimes h^* \otimes h_1^* \otimes \cdots \cdots \otimes h_{I-1}^*(\theta)$$

Assume there is not noise, then $y(\theta) = x(\theta) \otimes h(\theta)$,

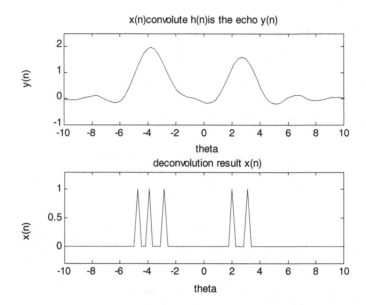

Fig. 53.3 The convolution echo y(n) and deconvolution result x(n) without noise

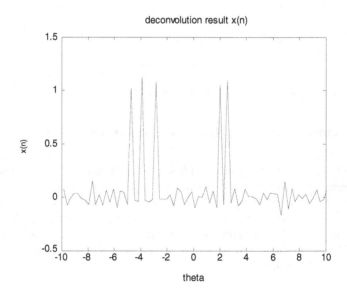

Fig. 53.4 The convolution echo y(n) and deconvolution result x(n) with SNR = 50 dB

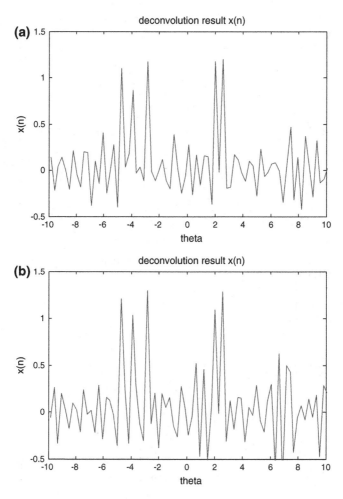

Fig. 53.5 The convolution echo y(n) and deconvolution result x(n) with **a** SNR = 40 dB, **b** SNR = 35 dB

$$y_I = y \otimes h^* \otimes h_1^* \otimes \cdots \cdots \otimes h_{I-1}^*(\theta)$$
$$= x(\theta) \otimes h \otimes h^* \otimes h_1^* \otimes \cdots \cdots \otimes h_{I-1}^*(\theta)$$
$$= Ax(\theta) \otimes \delta(\theta)$$

In fact, noise is excited, then $y(\theta) = x(\theta) \otimes h(\theta) + n(\theta)$,

$$y_I = y \otimes h^* \otimes h_1^* \otimes \cdots \cdots \otimes h_{I-1}^*(\theta)$$
$$= \{ x(\theta) \otimes h(\theta) + n(\theta) \} \otimes h^* \otimes h_1^* \otimes \cdots \cdots \otimes h_{I-1}^*(\theta)$$
$$= Ax(\theta) \otimes \delta(\theta) + n(\theta) \otimes h^* \otimes h_1^* \otimes \cdots \cdots \otimes h_{I-1}^*(\theta)$$

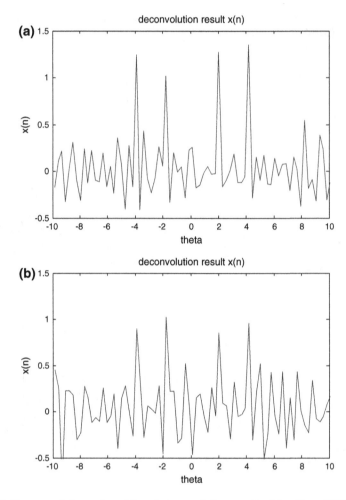

Fig. 53.6 The convolution echo y(n) and deconvolution result x(n) with **a** SNR = 35 dB, **b** SNR = 30 dB

Then define the part $h^{-1}(\theta) = h^* \otimes h_1^* \otimes \cdots \otimes h_{I-1}^*(\theta)$ as the inverse of $h(\theta)$. In cyclic convolution, convolution operation does not make the sequence's length growth, A can be obtained by h_I. Cyclic convolution also is easier than linear convolution in the hardware implementation because of the sequence of fixed length. This approach can distinguish the targets within the same beam width from the mixed echo signal.

53.4 MATLAB Simulation Result

In the following paragraph, the results using the cyclic convolution techniques are presented. The parameter for the simulation is as the follow:

Pulse Frequency: 2,000 HZ; Antenna Scanning Speed: $30°/s$; Scanning Angle Range: $-10° \sim 10°$. Antenna pattern function is sinc function. Zero power beam width is three degree. Target amplitude is one (Figs. 53.2, 53.3, 53.4, 53.5).

When the SNR < 35 dB, it is difficult to resolve two targets that just have a $0.5°$ interval. But SNR can be lower than 35 dB with the interval increase, the targets' interval should not larger than the beam width $3°$. In the following, the interval is $2°$, SNR can be low down to 30 dB. $x(n) = [-4 \ -2 \ 2 \ 4]$ (Fig. 53.6).

From the simulation results it can be seen that the signal processing method of generalized inverse filtering can resolve the targets in the same beam, and it also can distinguish the targets clearly when there is noise in the receiving echo.

53.5 Conclusion

In this study, the application of generalized inverse filtering method to deconvolution achieves the goal of azimuth super resolution. It can clearly distinguish the mixed targets in the same beam. Cyclic convolution instead of linear convolution can fix the number of the sequence which is convenient for hardware implementation and is easier for programming. But fix the length may increase the operation when the length of the echo sequence and the antenna pattern function sequence get a large difference, this problem need a further study. Numerical simulations are conducted and the results show the consistency with the application requirement. This validates our approach.

Acknowledgments This work is partially supported by the Fundamental Research Funds for the Central Universities (No. ZYGX2010J022) and the National Natural Science foundation of China (No. 61139003).

References

1. Ruidong L, She S, Hongtai Z (2009) Analysis on the affection of noise in radar super resolution through deconvolution. In: Radar conference, IET international, 20–22 Apr 2009
2. Xun L, Chuanhua W, Shimin W (2000) Deconvolution improve azimuth resolution. Ship Electron Eng 20(5):50–52 (in Chinese)
3. Yiyuan D, Jianyu Y, Weihua Z (1993) Improvement of angular resolution of real aperture radar via generalized inverse filtering. Acta Electronica Sinica 21(9):15–19 (in Chinese)
4. Fannian K (1985) A new method of deconvolution in the time domain. Acta Electronica Sinica 13(4):9–14 (in Chinese)
5. Mei L, Zaosheng C, Yinong L (1993) Deconvolution algorithm based on generalized inverse. J East China Univ Sci Technol 21(9):15–19 (in Chinese)

Chapter 54
Tracking Threshold Analysis of a Kind of FLL Based on Kalman Filter

Jianhui Wang, Shaojie Ni, Xiaomei Tang and Gang Ou

Abstract In this paper, the tracking threshold of a kind of FLL based on kalman filter related to integral time are discussed. The analyse and simulation result indicates that integral time can not be too long for the sake of reducing the tracking threshold. Shorter integral time can produce wider pulling in frequency range when entering tracking state from acquisition process, so it can reduce the required precision of the Doppler frequency's estimate in acquisition.

Keywords FLL · EKF · Tracking threshold

54.1 Introduction

Generally, PLL is adopted to track the phase of carrier in GNSS receiver. But when SNR is low, PLL suffered to lose of lock or cycle slipping. Compared to PLL, the FLL has a wider frequency discriminating range. So FLL can track carrier phase robustly. FLL based on EKF is a kind of loop which can adaptively change noise bandwidth, and this kind of loop has better performance than traditional FLL especially under high dynamic and weak signal environments. References [1, 2] analyzed the structure and performance of traditional FLL. Reference [3] presented two new frequency tracking loops, ODAFC and FEKF, based on traditional FLL, and these two presented loops has lower thermal noise tracking threshold than traditional FLL. Reference [4] studied the general principle

J. Wang (✉) · S. Ni · X. Tang · G. Ou
Satellite Navigation Research and Development Center, National University
of Defense Technology, Changsha, China
e-mail: wjh0369@163.com

W. Lu et al. (eds.), *Proceedings of the 2012 International Conference on Information*
Technology and Software Engineering, Lecture Notes in Electrical Engineering 210,
DOI: 10.1007/978-3-642-34528-9_54, © Springer-Verlag Berlin Heidelberg 2013

of FLL based on EKF. Reference [5] put forward an algorithm based on EKF which has good performance in high dynamic and low SNR environments.

In a FLL based on EKF, increasing integral time can improve SNR, but the tracking threshold is not proportional to integral time. This paper analyzed the relation of tracking threshold and integral time.

54.2 The Model of FLL Based on EKF

The received signal expression of the ith integral is

$$r(t) = \cos(2\pi(f_{IF} + f_i)(t + t_i) + \varphi_0) + n(t + t_i) \qquad (54.1)$$

where f_{IF} is intermediate frequency, and f_i is the Doppler frequency of the ith integral. φ_0 and t_i are starting phase and time of the ith integral. And $n(t)$ is white noise which has one-sided power spectrum density N_0.

In order to simplify the analysis, we use a FLL with two NCOs. Figure 54.1 shows the structure of FEKF with two NCOs. NCO1 is normal, and NCO2 has two ports (one is for frequency update, and the other is for phase update). At the same integral and dump period, the phase of NCO2 at the end is equal to the phase of NCO1 at the start. This structure can ensure continuous phase of the two signal used for frequency discriminating.

The in-phase and quadra-phase outputs of NCO1 are

$$\hat{r}_{1I}(t) = 2\cos\left(2\pi\left(f_{IF} + \hat{f}_{i|i-1}\right)(t + t_i)\right) \qquad (54.2)$$

$$\hat{r}_{1Q}(t) = 2\sin\left(2\pi\left(f_{IF} + \hat{f}_{i|i-1}\right)(t + t_i)\right) \qquad (54.3)$$

where $\hat{f}_{i|i-1}$ is the Doppler frequency estimated of the received signal in a integral and dump period with starting time t_i.

The outputs of integral and dump 1 and integral and dump 2 are

$$I_{1,i} = \frac{\sin(\pi_i T)}{\pi \Delta f_i T} \cos\left[2\pi_i\left(t_i + \frac{T}{2}\right) - \varphi_0\right] \\ + n_{1,I}(i) \qquad (54.4)$$

$$Q_{1,i} = \frac{\sin(\pi_i T)}{\pi \Delta f_i T} \sin\left[2\pi_i\left(t_i + \frac{T}{2}\right) - \varphi_0\right] \\ + n_{1,Q}(i) \qquad (54.5)$$

where $\Delta f_i = \hat{f}_{i|i-1} - f_i$. $n_{1,I}(i)$ is the output noise of integral and dump 1, and $n_{1,Q}(i)$ is the output noise of integral and dump 2. T is the integral time.

Similarly, the outputs of integral and dump 3 and integral and dump 4 are

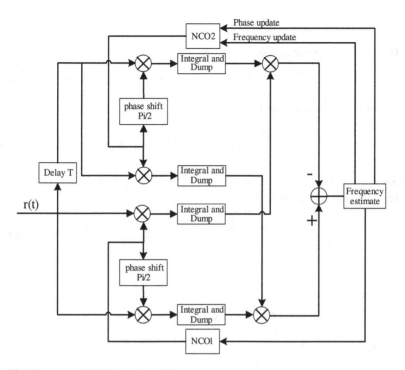

Fig. 54.1 Structure of FLL based on EKF

$$I_{2,i} = \frac{\sin(\pi \Delta f_i' T)}{\pi \Delta f_i' T} \cos\left[2\pi \Delta f_i'\left(t_i - \frac{T}{2}\right) - \varphi_0\right] + n_{2,I}(i) \qquad (54.6)$$

$$Q_{2,i} = \frac{\sin(\pi \Delta f_i' T)}{\pi \Delta f_i' T} \sin\left[2\pi \Delta f_i'\left(t_i - \frac{T}{2}\right) - \varphi_0\right] + n_{2,Q}(i) \qquad (54.7)$$

Where $\Delta f_i' = \hat{f}_{i|i-1} - f_{i-1} \approx \Delta f_i$. $n_{2,I}(i)$ is the output noise of integral and dump 3, and $n_{2,Q}(i)$ is the output noise of integral and dump 4.

The output of cross product frequency discriminator is

$$Q_{1,i}I_{2,i} - I_{1,i}Q_{2,i} \approx \frac{\sin^2(\pi \Delta f_i T)}{\pi^2 \Delta f_i^2 T^2} \sin(2\pi \Delta f_i T) + n'(i) \qquad (54.8)$$

The model of FEKF used in this paper is constructed as follows.
The state equation is

$$f_i = f_{i-1} + q_i \qquad (54.9)$$

Where q_i is a random value. Its expectation is zero, and its variance is σ_q^2.

The prediction of Doppler frequency of the ith integral and dump period is

$$\hat{f}_{i|i-1} = \hat{f}_{i-1} \qquad (54.10)$$

The prediction error covariance matrix is

$$P_{i|i-1} = P_{i-1} + \sigma_q^2 \qquad (54.11)$$

The observation equation is

$$Y_i = \frac{\sin^2(\pi \Delta f_i T)}{\pi^2 \Delta f_i^2 T^2} \sin(2\pi \Delta f_i T) + n'(i) \qquad (54.12)$$

Where $n'(i)$ is random noise. Its expectation is zero, and its variance is $\sigma_{n'}^2$.
The observation matrix is

$$H = \left. \frac{\partial h(f_i)}{\partial f_i} \right|_{f_i = \hat{f}_{i|i-1}} = -2\pi T \qquad (54.13)$$

Where

$$h(f_i) = \frac{\sin^2(\pi \Delta f_i T)}{\pi^2 \Delta f_i^2 T^2} \sin(2\pi \Delta f_i T) \qquad (54.14)$$

The Kalman gain is

$$K_i = P_{i|i-1} H^T \left[\sigma_{n'}^2 + H P_{i|i-1} H^T \right]^{-1} \qquad (54.15)$$

The estimation of the Doppler frequency is

$$\hat{f}_i = \hat{f}_{i|i-1} + K_i \left(Y_i - h(\hat{f}_{i|i-1}) \right) \qquad (54.16)$$

The estimation error covariance matrix is

$$P_i = (I - K_i H) P_{i|i-1} \qquad (54.17)$$

54.3 The Relation of Tracking Threshold and Integral Time

54.3.1 The Stable Estimate Error Covariance Matrix

After a long time, the FEKF will arrive at a stable state. Let K, P_1 and P_2 represent the stable Kalman gain, stable prediction error covariance matrix and stable estimation error covariance matrix. From the model of FEKF, we get

$$P_1 = P_2 + \sigma_q^2, \quad K = \frac{P_1 H}{\sigma_n^2 + H^2 P_1}, \quad P_2 = (1 - KH) P_1$$

From the three expressions above, the expression of P_1 can be derived as follows.

$$P_1 = \frac{\sigma_q^2}{2}\left(1 + \sqrt{1 + \frac{4\sigma_{n'}^2}{H^2\sigma_q^2}}\right) \tag{54.18}$$

Where $H = -2\pi T$.

Now we have to derive the expression of $\sigma_{n'}^2$. The received noise can be written as follows

$$n(t) = n_c(t)\cos\omega_{IF}t - n_s(t)\sin\omega_{IF}t \tag{54.19}$$

Where $n_c(t)$ and $n_s(t)$ are the in-phase and quadra-phase components of $n(t)$. They are orthogonal, and they have the same autocorrelation function.

The two-sided power spectrum density of $n(t)$ is $N_0/2$, so the $n_c(t)$ and $n_s(t)$ have baseband two-sided power spectrum density N_0. The relation of CNR (carrier power to noise density ratio) and N_0 is

$$N_0 = \frac{1}{2CNR} \tag{54.20}$$

The in-phase and quadra-phase local repeated carriers are $2\cos(\omega_{IF}t + \varphi)$ and $2\sin(\omega_{IF}t + \varphi)$ respectively. After the multiplication of the received noise and the local repeated carriers, the outputs of in-phase and quadra-phase are (the high frequency components are ignored)

$$n_I(t) = n_c(t)\cos\varphi + n_s(t)\sin\varphi \tag{54.21}$$

$$n_Q(t) = n_c(t)\cos\varphi - n_s(t)\sin\varphi \tag{54.22}$$

where φ is a constant phase error.

$n_I(t)$ and $n_Q(t)$ are orthogonal, and they have the same autocorrelation function as $n_c(t)$ or $n_s(t)$. So $n_I(t)$ and $n_Q(t)$ have zero expectation, and their two-sided power spectrum density is N_0. $n_{1,I}(i)$ is the integral of $n_I(t)$ over a period T, and $n_{1,Q}(i)$ is the integral of $n_Q(t)$ over a period T. The bandwidth of integral and dump filter is $1/T$. So $n_{1,I}(i)$ and $n_{1,Q}(i)$ have zero expectation, and their variance is N_0/T. And the following two equations hold approximately.

$$n_{2,I}(i) = n_{1,I}(i-1) \tag{54.23}$$

$$n_{2,Q}(i) = n_{1,Q}(i-1) \tag{54.24}$$

The expression of $n'(i)$ is

$$n'(i) = \left(n_{1,Q}(i) - n_{2,Q}(i)\right) \cos \Delta\varphi$$
$$+ \left(n_{2,I}(i) - n_{1,I}(i)\right) \sin \Delta\varphi \qquad (54.25)$$
$$+ n_{1,Q}(i)n_{2,I}(i) - n_{1,I}(i)n_{2,Q}(i)$$

$n_{1,I}(i)$ and $n_{2,I}(i)$ are integrals of $n_I(t)$ over different integral period, so they are independent. Similarly, $n_{1,Q}(i)$ and $n_{2,Q}(i)$ are independent too. Based on the analysis above, we get the expectation and variance of $n'(i)$ as follows.

$$E[n'(i)] = 0, \quad \sigma_{n'}^2 = \frac{2N_0}{T}\left(1 + \frac{N_0}{T}\right)$$

Considering (54.20), we get

$$\sigma_{n'}^2 = \frac{1}{TCNR}\left(1 + \frac{1}{2TCNR}\right) \qquad (54.26)$$

Combining (54.18) and (54.26), we get

$$P_1 = \frac{\sigma_q^2}{2}\left(1 + \sqrt{1 + \frac{2TCNR + 1}{2CNR^2\pi^2T^4\sigma_q^2}}\right) \qquad (54.27)$$

54.3.2 The Tracking Threshold

The rule of thumb for tracking threshold is that the 3-sigma must not exceed one-fourth of the pull-in range of the FLL discriminator [6]. The pull-in range of the FLL discriminator is $1/T$ Hz. We can derive the tracking threshold from the following in equation.

$$P_1 = \frac{\sigma_q^2}{2}\left(1 + \sqrt{1 + \frac{2TCNR + 1}{2CNR^2\pi^2T^4\sigma_q^2}}\right) < \left(\frac{1}{4T}\right)^2 \qquad (54.28)$$

Solve the above in equation, and the CNR tracking threshold is as follows

$$CNR > \frac{T + \sqrt{T^2 + ab}}{ab} \qquad (54.29)$$

The expressions of a and b are

$$a = 2\pi^2T^4\sigma_q^2, \quad b = \left(\frac{1}{8\sigma_q^2T^2} - 1\right)^2 - 1$$

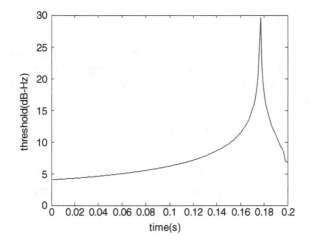

Fig. 54.2 Relation of tracking threshold and integral time

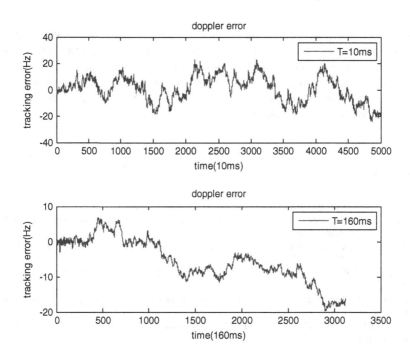

Fig. 54.3 Frequency tracking of different integral times

When σ_q^2 is constant, the tracking threshold is dependent on integral time T. Figure 54.2 shows the relation of the tracking threshold and the integral time, where σ_q^2 equals to 2 Hz^2.

54.4 Simulations

Long integral time can improve the SNR, but will depress the pull-in range of frequency discriminator. So the tracking threshold may be increased. To verify the analysis above, Fig. 54.3 shows frequency tracking of two different integral times with the same CNR and σ_q^2.

$$CNR = 10 \text{ dB-Hz}, \ \sigma_q^2 = 1$$

In the upper figure, the integral time is 10 ms, and the pull-in range is 100 Hz. The tracking error is approximately in the $[-20, 20 \text{ Hz}]$, and the loop maintains lock. In the lower figure, the integral time is 160 ms, and the pull-in range is 6.25 Hz. The tracking error exceeds the pull-in range, and the loop loses lock.

54.5 Conclusion

The relation of tracking threshold and integral time of FEKF was analyzed in this paper. The analysis shows that the tracking threshold is not proportional to integral time. In order to decrease the threshold, the integral time can not be too long. When the CNR is high enough, we can prolong the integral time to improve the tracking performance. On the other hand, when the CNR is low, we should adopt short integral time to maintain the loop locked.

References

1. Natali FD (1984) AFC tracking algorithms. IEEE Trans Commun 32(8):1064–1070
2. Natali FD (1986) Noise performance of a cross-product AFC with decision feedback for DPSK signals. IEEE Trans Commun 34(3):303–307
3. Aguirre S, Hinedi S (1989) Two novel automatic frequency tracking loops. IEEE Trans Aerosp Electron Syst 25(5):749–760
4. Barbara F (1996) Design of an extended Kalman filter frequency tracker. IEEE Trans Signal Process 44(3):739–742
5. Zhang H (2010) A novel frequency tracking algorithm in high dynamic environments. WCNIS, p 31–34
6. Kaplan ED (2005) Understanding GPS—principles and applications, 2nd edn. Artech House, Boston, p 192

Chapter 55
A Modified Cross-High-Order Spectral Estimation Approach Based on Rotational Invariance Technique in Hybrid Colored Noises

Yuling Gao and Qing Huang

Abstract In signal processing, a frequently encountered problem is harmonic retrieval in additive colored noise. Based on the theory of cross-power spectrum and higher order statistics, a modified cross-high-order spectral estimation approach based on rotational invariance technique is developed in this paper. In this approach the signal parameters are directly determined by eigenvalue without searching procedure inherent in all previous cross-high-order spectral estimation methods. Simulation results indicate that spectral density curve is smooth without false peaks existence. This approach is ideally suited for harmonic retrieval in additive colored noise and short data conditions, and is also accurate to estimate signal parameter in hybrid colored noises.

Keywords Harmonic retrieval · Cross-high-order spectral · Rotational invariance technique · False peaks

55.1 Introduction

The harmonic retrieval problem arises in various areas such as radar, sonar, geophysics [1]. Harmonic retrieval in additive colored noise is an important research topic [2]. In recent years, the cross-high-order spectral technique is widely used in signal processing, such as cross-high-order spectral QR decomposition approach [3], cross-high-order spectral SVD approach [4], but there is spectral peaks bias in these methods and stability is not high. Cross-high-order spectral MUSIC approach [5] can well inhibit the independent noise in each

Y. Gao (✉) · Q. Huang
Department of Electronic Information Nanjing College of Information Technology,
Nanjing, China
e-mail: yuling_gao2010@126.com

W. Lu et al. (eds.), *Proceedings of the 2012 International Conference on Information Technology and Software Engineering*, Lecture Notes in Electrical Engineering 210, DOI: 10.1007/978-3-642-34528-9_55, © Springer-Verlag Berlin Heidelberg 2013

channel, but need to search for spectral peaks throughout the frequency domain and exist large computing. For this situation, based on the theory of cross-power spectrum and higher order statistics, a modified cross-high-order spectral estimation approach based on rotational invariance technique is developed in this paper. Simulation results indicate that this approach is ideally suited for harmonic retrieval in additive colored noise and short data conditions, and is also accurate to estimate signal parameter in hybrid colored noises. High resolution of spectral estimation and stability are realized in this method, and estimation accuracy has little change with SNR decreased. Comparison with literature [6], spectral density curve is smooth without false peaks existence, and side lobe level is very low, therefore the signal is conducive to be extracted.

55.2 Theoretical Basis

55.2.1 Signal Model

Assumption the zero mean sinusoid random processing with additional colored noises, the next equation is described as follow,

$$x(n) = \sum_{i=1}^{q} \alpha_i \exp[j(w_i n + \phi_x)] + \xi_x(n) + \eta_x(n) \tag{55.1}$$

$$y(n) = \sum_{i=1}^{q} \beta_i \exp[j(w_i n + \theta_i + \phi_y)] + \xi_y(n) + \eta_y(n) \tag{55.2}$$

Where, α_i and β_i are amplitudes of complex harmonic signals; ω_i is the frequency of the harmonic signal; θ_i is the difference of phase among harmonic signals; ϕ_x and ϕ_y are initial phases of harmonic signals, and are distributed within $[0, 2\pi]$; q is the quantity of sinusoid harmonic (assumption q is known); ξ_x, ξ_y and η_x, η_y are zero mean colored noises with unknown spectral density, where, ξ_x and ξ_y are uncorrelated Gauss or non-Gauss noises, η_x and η_y are correlated Gauss noises, ξ_x, ξ_y and η_x, η_y are independent.

55.2.2 Cross-Fourth-Order Cumulant Matrix

Cross-fourth-order cumulant defined by Eqs (55.1) and (55.2) is as followed:

$$
\begin{aligned}
c_{xyyy}(m) &= cum\{x(n), y(n+m), y^*(n+m), y^*(n+m)\} \\
&= -\sum_{i=1}^{q} \alpha_i \beta_i^3 \exp[-j(w_i m + \theta_i)]
\end{aligned} \tag{55.3}
$$

p-order cross-fourth-order cumulant matrix is constructed as followed$(p\rangle\rangle q)$:

$$C_{xyyy} = \begin{bmatrix} c(0) & c(-1) & \cdots & c(-p+1) \\ c(1) & c(0) & \cdots & c(-p+2) \\ \vdots & \vdots & \ddots & \vdots \\ c(p-1) & c(p-2) & \cdots & c(0) \end{bmatrix} \tag{55.4}$$

Where,$c(i) = c_{xyyy}(i) = c_{xyyy}(i,0,0,0)$ $(i = 0, \pm 1, \cdots, \pm(p-1))$

55.3 The Modified Cross-High-Order Spectral Estimation Approach Based on Rotational Invariance Technique

55.3.1 Rotational Invariance Technique

Suppose that vector Y is obtained by rotating vector X, and this rotation keeps the invariance of X and Y corresponding to the signal subspace, witch is the basic idea of the rotational invariance. So a suitable matrix beam needs to be constructed in this paper, and the signal subspace spanned by matrix beam is the same subspace.

55.3.2 The Establishment of Matrix Beam

Assumption,

$$W_1 = \begin{bmatrix} 0 & & & & \\ 1 & \ddots & & 0 & \\ & \ddots & \ddots & & \\ 0 & & \ddots & & \\ & & & 1 & 0 \end{bmatrix} \quad W_2 = \begin{bmatrix} 1 & & & \\ & \ddots & & 0 \\ & 0 & \ddots & \\ & & & 1 \end{bmatrix}$$

Definition,

$$C_{xyyy1} = C_{xyyy}W_1 = FPF^H \tag{55.5}$$

$$C_{xyyy2} = C_{xyyy}W_2 = FP\Phi F^H \tag{55.6}$$

Where, $F = [F_1, F_2, F_3, \cdots, F_q]$; $F_i = [1, e^{jw_i}, e^{j2w_i}, \cdots, e^{j(p-1)w_i}]^T$ $P = diag$ $[-\alpha_1\beta_1^3, -\alpha_2\beta_2^3, \cdots, -\alpha_q\beta_q^3]\Phi = diag[e^{j\theta_1}, e^{j\theta_2}, e^{j\theta_3}, \cdots, e^{j\theta_q}]$

Φ is rotational operator. $\{C_{xyyy1} \; C_{xyyy2}\}$ is the matrix beam constructed in this paper, which meets the general structure of the matrix beam and generalized eigenvalue of the matrix beam$\{C_{xyyy1} \; C_{xyyy2}\}$ can be obtained easily byC_{xyyy}. The modified approach determines the rotation factor by solving the generalized eigenvalue of the matrix beam $\{C_{xyyy1}\ C_{xyyy2}\}$, then obtains the harmonic frequency.

Theorem *We define Γ as generalized eigenvalue matrix of the matrix beam, then generalized eigenvalue matrix Γ maintain the information of the rotational operator Φ [7].*

$$\Gamma = \begin{bmatrix} \Phi & 0 \\ 0 & 0 \end{bmatrix} \tag{55.7}$$

55.3.3 The Modified Cross-High-Order Spectral Estimation Approach

According to above analysis, we know $\{C_{xyyy1}\ C_{xyyy2}\}$ has a further contact with C_{xyyy} in addition to meeting the general structure of the matrix beam. Thus generalized eigenvalue problem of the matrix beam transforms into the q × q square matrix eigenvalue problem expressed by the characteristic data of the C_{xyyy}.

The singular value decomposition matrix of C_{xyyy} is shown as followed,

$$C_{xyyy} = U \begin{bmatrix} \sum & 0 \\ 0 & 0 \end{bmatrix} V^H \tag{55.8}$$

Where, the column vectors of matrix U and V are the left and right singular rectors of C_{xyyy}, $\sum$ is defined as follow,

$$\sum = diag[\sigma_1, \sigma_2, \cdots, \sigma_q]$$

Where $\sigma_i(i = 1, \cdots, q-1)$ is the non-zero singular value of C_{xyyy}, and $\sigma_i \geq \sigma_{i+1}(i = 1, \cdots, q-1)$.

Decompose the singular vector matrix V as follow,

$$V = [V_1, V_2]$$

Where,

$$V_1 = [v_1, v_2, \cdots, v_q]$$
$$V_2 = [v_{q+1}, v_{q+2}, \cdots, v_p]$$

Similarly, decompose the singular vector matrix U as follow,

$$U = [U_1, U_2]$$

Where,

$$U_1 = [u_1, u_2, \cdots, u_q]$$
$$U_2 = [u_{q+1}, u_{q+2}, \cdots, u_p]$$

Similar to the correlation function, $u_i, v_i (i = 1, 2, \cdots, q)$ is called signal eigenvectors, space consisted of U_1, V_1 is called signal eigenvector space, $u_i, v_i (i = q+1, q+2, \cdots, p)$ is called noise eigenvectors, space consisted of U_2, V_2 is called noise eigenvector space.

Thus, C_{xyyy} can be transformed into the following form,

$$C_{xyyy} = U_1 \Sigma V_1^H + U_2 \Sigma V_2^H$$

Now matrix beam $\{C_{xyyy1} \ C_{xyyy2}\}$ can be transformed into the equivalent form of the signal space,

$$C_{xyyy1} = U_1 \Sigma V_1^H W_1 = U_1 \Sigma S_{11}^H \tag{55.9}$$

$$C_{xyyy2} = U_1 \Sigma V_1^H W_2 = U_1 \Sigma S_{12}^H \tag{55.10}$$

By (9, 10) two equations, we come to the conclusion obviously: the generalized eigenvalue of matrix beam $\{C_{xyyy1} \ C_{xyyy2}\}$ is equivalent to the generalized eigenvalue of matrix beam $\{S_{11} = W_1^T V_1 \ S_{12} = W_2^T V_1\}$, that is equivalent to the eigenvalue of following matrix,

$$\Psi = [S_{11}^H S_{11}]^{-1} S_{11}^H S_{12}$$

If we find the generalized eigenvalue, then calculate the signal frequency.

55.4 Numerical Simulations

55.4.1 Simulation Model

The simulation model can be defined as,

$$x(n) = \cos(2\pi f_1 n + \pi/4) + \cos(2\pi f_2 n) + \xi_x(n) + \eta_x(n)$$

Where, $f_1 = 0.17\,\text{Hz}, f_2 = 0.19\,\text{Hz}$, $\xi_x(n), \eta_x(n)$ are zero mean, stable colored noised with unknown spectral density and are independent each other. They are produced by white noises with independent each other passed two filters that have the same type. The transfer function of the filter is below,

$$H(z) = \frac{k(1 + 2z^{-2} - 1.23z^{-4})}{1 - 1.637z^{-1} + 2.237z^{-2} - 1.307z^{-3} + 0.641z^{-4}}$$

The curve of the power spectral density is shown in Fig. 55.1.

Fig. 55.1 Power spectral density

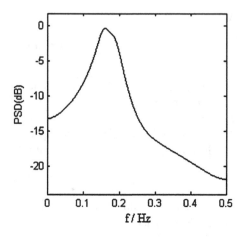

Fig. 55.2 The modified method

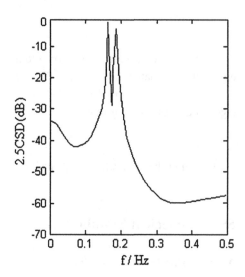

If $\eta_x(n) = \eta_y(n)$, then $\eta_x(n)$ and $\eta_y(n)$ are dependent, then, $T = 50T_e.\xi_y(n) = \xi_x(n + m)$, where $m \geq (T + pT_e)/T_e$. And $m = 40$, then $\xi_x(n)$ and $\xi_y(n)$ are irrelevant. However, sinusoid signal is period correlate. Thus,

$$y(n) = \cos\left(2\pi f_1(n + m) + \frac{\pi}{4}\right) + \cos(2\pi f_2(n + m)) + \xi_y(n) + \eta_y(n)$$

$$= \cos\left(2\pi f_1 n + \frac{\pi}{4} - 1.256\right) + \cos(2\pi f_2 n - 2.512) + \xi_x(n + m) + \eta_x(n)$$

Table 55.1 The statistic results of frequency estimation

True value	$f_1 = 0.17$ Hz	$f_2 = 0.19$ Hz
The modified approach	$0.1689 \pm 2.674\text{E} - 04$	$0.1903 \pm 3.223\text{E} - 04$
ESPRIT	$0.1708 \pm 3.183\text{E} - 04$	$0.1911 \pm 5.342\text{E} - 04$

Data are mean $\pm$S.D

Fig. 55.3 ESPRIT method

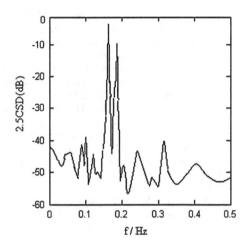

55.4.2 Simulation Results

The length of data is 1,024, SNR of two sinusoidal harmonics in $x(n)$ and $y(n)$ is also -10 dB. The order of extended function matrix is 80, 20 independent simulations are done. The spectral estimation curve of the modified method based on rotational invariance technique is shown in Fig. 55.2. The statistic results of sinusoid signals frequencies are shown in Table 55.1.

In order to compare with the new approach, we still use the mathematical model shown above, and get the spectrum estimation by cross-high-order spectral ESPRIT method [6], which is shown in Fig. 55.3.

The simulation results show that when SNR are -10 dB, the modified approach in Fig. 55.2 shows that spectral density curve is smooth without false peaks. The methods have good resolving power and performance. Cross-higher-order spectral ESPRIT approach in Fig. 55.3 distinguishes the frequencies of two sinusoid signal, but there are some false peaks, and side lobe level is higher.

55.5 Conclusions

Using the modified approach based on rotational invariance technique can eliminate almost the false peaks of spectral estimate and has high resolution. The spectral estimation performances of this approach are better than the ESPRIT method.

Acknowledgments This work was supported financially by the project of NanJing College of Information Technology Science Foundation (No.YKJ10-005).

References

1. Vossen RV, Naus HWL, Zwamborn APM (2010) High-resolution harmonic retrieval using the full fourth-order cumulant. Signal Process 90:2288–2294
2. Wang FS, Li HW, Li R (2010) Harmonic signals retrieval approach based on blind source separation. Circuits Syst Signal Process 29:669–685
3. Gao YL, Huang Q, Yu BM et al (2012) False peaks suppression based on cross-high-order spectral QR decomposition approach. Adv Mater Res 403–408:177–181
4. Gao YL, Kang XT, Shi YW (2006) Harmonic retrieval embedded in hybrid colored noises: cross-high-order spectral SVD-TLS approach with cleared false peaks. SCI-Tech Inf Dev Econ 16:143–150 (in Chinese)
5. Yang SY, Li HW (2007) Estimating the number of harmonics using enhanced matrix. IEEE Signal Process Lett 14(2):137–140
6. MA Y, Shi YW, Sun WT (2002) Signal frequency estimation in colored noise cross fourth-order cumulant on based ESPRIT-SVD method. J Jilin Univ (Inf Sci Edn), 2,30–32,38 (in Chinese)
7. Li XJ, Li HF (2005) Weighted cross—spectral ESPRIT method for harmonic retrieval in colored noise. Electr Mach Control 11:575–578 (in Chinese)

Chapter 56
A Simplified DBF Channels Calibration Algorithm and Implementation

Jiangxian Song, Chaohai Li, Wei Xia, Lilin Liang and Zishu He

Abstract Digital beam-forming (DBF) technology is one of the popular radar technologies. The performance of DBF is greatly affected by the consistency of receiving channel. In this paper, a simplified algorithm is proposed to calibrate the inconsistency of the channels, and an efficient method to implement the proposed algorithm based on FPGA (Field Programmable Gate Array) and DSP (Digital Signal Processor) is introduced. Simulation and experiments results are given to demonstrate the efficiency of the algorithm.

Keywords Channels · Calibration · DBF · Implementation

56.1 Introduction

In comparison with analog beam-forming method, digital beam-forming is more flexible and much easier to reduce side lobe and form multiple beams. Therefore DBF is more and more popular in the fields of mobile communication, radar and satellite communication. However the existence of Mutual Coupling, amplitude, phase and I/Q branches quadrature errors of each channel are important facts that can cause the channels inconsistency, which as a result will lead to the decrease in the antenna array's performance of gain, side lobe level and beam-steering accuracy. Recent interest of digital beam-forming has been focused on eliminating those errors through calibrating channels.

J. Song (✉) · C. Li · W. Xia · L. Liang · Z. He
School of Electronic Engineering, University of Electronic Science
and Technology of China Chengdu, China
e-mail: songjiangxian@126.com

W. Lu et al. (eds.), *Proceedings of the 2012 International Conference on Information Technology and Software Engineering*, Lecture Notes in Electrical Engineering 210, DOI: 10.1007/978-3-642-34528-9_56, © Springer-Verlag Berlin Heidelberg 2013

Fig. 56.1 Basic structure of a GNSS receiver

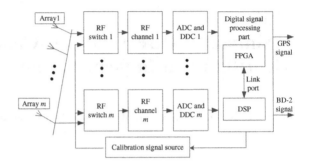

Calibrating amplitude-phase errors of DBF transmitter based on maximum likelihood estimation [1] and synchronous orthogonal codes [2] have been proposed. Combined optimization method (COM) [3] is developed to minimize element pattern errors in specified narrow angle and keep global errors being suboptimum. The method proposed in [4] is to generate calibration signal and inject them into each channels. Also there are many other methods being proposed in [5–7], however they can only calibrate some certain errors. Those algorithms are complex and difficult to implement in practical engineering. This paper proposes a simplified, low computational-complex algorithm, along with an easy and efficient way to implement it in practical project based on FPGA (Field Programmable Gate Array) and DSP (Digital Signal Processor).

In the following section, the basic structure of GNSS receiver in our practical project is introduced, along with the general algorithm and an optimized algorithm to calibrate channels are presented. Simulation results with MATLAB are shown in Sect. 56.3, and then an efficient way to realize the optimized algorithm in project based on FPGA and DSP is proposed. Finally some conclusions are presented in Sect. 56.4.

56.2 Calibration Algorithm Optimization

56.2.1 The Structure of GNSS Receivers

Consider the GNSS receiver of array elements. It covers GPS and BD-2 frequency band, the system can form two digital beams at the same time and it has the ability to calibrate channel errors by itself as well. The basic structure of system is shown in Fig. 56.1. Each channel includes an antenna array element, RF (Radio Frequency), DDC (Digital Down Converter) and ADC (Analog to Digital Converter). A calibration signal is generated by calibration signal source, which is split out sixteen ways and injected to each channel. RF switches choose which signal is to be injected to channels. FPGA and DSP implement channels calibration and digital beam-forming.

Fig. 56.2 Channels calibration principle

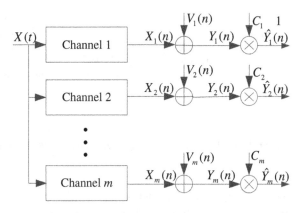

56.2.2 General Channel Calibration Algorithm

In the DBF system, beams are formed in the digital baseband, the inconsistency of receiver channels will reflect in the transfer functions finally. So we can only calibrate the whole channels' transfer functions rather than each part of the receiver channels. As usual, transfer functions of channels depend on the working frequency, while in narrow-band system, it can be considered that the amplitude and phase of channels change little or even invariable with different frequencies, so we can only calibrate the central frequency. For the case of practical radar applications, we can handle it as in the narrow-band situation.

The calculation method of channels calibration coefficients is shown in Fig. 56.2. Take the first channel as the reference channel. Assuming calibration signal is a single mono-frequency signal, and then injects it into each of those channels. The channel calibration coefficients can be given as

$$C_i = \frac{Y_1(n)}{Y_i(n)} \qquad (i = 2, 3, \cdots, m) \tag{56.1}$$

where $Y_i(n) = Y_i^I(n) + jY_i^Q(n)$ is the complex-valued baseband signal of the ith channel with noise. $Y_i^I(n)$ is the I branch data and $Y_i^Q(n)$ is the Q branch data. The calibrated signal can be represented as

$$\hat{Y}_i(n) = C_i Y_i(n) = Y_1(n) \tag{56.2}$$

We assume the input signal as $X(t) = A_0 \cos(2\pi f_0 t + \varphi)$, then after DDC $X_i(n) = X_i^I(n) + jX_i^Q(n)$ is the complex-valued baseband signal of the ith channel without noise. In practical project we can express

$$\begin{aligned} X_i^I(n) &= A_i e^{j\varphi_i}(\cos \omega n) \\ X_i^Q(n) &= A_i e^{j\varphi_i}(\sin \omega n) \end{aligned} \tag{56.3}$$

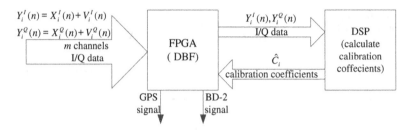

Fig. 56.3 Block diagram of channels calibration in FPGA and DSP

where $X_i^I(n)$ is the I branch data and $X_i^Q(n)$ is the Q branch data. The signal $V_i(n) = V_i^I(n) + jV_i^Q(n)$ is additive white Gaussian noise of the ith channel. It will inevitably lead to errors in the calibration coefficients with only one snapshot because of noise. So we average $C_i(n)$ which is acquired from experiments to get high precision estimation of calibration coefficients $\hat{C}_i$. In practical project we calculating $\hat{C}_i$ expectation by a sample mean

$$E\{\hat{C}_i\} = E\left\{\frac{1}{N}\sum_{n=1}^{N} C_i(n)\right\} = E\left\{\frac{1}{N}\sum_{n=1}^{N} \frac{Y_1(n)}{Y_i(n)}\right\} \tag{56.4}$$

Simplify Eq.(56.4) we can obtain that

$$E\{\hat{C}_i\} = \frac{A_1}{A_i} e^{j(\varphi_1 - \varphi_i)} \tag{56.5}$$

56.2.3 Algorithm Optimization

It is known that FPGA can implement fast parallel operations and DSP can achieve high precision floating point operations. For fast realization of channels calibration and high precision estimation of coefficients, we use FPGA to achieve I/Q data and implement the channels calibration [i.e. Eq. (56.2)], while DSP to calculate the calibration coefficients [i.e. Eq. (56.4)]. The block diagram of channels calibration in FGPA and DSP is shown in Fig. 56.3. It is obvious that the data quantity FPGA transmits to DSP will increase times along with the average data. So it's quite necessary to simplify the algorithm to improve calculate speed and reduce the occupancy rate of the hardware.

We can firstly get I/Q branches averaged data respectively

$$\bar{Y}_i^I(n) = \frac{1}{N}\sum_{n=1}^{N}(X_i^I(n) + V_i^I(n))$$

$$\bar{Y}_i^Q(n) = \frac{1}{N}\sum_{n=1}^{N}(X_i^Q(n) + V_i^Q(n))$$

(56.6)

Then calculate channels calibration coefficients estimation $\hat{\bar{C}}_i$ according to I/Q branches average data $\bar{Y}_i^I(n)$ and $\bar{Y}_i^Q(n)$. Assuming the noise signal $V_i(n)$ is also additive white Gaussian noise of the ith channel and calculates calibration coefficient estimation $\hat{\bar{C}}_i$ expectation

$$\mathrm{E}\left\{\hat{\bar{C}}_i\right\} = \mathrm{E}\left\{\frac{\frac{1}{N}\sum_{n=1}^{N}Y_1(n)}{\frac{1}{N}\sum_{n=1}^{N}Y_i(n)}\right\} = \mathrm{E}\left\{\frac{\bar{Y}_1(n)}{\bar{Y}_i(n)}\right\}$$

(56.7)

Simplify Eq. (56.7) can be represented as Eq. (56.8):

$$E\{\hat{C}_i\} = E\left\{\frac{1}{N}\sum_{n=1}^{N}C_i(n)\right\} = E\left\{\frac{1}{N}\sum_{n=1}^{N}\frac{Y_1(n)}{Y_i(n)}\right\}$$

(56.8)

Compare Eqs. (56.4) and (56.7), it needs N times complex numbers division and $N-1$ times addition to accomplish calculation of channels coefficients by Eq. (56.4), however it only needs $2N-2$ times addition and once division by Eq. (56.7), it is obvious that the computation has been greatly reduced. Then Compare Eqs. (56.5) and (56.8), we can conclude that the simplified channels calibration algorithm is an unbiased version of the general algorithm.

56.3 Performance Analysis with Implementation Results

56.3.1 Simulation Results with MATLAB

Assuming that the array elements $m = 16$, the first channel signal is $X_1(n) = e^{j2\pi fn}$, the second channel is $X_2(n) = 0.8e^{j2\pi fn-0.05\pi}$, the noise is white Gaussian noise. We implement 500 times independent repeated experiments in MATLAB under different SNRs (Signal-to-Noise Ratios) and normalized digital frequency independently to obtain variance

$$E\{\hat{C}_i\} = E\left\{\frac{1}{N}\sum_{n=1}^{N}C_i(n)\right\} = E\left\{\frac{1}{N}\sum_{n=1}^{N}\frac{Y_1(n)}{Y_i(n)}\right\}$$

(56.9)

Table 56.1 MATLAB simulation results

SNR (dB)	Normalized digital frequency f	General algorithm variance $D\{\hat{C}\}$	Simplified algorithm variance $D\{\hat{\hat{C}}\}$
35	0.00001	0.0027	0.0027
35	0.0001	0.0027	0.0027
35	0.001	0.0027	0.0039
35	0.01	0.0027	0.0067
45	0.00001	0.0009	0.0009
45	0.0001	0.0009	0.0009
45	0.001	0.0009	0.0013
45	0.01	0.0009	0.0022

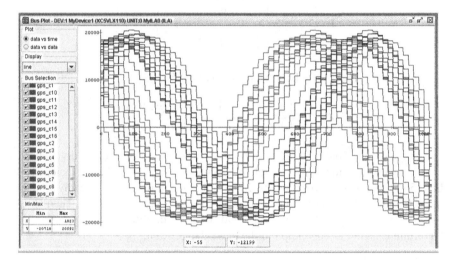

Fig. 56.4 I/Q branches data before calibration

Experiments results are shown in Table 56.1. It is obvious from Table 56.1 that both algorithms almost have the same variance through all different SNRs when normalized digital frequency is less than 0.0001. That means the two algorithms can acquire the same performance under certain conditions.

56.3.2 Implementation Results

We chose simplified algorithm to implement channel calibration coefficients calculation. The main steps are stated as follows, firstly utilize Eq. (55.6) to implement calculation of I/Q branches average separately in FPGA, which realizes division by shift bits, it can effectively reduce the resource costs and improve operation speed. Secondly FPGA transmits I/Q branches average data to DSP

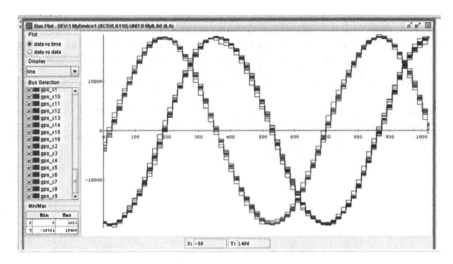

Fig. 56.5 I/Q branches data after calibration

through link-port, where DSP performs 32 bit complex data division to estimate the calibration coefficients. Finally DSP transmits back to FPGA, which stores and realizes channels calibration in time.

Experiments results acquired by FPGA are shown in Figs. 56.4 and 56.5. We have done forty times tests, the signal phase errors after calibration in each test are less than ±3° and amplitude relative errors are less than 0.01. The system is of high performance.

56.4 Conclusion

A simplified method of calibrating channels inconsistency has been proposed, which can significantly reduce the resource costs and improve operation speed. An efficient implementation of it in practical project based on FPGA and DSP has been presented as well. Experiments results in project and MATLAB indicate that this channels calibration method is of low costs, high accuracy and great performance. It is of high practical value.

References

1. Gu J, Liang G, Gong W et al (2009) Channel calibration of amplitude-phase errors of DBF transmitter based on maximum likelihood estimation. International conference on networks security wireless communications and trusted computing (April, Wuhan), 2:2378–2381
2. Oodo M, Miura R (1999) A remote calibration for DBF transmitting array antennas by using synchronous orthogonal codes. Antennas and propagation society international symposium IEEE (Jul, Orlando), 2:1428–1431

3. Yan W, Xu SJ (2003) Mutual coupling calibration of DBF array with combined optimization method. IEEE Trans Antennas Propag 51:2947–2952 Oct
4. Nuteson TW The Aerospace and Chantilly et al (2002) Digital beam forming and calibration for smart antennas using real-time FPGA processing. Microwave symposium digest, 2002 IEEE MTT-S International (Jun, Seattle), 1:307–310
5. Qu WJ, Yang C (1992) DBF system channel calibration technology. J UEST China 21:593–601 (n Chinese)
6. Chen L, Zuo Y, Zhu L et al (2006) FPGA based calibration method for channel imbalance of array antenna. Electron Opt Control 13:1671–1673 (in Chinese)
7. Wang HJ, Wei G (2008) Channel inconsistent calibration algorithm for array antenna. Comput Eng Appl 44(27):16–17 (in Chinese)

Chapter 57
Parameter Estimation of Chirp Signals Based on FRFT Under the Low SNR

Wei Yang and Yaowu Shi

Abstract We present a new method of estimating the Chirp signal parameter based on fractional Fourier transform (FRFT). The slope of the Chirp single is firstly estimated using FRFT according to its expectation. By redefining the variable meaning of the FRFT and the improved FRFT, the time-delay of multiple Chirp signals is estimated. The validity of the methods is demonstrated through simulations, and the proposed methods can achieve both high accuracy and less calculation.

Keywords Fractional fourier transform · The expectation · Multiple chirp signals

57.1 Introduction

Detection and parameter estimation are significant issues in digital signal processing [1] such as radar [2], underwater acoustics [3], speech processing [4], and array processing [5], etc. However, most of the existing detection and estimation methods [6, 7] are based on the maximum likelihood estimation (MLE) principle.

This work is supported in part by the National Natural Science Foundation of China under the Grant 51075175

W. Yang (✉) · Y. Shi
School of Communication Engineering, Jilin University, Changchun, China
e-mail: weiyang09@mails.jlu.edu.cn

Y. Shi
e-mail: shiyw@jlu.edu.cn

W. Lu et al. (eds.), *Proceedings of the 2012 International Conference on Information Technology and Software Engineering*, Lecture Notes in Electrical Engineering 210, DOI: 10.1007/978-3-642-34528-9_57, © Springer-Verlag Berlin Heidelberg 2013

The MLE-based methods achieve Crame-Rao low bound, but it suffers from the huge computational burden. To improve the methods, the time–frequency analysis has also attracted considerable attention, by which many problems are studied jointly in time–frequency plane, rather than separately. In common case, the Wigner-Ville distribution (WVD) [8, 9] is used to estimate the instantaneous frequency. When dealing with multicomponent, the cross terms caused by WVD make it difficult to interpret WVD image. The fractional Fourier transform (FRFT) [10] can reduce the computational burden by converting the two dimension problem to one dimension problem. Moreover, it is better to centralize the energy of Chirp signals in different angles and can directly estimate the parameter of Chirp signals.

In practise, the FRFT methods are always confronted with deciding the value of the rotation angle, which can reduce to one dimension problem instead of time–frequency plane. In this paper, two-fold works are employed to bypass this drawback: (1) combining the FRFT with the expectation form, the estimation value is obtained by searching the maximum power; (2) fixing the angle at the certain value so as to easily find the maximum energy of the multiple Chirp signals. Due to the expression of the FRFT kernel, we obtain a polynomial for the Chirp signals with less calculation through simplifying the equations. We present the analysis of output SNR, and simulations verify the validity of our proposed method.

57.2 Parameter Estimation with the FRFT

57.2.1 The FRFT of Chirp signals like Expectation

Given a Chirp signal $f(t)$, the FRFT is defined as [11]

$$F_\alpha(u) = \int_{-\infty}^{\infty} f(t)K(u,t)dt = A_\alpha \int_{-\infty}^{\infty} f(t)\, e^{j\frac{u^2+t^2}{2}\cot\alpha - jut\csc\alpha}dt, \alpha \neq 2n\pi, 2n\pi + \pi,$$

(57.1)

where $K(u,t)$ is the kernel, and $A_\alpha = \sqrt{(1 - j\cot\alpha)/2\pi}$. Moreover, when $\alpha = 2n\pi$ and $\alpha = 2n\pi + \pi$, $F_\alpha(u)$ equal to $f(t)$ and $f(-t)$, respectively. The FRFT of the finite Chirp signals is defined as

$$F_\alpha(u) = A_\alpha \int_0^T e^{\pi jkt^2 + 2\pi ft}e^{j\frac{u^2+t^2}{2}\cot\alpha - jut\csc\alpha}dt,$$

(57.2)

Where T is the Chirp time duration, B is the Chirps bandwidth, $k = B/T$ is the Chirp rate, and f is the carrier frequency. After simplifying, $F_\alpha(u)$ is expressed as

$$F_\alpha(u) = A_\alpha e^{j\frac{u^2}{2}\cot\alpha} \int_0^T e^{j(\pi k + (\cot\alpha)/2)t^2} e^{(2\pi f - ju\csc\alpha)t} dt. \tag{57.3}$$

From (57.3), we consider the definition of Chirp signal expectation

$$E(t) = \int_0^T e^{j\pi k t^2} e^{2\pi f t} dt, \tag{57.4}$$

where (57.4) is convergation, $\int_0^T e^{j\pi k t^2} e^{2\pi f t} dt \le \int_0^T 1 dt$, meanly $E(t) \le T$. Only when k and f equal to 0, we can obtain $E(t) = T$. Similarly, when $\pi k + \cot\alpha/2 = 0$, and $2\pi f - ju\csc\alpha = 0$, (57.3) is simplified to

$$F_\alpha(u) = T\sqrt{\frac{1 - j\cot\alpha}{2\pi}} \; e^{j\frac{u^2}{2}\cot\alpha}. \tag{57.5}$$

Substituting $\cot\alpha = -2\pi k$ into (57.5), we can obtain

$$F_\alpha(u) = Te^{-ju^2\pi A}\sqrt{\frac{1}{2\pi} + jA}. \tag{57.6}$$

The Chirp rate can be estimated through (57.6), once the maximum of $F_\alpha(u)$ is obtained by searching for the maximum value of Chirp amplitude. From the upper procedure, we can see this method is very easy to be put into practice, and the time delay estimation in the next subsection is based on the prior information.

57.2.2 The FRFT of Multiple Chirp Signals with Different Time Delay

In some situations, it is a big issue to analyze the time-delay of multiple Chirp signals because the received signals are always delayed. Here, the FRFT is used to analyze the time-delay of multiple Chirp signals. The multiple component Chirp signals are given by

$$s(t - \tau) = \sum_{i=0}^{N} e^{j2\pi(f_i(t-\tau_i) + \frac{1}{2}m_i(t-\tau_i)^2)}, \tag{57.7}$$

where f_i and m_i are the initial frequency and slope of the i-th corresponding Chirp signal. So the FRFT of the delayed multiple Chirp signals is written as

$$F_\alpha(u) = \int_0^T A_\alpha \left(\sum_{i=0}^{N} e^{j2\pi(f_i(t-\tau_i) + \frac{1}{2}m_i(t-\tau_i)^2)} \right) e^{j\frac{u^2 + t^2}{2}\cot\alpha - jut\csc\alpha} dt, \tag{57.8}$$

According to the time–frequency distribution of Chirp signal and the property of the linear transform, we can get the k_i definition, expressed as

$$k_i = -\cot\alpha_i. \tag{57.9}$$

Using $u = \tau$, (57.8) is denoted as Use a long dash rather than a hyphen for a minus sign. Punctuate equations with commas or periods when they are part of a sentence, as in

$$\sum_{i=0}^{N} F_{\alpha_i}(\tau) = \sum_{i=0}^{N} \left[\int_0^T A_{\alpha_i} e^{j2\pi(f_i(t-\tau_i) - \frac{1}{2}\cot\alpha_i(t-\tau_i)^2)} e^{j\frac{t^2+\tau^2}{2}\cot\alpha_i - j\tau\tau\csc\alpha_i} dt \right].$$

(57.10)

In (57.10), the time-delay parameter of the Chirp signals could be determined by

$$\tau = \sum_{i=0}^{N} \max|F_{\alpha_i}(\tau)|.$$

(57.11)

Finally, we can obtain the time-delay using (57.11). The significant advantage of this method is that the energy of $F_{\alpha_i}(\tau)$ is focused by the Chirp rate, which is feasible to estimate the time-delay.

57.2.3 The Signal to Noise Ratio Analysis

The SNR of output multiple Chirp signals in the FRFT can be given as [12]

$$SNR_{out} = \frac{|F_\alpha(\tau)|^2}{P_{noise}},$$

(57.12)

where $F_\alpha(\tau)$ is the FRFT of the multiple Chirp signals $s(t - \tau)$. P_{noise} is the power of the additive Gaussian noise $n(t) = A \exp\left[-j(t - \mu)^2 / \sigma^2\right]$, and also equals to the power of the Gausssion noise in the FRFT domain according to the Parseval theorem. Taken the i-th Chirp signal as illustration, the $|F_{\alpha_i}(\tau)|^2$ in the FRFT plane is

$$F_{\alpha_i}(\tau) = \int_0^T A_{\alpha_i} e^{j2\pi(f_i(t-\tau_i) - \frac{1}{2}\cot\alpha_i(t-\tau_i)^2)} e^{j\frac{t^2+\tau^2}{2}\cot\alpha_i - j\tau\tau\csc\alpha_i} dt.$$

(57.13)

Due to $m_i = -\cot\alpha_i$ and Talor series, (57.13) is simplified as

$$F_{\alpha_i}(\tau) = \frac{A_\alpha}{2\sqrt{j\cot\alpha_i(\frac{1}{2} - \pi)}} \exp\left[-j2\pi f_i\tau_i + j\frac{\tau_i^2}{2}\cot\alpha_i - j\pi\tau_i^2\cot\alpha_i + \frac{(2\pi f_i + 2\pi\tau_i\cot\alpha_i - \tau_i\csc\alpha_i)^2}{2jC\cot\alpha_i}\right]$$
$$\cdot \sqrt{-\pi}\left\{-erf\left[\frac{-2\pi f_i - 2\pi\tau_i\cot\alpha_i + \tau_i\csc\alpha_i - C\cot\alpha_i T}{\sqrt{2jC\cot\alpha_i}}\right] + erf\left[\frac{(-2\pi f_i - 2\pi\tau_i\cot\alpha_i + \tau_i\csc\alpha_i)}{\sqrt{2jC\cot\alpha_i}}\right]\right\}.$$

(57.14)

where $C = 1 - 2\pi$. In the original time–frequency plane, the signal power is expressed as

$$F(\tau) = \int_0^T e^{j2\pi(f_i(t-\tau_0)-\frac{1}{2}\cot\alpha(t-\tau_0)^2)} dt$$

$$= \frac{\exp\left(j\pi\tan\alpha f_i^2\right)}{2\sqrt{j}\cot\alpha} \cdot \left[-erf\left(\frac{-\pi f_i - \pi\tau_0\cot\alpha + \pi T\cot\alpha}{\sqrt{-j\pi\cot\alpha}}\right) \right. \tag{57.15}$$

$$\left. +erf\left(-\frac{\pi f_i + \pi\cot\alpha\tau_0}{\sqrt{-j\pi\cot\alpha}}\right)\right]$$

While $erf(x)$ is non-elementary function, we represent $F_\alpha(\tau)/F(\tau)$ using Taylor series, given by

$$e = \frac{T^3(-C^2\cot^2\alpha) + 3T^2C\cot\alpha(-2\pi f - 2\pi\tau\cot\alpha + \tau\cot\alpha) + 6jCT\cot\alpha - 3T(-2\pi f - 2\pi\tau\cot\alpha + \tau\cot\alpha)^2}{T^3(\pi^2\cot^2\alpha) + 3T^2\pi\cot\alpha(-\pi f - \pi\tau\cot\alpha) + 3jT\pi\cot\alpha + 3T(-\pi f - \pi\tau\cot\alpha)^2}$$

$$\tag{57.16}$$

From Eq. (57.16), we can obtain the output SNR in the FRFT is dependent on τ, α. The output SNR does not increase substantially with T, which one can know that our method is very effective.

57.3 Simulation Results

In this section, we evaluate the performance of the proposed method by the slope of multiple Chirp signals. The simulation of two Chirp signals with the same time delay is implemented by considering of $f_1 = 6$, $f_2 = 0.6$, $m_1 = -\cot 30°$ and $m_2 = \cot 30°$. At $\tau = 0.02$, the simulation results in noiseless and zero mean Gaussian noise are shown in Fig. 57.1a and b, respectively. One can see that the two figures show very similar profiles and the sharp peaks of the FRFT is at the correct angle, which means our method can estimate the time delay successfully.

In the simulation of the multiple Chirp signals with different time delays, the parameters of two corresponding Chirps signals are $f_1 = -0.8$ Hz, $m_1 = -\cot 20°$, and $f_2 = 0.56$ Hz, $m_2 = -\cot(-30°)$, Under $SNR = 0$ dB, the estimation result is $\tau_1 = 0.201$ and $\tau_2 = 0.0396$. The FRFT of the delayed multiple Chirp signals at $\tau = 0.02$ with zero mean Gaussian noise and without noise are shown in Fig. 57.2a and b, respectively. From Fig. 57.2a, one can exactly estimate the value of the time delay. Compared to Fig. 57.2a, under the condition of $SNR = -5$ dB (Fig. 57.2b), we can still obtain the delayed time, no matter its value already slightly deviate from the real value.

Both Figs. 57.1 and 57.2 show that the algorithm works well and the time-delay estimation is not much affected by the Gaussian noise. The high accuracy of the estimation indicates that the proposed the time-delay estimation of the FRFT is effective.

Fig. 57.1 FRFT of the single
signal with same time delay.
a Without noise; **b** with zero
mean Gaussian noise

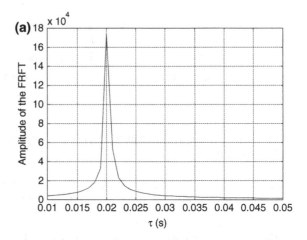

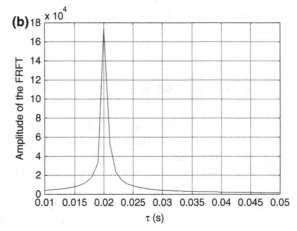

In Fig. 57.3, the Root mean square error (RMSE) [13] of the time delay is evaluated in 100 Monte Carlo runs, under the background of the zero mean independent Gaussian noise. $RMSE = \sqrt{\frac{1}{N}\sum_{i=1}^{N}\left(\hat{\theta}_i - \theta\right)^2}$, $\hat{\theta}_i$, is the estimated value, θ is the real value. For comparison of the performance, Fig. 57.3 also shows the comparison with ML and CRLB. At the low SNR range, our method show better accuracy because of the special angle. With the increasing of SNR, the three methods are approaching identical tendency. Therefore, our method is very applicable to estimate the time-delay.

Fig. 57.2 FRFT of the
multiple signals with different
time delay. **a** Without noise;
b with zero mean Gaussian
noise

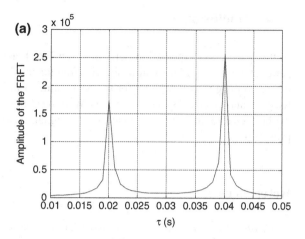

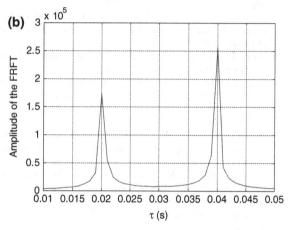

Fig. 57.3 RMSE of the SNR

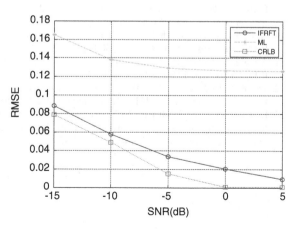

57.4 Conclusion

In this paper, the expectation definition and FRFT are connected to estimate the parameter of the single Chirp signal, simultaneously. The value of time-delay estimation is got by searching for the peak value of the FRFT. The method is very feasible and needs less computation. Moreover, the very high accuracy of our method enable its wide applications in signal analysis, especially geological prospecting.

References

1. Cadzow JA (1982) Spectral estimation: an over determined rational model equation approach. IEEE Process pp 907–939
2. Guohua W, Siliang W, Mao E (2003) Differential evolution for target motion parameter estimation. In: Proceeding of the 2003 international conference on neural network and signal processing, pp 563–566
3. Prior MK (2009) Estimation of K-distribution shape parameter from sonar data: sample size limitations. IEEE J Oceanic Eng 34(1):45–50
4. Chan Y, Riley J, Plant J (1980) A parameter estimation approach to time-delay estimation and signal detection. IEEE Trans Acoust Speech Signal Process 8–16
5. Krim H, Viberg M (1996) Two decades of array signal processing research. IEEE Signal Process Mag 13(4):67–94
6. Abatzoglou T (1986) Fast maximum likelihood joint estimation of frequency and frequency rate. In acoustics, speech, and signal processing, IEEE international conference on ICASSP '86
7. Besson O, Ghogho M, Swami A (1999) Parameter estimation for random amplitude chirp signals. IEEE Trans Signal Process 47(12):3208–3219
8. Kay S, Boudreaux-Bartels G (1985) On the optimality of the Wigner distribution for detection. In: IEEE international conference on acoustics, speech, and signal processing, ICASSP '85
9. Barbarossa S (1995) Analysis of multicomponent LFM signals by a combined Wigner-Hough transform. IEEE Trans Signal Process 43(6):1511–1515
10. Almeida LB (1994) The fractional Fourier transform and time-frequency representations. IEEE Trans Signal Process 42(11):3084–3091
11. Almeida LB (1994) The fractional fourier transform and time-frequency representations. IEEE Trans Signal Process 11:3084–3091
12. Tao Ran, Li X-M, Li Y-L, Wang Y (2009) Time-delay estimation of chirp signals in the fractional fourier domain. IEEE Trans Signal Process 7:2852–2855
13. De Clercq EM, De Wulf RR (2007) Estimating fuzzy membership function based on RMSE for the positional accuracy of historical maps. International workshop on the analysis of multi-temporal remote sensing images, pp 1–5

Chapter 58
The Study on Frequency Agility in Radar Combining Pulse-to-Pulse Frequency Agility and Doppler Processing

Xie Feng, Liu Feng and Ye Xiaohui

Abstract In this paper, an OFDM pulse burst waveform is investigated that enables combining pulse-to-pulse frequency agility and Doppler processing. The number of subcarriers for the baseband OFDM has been evaluated first. The use of a single subcarrier gives the best resolution performance at the cost of very poor spectrum occupancy. The extreme case of a maximum number of subcarriers is a good trade-off. It has been used to analyze three patterns for frequency agility.

Keywords OFDM · Frequency agility · Doppler

58.1 Introduction

Pulse-to-pulse frequency agility or frequency hopping forces the jammer to spread its power over a wide band-width. It reduces the jammer power density the radar has to face compared with what it would be if all the jammer power were concentrated within the narrow bandwidth of a fixed-frequency radar receiver [1]. Unfortunately, not only the RF analog front end suffers from more complexity but also conventional Doppler processing becomes impossible [1, 2]. Therefore either non-trivial technique, other than discrete fourier transform (DFT) is needed or the concept of frequency agility should be reviewed to trade off the use of multiple frequencies and the opportunity for Doppler processing. In this regard, some compromises have already been pro-posed. The idea of overlapping bursts, i.e. a burst not from adjacent pulses with different RF frequencies, but from pulses with

X. Feng (✉) · L. Feng · Y. Xiaohui
School of Electronic Engineering, Naval University of Engineering, 430033 Wuhan, China
e-mail: whxiefeng@163.com

W. Lu et al. (eds.), *Proceedings of the 2012 International Conference on Information Technology and Software Engineering*, Lecture Notes in Electrical Engineering 210, DOI: 10.1007/978-3-642-34528-9_58, © Springer-Verlag Berlin Heidelberg 2013

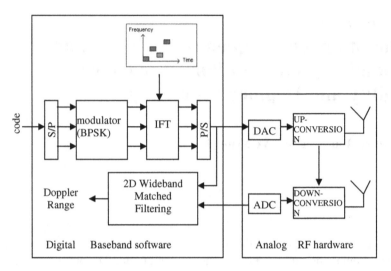

Fig. 58.1 OFDM agile radar transceiver block-scheme

the same RF has been implemented in Eagle [3]. A different technique based on burst-to-burst agility has been used in Senrad to perform Doppler processing [4].

In the past, most of the effort put on Multi-carrier (MC) signals was devoted to serve the communication community [5]. Recently, the radar community has also started to consider them as the possible new waveform generation. In [6, 7] performances achieved with the so-called Multi-carrier phase-coded (MCPC) signals are widely studied.

In this paper, MC waveforms are studied that can easily exhibit wideband characteristics, desirable for high-resolution radar performance yet at the same time, provide for ease of system design and data processing. Their remarkable structure, essentially a linear combination of narrowband signals makes the opportunity for RF agility straightforward. Indeed, selection of the desired sub-bands at any instant of time can be controlled digitally in the transmitter at the IFFT block, referring to the transceiver block-scheme in Fig. 58.1. Similarly, they can mitigate narrowband jamming/interference by simply turning off certain sub-bands. The original OFDM scheme that allows controlling the transmission of multiple frequencies simultaneously could enable both concepts of pulse-to-pulse agility and Doppler processing to coexist in the same system. To the authors' best knowledge this approach has not been proposed earlier.

In Sect. 58.2, our model for the simulations is described. It tackles the waveform design and introduces the number of subcarriers as a key parameter. Hints for adapted Doppler processing are also given. Results of the simulations are presented in Sect. 58.3. The time/frequency structure of the OFDM pulse when no frequency agility is present is inspected first, before different patterns for frequency agility are included. Preliminary results on Doppler processing are finally introduced.

Table 58.1 Basic specifications that determine the waveform design for short-range radar

Radar carrier frequency, f_c	10	(GHz)
Minimum speed, v_{min}	0.5	(m.s^{-1})
Minimum range, d_{min}	150	(m)
Maximum unambiguous speed, v_{unmax}	150	(m.s^{-1})
Maximum unambiguous range, d_{unmax}	15	(km)
Bandwidth of the signal, B_w	300	(MHz)
Range resolution, δR	0.5	(m)
Maximum speed, v_{max}	300	(m.s^{-1})

58.2 Simulation Model

The radar waveform considered is a pulse burst or a train of pulses, and the target model is a point target. The propagation channel is the perfect channel with no multiple paths. The Doppler Effect, observed when an electromagnetic wave at a frequency f_c is reflected by a moving target fleeing the radar with a constant radial velocity $v_r \ll c$ implies a frequency shift (e.g. [1]):

$$f_{Doppler} = \frac{-2v_r f_c}{c} \tag{58.1}$$

where c is the speed of light. Basic specifications that determine the waveform design for short-range radar are defined in Table 58.1. (From now on, absolute values are considered for the speed and Doppler, i.e. $v \equiv |v|$).

The minimum detectable speed v_{min} implies the Doppler resolution, i.e. a minimum Doppler shift, f_{Dmin}. Accordingly, the observation time T_{obs}, i.e. the duration of the burst, is given by:

$$T_{obs} \geq \frac{1}{f_{Dmin}} = \frac{c}{2v_{min}f_c} = 30mc \tag{58.2}$$

The minimum detected range d_{min} implies the pulse duration T_{pulse} given by:

$$T_{pulse} \leq \frac{2d_{min}}{c} = 1\mu s \tag{58.3}$$

The maximum unambiguous speed v_{unmax} and the maximum unambiguous range d_{unmax} limit the pulse repetition time (PRT), as follows:

$$\frac{2d_{unmax}}{c} \leq PRT \leq \frac{c}{2v_{unmax}f_c} \tag{58.4}$$

With the values presented in Table 58.1, (58.4) enables PRT to take one single value: 100 µs. If another value is selected then, there will be ambiguities either in the speed or in the range domain. Especially, our scenario claims a maximum speed $v_{max} = 300\,m\,s^{-1} > v_{unmax}$ therefore ambiguities in the speed domain will occur. Figure 58.2 gathers all these waveform parameters in a graph.

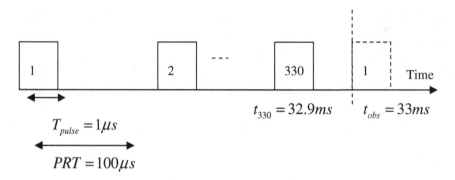

Fig. 58.2 Pulse burst

The ambiguity function (AF) is used to assess the performance of radar signals. Although the narrowband approximation suffices our scenario with low speeds, its wideband version is considered for the sake of completeness and future work [8]. It correlates a reference signal x (t, 1) with the wideband model of the received signal x (t−τ, s), i.e. a delayed and scaled version of the reference signal, as follows (e.g. [9]):

$$AF(\tau, s) = \int\limits_{-\infty}^{\infty} x(t,1) \cdot x * (t - \tau, s) \cdot \frac{1}{\sqrt{s}} dt \qquad (58.5)$$

where x(t, s) is given later in Eq. (58.10) and s is the scaling factor (caused by the Doppler effect) given by:

$$s = \frac{c + v_r}{c - v_r} \approx 1 + 2\frac{v_r}{c} \qquad (58.6)$$
$$v_r \ll c$$

The above simplification is true in radar but not in sonar where the propagation speed c is much lower.

The AF is based on the matched filter, and actually applies the echo reflected from a point target. At the end, the capability of resolving targets along the two dimensions, Doppler (scale) and range (delay) is derived.

It is obvious in Eq. (58.1), that the Doppler frequency shift has a linear dependency with the carrier frequency f_c. Therefore, a different frequency shift $\Delta f_{Doppler}$ appears for a different f_c. However, the difference in $\Delta f_{Doppler}$ is not visible if the signal is observed over a time interval $\Delta t \ll 1/\left|\Delta f_{Doppler}\right|$.

In the narrowband assumption we ignore this difference and consider that $\Delta f_{Doppler}$ is the same for all frequencies. Calling f_{min} and f_{max} the minimum and maximum frequencies, respectively, of the transmitted signal, the difference in frequency shift is:

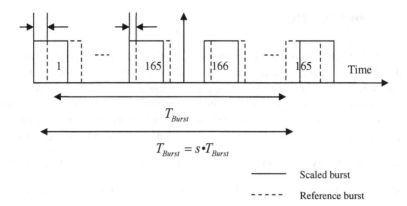

Fig. 58.3 Pulse burst with the scaling effect and no delay (the target flees the radar), $v_r = 150 \text{ m s}^{-1}$ and $s-1 = 10^{-6}$

$$\Delta f_{Doppler} \leq (2v_r/c) \cdot (f_{max} - f_{min}) \tag{58.7}$$

With the maximum target speed v_{max}

$$\Delta f_{Doppler} \leq (2v_{max}/c) g B_w \tag{58.8}$$

Where $B_w = f_{max} - f_{min}$ is the bandwidth of the signal. Hence, the narrowband hypothesis is valid for

$$\Delta t \ll c/(2v_{max}B_w) \tag{58.9}$$

The bandwidth and maximum speed in our scenario (Table 58.1) give $\Delta t \ll 1.7 \text{ ms}$

Obviously, the narrowband assumption is violated during the whole pulse burst but not during one pulse [9]. Thus, the Doppler scaling will result in a stretching or compressing of the PRT but its effect will be negligible on T_{pulse} that is short enough.

With the reference time taken at half the pulse burst, Fig. 58.3 shows and exaggerates the effect of scaling on PRT, while T_{pulse} is left unchanged.

Even though the resolution analysis will focus on small Doppler shifts where the narrowband AF is a good approximation, the hereby concept [9], also true with high velocities, will be preferred. The reference signal in our model is given by:

$$x(t,s) = \sum_{m=-N_{pulse}/2}^{N_{pulse}/2-1} u_m \left(t - (2m+1) \cdot s \cdot \frac{PRT}{2} \right) \tag{58.10}$$

where $N_{pulse} = 330$ and fsu_m is an OFDM waveform.

58.3 Simulation Results

The OFDM design options are evaluated first. In the second step, the best option is used and the resolution performance of three distinct patterns for frequency agility is assessed. Preliminary results on Doppler processing are also provided.

To compute the speed resolution, the narrowband approximation can be used since we look at the AF around the central peak, where the speed is low. At zero-delay, the received echo is $e^{j\omega_D t} \cdot x(t, 1)$. Adjacent pulses being separated by PRT imply the ambiguity function at zero-delay to be proportional to $\sum_{p=0}^{P-1} (e^{j\omega_D t PRT})^p$, where P is the number of pulses. One recognizes a geometric series, which simplifies into:

$$AF_{zero-delay} = \frac{\sin(\pi f_D T_{obs})}{\sin(\pi f_D PRT)}$$

The first null is for $f_D = 1/T_{obs}$. Referring to Eq. (58.1), it corresponds to $v_r = c/(2f_c T_{obs})$. The expected resolution in the scale domain is therefore:

$$(s-1)_{res} = 1/(f_c T_{obs})$$

When there is only one carrier, we only see the effect of phase coding. With a delay larger than the symbol duration Ts, the symbols we match are not correlated and the AF is close to zero. When the delay is smaller than Ts, only part of the symbol is matched with the same symbol in the reference sequence. The duration of this part is equal to $T_s - |\tau|$ and we expect the AF to be proportional to this value. Thus the shape of the AF is a triangle and the range resolution is $T_s = 1/B_w$.

If an OFDM baseband signal b (t) is transmitted $b(t) = \sum_{k=1}^{K} c_k e^{j2\pi kt/T_s}$, the autocorrelation function is

$$ACF(\tau) = \int \sum c_k(t)c_k \times (t+\tau)e^{-j2\pi kt/T_s} dt$$

For small delays τ, compared to T_s, ACF (τ) is simply the sum of a geometric sequence:

$$\sum e^{-j2\pi k\tau/T_s} = \frac{\sin(\pi K\tau/T_s)}{\sin(\pi \tau/T_s)}$$

The first null is for $\tau = T_s/K$ where K is still the number of subcarriers.

Because of the shape of the sinc (wide mainbeam) the range resolution is $\tau = 1.2 \cdot T_s/K = 1.2/B_w$. It is slightly larger than for the single carrier case but again it does not depend on the symbol duration T_s.

Note that when the delay is longer than T_s, in the intermediate case, the effect of the phase coding will appear and lower the sidelobes as seen in the zoom window in Fig. 58.4.

Fig. 58.4 AF for intermediate case with K = 10 and L = 30

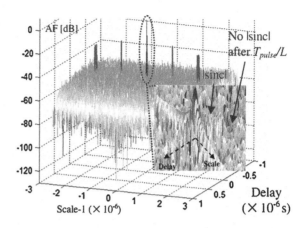

58.4 Conclusion

The random spread subcarriers pattern provides the best AF in terms of sidelobe levels and speed ambiguities compared to the random grouped subcarriers and repeated Costas grouped subcarriers. However, it does not benefit from any instantaneous bandwidth reduction as the other two cases. Finally, it was shown that OFDM pulse burst supports a Doppler processing per subcarrier when the Doppler is low. The Doppler spectrum of the estimate gains in SNR.

References

1. Skolnik MI (2010) Introduction to radar systems. Mc Graw-Hill, New York
2. Edde B (2009) Radar: principles, technology, applications. Prentice Hall PTR, Upper Saddle River
3. Oderland I, Nordlöf S, Leijon B (1990) EAGLE: high accuracy 35 GHz tracking radar. In: Proceedings of RADARCON 1990, pp 461–466
4. Skolnik MI, Linde G, Meads K (2010) Senrad: an advanced wideband air-surveillance radar. IEEE Trans Aerosp Electron Syst 37(4):1163–1175
5. Hara S, Prasad R (2003) Multicarrier techniques for 4G communications. Artech House, London
6. Levanon N, Mozeson E () Radar Signals. John Wiley & Sons, Inc., 2004
7. Levanon N, Mozeson E (2002) Multicarrier radar signal-pulse train and CW. IEEE Trans Aerosp Electron Syst 38(2):707–720
8. Lellouch G, Pribic R, van Genderen P (2008) Wideband OFDM pulse burst and its capabilities for doppler processing in radar. Submitted to RADAR2008, Adelaide, Sept 2008
9. Rihaczek AW (2010) Inc principles of high-resolution radar. Artech House, London

Chapter 59
An Improved Collision Avoidance Method for IEEE 802.11 DCF with Information-Sharing-List

Weizhen Tian, Mingrui Guo and Zhixin Li

Abstract In wireless networks, stations need to share the limited bandwidth of the wireless channel with each other, inevitably leading to packets collisions. To enhance DCF, we propose a new collision avoidance (CA) method based on information sharing list (ISL). During the RTS/CTS exchange, the backoff value piggybacked in CTS (new CTS) frame will be broadcasted to other stations, stored in their lists in order to avoid the re-use of the same values in next backoff stage. The results show that the improved CA mechanism can acquire better performance in throughput and delay than legacy DCF as the network load stabilizes.

Keywords CSMA/CA · ISL · Backoff · Contention window · CTS

59.1 Introduction

With the development of mobile communication technology, the Wireless Local Area Network (WLAN) has been widely used. From 1997 to date, IEEE 802.11 working group has released a series of wireless LAN standards, providing physical layer and Medium Access Control (MAC) layer technology about wireless LAN. MAC layer access mechanism includes Distributed Coordination Function (DCF) and Point Coordination Function (PCF). PCF is adopted in infrastructure networks

W. Tian (✉) · M. Guo · Z. Li
Beijing Institute of Technology, No. 5 Yard, Zhong Guan Cun South Street, 4#413 Room, 4# Teaching Building, Haidian District, Beijing, China
e-mail: twz-flyaway@163.com

W. Tian
Department of Command Tactical Information System, Xi'an Commun. Institute, Xi'an, China

W. Lu et al. (eds.), *Proceedings of the 2012 International Conference on Information Technology and Software Engineering*, Lecture Notes in Electrical Engineering 210, DOI: 10.1007/978-3-642-34528-9_59, © Springer-Verlag Berlin Heidelberg 2013

and uses polling mechanism. DCF is used in ad-hoc networks and operates by Carrier Sense Multiple Access with Collision Avoidance (CSMA/CA) which is composed of physical and virtual carrier sense and Binary Exponential Backoff (BEB) [1].

In CSMA/CA mechanism, the stations tending to send packets need to compete with others before capturing the channel successfully. Therefore, collisions will increase with the increase of station numbers. In order to improve the backoff mechanism, a great amount of work has been done. Multiplicative Increase Linear Decrease (MILD) proposed in [2] can improve the fairness of nodes, but linear decrease is too conservative, especially when the number of active stations is small. In [3–6], Contention Window (CW) adjusts as the number of stations estimated by acquiring information of collision probability or cross-layer packet-switching, but the realization is very complex. The study above can bring about some improvements although, but the collision problem caused by different stations selecting the same random number in one competing stage still exists. In [7], newly-joined stations select a larger random value between 0 and CW to avoid collision with old stations. In order to minimize the collisions, this paper proposes an improved CSMA/CA based on Information Sharing List (ISL), stations that have packets to send will select a random value that is not equal to the values in its own ISL. As the network running, the information in the lists will cover all the stations within the scope, the collisions will seldom occur.

59.2 IEEE 802.11 DCF

The standard defines a basic two-way handshaking and an optional four-way handshaking medium access mechanism, the latter is only used in the case that an (MAC Service Data Unit) MSDU size exceeds the (MAC Protocol Data Unit) MPDU payload size. The four-way handshaking mechanism adds (Request to Send/Clear to Send) RTS/CTS before DATA/ACK to achieve channel reservation. RTS and CTS frames both consists a Duration field that denotes the period of the time the channel remain busy.

If a station has packet to send and finds out the channel is busy after a DIFS, the station will activate the backoff mechanism and select a random number in [0, CW_{min}], by this way, the backoff time before access the channel is Backoff_Time:

$$Backoff\_Time = Random(CW) \times slot\_time \qquad (59.1)$$

The backoff timer is decreased as long as the medium is sensed as idle, frozen when a transmission is detected on the channel, resumed when the medium is sensed as idle again for more than a DIFS interval. Only when the backoff timer reaches zero, the station transmits its packet. If two or more stations select the same backoff random number at the beginning, they will complete the backoff at the same time; they transmit RTS simultaneously so that collision will occur. At

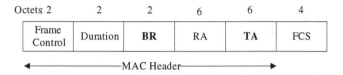

Octets 2	2	2	6	6	4
Frame Control	Duration	**BR**	RA	**TA**	FCS

◄─────────────────MAC Header─────────────────►

Fig. 59.1 CTS' frame format

this point, the CW of the two stations will both double and they must re-select a backoff number, the CW will increase exponentially until packets are transmitted successfully, but the CW has a maximum value CW_{max}.

59.3 Proposed Mechanism of CSMA/CA

59.3.1 CTS' Frame Format

The frame format for the CTS' is as defined in Fig. 59.1. The newly added BR filed occupying 2 bytes records the current remaining backoff value of the station. This article uses the 802.11b standard, the maximum CW 1,024 is small than 2^{16} and cannot lead to overflow. The TA field is the address of the STA transmitting this CTS' frame. The TA and BR filed will be written as a pair in the ISL of the stations that receive CTS'. With the advance of time slots, the BR value in the list will be decreased like the NAV. When the BR value declines to 0, it indicates the station with responding TA get the chance to transmit, then, the BR value will update after a new random number is selected. At the very beginning, the ISL may consists a few information in that not all stations have right to send CTS', but the lists will cover most of the active stations as the network status becomes stable.

59.3.2 Handshake Procedure

The improved CSMA/CA mechanism is illustrated by an example with the four-way handshaking. The update of ISL and the application of CTS' when used in random number selecting are displayed in Figs. 59.2, 59.3. We make some assumptions in this procedure: (1) the channel is idle with no transmission errors and no hidden terminal. (2) Stations always have packets to send. (3) The destination of packets is random. We also assume the example communication pairs are A to B, C to A. Figure 59.2 shows part of the procedure. The handshake details of s1 in Fig. 59.3 are elaborated below by dividing it into three stages.

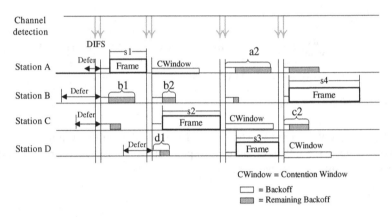

Fig. 59.2 Improved backoff procedure

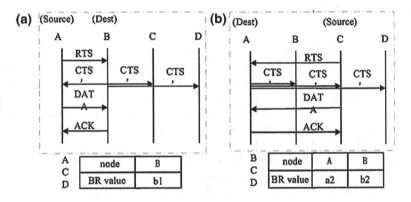

Fig. 59.3 Handshake between **a** A, B. **b** C, A and ISL of other stations

1. After a DIFS, four stations collide with each other while they send out RTS to reserve the channel at the same time. Then, they all select random numbers to begin backoff.
2. Station A selected the smallest random number, so it acquires the opportunity to send packets first. The destination address is B, so it sends RTS to B. After receiving RTS from A successfully, B replies A with CTS', other stations within the communication range will all receive CTS' and extract not only the duration value as NAV, but also TA and BR information that will be recorded in the ISL in each station, as shown in the right part of Fig. 59.2.
3. DATA/ACK handshaking between A and B is followed.

After s1, station D begins its activity and it will choose a random number other than b1 that is already in its ISL to avoid collision with B. A will select a new random number between 0 and CW_{min} to start the next round of channel acquiring.

After a while, station C's counter expires and will have a four-way handshake with destination A, as shown in Fig. 59.3. After this process, the lists of station B, C, D are all updated as shown in the right part of Fig. 59.3. After handshaking, station C will select random number that is not equal to a2 and b2 already in the list. The collisions caused by selecting the same value are eliminated successfully.

59.4 Performance Analysis

The factors that affect the network performance are station numbers, packet size, burst of traffic, channel conditions and etc. We evaluate the capacity and delay performance of the proposed CSMA/CA in this section.

59.4.1 Capacity

Suppose that all stations are active and always have packets to send. Capacity analysis is put forward by Cali in [8]. t_v is the average time required to transmit a packet under DCF basic mechanism. Then, the channel capacity ρ is defined as the ratio of the average packet length with the average time to send a packet:

$$\rho = \frac{\bar{l}}{t_v} = \frac{\bar{l}}{2 \times \tau + m + t_{SIFS} + t_{ACK} + t_{DIFS}} \tag{59.2}$$

$\bar{l}$ is the average packet length dependent on the traffic setting. τ is the propagation delay and m is the transmission delay. Under the RTS/CTS/DATA/ACK mechanism, the expression of t_v is revised as (59.3):

$$t_v = 2 \times \tau + m + 3 \times t_{SIFS} + t_{ACK} + t_{DIFS} + t_{RTS} + t_{CTS'} \tag{59.3}$$

After the additional information sharing mechanism, the handshaking mechanism is RTS/CTS'/DATA/ACK, so the capacity is modified in (59.4):

$$\rho' = \frac{\bar{l}}{t_v + \Delta t} \tag{59.4}$$

Δt is the time for transmitting the additional 8 bytes in CTS'. Although the overhead increases, it will have little impact on the overall system throughput and network capacity.

The analysis above does not take the collisions into account, from Cali's paper [8]; the formula should be Eq. (59.5) if considering collisions:

$$t'_v = E[N_c]\{E[Coll] + \tau + t_{DIFS}\} + E[Idle\_p] \cdot (E[N_c] + 1) + E[t_v] \tag{59.5}$$

Table 59.1 Parameter values used in the simulations

Parameter	Value	Parameter	Value
Packet payload	512, 4,348, 8,184 bits	Propagation delay	$1\mu s$
MAC header	224 bits	Slot Time	$20\mu s$
PHY header	192bits	SIFS	$10\mu s$
ACK	304bits	DIFS	$50\mu s$
RTS	352bits	CW_{min}	31
CTS	304bits	CW_{max}	1,023
Channel bit rate	1Mbps	Data rate	2 Mbps

N_c is the number of collisions within a transmission, the value can become smaller under the information sharing mechanism, especially when the network status becomes stable. The ideal value of t_v is smaller than t'_v. Then, we can conclude that the capacity of the proposed scheme increases in the condition that the average length of the packets remains the same.

59.4.2 Delay

We use Erlang C formula to analyze the delay in this paper.

The application of Erlang C needs some assumptions: (1) the data arrival rate follows Poisson distribution. (2) Packets that need to be sent out can wait without dropping except for collision. The basic formula of Erlang C is (59.6):

$$P(> 0) = \frac{\frac{A^N}{N!} \cdot \frac{N}{N-A}}{\sum_{x=0}^{N-1} \frac{A^x}{x!} + \frac{A^N}{N!} \cdot \frac{N}{N-A}} \tag{59.6}$$

A is Offered Traffic; the unit of A is erlang. N is the number of active stations in the network. P > 0 denote: probability of Delay >0. If used in CSMA/CA with above assumptions, (59.7) can be deduced as the average delay of the system. The estimation of N can be acquired through the statistics of ISL in each station.

$$D = P(> 0) \cdot \frac{t_v}{N - A} \tag{59.7}$$

59.5 Simulation Results

In this section, we use OPNET to evaluate the performance of proposed mechanism including global delay, throughput, retransmission attempts, and compare the results with legacy DCF, Table 59.1 shows some relevant parameters in the

Fig. 59.4 Global delay
versus number of stations

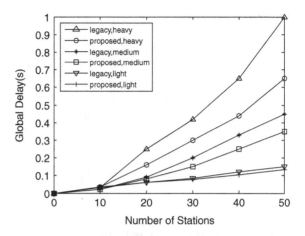

Fig. 59.5 Throughput versus
number of stations

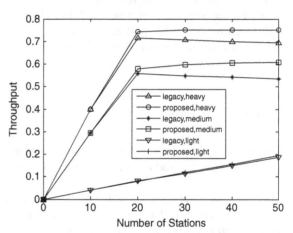

simulations. To get a comprehensive view of the proposed scheme, we do simulation with different payloads that denote light/medium/heavy traffic respectively.

Figure 59.4 shows the global delay of the network with different stations,
different schemes and different traffic sizes. The X-axis is the number of stations.
The delay becomes larger as station number increases, because of more stations
attempting to transmit that result in more collisions. Considering we have a fixed
station number, the result is that: when the traffic is heavy, the proposed scheme
performs better due to fewer collisions under the information sharing mechanism
between stations; and the effect is not so obvious under light traffic conditions.

We can have the following observations in Fig. 59.5. First, as the number of
stations increases, the throughput tends to be saturated if the traffic is medium or
heavy, but increasing all the time under light traffic. Second, the proposed scheme
can get a higher throughput (percentage of the bandwidth) than the legacy CSMA/
CA since it decreases collisions and retransmission and can get full utilization of
the channel.

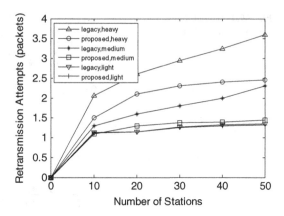

Fig. 59.6 Retransmission attempts versus number of station

Figure 59.6 depicts the retransmission attempts that per packet need to try before transmitting the packet successfully. The results in Figs. 59.5 and 59.6 are both based on the decrease of retransmission attempts. The proposed scheme not only makes the value reduced, but also obtains a steady state of collisions as the number of stations increases.

59.6 Conclusions

In this research, we first set forth the drawbacks of legacy CSMA/CA that leads amount of collisions and bad performance, especially when the number of stations is large.

We proposed an improved Collision Avoidance method with information sharing list storing the backoff values of other stations. The capacity performance are analyzed through Cali' theory, and delay performance with Erlang C formula. From the simulation result, we can conclude the new proposed mechanism can be used with almost no extra overhead. There only shortcoming we need to overcome in the future is the delay jitter the proposed scheme has brought out.

References

1. IEEE P802.11 (1997) Standard for wireless LAN medium access control (MAC) and physical layer (PHY) specifications
2. Bharghavan V et al (1994) MACAW: a media access protocol for wireless LAN's, In: Proceedings of ACM SGCOMM'94, 24(4):212–225. doi:10.1145/190809.190334
3. Qian W, Dongfeng Y (2010) An adaptive backoff algorithm for IEEE 802.11 DCF with cross-layer optimization, 6th international conference on wireless communications networking and mobile computing (WiCOM). doi:10.1109/WICOM.2010.5601153. (in Chinese)
4. Bianchi G, Fratta L, Oliveri M (1996) Performance evaluation and enhancement of the CSMA/CA MAC protocol for 802.11 wireless LANs, PIMRC'96, 2:392–396

5. Zhao L-Q, Fan C-X (2003) ADCF: a simple algorithm of adaptive IEEE 802.11 DCF. J Circ Syst 8(4):100–102 (in Chinese)
6. Bianchi G, Tinnirello I (2003) Kalman filter estimation of the number of competing terminals in an IEEE 802.11 network. INFOCOM 2:844–952
7. Weinmiller J et al (1996) Analyzing and tuning the distributed coordination function in the IEEE 802.11 DFWMAC draft standard. In Proceedings of MASCOT
8. Kuo W-K, Jay Kuo C-C (2003) Enhanced backoff scheme in CSMA/CA for IEEE 802.11, Vehicular technology conference 2003, 5:2809–2813

Chapter 60
The Evaluation System for Cloud Service Quality Based on SERVQUAL

Hao Hu and Jianlin Zhang

Abstract Cloud computing is now playing an important role in resource sharing. Especially in business areas, many companies provide computing resources as a service to their customers. However, few researches focus on the measurement of cloud service quality, especially the external metrics for PaaS and IaaS service. By analyzing researches on cloud computing and summarizing the indicators of SERVQUAL in information system, a new cloud service quality evaluation system is proposed. Five dimensions (rapport, responsiveness, reliability, flexibility, security) from SaaS-Qual and their external metric indicators are introduced into this system. Moreover the paper presents the application of the evaluation system for cloud service pricing method.

Keywords Cloud computing · Service quality · Evaluation system · SERVQUAL · Pricing method

60.1 Introduction

Cloud computing is emerged as a computing resource sharing method in recent years. It proposed the pay-as-you-go model, which improves the utilization of funds for numerous enterprises in the process of informatization and reduces the pre-construction costs of hardware and software facilities. Once the cloud computing resources are subscribed based on enterprises' demand, they are able to

H. Hu (✉) · J. Zhang
School of Management, Zhejiang University, Yuhangtang Road No. 388,
Hangzhou, Zhejiang, China
e-mail: howl.hu@gmail.com

W. Lu et al. (eds.), *Proceedings of the 2012 International Conference on Information Technology and Software Engineering*, Lecture Notes in Electrical Engineering 210, DOI: 10.1007/978-3-642-34528-9_60, © Springer-Verlag Berlin Heidelberg 2013

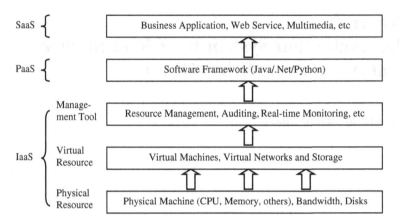

Fig. 60.1 Architecture of cloud service

conduct their own business. Many scholars approved that cloud computing as a service to provide computing resources, and proposed X as a Service (XaaS) to describe the service. But cloud computing involves many kinds of resources, which is based on multi-tenancy support and several technologies. More and more inappropriate description had been appeared and confused the understanding of cloud service, such as IPaaS, NaaS and so on. Hence Mell and Grance [1] unified the definition and architecture of cloud computing, and proposed the three-tier service model: SaaS, PaaS and IaaS. As shown in Fig. 60.1, each tier which is based on the same technologies provides resources and technology support for upper tier, and provides service to customers in different way [2].

Recently, because of high efficiency and good popularization of cloud computing, more and more companies put their information systems into cloud service providers' data center. But they must ensure their service quality to hold the customers with the growth of customers and the cancellation of popularization strategies. Gronroos [3] proposed service quality model and considered that service quality is the gap between customers' expectation and their perception of service. Using insights from this study as a starting point, many scholars evaluated the perceived service quality of customers by constructing scale and then guided enterprises' improvement. But the measurement of scale doesn't have real-time operability, while the results are subjective. Therefore, we developed the index system in SaaS-Qual [4] based on previous study, which focus on PaaS and IaaS, and using the external metrics to measure service quality.

The remainder of this paper is organized as follows. In the next section, we develop the theoretical basis for this work and introduce five dimensions into evaluation system. Then we detail each dimension and develop the external metrics. To end up, we highlight their implication for pricing and point out limitation.

60.2 Related Works

The earliest evaluation system of service quality named SERVQUAL was proposed by Parasuraman et al. [5]. They conducted empirical studies and put forward five dimensions (tangibles, reliability, responsiveness, assurance, empathy) of enterprises' service quality based on customers' perspectives. Keetinger and Lee [6] introduced SERVQUAL into the area of information system. They developed its contents and merged original five dimensions into four ones (a combined dimension of assurance and empathy, named rapport). Meanwhile they proposed Zone of Tolerance (ZOT) theory, which suggested that customers would be satisfied in a range of service performance. They named upper limit as desired service, to which the service a customer believes can be delivered, and lower limit as adequate service, which the service a customer consider acceptable. Gefen [7] introduced SERVQUAL into online information system from offline system and merged five dimensions into three ones, who combined reliability, responsiveness, assurance as one dimension. Parasurman et al. [8] refreshed their own SERVQUAL in electronic service setting and renamed E-S-Qual, including efficiency, system availability, fulfillment, and privacy, which provided a new direction for latter studies.

With the development of electronic commerce, more service providers will offer information system in B/S structure instead of C/S, so several researchers applied previous findings to Application Service Provider (ASP). Ma et al. [9] proposed ZOT-based ASP-Qual scale to measure the service quality, and she added security into evaluation system based on SERVQUAL. Benlian et al. [4] emphasized that flexibility (such as elastic resources, pay-as-you-go model) is the core competitiveness of cloud service after comparing SaaS with ASP. Then he proposed SaaS-Qual to capture the specifics in cloud environment, which contained rapport, reliability, responsiveness, flexibility, features, and security.

60.3 Cloud Service Quality Evaluation System

The original SaaS-Qual was based on customers' perception to evaluate service quality of cloud service providers. As the reason that SaaS mainly provides software to customers and its hardware facilities are transparent to users, it is appropriate to use customers' perception to measure service quality in SaaS. But in IaaS and PaaS model, users prefer to make use of the hardware facilities in providers' data center to develop, test, or host their applications and systems. Only the customers' perception data cannot capture the complete status of cloud services. Furthermore, such as the dynamic expansion of resources, data migration and image backup, users don't have access to these in SaaS. However, such as the specifics of user-friendly, interface layout are not important in IaaS and PaaS. Therefore, we combined with the external measurement of cloud service quality to

Fig. 60.2 Cloud service quality evaluation system

capture service specifics of IaaS and PaaS by objective indicators. Meanwhile, we assumed that the customers' expectation to service quality will be written down in the Service Level Agreement (SLA). The customers can choose different level according to the indicators in our study and decide their desired service and adequate service based on ZOT theory. Then service providers can determine customers' requirement to the cloud service effectively by external measurement.

Comparing SaaS with PaaS and IaaS, we found they have almost same cloud technology foundations (such as virtualization, distributed computing). According to Fig. 60.1, they are just different services based on the same resources. For example, SaaS transforms cloud computing resources into software applications. PaaS provides application deployment environment for SaaS. IaaS maintains the physical environment where applications and platforms exist. SaaS is based on the resources and technologies provided by PaaS and IaaS. Therefore, we use rapport, reliability, responsiveness, flexibility, and security mentioned in SaaS-Qual to build the cloud service quality evaluation system for describing IaaS and PaaS, as shown in Fig. 60.2. The dimension of features in original SaaS-Qual was used to describe perceptive specifics and function. We don't add it into the system because it can't be measured by external metrics.

60.3.1 Rapport

Rapport was proposed by Keetinger and Lee who emphasized that assurance and empathy in SERVQUAL should be combined as one indicator, named as rapport, which describes the service quality of technology support and customer care made by cloud service providers. It contains training system (Ra1), customized service (Ra2), and customer support (Ra3). Customized service includes international customer service and personalized information pushing. Customer support includes real-time counseling, remote assistance or any other methods to help customers.

Ra1 is measured by training hours. Ra2 is measured by information pushing frequency. Ra3 is measured by customers' selection results in SLA, which means the service quality can be identified directly by comparing the selection results with reality.

60.3.2 *Responsiveness*

Responsiveness describes the capacity of service providers to ensure service availability and normal performance. It contains dynamic scalability response (Rb1), disaster recovery time (Rb2), technology support availability (Rb3). When customers need more resources than they subscribed in SLA, cloud service should extend computing capacity in response for users' requirements conducted by automatic detection or manual request. Technology support means providers should response for customers' confusion and request.

Rb1 and Rb2 are measured by response time decided in SLA. Rb3 is measured by non-support times in commitment time.

60.3.3 *Reliability*

Reliability describes the capacity of service providers to provide cloud service correctly for users in time. It contains elastic service availability (Rc1), service accuracy (Rc2), and budget control (Rc3). Each customer has different using time for service. For example, small companies probably want to close the service in midnight to save capital. But large companies need 24×7 service available to meet customers' demands. Thus service providers need to make service available in commitment time. Meanwhile, strengthening recovery strategies and systems operation to keep service accuracy is one of the most important purposes for service management. Furthermore, budget control means providing users with budget overrun warning, list of costs and budget-use planning services, which ensures customers' funds utilization efficient.

Rc1 is measured by available uptime percentage: available uptime percentage = available uptime/total time. Rc2 is measured by the frequency of failure. Rc3 is measured by customers' selection results in SLA.

60.3.4 *Flexibility*

Flexibility describes the capacity that service providers support users to change default parameters flexibility. It contains adjustment of multi-client access (Rd1), extra resources allocation (Rd2), data migration (Rd3). The number of end-users isn't confirmed before the companies begin to accept cloud service, so service providers need adjust the number of user-access dynamically. As the important features of cloud service, scalability should be guaranteed for offering extra resources to meet users' dynamic demand. Furthermore, providers need to meet users' requirement to migrate data between different operating systems.

Rd1 is measured by multi-client variation balance. Rd2 is measured by coverage of resources: coverage of resources = amount of allocated resources/ total amount of requested resources. Rd2 is just for IaaS because users can't access to the configuration of computing resources directly in PaaS. Rd3 is measured by amount of data migration. Rd3 is for PaaS because the store strategy of IaaS is based on image, which is hard to migrate between them.

60.3.5 Security

Security describes the systematic protection methods adopted by service providers to avoid data missing and system collapse. It contains data backup (Re1), fault recovery strategy (Re2), regular security audit (Re3), anti-virus tool (Re4), data secrecy (Re5), and access control (Re6). Data backup means image backup in IaaS, and database backup in PaaS. Security audit will give the providers chances to find defects in management, thus avoid losses effectually. Data secrecy contains penalty measure made for data leakage between users and providers. Furthermore, access control will avoid the interference of unrelated persons for the use of cloud service.

Re1 is measured by frequency of backup. Because each cloud service providers has different security strategy, rest of indicators cannot be measured by unified standard [10]. Therefore, the remainder is measured by customers' selection results in SLA.

The cloud service quality evaluation system is shown in Table 60.1.

60.4 Pricing Method Based on Evaluation System

The service value created by reasonable price is critical factor for customers to adopt cloud service. Service providers not only charge fee based on customers' usage amount, but also offer efficient service quality to maintain their value. Bhargava et al. [11] compared standard price with pricing based on threshold, the latter will bring more profits and fairness. They found that customers would like to pay more to obtain more specific and superior service. Therefore, we proposed that service providers define the specific services and their level of service quality according to our evaluation system in customers' SLA. They can divide service quality into different levels and each level is priced at a reasonable price. Customers decided their level based on their preference and actual demand before subscribing service. For example, cloud service providers divide second grade indicators in our evaluation system into several levels according to their external metrics and each level has its own reasonable price based on the cost of management and resources. If the customers do not care about service quality, their level will be set only the lowest one. Higher level means more superior

Table 60.1 Cloud service quality evaluation system

First grade indexes	Second grade indexes	External metric
Rapport (Ra)	Training system (Ra1)	Training hours
	Customized service (Ra2)	Information pushing frequency
	Customer support (Ra3)	Selection results in SLA
Responsiveness (Rb)	Dynamic scalability response (Rb1)	Response time
	Disaster recovery time (Rb2)	Response time
	Technology support availability (Rb3)	Non-support times in commitment
Reliability (Rc)	Elastic service availability (Rc1)	Available uptime percentage
	Service accuracy (Rc2)	Failure frequency
	Budget control (Rc3)	Selection results in SLA
Flexibility (Rd)	Multi-client access adjustment (Rd1)	Multi-client variation balance
	Extra resources allocation (Rd2)	Coverage of resources[a]
	Data migration (Rd3)	Amount of data migration[b]
Security (Re)	Data backup (Re1)	Image backup frequency[a]
		Database backup frequency[b]
	Fault recovery strategy (Re2)	Selection results in SLA
	Regular security audit (Re3)	Selection results in SLA
	Anti-virus tool (Re4)	Selection results in SLA
	Data secrecy (Re5)	Selection results in SLA
	Access control (Re6)	Selection results in SLA

Notes
[a] means the metric in IaaS
[b] mean the metric in PaaS.
Selection results in SLA means the metric can be identified directly by comparing the selection results with reality

quality. When providers cannot satisfy the level of quality in SLA, they must compensate customers for their loss. Thus customers will have more specific and flexible choices and providers will prefer to increase the cloud service value to ensure the quality which customers care about.

We can't apply SERVQUAL and SaaS-Qual to the commercial pricing directly because they are all based on customers' perception. But the measurement based on external metrics makes quality parameters acquisition more objective and real-time, so the pricing method based on service quality will be more fair and feasible. Considering each quality indicator has different effects on different users [12], we suggested that pricing methods should be depended on different target customers according to reality. Therefore, we should determine our target customers by market investigation and adopt analysis hierarchy process (AHP) to confirm the price of service quality for them. Firstly, we distribute questionnaire about cloud service quality evaluation to several experts. After acquiring adequate data, we can construct judgment matrixes and ensure them available by consistency check. Then we can decide price according to specific weight. The chosen experts should be representative and comprehensive, so we can select the researchers as the experts from universities and companies in cloud computing area.

60.5 Conclusion

Cloud service is emerged as an effective data service paradigm. Service providers can acquire real-time data from logs stored in data center to describe their service quality. Combined with external metrics, it will be objective to measure the efficiency of cloud service and the requirement of customers. We believed that this evaluation system will be meaningful for cloud service providers because it provides a method to evaluate the cloud service in quantitative manner.

In this paper, a comprehensive evaluation system of cloud service quality was summarized from various references about SERVQUAL. We proposed external metrics combined with the application of SLA to measure users' requirement of service quality. Moreover, we suggested the pricing method based on our evaluation system to capture customers' preference and satisfy their requirement for service quality. But, as the development of cloud computing, the effect of several technologies and resources will be weaken, and different target customers probably will bring the variation of weight. Hence we should adjust the service price according to the specific target customers in practice.

References

1. Mell P, Grance T (2009) The NIST definition of cloud computing. National Inst Stand Technol 53(6):50–52
2. Jing SY, Ali SS, She K (2011). State-of-the-art research study for green cloud computing. J Supercomputing 56:1–24
3. Gronroos C (1982) Strategic management and marketing in the service sector. Swedish School of Economics and Business Administration, Helsingfors
4. Benlian A, Koufaris M, Hess T (2011) Service quality in software-as-a-service: developing the SaaS-Qual measure and examining its role in usage continuance. J Manag Inf Syst 28(3):83–126
5. Parasuraman A, Zeithaml V, Berry L (1988) SERVQUAL: a multiple-item scale for measuring consumer perceptions of service quality. J Retail 64(1):12–40
6. Kettinger W, Lee C (1997) Pragmatic perspectives on the measurement of information systems service quality. MIS Quart 21(2):223–240
7. Gefen D (2002) Customer loyalty in e-commerce. J Assoc Inf Syst 3(1):27–51
8. Parasuraman A, Zeithaml V, Malhotra A (2005) E-S-QUAL: a multiple-item scale for assessing electronic service quality. J Serv Res 7(3):213–233
9. Ma Q, Pearson J, Tadisina S (2005) An exploratory study into factors of service quality for application service providers. Inf Manage 42(8):1067–1080
10. Liu X, Yuan D, Zhang G (2012). The design of cloud workflow systems. Springer, New York, pp 47–49
11. Bhargava HK, Sun D (2008) Pricing under quality of service uncertainty: market segmentation via statistical QoS guarantees. Eur J Oper Res 191(3):1189–1204
12. Chandrasekar A, Chandraskar K, Mahadevan M (2012) QoS monitoring and dynamic trust establishment in the cloud advances in grid and pervasive computing. Lect Notes Comput Sci 7296:289–301

Chapter 61
The Design and Realization of a High Speed Satellite Network Management Agent

Chuanjia Fu and Zhimin Liu

Abstract This paper details the design and realization of a high speed satellite network management agent. It covers the analysis, architecture, and process of this software. The first part of this paper discusses main components of the satellite network management system. In order to realization this project, we design a Simple Talk Control Protocol for the network control process, and introduce the Simple Network Management Protocol (SNMP) and ROHC protocol to achieve the goal of reliable management and rate improvement. Finally, we set up the test platform in laboratory and design a series of tests to confirm the design has achieved it object.

Keywords Satellite management agency · SNMP · Embedded · ROHC · Simple talk control protocol

61.1 Background and Software Architecture Design

With the increasing importance of embedded technology, the development of embedded systems for all uses are becoming more and more widely. This paper will introduce a network management agent design, which is realized in a Single Channel Per Carrier (SCPC) satellite system [1].

The Satellite management system we design is based on "Center/Agent" architecture [2]. Therefore, those methods and structure design we used also can be

C. Fu (✉) · Z. Liu
School of Software and Microelectronics, Peking University, No.5 Yiheyuan Road Haidian District, Beijing 100871, People's Republic of China
e-mail: chuanjiafu@gmail.com

Z. Liu
e-mail: lzmpku@126.com

W. Lu et al. (eds.), *Proceedings of the 2012 International Conference on Information Technology and Software Engineering*, Lecture Notes in Electrical Engineering 210, DOI: 10.1007/978-3-642-34528-9_61, © Springer-Verlag Berlin Heidelberg 2013

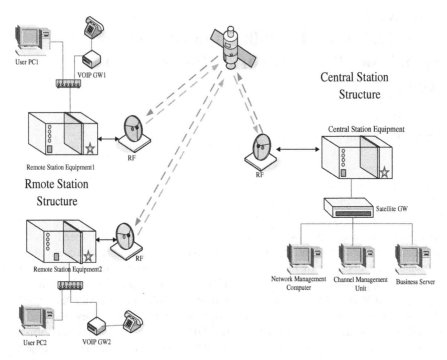

Fig. 61.1 Satellite communications system network structure

transplant to other similar systems, such as remote control system and telemedicine system etc. This SCPC satellite system is composed of a network management computer, business server, a channel management unit, and multiple remote stations and central station equipment etc. The network structure is as show in Fig. 61.1.

In order to manage the satellite communication system described above, we designed a series of network management software. The network management software is composed two parts, center and agent. This article is mainly focus on agent software design, which is based on traditional SNMP agent and cooperated with ROHC to improve data rate. In order to achieve effective management we design the Simple Talk Control Protocol. The Software structure is show as Fig. 61.2.

The detailed design and realization of the Simple Talk Control Protocol, SNMP agent and ROHC will be described as following.

61.2 The Simple Talk Control Protocol Design

The Simple Talk Control Protocol we design between the network manager and agent is reference to the Radio Resource Control (RRC) of the 3GPP (The 3rd Generation Partnership Project) [3]. The Simple Talk Control Protocol we design

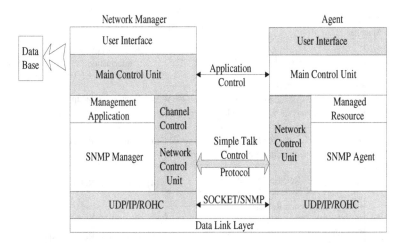

Fig. 61.2 Software structure

including mechanisms of system information broadcast process, connection setup process, connection release process, connection re-establishment process and failure reporting process etc. Their processes are show as Fig. 61.3.

- Broadcast of System Information is used to inform the new join remote station about current system signaling channel parameters.
- Connection Setup Process is used to setup traffic channel.
- If no traffic channel resource is available or Connection Request is invalid, Connection Request will be reject.
- Connection Release Process is used to release traffic channel resource.
- If connection breakdown abnormally Connection Re-establishment Process will be used to re-establish connection.
- Error Report is used to report breakdown information.

61.3 SNMP Module

Simple Network Management Protocol (SNMP) was first defined by the Internet Engineering Task Force (IETF) in 1989. Since then, SNMPv2 has become an industry standard for controlling networking devices from a single management application. For information on the SNMPv2 standard, refer to RFC 1098 [4]. SNMPv2 is an implementation of the "client/server" relationship. The client application, called the SNMPv2 agent, running on a remote station device. The database controlled by the SNMPv2 agent is Management Information Base (MIB), which is the definition of the data, or objects, that are stored in the agent for the manager to access [5]. The standard MIB's structure is represented by a tree. There are three components of the tree.

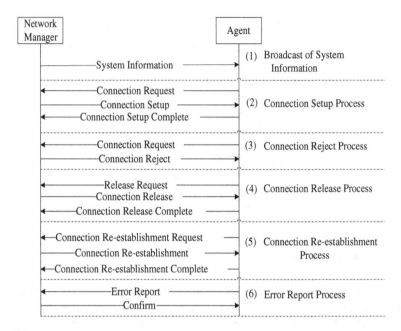

Fig. 61.3 Simple talk control protocol

- Root: The unnamed root of the tree contains a set of characters common to all MIB objects.
- Subtree: The sub-tree contains a subset of the information available in the root.
- Leaf: The leaf is a sub-tree with no additional sub-trees in its domain.

The structure of MIB is showing in Fig. 61.4.

Enterprises subtree is used to permit parties providing networking subsystems to register models of their products. So we register our networking subsystem is under this subtree in laboratory environment only for study. Our OID table is showing as Table 61.1.

Information is passed between layers in the form of packets, known as protocol data units (PDUs). The SNMP PDUs support five commands: GetRequest, GetNextRequest, GetResponse, SetRequest, and Trap. PDUs are used between the Agent and the Manager to pass information.

61.4 ROHC Module

As we know, RObust Header Compression (ROHC) itself shows around 20 % achieved rate improvement on a line of wireless link [6]. Therefore, the ROHC is used in our system to reduce the header overhead, packet loss rate and improve interactive responsiveness. Header compression with ROHC can be characterized

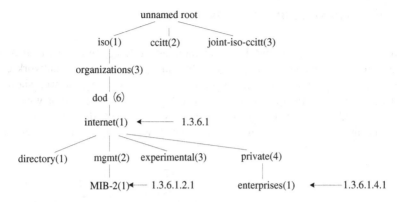

Fig. 61.4 MIB structure

Table 61.1 OID definition

NO.	OID	Name	Type
1	1.3.6.1.4.1.1.1.1.0	Receive frequency	int
2	1.3.6.1.4.1.1.1.5.0	Send frequency	int
3	1.3.6.1.4.1.1.3.2.0	Device IP	unsigned long
…	…	…	…

as an interaction between two state machines, one compressor machine and one decompressor machine, each instantiated once per context. The compressor and the decompressor have three states each, which in many ways are related to each other [7].

• Compressor States and logic

For ROHC compression, the three compressor states are the Initialization and Refresh (IR), First Order (FO), and Second Order (SO) states. The compressor starts in the lowest compression state (IR) and transits gradually to higher compression states. The transition logic for compression states is based on three principles: the optimistic approach principle, timeouts, and the need for updates.

• Decompressor States and logic

The decompressor also includes three states: No Context, Static Context and Full Context. The decompressor starts in its lowest compression state, "No Context" and gradually transits to higher states. The decompressor state machine normally never leaves the "Full Context" state once it has entered this state. For information on the ROHC standard, refer to RFC 3095 [7].

61.5 Experimental and Results

In order to show that the network management agent design fulfilled all functions required for the network management system and generates better network condition, we have performed a series of experiments on the platform show in Fig. 61.5. The test platform consist a Network Manager Computer (runs windows), four embedded devices (run Linux) and necessary network devices (Multi-band Combiner and multiplexer Gateway etc.,). The network management system runs

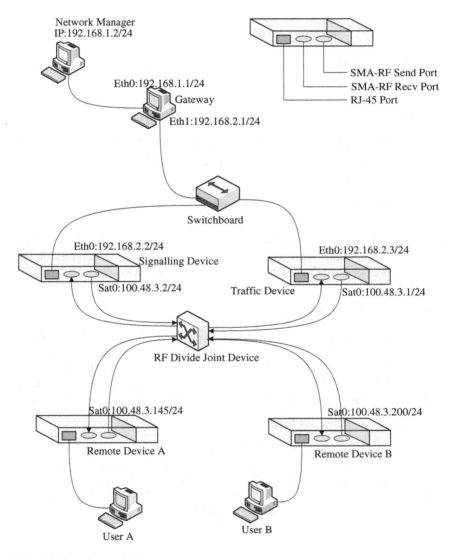

Fig. 61.5 Experiment platform

Table 61.2 Test result

No.	Bandwidth	System delay (ms)	Package lose (%)
1	100 kHz	150	0.012
2	800 kHz	45	0.009
3	2 MHz	13	0.005

on the Network Manager Computer, agent program runs on the four embedded devices.

On the platform we can setup connection and release connection between any Remote devices and the Traffic Device. Additionally, the Network Management Computer sends SNMPv2 GetRequest every 20 s, to get real time parameters of each device. Moreover, to make the experiments as close to real world scenarios as possible, we generate a typical workload on the experimental network, and record system control time delay and package lose on different traffic bandwidths condition (Table 61.2).

61.6 Conclusion

The main contribution of this paper is to present techniques and methods on the network management agent design, which are proved possessing practical significance. First, the Network Control Protocol described in this experiment is suitable for simple network control used in other condition, such as remote control and remote monitoring. Furthermore, in order to increase the efficiency of the network we used robust header compression (ROHC) protocol on each of the embedded devices. Based on the platform described in Fig. 61.5, experiments show that the ROHC decreases the time delay by 16.2 % on average.

References

1. Guffee CO, Jr. Voice of America Weinreich D (1995) Satellite audio distribution on a video transponder using digital compression and band-edge SCPC technologies, (digital satellite communications, tenth international conference on date of conference), vol 2, pp 390–397
2. Adler RM (1995) Distributed coordination models for client/server computing (Journals and Magazines)
3. 3GPP TS 25.331 v1.0.0 (1999) RRC protocol specification (3rd Generation Partnership Project)
4. Case JD, Fedor M, Schoffstall ML, Davin C (1989) A simple network management protocol (SNMP), (RFC1098)
5. Teegan HA (1996) Distributed performance monitoring using SNMPv2 (network operations and management symposium, IEEE), pp 616–619

6. Jung S, Hong S, Park P (2006) Effect of RObust Header Compression (ROHC) and packet aggregation on multi-hop wireless mesh networks (computer and information technology, CIT '06. The sixth IEEE international conference), p 91

7. Bormann C, Burmeister C, Degermark M, Fukushima H, Hannu H, Jonsson LE, Hakenberg R, Koren T, Le K, Liu Z, Martensson A, Miyazaki A, Svanbro K, Wiebke T, Yoshimura T, Zheng H (2001) RObust Header Compression (ROHC):framework and four profiles: RTP, UDP, ESP, and uncompressed, (RFC 3095)

Chapter 62
Simulation and Optimization of Via in Multi-Layer PCB

Jia Lu and Chunrong Zhang

Abstract The discontinuity of via is one of the key factors in multi-layer PCB design. As frequency increases, via can cause reflection that seriously affects system performance and the integrity of signal. In this paper, the full-wave HFSS simulation software is used for modeling and simulation of multi-layer PCB via. The size of via on signal integrity is analyzed and the influence of back-drilling to stub is researched. The simulation results can be practical guidance for multi-layer PCB design.

Keywords Multi-layer PCB Via · Transmission · Characteristics · Back-drilling

62.1 Introduction

Microwave communication equipments tend to be small, low cost and high performance because of the rapid development of modern electronic industrial technology. Multi-layer PCB board is widely used with its unique technical advantages in the manufacturing of electronic equipment. The market share of multi-layer PCB board is gradually expanding, and processing costs also decreased. Vias are often used in high-speed multilayer design. Electrical connection between the layers of the printed circuit board panels or between components and alignment is achieved through vias [1]. As the continuous improvement of signal speed and wiring density, many factors will affect the signal integrity, and via is one of the key factors affecting the quality of signal transmission.

J. Lu (✉) · C. Zhang
Xi'an Electronic Engineering Research Institute, 710100 Xi'an, China
e-mail: lujia206@126.com

W. Lu et al. (eds.), *Proceedings of the 2012 International Conference on Information Technology and Software Engineering*, Lecture Notes in Electrical Engineering 210, DOI: 10.1007/978-3-642-34528-9_62, © Springer-Verlag Berlin Heidelberg 2013

Fig. 62.1 The types of via

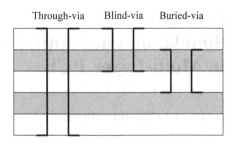

Fig. 62.1 The types of via

A number of factors will affect the parasitic parameters of via, such as the size of pad and anti-pad, coating metal, number of layers, etc. This letter will use ANSOFT HFSS 3D electromagnetic simulation software to establish the precise via model, then simulation and analysis of the effect of the impact of via.

62.2 Descriptions About Via

62.2.1 The Types of Via

Via is divided into three categories: through-via, blind-via and buried-via from the manufacturing process [2]. Blind-via is used to connect surface and inner of the printed circuit board, which has a certain depth. The depth of via is usually not more than a certain ratio (aperture) (Fig. 62.1).

Buried-vias connect the inner layer of the printed circuit board which do not extend to the surface of the circuit board. Both of the above two types of vias are only through part of the layers in PCB board. The third type is called through-via that through the entire circuit board, which can be used for internal interconnect or components' installing. Blind and buried vias help to increase the density of the multi-layer PCB board. For the reason of complex manufacturing process and high cost, they are mainly used for the production of flexible board (FPC) and frequency (RF) board that require a higher signal quality radio. Most of PCBs use through-via as its easy manufacturing process and low cost.

62.2.2 Transmission Characteristics of Via

Via connects the different layers of PCB, its structure includes barrel, pad and anti-pad [3]. Vias through the ground plane will produce the parasitic capacitance, expressed as follows:

$$C = 1.41\varepsilon_0\varepsilon_r TD_1/(D_2 - D_1) \tag{62.1}$$

Fig. 62.2 Schematic
diagram of via

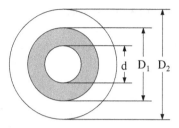

Where ε_r is the dielectric constant, D1 is pad diameter, D2 is anti-pad diameter, T is the thickness of the media. The influence of the parasitic capacitance is particularly important for digital circuits that should be designed to minimize delay caused by the transmission signal (Fig 62.2).

Via also has parasitic series inductance that will reduce the effect of supply bypass capacitor filtering. Series inductance is mainly determined by the the via depth h and the diameter via d, expressed as follows:

$$L = 5.08h \left[\ln \left(\frac{4h}{d} \right) + 1 \right] \tag{62.2}$$

As an uncontinuous structure, via will make transmission characteristics poorer for its parasitic capacitance and inductance. The performance of via is determined by many complex factors. The coupling between vias and traces or the pad and anti-pad, as well as the impact of a strong electric field at the existence of the anti-pad are all influencing factors. Different sizes of via, pad and anti-pad will lead to impedance changes, which directly affect the integrity of the signal transferred through vias.

62.3 Simulation of Via

62.3.1 Modeling of Via

The simulation results that modeling analysis of via in the circuit using the traditional way (using the equivalent circuit or lumped elements to model) are not able to achieve good accuracy. However, full-wave analysis using the method of the electromagnetic field can solve the problem of accurate modeling and simulation. This letter uses ANSOFT HFSS 3D electromagnetic simulation software to simulate vias in multi-layer PCB [4], scanning frequency range s from 0 to 10 GHz.

The model shows in Fig. 62.3, the PCB has six layers, size of which is 10×10 mm, $\varepsilon_r = 4.4$. This via connects the top and bottom. Layer 2 and 5 are ground and the remaining layers are used for signal transmission. In this way, each signal layer has a reference plane which can mask interference from layers and external. It is conducive to the return of the high frequency components of the

Fig. 62.3 HFSS model via in multi-layer PCB

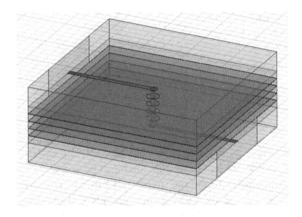

signal. The actual manufacturing process of via is drilling, so the section of via, pad and anti-pad are circular.

Study impacts that size of vias, pads and anti-pad, as well as the dielectric thickness on transmission characteristics. The simulation results are reflected in the form of S-parameters. The network parameters S is on the basis of the relationship between incident and reflected wave which is suitable for microwave circuit analysis.

62.3.2 Transmission Characteristics on the Change of Via Radius

Figure 62.4 shows the S11 and S21 simulation curves in the conduction that via radius changes from 0.3 to 0.6 mm. The dielectric thickness is 0.4 mm, pad radius is 0.6 mm and anti-pad radius is 0.8 mm.

62.3.3 Transmission Characteristics on the Change of Pad Radius

Figure 62.5 shows the S11 and S21 simulation curves in the conduction that pad radius changes from 0.4 to 0.7 mm. The dielectric thickness is 0.4 mm, via radius is 0.3 mm and anti-pad radius is 0.8 mm.

62.3.4 Transmission Characteristics on the Change of Anti-Pad Radius

Figure 62.6 shows the S11 and S21 simulation curves in the conduction that anti-pad radius changes from 0.5 to 0.8 mm. The dielectric thickness is 0.4 mm, via radius is 0.3 mm and pad radius is 0.5 mm.

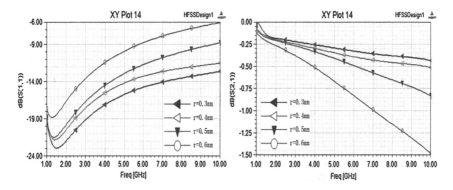

Fig. 62.4 S11 and S21 simulation curves on the change of via radius

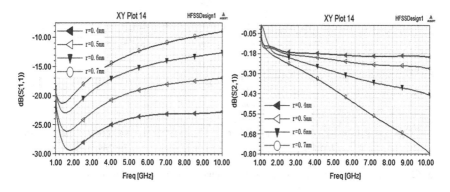

Fig. 62.5 S11 and S21 simulation curves on the change of pad radius

62.3.5 Transmission Characteristics on the Change of Dielectric Thickness

Figure 62.7 shows the S11 and S21 simulation curves in the conduction that dielectric thickness changes from 0.4 to 0.7 mm. The via radius is 0.3 mm, pad radius is 0.5 mm and anti-pad radius is 0.7 mm.

Thus, when the dielectric thickness, pad and anti-pad size is unchanged, S11 parameter is proportional to the via radius and S21 is inversely proportional to it. As the frequency increases, the transmission characteristics above are more significant. From Figs. 62.5, 62.6 and 62.7, when one of other sizes changes, we can find similar characteristics.

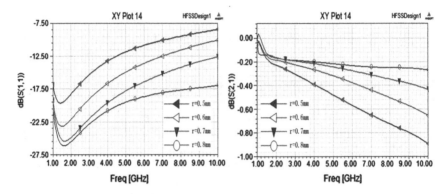

Fig. 62.6 S11 and S21 simulation curves on the change of anti-pad radius

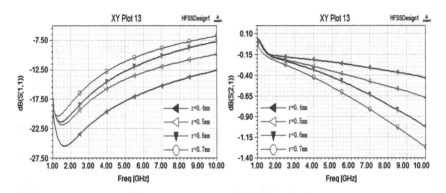

Fig. 62.7 S11 and S21 simulation curves on the change of dielectric thickness

62.4 Optimization of Via

In a multi-layer PCB design, when the signal transmitted from the top to an internal layer, it will produce a redundant vias stub if we use through-via which greatly affects the quality of signal transmission. Though the use of the blind-via can avoid vias stub, but it's complex and costly. Therefore, the study that the optimization of via stub will help the designer to balance cost and performance.

To solve this problem, a technology called back-drilling is used [5]. Back-drilling is a process to removed the unused portion of through-via.

Combined with the HFSS model of Fig. 62.3, signal transfers from top to layer 3 through the via, and a via stub appears as showed in Fig. 62.8. We can see resonance [6] in higher frequency through the S11 and S21 simulation curves, which will greatly affect the transmission characteristics.

After the process of back-drilling, as showed in Fig. 62.9, the resonant phenomenon almost disappears from the simulation curve. From Fig. 62.10, though the via stub does not export signal, but it has relatively strong electric field

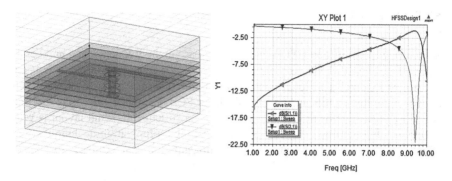

Fig. 62.8 Model and simulation results before back-drilling

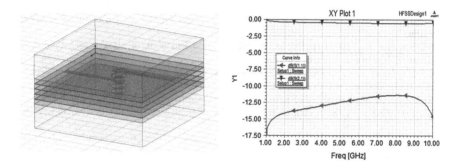

Fig. 62.9 Model and simulation results after back-drilling

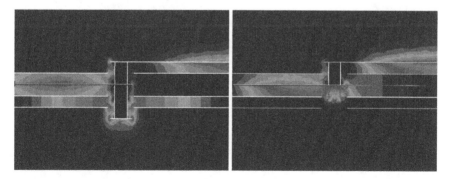

Fig. 62.10 Electric field of via before and after back-drilling

distribution. It will form an additional LC slip, which can lead to more impedance discontinuities and corresponding resonance. After back-drilling, with the removal of the via stub, the electric field distribution that has nothing to do with the signal transmission will disappear.

62.5 Conclusion

In the design of multi-layer PCB, with the increase in the frequency and number of layers. The impact of via parasitic parameters on the transmission characteristics can not be ignored. This letter contains the simulation and optimize of via which has some guiding significance with multi-layer PCB design. In order to reduce the adverse effects of vias, the rational design of via structure size and layout is very important. In addition, the use of back-drilling optimization techniques can effectively reduce the resonant to improve transmission characteristics in via.

References

1. Bogatin Eric (2004) Signal integrity: simplified. Publishing House of Electronics Industry, Beijing
2. Johnson Howard (2004) High-speed digital design: a handbook of black magic. Publishing House of Electronics Industry, Beijing
3. Li M-Y, Guo C-Q (2009) Simulation and analysis of via in high speed circuit design. Electron Instrum Customer, pp 105–107 (in Chinese)
4. Ansoft Corporation (2005) HFSS full book. Ansoft Corporation
5. Liu Y-M, Cao Y-S (2008) Analysis and simulation of via in high-speed PCB. Comput Eng Des 713–715 (in Chinese)
6. Kim Y-W (2005) A new via hole structure of MLB (multi-layered printed circuit board) for RF and high speed systems. IEEE, pp 1378–1382

Chapter 63
On-Chip Bus Architecture Performance Estimation for a Microprocessor Based on Queuing Model

Xuexiang Wang, Zhi Qi and Peng Cao

Abstract Higher performance of interconnect of on-chip module is needed because of the improvement of the system-on-chip integration and application requirements. On-chip bus is the main interconnect structure used in the current system-on-chip, it makes a target of high performance, low power dissipation and low hardware spending. In order to make the best choice of the on-chip bus structure, performance estimation is a quite significant and obligatory work. This paper proposes a method for estimating the on-chip bus structure performance.

Keywords On-chip bus · Microprocessor · Performance estimation · Queuing model

63.1 Introduction

Expanding continuously along with the embedded application, more and more various components and RISC core were integrated into a single chip to form a microprocessor. The on-chip bus plays a key role in the microprocessor by enabling efficient integration of heterogeneous components. The on-chip bus structure indicates the basic characteristic of on-chip bus, such as single or hierarchy, bidirectional or unidirectional, share or point to point, asynchronous or synchronous etc.

X. Wang (✉) · Z. Qi · P. Cao
National ASIC System Engineering Research Center, Southeast University, Nan Jing
210096, P.R. China
e-mail: wxx@seu.edu.cn

W. Lu et al. (eds.), *Proceedings of the 2012 International Conference on Information Technology and Software Engineering*, Lecture Notes in Electrical Engineering 210, DOI: 10.1007/978-3-642-34528-9_63, © Springer-Verlag Berlin Heidelberg 2013

It is effective for performance estimation based on the analytical model and specially there is many previously related works have done some research based on the queuing network theory. For example, some research adopt the classical queuing network model based on Markov chain [1], some adopt the mean value analysis (MVA) method [2] or other mixed methods [3]. Onyuksel used a closed queuing network model such as M/G/1/N model for performance analysis of a single-bus multiprocessor system [4]. Bodnar and Liu used the stochastic hierarchy queuing network model to analyzing the single-bus, coupling architecture for multiprocessor system [5]. Mahmud analyzed a asynchronous hierarchy bus structure based on the closed queuing network model [6] and he also finished the performance evaluation according to the proposed hierarchy bus architecture [7]. Zhang applied analytical model based on simulation method [8]. Compared to the simulation model, performance estimation based on the analytical model can educe the estimation results more quickly.

In this paper, the queuing network models are proposed according to a RISC microprocessor. These models are applied to analyze bus structure, which impact the on-chip bus performance remarkably. The results of the queuing network model analysis were found to be valuable for the architecture choice.

In Sect. 63.2, we put forward the queuing network model for the on-chip bus structure. Section 63.3 describes the evaluating results for two kinds of structures. Finally we give a conclusion in Sect. 63.4.

63.2 On-Chip Bus Structure and Queuing Network Model

In this section, we first present a overview of the bus structure of REMUS [9], including the single-layer bus architecture and two-layer bus architecture. Next, we bring forward the corresponding queuing network models respectively according to the single-layer and two-layer bus architecture.

63.2.1 Bus Structure

Structure is the basic element of the on-chip bus communication scheme. In terms of the bus interface, it can be divided into the standard bus structure such as AMBA, Core-Connect etc. and the wrapper-based bus structure such as Virtual component Interface (VCI), Open Core Protocol (OCP). Some bus architectures uses the synchronous bus transfer mode and the others use the asynchronous bus transfer mode (MARBLE). And the bus structure can also be divided into two parts: the shared architecture and the point-to-point architecture (CoreFrame).

AMBA is flexible bus architecture, and it can be organized in many kinds of fashion. According to the microprocessor, firstly, a single bus structure was introduced and then we develop a two-layer bus architecture still based on AMBA.

Figure 63.1 show the single-layer shared bus architecture used in REMUS. In the system, including one layer system bus (AHB).

Because more and more various components were integrated into REMUS, the bus bandwidth has becoming the bottleneck of the whole system. The hierarchical system bus structure based on AMBA is called multi-layer AHB system. The elementary transfer still carry through base on AHB transfer protocol and a more complex exchange matrix was used for realizing the parallel transfers between the masters and the slaves. The block diagram of two-layer system is shown in Fig. 63.2.

63.2.2 Structure Queuing Network Model

If we view on-chip bus as a server, and view all masters in the system as customers, the system could be viewed as a queuing system. In our real system, LCDC corresponds to component C_1, MMA corresponds to C_2, DMAC corresponds to C_3, and RISC is C_4.

Fig. 63.1 One-layer AMBA bus architecture

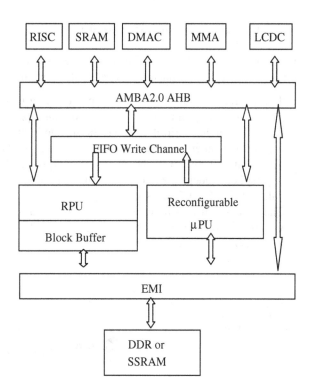

Fig. 63.2 Two-layer AMBA
bus architecture

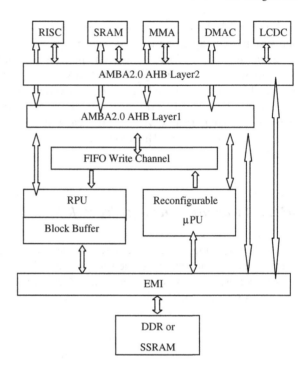

63.2.2.1 Single-Layer Bus Architecture

Figure 63.3 shows the single-layer bus queuing system. The following assumptions are made for the queuing model of the system.

- The queuing network model is an open queuing network system.
- The arbitration scheme is no preemptive static priority arbitration; component C1 has the highest priority, C2 has the second highest priority, and so on. There are four customer streams in the queuing system relative to the real system.
- The duration between the beginning of a request and the generation of the next request is an independent, exponentially distributed random variable. The mean value of C1, C2, C3 and C4 are $1/\lambda_1$, $1/\lambda_2$, $1/\lambda_3$, and $1/\lambda_4$.
- The duration of a transaction initiated by a master is an independent random variable. And the mean value of C1, C2, C3 and C4 are $1/\mu_1$, $1/\mu_2$, $1/\mu_3$, and $1/\mu_4$.

Fig. 63.3 Single-layer bus
queuing network model

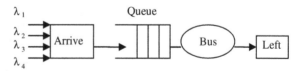

Then the mean wait time of C_i is Tw_i, which can be obtained by analyzing the queuing network model. The arriving intensity λ for the shared queuing is equal to the addition of all components' arriving intensity. Thus,

$$\lambda = \lambda_1 + \lambda_2 + \lambda_3 + \lambda_4 \tag{63.1}$$

$$Tw_i = E[W^1] + E[W^2] + E[W^3] \tag{63.2}$$

And we define K_i is the square of the variation coefficient of service time of C_i. Then the mean request waiting time can be obtained from three parts.

In Eq. (63.2), $E[W^1]$ denotes the mean waiting time arises from current component's transaction (service) when C_i requests. $E[W^2]$ denotes the mean waiting time arises from the higher priority components' transaction in the waiting queuing when C_i requests. And $E[W^3]$ denotes the mean waiting time arises from the higher priority components' transaction when C_i is in the waiting queuing. And the 1-stage and 2-stage moments of the service time distribution of C_i is,

$$E[S_i] = \sum_{j=1}^{i} \frac{\lambda_j}{\lambda} \cdot \frac{1}{\mu_j} \tag{63.3}$$

$$E[S_i^2] = \sum_{j=1}^{i} \frac{\lambda_j}{\lambda} \cdot \frac{K_j + 1}{\mu_j^2} \tag{63.4}$$

Then we can get the results as below,

$$E[W^1] = \hat{\rho}_r \cdot \frac{E[S_i^2]}{2E[S_i]} = \frac{1}{2} \sum_{j=1}^{r} \left(\lambda_j \cdot \frac{K_j + 1}{\mu_j^2} \right) \tag{63.5}$$

$$E[W^2] = \sum_{j=1}^{i} \left(\rho_j \cdot Tw_j \right) \tag{63.6}$$

$$E[W^3] = \hat{\rho}_{i-1} \cdot Tw_i \tag{63.7}$$

$$Tw_i = \frac{\sum_{j=1}^{r} \left(\lambda_j \cdot \frac{k_j + 1}{\mu_j^2} \right)}{2(1 - \hat{\rho}_{i-1})(1 - \hat{\rho}_i)} \tag{63.8}$$

Where

$$\rho_i = \frac{\lambda_i}{\mu_i}, \hat{\rho}_i = \sum_{j=1}^{i} \rho_j \tag{63.9}$$

63.2.2.2 Two-Layer Bus Architecture

The two-layer bus queuing system is shown in Fig. 63.4. Some assumption also are made for the queuing model of the system.

- The queuing network model is also modeled as an open queuing network system.
- The arbitration scheme is also no preemptive static priority arbitration; The priority order of layer1 queuing system is 1-2-4, and the order of layer2 is 2-3-4.
- The duration between the beginning of a request and the generation of the next request is an independent, exponentially distributed random variable. The mean value of C_1, C_2, C_3 and C_4 are $1/\lambda_1$, $1/\lambda_2$, $1/\lambda_3$, and $1/\lambda_4$. And the probability of accessing layer1 and layer2 queuing of C_2 is q_{21} and q_{22}. Similarly, the probability of C_4 is q_{41} and q_{42}.
- The duration of a transaction initiated by a master is an independent random variable. And the mean value of C_1, C_2, C_3 and C_4 are $1/\mu_1$, $1/\mu_2$, $1/\mu_3$, and $1/\mu_4$. At the same time, the mean transaction time of C_2 and C_4 respectively in layer1 and layer2 queuing is $(1/\mu_{21}, 1/\mu_{22})$ and $(1/\mu_{41}, 1/\mu_{42})$, their relationship is,

$$\frac{1}{\mu_2} = q_{21} \cdot \frac{1}{\mu_{21}} + q_{22} \cdot \frac{1}{\mu_{22}} \tag{63.10}$$

$$\frac{1}{\mu_4} = q_{41} \cdot \frac{1}{\mu_{41}} + q_{42} \cdot \frac{1}{\mu_{42}} \tag{63.11}$$

Because of addition of the exponential distribution, the queuing system in Fig. 63.4 could be divided into two independent queuing systems: layer1 queuing system and layer2 queuing system.

In layer1 queuing system, including three customer queues, whose arriving rates are $\lambda_1, q_{21}\lambda_2$, and $q_{21}\lambda_4$ mean transaction time are $1/\mu_1$, $1/\mu_{21}$ and $1/\mu_{41}$, the square of the variation coefficient of service time are K_1, K_{21} and K_{41}. Then three components' mean wait time can be obtained,

Fig. 63.4 Two-layer bus queuing network model

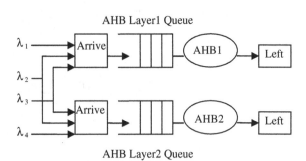

AHB Layer1 Queue

AHB Layer2 Queue

$$Tw_1 = \frac{Q}{1 - \rho_{11}} \tag{63.12}$$

$$Tw_{21} = \frac{Q}{(1 - \rho_{11})(1 - \rho_{11} - \rho_{12})} \tag{63.13}$$

$$Tw_{41} = \frac{Q}{(1 - \rho_{11} - \rho_{12})(1 - \rho_{11} - \rho_{12} - \rho_{13})} \tag{63.14}$$

Where

$$Q = \frac{1}{2} \left[\lambda_1 \frac{K_1 + 1}{\mu_1^2} + q_{21}\lambda_2 \frac{K_{21} + 1}{\mu_{21}^2} + q_{41}\lambda_4 \frac{K_{41} + 1}{\mu_{41}^2} \right] \tag{63.15}$$

And

$$\rho_{11} = \frac{\lambda_1}{\mu_1}, \quad \rho_{12} = \frac{q_{21}\lambda_2}{\mu_{21}}, \quad \rho_{13} = \frac{q_{41}\lambda_4}{\mu_{41}} \tag{63.16}$$

Similarly, the layer2 queuing system can be evaluated, whose parameters are $(q_{22}\lambda_2, \lambda_3, q_{42}\lambda_4)$, $(1/\mu_{22}, 1/\mu_3, 1/\mu_{42})$ and (K_{22}, K_3, K_{42}), the results as below,

$$Tw_{22} = \frac{Q}{1 - \rho_{21}} \tag{63.17}$$

$$Tw_3 = \frac{Q}{(1 - \rho_{21})(1 - \rho_{21} - \rho_{22})} \tag{63.18}$$

$$Tw_{42} = \frac{Q}{(1 - \rho_{21} - \rho_{22})(1 - \rho_{21} - \rho_{22} - \rho_{23})} \tag{63.19}$$

Where

$$Q = \frac{1}{2} \left[q_{22}\lambda_2 \frac{K_{22} + 1}{\mu_{22}^2} + \lambda_3 \frac{K_3 + 1}{\mu_3^2} + q_{42}\lambda_4 \frac{K_{42} + 1}{\mu_{42}^2} \right] \tag{63.20}$$

And

$$\rho_{21} = \frac{q_{22}\lambda_2}{\mu_{22}}, \quad \rho_{22} = \frac{\lambda_3}{\mu_3}, \quad \rho_{23} = \frac{q_{42}\lambda_4}{\mu_{42}} \tag{63.21}$$

Especially, the mean wait time of C_2 and C_4 could be,

$$Tw_2 = q_{21}Tw_{21} + q_{22}Tw_{22} \tag{63.22}$$

$$Tw_4 = q_{41}Tw_{41} + q_{42}Tw_{42} \tag{63.23}$$

Table 63.1 Components' parameters 1

Device	$1/\lambda_i$	$1/\mu_i$	$\delta^2(\mu_i)$
LCDC(1)	200.22	41.326	0.012
MMA(2)	214.78	26.01	189.75
DMAC(3)	9997.643	32.04	0.1178
RISC(4)	5.3659	3.82	5.58

63.3 Experimental and Results

Our aim in this chapter is to provide a better choice of the bus structure according to REMUS. We make performance estimation based on single-layer and two-layer bus structure.

Performance estimation comparison between single-layer and two-layer bus structure was finished based on the same background application and the input parameters are list in Table 63.1. And according to MMA and RISC, they are connected both tow system buses in the two-layer bus architecture. Their parameters are $(q_{21}, q_{22}) = (0.543, 0.457)$, $(q_{41}, q_{42}) = (0.0806, 0.9194)$; $(1/\mu_{21}, 1/\mu_{22}) = (16.06, 37.82)$, $(1/\mu_{41}, 1/\mu_{42}) = (2.64, 3.92)$; $(\sigma^2(\mu_{21}), \sigma^2(\mu_{22})) = (177.63, 204.14)$, $(\sigma^2(\mu_{41}), \sigma^2(\mu_{42})) = (4.38, 5.69)$.

Then the estimation results are shown in Fig. 63.5, and we find that the mean wait time of LCDC, MMA and DMA in the two-layer bus structure decrease respectively 39.1, 59.4 and 77.7 % relative to the single-layer bus structure based on the same application.

63.4 Conclusion

This chapter proposes a method for estimating the performance of a microprocessor called REMUS based on queuing network model. To take analyze structure and arbitration, which are the most important two aspects of the on-chip bus, provide a better choice of the bus structure and arbitration algorithm. The results indicate that the two-layer bus structure is better than others in terms of the system

Fig. 63.5 Single-layer and two-layer bus estimation results

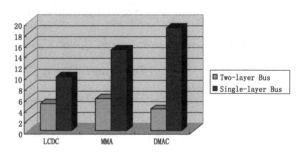

performance. With the help of the proposed estimating method, bus structure can be determined fast and accurately for the system designer.

Acknowledgments This work was supported by the National Natural Science Foundation of China (grant no.61204023 and grant no.61203251).

References

1. Marsan M A, Ballbo G, Conte G (1982) Comparative performance analysis of single bus multiprocessor architectures. IEEE Trans Comput C-31(12):1179–1191
2. Goyal A, Agerwala T (1984) Performance analysis of future shared storage systems. IBM J Res Devel 28(1):95–108
3. Tsyei TF, Vernon MK (1992) A multiprocessor bus design model validated by system measurement. IEEE Trans Parallel Distrib Syst 3(6):712–727
4. Onyuksel IH, Irani KB (1990) Markovian queueing network models for performance analysis of a single-bus multiprocessor system. IEEE Trans Comput 39(4):975–980
5. Bodnar BL, Liu AC (1998) Modeling and performance analysis of single-bus tightly-coupled multiprocessor. IEEE Trans Comput 38(3):464–470
6. Mahmud SM (1990) Performance analysis of asynchronous hierarchical-bus multiprocessor systems using closed queuing network models. Proc IEEE Int Symp Circuits Syst 4:2689–2692
7. Mahmud SM (1994) Performance analysis of multilevel bus networks for hierarchical multiprocessors. IEEE Trans Comput 43(7):789–805
8. Zhang LQ, Chardhary V (2002) On the performance of bus interconnection for SoCs. In: Proceedings of the 4th workshop on media and stream processors, pp 5–14
9. Zhu M, Liu L, Yin S et al (2010) A cycle-accurate simulator for a reconfigurable multi-media system. IEICE Trans Inf Syst 93:3210–3302

Chapter 64
Study on Time Delay Estimation Algorithm in Cable Monitoring Based on Wavelet Theory

Jiawen Zhou and Ying Liu

Abstract On the Basis of the principle of telecommunication cable fault detection using pulse reflection method, we proposed a generalized related transmission delay estimation method for reflected pulse. This method which bases on wavelet de-noising could estimate the transmission delay parameters effectively in condition of low SNR and theoretically overcomes the following limitations: small diameter of local cables, large attenuation to the test pulse signal and infeasibility for local cable fault detection. Simulation results have demonstrated the feasibility and effectiveness of this method.

Keywords Telecommunication cable · Monitoring · Wavelet · Time delay estimation

64.1 Introduction

In the modern information society, customers' demands for communication service quality are getting much higher that makes it of great significance for telecom operators to keep telecommunication lines smooth. As an important carrier for information transmission, telecommunication cables undertake the task of transferring relay information, local calls and other kinds of communication services.

Z. Jiawen (✉) · L. Ying
School of Electrical and Information Engineering,
Beijing Jiaotong University, Beijing 100044, China
e-mail: 11120208@bjtu.edu.cn

L. Ying
e-mail: liuying@bjtu.edu.cn

W. Lu et al. (eds.), *Proceedings of the 2012 International Conference on Information Technology and Software Engineering*, Lecture Notes in Electrical Engineering 210, DOI: 10.1007/978-3-642-34528-9_64, © Springer-Verlag Berlin Heidelberg 2013

Fig. 64.1 Inflatable cable pressure measurement diagram

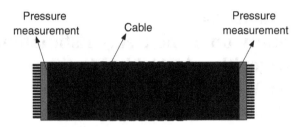

Among all kinds of telecommunication cable faults, disconnection breakdowns caused by anthropogenic and natural factors are more common. If there are any faults with the telephone lines, the telecom department needs to repair them as soon as possible and try their best to reduce the disable time in order to meet the needs of the customers. So it is crucial for telecom department to locate the position of the breakdowns timely and accurately so as to keep telecommunication lines smooth. However, it usually takes a long time to get some of the telecommunication lines back to work mostly because telecom department failed to locate the position of the breakdowns accurately enough. The main reason for this failure is that the test methods and measuring precision of the test instruments can't achieve ideal state.

Cable fault detection methods basically include air pressure method, capacitance method and pulse reflection method at present. If there happens to be any breakdowns in the cable filled with hydrogen, air pressure method could locate the position of the breakdown by analyzing the test points' hydrogen pressure changes, as is shown in Fig. 64.1. With this method, the accuracy of the measuring results mainly depends on the precision of the hydrogen sensor.

Capacitance method mainly makes use of the character that if there's a disconnection, the cable could be equivalent to a capacitor in electrical characteristics and the relationship between the length and the equivalent capacitance of the cable is linear, as is shown in Fig. 64.2. So the position of the breakdown could be determined by testing the size of the equivalent capacitance. Obviously, the accuracy of the cable fault location based on capacitance method depends on the measurement precision of the cable capacitance [1–3].

Pulse reflection method determines the state (such as open circuit, short circuit, etc.) and the location of the breakdown by sending test pulse at the test point and then analyzing the reflected time and the state of the pulse reflected by the breakdown [4, 5], as is shown in Fig. 64.3.

Compared with the other two methods, the structure of the test circuit for pulse reflection method is more complex, while the measurement precision is

Fig. 64.2 Open circuit cable equivalent to capacitance diagram

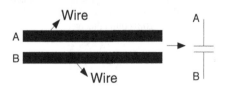

Fig. 64.3 Relationship
between the reflected pulse
and the emission pulse
diagram

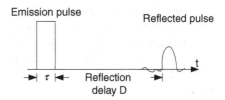

Emission pulse   Reflected pulse

accordingly much higher. Back to the 60s, pulse reflection method has already
been used in long-distance lines which greatly improved the accuracy of the
circuitry fault detection and reduced the disable time [3–5]. However, this
advanced method couldn't be used for the faults detection of local calls during a
long time. This is mainly because that local cable is voice-frequency cable whose
diameter is too small and therefore leads to large attenuation to the test pulse
signal. In order to weaken the influence brought by the cable to the test pulse and
the test noise, an efficient method measuring time delay for cable reflected pulse is
proposed in this paper. This method is based on wavelet de-noising theory which is
a kind of modern information processing theory.

64.2 Pulse Time Delay Estimation Method Based on Wavelet Transform

64.2.1 Correlation Estimate Method

Supposing that x(n) and y(n) are respectively the sample value of the emission
pulse signal and the reflected pulse signal. The measuring model for the time delay
parameters of the reflected pulse is

$$\begin{cases} x(n) = s(n) + n_1(n) \\ y(n) = \beta s(n - D) + n_2(n) \end{cases} \tag{64.1}$$

$$n = 0, 1, \ldots, N - 1$$

N is the length of the sampling data for the testing signal. Supposing that $n_1(n)$
and $n_2(n)$ are smooth, zero-mean, unrelated Gaussian white noise which is independent of $s(n)$. Usually, the power of $n_1(n)$ is small, while the power of $n_2(n)$
depends on the properties of the cable and the distance between the breakdown and
the test point. In this chapter, the SNR of the reflected pulse refers to the ratio of
maximum instantaneous power for the reflected pulse, i.e., $\beta s(n-D)$ and the
average power of the noise, i.e., $n_2(n)$. Here β is the attenuation factor, $\beta \in (0, 1)$;
D is the transmission delay of the reflected pulse.

In order to make it easier for signals to transfer in cables with different bandwidth characteristics, emission pulse $s(n)$ could be narrow-pulse shown in Fig. 64.3 and could also be other kinds of waveforms.

With this method, the estimation process of the transmission delay D for the reflected pulse is the process of estimating parameter D from finite-length $x(n)$ and $y(n)$.

The estimation process for the transmission delay τ based on correlation estimate method is shown as follows.

$$r_{xy}(\tau) = E\{x(n)y(n-\tau)\} = r_{ss}(\tau - D)(-\infty < \tau < +\infty) \qquad (64.2)$$

Including:

$$r_{ss}(\tau) = E\{s(n)s(n+\tau)\} \qquad (64.3)$$

Symbol $E\{\cdot\}$ represents mathematical expectation.

$r_{xy}(\tau)$ reaches peak at $\tau = D$. The estimation result for time delay τ is $\hat{D} = \arg\max_{\tau}\{r_{xy}(\tau)\}$.

In practical application, correlation estimate method can't get effective estimation results for reflected pulse transmission delay D in condition of low SNR.

64.2.2 Correlation Estimate Method Based on Wavelet De-Noising

There are a variety of wavelet de-noising methods [6–8]. The threshold method proposed by Mr. Donoho is introduced here in this chapter [9, 10]. This method considers that the wavelet coefficients of general signal contain abundant information. The number of large amplitudes isn't that much while the wavelet coefficients corresponding to the noise are identically distributed and the amplitude of the coefficients is rather even. So, among many of the wavelet coefficients, we set coefficients with smaller absolute values to zero and use hard-threshold or soft-threshold method to retain or expand coefficients with larger absolute values, in order to get the wavelet coefficients of the estimated signal. Then reconstruct the signal directly with the estimated wavelet coefficients, so as to achieve the purpose of de-noising. This reflected pulse time delay estimation method which is on the basis of wavelet threshold de-noising is shown in Fig. 64.4.

As is shown in Fig. 64.4, the first step is wavelet de-noising for the emission signal x(n) and the reflected signal y(n). Essentially, this process optimizes the minimum mean square error because the source signal $\hat{s}(n)$, recovered from the noise, can be as smooth as $s(n)$ to a large extent, so the mean square error between them can attain minimum. Taking related operations with signal $\hat{s}(n)$ and $\hat{s}(n-D)$ and then getting time delay estimation. This method can be seen as a special

Fig. 64.4 Time delay estimation method based on wavelet de-noising diagram

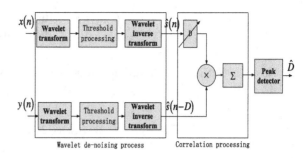

generalized related method which can be applied to the situation of dealing with related noises.

Basic procedure of correlation time delay estimate method which is based on wavelet de-noising is as follows.

1. First of all, take wavelet transform respectively with signal x(n) and y(n) who contain the Gaussian white noise and get two groups of wavelet coefficients;

$$Wx(j,k) = 2^{-\frac{j}{2}} \sum_{n=0}^{N-1} x(n)\psi\left(2^{-j}n - k\right) \tag{64.4}$$

$$Wy(j,k) = 2^{-\frac{j}{2}} \sum_{n=0}^{N-1} y(n)\psi\left(2^{-j}n - k\right) \tag{64.5}$$

2. Select suitable wavelet basis function $\psi(n)$;
3. Select the threshold λ_x and λ_y [9, 10];

$$\lambda_x = \sigma_x \sqrt{\frac{2\log(N)}{N}} \tag{64.6}$$

$$\lambda_y = \sigma_y \sqrt{\frac{2\log(N)}{N}} \tag{64.7}$$

σ_x and σ_y respectively stand for the power of the signals.

4. Take threshold process according to Eqs. (64.3, 64.4) and get $\hat{W}x(j,k)$ and $\hat{W}y(j,k)$;

$$\hat{W}x(j,k) = \begin{cases} Wx(j,k), & |Wx(j,k)| \geq \lambda_x \\ 0 & |Wx(j,k)| < \lambda_x \end{cases} \tag{64.8}$$

$$\hat{W}y(j,k) = \begin{cases} Wy(j,k), & |Wx(j,k)| \geq \lambda_y \\ 0 & |Wx(j,k)| < \lambda_y \end{cases} \qquad (64.9)$$

5. Take wavelet inverse transform and get de-noising signal estimation $\hat{s}(k)$ and $\hat{s}(k-D)$;

$$\hat{s}(n) = \sum_{j,k} Wx(j,k)\psi\left(2^{-j}n - k\right) \qquad (64.10)$$

$$\hat{s}(n-D) = \sum_{j,k} \hat{W}y(j,k)\psi\left(2^{-j}n - k\right) \qquad (64.11)$$

6. Calculate the cross-correlation function.

$$\hat{R}_{\hat{s}\hat{s}}(\tau) = \frac{\beta}{N} \sum_{k=0}^{N-\tau-1} \hat{s}(k)\hat{s}(k - D + \tau),$$

$$\tau = 1 - N, 2 - N, \ldots, 0, 1, \ldots N - 1 \qquad (64.12)$$

The peak value of $\hat{R}_{\hat{s}\hat{s}}(\tau)$ is the estimated value $\hat{D}$ for the reflected pulse's transmission delay parameter D, that is to say, the estimated value for parameter D is $\hat{D} = \arg\max_{\tau}\{\hat{R}_{\hat{s}\hat{s}}(\tau)\}$.

64.3 Simulation Experiment

In the process of simulation experiment, supposing that the length of the cable is L kilometers, so the maximum delay time for transmission reflection is $T_L = \frac{L}{c}$, c is the test pulse signals' propagation velocity in cable and the size of T_L is proportional to the length of the cable. At this time, the period of the test pulse T must be larger than T_L.

Experiment 1: Supposing that the test pulse is rectangular narrow pulse, pulse width is 1 ms, sampling period of the signal is Ts = 0.01 ms and the transmission delay for the test pulse reflected to the emission point which has passed the breakdown is D = 8T s. The SNR for signal x(n) is 20 dB and for signal y(n) is 0 dB. The estimation result for the reflected pulse transmission time parameter D which is based on the traditional correlation estimate method is shown in Fig. 64.5. The peak value is the estimation result for D. It is obvious that parameter D can be estimated effectively. When the SNR for signal y(n) is −10 dB, the estimation result for the reflected pulse transmission time parameter D which is based on the traditional correlation estimate method is shown in

Fig. 64.5 Estimation result
for reflected signal time delay
D basing on correlation
estimate method while
SNR2 = 0 dB

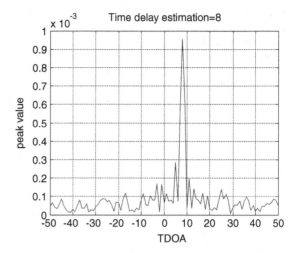

Fig. 64.6 Estimation result
for reflected signal time delay
D basing on correlation
estimate method while
SNR2 = −10 dB

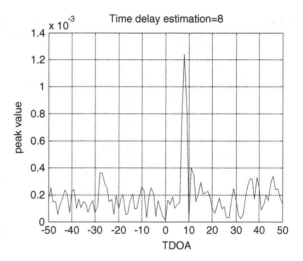

Fig. 64.6. As we can see, in condition of low SNR, the reflected pulse transmission
time parameter D can't be accurately estimated.

Experiment 2: Supposing that the test pulse is rectangular narrow pulse, pulse
width is 1 ms, sampling period of the signal is Ts = 0.01 ms and the transmission
delay for the test pulse reflected to the emission point which has passed the
breakdown is D = 8Ts. The SNR for signal x(n) is 20 dB and for signal y(n) is
−10 and −20 dB respectively. The estimation results for the reflected pulse
transmission time parameter D which is based on wavelet de-noising method are
shown in Figs. 64.7 and 64.8. Obviously, the method proposed in this chapter can
estimate the transmission time parameters efficiently in condition of low SNR.

Experiment 3: Supposing that the test pulse is carrier pulse, pulse width is 1 ms,
the frequency for the cosine carrier is 2 kHz, sampling period of the signal is

Fig. 64.7 Estimation result
for rectangular narrow pulse
signal time delay D basing on
wavelet de-noising
correlation estimate method
while SNR2 = −10 dB

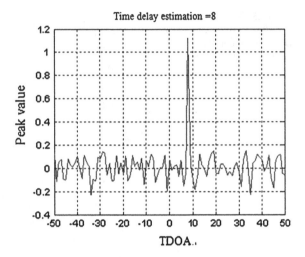

Fig. 8 Estimation result for
rectangular narrow pulse
signal time delay D basing on
wavelet de-noising
correlation estimate method
while SNR2 = −20 dB

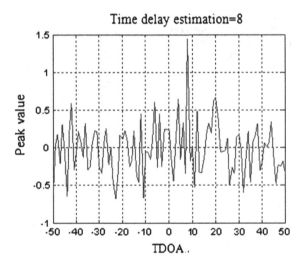

Ts = 0.01 ms and the transmission delay for the test pulse reflected to the emission point which has passed the breakdown is D = 8Ts.The SNR for signal x(n)is 20 dB and for signal y(n)is −10 and −20 dB respectively. The estimation results for the reflected pulse transmission time parameter D which is based on wavelet de-noising method are shown in Figs. 64.9 and 64.10. As we can see, there is an obvious improvement with the estimation result.

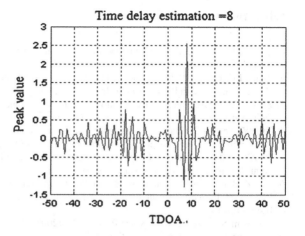

Fig. 9 Estimation result for carrier pulse signal time delay D basing on wavelet de-noising correlation estimate method while SNR2 = −10 dB

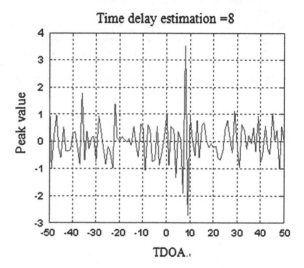

Fig. 10 Estimation result for carrier pulse signal time delay D basing on wavelet de-noising correlation estimate method while SNR2 = −20 dB

64.4 Conclusion

In this chapter, wavelet de-noising theory is exploited in the process of cable fault detection using pulse reflection method. On condition of large attenuation to reflected signal and low SNR at the test point, the transmission delay for reflected pulse can be estimated accurately. This method provides reliable theory basis to realize accurate positioning for the cable breakdowns and also broaden the application scope for pulse reflection method in cable fault detection. Simulation results have demonstrated the feasibility of this method.

Acknowledgments This paper is supported by National Natural Science Foundation of China (Grant No. 61172130).

References

1. Chai Z, Liu GZ, Che XG (2006) Development on communication cable monitor. Electron Meas Instrum J 20(2):93–97, (in Chinese)
2. Zhang LC, Liu Y (1999) Design and realization of cable fault centralized monitoring system. Chang Chun Post and Telecommun Inst J 17(4):35–39, (in Chinese)
3. Zhang JY Application of activeX automation techniques in cable monitoring system. J Jilin Univ (Information science edition) 21(3), (in Chinese)
4. Xu BY et al. (2001) Barrier testing technique for communication cable. Beijing University of Posts and Telecommunications Press, Beijing, 2001, (in Chinese)
5. He ZY (2001) Design of S-C microcomputer control system for carrier detection of cable cut point. Electron Autom 3:6–8, (in Chinese)
6. Hao Z, Blackburn TR, Phung BT, Sen D (2007) A novel wavelet transform technique for on-line partial discharge measurements. 1. WT de-noising algorithm. IEEE Trans Dielectr Electr Insul 14:3–14
7. Seramani S, Taylor EA, Seekings PJ, Yeo KP (2007) Wavelet de-noising with independent component analysis for segmentation of dolphin whistles in a noisy underwater environment OCEANS 2006—Asia Pacific, pp 1–7
8. Wenqing H, Yuxing D (2006) Adaptive thresholding denoising algorithm based on cross-validation 2006 International conference on communications. Circ Syst Proc 1:276–279
9. Xu C, Zhao RZ, Gan XB (2004) Wavelet analysis applied algorithm. Science Press, Beijing, 200, (in Chinese)
10. Donoho DL (1995) De-noising by soft-thresholding. IEEE Trans IT 41(3):613–616

Chapter 65
Design and Implementation of a Security System for Mobile Terminals' Digital Content

Honglei Liu, Haihong E and Danzhi Wang

Abstract With the continuous development of mobile devices, the digital content on users' mobile terminals is facing more security challenges. This paper attempts to realize a system that allows users to backup their personal data securely, by taking the measure of combining digital signature with digital envelope. In the end, it verified the efficiency and reliability of the research through a complete workflow. Experiment proves that the system can effectively guarantee the security of digital content on mobile terminals.

Keywords Mobile terminals · Digital content · Data encryption · Digital signature

65.1 Introduction

With the processing capabilities of mobile terminals continuously enhanced, more and more business can be carried out on them. Moreover, because of their mobility compared with personal computers, growing number of people have mobile phones [1]. However, as the digital content and business dependent on mobile

H. Liu · H. E (✉) · D. Wang
PCN&CAD Center, Beijing University of Posts and Telecommunications,
Beijing 100876, China
e-mail: ehaihong@bupt.edu.cn

H. Liu
e-mail: liu19880@163.com

D. Wang
e-mail: dzwang@bupt.edu.cn

W. Lu et al. (eds.), *Proceedings of the 2012 International Conference on Information Technology and Software Engineering*, Lecture Notes in Electrical Engineering 210, DOI: 10.1007/978-3-642-34528-9_65, © Springer-Verlag Berlin Heidelberg 2013

terminals is becoming more varied, the security issues of mobile terminals have gained great attention. It will directly affect the users' experience if you cannot provide secure services [2].

Digital content refers to products or services integrated from images, text, videos, voice and so on, by high-tech tools and information technology [3]. It includes six important elements: text, images, audios, videos, graphics and software.

At present, the risks which mobile terminals are faced with are as follows: (1) Eavesdropping, which means obtaining communicating or stored information illegally, considering the fact that anyone can monitor user's data in the air through the digital wireless interface of scanning equipment now [4]. (2) Masquerade, which means attackers can disguise as legitimate servers or communicators to gain access to sensitive data [5]. (3) Tampering, which means attackers can destroy the integrity of data by modifying, inserting, reordering the information illegally. (4) Repudiation, which is divided into receiving repudiation and source repudiation. The former refers that the receiver denies he has received the information, and the latter refers that the sender denies he has sent the information [6]. (5) Unauthorized access of resources, which refers to the illegal use of banned resources, or unauthorized use of encryption terminal [7].

For personal users, the digital content on their mobile terminals is also faced with the same risks, and there is also a risk of data loss [8]. Therefore, how to safely backup your personal digital content has become a hotspot at present.

In PC end, we mainly use PKI technology to ensure the security and confidentiality of data during transmission [9]. Similarly, we can use PKI's theory of digital envelope and digital signature to backup the information on mobile terminals as well.

65.2 Overseas and Domestic Research Status

In order to backup the digital information on mobile terminals safely and quickly, the mobile terminal researchers at home and abroad as well as major Internet companies are exploring and introducing many products.

HuaXinAnChuang Company has developed an end-to-end encryption product to ensure the security of mobile phone text messages during storage and transmission—BaiMiXin, which adopts the next generation of asymmetric encryption technology—IBC technology. It can ensure that the SMS is secure once it leaves the phone, whether it is in the air during the transmission, or stored in the operators SMS gateway. Even if the messages are intercepted, others cannot view the content without the correct decryption key. IBC technology uses the recipient's mobile phone number as the public key [10], so we don't need to exchange a public key online or agree on a key offline, which makes the product simple and secure.

Cloak SMS Free, a product of the Hamish Medlin Company, is used to encrypt the short messages between mobiles of android operating system. It adopts the symmetric key encryption technology, using AES as the encryption algorithm. Before sending a message, the sender needs to enter a previously defined password as the key, while the recipient needs to enter a same password to decrypt the message. Unfortunately, if the recipient forgets the password, he cannot decrypt the message any more.

Text Secure is an open-source android application used to encrypt SMS. This application uses ECC algorithm to generate the asymmetric key, and uses 128 bit AES algorithm to encrypt the data and 256 bit HMAC–SHA1 algorithm to generate a digital summary for data validation [11]. Besides, this application provides the function to export and import SMS. The main drawback of the application is that it can only be used for SMS.

In addition, domestic Internet companies' products such as Qihoo 360s 360 Mobile Assistant, ZhuoYiXunChang's pea pods assistant and so on all provide a backup function of the digital content. But all of these products don't have encryption capabilities, only simply transfer data to physical devices. It can be seen that present digital content backup for mobile terminals is mainly in the form of plaintext transfer, and encryption algorithm is mostly symmetric, such as DES, IDEA, AES, etc., [12]. While sending encrypted messages, the sender needs to tell the password to the recipient, through phone, SMS and so on. While receiving the message, the recipient needs to enter the password to decrypt it. In this static password mode, password consultation is completely in artificial way, and using the same password for a long time can bring security implications. In addition, some people have tried to download key based on SIM card [13]. But it's not conducive to the expansion in the late stage as its storage space is too small. Based on the above, this paper applies the concept of digital envelope, digital signature of PC-side's data encryption technology to mobile devices' content security, designing and implementing a security system for mobile terminals' digital content.

65.3 Design of the System

The system adopts a combination of symmetric key and asymmetric key to securely backup personal mobiles' digital content. Encryption process is divided into three steps: Firstly, mobile devices generate the symmetric key and encrypt the digital content. Secondly, the key is encrypted by the public key generated by the server. Finally, after finishing digital signature to message digest, the data packet is uploaded to the server. As for the decryption process, firstly, the user needs to download the encrypted data packet from server, then decrypts the symmetric key using personal private key, and verifies the digital signature using his public key to see if the file has been tampered with. If not, he can use the symmetric key to decrypt the encrypted data and obtain the plaintext in the end.

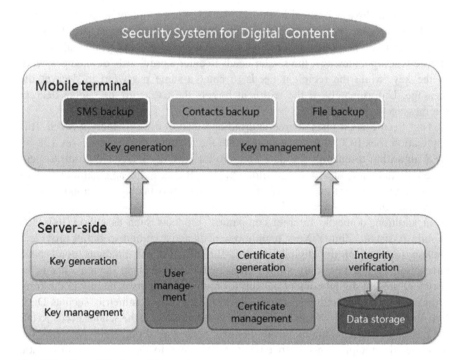

Fig. 65.1 The architecture of the system

65.3.1 The Overall Architecture

Figure 65.1 illustrates the architecture of the system.

The system can be divided into two parts: the mobile terminal and the server-side. The mobile terminal provides the interfaces of data encryption, data decryption, key generation and key management for users to backup the SMS, contacts and files. The server-side is mainly for certificate management and data content management. The certificate management module provides users with an interface to generate digital certificate and private key. The data content management interface is mainly used to verify the integrity of data and store encrypted data packet.

65.3.2 Module Design

65.3.2.1 Mobile Terminal Interfaces

- SMS backup

This module is to provide users with an SMS backup interface. After selecting the SMS backup option in the menu, the system will automatically validate the user's information. If it is verified, the short message will be encrypted and uploaded to the server. If not, the user will need to manually enter his personal information beforehand. When the user wants to restore the backup data, he just needs to download the encrypted files from the server.

- Contacts backup

This module is similar to the SMS backup module, after selecting this option, encrypted data is uploaded to the server if the user is verified. While restoring, the user needs to download the appropriate data from the server.

- File backup

This module helps to backup files. Firstly, the user needs to select the object file. Then, after validating the user's information, encrypted file is uploaded to the server. For decryption, the user only needs to download the corresponding file.

- Key generation

Due to the performance of mobile terminals, this system chooses to use symmetric key to encrypt digital content. This interface adopts the AES algorithm to generate a symmetric key, and produces a new key each time when users backup their data.

- Key management

The user's key file is stored in the mobile terminal's memory card and he can obtain it by calling the operation interface in the bottom.

65.3.2.2 Server-Side Interfaces

The server-side is mainly responsible for certificate management, key management, user management and digital content management.

- Key generation

We use this module to generate asymmetric keys. Compared with the symmetric key, it's of higher security, at the cost of high complexity of algorithm.

- Key management

The asymmetric keys generated by the above interface are stored in a key store, which records users' keys and provides CRUD functions to them.

- Certificate generation

The module is to generate digital certificate by underlying java interface. After entering personal information, it will produce the digital certificate according to the personal information authentication.

- Certificate management

We use this module to maintain the information of certificates produced by the server. It also provides an interface for users to view and update their certificates. Besides, by setting aside invalid certificates information, it can improve the system's security.

- User management

When using this system for the first time, the user is required to register an account, which is bound with user information. Only authenticated users are able to upload the encrypted files.

- Integrity verification

This module is to ensure data security, to prevent data from being tempered with or loss of information. The server verifies the signature of encrypted packet, if it's not verified, the server will prompt a message that the data is incomplete.

- Data storage

This module is mainly used to save the users' encrypted data to the MySQL database server. The server records all the versions of the backup information, and the user can choose the particular version which to restore. Besides, the database system provides timing backup function to ensure no data loss.

65.4 Workflow of the System

65.4.1 Workflow of the System

Figure 65.2 shows the workflow of the system.

1. Generate the symmetric key and encrypt the data;
2. Generate message digest and sign it with private key;
3. Encrypt symmetric key with public key;
4. Decrypt the symmetric key which has been encrypted, using private key;
5. Decrypt the data which has been encrypted, using the symmetric key obtained in the former step;
6. Verify the signature using public key.

65.4.2 Experimental Data from the System

According to the definition of the mobile terminal's digital content, the objective of the test is mainly documents, audios, videos, software and other digital content, and test steps are as follows:

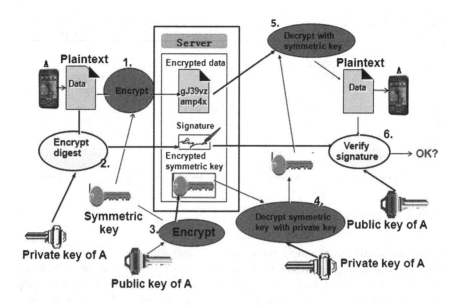

Fig. 65.2 Workflow of the system

1. Test environment

Mobile terminal: Samsung Galaxy I9100, Android2.3 operating system
Server-side: Windows xp, Tomcat6
Network environment: wifi of 2M bps bandwidth.

2. Test results

The results are shown in Table 1.1 through encrypting and decrypting audios, videos, documents, applications, SMS, contacts and other digital content. And the backup time refers to the sum of the encryption time and upload time.

3. Test conclusions

The results are in line with expectations as a whole, which doesn't exceed the users' time tolerance for network access. We can hardly feel any difference while encrypting small digital content, though there is some delay in the encryption process of large files. However, there is little difference in time of encrypted backup process compared with non-encrypted backup process by 91 Phone Assistant and 360 Mobile Assistant using the same data file. Decryption process, as a whole, takes much shorter time than the encryption process. This is because the system needs to encrypt the data first and then upload it to the server-side for backup, in addition to subject to the limitations of the encryption algorithm, it's also affected by the network environment.

Table 65.1 Test result

File format	File size	Backup time(ms)	Decrypt time(ms)
Doc	30 K	722	285
Doc	300 K	758	307
Mp3	8.3 M	2,315	1,023
Jpg	1.83 M	986	569
3gp	27 M	5,675	2,086
Pdf	1.28 M	1,024	478
Exe	9.2 M	1,047	532
SMS	124	532	82
Contracts	267	876	229

65.5 Conclusion

Mobile terminals are playing a more and more important role in peoples' digital life, as the continuous development and the tremendous advantage of mobility compared to the PC. In the future, the mobile terminal is not just a tool for calling and sending messages, more is to act as a mobile computer. Therefore, it's necessary to control the transmission and storage of the digital content. However, due to the particularity of mobile terminals and insecure factors such as theft and masquerade, it's difficult to guarantee the security of digital content during transmission and storage process. On the other hand, how to ensure the security of digital content after the backup process is equally important. The security system achieved by this paper aims to solve these problems from the perspective of digital content encryption, authentication, data storage, data integrity verification and so on. Practice shows that the system can effectively guarantee the security of digital content on mobile terminals. It will be of great significance in both theory and practical application.

Acknowledgments This work is supported by the National Key project of Scientific and Technical Supporting Programs of China (Grant No. 2009BAH39B03); the National Natural Science Foundation of China (Grant No. 61072060); the National High Technology Research and Development Program of China (Grant No. 2011AA100706);the Research Fund for the Doctoral Program of Higher Education (Grant No. 20110005120007); the Fundamental Research Funds for the Central Universities(2012RC0205); the Co-construction Program with Beijing Municipal Commission of Education; Engineering Research Center of Information Networks, Ministry of Education.

References

1. Anurag K et al (2008) Wireless networking. Elsevier Morgan Kaufmann Publishers, San Francisco
2. Balachandran K, Rudrapatna A (2009) Performance assessment of next-generation wireless mobile sys tems. Bell Labs technical Jornal, 33–58

3. Wang M, Fan K, Li X et al (2008) A Digital TV Content Protection System. Chin J Electron 17(3):427–431
4. Maggi F, Volpatto A, Gasparini S et al. (2010). Don't touch a word! a practical input eavesdropping attack against mobile touchscreen devices. Politecnico di Milano, Tech. Rep. TR-2010:58–59
5. Ashley P, Hinton H (2001) Wired versus wireless security. The internet, WAP and mode for e-business. In proceedings of annual computer security application conferences
6. Dunkelman O, Keller N, Shamir A (2010) A Practical-Time Attack on the A5/3 Cryptosystem Used in Third Generation GSM Telephony. Cryptology ePrint Archive, Report 2010/013
7. Yang Y (2008) An integrated security framework for mobile business application. WiCOM
8. Zhou H-Q (2008) PKI-based E-Business Security System. 3rd international conference on innovative computing information and Control, 18–20
9. Wenbo H.(2009) Smock: A scalable method of cryptographic key management for mission-critical networks. IEEE Transactions on Information Forensics and Security, 4(1):144–150
10. Cheng Z (2011) Application of WPKI in Mobile business Security. China management informationization: 20–21
11. Haleem M, Chetan C, Chandramouli R, Subbalakshmi KP (2007) Opportunistic encryption: A trade-off between security and throughput in wireless networks. IEEE Trans Dependable Secure Comput 4(4):313–324
12. Monares A, Ochoa SF, Pino JA et al. (2010) Mobile computing in urban emergency situations: Improving the support to firefighters in the field. Expert systems with applications, 1–13
13. Caliskan D (2011) An application of RSA in data transfer. Application of Information and Communication Technologies (AICT), 1–4

Chapter 66
Kane's Approach to Modeling Underwater Snake-Like Robot

Ke Yang, Xu Yang Wang, Tong Ge and Chao Wu

Abstract The paper presents a detailed methodology for dynamic modeling of underwater snake-like robot using Kane's dynamic equations. This methodology allows construction of the dynamic model simply and incrementally. The generalized active forces and the generalized inertia forces were deduced. The forces which contribute to dynamics were determined by Kane's approach. The model developed in this paper includes gravity, control torques and three major hydrodynamic forces: added mass, profile drag and buoyancy. The equations of hydrodynamic forces were deduced. The methodology provides a direct method for incorporating external environmental forces into the model. The resulting model is obtained in closed form. It is computationally efficient and provides physical insight as to what forces really influence the system dynamics. The dynamic model provides a framework for modern model-based control schemes.

Keywords Underwater snake-like robot · Dynamic modeling · Kane's dynamic equations · Generalized active forces · Generalized inertia forces · Hydrodynamic forces

K. Yang (✉) · X. Y. Wang · T. Ge · C. Wu
Underwater Engineering Institute, Shanghai Jiao Tong University, Shanghai, China
e-mail: yjs2yangke@163.com

X. Y. Wang
e-mail: wangxuyang@sjtu.edu.cn

T. Ge
e-mail: tongge@sjtu.edu.cn

C. Wu
e-mail: wuchaorr@sjtu.edu.cn

W. Lu et al. (eds.), *Proceedings of the 2012 International Conference on Information Technology and Software Engineering*, Lecture Notes in Electrical Engineering 210, DOI: 10.1007/978-3-642-34528-9_66, © Springer-Verlag Berlin Heidelberg 2013

66.1 Introduction

There is a need for robots that move on and within complex material such as sand, rubble, and loose debris. For example, such robots could locate hazardous chemical leaks, function as self-propelled inspection devices, and search for victims in disaster sites [1]. Inspired by biological snake locomotion, snake robots carry the potential of meeting the growing need for robotic mobility in unknown and challenging environments [2]. They may push against rocks, branches, or other obstacles to move forward more efficiently [3]. Snake robots typically consist of serially connected joint modules that are capable of bending in one or more planes. The many degrees of freedom of snake robots make them difficult to control but provide traversability in irregular environments that surpasses the mobility of the more conventional wheeled, tracked, and legged forms of robotic [3].

To achieve better performance on adaptive dynamic walking, biologically inspired approaches or biological concepts based control are introduced to various robot control system [4, 5]. It has been widely recognized that computer modeling and simulation can provide valuable contributions to the research of biological concepts based control [6, 7]. In this paper, we establish an underwater snake-like robot model using Kane's dynamic equations. Kane's approach provides a direct method for incorporating external environmental forces into an underwater snake-like robot model. The resulting model can provide physical insight as to what forces really influence the dynamic model.

66.2 Kinematic Analysis

In this section we will develop expressions for the linear and angular velocities, and linear and angular accelerations for underwater snake-like robot. The Link 1 is a rigid mass with 3 translational and 3 rotational degrees of freedom. The other Link j has one degree of freedom which allows rotation about z^j.

66.2.1 Coordinate System

Consider a fixed inertia frame, {I} as in Fig. 66.1. Axes x^I and y^I are horizontal.

At the Link j, frame {j} is attached. The superscript on the right side of any quantity will denote the rigid body which it refers to. The superscript on the left will denote the reference frame with respect to which the quantity is expressed. When omitted, frame {I} is assumed.

Each frame is related to the fixed frame through a rotation matrix ${}_j^I C$, indicating transformation of vectors in frame {j} to the inertia frame {I}. The robot is

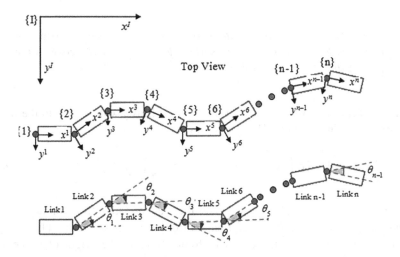

Fig. 66.1 Frames assignment on the underwater snake-like robot

supposed to have $n-1$ rotational joints (Fig. 66.1). Define the generalized speeds of the system as:

$$\dot{q}=\left[v_x^1 \ v_y^1 \ v_z^1 \ \omega_x^1 \ \omega_y^1 \ \omega_z^1 \ \dot{\theta}_1,\ldots,\dot{\theta}_{n-1}\right]^T \tag{66.1}$$

where, $v_o^1 = \left[v_x^1 \ v_y^1 \ v_z^1\right]^T$ is the linear velocity of the origin of frame $\{1\}$ with respect to the frame $\{I\}$. $\omega^1 = \left[\omega_x^1 \ \omega_y^1 \ \omega_z^1\right]^T$ is the angular velocity of Link 1 with respect to the frame $\{I\}$. $\left(\dot{\theta}_1,\ldots,\dot{\theta}_{n-1}\right)$ are joint speeds of the underwater snake-like robot.

66.2.2 Position Vector

The position vector of the CM (center of mass) of Link 1 with respect to the frame $\{I\}$ expressed in the frame $\{I\}$ is given by:

$$p_{cm}^1 = p_o^1 + {}_1^I C^1 p_{cm}^1 \tag{66.2}$$

where, $p_o^1 = [x^1 \ y^1 \ z^1]^T$ is the position vector of the origin of frame $\{1\}$ with respect to the frame $\{I\}$. ${}^1 p_{cm}^1$ is the position vector from the origin of frame $\{1\}$ to the CM of Link 1. ${}_1^I C$ is the transformation from frame $\{1\}$ to frame $\{I\}$.

$$\begin{aligned}
{}_1^I C &= {}_1^I C^T \\
&= \begin{bmatrix}
\cos\theta\cos\psi & -\cos\phi\sin\psi+\sin\phi\sin\theta\cos\psi & \sin\phi\sin\psi+\cos\phi\sin\theta\cos\psi \\
\cos\theta\sin\psi & \cos\phi\cos\psi+\sin\phi\sin\theta\sin\psi & -\sin\phi\cos\psi+\cos\phi\sin\theta\sin\psi \\
-\sin\theta & \sin\phi\cos\theta & \cos\phi\cos\theta
\end{bmatrix}
\end{aligned}$$
(66.3)

The position vector of the CM of an arbitrary Link j with respect to the frame {I}, expressed in the frame {I} is given by:

$$p_{cm}^j = p_o^1 + {}_1^I C\left(\sum_{j=2}^{j} \left({}_{i-1}^1 C^{i-1}p_o^i\right) + {}_j^1 C^j p_{cm}^j\right) \quad (j = 2,\ldots,n)$$
(66.4)

where, ${}_{i-1}^1 C = {}_1^{i-1} C^T = {}_1^1 C_2^1 C_3^2 C \ldots {}_{i-1}^{i-2} C$ is the transformation from frame {i−1} to frame {1}. ${}_1^1 C = I_{3\times3}$ is the transformation from frame {1} to frame {1}. ${}^{i-1}p_o^i$ is the position vector from the origin of frame {i-1} to the origin of frame {i}, expressed in frame {i−1}. ${}_j^1 C = {}_1^j C^T = {}_2^1 C_3^2 C \ldots {}_j^{j-1} C$ is the transformation from frame {j} to frame {1}. ${}^j p_{cm}^j$ is the position vector from the origin of frame {j} to the CM of Link j, expressed in frame {j}.

$$\substack{j-1 \\ j}C = {}_{j-1}^j C^T = \begin{bmatrix}
\sin\theta_{j-1} & 0 & \cos\theta_{j-1} \\
0 & 1 & 0 \\
-\cos\theta_{j-1} & 0 & \sin\theta_{j-1}
\end{bmatrix}$$
(66.5)

66.2.3 Angular Velocity and Linear Velocity

The angular velocity of an arbitrary Link j with respect to the frame {I}, expressed in the frame {j} is given by:

$$\begin{aligned}
{}^I\omega^j &= \omega^1 + {}_1^I C\sum_{i=2}^{j} \dot{\theta}_{i-1} \quad (j = 1,\ldots,n) \\
{}^j\omega^j &= {}_I^j C^I\omega^j \quad (j = 1,\ldots,n)
\end{aligned}$$
(66.6)

where, ${}_I^j C = {}_1^j C_I^1 C$ is the transformation from frame {I} to frame {j}. $\dot{\theta}_{i-1}$ is the angular velocity of Link i with respect to Link i−1, expressed in frame {i−1}.

The linear velocity of the CM of an arbitrary Link j with respect to the frame {I}, expressed in the frame {j} is given by:

$$\substack{j\\}v_{cm}^j = {}_I^j C\frac{dp_{cm}^j}{dt} \quad (j = 1,\ldots,n)$$
(66.7)

66.2.4 Angular Acceleration and Linear Acceleration

The angular acceleration of an arbitrary Link j with respect to the frame {I}, expressed in the frame {j} is given by:

$$^j\alpha^j = {}^j_I C \frac{d^I \omega^j}{dt} \quad (j = 1, \ldots, n) \tag{66.8}$$

The linear acceleration of an arbitrary Link j with respect to the frame {I}, expressed in the frame {j} is given by:

$$^j a^j_{cm} = {}^j_I C \frac{d^2 p^j_{cm}}{dt^2} \quad (j = 1, \ldots, n) \tag{66.9}$$

66.3 Dynamic Analysis

In this section we will develop a dynamic model for an underwater snake-like robot. The task involves applying Kane's dynamic equations to a system composed of n Links. The technique requires that we develop expressions for the generalized inertia forces of the underwater snake-like robot, and expressions for all of the generalized active forces which act on the robot.

66.3.1 Gravity, Buoyancy and Inertia Forces

Gravity can be treated as a generalized active force which acts at the CM of an arbitrary Link j.

$$G^j = m^j g \quad (j = 1, \ldots, n) \tag{66.10}$$

The generalized active force due to gravity is given by the following:

$$(F_r)_G = \sum_{j=1}^{n} \left(\frac{\partial^j v^j_{cm}}{\partial \dot{q}_r} \cdot G^j \right) \quad (r = 1, \ldots, n+5) \tag{66.11}$$

where, $\dot{q}_r$ is the *rth* generalized speed.

The buoyancy is proportional to the mass of the fluid displaced by the Link j and acts through the center of buoyancy of the Link j.

$$B^j = -\rho V^j g \quad (j = 1, \ldots, n) \tag{66.12}$$

The generalized active force due to buoyancy is given by the following:

$$(F_r)_B = \sum_{j=1}^{n} \left(\frac{\partial^j v_{cb}^j}{\partial \dot{q}_r} \cdot B^j \right) \quad (r = 1, \cdots, n+5)$$

$$^j v_{cb}^j = {}^j v_{cm}^j + {}^j \omega^j \times {}^j p_{GB}^j$$

(66.13)

where, $^j v_{cb}^j$ is the linear velocity of the center of buoyancy of Link j, with respect to frame {I}, expressed in frame {j}. $^j p_{GB}^j$ is the position vector from center of mass to center of buoyancy of Link j, expressed in frame {j}.

The generalized inertia force of the system requires that we develop expressions for the inertia force and torque of Link j in the system. The inertia force of an arbitrary Link j, is given by the following:

$$R_j^* = -m^j \, {}^j a_{cm}^j \quad (j = 1, \ldots, n)$$

(66.14)

The inertia torque of an arbitrary Link j is given by the following:

$$T_j^* = -{}^j \alpha^j \cdot I^j - {}^j \omega^j \times I^j \cdot {}^j \omega^j \quad (j = 1, \ldots, n)$$

(66.15)

where I^j is the central inertia matrix of Link j.

The generalized inertia force for the system is given by the following:

$$F_r^* = \sum_{j=1}^{n} \left(\frac{\partial^j \omega^j}{\partial \dot{q}_r} \cdot T_j^* + \frac{\partial^j v_{cm}^j}{\partial \dot{q}_r} \cdot R_j^* \right) \quad (r = 1, \cdots, n+5)$$

(66.16)

66.3.2 Hydrodynamic Forces

Two types of hydrodynamic effects were considered in the dynamical model of the underwater snake-like robot. The first one consists of the forces and moments induced on the robot due to drag, and the second one is the effect of added mass and moment of inertia. To include the hydrodynamic effects, we assumed the Links as rectangular boxes.

66.3.2.1 Drag Force

The drag forces are mainly determined by flow turbulence, body shape, and skin roughness.

$$F_d^j = -\frac{1}{2} \rho \left\{ \begin{array}{c} C_{dx}^j A_{px}^j \left| {}^j v_x^j \right| {}^j v_x^j \\ C_{dy}^j A_{py}^j \left| {}^j v_y^j \right| {}^j v_y^j \\ C_{dz}^j A_{pz}^j \left| {}^j v_z^j \right| {}^j v_z^j \end{array} \right\} \quad (j = 1, \ldots, n)$$

(66.17)

where A_{px}^j, A_{py}^j and A_{pz}^j are the projected areas to the x^j, y^j and z^j axes. $^jv_x^j$, $^jv_y^j$ and $^jv_z^j$ are the relative translational velocities in x^j, y^j and z^j directions respectively. C_{dx}^j, C_{dy}^j and C_{dz}^j are drag coefficient.

The hydrodynamic moments are calculated by integrating the infinitesimal moment components due to the drag force around the body. When the body is rotated about z^j with an angular velocity ω_z^j, the drag forces from x^j and y^j directions will induced a hydrodynamic moment. The infinitesimal drag forces in x^j and y^j directions and their resulting moment components are expressed as:

$$
\begin{aligned}
dF_{dx}^j &= -\frac{1}{2}C_{dx}^j\rho\left(h_j dy^j\right)\left|v_x^j\right|v_x^j \\
dF_{dy}^j &= -\frac{1}{2}C_{dy}^j\rho\left(h_j dx^j\right)\left|v_y^j\right|v_y^j \\
dM_{dz,x}^j &= dF_{dx}^j y^j = -\frac{1}{2}C_{dx}^j\rho\left(h_j dy^j\right)\left|^j\omega_z^j y^j\right|\left(^j\omega_z^j y^j\right)y^j \\
dM_{dz,y}^j &= dF_{dy}^j x^j = -\frac{1}{2}C_{dy}^j\rho\left(h_j dx^j\right)\left|^j\omega_z^j x^j\right|\left(^j\omega_z^j x^j\right)x^j
\end{aligned}
\tag{66.18}
$$

where, dF_{dx}^j and dF_{dy}^j are the infinitesimal drag force components in x^j and y^j directions, and $dM_{dz,x}^j$ and $dM_{dz,y}^j$ are the resulting moment components (acting in z^j direction) due to these forces. Hence the total moment M_{dz}^j is computed as follows:

$$
M_{dz}^j = 2\int_0^{L_j/2} dM_{dz,y}^j + 2\int_0^{w_j/2} dM_{dz,x}^j = -\frac{\rho}{64}\left(C_{dx}^j w_j^4 + C_{dy}^j L_j^4\right)h_j\left|^j\omega_z^j\right|^j\omega_z^j
\tag{66.19}
$$

Likewise, the moment acting in the y^j and x^j directions are computed as follows:

$$
\begin{aligned}
M_{dy}^j &= -\frac{\rho}{64}\left(C_{dx}^j h_j^4 + C_{dz}^j L_j^4\right)w_j\left|^j\omega_y^j\right|^j\omega_y^j \\
M_{dx}^j &= -\frac{\rho}{64}\left(C_{dy}^j h_j^4 + C_{dz}^j L_j^4\right)L_j\left|^j\omega_x^j\right|^j\omega_x^j
\end{aligned}
\tag{66.20}
$$

where, L_j, w_j and h_j are respectively, length, width and height (in x^j, y^j and z^j directions) of Link j.

The generalized active force due to the drag force and torque for the system is then given by:

$$
(F_r)_d = \sum_{j=1}^n\left(\frac{\partial^j\omega^j}{\partial\dot{q}_r}\cdot M_d^j + \frac{\partial^jv_{cm}^j}{\partial\dot{q}_r}\cdot F_d^j\right) \quad (r=1,\ldots,n+5)
\tag{66.21}
$$

where, $M_d^j = \begin{bmatrix} M_{dx}^j & M_{dy}^j & M_{dz}^j \end{bmatrix}^T$ is drag torque.

66.3.2.2 Add Mass

When a submerged body is accelerated through a fluid, the driving force must overcome not only the drag force but also the body's inertia and the inertia of the fluid accelerated by the body. When a body undergoes angular accelerations, the surrounding fluid may also be entrained and accelerated. The added inertia forces and moments acting on the body due to the motion of the surrounding fluid were expressed as:

$$
F_a^j = -C_a^j \rho V^j \, {}^j a_{cm}^j
$$
$$
M_a^j = -C_A^j \left\{ \begin{array}{c} I_{xxf} \, {}^j\alpha_x^j \\ I_{yyf} \, {}^j\alpha_y^j \\ I_{zzf} \, {}^j\alpha_z^j \end{array} \right\}
\tag{66.22}
$$

where C_a^j is the added mass coefficient and ρV^j is the mass of the fluid with the same volume of the Link j. C_A^j is the added moment inertia coefficient and I_{xxf}, I_{yyf} and I_{zzf} are the center inertia moment of the fluid with the same volume of the Link j. ${}^j\alpha_x^j$, ${}^j\alpha_y^j$ and ${}^j\alpha_z^j$ are angular accelerations in x^j, y^j and z^j directions.

The generalized inertia force due to the added mass for the entire system is then given by the following:

$$
(F_r^*)_{AM} = \sum_{j=1}^{n} \left(\frac{\partial^j \omega^j}{\partial \dot{q}_r} \cdot M_a^j + \frac{\partial^j v_{cm}^j}{\partial \dot{q}_r} \cdot F_a^j \right) \quad (r = 1, \cdots, n+5)
\tag{66.23}
$$

66.3.3 Control Forces

To make this model complete, we will consider a control input for each of the joints. The n − 1 joints control for the robot can be expressed as an arbitrary Link j acted by torque T^j. The generalized active force for all of the control inputs of the system is then expressed as:

$$
(F_r)_{control} = \sum_{j=2}^{n} \left(\frac{\partial^j \omega^j}{\partial \dot{q}_r} \cdot T^j \right) \quad (r = 1, \ldots, n+5)
\tag{66.24}
$$

66.3.4 System of Dynamic Equations

Having developed all of the generalized inertia forces and generalized active forces for an underwater snake-like robot, the equations of motion are found to obtain the following dynamic model:

$$F_r^* + \left(F_r^*\right)_{AM} + (F_r)_G + (F_r)_B + (F_r)_d + (F_r)_{control} = 0 \quad (r = 1, \ldots, n+5) \quad (66.25)$$

66.4 Conclusion

Development of a robust autonomous underwater robotic vehicle is a key element to the exploitation of marine resources. An accurate dynamic model is important for both controller design and mission simulation, regardless of the control strategy employed. In this paper a complete model for an underwater snake-like robot has been constructed, consistent of a set of dynamic equations and a set of constraint equations. The model is formulated following Kane's approach which is advantageous to conventional approaches. Kane's method provides a straightforward approach for incorporating external forces into the model. External hydrodynamic forces consisted in this model include: buoyancy, profile drag, added mass. Other external forces include gravity, joint torques. The resulting set of dynamic equations is a closed form representation of the complete system and provides the physical insight necessary to study the behavior of the total system.

References

1. Maladen RD, Ding Y, Umbanhowar PB, Goldman DI (2011) Undulatory swimming in sand: experimental and simulation studies of a robotic sandfish. Int J Rob Res 30(7):793–794
2. Liljeback P, Pettersen KY, Stavdahl O, Gravdahl JT (2010) Hybrid modeling and control of obstacle-aided snake robot locomotion. IEEE Trans Rob 26(5):781
3. Transeth AA, Leine RI, Glocker C (2008) Snake robot obstacle-aided locomotion: modeling, simulation, and experiments. IEEE Trans Rob 24(1):88–103
4. Lewinger WA, Quinn RD (2011) Neurobiologically-based control system for an adaptively walking hexapod. Ind Rob 38(3):258–263
5. Scrivens JE, De Weerth SP, Ting LH (2008) A robotic device for understanding neuromechanical interactions during standing balance control. Bioinspiration Biomimetics 3(2):2–5
6. Schmitt J, Clark J (2009) Modeling posture-dependent leg actuation in sagittal plane locomotion. Bioinspiration Biomimetics 4(4):13–16
7. Gorner M, Wimbock T, Hirzinger G (2009) The DLR Crawler: evaluation of gaits and control of an actively compliant six-legged walking robot. Ind Rob 36(4):344–351

Chapter 67
Temperature and Strain Characteristic Analysis of Fiber Bragg Grating Sensor

Bing Bai, Yubin Guo, Xiaohui Zhao, Jiayu Huo and Tiegang Sun

Abstract A novel method is proposed to simulate temperature and strain characteristic of fiber Bragg grating (FBG) sensor with OptiGrating software, whose temperature characteristic is investigated experimentally with FBG sensing system based on broadband source one step further, both the simulation and experiment results show that the Bragg wavelength shift is linearly proportional to uniform temperature distribution. The parameters of FBG sensing system based on fiber ring laser have been optimized for good sensing performance, high signal-to-noise ratio (SNR) and powerful multiplexing capacity indicate its effectiveness and reliability in the application of long-haul sensing system.

Keywords Fiber bragg grating sensor · FBG sensing system · Fiber ring laser

67.1 Introduction

With the rapid development of photonic technology, optical fiber sensors play an important role in detecting outside information of nature. The fiber Bragg grating sensors have enormous potential in the application field of fiber-optic sensors as the result of immunity to electromagnetic interference, high sensitivity, small size, light weight and low cost [1–5]. The trial and error approach during fiber optical grating manufacturing process is extremely slow, expensive and unreliable, an

B. Bai · Y. Guo (✉) · X. Zhao · J. Huo · T. Sun
College of Communication Engineering, Jilin University, 5372 Nanhu Road, ChangChun, Jilin province, China

T. Sun
e-mail: suntieg@163.com

W. Lu et al. (eds.), *Proceedings of the 2012 International Conference on Information Technology and Software Engineering*, Lecture Notes in Electrical Engineering 210, DOI: 10.1007/978-3-642-34528-9_67, © Springer-Verlag Berlin Heidelberg 2013

appropriate fiber optical grating design and modeling software will help it develop towards an optimized production.

The essence of fiber grating sensor signal demodulation system is real-time monitoring of reflection spectrum from the sensing nodes, and analyses Bragg wavelength shift due to outside environment through the signal processing technology, and how to identify and measure a bunch of different wavelengths among reflected light is essential to wavelength demodulation and sensing grating addressing [6–9]. In order to realize Bragg wavelength shift detection, several optical fiber sensor signal interrogation techniques according to FBG sensor application situations have been proposed, including interrogation schemes based on a matched grating filter [10–12], an edge filter [13], a tunable F–P filter [14], an unbalanced M-Z interferometer [15], an unbalanced Michelson interferometer [16, 17], etc.

In this paper, OptiGrating software is adopted to simulate Bragg wavelength shift due to uniform temperature distribution and uniform strain distribution of the FBG sensor and its temperature characteristic is investigated based on broadband source/narrowband filter configuration. Operating principle is analyzed theoretically and sensing performance is investigated experimentally in fiber sensing system based on fiber ring laser [18, 19].

67.2 Fiber Bragg Grating Sensor

Simulation principle of OptiGrating is based on mode coupling theory, which is calculated by the Transfer Matrix Method, the Bragg wavelength and the FWHM bandwidth of the fiber grating can be given by:

$$\lambda_B = 2n_{eff}\Lambda \tag{67.1}$$

$$\Delta\lambda_{FWHM} = \lambda_B S \times \sqrt{\left[\left(\frac{\partial n}{2n}\right)^2 + \left(\frac{\Lambda}{L}\right)^2\right]} \tag{67.2}$$

Where the Bragg wavelength λ_B is 1549.38 nm, grating period Λ is 0.5345626 μm, and the length of fiber grating L is 50,000 μm, $S = (k^2 - \partial\beta^2)^{1/2}$, ∂n is the maximum range of refractive index.

The fiber Bragg grating shape is sinusoidal, average index is uniform. There is no chirp in the grating period, it is apodized by Gaussian function, and index modulation value is 0.00038. In addition, the sensors option needs to be checked in the grating manager dialog box. The FBG sensor is a narrowband optical filter, incident light who meet the Bragg conditions would be reflected back into the fiber in the opposite direction, transmission and reflection spectrum of FBG sensor in the free condition is shown in Fig. 67.1, its FWHM bandwidth is 0.30 nm.

Temperature and strain will change the grating refractive index and the grating period, spectral response of the FBG sensor changes when applied temperature and

Fig. 67.1 Transmission and reflection spectrum of fiber Bragg grating sensor

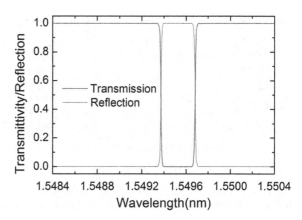

strain distributions change. Refractive index change and grating period change caused by the temperature and strain contribute to the Bragg wavelength shift $\Delta\lambda_{BT}$ and $\Delta\lambda_{B\varepsilon}$.

$$\frac{\Delta\lambda_{BT}}{\lambda_B} = (\alpha + \xi) \cdot \Delta T \tag{67.3}$$

$$\frac{\Delta\lambda_{B\varepsilon}}{\lambda_B} = \left(1 - \frac{n^2}{2[P_{12} - v(P_{11} - P_{12})]}\right)\Delta\varepsilon \tag{67.4}$$

Where thermal-expansion coefficient α is 5.5e − 007/°C, thermal-optic coefficient ξ is 8.3e − 006/°C, strain-optical tensor P_{11} and P_{12} entered are 0.121 and 0.27 respectively, and Poisson ratio v is 0.17.

Uniform temperature distribution and uniform strain distribution are applied to the FBG sensor respectively, Bragg wavelength shift will be detected due to the temperature change or strain change, the problem of cross sensitivity is not included within the scope of the paper. The temperature varies from 25 to 45 °C with a step of 1 °C, Fig. 67.2 a shows Bragg wavelength shift due to uniform temperature change. The strain increases from 0 to 150 με with a step of 5 με Fig. 67.2b shows Bragg wavelength shift due to uniform strain change. As can be seen in Fig. 67.2, the Bragg wavelength shift is linear with uniform temperature change or uniform strain change, Bragg wavelength moves to longer wavelength as temperature increases or strain increases. The estimated linear regression models are stated: y = 0.01367*x + 1,549.03887 and y = 0.00101*x + 1549.37963, whose correlation coefficients are 0.99979 and 0.99928 respectively, temperature sensitivity obtained is 13.67 pm/°C and strain sensitivity obtained is 1.01 pm/με.

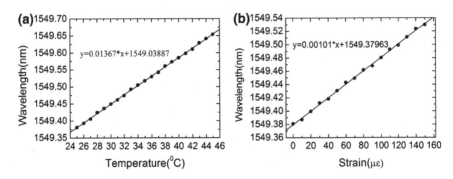

Fig. 67.2 **a** Relation between Bragg wavelength shift and uniform temperature distribution. **b** Relation between Bragg wavelength shift and uniform strain distribution

67.3 FBG Sensing System

67.3.1 Broadband Source

The schematic diagram of the basic FBG sensing system configuration based on broadband source (BBS) is shown in Fig. 67.3. The light source for FBG sensing system is a broadband source, narrowband scanning filter (SF) spectrum convolutes with reflection spectrum of FBG sensor. The maximum convolution results reach when reflection spectrum of FBG sensor and SF spectrum match, a measured reflection curve will be obtained, Bragg wavelength relates to outside temperature information.

In the experiment, the free termination have been immersed in refractive index matching gel (IMG) to avoid undesired reflections [20], and isolator (ISO) is introduced to protect BBS. In addition, the resolution of temperature cabinet is limited to 1 °C. The relation between the Bragg wavelength shift and temperature distribution is measured based on broadband source/narrowband filter configuration. According to Bragg wavelength shift and temperature change data, a measured reflection curve is shown in Fig. 67.4. The estimated linear regression model is stated: $y = 0.01313*x + 1549.04699$, whose correlation coefficient is 0.99515, temperature sensitivity obtained is 13.13 pm/°C. The simulation and experimental results are consistent with each other, which prove that the Bragg wavelength is linear with the temperature change.

Fig. 67.3 FBG sensing system based on broadband source

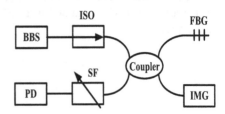

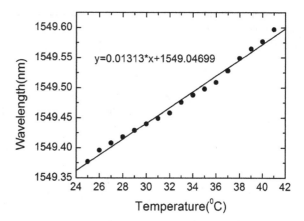

Fig. 67.4 Relation between Bragg wavelength shift and uniform temperature distribution

67.3.2 Fiber Ring Laser

Based on the simulation and experiment investigation on FBG sensor's sensing characteristic, a FBG sensing system scheme based on fiber ring laser is designed, as shown in Fig. 67.5. In order to choose the lasing wavelength in the waveband 1526.438–1564.679 nm, a dielectric thin-film filter (DTF) is introduced. The erbium doped fiber (EDF) makes up one end of the ring cavity while FBG sensors composes the other end. The Bragg wavelengths are different from each other, but they are all located in the gain bandwidth of EDF. Fiber sensing system chooses a lasing wavelength when the transmission window of DTF matches the wavelength of FBG sensor's reflection spectrum from remote sensing node, then real-time monitoring for Bragg wavelength shift has been realized.

The parameters of FBG sensing system are optimized in order to improve sensing performance, then 1.8 m-long EDF is chosen as gain medium whose peak absorption coefficient is 24 dB/m at 1,530 nm. 2×2 coupler with split ratio of 90:10 is incorporated into the ring cavity instead of 1×2 coupler and circulator, which decreases system loss, complexity and cost. A FBG sensor is located at

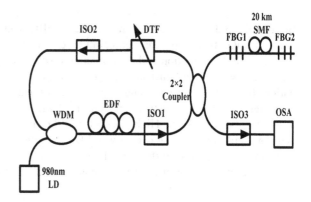

Fig. 67.5 FBG sensing system based on fiber ring laser

Fig. 67.6 A 20 km FBG
sensor interrogation signal
spectrum

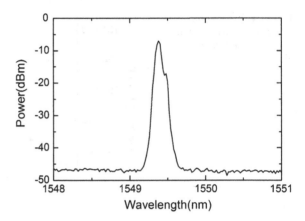

20 km sensing node, interrogation signal taken from 90 % output of 2×2 coupler is measured by optical spectrum analyzer (OSA). As can be seen from Fig. 67.6, the relative peak power is -7.03 dBm for a pump power of 100 mW, the side mode suppression ratio (SMSR) is lager than 38 dB, and 3 dB bandwidth is less than 0.08 nm.

The FBG sensing system based on broadband source suffers from a low SNR as result of the low power spectrum intensity of broadband source, and as is known that the SNR declines as the distance increases. However, FBG sensing system based on fiber ring laser enhances the SNR and provides a higher output power with a narrow spectral width. Even though the back-reflected signal degrades along 20 km fiber link, the SNR of the interrogation signal remains high, which indicates its demodulation performance is good. Because of the wide tuning range of DTF, FBG sensing system based on fiber ring laser is capable of supporting 18 FBG sensors with 2 nm per bandwidth interval at least, which means its multiplexing capacity is powerful.

67.4 Conclusion

Both strain and temperature characteristic of FBG sensor are simulated incorporating OptiGrating software, temperature characteristic is investigated experimentally. Both simulation and experimental results are consistent with each other, the Bragg wavelength shift is linearly proportion to uniform temperature distribution. Based on simulation and experimental research on FBG sensor's sensing characteristic, we demonstrate an optimized FBG sensing system configuration based on fiber ring laser. High SNR of a 20 km FBG sensor interrogation signal and powerful multiplexing capacity of FBG sensing system indicate good sensing performance, which have theoretical significance to the high precision, long-haul efforts and extremely sensitive fiber sensing market development in future.

References

1. Bazzo JP, Lukasievicz T, Vogt M et al (2012) Thermal characteristics analysis of an IGBT using a fiber Bragg grating. Opt Laser Eng 50:99–103
2. Childs P, Wong ACL, Yan BB et al (2010) A review of spectrally coded multiplexing techniques for fibre grating sensor systems. Meas Sci Technol 21:094007
3. Luyckx G, Voet E, Lammens N et al (2011) Strain measurements of composite laminates with embedded fibre Bragg gratings: criticism and opportunities for research. IEEE Sens J 11:384–408
4. Liu QW, Tomochika T, Zuyuan H (2011) Realization of nano static strain sensing with fiber Bragg gratings interrogated by narrow linewidth tunable lasers. Opt Express 19(21):20214
5. Wang YP, Wang M, Huang XQ (2011) Spectral characterization of polarization dependent loss of locally pressed fiber Bragg grating. Opt Express 19(25):25535–25544
6. Li RM, Yu YL, Shum P (2011) Addressing fiber Bragg grating sensors with wavelength-swept pulse fiber laser and analog electrical switch. Opt Commun 284:1561–1564
7. Liu WJ, Li M, Wang C (2011) Real time interrogation of a linearly chirped fiber Bragg grating sensor. J Lightwave Technol 29(9):1239–1247
8. Wang C, Yao JP (2011) Ultrafast and ultrahigh-resolution interrogation of a fiber Bragg grating sensor based on interferometric temporal spectroscopy. J Lightwave Technol 29(19): 2927–2933
9. Yang MW, Liao CR (2011) Fiber Bragg grating with micro-holes for simultaneous and independent refractive index and temperature sensing. IEEE Photon Technol Lett 23(20):1511–1513
10. Jiang BQ, Zhao JL, Qin C et al (2011) An optimized strain demodulation method based on dynamic double matched fiber Bragg grating filtering. Opt Laser Eng 49:415–418
11. Zhan YG, Yu MH, Pei JC et al (2010) A linearity interrogation technique with enlarged dynamic range for fiber Bragg grating sensing. Opt Commun 283:3428–3433
12. Zou HB, Liang DK, Zeng J (2012) Dynamic strain measurement using two wavelength-matched fiber Bragg grating sensors interrogated by a cascaded long-period fiber grating. Opt Laser Eng 50:199–203
13. Navneet SA, Kaler RS (2011) Fiber Bragg grating interrogator using edge filtering technique with microbend loss error mitigation. Opt 122(9):796–798
14. Kuo ST, Peng PC, Kao MS (2011) Tunable erbium-doped fiber ring laser with signal-averaging function for fiber-optic sensing applications. Laser Phys 21(1):188–190
15. Liao CR, Wang Y, Wang DN et al (2010) Fiber in-line Mach-Zehnder interferometer embedded in FBG for simultaneous refractive index and temperature measurement. IEEE Photon Technol Lett 22(22):1686–1688
16. Amaral LMN, Frazão O, Santos JL et al (2011) Fiber-optic inclinometer based on taper Michelson interferometer. IEEE Sens J 11(9):1811–1814
17. Min JW, Yao BL, Gao P (2010) Parallel phase-shifting interferometry based on Michelson-like architecture. Appl Opt 49(34):6612–6616
18. Tsuda H (2010) Fiber Bragg grating vibration-sensing system, insensitive to Bragg wavelength and employing fiber ring laser. Opt Lett 35(14):2349–2351
19. Gan JL, Hao YQ, Ye Q et al (2011) High spatial resolution distributed strain sensor based on linear chirped fiber Bragg grating and fiber loop ringdown spectroscopy. Opt Lett 36(6): 879–881
20. Quintela MA, Herrera RAP, Canales I et al (2010) Stabilization of dual-wavelength erbium-doped fiber ring lasers by single-mode operation. IEEE Photon Technol Lett 22(6):368–370

Chapter 68
A Fuzzy Neartude Based Approach to Simulate the Two-Compartment Pharmacokinetic Model

Lei Zhao, Xinling Shi, Yufeng Zhang and He Sun

Abstract The estimation of parameters of compartmental models is very important in pharmacokinetic (PK) modeling which is sometimes rather hard. One of the suitable ways is to represent the parameters as fuzzy sets. Using the fuzzy neartude as an efficient tool the influence of each of the four parameters i.e. oral bioavailability, absorb rate constant, the central volume of distribution and the clearance on the concentration time profile was simulated. The simulation results demonstrate that fuzzy neartude-based approach is a useful tool to predict the plasma drug concentration during the early stage of drug discovery when the information for some parameters is limited, vague and semi-quantitative.

Keywords Fuzzy neartude · Simulation · Prediction of concentration · Two-compartment model

68.1 Introduction

Compartmental models are classical models used in pharmacokinetic (PK) research to represent drug distribution and elimination in the body [1]. One of the most important parts of the PK modeling is the estimation of parameters. However, the information accumulated during early stage drug development is quite often to

L. Zhao (✉) · X. Shi · Y. Zhang · H. Sun
Yunnan University, 310 room, Huaizhou Building 121 Street, Kunming, China
e-mail: ynzhaolei@foxmail.com

L. Zhao
School of Sciences, Department of Physics, Kunming University of Science
and Technology, Kunming, China

W. Lu et al. (eds.), *Proceedings of the 2012 International Conference on Information Technology and Software Engineering*, Lecture Notes in Electrical Engineering 210, DOI: 10.1007/978-3-642-34528-9_68, © Springer-Verlag Berlin Heidelberg 2013

be incomplete and insufficient [2]. That means it is hard to estimate the model parameters. A suitable way is to represent the parameters as fuzzy sets. By now, there have been four main existing methods for implementing fuzzy arithmetic, conventional fuzzy arithmetic, vertex method, transformation and optimization methods [3].

Fuzzy sets theory (FST) has been widely applied in various fields of science in the presence of vague, limited information. Nestorov et al. [4] developed a methodology of fuzzy sets for the prediction of in vivo hepatic clearance to account for the existing uncertainty and variability. Gueorguieva et al. [5] used FST to predict diazepam pharmacokinetics based on a whole body physiologically based model. FST is also used to predict pain relief of migraine in patients following naratriptan oral administration [6].

In this paper, we used the fuzzy neartude-based approach to simulate the plasma drug concentration in a two-compartment model with oral administration. Based on the selected model, the influence of each of the four parameters i.e. oral bioavailability (F), absorption rate constant (K_a), the central volume of distribution (V_c) and the clearance (CL) on the concentration time profile was simulated. The simulation results demonstrate that fuzzy approach is a useful tool to predict the plasma drug concentration during the early stage of drug discovery when the information for some parameters is limited, vague and semi-quantitative.

68.2 Methods

68.2.1 Fuzzy Number

Under the fuzzy sets theory, model parameters are represented as fuzzy sets or fuzzy number. A fuzzy number is defined by an interval (support) and a membership function [5]. The support of a fuzzy set is the region of elements of a universe of discourse. The membership function maps each element in the universe of discourse to a membership value between 0 and 1. The more the object belongs to the fuzzy set, the higher the degree of membership [7].

Generally, the function itself can be arbitrary shape. There are several kinds of membership function that commonly used to represent fuzzy number, e.g. triangular membership function, trapezoidal membership function, bell-shaped membership function, and so on.

68.2.2 Fuzzy Neartude

Fuzzy neartude is proposed as a measure that indicates the degree of closeness between two fuzzy sets. Fuzzy neartude encompasses a family of compatibility measures based on different distance metrics between the corresponding fuzzy sets [7].

The neartude of two fuzzy sets is defined as

$$\sigma(\underset{\sim}{A}, \underset{\sim}{B}) = 1 - d(\underset{\sim}{A}, \underset{\sim}{B}) \tag{68.1}$$

where $d(\underset{\sim}{A}, \underset{\sim}{B})$ is the distance, e.g. Hamming distance, Euclidean distance.

The normalized Hamming distance is defined as

$$d_H\left(\underset{\sim}{A}, \underset{\sim}{B}\right) = 1/n \sum_{i=1}^{n} \left| \underset{\sim}{} (x_i) - \underset{\sim}{B}(x_i) \right| \tag{68.2}$$

And the normalized Euclidean distance is defined as

$$d_O(\underset{\sim}{A}, \underset{\sim}{B}) = \sqrt{1/n \sum_{i=1}^{n} \left(\underset{\sim}{A}(x_i) - \underset{\sim}{B}(x_i) \right)^2} \tag{68.3}$$

68.2.3 PK Model

A two-compartment model with first-order rate process for the oral route is presented in Fig. 68.1 [8] where X_0 is the oral dose, F the oral bioavailability, K_a the absorption rate constant, K_{12} the transfer rate constant from the central compartment to the peripheral compartment, K_{21} the transfer rate constant from the peripheral compartment to the central compartment, and K_{10} is the elimination rate constant.

According to Fig. 68.1, plasma concentrations in the central compartment are described in the following manner [1];

$$C(t) = (X_0 F K_a)/V_c \cdot (L \cdot e^{-\alpha t} + M \cdot e^{-\beta t} + N \cdot e^{-K_a t}) \tag{68.4}$$

where V_c is the central volume of distribution and L, M, and N are given as bellow;

$$L = (K_{21} - \alpha)/[(K_a - \alpha)(\beta - \alpha)] \tag{68.5}$$

Fig. 68.1 A two-compartment model of oral administration

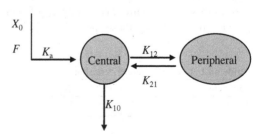

$$M = (K_{21} - \beta)/[(K_a - \beta)(\alpha - \beta)] \qquad (68.6)$$

$$N = (K_{21} - K_a)/[(\beta - K_a)(\alpha - K_a)] \qquad (68.7)$$

where α and β are the hybrid rate constants associated with distribution and elimination phases respectively and they were calculated as follows;

$$\alpha \cdot \beta = K_{10} \cdot K_{21} \qquad (68.8)$$

$$\alpha + \beta = K_{10} + K_{21} + K_{12} \qquad (68.9)$$

The elimination rate constant K_{10} was calculated by

$$K_{10} = CL/V_c \qquad (68.10)$$

where CL is the clearance.

68.2.4 Simulation

A two-compartment model with first-order process for oral administration was selected in the simulation. All the simulations were performed by using MATLAB software. The mean and standard deviation of the parameters demonstrated in Table 68.1 was taken from the literature [9]. For another two parameters the oral dose X_0 and the oral bioavailability F were set as 30 mg and 0.6, respectively.

In the fuzzy approach, the model parameters were presented as fuzzy sets or fuzzy numbers as illustrated in Fig. 68.2. All of the parameters are assigned trapezoidal-shaped membership functions with different support which are shown in Fig. 68.2. The fuzzy number for the bioavailability (F) is given in the upper left panel with the most typical values are within the closed interval [0.4 0.8]. The fuzzy number for the absorption rate constant (K_a) is given in the upper right panel with the most typical values are within the closed interval [1.6 4.2]. The fuzzy number for the central volume of distribution (V_c) is given in the lower left panel with most typical values in the closed interval [0.75×10^6 1.3×10^6]. The fuzzy number for the clearance (CL) is given in the lower right panel with most typical values in the closed interval [0.5×10^5 1.15×10^5]. All of the reference mean values have a membership degree of 1 in its membership function respectively.

According to the membership function, seven different values of each parameter were selected with several degree of membership, so there was total 7^4 i.e. 2,401 subjects. Using these individual concentrations a data table with F, K_a, V_c,

Table 68.1 Literature values describing the two-compartment pharmacokinetics model

Parameter	V_c (L)	CL (L/h)	K_{12} (h^{-1})	K_{21} (h^{-1})	K_a (h^{-1})
Value[a]	961(167)	75(13.9)	$0.126(2.14 \times 10^{-4})$	$0.062(4.32 \times 10^{-5})$	2.6(0.9)

[a] Mean(SD)

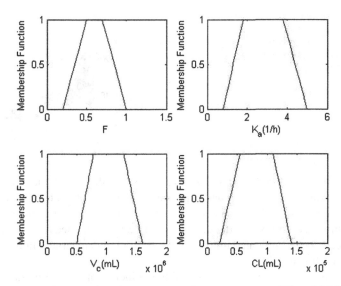

Fig. 68.2 Assigned fuzzy numbers for parameters of two-compartment PK model following oral administration, i.e. *upper left panel*, bioavailability (*F*), *upper right panel*, the absorption rate constant (*K_a*), *lower left panel*, the central volume of distribution (*V_c*) and *lower right panel*, clearance (*CL*)

CL as the characteristic value was established. It was stored as a matrix of 2,401 × 4. According to (68.2), the neartude between each individual concentration and the mean concentration was calculated.

Based on the reference mean value and standard deviation, twenty concentration data sets of simulated subjects were generated stochastically by the MATLAB program used to be compared with the fuzzy-valued concentration in each simulation.

68.3 Results and Discussion

Based on the selected model, the influence of each of the four parameters on the concentration time profile was simulated. The predicted fuzzy-valued plasma drug concentrations varying along with time t (t = 0–12 h) are shown in Fig. 68.3. The color of the line represents the neartude between each individual concentration and the mean concentration. The red line is with the largest degree of proximity and the color of blue represents the smallest degree.

The profiles in Fig. 68.3 demonstrate the influence of bioavailability (*F*) on the concentration (*C*). For the relationship between *F* and *C* is linear, the range of variation of concentration along with the whole time axes is wide. The C_{max} i.e. the peak concentration varies from 7.556 to 22.68 ng/mL and the concentration at 12 h varies from 1.83 to 5.49 ng/mL with the *F* varying from 0.3 to 0.9. While the

Fig. 68.3 Predicted
fuzzy-valued plasma drug
concentration time profiles
with the fuzzy numbers of
bioavailability (F)

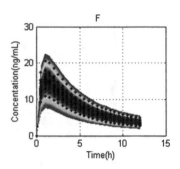

fuzzy numbers F with 0.3 and 0.9 have the same degree of membership, the
bottom line and the top line have the same smallest neartude with the mean
concentration profile. The concentration of subjects with F having the largest
membership accordingly has the largest neartude with the mean concentration. It
can be seen that the simulated individual concentration data largely falls within the
region of high neartude.

The profiles in Fig. 68.4 demonstrated the influence of absorption rate constant
(K_a) on the concentration. K_a affects both the C_{max} and T_{max} i.e. the time required
to reach the peak concentration. With the increasing K_a C_{max} increases and T_{max}
decreases. When K_a has the largest degree of membership, the corresponding
concentration profile is the closest to the mean concentration. That means it has the
red color. Similarly the blue line is with the lowest degree of neartude. It can be
seen that the simulated individual concentration data largely falls within the region
of high neartude.

The profiles in Fig. 68.5 demonstrated the influence of the central volume of
distribution (V_c) on the concentration. While V_c just affects C_{max}, the T_{max} of the
concentration profiles does not vary. The top concentration profile with the lowest
neartude i.e. with the blue color is generated based on V_c of 560 L and the bottom
concentration profile is generated based on V_c of 1,510 L. It can be seen that with
the increasing V_c C_{max} decreases and the simulated individual concentration data
also largely falls within the region of high neartude.

The profiles in Fig. 68.6 demonstrated the influence of the clearance (CL) on
the concentration. While CL just affects the phase of elimination, so the concentration time profiles before T_{max} are the same.

Fig. 68.4 Predicted
fuzzy-valued plasma drug
concentration time profiles
with the fuzzy numbers of
absorption rate constant (K_a)

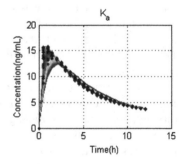

Fig. 68.5 Predicted
fuzzy-valued plasma drug
concentration time profiles
with the fuzzy numbers of
central volume of
distribution (V_c)

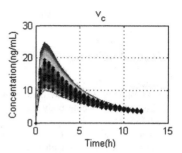

Fig. 68.6 Predicted
fuzzy-valued plasma drug
concentration time profiles
with the fuzzy numbers of
clearance (CL)

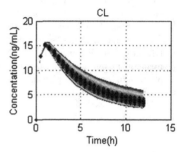

The variation of concentration with each of the four parameters can be seen clearly in Figs. 68.3, 68.4, 68.5, 68.6. It also can be seen that all of the simulation results are accordance with the characteristic of PK in the real situation.

68.4 Conclusion

Fuzzy set method is a useful tool to predict the plasma drug concentration profile during the early stage of drug discovery when the information for some parameters is limited, vague and semi-quantitative. Although the fuzzy approach was specifically used to predict pharmacokinetics, it also can be applied to pharmacodynamics.

Acknowledgments This work was supported by the Natural Science Foundation of China (No. 61261007) and Natural Science Foundation of Yunnan Province (2009CD016).

References

1. Shargel L, Wu-Pong S, Yu ABC (2005) Applied biopharmaceutics and pharmacokinetics. McGraw-Hill Companies, New York
2. Seng K-Y, Vicini P, Nestorov IA (2006) A fuzzy physiologically based pharmacokinetic modeling framework to predict drug disposition in humans. Conf Proc IEEE Eng Med Biol Soc 1:5037–5040

3. Seng K-Y, Nestorov I, Vicini P (2007) Simulating pharmacokinetic and pharmacodynamic fuzzy-parameterized models: a comparison of numerical methods. J Pharmacokinet Pharmacodyn 34:595–621
4. Nestorov I, Gueorguieva I, Jones HM et al (2002) Incorporating measures of variability and uncertainty into the prediction of in vivo hepatic clearance from in vitro data. Drug Metab Dispos 30:276–282
5. Gueorguieva I, Nestorov I, Rowland M (2004) Fuzzy simulation of pharmacokinetic models: case study of whole body physiologically based model of diazepam. J Pharmacokinet Phar 31:185–213
6. Gueorguieva I, Nestorov I, Aarons L, Rowland M (2005) Uncertainty analysis in pharmacokinetics and pharmacodynamics: application to naratriptan. Pharm Res 22:1614–1626
7. Cross V, Sudkamp T (2002) Similarity and compatibility in fuzzy set theory. Physica, Heidelberg
8. Gabrielsson J, Weiner D (2000) Pharmacokinetic and pharmacodynamic data analysis: concepts and applications, 3rd edn. Swedish Pharmaceutical Press, Stockholm
9. Dickinson GL, Rezaee S, Proctor NJA et al (2007) Incorporating in vitro information on drug metabolism into clinical trial simulations to assess the effect of CYP2D6 polymorphism on pharmacokinetics and pharmacodynamics: dextromethorphan as a model application. J Clin Pharmacol 47:175–186

Chapter 69
A Three-Dimensional Role Based User Management Model in Web Information Systems

Jiangfeng Li and Chenxi Zhang

Abstract Web application plays an important role in compute application. User management system is a significant part in web application system, especially in a large scale software system. In the user management models of existed web application systems, role is a one-dimensional space. It contains only a set of authorities, and lacks of the control of valid scopes and permission time. A three-dimensional role is presented. It is a set of three-dimensional vectors composed of authorities, scopes, and permission time. Moreover, a three-dimensional role based user management model is proposed. Analysis in mathematical approach proves that the model is independent of business requirements in information systems. It results in that the model is suitable for the user management subsystem of every business information system. In addition, verification on information systems shows that the model performs well in developing user management system and can satisfy requirements of various applications in web information systems.

Keywords Three-dimensional role · Authority · Scope and permission time · User management · Web application

The research is supported by Program for Young Excellent Talents in Tongji University (Grant NO. 2010KJ061).

J. Li (✉) · C. Zhang (✉)
School of Software Engineering, Tongji University, Shanghai 201804, China
e-mail: zhangcx2000@163.com

W. Lu et al. (eds.), *Proceedings of the 2012 International Conference on Information Technology and Software Engineering*, Lecture Notes in Electrical Engineering 210, DOI: 10.1007/978-3-642-34528-9_69, © Springer-Verlag Berlin Heidelberg 2013

69.1 Introduction

Recently, it is a tendency that enterprises administer their business using special management information system. A large scale management information system is user-oriented, and it has various users. Users who belong to different department have different identities and duties. So, it is necessary for the system to have a function that, system is able to entitle users in multi-level. An approach that seems to be useful in this field is the exploitation of the abstraction concept of roles [1]. Role Based Access Control (RBAC) [2] has been widely applied to authorize certain users to access certain data or resources within complex systems [3]. In the role based access control model [4–6], role is a collection of access rights or permission. An approach for extending the role based access control model is presented to support activities that demand runtime mutability in their authorization attributes [7]. Such activities cannot be subdivided in a set of subtasks executed sequentially neither can be accomplished by a single role. In the role models for agent systems, role is a set of computation specification and function to realize relationship between agents [8].

In the existed enterprise systems [9–12], role is only a set of authorities. It is a conception of one-dimensional space. However, users participate in an enterprise system as a particular identity. It results in that roles of different users have different authorities, and each authority is valid in a relevant scope during its permission time. The chapter presents a three-dimensional role which is a conception of three-dimensional space. A user management model based on the three-dimensional role can satisfy requirements of modern application systems and large scale systems.

In Sect. 2, some basic conceptions of three-dimensional role are defined. Section 3 proposes a role based user management model. Section 4 analyzes the user management model, and verifies it in web information systems. Conclusion is presented in Sect. 5.

69.2 Basic Conceptions

In an enterprise system, a user's role is not invariable. It varies dynamically in term of the changing of enterprise activities. So, the system should grant multiple roles dynamically with the development of enterprise. In our definition, role is combination of authority, scope and permission time. It is a conception of three-dimensional space, called three-dimensional role. We introduce the three-dimensional role in a mathematical approach as follows.

Definition 1 (*Basic Authority*) Basic Authority is an operation authority, which is an atomic operation and cannot be divided any more. Let B. is the set of basic authority.

$$B = \{b_1, b_2, b_3, \cdots, b_n\}$$

where b_i is the ith basic authority.

Definition 2 (*Authority*) Authority is a union of finite basic authorities. Let p and P are an authority and the set of all authorities respectively.

$$p = \bigcup_i b_i \, (b_i \in B, 1 \leq i \leq |B|)$$

$$P = \{p_1, p_2, p_3, \cdots, p_n\}$$

where $|B|$ is the number of elements in set B, and p_i is the ith authority.

Definition 3 (*Scope*) Scope is an operation place where the operation is valid in an information system. The scope is related to an operation. Let S is the set of scopes.

$$S = \{s_1, s_2, s_3, \cdots, s_n\}$$

where s_i is the ith scope.

Definition 4 (*Permission Time*) Permission Time is the period of time when an operation is operated in an information system. It is a closed interval from starting time to ending time. Let t and T are a permission time and the set of all permission time respectively.

$$t = [t_{begin}, t_{end}]$$

$$T = \{t_1, t_2, t_3, \cdots, t_n\}$$

where t_{begin} is the time when the operation starts, t_{end} is the time when the operation ends, t_i is the ith permission time.

Definition 5 (*Role*) Role is a set of finite three-dimensional vectors composed of authorities, scopes, and permission time.

Let r and R are a role and the set of all roles respectively.

$$r = \{ <p_i, s_i, t_i> \,|(p_i \in P, s_i \in S, t_i \in T, 1 \leq i \leq |P|)\}$$

$$R = \{r_1, r_2, r_3, \cdots, r_n\}$$

where $|P|$ is the the number of elements in set P, S is the set of all scopes, T is the set of all permission time, r_i is the ith role.

Definition 6 (*User Group*) User Group which is composed of finite roles is a set of unions of finite roles. Let g and G are a user group and the set of all user groups respectively.

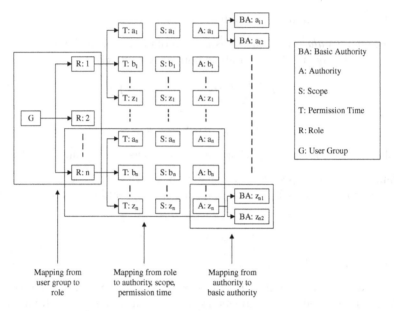

Fig. 69.1 Relationships among user group, role, authority and basic authority

$$g = \bigcup_i r_i \ (r_i \in R, 1 \leq i \leq |R|)$$

$$G = \{g_1, g_2, g_3, \cdots, g_n\}$$

where $|R|$ is the number of elements in set R, g_i is the ith user group.

From the definitions above, we could find that a member in a user group has some roles. A role has authorities in relevant scopes during the period of permission time. An authority comprises a few basic authorities. Figure 69.1 shows the relationship among user group, role, authority and basic authority.

69.3 Role Based User Management Model

A Role Based User Management (RBUM) model is proposed according to the three-dimensional role defined in Sect. 2. There are four modules in RBUM model, User Account Module, Power Validation Module, Power Configuration Module, Apply and Permit Interface. Apply and Permit Interface is an interface connecting User Account Module and Power Configuration Module. Figure 69.2 is the structure of RBUM model.

Fig. 69.2 Structure of
RBUM model

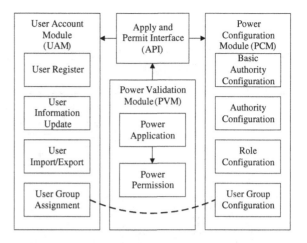

1. *User Account Module*

The User Account Module (UAM) contains the operation that users register to get accounts and update their personal information. Every user belongs to a User Group which is configured in the Power Configuration Module (PCM).

2. *Power Configuration Module*

Power Configuration Module (PCM), the key module in RBUM model, deals with the operation of power configuration management involving configuration of Basic Authority, Authority, Roles, and User Groups.

3. *Power Validation Module*

Power application and power permission are two key operations in Power Validation Module (PVM).

Power Application: a user previously applies for a user group that it wants to join in.

Power Permission: an administrator permits a user's application to join in an user group, after verifying the information of the applicant. In addition, an administrator can remove a user from any user group.

4. *Apply and Permit Interface*

Apply and Permit Interface (API), supported by the PVM, is an interface connecting UAM and PCM.

System operators, administrators, and users have their own duties in the RBUM model. The system operators and administrators work at different steps in the processes of configuring the user group. A registered user needs to be assigned to a user group. Figure 69.3 shows the workflow of user group configuration and assignment in the RBUM model.

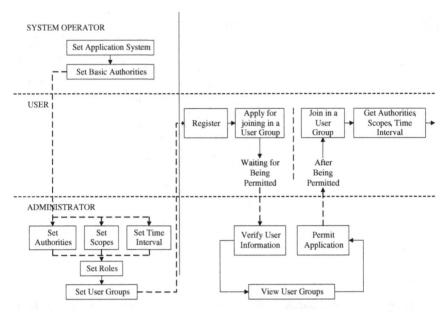

Fig. 69.3 Workflow of RBUM model

69.4 Analysis and Verification

69.4.1 Analysis in Mathematics Approach

We use theorems to manifest the relationship between user groups in the RBUM model.

Theorem 1 *Let G is the set of user groups, $\forall g, g' \in G$, there are three relationships between the two user groups g and g'.*

1. Coordinate relationship, if $g \cap g' = \emptyset$
2. Inclusive relationship, if $g \subseteq g'$
3. Mixed relationship, $g \cap g' \neq \emptyset$, $g \nsubseteq g'$ and $g' \nsubseteq g$

Proof $\forall g, g' \in G$, we have the formulations

$$g = \bigcup_i r_i | (r_i \in R, 1 \leq i \leq |R|), \ g' = \bigcup_j r'_j | (r'_j \in R', 1 \leq j \leq |R'l|)$$

$\square$

Since the sets are selected randomly, there are three possibilities below:

1. $\forall 1 \leq i \leq |R|$, $\forall 1 \leq j \leq |R'|$, such that, $\cup_i r_i \neq \cup_j r'_j$ therefore, $g \cap g' = \emptyset$, so g and g' are coordinate;
2. $\forall 1 \leq i \leq |R|$, $\exists 1 \leq j_i \leq |R'|$, such that, $\cup_{j_i} r'_{j_i} = \cup_i r_i$ therefore, $g \subseteq g'$, so g and g' are inclusive;
3. $\exists 1 \leq i_0 \leq |R|$, $\exists 1 \leq j_0 \leq |R'|$, such that, $\cup_{i_0} r'_{i_0} = \cup_{j_0} r'_{j_0}$ therefore, $g \cap g' \neq \emptyset$,

   $\exists 1 \leq i_0 \leq |R|$, $\forall 1 \leq j \leq |R'|$, such that, $\cup_j r'_j \neq \cup_{i_0} r_{i_0}$ therefore, $g \nsubseteq g'$,

   $\exists 1 \leq j_0 \leq |R'|$, $\forall 1 \leq i \leq |R|$, such that, $\cup_i r_i \neq \cup_{j_0} r'_{j_0}$, therefore, $g' \nsubseteq g$.

   So g and g' are mixed.

Summarizing the three possibilities above, we can find that coordinate relation, inclusive relation and mixed relation are three relations between two user groups.

Theorem 2 *Let G is the set of user groups, for any g, $g' \in G$,*

$$g = \left\{ \bigcup_i r_i | (r_i \in R, 1 \leq i \leq |R|) \right\}$$

$$g' = \bigcup_j r'_j | (r'_j \in R', 1 \leq j \leq |R'|)$$

Let P and P' are the authorities of g and g', S and S' are the scopes of g and g', T and T' are the permission time of g and g', respectively.

If $g \subseteq g'$, then $P \subseteq P'$, $S \subseteq S'$ and $T \subseteq T'$.

Proof Since $g \subseteq g'$, so that $\forall 1 \leq i \leq |R|$, $\exists 1 \leq j_i \leq |R'|$, we have the formulation□

$$r'_{j_i} = r_i$$

Noting that

$$r_i = \{ <p_i, s_i, t_i > | (p_i \in P, s_i \in S, t_i \in T, 1 \leq i \leq |P|) \}$$

$$r'_{j_i} = \left\{ \left\langle p'_{j_i}, s'_{j_i}, t'_{j_i} \right\rangle | \left(p'_{j_i} \in P', s'_{j_i} \in S', t'_{j_i} \in T', 1 \leq j_i \leq |P'| \right) \right\}$$

This implies that

$$<p'_{j_i}, s'_{j_i}, t'_{j_i} > = <p_i, s_i, t_i >$$

Therefore $\forall 1 \leq i \leq |R|$, $\exists 1 \leq j_i \leq |R'|$, such that, $p'_{j_i} = p_i$, $s'_{j_i} = s_i$, $t'_{j_i} = t_i$ So $P_1 \subseteq P_2$, $S_1 \subseteq S_2$, and $T_1 \subseteq T_2$.

The theorem above indicates that, in two user groups with inclusive relationship, the authorities, scopes, and permission time of the sub-group are no more than those of the super-group.

69.4.2 Verification on Management System

For verifying the RBUM model, we designed and implemented some systems, such as China School Resource System, Shanghai Fine Curriculum Network System. In the systems, user group is changed dynamically with the variation of business requirements. Users work in their own scopes during their own time period by assigning different authorities, scopes, and permission time to each role.

69.5 Conclusion

The chapter presented a Three-Dimension Role, which is a set of three-dimensional vectors, forming as <authority, scope, permission time>. Users who have the same role own the same authorities, scopes, and permission time. According to the three-dimensional role, we proposed a Role Based User Management (RBUM) mode. It is verified in mathematic approach that RBUM model is independent of business requirements in information systems. In addition, the three-dimensional RBUM model is used effectively in developing a user management system for the particular web information system.

References

1. Venkatesh V, Thong JY, Chan FY, Hu PJ, Brown SA (2011) Extending the two-stage information systems continuance model: incorporating UTAUT predictors and the role of context. Inform Sys J 21(6):527–555
2. Feng X, Ge B, Sun Y et al (2010) Enhancing role management in role-based access control. In: Proceedings of 3rd IEEE international conference on broadband network and multimedia technology, 677–683
3. Raje S, Davuluri C, Freitas M, Ramnath R, Ramanathan J (2012) Using ontology-based methods for implementing role-based access control in cooperative systems. In: Proceedings of the ACM symposium on applied computing, 763–764
4. Kabir ME, Wang H, Bertino E (2012) A role-involved purpose-based access control model. Inform Sys Frontiers 14(3):809–822
5. Kim S, Kim D-K, Lu L (2011) A feature-based approach for modeling role-based access control systems. J Sys Softw 84(12):2035–2052
6. Mohammed SA, Yusof MM (2011) Towards an evaluation method for information quality management of health information systems. In: Proceedings of the 2nd international conference on information management and evaluation, 529–538
7. Liqing L, Rong Y, Hai L, Xudong L (2011) Research of extended RBAC model on permission control in WEB information system. In: Proceedings of 3rd IEEE international conference on communication software and networks, 359–362
8. Adam E, Berger T, Sallez Y (2011) Role-based manufacturing control in a holonic multi-agent system. Int J Prod Res 49(5):1455–1468
9. Franqueira NL, Wieringa J (2012) Role-Based Access Control in Retrospect. Computer 45(6):81–88

10. Guodong C (2010) System analyses to restructure resources management information system. In: Proceedings of 2nd conference on environmental science and information application technology, 597–600
11. Petch-urai O, Sombunsukho S (2010) A Development of supply management information system department of computer and information technology. In: Proceedings of 2nd international conference on computer technology and development, 402–404
12. Helil N, Kim M, Han S (2011) Trust and risk based access control and access control constraints. KSII Trans Internet Inform Sys 5(11):2254–2271

Chapter 70
Key Management for Wireless Hierarchical Multicast

Jingjing Zheng, Chunyu Zhao and Hongyun Wang

Abstract A new design scheme of key management tree is proposed for multiple levels multicast communication in wireless network. The network key tree matches with the actual network topology in the scheme and the key nodes in the trees only need to transfer its associated key information. At the same time using the partial order relation between data resources and user ordering in multicast service, users can be grouped; those who have the same ordering relations are in the same group. In this scheme multiple key trees are unified into a single key tree. The amount of the keys stored in key management center (KDC) is calculated and key update process is offered. The simulation results show that the scheme effectively reduces the redundancy key amount and the communication overload produced by the key updating.

Keywords Logical key tree · Topology · Partial order · Multicast

70.1 Introduction

Because in a large group communication security network, it is impossible that only a multicast data stream is present. The previous method needs set a key management tree for each multicast data stream. It generates a large number of redundant key [1]. From access control on the network, different users may have different access levels; therefore, a key management tree of the multilayer data flows will decrease the number of keys.

J. Zheng (✉) · C. Zhao · H. Wang
HuangHe S&T College, Zhengzhou, Henan province 450005, People's Republic of China
e-mail: waterjing@126.com

W. Lu et al. (eds.), *Proceedings of the 2012 International Conference on Information Technology and Software Engineering*, Lecture Notes in Electrical Engineering 210, DOI: 10.1007/978-3-642-34528-9_70, © Springer-Verlag Berlin Heidelberg 2013

For example, in wireless multicast service, it may contain a large number of data stream, such as news, weather, financial and others. The order quantity of users who belong to different levels is not the same, some users may only order news, and others may order two of them. The users who order the same data stream are divided into a group, and the data stream which the same users order is divided into a group. Thus, it is a partial order relation between the user group and the data group. The partial order relation can be used to constitute a multilevel key management scheme [2].

In wireless network, the multicast network model is composed of three parts: SH (super host), BS (base station), MH (mobile host). SH is connected to BS through a wired connection, BS and MH are wireless connection. Wireless network also has the problem of multilayer multicast and different access levels of users. However, the wireless network and wired network model are very different, the users move back and forth between different cells in wireless network, so the scheme of the cable network cannot be applied directly to wireless network. In wireless network, the key management issues of multilayer multicast information flow are as follows [3–5]:

- The grouping problem of the users which accept the same data stream in the same BS.
- Whether SH knows these services ordered by the users in each BS.
- Whether BS knows the service data flow of the users in it.
- When a user moves from one BS to another BS, whether the BS has its service flow. If the service flow is existing, it can directly add the corresponding service group, and the private key is distributed, don't need the key update. If not, add it.
- For all SH, a sub tree can be designed base on the partially ordering relation between users and data group.

70.2 Instruction Method of Hierarchical Key Tree in Wireless Network Tree

A scheme of key management mode of multiple hierarchical multicasts in wireless cellular network is proposed.

We use a typical cellular network model. The network contains SH (super host), BS (base station) and the users. The corresponding key management tree of topology matching is shown in Fig. 70.1. In this diagram the users who are in the same BS are viewed as a subgroup. If the users are the neighbors in the network they are also neighbors in the actual physical location. Firstly, assuming that only the SH is responsible for the management of BS and the keys which protect the multicast communication, the users evenly distribute under BS. The number of users is huge; the users in the same BS are viewed as a subgroup and contain all of

Fig. 70.1 The subtree for
users in each BS

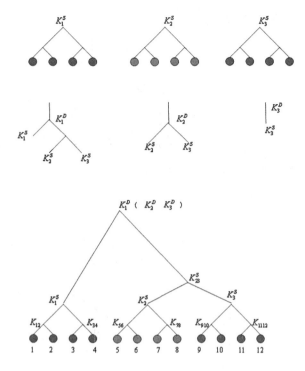

the data group and service group. BS knows whether a reset key message is
necessary for its group members [6].

The construction steps of the key graph are as follows:

Construct subtree for the multi-layered multicast users in each BS, called user
subtree. Different encryption keys of each data stream are persevered in the BS sub
tree's root. If contains the number of K data stream, the BS contains K different
keys, and BS has each data stream key. The construction method needs structure a
tree for every service group S_j. If leaf node is the user in S_j, the root node is
represented by the key K_j^s. It needs structure a tree for each data stream, the root is
K_m^D, the leaf node is $\left\{ K_j^s, \forall i : t_m^i = 1 \right\}$. Finally the data group sub tree's leaves are
linked to the root node of service group subtree [1].

- Construct a subtree for all BS in SH, called BS subtree.
- Link the leaf of the BS subtree to the root of the user subtree, called key tree.

Key tree construction process as shown below:
Step one: (Fig. 70.1)
Step two: (Fig. 70.2)
Step three: (Fig. 70.3)

In the Fig. 70.2, a leaf node represents a user, the users in the same branch is in
the same service group S_i. In which, K_{bs_i} represents each data stream's encryption

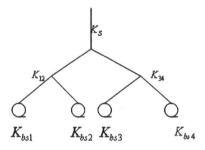

Fig. 70.2 The subtree for all BS in SH

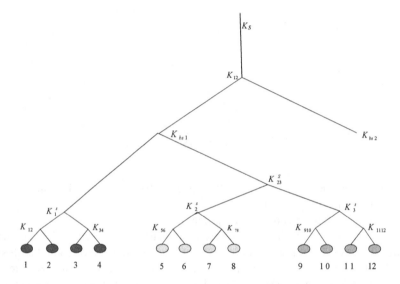

Fig. 70.3 Link the leaf of the BS subtree to the root of the user subtree

key. We define that the keys have the same number as the data flows in each BS. Each key represents a data stream; each data stream is encrypted by different keys in the BS. Each user saves the key of data stream that the user has ordered.

70.3 Data Analysis

In order to analyze the superiority of this hybrid key management tree, we calculate the total number of key stored in the key management center (KDC) in the two modes.

We assume that all the key management trees are the balanced trees. Suppose that the tree's height is $L_0 + 1$ and the degree is d.

The char n is the number of users. $R_d(n)$ is all the keys of the key management tree, $F_d(n)$ is the length of branches of the key tree. Then, for a balanced tree, $F_d(n) = L_0$ or $L_0 + 1$, in which $L_0 = \log_d^n$.

The following is the analysis of storage. When the trees are the independent key trees, all keys of the M key management trees are stored in KDC. Because SH and BS construction method are the same for every key management tree. Then the number of keys in the branches is $d^{L_0} - \left\lceil \frac{n - d^{L_0}}{d-1} \right\rceil$ when $F_d(n) = L_0$.

The number of keys in the branches is $n - d^{L_0} + \left\lceil \frac{n - d^{L_0}}{d-1} \right\rceil$ when $F_d(n) = L_0 + 1$.

So the total number of keys $R_d(n)$ is expressed in the following:

$$R_d(n) = n + 1 + \frac{d^{L_0} - 1}{d - 1} + \left\lceil \frac{n - d^{L_0}}{d - 1} \right\rceil \tag{70.1}$$

Using $\frac{n - d^{L_0}}{d-1} \le \left\lceil \frac{n - d^{L_0}}{d-1} \right\rceil < \frac{n - d^{L_0}}{d-1} + 1$, the below equation holds:

$$\frac{dE[n] - 1}{d - 1} + 1 \le E[R_d(n)] < \frac{dE[n] - 1}{d - 1} + 2 \tag{70.2}$$

In addition, because $\log_d^n$ is a concave function and $\lfloor \log_d n \rfloor \le \log_d n$, the Eq. 70.3 holds:

$$E[F_d(n)] \le E\left[\log_d^n\right] + 1 \le \log_d E[n] + 1 \tag{70.3}$$

Using the above (70.2) and (70.3) equation, we analysis of the key storage quantity. In the scheme of the independent key tree, the key management center needs to store the entire keys of the key tree, $n(D_m)$ is the number of the users in the m key tree. So we calculate the key storage capacity in KDC in separation scheme when the total number of the data flow is M is expressed in the following:

$$R_{KDC}^{ind} = \sum_{m=1}^{M} E[R_d(n(D_m))] \tag{70.4}$$

In the scheme of the hybrid key management tree, in general, each BS contains all data flows, and the number of users is equal. The total number of BS is expressed by S_{bs}. When the total number of users is n, the number of user of each BS is n/S_{bs}. So the key amount storied in KDC is expressed in the following:

$$R_{KDC}^{mg} = \sum_{k=1}^{N_{bs}} E[R_d(BS_k)] + E[R_d(N_{bs})] + (K - 1)N_{bs} \tag{70.5}$$

The first part of the above equation is the total sum of mathematical expectation of all key of the user trees from the first to the K BS. The second portion represents the mathematical expectation of the number of key of the user subtrees. The last portion represents the number of key of data stream in each BS node.

In which, suppose that the number of users is equal to n_0 in each service group. When $n_0 \to \infty$, the (70.4) and (70.5) equation can be simplified to (70.6) and (70.7).

$$R_{KDC}^{ind} = \sum_{m=1}^{M} E[r_d(m \cdot n_0)] \tag{70.6}$$

$$R_{KDC}^{mg} \leq M \cdot E[r_d(n_0)] + (M-1)N_{bs} \tag{70.7}$$

From the above formula, when $n_0 \to \infty$, we get the Eq. 70.8:

$$R_{KDC}^{ind} \sim O\left(\frac{d}{d-1} \cdot \frac{M(M+1)}{2} n_0\right) \tag{70.8}$$

$$R_{KDC}^{mg} \sim O\left(\frac{d}{d-1} \cdot M \cdot n_0\right) \tag{70.9}$$

The above equation displays that the key storage quantity in the user terminal and key management center relative reduction.

70.4 Performance Analysis

In the section, we compare the scheme with the independent tree mode (In depend-tree) in the wireless network and analysis the experimental data in different application mode.

We use statistical dynamic membership to simulate system. Firstly we set the number of users at most 4,000 peoples, M is 4 and the degree is 3. In the Fig. 70.4 the linear represents the variation of the number of keys stored in KDC when the number of the users increases under Mg scheme, the star is the number of keys stored in KDC under the Ind program. With the increase of the number of users, the number of keys stored in KDC under Mg scheme is less than Ind scheme.

In the following we analyze the amount of the keys stored in the key management center (KDC) in two different modes when the number of data groups is different.

Because the number of the data groups can be changed at any time. That is, the service layer is changed. In Fig. 70.5 we compared the number of the keys that the key management center needs to save under the different layers, in which N_0 is 1,500 and the degree is 3.

In Fig. 70.5, the simulation results indicate that in independent mode the total number of keys in KDC is almost a linear growth with the increasing of the number of data group. But under Hybrid mode the total number of keys in KDC almost changes very little. This means that the Mg scheme is obviously reduced the redundancy of the key caused by the changes of the data group.

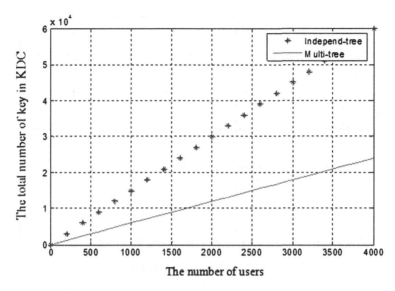

Fig. 70.4 The comparison of the number of keys in KDC

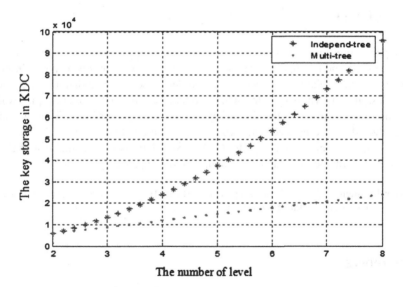

Fig. 70.5 The key storage in KDC when the layer is different

In Fig. 70.6, the simulation results indicate that when the data group layers are different the storage volume ratio is linear growth along with the data group increases. The storage volume under Ind mode in KDC is several times larger than the storage capacity under Mg mode.

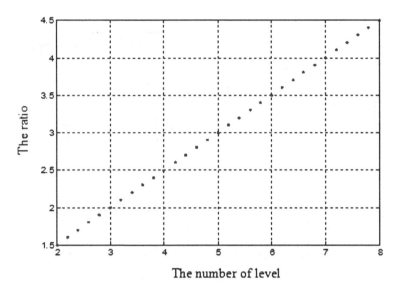

The number of level

Fig. 70.6 The storage volume ratio in the two modes

70.5 Conclusion

This paper presents a hybrid key tree construction scheme of multiple levels service group and key update process in wireless network. In this manner, through the partial order relation between the user ordering service and data resources, we have designed a hybrid key tree which contains all the data stream replaces the scheme that requires constructing a key tree for each data stream. Then we analyzed the difference between the two schemes, included the total number of keys stored in the key management center (KDC). Simulation experimental results show that the performance by our way is significantly improved.

Acknowledgments This work was supported by Huang He Science and Technology College' research projects under Grant KYZR201204.

References

1. Sun Y, Trappe W, Ray Liu KJ (2003). Topology-aware key management schemes for wireless multicast. In: Proceedings of IEEE GLOBECOM. San Francisco, CA, pp 1471–1475
2. Sun Y, Trappe W, Ray Liu KJ (2004) A scalable multicast key management schemes for heterogeneous wireless networks. IEEE/ACM Trans Networking 12(4):653–666
3. Xu M-W, Dong X-H, Xu K (2004) A survey of research on key management for multicast. J Software 200415(1):141–150 (in Chinese)
4. Decleene B, Dondeti L, Griffin et al (2001) Secure group communications for wireless networks. In: Proceedings of IEEE MILCOM 2001. Mclean, VA, pp 113–117

5. Balenson D, McGrew D, Sherman A (2003) Key management for large dynamic groups: one-way function trees and amortized initialization. IEEE Trans Software Eng 29(5):444–458
6. Haney H, Muckenhirn C, Rivers T (1997) Group key management protocol (GKMP) specification. RFC2093, Internet Engineering Task Force

Chapter 71
Novel Network Protocol for WSN Based on Energy and Distance

Haiyan Wu and Nana Gong

Abstract Wireless sensor network (WSN) is a special type of network formed from a large amount of wireless sensor nodes with energy, memory and processing constraint factors. Utilization of energy and enhancing network lifetime is a challenging task. The proposed cluster based routing algorithm selects cluster heads on the basis of two critical factors: the node's energy and distance to base station (BS). Cluster heads transmit data to the BS by single-hop or multi-hop due to the distance to BS. Computer simulation shows that the proposed algorithm performs with lower energy consumption and longer life time than LEACH.

Keywords WSN · Cluster routing protocol · Multi-hop · Distance

71.1 Introduction

In recent year wireless sensor network (WSN) becomes an important topic for researchers. Such network can be used to monitor physical or environmental conditions, such as temperature, sound, vibration, pressure, motion or pollutants, and to cooperatively pass their data through the network to BS. Sensor nodes are spatially distributed autonomous and hold small amount of power, hence it is necessary for communication protocols to save nodes' energy consumption.

A large number of existing routing protocols partition WSN into clusters. Each cluster has a cluster head (CH) that acquires the sensed data from cluster member

H. Wu (✉) · N. Gong
School of Information and Engineering, Huanghe Science and Technology College,
Zhengzhou, China
e-mail: wuhy@haut.edu.cn

W. Lu et al. (eds.), *Proceedings of the 2012 International Conference on Information Technology and Software Engineering*, Lecture Notes in Electrical Engineering 210, DOI: 10.1007/978-3-642-34528-9_71, © Springer-Verlag Berlin Heidelberg 2013

nodes, aggregates and forwards it to other cluster heads or to the base station. Low-Energy Adaptive Clustering Hierarchy (LEACH) [1] was the first cluster based protocol. LEACH has introduced a CH selection and randomized rotation technique. Through simulation results, it has been proved that the network life time increases as well as better load-balancing issue than flat routing algorithm. PEGASIS [2] use chain topology with data fusion to reduce energy consumption. However, nodes far away from the base station exhaust their energy rapidly and would die soon. HEED [3] elects cluster heads based on residual energy and node connectivity. But HEED increases the control packages. TEEN [4] reduces the information packages that cluster members send to CH through adjusting two thresholds, yet it is difficult to select the thresholds.

This paper addresses the advanced method for cluster head selection based LEACH. To appropriately select a cluster-head for each cluster, the algorithm takes into account each node's residual energy, and distance from node to the BS. Communication for intra-cluster adopts model that CHs transfer packages to BS directly or multi-hop to BS via relaying CHs. From the simulation results we can see the advanced protocol extends the life time of the network.

71.2 Energy and Channel Models

We make some assumptions about the sensor nodes and the underlying network model. For the sensor nodes, we assume that all nodes can transmit with enough power to reach the BS if needed, that the nodes can use power control to vary the amount of transmit power, and that each node has the computational power to support different MAC protocols and perform signal processing functions [5].

The proposed algorithm uses a simple propagation loss model [5] to compute the energy consumption of radio transmission. According the model, both the free space and the multi-path fading channels are used, depending on the distance between the transmitter and receiver. The energy dissipation for transmitting a k-bit message to a receiver at a distance d is:

$$E_{Tx}(k,d) = kE_{elec} + k\varepsilon_{fs}d^2, \quad d < do \tag{71.1}$$

$$E_{Tx}(k,d) = kE_{elec} + k\varepsilon_{amp}d^4, \quad d \geq do \tag{71.2}$$

To receive this message, the radio expends energy amounting to:

$$E_{Rx}(k,d) = kE_{elec} \tag{71.3}$$

E_{elec} is the radio dissipation, $\varepsilon_{fs}d^2$ and $\varepsilon_{amp}d^4$ is transmit amplifier dissipation, they depend on the transmission distance.

71.3 The Proposed Routing Mechanism

The advanced algorithm organizes nodes into local clusters. In every cluster, one cluster head (CH) receives data from other cluster members and performs data aggregation, then transmits data through direct or multi-hop pattern to the remote BS. All process based on round and each round consists of three phases: cluster head selection; cluster formation; data transmission and cluster head rotation.

71.3.1 Cluster Head Selection

In every round, each node itself decides whether or not to become a cluster head. At first if node i hasn't work as CH in recent $1/p$ rounds, the node i chooses a random number between 0 and 1. If the number is less than a threshold T(n), the node i becomes a cluster-head for the current round. CHi is chosen using the threshold T(n) [6] as below:

$$T(n) = \begin{cases} \frac{P}{1-P[r \bmod (1/P)]} \times \left[w_1 \frac{E_{res}}{E_{init}} + (1-w_1) \frac{d_{max}-d_i}{d_{max}-d_{min}} \right] & n \in G \\ 0 & \text{otherwise} \end{cases} \quad (71.4)$$

P is the desired ratio of cluster heads, r is the current round and G is the set of nodes that has not been cluster heads in recent $1/P$ rounds. d_i is the distance from node i to BS. d_{max} is the maximize distance from node to BS, d_{min} is the minimize distance from node to BS. w_1 is a constant.

All CHs broadcast their status to the other sensors in the network. Then the advanced protocol acts on Clusters formation in the same way as the LEACH [1] protocol.

71.3.2 Communication Routing Mechanism

In LEACH protocol, the CH has all the data from the nodes in its cluster and aggregates the data, then directly transmits the compressed data to the base station. Since CHs far away from BS spend higher energy than those nearer to BS.

In mechanism described, CHs far away from BS route data through intermediate CHs which meet the requirements. If the distance from CHi to BS is greater than do, then CHi should need certain relay node. The relay node should have short distance to source node and destination node. Figure 71.1 shows if source node CH0 sends data to destination node CH4, CH1 as the relay node is a fine choice [7]. Meanwhile we consider residual energy factor to devise the balance of the energy consumption between nodes.

Fig. 71.1 Choice of the relay node

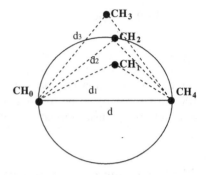

Relay candidate set consists of the satisfied CHs that meet the following conditions. The protocol will choose candidate with small serial number when other conditions are the same.

$$candidate = \left\{ CH_j \middle| d_{j\_BS} \le do \cup (d^2_{i\_j} + d^2_{j\_BS}) < d^2_{i\_BS} \cup E_{jres} > E_{ires}, CH_j \in CH \right\}$$

(71.5)

Where d_{i_BS} is the distance from BS to CH_i, and d_{j_BS} is the distance from BS to CH_j, d_{i_j} is the distance between CH_i and CH_j. E_{ires} is the residual energy of CH_i, and E_{jres} is the residual energy of CH_j.

71.4 Performance Evaluation

71.4.1 Simulation Environment

We randomly and uniformly deploy 50 nodes distributed over 100×100 m network area using MATLAB programming language. The BS is located at (50, 0). The data packet size is 4,000 bits and control packet size is 100 bits. Initial energy of each node is 0.5 J, E_{elec} is 50 nJ/bit, ε_{fs} is 10 pJ/bit/m^2, ε_{amp} is 0.0013 pJ/bit/m^4 and $do = (\frac{E_{fs}}{E_{amp}})^{1/2}$ m. The initial number of cluster equals to 5 % of the total live nodes. This simulation runs for 2,000 rounds.

71.4.2 Simulation Results and Discussion

In our simulation, we focus on comparing three metrics of network performance: number of nodes alive, distribution of dead nodes and network energy consumption among LEACH and Advanced-LEACH.

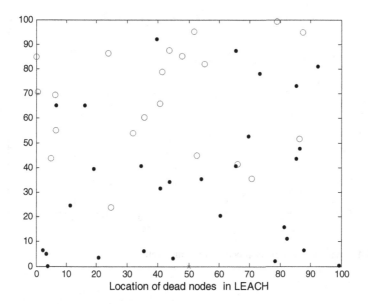

Fig. 71.2 Location of dead nodes exceeds 20 in LEACH

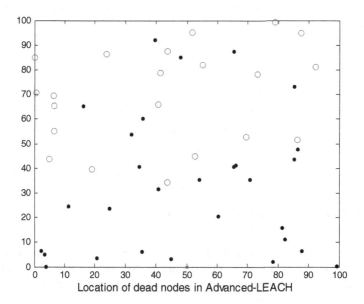

Fig. 71.3 Location of dead nodes exceeds 20 in advanced-LEACH

In Figs. 71.2 and 71.3 we simulated a case to reflect the location of dead nodes when the number of dead nodes is over 20. In Figure red "o" represents dead node and black "·" represents survival node. With the assumption that the BS is below

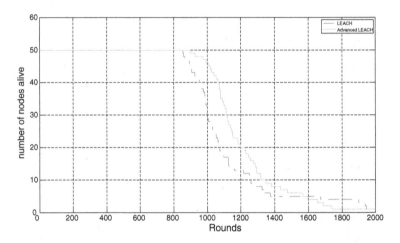

Fig. 71.4 Live nodes comparison between routing protocols

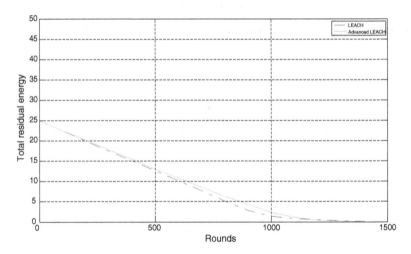

Fig. 71.5 Total energy consumption versus rounds

each subfigure, nodes which is farther from the BS will die off sooner than the other. Compared with LEACH, dead nodes in Advanced-LEACH spread more widely rather than concentrate in a small area, and thus the result show the better energy balance in the latter.

Figure 71.4 shows number of nodes alive as rounds proceed for two schemes. The CHs farther from BS would consume more energy to send aggregated data in single-hop way. Consequently, it would drain its energy very fast. To maintain the load balance, in the proposed scheme multi-hop communication is considered when CHs farther from BS send data to BS. The distance-energy factor is considered in the selection of relay nodes. The result tells that Advanced-LEACH compared with LEACH prolongs the lifetime of network with a gain of 14 %.

Figure 71.5 demonstrates the total residual energy in the whole system versus rounds for LEACH and Advanced-LEACH. This metric in Advanced-LEACH is higher than LEACH with a gain of 2 %.The result shows the proposed algorithm is more energy efficient than LEACH.

71.5 Conclusion

In this paper, we proposed Advanced-LEACH, an algorithm to select appropriate cluster-heads and better communication routing mechanism. This algorithm considers some critical parameters including nodes' remaining energy, location to achieve optimal energy efficiency and energy balance. Our simulation results prove that in the same conditions the proposed algorithm outperforms LEACH in terms of both network energy efficiency and energy balance.

Acknowledgments This work was supported by Science and Technology Innovation Team of IOT in Zhengzhou under Grant 112PCXTD343, and Major research projects in Henan Province Science and Technology Agency under Grant 20110992.

References

1. Heinzelman WR, Chandrakasan A, Balakrishnan H (2000) Energy efficient communication protocol for wireless micro sensor networks. In: Proceedings of Hawaii international conference on system sciences, pp 1–10
2. Lindsey S, Raghavendra CS (2002) PEGASIS: power efficient gathering in sensor information systems. In: Proceedings of IEEE aerospace conference, vol 3, pp 1125–1130
3. Younsis O, Fahmy S (2004) Distributed clustering in Ad hoc sensor networks: a hybrid, energy-efficient approach. In: Proceedings of 13th joint conference on IEEE computer and communications societies, INFOCOMM
4. Manjeshwar A, Agrawal DP (2001) TEEN: a protocol for enhanced efficiency in wireless sensor networks. In: Proceedings of the 1st international workshop on parallel and distributed computing issues in wireless networks and mobile computing (with IPDPS'01)
5. Heinzelman WB, Chandrakasan AP, Balakrishnan H (2002) An application–specific protocol architecture for wireless microsensor networks. In: IEEE transaction on wireless communications. IEEE Press, New York 1(4):660–670
6. Chen XJ, Li XY (2009) Research and improvement of LEACH protocol in wireless sensor network. J Comput Appl 29(12):3241–3243 (in Chinese)
7. Zhou Y, Shi R, Zhou Y (2011) Multi-hop routing algorithm based on uneven clustering for WSN. J Comput Appl 28(2):642–644 (in Chinese)

Chapter 72
Strategies for Improving the Convergence Rate of High Accuracy Surface Modeling

Na Zhao and Tianxiang Yue

Abstract To improve the convergence rate of high accuracy surface modeling (HASM) method, we give three preconditioned conjugate gradient methods: incomplete Cholesky decomposition conjugate gradient method (ICCG), symmetric successive over relaxation-preconditioned conjugate gradient method (SSORCG), and polynomial preconditioned conjugate gradient method (PPCG). Besides, we give adequate consideration to storage schemes of the large sparse matrix and optimize the performance of sparse matrix–vector multiplication. The cost of the computation is also considered in each iteration. Numerical tests show that ICCG has the fastest convergence rate of HASM followed by SSORCG method. PPCG method has the lowest convergence rate than others since the sparse matrix–vector multiplication has to be formed three other times in every iteration; however, it is efficient on vector and parallel computers.

Keywords Preconditioned conjugate gradients methods · Incomplete cholesky factorization · Successive over-relaxation algorithm · Polynomial preconditioner · HASM · Surface modeling

72.1 Introduction

The methods of surface modeling have been experienced rapid development since the first digital elevation model was developed and the error problem has become a major concern since the 1960s [1–3]. To solve the error problems produced by

N. Zhao · T. Yue (✉)
Institute of Geographical Sciences and Natural Resources Research,
Chinese Academy of Sciences, 100101 Beijing, China
e-mail: yue@lreis.ac.cn

W. Lu et al. (eds.), *Proceedings of the 2012 International Conference on Information Technology and Software Engineering*, Lecture Notes in Electrical Engineering 210, DOI: 10.1007/978-3-642-34528-9_72, © Springer-Verlag Berlin Heidelberg 2013

geographical information systems [4], high accuracy surface modeling (HASM) method was developed by Yue et al. [3, 5–9].

According to differential geometry, a surface is uniquely defined by the first and the second fundamental coefficients [10]. The whole calculation process of HASM can be divided into three stages: deriving finite difference approximations to differential equations, establishing the sampling point equations and solving the algebra equations. Numerical tests show that the accuracy of HASM is higher than some classical surface modeling methods [11].

To solve the time-consuming calculation problem, Yue et al. [9, 12] gave the optimum formulation of HASM. However, several problems need to be settled, such as the data storage issue and low computational speed problem. The purpose of this paper is to improve the convergence rate of HASM by using PCG, a numerical code which uses the preconditioned conjugate gradient method to solve the matrix equations produced by HASM. Three preconditioned strategies are introduced and we find that both ICCG and SSORCG methods perform better than the existing methods. Because of the selection of coefficients in PPCG, PPCG method has a lower convergence rate than the methods analyzed. Among various parallel programming environments, parallel degree for SSORCG method is increased while the sample points using red–black ordering method. However, in this case, the convergence rate of the SSORCG method is decreased and thus the speedup ratio is unsatisfied. PPCG method is efficient on vector and parallel computers since we only form the sparse matrix–vector multiplication in every iteration and this can be realized effectively when we select the optimal coefficients.

72.2 Preconditioned Conjugate Gradient Methods

In this work, selected numerical methods are presented for solving the matrix equations that arise when the finite difference method is applied. The finite difference model produces a set of linear equations which can be expressed in matrix notation as:

$$Ax = b, \tag{72.1}$$

where $A \in R^{n \times n}$ is symmetric and positive definite.

Linear equations such as Eq. (72.1) can be solved using direct or iterative methods. In most direct methods the matrix A is factored exactly and the true solution is obtained by existing one backward and one forward substitution. In most iterative methods, an initial estimate of the solution is refined iteratively using an approximately factored matrix. Direct methods are for the small size problems, and iterative methods are more efficient for large problems and require less computer storage [13]. The conjugate gradient method (CG) [14] is an iterative method which can be used to solve matrix equations if the matrix is symmetric and positive definite. CG method, however, works well on matrices that are either well conditioned or have just a few distinct eigenvalues. And the matrix A in HASM is ill-conditioned. Then we further introduce the preconditioned conjugate gradient method which has a fast convergence rate through modifying the condition of matrix A.

Algorithm 1 *Preconditioned Conjugate Gradients (PCG)* [15]:

Given an initial value x_0, let $k = 0$, $r_0 = b - Ax_0$,
while ($r_k \neq 0$)

$$\text{solve} \quad \begin{matrix} Mz_k = r_k \\ k = k + 1 \end{matrix}$$

If ($k = 1$) $p_1 = z_0$
else

$$\beta_k = r_{k-1}^T z_{k-1} / r_{k-2}^T z_{k-2}; \ p_k = z_{k-1} + \beta_k p_{k-1}$$

end

$$\alpha_k = r_{k-1}^T z_{k-1} / p_k^T A p_k; \ x_k = x_{k-1} + \alpha_k p_k; \ r_k = r_{k-1} - \alpha_k A p_k$$

end

$$x = x_{k-1}$$

The major computational cost is $Mz_k = r_k$. The choice of a good preconditioner M can have a dramatic effect upon the convergence rate. Next, we will find different matrix M to improve the computational speed of the algorithm above and to reduce the computational cost of $Mz_k = r_k$.

72.2.1 Incomplete Cholesky Preconditioners

The linear system Eq .(72.1) can be converted into the following system:

$$\bar{A}x = \bar{b} \tag{72.2}$$

where $\bar{A} = MA$, $\bar{b} = Mb$. If $k(\bar{A}) < k(A)$ ($k(A)$ denote the condition of A), then the computational speed of system Eq .(72.2) is faster than Eq .'(72.1) by using the same algorithm.

The incomplete Cholesky preconditioner (ICCG) has been very popular [16]. Suppose the incomplete Cholesky factorization of A is $A = M + R = LL^T + R$, where L is a lower triangular matrix with the property that $M = LL^T$ is close to A and L has the same sparsity structure with A. The structure of L can be controlled by the matrix R. Consider that there exist many zero elements of R and the nonzero elements value of R are small in actual computation. In this paper, we consider the non-filled Cholesky factorization of A, that is, the position of L is zero if the corresponding position of A is zero. The code of this algorithm is as follows:

```
for k = 1 : n
    A(k,k) = √A(k,k)
    for i = k + 1 : n
        if A(i,k) ≠ 0
            A(i,k) = A(i,k)/A(k,k)

        end
    end
    for j = k + 1 : n
        for i = j : n
            if A(i, j) ≠ 0
                A(i, j) = A(i, j) − A(i, k) A(j, k)

            end
        end
    end
end
```

Since $k(A) \geq 1$, it is easy to find that $k(M^{-1}A) < k(A)$. In the process of calculation, to take advantage of the large number of zeros in matrices A and L, we employ the block compressed sparse row (BCSR) format [17]. $Mz_k = r_k$ can be transformed to $L^T z_k = y_k$ and $Ly_k = r_k$. The computational cost is $O(n^2)$.

72.2.2 Symmetric Successive Over-Relaxation (SSOR) Preconditioner

The major computational cost of HASM lie in the sparse matrix–vector multiplication and the matrix inverse. The computational cost of matrix inverse is larger than matrix–matrix multiplication. SSOR preconditioning can directly compute the inverse matrix of M which reduces the computational cost.

In terms of matrix splitting, if $A = L + D + L^T$, where D is diagonal and L is strictly lower triangular, SSOR preconditioner is given by [17]:

$$M = KK^T, \quad K = \frac{1}{\sqrt{2-\omega}}(\frac{1}{\omega}D + L)(\frac{1}{\omega}D)^{\frac{-1}{2}}, 0 < \omega < 2, \tag{72.3}$$

The inverse of K is $K^{-1} = \sqrt{2-\omega}(\frac{1}{\omega}D)^{\frac{1}{2}}(I + \omega D^{-1}L)^{-1}\frac{1}{\omega}D^{-1}$. Denote $\overline{D} = \frac{1}{\omega}D$, since $(I + \overline{D}^{-1}L)^{-1} = I - \overline{D}^{-1}L + (\overline{D}^{-1}L)^2 - \ldots$, we have $K^{-1} \approx \sqrt{2-\omega}\overline{D}^{\frac{1}{2}}(I - \overline{D}^{-1}L)\overline{D}^{-1} = \sqrt{2-\omega}\overline{D}^{\frac{-1}{2}}(I - \overline{D}^{-1}L) \equiv \overline{K}$. Then the approximate inverse matrix of A is $\overline{M} = \overline{K}^T \overline{K}$ and $Mz_k = r_k$ in Algorithm 1 can be changed into $z_k = \overline{M}r_k$ which can be further converted into $y_k = \overline{K}r_k$, $z_k = \overline{K}^T y_k$. The space complexity is $O(n)$ as we can implement this algorithm explicitly.

72.2.3 Polynomial Preconditioner

Given a symmetric system:

$$Ax = b$$

the principle of polynomial preconditioning, which goes back to Rutishauser [18], consists in solving the preconditioned system:

$$s(A)Ax = s(A)b,$$

where $s(A)$ is a polynomial and approximate to A^{-1}. Many polynomial preconditionings have been suggested. Among them the simplest polynomial suggested by Dubois et al. [19] is as follows: $A^{-1} = \left(\sum_{i=0} \left(I - M^{-1}A \right)^i \right) M^{-1}$, where $A = M - N$. In fact, from the Neumann series, we have $(I - A)^{-1} = \sum_{i=0} A^i$ and $A^{-1} = \sum_{i=0} (I - A)^i$. Then we can approximate A^{-1} by using a truncated Neumann series. That is $s(A)^{-1} \approx c_0 I + c_1 A + c_2 A^2 + c_3 A^3$. Therefore, $s(A)^{-1} r_k$ in the PCG algorithm can be written as $s(A)^{-1} r_k = c_0 r_k + c_1 A r_k + c_2 A^2 r_k + c_3 A^3 r_k$. Here, $c_j, j = 0, 1, 2, 3$ are coefficients chosen to optimize convergence. Note that at each iteration we simply form the product $s(A)^{-1} r_k$, $s(A)^{-1}$ need not actually be calculated and stored. This can be accomplished as follows:

$$z_1 = c_2 r_k + A r_k; \ z_2 = c_1 r_k + A z_1; \ s_k = c_0 r_k + A z_2$$

We can see that the powers of A are never formed explicitly. This will efficient on vector and parallel computers. However, the calculation of coefficients $c_j, j = 0, 1, 2, 3$ is a difficult problem. In this research, we use the method described by Saad [20] to compute these coefficients as: $c_0 = \frac{-15}{32} g^3$, $c_1 = \frac{27}{16} g^2$, $c_2 = \frac{-9}{4} g - 15$, where g is the upper bound on the largest eigenvalue of A. The matrix A in HASM is large and sparse, and the largest eigenvalue can be computed by iterative methods such as power method. Here, we estimate the maximum eigenvalue as the largest sum of the absolute values of the components in any row of A.

72.3 Numerical Tests

In this section, we employ Gaussian synthetic surface to validate the efficiency of PCGs for solving HASM. The formulation of Gaussian synthetic surface is expressed as:

$$f(x,y) = 3(1-x)^2 e^{-x^2-(y+1)^2} - 10\left(\frac{x}{5} - x^3 - y^5\right)e^{-x^2-y^2} - \frac{e^{(x+1)^2-y^2}}{3}$$

The computational area is $\{(x,y)|-3 < x < 3, \quad -3 < y < 3\}$, and $-6.5510 < f(x,y) < 8.1062$. For HASM, the process of solving the linear system Eq. (72.2) is named inner iteration, while updating the right vector $\bar{b}$ is termed outer iteration. The first test is that we fix the sampling interval (m = 4), outer iterative times (1time) and inner convergence criteria ($\|f^{n+1} - f^n\|_2 < 10^{-6}$). We change the grid number of computational area to compare the computational efficiency of PCG for different preconditioned strategies in HASM. Meanwhile, we present the relationship between the computational time and the grid numbers for the ICCG preconditioner and the diagonal preconditioner. In this paper, the diagonal preconditioner denote the preconditioner is a diagonal matrix whose diagonal elements are the diagonal elements of A. Numerical results are showed in Table 72.1 and Fig. 72.1. Here, S in Table 72.1 denotes the diagonal preconditioned conjugate-gradient method (DCG) while T denotes the tridiagonal preconditioned conjugate-gradient method (TCG). We can see that the computational time of ICCG and SSORCG methods are less than DCG and TCG methods. The difference against the computational time of ICCG and DCG has good linear relationship with the grid numbers, that is, the efficiency of ICCG becomes sharper with the number of grids increasing.

Results show that when we apply the PCG method in HASM, ICCG and SSORCG methods are more effective in computational time than former preconditioned conjugate gradient methods. Furthermore, although PPCG method has the lowest computational speed, it will be effective in the parallel environment.

The second test is that we fix the sampling interval (m = 4), grid numbers (101 × 101) and outer iterations (5 times). We compare the computational accuracy of the different PCG methods. The third test is that we fix the sampling interval (m = 4), the grid numbers (501 × 501) and the inner iterative numbers

Table 72.1 Comparisons of computational efficiency with different PCGs

Number of grids	Compute time(s)					T(s)
	ICCG	SSORCG	PPCG	S	T	S-ICCG
101 × 101	0.0797	0.0922	0.3804	0.1178	0.2418	0.0381
301 × 301	0.6826	0.7896	4.6367	1.0882	1.7557	0.4056
501 × 501	1.9908	2.4962	15.748	3.6256	5.211	1.6348
1001 × 1001	8.1942	10.409	67.406	15.612	21.35	7.4173

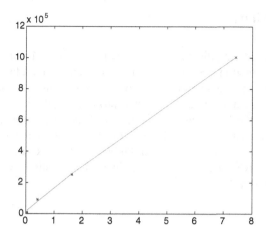

Fig. 72.1 Relationship between the time interval of DCG-ICCG and grid numbers

Table 72.2 Simulation accuracy with different inner-iteration numbers

Inner iteration	ICCG	SSORCG	S	T
5	2.3482	7.4335	6.4740	7.9532
10	0.0395	0.8129	4.7385	1.9689
20	0.000026	0.0145	0.4666	0.0876
50	0	0	0.000513	0.000044

Table 72.3 Simulation accuracy with different outer-iteration numbers

Outer iteration	ICCG	SSORCG	S	T
2	0.001163	0.017035	0.113418	0.094594
4	0.000138	0.002427	0.013710	0.007262
6	0.000064	0.001169	0.006030	0.003271
8	0.000042	0.000821	0.007483	0.002054

(10 times). We then compare the convergence accuracy under different outer iterative numbers. Results in Tables 72.2 and 72.3 show that when we apply PCG method in HASM, ICCG method is the fastest followed by SSORCG method.

Numerical tests above reveal that we can apply ICCG method and SSORCG method in HASM, and use PPCG method when we propose a parallel implementation of HASM.

72.4 Conclusions

In this paper, considering that the matrix equations of HASM is ill-conditioned, we present ICCG method and SSORCG method to improve the efficiency of HASM and consider the implementation details of the two methods regards of the computational time and data storage. We then propose a polynomial preconditioned method based on the Neumann series, and show the implement detail for PPCG which will be efficient in the parallel environment. Gaussian synthetic surface is used to validate these methods on DELL OP990 machine. Results show that ICCG method and SSORCG method are better than other existing PCG method when they are used in HASM.

References

1. Stott JP (1977) Surface modeling by computer. Thomas Telford Ltd for the Institution of Civil Engineers, London, pp 1–8
2. Crain IK (1970) Digital representation of topographic surface. Photogram Eng Remote Sens 11:1577
3. Yue TX, Du ZP (2001) A digital model for multi-sources information fusion. World Sci Technol R&D 23(5):1–4 (in Chinese)
4. Wise S (2000) GIS data modeling-lessons from the analysis of DTMs. Int J Geogr Inform Sci 14(4):313–318
5. Yue TX, Ai NS (1990) A mathematical model for cirque surface. Glaciol Cryopedology 12(3):227–234 (in Chinese)
6. Yue TX, Chen SP, Xu B, Liu QS, Li HG, Liu GH, Ye QH (2002) A curve-theorem based approach for change detection and its application to yellow river delta. Int J Remote Sens 23(11):2283–2292
7. Yue TX, Du ZP, Liu JY (2004) High precision surface modeling and error analysis. Prog Nat Sci 14(2):83–89 (in Chinese)
8. Yue TX, Du ZP (2005) High precision surface modeling: a core module of new generation CAD and GIS. Prog Nat Sci 15(3):73–82 (in Chinese)
9. Yue TX (2011) Surface Modeling: High Accuracy and High Speed Methods. CRC Press, New York
10. Henderson DW (1998) Differential geometry. Prentice-Hall, Inc., London
11. Yue TX, Du ZP (2006) High accuracy surface modeling and comparative analysis of its errors. Prog Nat Sci 16(8):986–991 (in Chinese)
12. Yue TX, Du ZP (2006) Numerical test for optimum formulation of high accuracy surface modeling. Geo-Inform Sci 8(3):83–87 (in Chinese)
13. Irwin R, Hornberger GM, Molz FJ (1971) Numerical methods in subsurface hydrology. Wiley, New York
14. Hestenes MR, Stiefel EF (1952) Methods of conjugate gradients for solving linear systems. J Res Natl Bur Stand 49:409–436
15. Golub GH, Van Loan CF (2009) Matrix computations. Posts and Telecom press, Beijing
16. Helfenstein R, Koko J (2012) Parallel preconditioned conjugate gradient algorithm on GPU. J Comput Appl Math 236(15):3584–3590
17. Yuan E, Zhang YQ, Liu FF, Sun XZ (2009) Automatic performance tuning of sparse matrix-vector multiplication: implementation techniques and its application research. J Comput Res Dev 7(46):1117–1126 (in Chinese)

18. Rutishauser H (1959) Theory of gradient methods. In refined iterative methods for computation of the solution and the eigenvalues of self-adjoint boundary value problems. Institute of Applied Mathematics, Zurich, Basel-Stuttgart, pp 24–49
19. Dubois PF, Greenbaum A, Rodrique GH (1979) Approximating the inverse of a matrix for use in iterative algorithms on vector processors. Computing 22:257–268
20. Saad Y (1985) Practical use of polynomial preconditioning for the conjugate gradient method. SIAM J Sci Stat Comput 6:865–881

Chapter 73
Course-keeping Autopilot with RRD Based on GA-PD Adaptive Optimization

Lijun Wang and Sisi Wang

Abstract In order to develop a simple PD controlled autopilot system with rudder roll damping, the course-keeping controller and rudder roll damping controller are designed and adaptively optimized based on GA-PD method. The parameters of controllers are to be optimized by generic algorithm when the detected standard deviation of the course error or roll angle exceeds the set threshold values. The simulation tests are done on a nonlinear ship motion model, and results show that the controllers achieved good disturbance rejection near the ship's natural roll frequency band, and stable performance of course-keeping and changing, with strong adaptivity to sea states.

Keywords Rudder roll damping · Generic algorithm · PD control · Adaptive

73.1 Introduction

The motion control of under-actuated ships is one of current research topics to be overcome urgently. The ship occurs big amplitude motions of rolling and pitching in strong sea disturbances, of which the rolling movement has great influence on ship stability and the safety of cargo lashing. Consequently, the research work on the coordinate controlling of course-keeping and rudder roll damping (RRD) with no more input variable does great good to the research and development of the ship autopilot and the safety of sea going ships in heavy seas. It is well known that the ship system is of large inertia, large time delay and nonlinearity. At the same time,

L. Wang (✉) · S. Wang
Navigation College, Guangdong Ocean University,
43#, Middle Haibing Road, Zhanjiang, China
e-mail: 123wanglijun@163.com

W. Lu et al. (eds.), *Proceedings of the 2012 International Conference on Information Technology and Software Engineering*, Lecture Notes in Electrical Engineering 210, DOI: 10.1007/978-3-642-34528-9_73, © Springer-Verlag Berlin Heidelberg 2013

the parameter or construction perturbation of ship model can be caused by heavy weather navigation operating mode, therefore, the parameter adaptivity of control system directly affect the effectively of the rudder control system. In the past research work, a composite control system of RRD and stabilizing fin [1] was brought up in LQR method. The robust controller designing method was studied based on the loop shaping algorithm [2–4]. A complex RRD controller was developed using Neural network and internal model control [5]. The paper brings up course-keeping and RRD controllers based on simple PD control which has good engineering application. Further work has been done on the adaptive optimization of the system parameters by generic algorithm (GA). The simulation tests are done on a nonlinear ship motion model, and results show that the controllers achieved good disturbance rejection near the ship's natural roll frequency band, and stable performance of course-keeping and changing, with strong adaptivity to sea states.

73.2 The Adaptive Optimization Principle of Course-Keeping Autopilot with RRD

The ship autopilot system can achieve good reduction of rolling moments caused by sea disturbances in addition to course-keeping and course-changing, the schedule of which is indicated in Fig. 73.1. There are two control rings in the system: the course loop and the RRD loop.

As shown in Fig. 73.1, ϕ_d is the set roll angle, whose set value is 0; ψ_d stands for set course; K_{yaw} and K_{roll} are the PD course controller and RRD controller separately, whose parameters are adaptively optimized using GA; D_{waves} stands for the rolling and yawing moments caused by waves.

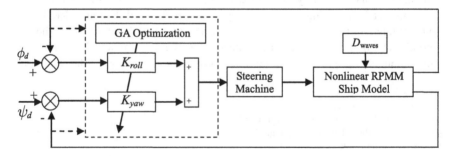

Fig. 73.1 The adaptive optimization principle of course-keeping autopilot with RRD

73.2.1 PD Course Controller

The structure of traditional PD course controller is indicated as:

$$K_{yaw} = K_{p1}e + K_{d1}\dot{e} \tag{73.1}$$

Where the course deviation $e = \psi_d - \psi$, and is ψ the output course; K_{p1}, K_{d1} are the proportional and differential parameters of the PD course controller, which can be obtained by trial and error or empirical formula.

73.2.2 PD Controller of RRD

The PD controller of RRD can be defined as:

$$K_{roll} = K_{p2}\sigma + K_{d2}\dot{\sigma} \tag{73.2}$$

Where the roll angle deviation $\sigma = \phi_d - \phi$, and ϕ is the output roll angle; K_{p2}, K_{d2} are the proportional and differential parameters of the PD controller of RRD.

The above PD controllers of course and RRD are of routine design. That is to say the outside disturbance is not taken into consideration and the parameters are invariable. In fact, the ship model is uncertain and the sea disturbance is time-variant, so the robustness and the adaptivity of PD controllers can not be guaranteed. In the chapter, the control accuracy of course and roll angle will be tested during the control process. Take the course deviation for example, the standard deviation of the course error $\sigma(\Delta\psi)$ is compared with the threshold τ, if $\sigma \leq \tau$, the course control accuracy meets the demand, on the contrary, the controller parameters need to be optimized to improve the course-keeping precision. The value of τ is set according to sea state.

73.3 The Nonlinear Ship Motion Mathematical Model

A nonlinear ship motion model based on roll planar motion mechanism (RPMM) was brought up in document [6], which can describe 4° of freedom motion: rolling, yawing, surging and swaying. Basic equations of motion are defined as:

$$
\begin{aligned}
m(\dot{u} - vr - x_G r^2 + z_G pr) &= X \\
m(\dot{u} + ur + x_G \dot{r} - z_G \dot{p}) &= Y \\
I_z \dot{r} + m x_G(\dot{v} + ur) &= N \\
I_x \dot{p} - m z_G(\dot{v} + ur) + \rho g \nabla \overline{GM}\phi &= K
\end{aligned}
\tag{73.3}
$$

Where, m is the ship mass, ∇ is the ship displacement, g is gravity acceleration, ρ is the water density, $\overline{GM}\phi$ is the ship righting arm, u, v are the surging and swaying speeds separately, r is yawing angular velocity, x_G, z_G are the distances from the ship barycenter to axes x and z of body coordinate system separately, I_x, I_z are the moments of inertia around axes x and z separately, X, Y, N and K are the hydrodynamic forces and moments of different axes, which can be indicated in nonlinear functions of ship motion variables and control inputs, and expressed in a 3rd Taylor series polynomial of nonlinear hydrodynamic constants.

Simulation study has been carried based on a container ship "SR 108". The main data of the ship is as follows: the Length is 175 m, the breadth is 25.4 m, the speed is 7.3 m/s, the loaded mean draft is 8.5 m, the block coefficient is 0.559, the full load displacement is 21,222 t, the rudder area is 33.04 m^2, the initial meta-centric height is 0.3 m. Detailed data and hydrodynamic force derivatives is provided in document [6], then the nonlinear ship motion model is realized with S-function in Matlab language.

73.4 The Adaptive Parameter Optimization Based on GA

73.4.1 The Steps of GA Optimization

Generic algorithm can simulate the natural evaluation by the basic operation, such as reproduction, crossover and mutation. The search procedure of GA is highly parallel, random and global adaptive. The detailed search procedure is as follows.

1. Improved genetic algorithm used two-dimensional code strategy is applied to initialize the population, which can improve the convergence speed and global searching ability;
2. The individuals are selected and evaluated according to the fitness function, which is defined to minimize the global variance between the output and the reference data, that is:

$$\min f(N) = \sum_{i=1}^{m} (U_i - U_{0i})^2 \tag{73.4}$$

Where, N is the population number, m is the number of samples, U_i is the ith output data, and U_{0i} is the ith reference data.

3. New populations are produced following the rules of reproduction, crossover and mutation according to the genetic probability. The crossover operator is combined with arithmetic cross and adjacent floating point crossover. The crossover rate P_c and the probability of mutation are adaptively adjusted according to the following rules:

$$\begin{cases} P_c^n = P_c^{n-1} - (P_c^0 - 0.6)/Q \\ P_m^n = P_m^{n-1} - (0.1 - P_m^0)/Q \end{cases} \tag{73.5}$$

Where, n is iterations, Q is the maximum number of generations, P_c^0, P_c^n are the initial value and the *nth* iterated value of crossover rate, P_m^0, P_m^n are the initial value and the *nth* iterated value of probability of mutation.

4. Repeat the steps 2 and 3 until the termination conditions are fulfilled, the best individual is recognized as the result of the optimization based on GA.

73.4.2 The Fitness Functions

When the ship is navigating in open waters, the course-keeping performance is more important for ship autopilot system. The variance of the sea state has direct influence to the course-keeping precision. In order to improve the performance, it is reasonable to minimize the course deviation and the steering gear wear. Then the fitness function can be defined with a minimum sum of the course error variance $\Delta\psi_i^2$ and the rudder angle variance δ_i^2.

$$\min : f_1(N) = \sum_{i=1}^{m} (\Delta\psi_i^2 + \delta_i^2) \tag{73.6}$$

If the ship is encountering severe sea state, the roll amplitude is increasing heavily to affect the safety of the ship and the cargo on it. In such case, the ship maneuvering strategy is to heading the wind and waves to avoid further rolling. Consequently, a course-keeping autopilot with RRD function does great good to the navigation safety in heavy seas. The fitness function for RRD can be defined to minimize the variance of roll angle $\Delta\phi_i^2$.

$$\min f_3(N) = \sum_{i=1}^{m} (\Delta\phi_i^2) \tag{73.7}$$

73.5 Numerical Simulations and Results

73.5.1 The Parameter Configuration

The sea waves are the main factor for ship roll moment. A linear state space model driven by white noise signal is used to simulate the wave induced ship motion. Assume: $y_w = x_{w2} = \dot{x}_{w1}$, the wave model is given as follows:

Table 73.1 The roll
reduction rates in different
wave encounter frequencies

ω_w (rad/s)	$\phi_{nrrd}(°)$	ϕ_{rrd} (°)	$R_{rrd}(\%)$	δ (°)
0.5	6.88	3.50	49.06	9.02
1.0	7.76	3.16	59.28	7.51
1.3	8.70	4.09	52.97	7.37

$$\begin{bmatrix} \dot{x}_{w1} \\ \dot{x}_{w2} \end{bmatrix} = \begin{bmatrix} 0 & 1 \\ -\omega_w^2 & -2\varsigma_w\omega_w \end{bmatrix} \begin{bmatrix} x_{w1} \\ x_{w2} \end{bmatrix} + \begin{bmatrix} 0 \\ K_w \end{bmatrix} W_n$$

$$y_w = \begin{bmatrix} 0 & 1 \end{bmatrix} \begin{bmatrix} x_{w1} \\ x_{w2} \end{bmatrix}$$

(73.8)

Where, ς_w is roll damping ratio, ω_w is the wave encounter frequency, K_w is the wave influence factor of roll moment, W_n is zero mean white noise process. In the simulation tests, the sea state is set to wind force 5 and moderate sea, ς_w is set to 0.1, ω_w is 1.0 rad/s, K_w is 0.0001.

During the parameter optimization process of course controller and RRD controller, the initial population size is 40, the stopping generation is 100, fitness functions are indicated as Eqs (73.6) and (73.7), the initial search domains of K_{p1}, K_{d1}, K_{p2}, K_{d2} are set as $[-100,100]$, after the preliminary tests, the search domains are set as $[0.05\ 0.5]$, $[-15\ -5]$, $[-30\ -20]$, $[-2\ -1]$. Then the final results are $K_{p1} = 0.15, K_{d1} = -10, K_{p2} = -24, K_{d2} = -1.25$.

Rudder servo system is considered to avoid saturating the steering machinery, the steering gear consists of two nonlinear parts, that is, the saturation and slew rate limitation. The rudder angles are assumed to be in the range $[-30°, 30°]$ and rudder rate limits are within $\pm 15°/s$.

73.5.2 Simulation Results

In order to have a reasonable evaluation of the performance of RRD controller, the roll angle reduction rate is define as follows:

$$R_{rrd} = \frac{\phi_{nrrd} - \phi_{rrd}}{\phi_{nrrd}} \times 100\ \%$$

(73.9)

Where ϕ_{nrrd}, ϕ_{rrd} are the standard deviation of roll angles when the RRD controller is turned off or on. The simulation results with different wave encounter frequencies are indicated in Table 73.1, in which the value of the rudder angle is the standard deviation when RRD is on.

As shown in Fig. 73.2 and Table 73.1, the course controller achieves good performance of course-keeping and course-changing in case of moderate seas and frequent using of rudder. The setting time of a 50° turn is 200 s. When RRD is off,

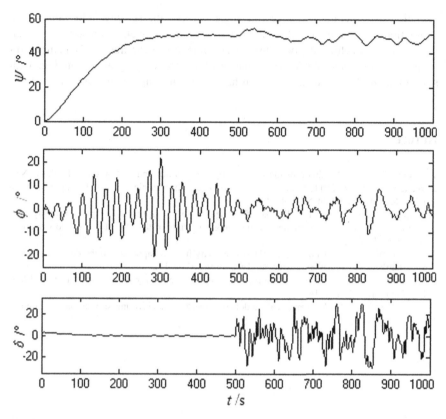

Fig. 73.2 Performance of the RRD autopilot during course-keeping and course-changing (50°). The RRD controller is active between $t = 500$ and $1,000$ s

the course deviation range is $\pm 1°$, or else $\pm 5°$. When the RRS controller is inactive, the rudder exhibits a typical steering characteristic. After the actuation of the RRS controller, the rudder angle varies with bigger amplitude and higher frequency. As can be seen from Fig. 73.2, the rudder angle never exceeded its magnitude and velocity limitations. When RRD controller is on, the roll reduction is about 50 % in a wide frequency range of sea waves.

73.6 Conclusion

In order to develop a simple PD controlled autopilot system with rudder roll damping, the course-keeping controller and rudder roll damping controller are designed and adaptively optimized based on GA-PD method. The parameters of controllers are to be optimized by generic algorithm when the detected standard deviation of the course error or roll angle exceeds the set threshold values. The simulation tests are done on a nonlinear ship motion model, and results show that the controllers achieved good disturbance rejection near the ship's natural roll frequency band, and stable performance of course-keeping and changing, with strong adaptivity to sea states, which has significant engineering significance.

References

1. Fossen TI (2012) Handbook of marine craft hydrodynamics and motion control. Wiley, NY
2. Wang L, Xianku Z (2011) Improved closed-loop gain shaping algorithm and its application in RRS. ICIC Express Lett 5(11):4215–4220
3. Raspa P, Torta E, Ippoliti G, Blanke M, Longhi S (2011) Optimized output sensitivity loop shaping controller for ship rudder roll damping. Preprints of the 18th IFAC World Congress, pp 13660–13665
4. Tzeng C-Y, Wu C-Y, Chu Y-L (2001) A sensitivity function approach to the design of rudder roll stabilization controller. J Mar Sci Technol 9(2):100–112
5. Alarcin F (2007) Internal model control using neural network for ship roll stabilization. J Mar Sci Technol 15(2):141–147
6. López E et al (2004) Mathematical models of ships for maneuvering simulation and control. J Marit Res 2:5–20

Chapter 74
The Research of Soft Handover Signaling for LTE System in High-Speed Railway Environment

Linlin Luan, Muqing Wu, Panfeng Zhou, Shiping Di and Shuyun Ge

Abstract Hard handover (HHO) algorithm has been used for LTE system according to 3GPP specifications. In this paper, a soft handover (SHO) signaling procedure is investigated to validate the feasibility of SHO algorithm for LTE system so as to solve high handover failure ratio in high-speed railway environment. The paper gives an overview of SHO signaling procedure, SHO algorithm and simulation approach for LTE system. The simulation is implemented to select different handover parameter sets for HOM (Handover hysteresis margin) and TTT (Time-to-Trigger), which is investigated the influence to Ping-Pong handover and radio link failure (RLF) in 350 km/h for the high-speed railway simulation scenario. The results suggest that SHO algorithm has better handover performance than A3 event HHO algorithm in 350 km/h.

Keywords Soft handover · Handover performance · High-speed railway · LTE

L. Luan (✉) · M. Wu
Broadband Communication Networks Lab, Beijing University of Posts
and Telecommunications, Postbox 214 Beijing, China
e-mail: barryluan@bupt.edu.cn

M. Wu
e-mail: wumuqing@bupt.edu.cn

P. Zhou · S. Di · S. Ge
Beijing National Railway Research and Design Institute of Signal
and Communication, Beijing, China
e-mail: 235zpf@crscd.com.cn

S. Di
e-mail: 433dsp@crscd.com.cn

S. Ge
e-mail: 408gsy@crscd.com.cn

W. Lu et al. (eds.), *Proceedings of the 2012 International Conference on Information Technology and Software Engineering*, Lecture Notes in Electrical Engineering 210, DOI: 10.1007/978-3-642-34528-9_74, © Springer-Verlag Berlin Heidelberg 2013

74.1 Introduction

Long-Term Evolution (LTE) is known as Evolved Universal Terrestrial Radio Access Network (E-UTRAN), which is a system currently under development within the 3rd Generation Partnership Project (3GPP) [1]. Orthogonal Frequency Division Multiple Access (OFDMA) and Single Carrier-Frequency-Division Multiple Access (SC-FDMA) have been used downlink and uplink respectively as radio access technology (RAT) for LTE system [2]. With the rapid development of China high-speed railway, there is a requirement for mobility support with high performance up to the speed of 350 km/h, even up to 500 km/h. With the velocity of the UE even higher, the handover will be frequent between the neighboring cells. Therefore, the handover performance becomes more important, especially for real-time service.

LTE system is consisted of three elements: evolved-NodeB (eNodeB), Mobile Management Entity (MME), and Serving Gateway (S-GW). The eNodeB performs all radio interface related functions such as packet scheduling and handover mechanism. MME manages mobility, user equipment (UE) identity, and security parameters. The Serving GW is the gateway which terminates the interface towards E-UTRAN [3]. This paper is focus on high-speed railway scenario, which deploys the eNodeB consisting of Base Band Unit (BBU) and Radio Remote Unit (RRU) along the railway line as demonstrated in Fig. 74.1.

Several studies have been published to investigate the HO performance for LTE recently. A generalized model has been proposed to analyze the impact of the propagation environment and the velocity on the handover performance of the UE for LTE system [4]. The proposed algorithm reduces the call dropping rate during handover procedure at the cell boundary of urban cells [5]. The improvement of

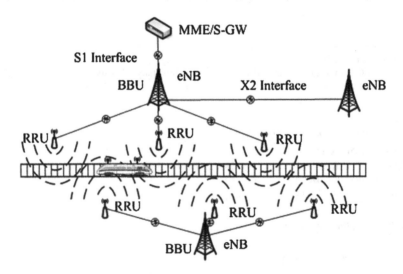

Fig. 74.1 E-NodeB deployment along railway line

LTE handover performance is studies through ICIC [6]. Two kinds of handover algorithms have been introduced in LTE system [7, 8]. And the simulation results show that these both algorithms have better handover performance.

The paper is organized as follows: In Sect. 74.2, overview of handover is introduced including two handover types (HHO and SHO) and two handover performance evaluation metrics (Ping-pong HO and RLF). The more detailed SHO signaling procedure for LTE system will be introduced in Sect. 74.3. SHO algorithm will be evaluated based on high-speed railway environment in Sect. 74.4. Then in Sect. 74.5, the simulation results are analyzed. Finally, conclusions are drawn in Sect. 74.6.

74.2 Overview of Handover

At present, there are two handover types in wireless communication systems: hard handover (HHO) and soft handover (SHO). Hard handover is a Break-Before-Connect (BBC) mode. Handover type in LTE system is hard handover according to 3GPP specification. It defines that the release of the connection with the source eNodeB should be finished and then the connection with the target eNodeB will be setup. Soft handover is a Connect-Before-Break (CBB) mode. It defines that a new radio link has been established with the target eNodeB before the release of the connection with the source eNodeB. Although it reduces the complexity of LTE network architecture, hard handover type may result in data being lost and higher handover failure ratio. To provide high data rate services in high-speed railway environment for LTE system, an efficient, self-adaptive and robust handover algorithm is required. Hence handover technique becomes an important research area since the proposition of LTE system by 3GPP in 2004 [9].

Soft handover mechanism has been applied to commercial market for WCDMA system in recent several years in the world. Compared to LTE network architecture, WCDMA network has a Radio Network Controller (RNC), which is a centralized controller to perform handover control for each UE to communicate with two or more cells simultaneously. So SHO is suitable for maintaining RRC active condition, preventing voice call drop and decreasing packets retransmission and being lost. However, this RNC network architecture adds the complexity of system and transmission latency. Since the UE connected two or more cells for SHO, this will waste more frequency resources. Although it is not suitable for common LTE system, SHO algorithm should be studied to apply for high-speed railway specialized network of LTE system. For some cell of high-speed specialized network, there are most two trains which move in the opposite direction. So we assume that there are enough sufficient frequency resources. Since SHO procedure has changed common HHO signaling procedure of LTE system completely, SHO signaling procedure should be investigated deeply. The detailed SHO signaling procedure will be introduced in Sect. 74.3.

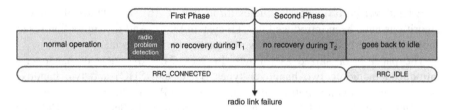

Fig. 74.2 Radio link failure procedure

Ping-pong handover is a very important handover failure mode for LTE system. We define that Ping-pong HO occurs more than one time when HO is implemented during the time the UE moves from the source eNodeB to the target eNodeB for HHO. For SHO, Ping-pong HO won't occur since the UE connects two neighboring eNodeBs at the same time during SHO implement process.

In Fig. 74.2, it shows RLF procedure as a whole [10]. When the radio link quality deteriorated serious enough in the first phase, a RLF procedure is triggered to lead interruption in communication service and a cell re-selection or handover procedure will be established. In principles, there are several factors to trigger RLF procedure: (1) HO delay is extensive enough; (2) there is high interference in the handover overlapping area; (3) when the UE is at the edge of coverage [11]. The UE always monitor the DL link quality via the measurement of cell-specific reference signal. The radio link quality has two quality thresholds. One is the threshold Q_{out}, which is the threshold of triggering RLF procedure; the other is the threshold Q_{in}, which is the threshold of cancelling RLF procedure. The threshold Q_{out} represents the level at which the DL radio link cannot be reliably received and corresponds to 10 % Block Error Rate (BLER) on the Physical Downlink Control Channel (PDCCH). The threshold Q_{in} shall correspond to 2 % Block Error Rate [12].

For SHO procedure, the UE connects with two neighboring cells simultaneously. We introduce a concept of *Active Sets* referred to WCDMA system. When the UE connects with a single cell, MME will put this cell id into active sets. After the UE begins SHO procedure and the new cell sends add path request to the MME, MME will add this new cell id into active sets. After the former cell sends delete path request to the MME, the former cell id is removed from active sets.

74.3 Soft Handover Signaling Procedure

In Fig. 74.3, it shows the more detailed SHO signaling procedure. We summary the main steps of SHO signaling procedure as follows:

1. The source eNodeB adopts the X2-based handover procedure via the intra-MME/S-GW.
2. The source eNodeB configures the UE measurement procedure preparing for SHO procedure triggering.

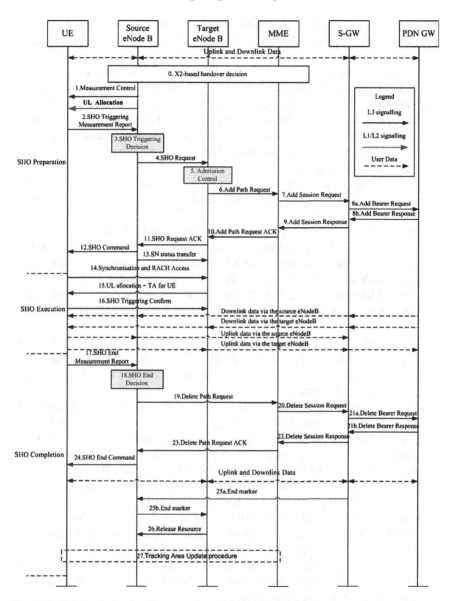

Fig. 74.3 Soft handover signaling procedure based on LTE system

3. The SHO procedure is triggered by the UE that sends a *SHO Triggering Measurement Report* to the source eNodeB including the reference symbols received power (RSRP) and the reference symbols received quality (RSRQ) of the source eNodeB and the target eNodeB when *SHO Triggering Measurement Criteria* is met.

4. The source eNodeB makes SHO Triggering Decision based on SHO Triggering Measurement Report and Radio Resource Management (RRM) information.

5. A *SHO Request* message is sent from the source eNodeB to the target eNodeB, which contains all the relevant bearer information for the target eNodeB .

6. The target eNodeB performs *Admission control* after receiving the *SHO Request* message from the source eNodeB. The target eNodeB can grant the resource for the radio access of the UE because the whole system resource will be prepared for the UEs (as trains) for high-speed railway specialized networks of LTE system.

7. An *Add Path Request* message is sent from the target eNodeB to the MME to inform that the target eNodeB should be added to *Active Sets* and will connect with the UE.

8. The MME sends an *Add Session Request* message to the S-GW.

9. The S-GW sends an *Add Bearer Request* message to the Packet GateWay (P-GW) and the P-GW replies an Add Bearer Response message to the S-GW.

10. The S-GW sends an *Add Session Response* message to the MME.

11. The MME sends an *Add Path Request ACK* message to the target eNodeB to confirm a new path added. And the cell id of the target eNodeB is added to *Active Sets*.

12. The target eNodeB saves the context of the UE, prepares L1/L2 for SHO preparation and responds to the source eNodeB with an *SHO Request ACK* message which provides information for the establishment of the new radio link (admission control).

13. The source eNodeB transfers *RRC Connection Reconfiguration* message to the UE in a *SHO command* message including *Mobility Control information*.

14. The *SN Status Transfer* message is sent from the source eNodeB to the target eNodeB to transport *uplink PDCP SN receiver status* and *downlink PDCP SN transmitter status* of the E-RABs (Evolved Radio Access Bearer).

15. The UE performs Synchronization to the target eNodeB and accesses the target eNodeB via *Random Access Channel* (RACH).

16. The target eNodeB responds with *UL allocation* and *timing advance* (TA).

17. A *RRC Connection Reconfiguration Complete* message is sent by the UE to confirm *SHO triggering procedure* after the UE accomplishes to access the target eNodeB. The C-RNTI of the *RRC Connection Reconfiguration Complete* message is verified by the target eNodeB. Until now, the target eNodeB can receive/send data from/to the UE.

18. When *SHO End Measurement Criteria* is met, the UE sends *SHO End Measurement Report* to the target eNodeB.

19. The source eNodeB makes *SHO End decision* based on *SHO End Measurement Report* and *Radio Resource Management* (RRM) information.

20. A *Delete Path Request* message is sent from the source eNodeB to the MME to inform that the source eNodeB has been changed by the UE.

21. The MME sends a *Delete Session Request* message to the S-GW.

22. The S-GW sends a *Delete Bearer Request* message to the P-GW and the P-GW replies a *Delete Bearer Response* message to the S-GW.
23. The S-GW sends a *Delete Session Response* message to the MME.
24. The MME sends a *Delete Path Request ACK* message to confirm the former path deleted. And the cell id of the source eNodeB has been removed from *Active Sets*.
25. The source eNodeB sends *RRC Idle Reconfiguration* message to the UE in a *SHO End command* message to disconnect the radio connection with the UE.
26. The S-GW disconnects the downlink data path to the source side. One or more "end marker" packets are sent on the old path to the source eNodeB and any User Plane resources can be released towards the source eNodeB by the S-GW.
27. The target eNodeB sends *UE Context Release* message to the source eNodeB and triggers the release of resource for the source eNodeB. The source eNodeB can flush its forwarded DL data buffer that is stored in for case of fallback after receiving the *UE Context Release* message.
28. *Tracking area update procedure* is implemented by E-UTRAN.

74.4 Soft Handover Algorithm

In Fig. 74.4, the flowchart of SHO algorithm is shown. At first, the UE performs SHO triggering measurement procedure. After it receives the measurements from the source and target eNodeB, the UE processes the data. When *SHO Triggering Measurement Criteria* is met, the UE sends the measurement report to the source eNodeB and the source eNodeB decides if SHO procedure should be triggered. It is shown in (1) when it continues to move in the handover overlapping area and *SHO End Measurement Criteria* is met, the UE sends the measurement report to the source eNodeB to disconnect the path with the UE. $Hyst_{SHO\_Trigger}$ is HO hysteresis margin (HOM) for SHO triggering procedure and $Hyst_{SHO\_End}$ is HOM for SHO completion procedure. $M_{S-eNodeB}$ is RSRP of the source eNodeB and $M_{T-eNodeB}$ is RSRP of the target eNodeB.

$$M_{S-eNodeB} - M_{T-eNodeB} \leq Hyst_{SHO\_Trigger} \tag{74.1}$$

$$M_{T-eNodeB} - M_{S-eNodeB} \geq Hyst_{SHO\_End} \tag{74.2}$$

In Table 74.1, it shows the deployment of eNodeB and S-GW. The list of eNodeB includes the following attributes: Evolved Cell Global Identifier (E-CGI), SHO reference point, eNodeB Number and S-GW Number. In this paper, SHO reference point is decided based on *SHO Triggering Measurement Criteria*. As soon as the movement direction of the UE in high-speed railway is decided, the deployment list of the whole eNodeBs is also confirmed. E-CGI is ordered according to the deployment of eNodeB. S-GW Number is shown to communicate

Fig. 74.4 The flowchart of
soft handover algorithm

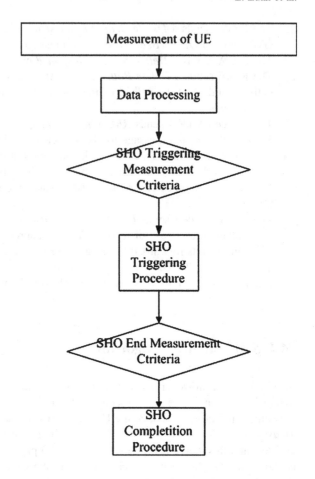

Table 74.1 E-NodeB/S-GW deployment list

E-CGI	SHO triggering reference point	eNodeB number	S-GW number
E-CGI(0)	(A0, B0)	eNodeB(0)	S-GW(N-1)
E-CGI(1)	(A1, B1)	eNodeB(1)	S-GW(N)
E-CGI(2)	(A2, B2)	eNodeB(2)	S-GW(N)
...	...	...	...
...	...	...	...
E-CGI(n)	(An, Bn)	eNodeB(n)	S-GW(N)
E-CGI(n + 1)	(An + 1, Bn + 1)	eNodeB(n + 1)	S-GW(N + 1)

with the related eNodeB Number. Because the target eNodeB has been confirmed
in advance, the source eNodeB can know which eNodeB the UE will trigger the
handover next to.

Table 74.2 Handover simulation parameter

Parameters	Values
UE number	1
eNodeB	2
Carrier frequency	2.6 GHz
Bandwidth	10 MHz
Antenna direction parameter	Horizontal 120/150°
	Vertical 65°
Distance to railway line	50 m
Path loss	130.699 + 34.786 logd
Shadowing standard deviation	8 dB
Shadowing correlation	0.5/1.0
Fast fading	Jakes model for the Rayleigh fading
Handover area length	500 m
Handover execution time	5 s
RRU transmitting power	46 dBm
$Hyst_{SHO\_Trigger}$	1 dB
$Hyst_{SHO\_End}$	1.0, 2.0, 3.0 dB
TTT	0 ~ 120 km/h: 0, 120, 240, 360, 480 ms
	120 ~ 250 km/h:0, 90, 180, 270, 360 ms
	250 ~ 350 km/h: 0, 60, 120, 180, 240 ms
Channel scenario	Mountain

In this paper, Ping-Pong HO and RLF are the most important key performance indicators (KPI) during SHO procedure of high-speed railway specialized network because they represents most of handover failure modes. In Table 74.2, it shows the simulation parameters of the system level simulation platform.

74.5 Soft Handover Simulation Analysis

74.5.1 Simulation Results of HHO

In Figs. 74.5, 74.6 and 74.7, the simulation results for HHO are shown when the speed of the UE is 350 km/h. The purpose of the simulation is to analyze the relation between handover performance and different TTT when HOM adopts the different values. Ping-pong HO ratio decreases with the growth of TTT and HOM, however RLF ratio increases inversely, shown in Figs. 74.5, 74.6 and 74.7. When HOM is 1 dB, RLF procedure doesn't occur and Ping-pong HO ratio begins to decline from 12.1 to 0.3 %, shown in Fig. 74.5. When HOM is 2 dB, Ping-pong HO begins to disappear and RLF ratio begins to rise from zero to 5.6 %, shown in Fig. 74.6. When HOM is 3 dB, RLF ratio becomes too big to meet the QoS of wireless communication. It is because RSRP and SINR of the source eNodeB will become so low that RLF procedure is easier to be triggered when the train moves

Fig. 74.5 Performance vs. different TTT when HOM is 1 dB for HHO

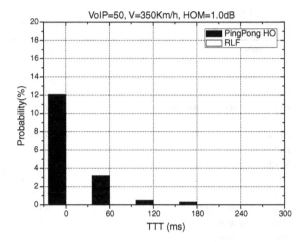

Fig. 74.6 Performance vs. different TTT when HOM is 2 dB for HHO

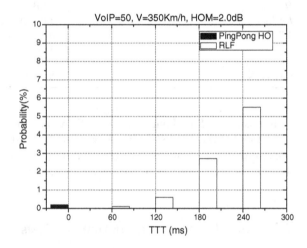

Fig. 74.7 Performance vs. different TTT when HOM is 3 dB for HHO

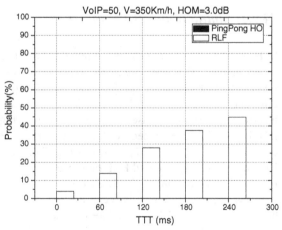

Fig. 74.8 Performance vs. different TTT when HOM is 1 dB for SHO

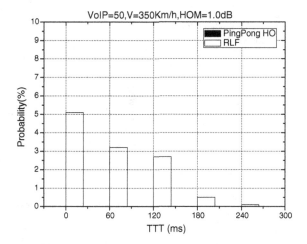

to the boundary of the source eNodeB. If RLF ratio is more than 2 %, system performance will be deteriorated, even results to communication outage. The feeling of the consumer is frequent call drop or poor call quality.

74.5.2 Simulation Results of HHO

In Figs. 74.8, 74.9 and 74.10 the simulation results for SHO are shown when the speed of the UE is 350 km/h. When HOM is 1 dB, RLF ratio decreases from 5.1 to 0.1 % with the growth of TTT, shown in Fig. 74.8. SHO algorithm forbids the UE to handover from the target eNodeB to the source eNodeB based on the deployment list of eNodeBs, even if the signal strength of the target eNodeB is worse

Fig. 74.9 Performance vs. different TTT when HOM is 2 dB for SHO

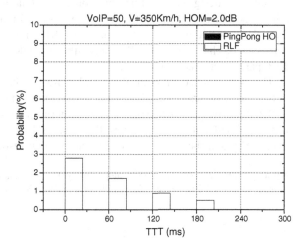

Fig. 74.10 Performance vs. different TTT when HOM is 3 dB for SHO

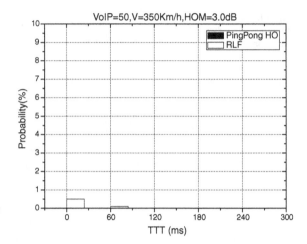

than the source eNodeB's. So this mechanism will add the probability of RLF procedure triggered. When HOM is 2 dB, RLF ratio declines from 2.8 % to zero, shown in Fig. 74.9. HOM is 3 dB, RLF ratio is lower than 0.5 %, shown in Fig. 74.10. According to aforementioned definition, RLF ratio is less than 2 % will sustain normal communication requirement. So HOM is 2 and 3 dB, handover performance of SHO is better than HHO's.

74.6 Conclusion

In this article, a SHO signaling procedure and algorithm have been proposed for high-speed specialized network of LTE, which may replace already existing A3 event HHO algorithm in high-speed specialized network of LTE. SHO signaling procedure is designed based on common LTE signaling procedure for high-speed railway scenario. According to the handover simulation results, it is clearly shown that SHO algorithm has gotten satisfying RLF ratio and have better handover performance than A3 event HHO algorithm via adopting the different HOM and TTT parameter sets in 350 km/h.

The performance of both HHO and SHO has been evaluated in high-speed railway environment. The simulation results show that the handover performance of SHO is extremely better than HHO's in term of Ping-pong ratio and RLF ratio. But the disadvantage of SHO is the cost of higher signaling implementation complexity and the requirement of higher system resources.

Acknowledgments The research is supported by The National Science and Technology Major Projects (No.2011ZX03001-007-03) and Beijing Natural Science Foundation (No.4102043).

References

1. Anas M, Calabrese FD (2007) Performance evaluation of received signal strength based hard handover for UTRAN LTE. IEEE 65th vehicular technology conference, VTC2007-Spring
2. Jansen T, Balan I (2010) Handover parameter optimization in LTE self-organizing networks. IEEE 72nd vehicular technology conference fall (VTC 2010-Fall)
3. Dimou K, Wang M (2009) Handover within 3GPP LTE: design principles and performance. IEEE 70th vehicular technology conference fall
4. Qureshi SUR, Nawaz SJ, Patway M (2010) The impact of propagation environment and velocity on the handover performance of LTE systems. International conference on wireless communications and signal processing (WCSP), pp 21–23
5. Aziz D, Sigle R (2009) Improvement of LTE handover performance through interference coordination. IEEE 69th vehicular technology conference, VTC Spring
6. Luan L, Wu M, Shen J, Ye J, He X (2012) Optimization of handover algorithms in LTE high-speed railway networks. JDCTA 6(5):79–87
7. Luan L, Wu M, Wang W, He X, Shen J (2012) A GPS-based handover algorithm in LTE high-speed railway networks. AISS 4(9):205–213
8. Lin CC, Sandrasegaran K (2011) Optimization of handover algorithms in 3GPP long term evolution system. 4th international conference on modelling, simulation and applied optimization (ICMSAO)
9. Sesia S, Baker M, Toufik I (2009) LTE-the UMTS long term evolution. Wiley, London
10. Myung HG (2008) Technical overview of 3GPP LTE. 18 May 2008
11. Anas M, Calabrese FD, Ostling PE (2007) Performance aanlysis of handover measurements and layer 3 filtering for Utran LTE. IEEE 18th international symposium on personal indoor and mobile radio communications, PIMRC
12. Rappaport TS (2009) Wireless communications: principles and practice. Pearson Education

Chapter 75
Track-Before-Detect Algorithm Based on Optimization Particle Filter

Huajian Wang

Abstract An improved track-before-detect (TBD) algorithm based on the Cubature particle filter is an efficient approach for weak target detection and track under low signal to noise radio environment. Under the framework of particle filter, the algorithm which combines the particle filter with cubature kalman filter (CKF) algorithm is presented to generate the important density function of particle filter. The simulation results demonstrate that the improved algorithm can provide stable and reliable detection as well as accurate tracking.

Keywords Weak target · Track-before-detect · Cubature particle filter

75.1 Introduction

The Track-Before-Detect (TBD) [1] is good for weak target detecting and tracking in low signal to noise radio (SNR) environment. The method uses the original measurement data of the sensor directly to promote SNR by accumulating target information for a period of time. At the same time the target is being joint detected and estimated, the detected results and target track are announcing.

In recent years, beasuce Particle filter (PF) [2] is not the limit of the nonlinear non-Gaussian problem. The Track-Before-Detect algorithm based on PF has been a lot of attention [3–8]. This is a crucial problem, which affect real-time and tracking accuracy of PF-TBD. In the Paper [5, 6], these described that the

H. Wang (✉)
Department of Information Engineering, Engineering University of China Armed Police Force, No. 710086 WuJing Road, Xi'An Shaanxi, China
e-mail: whj-winner@163.com

W. Lu et al. (eds.), *Proceedings of the 2012 International Conference on Information Technology and Software Engineering*, Lecture Notes in Electrical Engineering 210, DOI: 10.1007/978-3-642-34528-9_75, © Springer-Verlag Berlin Heidelberg 2013

algorithm, which Rao-Blackwellised particle filter-based track-before-detect algorithm for Over-the-horizon radar target. But The algorithm requires a state-space model for condition of linear Gauss model. In the Paper [7], it could avoid the sample impoverishment problem, but convergence problem can not be well guaranteed and operation time is long time. In the Paper [8], it described that the TBD algorithm based on gauss particle filtering avoided resampling and achieved relatively easy. But It required the posterior probability density for gauss distribution. So it's application was restricted.

In order to improve the accuracy and the algorithm running time relationship, this article proposed TBD method based on cubature particle filter. In the tracking phase, the algorithm uses CKF (cubature kalman filter) to construct of the particle filter importance proposal function. The simulation results demonstrate that the algorithm can be obtained under higher filtering accuracy in a small number of particles, meet the requirements of real-time control, and strong robustness.

75.2 Problem Description

75.2.1 State Model

Assuming that the target is in uniform motion in X–Y plane. System state equation is as follows.

$$X_k = FX_{k-1} + w_{k-1} \tag{75.1}$$

Where $X_k = [x_k, \dot{x}_k, y_k, \dot{y}_k, I_k]^T$, (x_k, y_k), $(\dot{x}_k, \dot{y}_k)$ and I_k expresses position, speed and strength of the target. w_k is a i.i.d process noise vector. Q_k is the process noise w_k covariance matrix.

State transition matrix is as follows.

$$F = \begin{bmatrix} F_x & & \\ & F_x & \\ & & 1 \end{bmatrix}, \ F_x = \begin{bmatrix} 1 & T \\ 0 & 1 \end{bmatrix} \tag{75.2}$$

The process noise covariance is as follows.

$$Q_x = \begin{bmatrix} Q_x & & \\ & Q_x & \\ & & q_2 T \end{bmatrix}, \ Q_x = \begin{bmatrix} \frac{q_1}{3} T^3 & \frac{q_1}{2} T^2 \\ \frac{q_1}{2} T^2 & q_1 T \end{bmatrix} \tag{75.3}$$

Where T is a sampling period. q_1 is Process noise power spectral density and q_2 is the target intensity noise power spectral density.

Definiting $E_k \in \{0, 1\}$, 1 is that the target is presence, 0 is that the target is not exist [4]. $\{E_k, k = 0, 1, 2 \ldots\}$ is submitted to Markov procedure.

Definiting $P_b = P\{E_k = 1 | E_{k-1} = 0\}$, $P_d = P\{E_k = 0 | E_{k-1} = 1\}$

Transfer probability matrix is as follows.

$$\Pi = \begin{pmatrix} 1 - P_b & P_b \\ P_d & 1 - P_d \end{pmatrix} \tag{75.4}$$

75.2.2 Measurement Model

Assuming that the sensor is scanning for $n \times m$ resolution cell in X–Y plane. Each resolution cell (i, j) corresponds to each the matrix region $\Delta_x \times \Delta_y$, $i = 1, 2, \ldots, n$, $j = 1, 2, \ldots, m$. The measured value of each resolution cell (i, j) in every moment is as follows.

$$z_k^{(i,j)} = \begin{cases} h_k^{(i,j)}(x_k) + v_k^{(i,j)} & \text{Target exist} \\ v_k^{(i,j)} & \text{Target not exist} \end{cases} \tag{75.5}$$

For the remote target, $h_k^{(i,j)}$ is the point spread function.

$$h_k^{(i,j)}(x_k) = \frac{\Delta_x \Delta_y I_k}{2\pi \sum^2} \exp\left\{ -\frac{(i\Delta_x - x_k)^2 + (j\Delta_y - y_k)^2}{2\sum^2} \right\} \tag{75.6}$$

Where $\sum$ is sensor's fuzzy degree, as the known quantity.

The measured value of k is $z_k = \{z_k^{(i,j)}, i = 1, 2, \ldots n, j = 1, 2, \ldots m\}$.

The all measured value in front of all time of k is as follows. $z_{1:k} = \{z_n | n = 1, 2, \ldots k\}$

75.2.3 Likelihood Function

Assuming that the every resolution cell is i.i.d. The likelihood function is as follows.

$$p(z_k | x_k, E_k) = \begin{cases} \prod_{i=1}^{n} \prod_{j=1}^{m} p_{s+N}(z_k^{(i,j)} | x_k) & E_k = 1 \\ \prod_{i=1}^{n} \prod_{j=1}^{m} p_N(z_k^{(i,j)}) & E_k = 0 \end{cases} \tag{75.7}$$

Where $p_N(z_k^{(i,j)})$ is the likelihood function of the resolution cell (i, j) noise. $p_{s+N}(z_k^{(i,j)} | x_k)$ is the likelihood function of the resolution cell (i, j) signal and noise. Because of each resolution element of the noises are independent, and the effects of target to measure the signal intensity is limit. Likelihood ratio is as follows.

$$L(Z_k|X_k, E_k) = \frac{p(Z_k|X_k, E_k)}{p(Z_k|X_k, E_k = 0)}$$

$$\approx \begin{cases} \frac{1}{M}\sum_{m=1}^{M} \prod_{i\in C_i(x_k^m)} \prod_{j\in C_j(x_k^m)} l(z_k^{(i,j)}|x_k^m) & E_k = 1 \\ 1 & E_k = 0 \end{cases} \tag{75.8}$$

Where

$$l(z_k^{(i,j)}|x_k^m) = \frac{p_{s+N}(z_k^{(i,j)}|x_k)}{p_N(z_k^{(i,j)})}$$

$$= \exp\left\{ -\frac{h_k^{(i,j)}(x_k^m)(h_k^{(i,j)}(x_k^m) - 2z_k^{(i,j)})}{2\sigma^2} \right\} \tag{75.9}$$

75.3 Track-Before-Detect Algorithm Based on Cubature Particle Filter

The problem of TBD is a mixed problem. What is the solution to the TBD problem is that the target state x_k and the joint posterior probability density $p(x_k, E_k = 1|Z_k)$ are estimated in k time. There is that the frame of Salmond is used.

The new algorithm that incorporates the latest observation into a prior updating phase develops the importance density function by CKF that is more close to posterior density in Nonlinear and non-Gauss [9, 10]. So a iteration steps used CKF to update particle is as follows

Step1 Calculation Cubature point

$$P_{k-1/k-1} = S_{k-1/k-1} S_{k-1/k-1}^T \tag{75.10}$$

$$X_{i,k-1/k-1} = S_{k-1/k-1}\xi_i + \hat{x}_{k-1/k-1} \tag{75.11}$$

Step2 One step prediction of state

$$X_{i,k/k-1}^* = f(X_{i,k-1/k-1}) \tag{75.12}$$

$$\hat{x}_{k/k-1} = \frac{1}{2n}\sum_{i=1}^{2n} X_{i,k/k-1}^* \tag{75.13}$$

$$P_{k/k-1} = \frac{1}{2n} \sum_{i=1}^{2n} X^*_{i,k/k-1} X^{*T}_{i,k/k-1} - \hat{x}_{k/k-1} \hat{x}^T_{k/k-1} + Q_{k-1} \tag{75.14}$$

Step3 Measurement update

$$P_{k/k-1} = S_{k/k-1} S^T_{k/k-1} \tag{75.15}$$

$$X_{i,k/k-1} = S_{k/k-1} \xi_i + \hat{x}_{k/k-1} \tag{75.16}$$

$$Z_{i,k/k-1} = h(X_{i,k/k-1}) \tag{75.17}$$

$$\hat{z}_{k/k-1} = \frac{1}{2n} \sum_{i=1}^{2n} Z_{i,k/k-1} \tag{75.18}$$

Step4 Estimated value

$$\hat{x}_{k/k} = \hat{x}_{k/k-1} + K_k(z_k - \hat{z}_{k/k-1}) \tag{75.19}$$

$$P_{kk} = P_{k/k-1} - K_k P_{zz,k/k-1} K^T_k \tag{75.20}$$

$$P_{zz,k/k-1} = \frac{1}{2n} \sum_{i=1}^{2n} Z_{i,k/k-1} Z^T_{i,k/k-1} - \hat{z}_{k/k-1} \hat{z}^T_{k/k-1} + R_k \tag{75.21}$$

$$P_{xz,k/k-1} = \frac{1}{2n} \sum_{i=1}^{2n} X_{i,k/k-1} Z^T_{i,k/k-1} - \hat{x}_{k/k-1} \hat{z}^T_{k/k-1} \tag{75.22}$$

$$K_k = P_{xz,k/k-1} P^{-1}_{zz,k/k-1} \tag{75.23}$$

Assuming that the posterior probability density of the joint target state (x_k, E_k) is $p(x_k, E_k|z_{1:k})$. So, the steps of the track-before-detect algorithm based on Cubature particle filter are as follows:

Step 1 Initialization

P_b is as appearance probability of targets. While $E_k = 1$, the prior distribution $p(x_0)$ samples from $\{x_0^{(i)}\}_{i=1}^N$.

$$\hat{x}_0^{(i)} = E[x_0^{(i)}] \tag{75.24}$$

$$\hat{P}_0^{(i)} = E[(x_0^{(i)} - \hat{x}_0^{(i)})(x_0^{(i)} - \hat{x}_0^{(i)})^T] \tag{75.25}$$

Step 2 Calculating the predicted particle targets exist variables

According to the transfer probability matrix Π and $\{E_{k-1}^n\}_{n=1}^N$, to calculate $\{E_k^n\}_{n=1}^N$

Step 3 Forecasting $\hat{x}_{k/k}$ and P_{kk} by CKF

For while "Newborn particle" of $E_{k-1} = 0, E_k = 1$, the prior distribution $p(x_0)$ samples from $\{x_0^{(i)}\}_{i=1}^N$.

For while "Survival particle" of $E_{k-1} = 1, E_k = 1, \hat{x}_k^{(i)}$ and $\hat{P}_k^{(i)}$ are estimated by CKF.

Step 4 The normalized weights

$$\{\omega_k^n = \tilde{\omega}_k^n / \sum_{n=1}^N \tilde{\omega}_k^n\}_{n=1}^N \tag{75.26}$$

Step 5 Resampling

If $\hat{N}_{eff} < \hat{N}_{tthreshold}$, weights is $\{\omega_k^n = \frac{1}{N}\}_{n=1}^N$.

Step 6 Statistics the number of $E_k = 1$, so the appearance probability of targets is as follows

$$\hat{p}_{E_k=1}(E_k = 1 | z_{1:k}) = \frac{N_{E_k=1}}{N} \tag{75.27}$$

Step 7 setting threshold p_T, if $\hat{p}_{E_k=1} > p_T$, this describes the target appears The target state estimation value is as follows

$$\hat{x}_k = \sum_{n=1}^N x_k^n \cdot \omega_k^n \tag{28}$$

75.4 Simulation Result

In order to verify the performance of the algorithm, there has 30 frame radar simulation data in 20×20 area. Every resolution cell is $\Delta_x \times \Delta_y = 1 \times 1, p = 2$. Radar located at the origin of coordinates. Scanning time of a frame is 1 s. Assuming that the target appears from 5 to 25 s for uniform motion. The initial location is (4.2,7.2), The initial velocity is (0.45,0.25). The target signal strength $I = 20$, $\Sigma = 0.7$. $P_b = 0.1$. $P_d = 0.1$.

In order to evaluate the effect of the target tracking, this algorithm compares to UPF-TBD which is introducted in paper [14]. To definine the root mean square error is as follows:

$$RMSE = \frac{1}{N} \sum_{i=1}^{N} \sqrt{(x_k - \hat{x}_k^i)^2 + (y_k - \hat{y}_k^i)^2} \qquad (75.30)$$

From Fig. 75.1, while the same number of particles ($N = 4,000$) is used, the performance of CPF-TBD is better than UPF-TBD in the latter tracking part in low SNR conditions(SNR = 6 dB). As shown in Table 75.1, position RMSE improves about 20 % and velocity RMSE improves about 23 %. At the same time, execution time is relatively less.

From Fig. 75.2 while the different number of particles is used, the performance of CPF-TBD is nearly as same as UPF-TBD in low SNR conditions(SNR = 6 dB). while particles number of CPF-TBD ($N_{CPF-TBD} = 3,000$) is only one third of UPF-TBD $N_{UPF-TBD} = 10,000$. As shown in Table 75.1, position RMSE only worsens about 4.3 % and velocity RMSE about 0.86 % while the execution time of CPF-TBD is relatively 38.2 % of UPF-TBD.

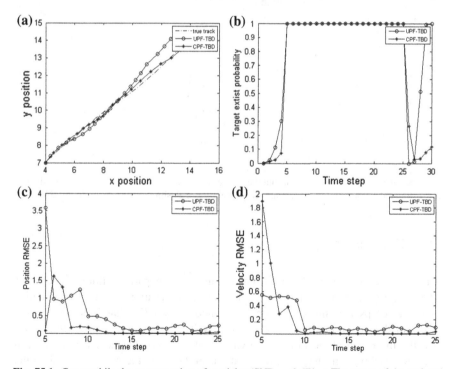

Fig. 75.1 Curve while the same number of particles (SNR = 6 dB). **a** The curve of the real and estimated values. **b** Target appearance probability. **c** Position RMSE. **d** Velocity RMSE

Table 75.1 The performance comparison in PF-TBD and CPF-TBD

Algorithm	Particles number	Position RMSE	Velocity RMSE	Execution time
UPF-TBD	4,000	4.618726	0.242404	2.2926
	10,000	3.892483	0.195826	5.1151
CPF-TBD	3,000	3.925910	0.204278	1.9563
	4,000	3.678081	0.184375	2.0038

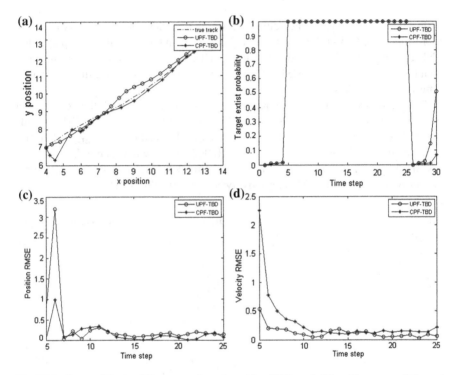

Fig. 75.2 Curve while the different number of particles (SNR = 6 dB). **a** The curve of the real and estimated values. **b** Target appearance probability. **c** Position RMSE. **d** Velocity RMSE

75.5 Conclusion

The track-before-detect algorithm based on Cubature particle filter is proposed. The new algorithm that incorporates the latest observation into a prior updating phase develops the importance density function by CKF that is more close to posterior density in nonlinear and non-Gauss. The simulation results demonstrate that the improved algorithm can provide stable and reliable detection as well as accurate tracking, and balances the relationship between the precision of the filter and the algorithm running time.

Acknowledgments This research work was supported by the foundation fund of China Armed Police Force No. WJY201114.

References

1. Kramer JDR, Reid WS (1990) Track-before- detect processing for an airbrone type radar. IEEE international conference on radar
2. Gordon NJ, Salmond DJ, Smith AFM (1993) Novel approach to nonlinear/non-Gaussian Bayesian state estimation. IEEE Proc-F 140(2):107–113
3. Salmond DJ, Birch H (2001) A particle filter for track-before-detect. In: Proceedings of the American control conference, pp 3755–3760
4. Rutten MG, Gordon NJ, Maskell S (2005) Recursive track-before-detect with target amplitude fluctuations. IEE Proc Radar, Sonar and Navig 152(5):345–352
5. Su HT, Wu TP, Liu HW et al (2008) Rao- Blackwellised particle filter based track- before-detect algorithm. IET Sig Process 2(2):169–176
6. Cuiyun LI, Hongbing JI (2009) Track-before- detection for multi weak targets based on Rao-Blackwellized particle filter. Opt Precis Eng 17(9):2342–2349
7. Ling F (2011) Fission reproduction particle filter-based track-before-detect algorithm. J Comput Appl 31(9):2581–2583
8. Ling F (2011) Gaussian particle filter-based track-before-detect method. Comput Eng Appl 47(23):121–123
9. Arasaratnam I, Haykin S (2009) Cubature kalman filter. IEEE Trans Autom Control 54(6):1254–1269
10. Zhang H (2006) Research for track-before-detect algorithm based on Bayesian filtering. Master degree thesis of Northwestern Polytechnical University, p 3
11. Wang X, Pan Q, Huang H, Gao A (2012) Overview of deterministic sampling filtering algorithms for nonlinear system. Control Decis 27(6):801–812
12. Sun F, Tang L (2011) Cubature particle filter. Syst Eng Electron 33(11):2554–2557

Chapter 76
End-to-End Delay Computation for Time Petri Nets

Li Pan

Abstract Time Petri nets are a powerful formal model for the specification and verification of real-time systems. End-to-end delays in task execution are a key requirement to real-time systems. In this paper, we propose a state class method based on time Petri nets for the computation of end-to-end delays. By introducing time constraints of fired transitions into firing domains of state classes, the computation of end-to-end delays can be performed synchronously with the evolution of state classes in the same firing rules. Therefore, the proposed method avoids the maintain of all visited state classes along an execution trace in the global set method, and is suitable for the on-the-fly verification of time properties based on time Petri nets.

Keywords End-to-end delay computation · State classes · Time Petri nets

76.1 Introduction

A Petri net is a powerful mathematical model for the description of distributed systems [1]. To support the specification and verification of time-critical systems, two main extensions of Petri nets in using time were developed: considering the time as duration or intervals, such as timed Petri nets [2] or time Petri nets (TPNs) [3]. It has been shown that TPNs are more general than timed Petri nets [4].

Suffering from infinity of the time domain, the state space of a TPN model is usually infinite. Hence contracting the infinite state space into a finite structure is

L. Pan (✉)
Hunan Institute of Science and Technology, 414006 Yueyang, China
e-mail: panli.hnist@gmail.com

W. Lu et al. (eds.), *Proceedings of the 2012 International Conference on Information Technology and Software Engineering*, Lecture Notes in Electrical Engineering 210, DOI: 10.1007/978-3-642-34528-9_76, © Springer-Verlag Berlin Heidelberg 2013

necessary. Berthomieu and Menasche first introduced the concept of state classes in [4]. A state class of a TPN represents all states reachable from the initial state by firing all feasible firing values corresponding to the same firing sequence.

An important limitation of the classical state class approach is not suitable to evaluate the end-to-end delay in an execution trace, which is a critical issue in the design and analysis of real-time systems [5]. The classical approach uses the relative time mode for state classes, thus the absolute (global) time value cannot be obtained efficiently through a state class graph.

In [5], a clock-stamped state class method based on the global time mode was proposed for the computation of global delays. However, this approach cannot determine the firability of transitions in some cases. Later, Xu et al. presented a combined approach based on both relative and absolute time modes to the analysis of schedulability in [6]. However, like [5], it disregards the timing inequalities on transition pairs, but retains only the firing intervals of individual transitions in relative domains. As a result, the firability problem remains unsolved.

In [7], Vicario presented a global set method for evaluating end-to-end delays of TPN models. A global set is a time inequality system that includes the relative firing domains of all classes along an execution trace. In the method, all visited firing domains must be saved until the end-to-end delays are calculated. Therefore, the method incurs high storage costs, and the computation of the global set cannot be performed synchronously with the evolution of state classes.

In this paper, we present an efficient state class method for the timing analysis of TPN models. First, we define the kernel of a state class as a set of firing intervals on enabled transition pairs to support the basic evolution of a time Petri net. Furthermore, fired transitions are introduced into state classes to compute end-to-end delays between reachable classes. Therefore, the computation of end-to-end delays can be performed synchronously with the evolution of state classes in the same firing rules. The proposed method avoids the maintain of all previous state classes in the global set method, and is thus suitable for the on-the-fly verification of time properties based on time Petri nets. The rest of the paper is organized as follows. In Sect. 76.2, we define the syntax and semantics of a TPN. Section 76.3 proposes an efficient state class approach to the evaluation of end-to-end delays for TPNs. Then, in Sect. 76.4, we illustrate an application of this approach to a flexible manufacturing system. Finally, Sect. 76.5 concludes the paper.

76.2 Time Petri Nets

Let R be the set of real numbers and R^+ the set of nonnegative real numbers. An *interval* is a connected subset of R. Formally, $I = [a, b]$ is an interval iff $I = \{x \in R \mid a \leq x \leq b\}$ where $a \in R$, $b \in R$ and $a \leq b$. The lower (upper) bound of an interval I is denoted by $\downarrow I$ ($\uparrow I$ respectively). The set of all (nonnegative) intervals is denoted as IR (IR^+). Let I_1 and I_2 stand for two intervals. Then the operations of I_1 and I_2 are defined as follows:

- $I_1 + I_2 = [\downarrow I_1 + \downarrow I_2, \uparrow I_1 + \uparrow I_2]$;
- $I_1 - I_2 = [\downarrow I_1 - \uparrow I_2, \uparrow I_1 - \downarrow I_2]$;
- $I_1 \cap I_2 = [\max\{\downarrow I_1, \downarrow I_2\}, \min\{\uparrow I_1, \uparrow I_2\}]$, if $I_1 \cap I_2 \neq \varnothing$.

A *time Petri net* is a 6-tuple $TPN = (P, T, Pre, Post, M_0, SI)$ where

- $P = \{p_1, p_2, \ldots, p_m\}$ is a finite nonempty set of *places*;
- $T = \{t_1, t_2, \ldots, t_n\}$ is a finite nonempty set of *transitions*;
- *Pre*: $P \times T \to N$ is the forward incidence matrix;
- *Post*: $P \times T \to N$ is the backward incidence matrix;
- $M_0 : P \to N$ is the initial marking;
- SI: $T \to IR^+$ is a mapping called static firing interval. For any $t \in T$, $SI(t)$ represents t's static firing interval relative to the time at which transition t begins enabled.

Let $Pre(t) \in N^P$ be the multi-set of input places of transition t, corresponding to the vector of column t in forward incidence matrix. *Post* (t) is defined similarly.

A transition t is *enabled* at marking M, if and only if

$$\forall p \in P, Pre(p, t) \leq M(p).$$

Let $En(M)$ be the set of transitions enabled at marking M. Let $Newly(M, t_f)$ denote the set of newly enabled transitions by firing t_f from M, which is defined by
$Newly(M, t_f) = \{t \in T \mid t \in En(M - Pre(t_f) + Post(t_f)) \wedge (t \notin En(M - Pre(t_f)) \vee (t = t_f))\}$.

A *state* of a $TPN = (P, T, Pre, Post, M_0, SI)$ is a pair $s = (M, f)$, where

- M is a marking;
- f is a global firing interval function. $\forall t \in En(M)$, $f(t)$ represents t's global firing interval in which each value is a possible firing time relative to the initial state.

Let $s_0 = (M_0, f_0)$ be the initial state of the TPN, where M_0 is the initial marking, and $\forall t \in En(M_0), f_0(t) = SI(t)$. The semantics of a TPN model can be characterized by a Labeled Transition System that is defined below.

A *labeled transition system* is a quadruple $L = (S, s_0, \sum, \to)$ where

- S is a finite set of states;
- $s_0 \in S$ is the initial state;
- Σ is a set of labels representing activities;
- $\to$ is the transition relation.

The *semantics* of a TPN is defined as a labeled transition system $L = (S, s_0, \Sigma, \to)$ such that

- $S = N^P \times (IR^+)^T$;
- $s_0 = (M_0, f_0)$;
- $\Sigma \subseteq T \times R^+$;
- $\to \; \subseteq S \times \Sigma \times S$, and $\forall t \in T, (M, f) \xrightarrow{(t_f, d)} (M', f')$ iff.

$$
\begin{cases}
Pre(t_f) \leq M & (1) \\
\forall t \in En(M), \downarrow f(t_f) \leq d \leq \uparrow f(t) & (2) \\
M' = M - Pre(t_f) + Post(t_f) & (3) \\
f'(t) = \begin{cases} [d+ \downarrow SI(t), d+ \uparrow SI(t)] & \text{if } t \in Newly(M, t_f) \\ [Max(\downarrow f(t), d), \uparrow f(t)] & \text{if } t \in En(M - Pre(t_f)) \end{cases} & (4)
\end{cases}
$$

From the definition of the transition relation, it is easy to see that (1) guarantees that t_f is enabled at M; (2) ensure that t_f is firable at state s. Namely, the transition can be fired within its global firing interval, and its firing time d is not higher than the latest firing time of any other enabled transition; (3) describes the standard marking transformation; and (4) calculates global firing intervals of all enabled transitions at the state s'. Specifically, persistently enabled transitions have their intervals at state s truncated to intervals beyond d, and newly enabled transitions are assigned their static intervals plus d.

76.3 State Classes with Fired Transitions

76.3.1 State Classes with Fired Transitions

Let t^i be the ith firing of transition t and $Fired(C)$ the set of fired transitions introduced into state class C. A *state class* with *fired transitions* of a $TPN = (P, T, Pre, Post, M_0, SI)$ is a pair $C = (M, D)$, where

- $M: P \rightarrow N$ is a marking;
- $D: T^* \times T^* \rightarrow IR$ is a *firing domain*, where $T^* = T \cup Fired(C)$, $D(t_i,t_j)$ represents the firing interval of transition t_j relative to transition t_i for all t_i, $t_j \in En(M) \cup Fired(C)$.

We can divide the firing domain of state class C into the following two parts: *the kernel domain* that is composed of firing intervals on any transition pair t_i and t_j where $t_i, t_j \in En(M)$, supporting the basic (state class) evolution of a TPN model; and *the computation domain* that includes the other firing intervals on any transition pair t_i and t_j where t_i or $t_j \in Fired(C)$, and is used to evaluate end-to-end delays of TPN models.

It is easy to show that $D(t_i, t_j) = -D(t_j, t_i)$. Especially, let $D(t_i, t_i) = 0$. The initial class $C_0 = (M_0, D_0)$ is defined as follows: M_0 is the initial marking; and $\forall t_i$, $t_j \in En(M_0) \cup \{t_0\}$, $D_0(t_i, t_i) = 0$ and $D_0(t_i, t_j) = SI(t_j) - SI(t_i)$ if $t_i \neq t_j$, where t_0 is defined as the firing transition of C_0 and $SI(t_0) = 0$.

Firable condition. A transition $t_f \in En(M)$ is *firable* at state class $C = (M, D)$, if and only if $\forall t_i \in En(M): \uparrow D(t_f, t_i) \geq 0$ (i.e., $\downarrow D(t_i, t_f) \leq 0$).

Firing rules. State class $C_{k+1} = (M_{k+1}, D_{k+1})$ reached from state class $C_k = (M_k, D_k)$ by firing transition $t_f \in Fr(C_k)$ can be computed as follows:

- Marking M_{k+1}: $M_{k+1} = M_k - Pre(t_f) + Post(t_f)$.
- Firing domain D_{k+1}: $\forall t_i,\ t_j \in En(M_{k+1}) \cup Fired(C_{k+1})$

1.

$$
D_{k+1}(t_i, t_f) = \begin{cases} 0 & \text{if } t_i = t_f \\ -SI(t_i) & \text{if } t_i \in Newly(M_k, t_f) \\ [\downarrow D_k(t_i, t_f), \operatorname{Min}\{\uparrow D_k(t_i, t) \mid t \in En(M_k)\}] & \text{otherwise} \end{cases}
$$

2.

$$
D_{k+1}(t_i, t_j) = \begin{cases} 0 & \text{if } t_i = t_j \\ D_{k+1}(t_i, t_f) + D_{k+1}(t_f, t_j) & \text{if } t_i, t_j \in Newly(M_k, t_f) \\ D_k(t_i, t_j) \cap (D_{k+1}(t_i, t_f) + D_{k+1}(t_f, t_j)) & \text{otherwise} \end{cases}
$$

The proposed method has been defined as a very general framework. The classical state classes can be obtained as particular subclasses of our method. If $Fired(C) = \{t_f\}$ for any class C, where t_f is C's generative transition, then C is turned to a classical state class given in [4, 7].

76.3.2 End-to-End Delay Computation

A *transition (firing) sequence* σ is a finite (or infinite) string consisting of symbols in T. The *empty sequence*, denoted by λ, is the sequence with zero occurrences of symbols. A sequence w is called a *prefix* of σ, if and only if there is a sequence v such that $\sigma = wv$. Let σ_i be the prefix of σ with length i.

Let SP denote a time span function. For any two reachable classes C_i and C_j, $SP(C_i, C_j)$ stands for the end-to-end delay from C_i to C_j. Especially, we set $SP(C_i, C_i) = 0$.

Theorem 1 *Let $\Pi = \{C_1, C_2, \ldots, C_n\}$ be the set of reachable state classes from the initial state class C_0 by firing transition sequence $\sigma = t_{r_1} t_{r_2} \ldots t_{r_n}$, then $\forall C_i, C_j \in \Pi$, $SP_n(C_i, C_j) = D_n(t_{r_i}, t_{r_j})$.*

Proof Because firing a transition takes no time, the global arrival times of C_i and C_j are just the firing times of transition t_{r_i} and t_{r_j}. Thus the time span between C_i and C_j is the firing interval of transition t_j relative to transition t_i. According to the definition of state classes, it follows that $SP_n(C_i, C_j) = D_n(t_{r_i}, t_{r_j})$. □

The global time stamp is an important performance indicator of real-time systems. It represents the global arrival time of a state class, or the execution time of a transition firing sequence. By introducing the zero transition at the initial state class, our method can efficiently evaluate the global times of state classes. Let ST_i be the global time stamp of class C_i.

Corollary 1 *Let $C_n = (M_n, D_n)$ be a reachable state class from the initial state class $C_0 = (M_0, D_0)$ by firing a transition sequence $\sigma = t_{r_1} t_{r_2} \ldots t_{r_n}$. Then $ST_n = SP_n(C_0, C_n)$.*

Proof The result is directly derived from Theorem 1. □

From the above corollary, it is easy to find that the global time stamp of a state class is in essence a special end-to-end delay between the state class and the initial state class, because the zero transition can be regarded as the firing transition of the initial state class.

76.4 Case Study

Flexible manufacturing systems provide a means to achieve better quality, lower cost, and smaller lead-time in manufacturing. Every part or partially finished product which enters and moves inside an FMS is called a part. A finished part that leaves the FMS is called a product. Each product is the result of a sequence of *processes* according to its technological requirements. Resource requirements are considered in an *operation*. An operation $O_{i,j,k}$ represents the jth process of the ith job type being performed at the kth machine.

Let us consider an FMS with three machines M_1, M_2 and M_3. The system handles two jobs J_1 and J_2 that have two processes each. Tables 76.1 and 76.2 show the job requirements and time constraints of operations. The TPN model of the FMS is given in Fig. 76.1.

Assume that the difference between the completion times of two successive operations on machine M_2 is not less than 3 time units. Three operations $O_{1,1,2}$, $O_{1,2,2}$ and $O_{2,2,2}$ on machine M_2 are represented by t_2, t_3 and t_7 respectively, so these transitions are also needed. Next, we analyze feasible schedules satisfying the above constraints by the proposed method.

The initial state class is $C_0 = (M_0, D_0)$ where
$M_0 = (1\ 1\ 1\ 1\ 0\ 0\ 1\ 0\ 0)$,

Table 76.1 Job requirements of the FMS

Process	Job 1	Job 2
1	M_1/M_2	M_1/M_3
2	M_2/M_3	M_2

Table 76.2 Times constraints of the operations

Operation	Time constraint
$O_{1,1,1}$	[1,3]
$O_{1,1,2}$	[2,4]
$O_{1,2,2}$	[2,4]
$O_{1,2,3}$	[3,3]
$O_{2,1,1}$	[4,4]
$O_{2,1,3}$	[5,6]
$O_{2,2,2}$	[2,4]

Fig. 76.1 TPN model of the FMS

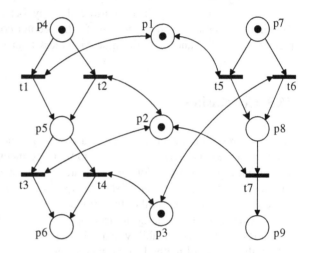

$$D_0 = \begin{array}{c} t_0 \\ t_1 \\ t_2 \\ t_5 \\ t_6 \end{array} \begin{pmatrix} 0 & [1,3] & [2,4] & [4,4] & [5,6] \\ & 0 & [-1,3] & [1,3] & [2,5] \\ & & 0 & [0,2] & [1,4] \\ & & & 0 & [1,2] \\ & & & & 0 \end{pmatrix}$$

According to the firable condition, $Fr(C_0) = \{t_1, t_2\}$. Firing t_2 will result in state class $C_1 = (M_1, D_1)$, where
$M_1 = (1\ 1\ 1\ 0\ 1\ 0\ 1\ 0\ 0)$,

$$D_1 = \begin{array}{c} t_0 \\ t_2^1 \\ t_3 \\ t_4 \\ t_5 \\ t_6 \end{array} \begin{pmatrix} 0 & [2,3] & [6,9] & [5,6] & [4,4] & [5,6] \\ & 0 & [4,6] & [3,3] & [1,2] & [2,4] \\ & & 0 & [-3,-1] & [-5,-2] & [-4,0] \\ & & & 0 & [-2,-1] & [-1,1] \\ & & & & 0 & [1,2] \\ & & & & & 0 \end{pmatrix}$$

Obviously, $ST_1 = SP_1(C_0, C_1) = D_1(t_0, t_2^1) = [2, 3]$, that is to say, the end-to-end delay between C_0 and C_1 is $[2, 3]$.

Firing t_3 at C_1 will reach state class $C_2 = (M_2, D_2)$, where
$M_2 = (1\ 1\ 1\ 0\ 0\ 1\ 1\ 0\ 0)$,

$$D_2 = \begin{matrix} t_0 \\ t_2^1 \\ t_3^1 \\ t_5 \\ t_6 \end{matrix} \begin{pmatrix} 0 & [2,2] & [4,4] & [4,4] & [5,6] \\ & 0 & [2,2] & [2,2] & [2,4] \\ & & 0 & [0,0] & [1,2] \\ & & & 0 & [1,2] \\ & & & & 0 \end{pmatrix}.$$

It is easy to see that $ST_2 = SP_2(C_0, C_2) = D_2(t_0, t_3^1) = [4, 4]$ and $SP_2(C_1, C_2) = D_2(t_2^1, t_3^1) = [2, 2]$. The end-to-end delay between C_1 and C_2 is $[2, 2]$. Therefore, schedule t_2t_3 does not satisfy the second constraint because the completion times of t_2 and t_3 on machine M_2 is less than 3 time units.

76.5 Conclusion

We have presented a state class method based on time Petri nets for the computation of end-to-end delays. In our method, timing inequalities on fired transitions are introduced into firing domains of state classes, and they are computed synchronously with the evolution of state classes in the same firing rules. Our method is more suitable for the on-the-fly verification than the global set method.

Currently, we are working on developing a new TPN model with mixed time semantics for the schedulability analysis of real-time systems [6, 8], and plan to extend the proposed method to the mixed semantics model.

Acknowledgments This work was supported by Hunan Provincial Natural Science Foundation of China (11JJ4058) and Scientific Research Fund of Hunan Provincial Education Department (11A041).

References

1. Murata T (1989) Petri nets: properties, analysis and applications. Proc IEEE 77(4):541–580
2. Ramchandani C (1974) Analysis of asynchronous concurrent systems by timed Petri nets. Project MAC, technical report 120, Massachusetts Institute of Technology
3. Merlin P, Farber DJ (1976) Recoverability of communication protocols: implication of a theoretical study. IEEE Trans Commun 24(9):1036–1043
4. Berthomieu B, Diaz M (1991) Modeling and verification of time dependent systems using time Petri nets. IEEE Trans Softw Eng 17(3):259–273
5. Wang J, Xu G, Deng Y (2000) Reachability analysis of real-time systems using time Petri nets. IEEE Trans Syst, Man, Cybernetics, 30(5):725–736
6. Xu D, He X, Deng Y (2002) Compositional schedulability analysis of real-time systems using time Petri nets. IEEE Trans Softw Eng 28(10):984–996

7. Vicario E (2001) Static analysis and dynamic steering of time dependent systems using time Petri nets. IEEE Trans Softw Eng 27(8):728–748
8. Wu N, Zhou MC (2010) A closed-form solution for schedulability and optimal scheduling of dual-arm cluster tools with wafer residency time constraint based on steady schedule analysis. IEEE Trans Autom Sci Eng 7(2):303–315

Chapter 77
Using the Reputation Score Management for Constructing FPFS System

Shufang Zhang and Jun Han

Abstract This paper has used the reputation score management for constructing a fair P2P file sharing system (abbr. FPFS system), the system design principle is simple and easy to realize, and every node entering into the P2P network obtains a certain reputation score, and obtains the corresponding resources reward according to the score. This paper has described the fair sharing strategies facing node network bandwidth and TTL, and these strategies can be used independently or be combined with other reputation score managements of P2P network. These two strategies have been discussed in the specific reputation score management system of P2P network Eigen Trust, and the test results indicate that: compared with a common P2P network, the fair sharing strategies of this chapter have faster file download speed and can decrease the network message communication amount during the process looking for resources.

Keywords Fair sharing · Reputation score P2P networks · Network bandwidth

77.1 Introduction

At present, the fairness in P2P network is a common problem, and most machines want to be not providers of resources but consumers of them. The research results indicate that 70 % Gnutella clients do not share any file. To construct a fair P2P

S. Zhang (✉)
School of Mechanical and Electronic Engineering, Suzhou University, 49 Middle Bianhe Road, Suzhou City, Anhui, People's Republic of China
e-mail: zhshf_sztc@126.com

J. Han
School of Information Science and Technology, University of Science and Technology of China, Hefei, China
e-mail: hanjun_sztc@126.coml

W. Lu et al. (eds.), *Proceedings of the 2012 International Conference on Information Technology and Software Engineering*, Lecture Notes in Electrical Engineering 210, DOI: 10.1007/978-3-642-34528-9_77, © Springer-Verlag Berlin Heidelberg 2013

network and promote the system efficiency, an effective method is to mainly consider the reputation mechanism of nodes [1], and to allow an entity to give its reputation score to another entity, and then the reputation can be used as a reference by other entities when they share resources, and another entity can obtain resources.

This chapter has putted forward that a P2P entity can be caused to enter into P2P network and to fairly use network resources by providing the incentive mechanism of reputation score to it, and it obtains the corresponding number of resources according to its reputation score. The obtaining of reputation score has referred to the reputation score management system of P2P network Eigen Trust [2]. Active participant nodes can obtain their priority according to network bandwidth or Time to Live (TTL) when they compete for the resources of other nodes in network. The design method of this chapter follows the principles of simplicity and easy realization, every node obtains a certain reputation score, and obtains the corresponding resources reward according to the score, then the behavior of free riding in a P2P network can be suppressed to some extent.

At present, the literature [3] also puts forward the calculation method of node reputation score based on value, the value is obtained during file upload, and files are downloaded according to the number of the value; another node reputation management is processed centrally on backstage [4], and there is also an independent calculation method of reputation score which does not depend on auxiliary nodes [5]. The method of this chapter is different from these methods, the calculation is based on the service quality of a node, such as rapid download time, and enhanced network bandwidth, it is mainly used to control nodes with low reputation score, and it can cause P2P network to exhibit fair file sharing through the incentive mechanism when a malicious node undermines. The two fair sharing strategies facing bandwidth and TTL have been putted forward, and they can be combined with some other reputation score managements. Finally, these two algorithms about reputation score management are proved to be feasible and efficient through test and performance analysis.

77.2 Fair Sharing Strategies

The aim of this chapter is to provide the incentive mechanism, which can cause nodes to fairly share the resources in P2P network and discard the nodes with free riding behavior or malicious behavior. But this system does not punish another kind of node, which is willing to contribute its network resources to the network, and does not download resources. This strategy uses the following two methods to calculate the reputation score of the active nodes entering into network: rapid download time and network bandwidth. Therefore, this chapter gives two kinds of fair sharing strategies in P2P network: one faces bandwidth, and the other faces TTL. It is worth noticing that these two kinds of strategies based on reputation score are very common, they can be applied to some other systems of reputation score management, and also can be used independently.

For example, if the nodes Peeri and Peerj download files from Peerk at the same time, the bandwidth of Peerk will be divided. The bandwidths obtained by Peeri and Peerj from Peerk are calculated according to the reputation score, i.e.

> *Bandwidth (peeri) = score(peeri)/(score(peeri) + score(peerj))*
> *Bandwidth (peerj) = score(peerj)/(score(peeri) + score(peerj))*

This is the condition that two nodes compete for network resources at the same time. If several nodes compete at the same time, the algorithm of the fair sharing strategy is as following:

> *Supposing the node Peeri has the available network bandwidth B, B should be distributed to several nodes Peerj (j = 1, 2, 3...n) which download files and data from Peeri. It can be described by pseudo codes as following:*
> *For each peerj download data from Peeri do*
> *{*
> *Bandwidth (peerj) = B * score (peerj)/(score (peer1)*
> *+score (peer2) +score (peer3) +score (peer n))*
> *}*
> *End*

To limit network hops, dynamic query or TTL is set out; therefore the fair sharing strategy facing TTL is also important at present. For every query request in Gnutella, TTL is set as 7 at present. The fair sharing strategy facing TTL putted forward in this chapter has set broader TTL for resources finding in network according to the reputation score. There are many ways to realize this aim, and a simple way is to define the maximum average TTL for every P2P entity (for example, High TTL = 10), and to define the minimum average TTL (for example, Low TTL = 5). The excitation of the node reputation score in this strategy is exhibited as the following algorithm.

> *If score (peer j) >=Average score then*
> *TTL (peer j) = High TTL;*
> *Else*
> *TTL (peer j) = Low TTL;*
> *End If*

In this strategy, the problem worthy of attention is that every node must know the average participant reputation score in network, and the calculation of the average participant reputation score may decrease the number of messages in network. Every node must know the participant score of other nodes in network, so the average participant reputation score of nodes must be known in advance in a specific reputation score management system of P2P network. The reputation score management systems such as Eigen Trust are easy to meet this requirement.

77.3 System Test and Its Performance Analysis

77.3.1 The Reputation Score Management System Eigen Trust

This section will mainly describe the test and performance analysis of the system. This fair sharing strategy has used the node reputation score, so the specific reputation score management system of P2P network Eigen Trust is chosen for test. In Eigen Trust, the reputation score of a P2P entity reflects its active extent to participate in network. In the experiments, the test environment is constructed according to the literature [6–11], a node submits application and responds to query according to a certain interest in every query cycle, it can share and download the files of other nodes after it gets the response, it can provide the nodes with free riding behavior and malicious nodes by way of simulation, and it can share some suspicious files.

Figure 77.1 shows the relation between the reputation score of every node and the number of upload files in 15 query cycles in the system Eigen Trust. X axis is the reputation score, and Y axis is the number of download files. The node reputation in Eigen Trust is closely related to the number of upload files, and those nodes which can provide many upload files have high reputation score. The correlation is 0.97, which means that the upload number of credible files and the node reputation score have very close relation.

In the fair sharing strategy facing TTL, the problem of average reputation score should be considered. The total score of network is 1 in the system Eigen Trust, so the average reputation score of every node can be set as 1/n, in which n is the number of all the nodes in P2P network. Here, it is supposed that a node knows or roughly knows the number n of all the nodes in P2P network. For a large-scale P2P network, the total

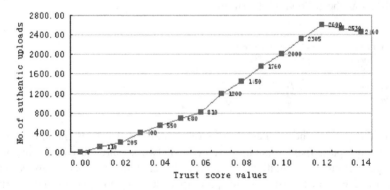

Fig. 77.1 The relation between reputation score and files download in Eigen Trust

Table 77.1 The average download speed of the fair sharing strategy facing bandwidth (its percentage compared with the total bandwidth speed)

P2P entity	Reputation score of Eigen Trust	Using the fair sharing strategy	Without the fair sharing strategy
0(A)	0.03012	67.41	47.98
1(B)	0.00483	55.14	47.98
2(C)	0.00201	61.44	47.98
3(D)	0.00121	48.33	47.98
4(E)	0	30	47.98
5(F)	0	30	47.98

number of nodes is unknown, and then the node reputation score can be directly set as 1, so every node can compare its reputation score with the average reputation score $1/n$ of the whole network. If the node reputation score is larger than the average reputation score, TTL can be set as 5. If else, TTL can be set as 0. In these conditions, a node does not need to clearly calculate the average participant reputation score of network, and then the problem posed earlier has been solved.

77.3.2 Test Facing Bandwidth

The reputation score management system of P2P network Eigen Trust can be used to test the fair sharing strategy of this chapter. First, the strategy facing network bandwidth is tested. This strategy attempts to obtain high reputation score through rapid download. The average speed of a node in P2P network is tested in the following two conditions: the fair sharing strategy with bandwidth and the common condition without fair sharing.

As is shown in Table 77.1, compared with a common P2P network, the average download speed of every user has very good performance through the combination of the bandwidth fair sharing strategy of this chapter with the reputation score management system Eigen Trust. A congested network is simulated, and the following congestion control scene is deployed during the test process: a P2P entity Peer i downloads files from Peer j, it competes for the network bandwidth resources of Peer j with other five nodes (from 0 to 4) at first, and the number of nodes is random. It is worth noticing that, if the fair sharing strategy is used in the simulated scene, the active nodes can compensate their own participant reputation score. The compensation amount is considerable, and the download speed of most nodes is larger than 48.33 %. For example, if the network bandwidth is 10 Mb/s, the download speed of the node A is: $10 \times 67.39\% = 6.739$ Mb/s.

Table 77.2 The test results of the fair sharing strategy facing TTL

P2P entity	Reputation score of eigen trust	Using TTL		Without the fair sharing strategy	
		The number of nodes (N_n)	The number of nodes in the TTL range (N_{ttl})	The number of nodes (NN_n)	The number of nodes in the TTL range (NN_{ttl})
The first group	<1/n	77	27	74	75
The second group	>1/n	23	98	26	75

77.3.3 Test Facing TTL

In the test of the fair sharing strategy facing TTL, an important problem is that whether the reputation score management of this chapter compensates those nodes with very high reputation score to let them have more time and chance for files download, and whether the free riding behavior can be avoided.

To accomplish the process above, 15 query cycles are chosen, and the algorithm of TTL setting is used in the test. The total number of nodes in P2P network is 100 (n = 100), and the default TTL is defined as 4, i.e. in a P2P network with 100 nodes, the search request of a node can reach 75 other nodes in 4 steps. In this condition, the average High TTL is defined as 5, and the average Low TTL is defined as 3.

In the test process, the P2P network is divided into two groups. Users in the first group have higher reputation score, i.e. the reputation score of every node is usually larger than the average reputation score 1/n in Eigen Trust. The reputation score of the users in the second group is smaller than the average reputation score 1/n. Table 77.2 shows the number of nodes of the first and second groups, and the corresponding average number of nodes which can be reached in the TTL range. The largest messages amount is given by:

$$Message\_Amount = Nn * Nttl + NNn*NNttl$$

The experimental result indicates that the adoption of the fair sharing strategy facing TTL can decrease the query load in network. With the fair sharing strategy facing TTL, when all the nodes in the network put out a query request, the largest messages amount is $77 \times 27 + 23 \times 98 = 4,333$. But with the common mode, the amount of network messages is about $74 \times 75 + 26 \times 75 = 7,500$. Moreover, the nodes with the free riding behavior can not be exclude from the network in this process because they can look for their destination node in the condition TTL = 3.

77.4 Conclusions

This chapter describes the two sharing strategies facing network bandwidth of nodes and facing TTL, which can construct fair sharing of P2P files. These two strategies have been tested in the specific reputation score management system of P2P network Eigen Trust, and the experimental results indicate that they can largely accelerate download time and decrease messages communication volume in network. These two sharing strategies based on reputation score have the advantages of simple design and easy realization. The next work is to consider more resources information (including software information and hardware information) in the calculation of reputation score and to research complicated algorithm of reputation score setting.

Acknowledgments The work is supported by the Nature Science Research Project of Suzhou University under Grant (2008yzk08, 2009yzk15, 2011cxy02), the National Natural Science Foundation of P.R.China (61174124, 61074033), and the Natural Science Foundation of Anhui Province Grant (KJ2012Z393,KJ2012Z397).

References

1. Li Y, Dai Y (2010) Research on trust mechanism for peer-to-peer network. Chin J Comput 3:390–405 (in Chinese)
2. Ma X, Geng J (2007) Survey of trust and reputation mechanism on P2P network. J Comput Appl 27(8):1935–1938 (in Chinese)
3. Huang Q, Song J, Liu W, Zhang J (2006) Survey of reputation system on peer-to-peer network. Mini Micro Syst 27(7):1175–1181 (in Chinese)
4. Feng Q, Dai Y (2007) Survey on trust mechinism for P2P network. Commun CCF 3(3):31–40 (in Chinese)
5. Chang J, Wang H, Yin G (2006) DyTrust: a time-frame based dynamic trust model for P2P systems. Chin J Comput 29(8):1301–1307 (in Chinese)
6. Audun J, Roslan I, Colin B (2007) A survey of trust and reputation systems for online service provision. Decis Support Syst 43(2):618–644
7. Artz D, Gil Y (2007) A survey of trust in computer science and the semantic web. Web Seman 5(2):58–71
8. Damiani E, di Vimercati DC, Paraboschi S, Samarati P, Violante F (2002) A reputation-based approach for choosing reliable resources in peer-to-peer networks. In: Proceedings of the 9th ACM conference on computer and communications security
9. Caballero A, Botia JA, Gomez-Skarmeta AF (2006) A new model for trust and reputation management with an ontology based approach for similarity between tasks. Multiagent Syst Technol 4196:172–183
10. Ruohomaa S, Kutvonen L, Koutrouli E (2007) Reputation management survey. In: Bob W (ed) Proceedings of the 2nd international conference on availability, reliability and security
11. De Alfaro L, Kulshreshtha A, Ian Pye B, Adler T (2011) Reputation systems for open collaboration. Commun ACM 54(8):81–87

Chapter 78
Design and Simulation of AUV Visual Servo Charging Manipulator

Zhang Lefang

Abstract A method of fast charging for remote AUV is presented in this chapter and a charging manipulator will be designed. The manipulator principle is expounded based on string robotics theory. A robotic visual servo system is built. A simulation system based on Puma560 and Scara is built with MATLAB/Simulink. The simulation results show that the manipulator based on Puma560 is simple and reliable. It has a high accuracy and little error, so it can better satisfy the practical application.

Keywords PUMA560 · SCARA · Visual servo · Charging manipulator · Design and simulation

78.1 Introduction

There are Two Types of Underwater unmanned vehicle (UUV) [1].One type is Remote operated vehicle (ROV) and the other is Autonomous underwater vehicle (AUV). ROV is provided with electric power by the mother ship. AUV carries energy and is controlled by itself. It has strong autonomy, good mobility, large voyage and can perform a variety of military tasks. The research and development of AUV is mainstream in current and future period of time, because it has a very good potential application prospect. However, due to the structure and load conditions, AUV can not carry much energy, which restrict its voyage and task cycle.

Z. Lefang (✉)
Xi'an Electronic City, Eurasia Road on the 1st, Information Engineering College
of Xi'an Eurasia University, Xi'an 710065 Shaanxi, China
e-mail: zhanglefang@126.com

W. Lu et al. (eds.), *Proceedings of the 2012 International Conference on Information Technology and Software Engineering*, Lecture Notes in Electrical Engineering 210, DOI: 10.1007/978-3-642-34528-9_78, © Springer-Verlag Berlin Heidelberg 2013

Fig. 78.1 Principle of charging

Charging Manipulator

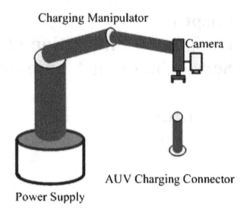

Camera

AUV Charging Connector

Power Supply

Therefore, the energy power problems is a bottleneck for the development of AUV research. So one of core technologies for AUV is effective energy supply.

Now, AUV can carry energy or generate energy by itself. The energy which AUV carries is mostly lead-acid battery, silver-zinc battery and lithium battery [2]. Recently, American has made an test which consider fuel cell as AUV power supply. At present, proton exchange membrane fuel cell technology (PEMFC) is more and more mature. The ratio of volume and power has exceeded 1 kw/L and the ratio of weight and power has exceeded 1 kw/kg [3]. Related organizations have presented the energy strategy which combine thermodynamic and electricity power. An aluminium-hydrogen peroxide power supply for AUV has been developed by Norwegian Defence Research Institute. The energy supply can make an AUV weighted 1t sail 1,200 nautical mile at the speed of 4 kn. Also, AUV can acquire energy from ocean or environment except the energy which it carries. For example, an underwater glide vehicle named GLIDER which based on ocean temperature differential power generation has been developed by Webber research company in the United States. Besides, an AUV named SAUVII has also been developed and it can come to the surface using solar energy. The methods referred above can solve the energy problem of AUV in a certain extent.

However, AUV can not carry much energy, so it can not sail too far. And the efficiency of ocean and solar energy is too low. Also, if using energy from environment, the structure of AUV must be changed. Therefore, this chapter presents a fast charging system for remote AUV. The system is similar to flight-refuel. AUV can get energy from the power supply device placed on the sea surface. The power supply device is similar to a buoy floating on the sea surface, and it has batteries and a charging manipulator as shown in Fig. 78.1. If AUV has not enough power, it will rise automatically, and protrude its charging connector. Then, the charging manipulator will capture the connector and connect reliably. At last, the AUV can get energy from the power supply device.

One of the key technology for the charging system is directing the manipulator to capture the AUV charging connector. At present, the manipulator can capture the connector through the technique of infrared, ultrasound, Bluetooth, radio

Fig. 78.2 Puma560
configuration

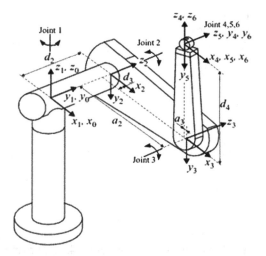

frequency or visual sensor. Infrared positioning technology is suitable only for short distance communication and easy interfered by fluorescent light etc., so it is limited in accurate location. Ultrasound positioning has higher accuracy but higher cost. Rest of the wireless location technology such as Bluetooth, radio frequency, etc., have poor stability and are interfered easily by noise. Visual sensor contains a large amount of information and has a wide range of application. So this chapter present that the manipulator can be directed to capture the charging connector through visual servo control [4–8].

Visual servo control is a technique which uses feedback information extracted from a vision sensor to control the motion of a robot [9]. In this chapter, we introduce the design principle of charging manipulator based on Puma560 and Scara robot. Then, we construct an image based visual servo control system (as shown in Fig. 78.2) with eye-in-hand configuration based on these two robot and make simulation analysis.

78.2 Design Principle of the Manipulator for Visual Servo Control

The charging manipulator is an important part of the visual servo system. In this chapter, we select and use series multiple joint robot as charging manipulator. Usually, this kind of robot has compact structure and occupies small volume. All joints of the robot are revolute. So this kind of robot is widely used in practice. Puma560 and Scara are typical robots of this kind [10]. Therefore, in this section we will introduce principle and configuration of Puma560 and Scara.

Puma560 is a typical series six joint robot and has six revolute joints as shown in Fig. 78.2, so it has six DOFS. Joint 1 is in the waist, Joint 2 is in the shoulder

Fig. 78.3 Scara
configuration

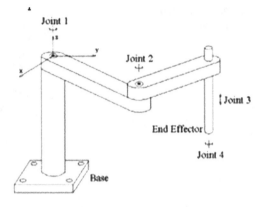

and Joint 3 is in the elbow. Joint 4, Joint 5 and Joint 6 are all in the wrist. Axes of Joint 2 and joint 3 are parallel and they are all vertical with the axis of Joint 1. Axes of the three joints in wrist intersect at a point. We can establish robot link coordinates [11] and describe the relationship of mechanism motion through D–H parameter [12]. Because of broad working space and flexible end-effector path, this robot has a wide range of application in many areas such as welding, spraying, mechanical processing, material carrying, manufacturing and so on.

We can also use Scara as manipulator besides Puma560. Selective compliance assembly robot arm (Scara) has three revolute joints and one prismatic joint (as shown in Fig. 78.3). Joint 1, Joint 2 and Joint 4 are revolute, Joint 3 is prismatic. Its arm was rigid in the Z-axis and pliable in the XY-axes, which allowed it to adapt to holes in the XY-axes [13]. By virtue of the Scara's parallel-axis joint layout, the arm is slightly compliant in the 'X–Y' direction but rigid in the 'Z' direction, hence the term: Selective Compliant. This is advantageous for many types of assembly operations, i.e., inserting a round pin in a round hole without binding. Because of particular shape, its work scope is similar to a sector.

78.3 Visual Servo Control Simulation

In the previous section, we have explained that we select Puma560 and Scara robot as the manipulator. In this section, we fix a camera at the end of the robot as shown in Fig. 78.1 to establish an image based visual servo control system using Puma560 and Scara robot respectively. Then, we simulate the system in MATLAB with Robotics Toolbox [14] and Machine Vision Toolbox [15]. These toolboxes are written by Australian scientist, Peter Corke. They including rich functions which are used in robotic research, such as robot kinematics and dynamics, trajectory planning, etc.

Fig. 78.4 Control system
structure

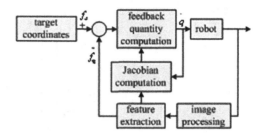

Fig. 78.5 Target model

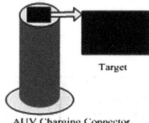

Fig. 78.6 Puma560's
simulation model

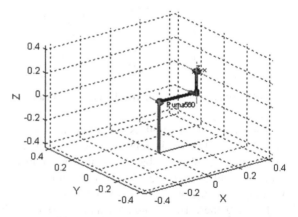

Figure 78.4 shows the structure of image based visual servo control system. We define the image feature error as $e(f) = f_d - f_g$ and the control volume as u. Then, we can establish a proportional feedback controller Eq. (78.1).

$$u = k \cdot J_i^{-1} e(f) \tag{78.1}$$

where k is a proportional coefficient.

The task is that the robot's end-effector is guided to reach the target position using feedback information extracted from a camera fixed at the end of the robot [5]. Figure 78.5 shows the target model. Assuming that the target is stationary, and its feature points are the four vertices of the rectangle as shown in Fig. 78.5. The coordinate matrix of the target in image plan is $f_d = [256, 456; 456, 456; 456, 256; 256, 256]$.

Figures 78.6 and 78.7 show Puma560 and Scaras' simulation model.

Fig. 78.7 Scara's simulation
model

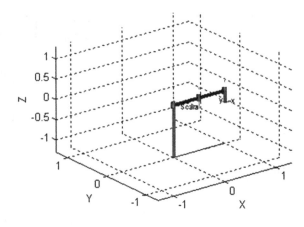

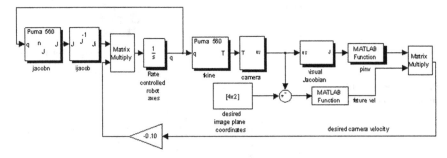

Fig. 78.8 Simulation model of visual servo system based on Puma560

The camera focal length is $f = 8$ mm and the size of the image plane in pixels is 512pix $\times$ 512pix (pix represents for a pixel).Then we can establish two simulation models of image based visual servo control system respectively based on Puma560 and Scara using MATLAB/Simulink. These two models are similar, so we only show the one based on Puma560 in Fig. 78.8.

Running these two simulation models respectively, we get the movement of robot's end-effector in image plan shown in Figs. 78.9 and 78.10. Also, we can get the image error and the speed of robot's end-effector as shown in Figs. 78.11, 78.12, 78.13 and 78.14.

In Figs. 78.8 and 78.9, we can see that both the Puma560 and the Scaras' end-effector are close to the target location (f_d). However, because of their different configuration, they show different trajectories. In Fig. 78.11, the image errors of all points of Puma560 are gradually decreasing and will eventually approach zero. But, in Fig. 78.12, the image errors of some feature points of Scara are gradually increasing within the same simulation time. Therefore, the algorithm presented in this chapter is appropriate for the visual servo control system based on Puma560. In Fig. 78.13, the speed of all feature points of Puma560's end-effector are gradually approaching zero. It explains that when the end-effector is close to the

Fig. 78.9 Movement of Puma560's end-effector in image plane

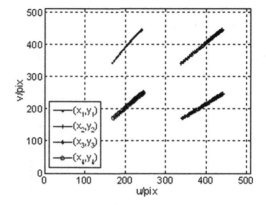

Fig. 78.10 Movement of Scara's end-effector in image plane

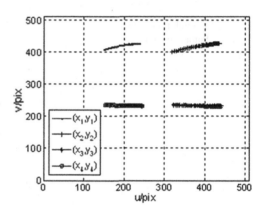

Fig. 78.11 Evolution of the image error of Puma560

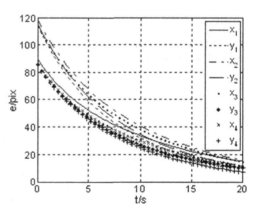

target, the robot is close to stability. However, the speed of some feature points of Scara's end-effector are gradually increasing (as shown in Fig. 78.14). So the evolution of speed is corresponding to that of image error. As a result, the Puma560's end-effector can be better close to the target than the scara's.

Fig. 78.12 Evolution of the
image error of Scara

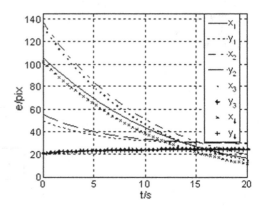

Fig. 78.13 Evolution of the
speed of Puma560's end-
effector

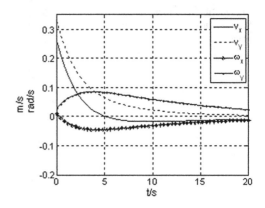

Fig. 78.14 Evolution of the
speed of Scara's end-effector

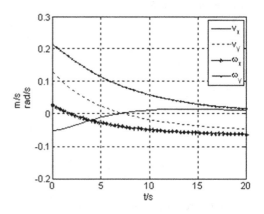

78.4 Conclusion

This chapter has presented a method of charging remote AUV based on the analysis of energy problems of remote AUV. And the key of this method is how to direct the charging manipulator to capture the charging connector. So, Puam560 and Scara are used as the charging manipulator. Then, the principles of Puma560 and Scara have been introduced and simulated model of the image based visual servo control system has been established using MATLAB/Simulink. The simulation of the visual control indicates that, according to the control algorithm developed in this chapter, Puam560 is suitable to use as the charging manipulator and will function effectively in practice.

References

1. Griffiths G (2003) Technology and applications of autonomous underwater vehicles. Ocean Sci Technol 2:342
2. Ishak D, Manap NAA, Ahmad MS, Arshad MR (2010) Electrically actuated thrusters for autonomous underwater vehicle. In: The 11th IEEE international workshop on advanced motion control, Nagaoka, 21–24
3. Lakeman JB et al (2006) The direct borohydride fuel cell for UUV propulsion power. J Power Sources 162:765–772
4. Piepmeier J, Murray GM, Lipkin H (2004) Uncalibrated dynamic visual servoing. IEEE Trans Robot Automat 20(1):143–147
5. Chen J, Dswson D, Dixon W, Behal A (2005) Adaptine homography-based visual servo tracking for fixed and camera-in-hand configurations. IEEE Trans Control Syst Technol 13(9):814–825
6. Corke P, Hutchinson S (2001) A new partitioned approach to image-based visual servo control. IEEE Trans Rob Autom 17(4):507–515
7. Yoshimi BH, Allen PK (1995) Alignment using an uncalibrated camera system. IEEE Trans Rob Autom 11(4):516–521
8. Espiau B, Chaumette F, Rives P (1992) A new approach to visual servoing in robotics. IEEE Trans Robot Automat 8(3):313–326
9. Hutchinson SA, Hager GD, Corke PI (1996) A tutorial on visual servo control. IEEE Trans Robot Automat 12(5):651–670
10. Colson J, Perreira ND (1983) Kinematic arrangements used in industrial robots. In: 13th industrial robots conference proceedings. April 1983
11. Manseur R, Doty KL (1988) A fast algorithm for inverse kinematic analysis of robot manipulators. Int J Robot Res 7(3):52–62
12. Denavit J, Hartenberg RS (1955) A kinematic notation for lower-pair mechanisms based on matrices. J Appl Mech 22:215–221 (June)
13. Westerland L (2000). The extended arm of man, a history of the industrial robot
14. Corke PI (2002) Robotics TOOLBOX for MATLAB(Release8). Manufacturing science and technology Pullenvale, Australia
15. Corke P (2005) The machine vision toolbox. IEEE Robot Autom Mag 12(4):16–25

Chapter 79
Study on the Tracking and Parameter Estimating in Passive Magnetic Detection

Junji Gao, Daming Liu and Guohua Zhou

Abstract Aiming at the tracking and parameter estimating of unknown moving magnetism objects, a particle filter algorithm combined with Hopfield network and magnetic potential tensor was brought out based on the submarine magnetic detection and magnetic defence of military port. This algorithm could effectively conquer the degeneration of particles, and ensure the speed and precision of tracking. The result of simulation and experiment shows the validity of this algorithm. This algorithm will have some significance in martial and civil field such as submarine magnetic detection, magnetic defence of military port, unexploded ordnance detection (UXO) and prospecting for ore.

Keywords Target tracking · Magnetic detection · Hopfield neural network · Localization by magnetic potential tensor · Particle filter

Target tracking is a process of modeling, estimating, and tracking to the moving objects. The detection to submarine by its surrounding magnetic field anomaly is an important antisubmarine means. There are many means of submarine magnetic detection, such as passive detection and active detection.

There are some studies on the tracking of immobile magnetic objects such as Unexploded Ordnance detection (UXO). The total field, the scalar field and the gradiometer data could be used in those problems. The conventional magnetic localization is inverse computing by least-squares approach [1]. There are study on the submarine magnetic detection from air in USA [2]. In recent years, wavelet and neural network were applied to magnetic detection. In [3], the UXO was localized by wavelet transform. Hopfield neural network was used in [4]. The

J. Gao (✉) · D. Liu · G. Zhou
Institute of Electrical Engineering, Naval University of Engineering, Liberation Road NO. 717, Wuhan, China
e-mail: blamer@sina.com

W. Lu et al. (eds.), *Proceedings of the 2012 International Conference on Information Technology and Software Engineering*, Lecture Notes in Electrical Engineering 210, DOI: 10.1007/978-3-642-34528-9_79, © Springer-Verlag Berlin Heidelberg 2013

Euler method was applied in [5] based on the measurement of gradiometer data. The Euler Deconvolution and Hilbert Transform were applied to compute the parameter of the magnetic object [6]. At the same time, there are few studies on the tracking and parameter estimating of unknown moving magnetic objects, which have great military and civil significance. Aiming at the passive magnetic detection, the approach on the tracking and parameter estimating of submarine was studied.

79.1 The Status Space Model of Magnetic Objects

The status space model is composed of status model which describing the movement information and measure model which describing the measure information.

The status vectors were defined as follows

$$\mathbf{X}(k) = [x(k), \ y(k), \ z(k), \ \theta(k), \ V(k), \ M_x(k), \ M_y(k), \ M_z(k)] \tag{79.1}$$

where $\mathbf{X}(k)$ is the status vector of time k, $[x(k), \ y(k), \ z(k)]$ is the relative position of the submarine center and the sensor at time k, $\theta(k)$ is the course of the submarine, $V(k)$ is the speed of the submarine, $[M_x(k), \ M_y(k), \ M_z(k)]$ is the equivalent magnetic moment of three components at time k.

The sampling cycle is T, if the T is short enough, there could be the hypothesis that the submarine is moving linearly in the same speed. So the status model is

$$\begin{cases} x(k) = x(k-1) + [V(k-1)\cos\theta]T; \ y(k) = y(k-1) + [V(k-1)\sin\theta]T; \\ z(k) = z(k-1); \ \theta(k) = \theta(k-1); \ V(k) = V(k-1); \\ M_x(k) = M_x(k-1); \ M_y(k) = M_y(k-1); \ M_z(k) = M_z(k-1). \end{cases} \tag{79.2}$$

In this problem, because the submarine is far away from the magnetic sensors, it could be equivalent to a magnetic dipole. The measure model is

$$H(k) = A(k)M(k) + w(k)$$

where $H(k)$ is the vectors of measured magnetic field, $\mathbf{w}(k)$ is Gauss noise. The observation matrix $A(k) = \frac{1}{r^5}\begin{bmatrix} 3x^2 - r^2 & 3xy & 3xz \\ 3xy & 3y^2 - r^2 & 3yz \\ 3xz & 3yz & 3z^2 - r^2 \end{bmatrix}$ where $x, \ y, \ z$ is the relative position of the magnetic dipole and the sensor, $r = \sqrt{x^2 + y^2 + z^2}$.

The status space model is

$$\begin{cases} X(k) = f(X(k-1), \varepsilon(k-1)) \\ H(k) = h(X(k), \omega(k)) \end{cases} \tag{79.3}$$

79.2 Model of Localization, Tracking and Parameter Estimating

Firstly, the tracking model by particle filter approach, Hopfield neural network and magnetic potential tensor were established respectively.

79.2.1 Particle Filter

Particle filter is the use of a sequential Monte Carlo based statistical signal processing method. In the Bayesian filtering technique, one attempts to construct an estimate of the posterior probability density function. The recursive propagation of the posterior density is only a conceptual solution that can be determined analytically only in a restrictive set of cases. When the analytical solution is intractable, a Monte Carlo based approach to recursive Bayesian filtering called the particle filter, which is one method that approximates the optimal Bayesian solution. In the Monte Carlo method, a set of random samples (particles) are drawn from a target distribution. In general, this distribution is unknown. We use the known distribution to denote a proposal distribution from which samples can be drawn. The main drawback of the conventional particle filter is that it uses transition prior as the proposal distribution. The transition prior does not take into account current observation data. To overcome this difficulty, the unscented Kalman filter (UKF) was proposed to generate better proposal distributions by taking into consideration the most recent observation. The particle filter based on the UKF is called unscented particle filter (UPF). UPF has better effect than particle filter. But UPF is sensitive to the decline of Signal-to-Noise. It can not be applied in the tracking of long distance.

79.2.2 Moments Estimation and Localization of Magnetic Objects Using Hopfield Neural Network

Hopfield neural network is a single layer recurrent network. The key idea of continuous Hopfield neural network optimization algorithm is that, by introducing Lyapunov energy function, topological structure of the neural network (in form of connection weight matrix) could correspond with problems to be solved (in form of objective function), which could be converted into evolutionary problems in dynamic system of neural networks. Simultaneously, the equilibrium state of the neural network should correspond with minimum point of the energy function. Finally, the process of searching the minimum point of the energy function could be equal as the evolutionary process of the neural network.

The magnetic moments estimation inverse problem can be expressed mathematically in the model:

$$Minf = \sum\nolimits_{i=1}^{n} (H_i - W_i M)^T (H_i - W_i M)$$

where, n is the number of sample points in magnetic anomaly detection, M is the magnetic moment vector ($M = M_x \boldsymbol{i} + M_y \boldsymbol{j} + M_z \boldsymbol{k}$), H_i is the observed magnetic flux density at the sample point $P(x_i, y_i, z_i)$, and W_i is the coefficient matrix with 3 rows and 3 columns, which depends on the source position (x_0, y_0, z_0), the observation point i and magnetic permeability of free space.

Construct the evolutionary equation of the Hopfield neural network

$$E = -\frac{1}{2} \sum_{i=1}^{n} \sum_{\substack{j=1 \\ j \neq i}}^{n} g_{ij} v_i v_j + \sum_{i=1}^{n} \theta_i v_i$$

where v_i is the ith output of the ith neuron, g_{ij} is connection weight of the ith neuron to the jth neuron, θ_i is the offset parameter of the ith, and n is the number of the neuron.

We choose objective function f as the energy function of Hopfield neural network. The magnetic moment vector has three components, so three neurons are needed here. Traditional gradient descent algorithm is adapted in each evolutionary step. We found out that the robustness is weak in our numerical simulation. In order to improve the robustness, evolutionary algorithm was corrected as follows.

Self-adaptive evolutionary algorithm [7]: $v_i^{k+1} = v_i^k - \tau \frac{\partial E}{\partial v_i} \frac{1}{\xi_i}$ where τ is learning coefficient, mostly $0 < \tau < 0.01$, i is code number of neurons, k is the step number of evolutionary process, the parameter ξ_i is given by $\xi_i = 3 \sum_{p=1}^{n} \left(\mathbf{w_p}(1, i) + \mathbf{w_p}(2, i) + \mathbf{w_p}(3, i) \right)$

For the localization, the areas in which the magnetic source may exist should be meshed into N elements. From definition of the neural network energy function, we can easily get the following relation $E_i \begin{cases} = E_k & if \ i = k \\ < E_k & if \ i \neq k \end{cases}$. That is, only if we apply the Hopfield neural network to estimate the magnetic moments at a correct element k, we can get the least neural network energy. It is indicated that the magnetic source position is corresponding with the least network energy element. It is obvious that, the effectiveness and accuracy of the source localization greatly depend on grid density. In order to provide a meaningful concept of the source localization, we take the weighted network energy as follows:

$$E_{Cell\ i} = E_i \left/ \sum_{j=1}^{n} H_j^T H_j \right. \quad i = 1, 2, \ldots, m$$

79.2.3 Localization by Magnetic Potential Tensor at Single Point

As concerned before, the submarine could be equivalent to a magnetic dipole. The magnetic field of the submarine at vector distance $\mathbf{r}$ is

$$\mathbf{H} = \frac{1}{4\pi} \frac{3(\mathbf{m} \bullet \mathbf{r}_0)\mathbf{r}_0 - \mathbf{m}}{r^3} \tag{79.4}$$

where $r = |\mathbf{r}|$ is the distance from the sensor to the object, $\mathbf{r}_0 = (\mathbf{r}/r)$ is the unit vector along $\mathbf{r}$. The magnetic potential tensor is [8, 9]

$$G = \begin{pmatrix} \partial H_x/\partial x & \partial H_x/\partial y & \partial H_x/\partial z \\ \partial H_y/\partial x & \partial H_y/\partial y & \partial H_y/\partial z \\ \partial H_z/\partial x & \partial H_z/\partial y & \partial H_z/\partial z \end{pmatrix} \quad \text{then } G\mathbf{r}_0 = -\frac{3}{r}\mathbf{H},$$

$$\text{so } \mathbf{r} = \begin{pmatrix} x \\ y \\ z \end{pmatrix} = -3G^{-1} \begin{pmatrix} H_x \\ H_y \\ H_z \end{pmatrix} \tag{79.5}$$

Equation (79.5) is the formula of localization by magnetic potential tensor. The magnetic moments is

$$\begin{bmatrix} M_x \\ M_y \\ M_z \end{bmatrix} = 2\pi r \begin{bmatrix} 3x^2 - 2r^2 & 3xy & 3xz \\ 3xy & 3y^2 - 2r^2 & 3yz \\ 3xz & 3yz & 3z^2 - 2r^2 \end{bmatrix} \begin{bmatrix} H_x \\ H_y \\ H_z \end{bmatrix} \tag{79.6}$$

The magnetic potential tensors could be gotten as long as five irrelevant quantities have been known. But the precision of magnetic potential tensors is confined by the measure precision and the measure distance.

79.2.4 Model of Tracking and Parameter Estimating in Passive Magnetic Detection

Through comparing those method concerned before, the approach of tracking was brought out which is a particle filter algorithm combined with Hopfield network and magnetic potential tensor. In that algorithm some optimized particles were offered by Hopfield neural network and magnetic potential tensor, the diversity of particles was improved, and the speed and precision of tracking could be ensured.

Considering of the measure information, measure model of 13 magnetic sensors was designed, which were laid in cross array (fig. 79.1).

The approach of tracking is as follows.

1. The first step: initialization $k = 1 \sim k\_before + k_1$.

- Before k_before, calculating the magnetic moments and position at each sampling point by Hopfield neural network and magnetic potential tensor.

  - When $k = 1$, the range of localization in Hopfield neural network is a large area where the submarine could be. The strategy of Hopfield neural network is that meshing the area into number A subareas, and calculating by Hopfield neural network, then ten subareas with least energy function (called type A

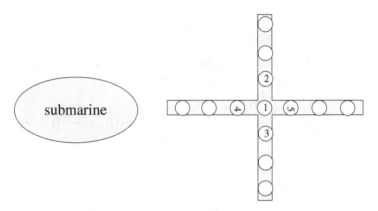

Fig. 79.1 Shelf of magnetic sensors

subarea) will be gotten. Then meshing type A subarea into number B subareas (called type B subarea), and searching by Hopfield network, an optimal type B subarea could be chosen from each type A subarea. Searching from the optimal type B subareas, one best type B subarea (called type C subarea) could be chosen. Meshing the type C subarea at gridding $2 \times 2 \times 2$ again and again, until reaching the satisfying precision, one best status vector will be chosen. Five symmetrical sensors were choosing from 13 sensors (Fig. 79.1) to calculate the status vectors of the object. Thus, we can get three status vectors. Take sensor 1 to sensor 5 as example, the magnetic potential tensors are

$$
\begin{cases}
\frac{\partial Hx}{\partial x} = \frac{Hx_5 - Hx_4}{2d}; \ \frac{\partial Hx}{\partial y} = \frac{\partial Hy}{\partial x} = \frac{Hy_5 - Hy_4}{2d}; \\
\frac{\partial Hx}{\partial z} = \overline{\partial x} = \frac{Hz_5 - Hz_4}{2d}; \\
\overline{\partial y} = \frac{Hy_3 - Hy_2}{2d}; \ \overline{\partial z} = \frac{\partial Hz}{\partial y} = \frac{Hz_3 - Hz_2}{2d}.
\end{cases}
$$

– When $k = 2 \sim k\_before$, the range of localization in Hopfield neural network is a symmetrical area whose center is the estimated submarine position at time $k-1$. The size of that area is decided by the estimated speed of the submarine. The area should contain the real time position of the submarine. And the strategy of Hopfield neural network is that meshing the area into number D subareas, and calculating by Hopfield neural network, then one subarea with least energy function (called type D subarea) will be gotten. Meshing the type D subarea at gridding $2 \times 2 \times 2$ again and again, until reaching the satisfying precision, one best status vector will be chosen.

• When $k = k\_before$, 4 status vectors will be generated. Averaging the k_before magnetic moments vectors gotten by those two methods, and adding the position vectors of time k_before, 8 initial status vectors will be gotten.

• When $k = k\_before + 1 \sim k\_before + k_1$, calculating the status vectors of each sampling point by those two methods, thus 21 status vectors could be gotten by Hopfield network and 3 status vectors could be gotten by magnetic potential tensor. Then updating the status vector ($x_k^i = f(x_{k-1}^i)$), and the new status vectors will be gotten by combining the updated vectors with the calculated vectors. Then the follow-up calculating should be carried out. Repeating those steps, and cumulating the vectors, thus $N_1 = 24\ k_1$ initial status vectors will be gotten, adding with the 8 vectors gotten at $k = k\_before$, there will be $N = 24\ k_1 + 8$ initial status vectors until time $k\_before + k_1$. The range of localization in Hopfield neural network is a symmetrical area whose center is the estimated submarine position at time $k-1$ (called type E subarea). And the strategy of Hopfield neural network is that meshing the type E subarea at gridding $2 \times 2 \times 2$ again and again, calculating by Hopfield neural network, one optimal status vector will be gotten at each iterative, and 20 status vectors could be gotten. Then choosing one best status vector from those 20 vectors, one vector group of 21 vectors will be gotten.

2. The second step: calculating the importance weights.

   When $k \geq k\_before + k_1$, calculating the status vectors of each sampling point by the two method (21 status vectors could be gotten by Hopfield network and 3 status vectors could be gotten by magnetic potential tensor), and new $N + 24$ particles will be gotten ($x_k^i \in q(x_k|B_{1:k})$, $i = 1, 2, \cdots, N + 24$), then calculating and updating the importance weights by the formula $\widehat{w}_k^i = w_{k-1}^i \frac{p(B_k|x_k^i)p(x_k^i|x_{k-1}^i)}{q(x_k^i|B_{1:k})}$.

   The noise of measurement abides by the Gauss formula.

3. The third step: eliminating 24 particles of least weights from the $N + 24$ particles, thus the total number of particles is always N. $x_k^{ii} \in q(x_k|B_{1:k})$.

4. The forth step: normalizing the importance weights by $\widehat{w}_k^i = w_{k-1}^i / \sum\limits_{i=1}^{N} w_k^i$.

5. The fifth step: re-sampling.

   For avoiding the degeneracy of particles, re-sampling by the method of remains errors will be applied. In which, the particles with low weights were eliminated and the particles with high weights were copied, thus N new particles $\tilde{x}_k^i$ could be gotten.

6. The sixth step: calculating the estimated values $\hat{x}_k = \sum\limits_{i=1}^{N} \widehat{w}_k^i x_k^i$, then going back to the second step.

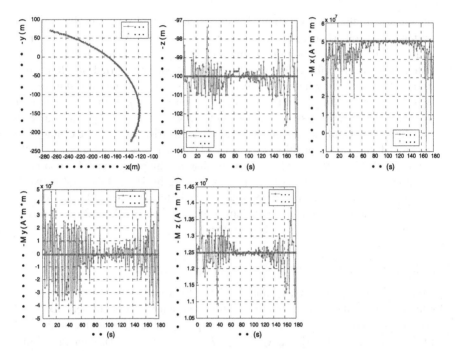

Fig. 79.2 Result of simulated experiment (position and magnetic moment)

79.3 Simulated Experiment and Magnetic Ship Model Experiment

For validate that algorithm, the simulated experiment and magnetic ship model experiment were designed as follows.

The shelf of magnetic sensors is as Fig. 79.1, the distance of adjacent sensor is 5 m. The magnetic object is a dipole of magnetic moments $(m_x, m_y, m_z) = (5 \times 10^7, -5 \times 10^5, 1.25 \times 10^7)$ A. m^2. The object is moving towards the sensors at speed $v = 2$ m/s with the initial position $(-150, -260, -100$ m) and initial course 60°. The change rate of course is 0.5°/s. The depth of object is fixed. The sampling period is 1 s. The time of moving is 200 s. The status noise is Gauss noise with the average 0 and variance 0.01. The measure noise is Gauss noise with average 0 and variance 1n T. The Signal-to-Noise is 200. The number of particles is 248, and $k\_before = 11$, $k_1 = 10$.

The result of simulated experiment is shown as Fig. 79.2.

From Fig. 79.2, it is shown that the tracking reached the steady state quickly, and the precision is high, the performance of that algorithm is good.

Magnetic ship model experiment was conducted based on the submarine magnetic detection and magnetic defence of military port. The measure equipment

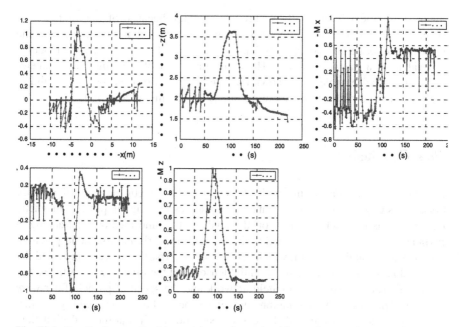

Fig. 79.3 Result of magnetic ship model experiment (position and magnetic moment)

was 7,100 MMS triaxial magnetic field measure system, whose resolving power is
1nT. The measure array is composed of 3 triaxial magnetic sensors. The optics
location system is TCA1800, whose precision is 0.001 m.

A shelf of magnetic sensors like Fig. 79.1 was made, which was laid upon the
ship model. Submarine model A was experiment object. The model was moving
on nonmagnetic car along the nonmagnetic railway. The magnetic field of submarine model was collected by measure system, and the position was collected by
TCA1800.

The origin of axis is down the center of two railways. The center sensor was
laid at the origin of axis. The distance of adjacent sensor is 0.2 m. The course of
ship model is magnetic north. The distance between the sensor plane and the ship
model is 2 m. The moving range of ship model is −12 to 12 m, the speed of ship
model is 0.1 m/s, and the sampling period is 1 s. There are 241 sampling points.
The Signal-to-Noise is 100.

The number of particles is 248, and $k\_before = 11$, $k_1 = 10$.

Considering the secrecy, the values of magnetic moments have been normalized, and the result of magnetic ship model experiment is as follows.

Form Fig. 79.3, it was shown that the algorithm reached the steady state at
about the 130th seconds. The error of the final position was increasing because the
Signal-to-Noise will increase with the increasing distance. Nevertheless, the error
of horizontal is less than 1.2 m, the error of vertical is less than 2.5 m, and the

magnetic moment is gradually steady. To sum up, because of the confining of space and magnetic sensors, the distance between the submarine model and the sensors can not be far enough, so it has a certain error that the submarine model is equivalent to a magnetic dipole. Therefore, the result of tracking and parameter estimating is acceptable. It is shown that the algorithm brought before is valid in the tracking and parameter estimating in passive magnetic detection.

79.4 Conclusion

Aiming at the submarine magnetic detection and magnetic defence of military port, the study on the tracking and parameter estimating in passive magnetic detection was conducted. Based on the analysis to some method on tracking, a particle filter algorithm combined with Hopfield network and magnetic potential tensor was brought out. In which, some optimal particles were provided by Hopfield network and magnetic potential tensor, and the diversity of particles was increased. The precision of Hopfield network and the speed of particle filter were utilized. The result of simulation and experiment shows the validity of that algorithm. This algorithm will have some significance in martial and civil field such as submarine magnetic detection, magnetic defence of military port, unexploded ordnance detection and prospecting for ore.

References

1. McFee JE, Das Y (1981) Determination of the parameters of a dipole by measurement of its magnetic field. IEEE Trans antennas propag Ap-29-2(3):282–287
2. Mcaulay AD (1977) Computerized model demonstrating magnetic submarine localization. IEEE Trans aerospace electron syst 13(3):246–254
3. Billings SD, Herrmann F (2003) Automatic detection of position and depth of potential UXO. In: UBC Geophysical Inversion Facility, 6339 Stores Rd, Vancouver, BC, V6T1Z4, Canada
4. Salem A, Ushijima K (2001) Automatic detection of uxo from airborne magnetic data using a neural network. Subsurf Sens Technol Appl 2(3):191–213
5. Salem A, Hamada T et al (2005) Detection of unexploded ordnance (UXO) using marine magnetic gradiometer data. Explor Geophys (36):97–103
6. Davis K, Li Y, Nabighian M (2005) automatic detection of UXO magnetic anomalies using extended Euler deconvolution. Automatic detection of UXO magnetic anomalies, 1–4
7. Zhou G, Xiao C, Liu S (2008) Magnetic moments estimation and localization of a magnetic object based on Hopfield neural network. In: The Proceeding of the 6th international conference on electromagnetic field problems and applications-I (21). San Antonio: TSI Press Series. 19–23
8. Nara T, Suzuki S, Ando S (2006) A closed-form formula for magnetic dipole localization by measurement of its magnetic field and spatial gradients. IEEE Trans magn 42(10):3291–3293
9. Changhan X, Huahui H (1997) Inverse problems in magnetic field measurement and the inverse transformation. J Huazhong Univ Sci Tech 25(8):81–82 (in chinese)

Chapter 80
Player Detection Algorithm Based on Color Segmentation and Improved CamShift Algorithm

Yanming Guo, Songyang Lao and Liang Bai

Abstract Aimed at the most proposed methods that detecting and tracking moving object cannot receive a good segmentation of the player in dynamic scenes, the chapter put forward an algorithm, based on color segmentation and the CamShift algorithm, to detect the tennis player. Firstly, apply a supervised clustering binary tree to automatically detect the target's area. Secondly, take the detected area as the initial tracking window, track the target by using the improved CamShift algorithm. On the basis of accurate tracking, the chapter raises a new extraction method of the player, taking use of the characteristic that mostly the court's color is simple. The method raises the thought that extract the player by the color feature, and it combines the CamShift algorithm with the difference in frame to overcome the tracking-loss problem. Experimental results show that our algorithm is effective in detecting the player in dynamic scenes.

Keywords Sports video · Color segmentation · CamShift · Player detection · Player tracking

80.1 Introduction

The accurate detection and tracking of the player in sports video are very important, it can not only provide necessary information for high-level processing like motion analysis, but also can provide effective technical support for the scientific training of player. However, owing to the frequent change of the sports video scenes, the accurate detection and tracking of the player is rather difficult.

W. Lu et al. (eds.), *Proceedings of the 2012 International Conference on Information Technology and Software Engineering*, Lecture Notes in Electrical Engineering 210, DOI: 10.1007/978-3-642-34528-9_80, © Springer-Verlag Berlin Heidelberg 2013

Fig. 80.1 Motion detection. **a** The moving areas. **b** The like-target areas. **c** The finial target area

Currently, the common detection methods include: interframe difference method [1], background subtraction method [2], statistical method [3], optical flow analysis [4]. These methods perform well in fixed scenes, and mostly used for monitoring, but seldom be applied in sports video.

The key issue for moving target's detection and tracking in dynamic scenes is to get rid of the impact which is brought by the scene's change, and to track the moving target instantly. At present, the common tracking methods [5] are based on model, region and character. The CamShift algorithm raised by literature [6] is robust in tracking the non-rigid object of irregular movement by taking advantage of the color features. However, the CamShift algorithm is a semiautomatic method, the initial tracking window should be set artificially, and it may result in the tracking-loss problem in a long-time tracking. Literature [7] raises a method by comparing the current pixel with the threshold, literature [8] raises a method by combining the CamShift algorithm with difference in frame, these two methods can automatically track the object, literature [9, 10] prevent the tracking-loss problem by predicting the trail next time. All of these methods can solve the tracking-loss problem in some extent. However, there is still no correct mechanism when the tracking lost in these methods.

The chapter improves the tracking process by combining the CamShift algorithm with difference in frame, and raises a player detection algorithm by combining the color segmentation and the improved CamShift algorithm; the algorithm performs well in tennis video.

80.2 Motion Detection

Motion detection, whose purpose is to detect the area which contains the player, aims to offer an initial tracking window to achieve the automatic tracking of the player, the steps are as follows:

1. Construct a decision-making binary tree, detect the video scenes by using interframe difference method, and divide the scene into stationary part and non-stationary part, the non-stationary part is the area of interest, covered with rectangular boxes, as Fig. 80.1a shows.

2. Taking into account the size of the player, we can make further processing of the interested area. Calculate the area of each rectangular box, set an area threshold to further divide the interested area into non-target areas, whose areas are below the area threshold, and like-target areas, whose areas are above, the like-target areas are preserved, as Fig. 80.1b shows.
3. Having the prior knowledge that the player wears a red T-shirt, we make a histogram statistics of the like-target areas, and choose the area that has the largest proportion of red as the finial motion region, as Fig. 80.1c shows.

By the motion detection of the video, we get the area which contains the player, this provide an initial window for the player tracking in the following text.

80.3 Player Tracking

The most common tracking methods are the MeanShift [11] algorithm and the CamShift [12] algorithm.

The MeanShift algorithm is the core of the CamShift algorithm, and it achieves the tracking purpose by searching the peak point in the probability distribution. The method lacks necessary updating mechanism, so it does not perform well during a long tracking. Besides, the size of the tracking window remains constant throughout the tracking process, also may result in the phenomenon of mismatch when the scale changes.

The CamShift algorithm is the improvement of the MeanShift algorithm, it can dynamically update the tracking window's size and position when iteratively using the MeanShift algorithm, which allows it to better track the non-rigid objects in irregular movement, and has a good noise immunity.

During operation, the initial center position of the next frame tracking window is the same as the center of the current tracking window, the size and the orientation can be calculated by the following formula [13]: suppose $I(x, y)$ is the pixel value of position (x, y) in the probability distribution, the value of x, y are in the range of the tracking window.

Calculate the second moment of the tracking window:

$$M_0 = \sum_x \sum_y I(x, y), \quad M_1 = \sum_x \sum_y x^2 I(x, y)$$
$$M_2 = \sum_x \sum_y y^2 I(x, y), \quad M_3 = \sum_x \sum_y xy I(x, y) \tag{80.1}$$

Suppose:

$$a = \frac{M_1}{M_0} - x_c^2, b = 2\left(\frac{M_3}{M_0} - x_c y_c\right), c = \frac{M_2}{M_0} - y_c^2 \tag{80.2}$$

(x_c, y_c) is the centroid of the tracking window, the length l and width w of the next tracking window are:

Fig. 80.2 The tracking effect
as scene switches, using
CamShift

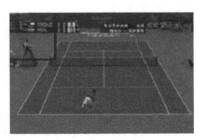

$$l = 2 * \sqrt{\frac{(a+c) + \sqrt{b^2 + (a-c)^2}}{2}}$$

$$w = 2 * \sqrt{\frac{(a+c) - \sqrt{b^2 - (a-c)^2}}{2}}$$

(80.3)

The orientation θ is gained by the following formula:

$$\theta = \frac{1}{2} \tan^{-1}\left(\frac{b}{a-c}\right)$$

(80.4)

Taking use of the CamShift algorithm to track the player can reduce the probability of loss significantly.

However, both the MeanShift algorithm and the CamShift algorithm are semiautomatic tracking algorithms, the initial tracking window needs to artificially set, and once the tracking-loss happens, the algorithm will obtain the color information in the wrong tracking window as the feature on which the following track based, thus making it difficult to relocate the correct target.

The effect of the CamShift algorithm when the scene switches is shown in Fig. 80.2:

Owing to this, a method of combining the CamShift algorithm and motion detection to track player is raised in the chapter, the method improves the traditional tracking method in two aspects: Firstly, it gets the initial tracking window automatically, but not semi-automatically; secondly, it proposes a correction mechanism in the tracking process, which ensures the effective target tracking.

Generally, when we use the CamShift algorithm to track, the tracking window will be the region which has the largest probability density, such as the player's coat region in this video. However, the region that motion detection gets is larger, which contains the whole player. Considering this, the chapter firstly use the motion detection method to gain the target region, and then judge whether the tracking window is in this region, if yes, we can assume that we have got the right tracking, the tracking will continue. If no, two situations may have occurred: (1) the tracking is not lost. The situation is caused by the shake of the camera or some, which make the motion detection cannot get the correct target region (2) the tracking is lost.

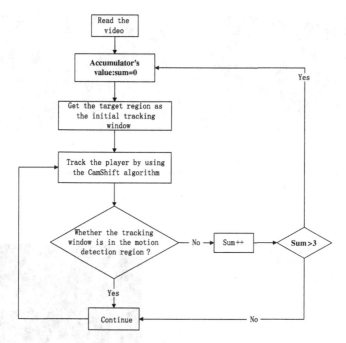

Fig. 80.3 Flow chart of the tracking method

To solve the problem of tracking lost, the chapter set an accumulator, the initial value is 0, if the tracking window is not in the motion detection region, the accumulator increases by 1. Generally speaking, if the situation is caused by the shaking of the camera or some, after one or two frames, the scene will be stabilized, so if more than three consecutive frames the tracking windows are not in the target regions, that is, the value of the accumulator exceed 3, we can consider that the tracking process has erred, we need to adjust the tracking window by taking the region that motion detection gained as the tracking window, and set the accumulator's value to 0.

The tracking flow chart of the player is shown in Fig. 80.3.

Using the algorithm mentioned above can not only obtain the correct initial window, but also ensure the real-time updates of the tracking process, thus it can solve the tracking-loss to a certain extent.

The effect of method which combines the CamShift algorithm and motion detection when the scene switches is shown in Fig. 80.4a.

In order to facilitate the use of the player's extraction method, the tracking window need to contain the whole player, so the chapter artificially expand the tracking window to the surrounding as the target region, based on the prior knowledge of the player's body size, as Fig. 80.4b shows.

It can be seen that, the method of combining the CamShift algorithm and motion detection can solve the tracking-loss problem caused by the factors such as

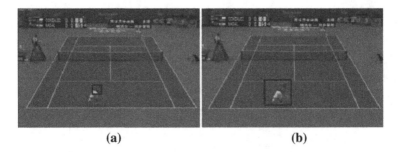

(a)   (b)

Fig. 80.4 The tracking effect as scene switches. **a** The improved CamShift algorithm. **b** The tracking window after extended

Fig. 80.5 The effect of the color-subtraction method

that the target itself is complex, the scene is complex, the camera shakes, to a certain extent, and receives a good effect in the practical application.

80.4 Player Extraction

Having got the correct target region, the detection range has decreased greatly.

Generally speaking, in the sports video, the color of the playground is simple, and makes a big difference with the player's clothing. Based on this feature, the chapter adopts a method to extract the player by color subtraction:

For the certain target region, get the main color by color histogram statistics, considering the impact of light, we can expand the pixel value by some points, and set the expanded color as the main color of the playground. Judge whether the pixel in the target region is in the range of the main color, if yes, then set the pixel white, else set black.

The effect of the method is shown in Fig. 80.5.

The figure shows that the method can accurately extract the player.

Table 80.1 The experimental result

Experimental video	Frame number	Precision ratio of motion detection (%)	Precision ratio of MeanShift algorithm (%)	Precision ratio of CamShift algorithm (%)	Precision ratio of new algorithm (%)
Group A	100	94	62	86	95
Group B	100	96	76	91	98

The detection algorithm proposed in the chapter could solve the problem which caused by factors such as light changes, scene switches, camera shakes, to a certain extent, and achieves a good effect in practice.

80.5 Experimental Studies and Results

Both the detection and tracking method proposed in the chapter can receive a good effect, the player's accurate tracking and extraction is based on the combination of the two aspects. The player's detection is consist of two parts: motion detection and the player's extraction. Motion detection is the initial part, it not only provides an initial tracking window for the tracking, but also provides a mechanism to correct the wrong tracking. The exact tracking is the foundation of the player's extraction, although the scene in the sports video may be complex, the region in the playground is simple, this feature ensures the success of the extraction method.

To better analyze the algorithm proposed in the chapter, we adopt formula (80.5) to measure the efficiency [14]:

$$\text{Precision ratio} = \frac{\text{the correct regions have been detected/tracked}}{\text{all of the regions have been detected/tracked}} \quad (80.5)$$

The system's experimental data is from Nadal, Federer and other's tennis video, divided into two groups: Group A and Group B. Group A is the video which has the scene switching, while Group B has not. Both the two groups set the player near the camera as the target to track and detect, the video scene may shake.

According to experimental result, we can get the data as Table 80.1 shows.

The experimental indicates that, whether the video has scene switching or not, the detection and tracking algorithm raised in the chapter can have a better performance in practical application.

80.6 Conclusion

The accurate detection and tracking of the player are the key steps in analyzing the sports video. On one hand, the chapter raises a mechanism to correct the tracking process when the tracking is lost, which is the mean innovation of this chapter. On the other hand, the chapter raises a new thought which combines the motion detection, player tracking and player extraction to make a good performance in handling the problems caused by factors such as scene switching, camera shaking and so on. However, the algorithm also has its limitations, the player's extraction relies too much on the accuracy of the player's tracking, and it has a high demand of the court, the algorithm requires the court is strictly simple, and the color of the court should make a difference with the color of the player's clothe. We will further study an algorithm which could overcome the shortages and could detect and track the player accurately in a complex scene in our next work.

Acknowledgments This work is supported by the National Natural Science Foundation of China under grants No. 60902094, and "12.5" Project of China under grants No. 40901040202.

References

1. Lipton AJ, Fujiyoshi H, Patil RS (1998) Moving target classification and tracking from real-time video[C]. WACV'98, Proceedings of the Fourth IEEE Workshop on Applications of Computer, Princeton, NJ, 8-14.
2. Jong BK, Hang JK (2003) Efficient region-based motion segmentation for a video monitoring system. Pattern Recogn Lett 24(1):113–128
3. Horprasert T, Harwood D, Davis LS (1999) A statistical approach for realtime robust background subtraction and shadow detection. In: Proceedings of IEEE frame rate workshop, Kerkyra, Greece, p 1–19
4. Horn B, Schunck B (1981) Determining optical flow. Artif Intell 17:185–203
5. Wang L, Hu WM, Tan TN (2002) The summary of human motion's visual analysis. Chinese J Comput, 25(3):225–237 (in Chinese)
6. Bradski GR, Clara S (1998) Computer vision face tracking for use in a perceptual user interface. Intell Technol J 2:1–15
7. Li G, Jiang J (2011) Object tracking method based on improved Camshift algorithm. Ind Control Comput 24(10):52–54 (in Chinese)
8. Chu HX, Song QC, Wang XF etc (2008) Target tracking algorithm based on Camshift algorithm combined with difference in frame. J Projectiles Rockets Missiles Guidance, 28(3):85–88 (in Chinese)
9. Li HH (2010) Improved CAMShift algorithm of target tracking. Mod Electron Tech 33(16):106–108 (in Chinese)
10. Wu HH, Zheng XS (2009) Improved and efficient object tracking algorithm based on CAMShift. Comput Eng Appl 45(27):178–180 (in Chinese)
11. Bradski G, Kaebler A (2005) Learning OpenCV. Qinghua University, Beijing
12. Boyle M (2001) The effects of capture conditions on the Camshift face tracker[R]. Alberta, Canada: Department of Computer Science, University of Calgary
13. Wu H, Hua QY (2008) The algorithm for athlete's detection and tracking in sport video. Comput Eng 34(19):230–235 (in Chinese)
14. Yilmaz A, Javed O, Shah M (2006) Object tracking: a survey. ACM assoc comput mach 38(4):1–45

Chapter 81
Reputation Mechanism Model Research Based on the Trusted Third-Party Recommendation

Guangqiu Huang and Weijuan Zhao

Abstract In this paper, by analyzing the existing reputation mechanism model and the key entities in the credit system, a reputation mechanism model based on the trusted third-party recommendation is put forward to deal with the phenomenon of honesty deficiency. Different from the current passive situation of the query agent understanding the target entity, the query agent will ask initiatively its third-party recommendation for the reputation value of the target entity in this model, which may reduce the malicious deception of the target entity to some extent. In addition, the model combined the different background factors in the query process, the recommended backgrounds and the time decay factors. It can make full use of the effective message from the third-party recommendation and provide more reliable credibility of the aggregate values.

Keywords Trusted · Third-party recommendation · Reputation model · Calculation of reputation value

81.1 Introduction

In recent years, the development of e-commerce is rather rapidly and become an important consumer in people's lives. However, because of the openness, dynamic, virtual and anonymous of the environment of e-commerce transaction,

G. Huang (✉) · W. Zhao
School of Management, Xi'an University of Architecture and Technology,
Xi'an 710055, China
e-mail: 735844454@qq.com

W. Zhao
e-mail: zwj592051078@126.com

W. Lu et al. (eds.), *Proceedings of the 2012 International Conference on Information Technology and Software Engineering*, Lecture Notes in Electrical Engineering 210, DOI: 10.1007/978-3-642-34528-9_81, © Springer-Verlag Berlin Heidelberg 2013

online trading frequently failed due to lack of the credibility and led to lots of constraints. So all kinds of experts and scholars went in for the research of e-commerce reputation mechanism and proposed some mathematical models such as trust measure model proposed by Marsh [1] and trust evaluation model by A. Adul-Rahman. Gradually some models like complain-based, Bayesian Network-based, social mechanisms-based and recommendation trust-based models [2] have been proposed and achieved a lot. In the domestic, also some well known universities like Peking University had invested in the research. The current reputation model involved trust description, trust calculation, recommended trust derivation and trust consolidated according to trust definition of Gambetta [3], on which the reputation models like BBK [4], Jøsang reputation model [5] and Poblano model [6] are based.

The existing models have provided some references to solve the problem of credit absence in e-commerce, and made some contribution to the development of the network economy. But with the development of online transactions, there were some problems: (1) The data of the reputation model were come from each other's transaction evaluation view, which would be likely to make some dishonest and malicious behavior; (2) Although the background of the third-party recommend view in some of the credibility model were involved, but it didn't give the specific model; (3) The existing model had established in the entire network information table, which cost too much space and time; (4) Timely and effective information is essential in today's society, but it wasn't be well used in current models.

There are some solutions to solve the problems in this paper: (1) Use the data of the trusted third-party agent's view as the main source of credibility data to prevent malicious fraud. (2) Use the background dissimilarity factor to differ the different trading partners and provide a more accurate credibility view; (3) Let every agent establish the relevant information of oneself only; (4) Establish a influence mechanism of unexpected factors and make full use of the timely information from the network to help the agents make decisions.

81.2 Reputation System Mechanism Model

Credibility had been widely used in various disciplines and systems. Scholars in various fields made different definitions. According to the definitions of key entities such as "the seeking agent", "the third-party recommended agent" and "the target entity", the Ref. [7] proposed the advanced reputation concept, on which this paper is based to discuss the reputation relation and its model.

81.2.1 The Credibility Relation Analysis

In real world, the credibility relations become more complex due to the participation of third-party agents. According to the definition, reputation relation was defined as a connection among the seeking agent, the third-party agent and the target entity. The seeking agent got the information of target entity by querying the third-party agent. The third-party recommended agent sent its recommended view to the seeking agent from the credit query. And the target entity established its relations with the other agent basing on the activities. The symbols were defined as follows:

The seeking agent x: $\otimes$ The third-party agent y:$\bigcirc$ The target entity z:$\bullet$

According to the definition and mission activities in real world, the reputation relations between the various entities of the reputation system was as follows (Fig. 81.1):

From the diagram, the seeking agent determined the target entity and make decisions. There was a credit relation between the third-party agent and the target entity. According to this relation the third-party agent sent its information or view to the target entity. Finally the seeking agent calculated all the recommended value and adjusted the value according to the unexpected factors mechanism.

In reputation system, there are three relations between various credit entities such as the recommended relation between the seeking agent and the third-party agent, the credit relation between the target entity and the third-party and the credit query relation between the query of reputation and the third-party agent.

81.2.2 Reputation Model Establish

81.2.2.1 The Model Framework Definition

In order to describe the recommended relation, the credit relation and the credit query relation better in the reputation system and establish the more accurate reputation model, the definition is as follows:

Fig. 81.1 The relations of trust entities network diagram

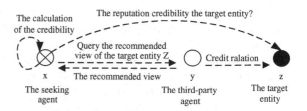

Definition 1 The credit system environment is described with a six-tuple:

$$SYS = (R, RS, RL, T, B, TR, OP)$$

In the formula, R represent the collection of all the credit entities including the Agent (short for A), the Product (P) and the Service (S); RS represent the collection of the credit query entities; RL represent the various relationship, expressed as: $RL = \{RL_1, RL_2, RL_3\} = \{$query of credit relation, recommended relation, the credit relation$\}$; T represent the time period that the trade have happened, expressed as: $T = \{t_1, t_2, ..., t_n\}$; B represent the background that the trade have happened, expressed as: $B = \{b_1, b_2, ..., b_n\}$; TR represent the recommended value or view in some time and background, and the value was described in the credit level of the Ref. [7], expressed as $TR = \{-1, 0, 1, 2, 3, 4, 5\}$; OP represent the way that the seeking agent get the recommended value, expressed as: $OP = \{$the first-hand view, the second-hand value, the third-hand value$\}$, and the third-hand view include all the view except the first-hand and second-hand view in this formula.

Definition 2 In the SYS, the query of credit relation RL_1 is described with a five-tuple as follows:

$$RL_1 = \{ <x, y, t, b, tr > \}$$

In the formula, $x \in A$ and $y \in RS$ and $x \neq y$ and $t \in T$ and $b \in B$ and $tr \in TR$, it expressed that: the third-party agent x get a credit query RS and return recommended value tr, and provide the time period t, background b and the delivery times op(op $= \{1,2,3\}$)w the trade happened.

Definition 3 In the SYS, the recommended relation RL2 is described with a five-tuple as follows:

$$RL_2 = \{ <x, y, t, b, tr > \}$$

In the formula, $x \in A$ and $y \in A$ and $x \neq y$ and $t \in T$ and $b \in B$ and $tr \in TR$, it expressed the seeking agent x's credit value tr to the third-party agent y in the certain time period t and certain background b.

Definition 4 In the SYS, the credit relation RL_3 is described with a five-tuple:

$$RL_3 = \{ <x, y, t, b, tr > \}$$

In the formula, $x \in A$ and y R and $x \neq y$ and $t \in T$ and $b \in B$ and $tr \in TR$, it expressed that the third-party agent x's credit value tr to the target entity y in the certain time period t and certain background b.

In the SYS, the different reputation relation value tr have the different meanings. In order to describe the model better, the $Rc(x, y, t, b)$ was be used to express the value of RL_1 relation, $Tc(x,y,t,b)$ to express the value of RL_2 relation, and $Rt(x,y,t,b)$ to express the value of RL_3 relation.

Definition 5 In the *SYS*, the credit agent is described with a three-tuple including kinds of reputation relation:

$$x = <\{rl_1\}, \{rl_2\}, \{rl_3\}>$$

In the formula, $rl_1(rl_1 \in RL_1)$ is a collection of query of credit relation which is relevant to x; $rl_2(rl_2 \in RL_2)$ is a collection of recommended relation which is relevant to x; $rl_3(rl_3 \in RL_3)$ is a collection of credit relation which is relevant to x.

Definition 6 In the *SYS*, the unexpected factor *Reb* of the target entity in a series trades is a three-tuple:

$$Reb = \{<z, n, detail>\}$$

In the formula, z express the unexpected behavior of the target entity in a series trade, and that's to say that the target entity is unhonest, n express the times of the behavior, and the *detail* express the detail situation of the unexpected behavior.

Definition 7 In the *SYS*, the new information *Newin* from the internet except in the normal trade is described with a two-tuple:

$$Newin = \{<z, c, detail>\}$$

In the formula, z express the target entity which is relevant to the collected information, c express the degree of the influence of the target entity's credit value, and *detail* express the detail situation of the information.

81.2.2.2 The Credibility Activities Process Description

The different reputation entities know each other in the internet by kinds of ways. And the way is like the process that the people understand things around us in the real life. They all gradually connect with each other in different methods, and know each other better further. Finally it will form a network of interpersonal relation. According the process and the former definitions, the credit entities activities timing diagram is as Fig. 81.2:

In the figure, x, y and z represent differently the seeking agent x, the third-party agent y, and the target entity z. $Rt(y, z, t, b)$ and $Rc(y, z, t, b)$ represent differently the credit value and the recommended credit value in the certain time period t and certain background b. $Tc(y, z, t, b)$ and $Tc'(y, z, t, b)$ represent differently the credibility values of the reputation view before and after the trade in the certain time period t and certain background b. T_{RA} and T'_{RA} represent differently the reputation measured value after calculated and the direct credit value after the transaction (Fig. 81.2).

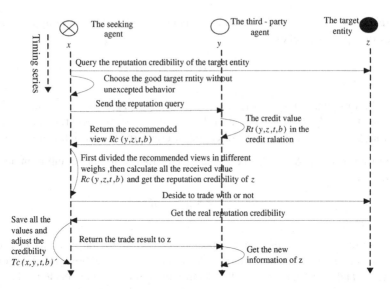

Fig. 81.2 Credit entities activities timing diagram

81.2.2.3 The Credibility Model Establish

After analysis of the credit relation, the factors must be considered to establish the credibility model. They are: (1) The time variability of the credibility views; (2) the background similarity between backgrounds the third-party provide the recommended view and that of the future trade; (3) How to deal with the weights of the reputation views received by different ways and the unexpected behaviors of the target entity; (4) The influence of the new information from the internet.

81.2.3 The Time Decay Factor

Obviously the longer the time period lasts, the lower the credibility value of the reputation view. And the time of the e-commerce credit is determined by the dynamic of the e-commerce. The credibility decay of the reputation views is as following formula [7] in this paper:

$$T(y_i, z, t, t_i) = e^{\frac{-(t-t_i)}{N}}$$

In this formula, N is the decay rate factor according to the different background, and the rate is becoming bigger as the value is becoming smaller.

81.2.4 Background Dissimilarity

In e-commerce activities, different backgrounds in the former and future trans-actions will influence the final credit of recommended views. The credit of the recommended view is higher in two similar backgrounds. However, the credibility is very low on the contrary, and maybe we should refuse the recommended views.

All the backgrounds of the different transactions are divided into n categories, expressed by $B = \{b_1, b_2, \ldots, b_n\}$. Different dissimilarity is digital, and marked by the number from 0 to M. The credit value of the background dissimilarity in two different backgrounds is as follows:

$$B(y_i, z, b, b_i) = 1 - \frac{|b_i - b|}{M}$$

In the formula, the value in similar background is 1. It expressed that the view in this background is fully adopted. If the value is 0, then it's hardly adopted.

81.2.5 The Weights Determined of Recommended Views

In the reputation query activities, the recommended views from the third-party agent are sent to the seeking agents in different ways. Some are sent directly and the other is indirectly. When the recommended views are across beyond three third-party agents, the recommended views should be treated as the third-hand views. The different situations of recommended views are as follows: (Fig. 81.3)

In this diagram, the recommended view is across only one third-party agent that have trade with the target agent directly (Fig. 81.4).

In this diagram, the recommended view is across two third-party agents and the last third-party agent have trade with the target agent directly (Fig. 81.5).

In this diagram, the recommended view is across three third-party agents and the last third-party agent have trade with the target agent directly.

By analysing the three situations of recommended views, we can get the following formula:

$$T = \sum_{i=\alpha, \beta, \gamma} \sum_{i=1}^{l} Tc(x, y_i, t, b) Rc(y_i, z, t_i, b_i) T(y_i, z, t, t_i) B(y_i, z; b, b_i)$$

In this formula, α, β and γ represent differently the different weighs of the recommended views, α, β, γ, $\in [0,1]$ and $\alpha + \beta + \gamma = 1$.

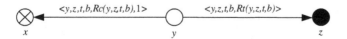

Fig. 81.3 The first view situation

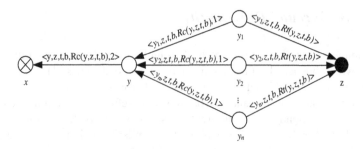

Fig. 81.4 The second view situation

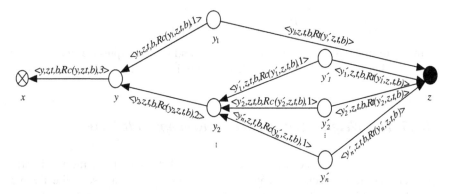

Fig. 81.5 The third view situation

81.2.6 Influence Mechanism of Unexpected factors

The unexpected factors are the unexpected behaviors and the new information from the internet in this paper. If the target entity have the unexpected behaviors, the entity will be added to the collection *Reb*, and *Reb* = {*reb1,reb2,...,rebn*}. Every *reb1(reb1* ∈ *Reb)* include the target entity z, the time n and the detail of the behavior. Also the new information that is relevant with the target entity z from the internet will be added to the collection *Newin*, and *Newin* = {*newin1, newin2,...,newinn*}. Every *newin1(newin1* ∈ *Newin)* include the target entity z, influence degree c and the detail. The final influence $\bar{c}$ is calculated as follows:

$$\bar{c} = f(c1, c2, \ldots, cn)$$

So the ultimate credit value of the reputation aggregate is as follows:

$$T_{RA} = T \times \bar{c}$$

81.2.7 Reputation Credibility Adjustment

The reputation credit will be changed after every transaction, and we should adjust the value according to the different results after the transaction. In the paper, the following formula was be used to adjust the reputation credit. The reputation credit after the transaction is T'_{RA}, the former calculated value is $Rc(y, z, t, b)$, and the η and ε are the parameters. If $|T'_{RA} - Rc(y, z, t, b)| > \varepsilon$, then

$$Rc'(y, z, t, b) = \eta \times Rc(y, z, t, b) - (1 - \eta) \times 5$$

If $|T'_{RA} - Rc(y, z, t, b)| < \varepsilon$, then

$$Rc'(y, z, t, b) = \eta \times Rc(y, z, t, b) - (1 - \eta) \times 5$$

81.3 Conclusion

This paper took the problem of credit absence in e-commerce for example and based on the existed reputation mechanism models. It analyzed the relations among all kinds of credit entities and the various problems in the previous models. At last it proposed a reputation mechanism model research based on the trusted third-party recommendation to solve all the problems.

References

1. Marsh S (1994) Formalising trust as a computational concept. PhD Thesis. University of Stirling, Seotland
2. Chen H, Yang et al (2004) Maze: a social peer-to- peer network, In: Proceedings of the international conference one-commerence for dynamic-business (CEC-EAST,04), Beijing, China
3. Abdul-Rahman A, Hailes S (2000) Supporting trust in virtual communities. In: Proceedings of 33rd Hawaii international conference on system sciences. Hawaii, pp 1769–1777
4. Beth T, Borcherding M, Klein B (1994) Valuation of trust in open net-work. In: Proceedings of European symposium on research in security (ES-ORICS).Springer, Brighton, pp 3–18
5. Jøsang AA (1998) Subjective metric of authentication. In: Quisquater J (ed) Proceedings of the ESORICS'98. Springer, Louvain-la-Neuve, pp 329–344
6. Chen R, Yeager W (2000) Poblano a distributed reputation model for peer-to-peernetworks. Technical report. Sun Microsystems, USA
7. Chen D, Zheng X, Gan H (2002) trust and reputation for service-oriented environment: technologies for building business intelligence and consumer confidence. ZHE-JIANG University Press, China, pp 53–62 (in Chinese)

Chapter 82
Non Antagonistic Integrated Combat Network Modeling

Yanbo Qi, Zhong Liu and Kun Sun

Abstract How to generate the topology model of non antagonistic integrated combat network is an important issue in combat analysis. A comprehensive construction algorithm was proposed to solve the problem. First a command network was developed by using an improved hierarchy network evolving method, and then new nodes was joined into it by growth and local priority connections, finally, the network model was analyzed according to the topology statistical parameters. The analyzing results show that when command level was increased, the command network topology doesn't change but the topology performance of the combat network is improved. This is in line with actual combat network, verifying the validity of the algorithm model.

Keywords Integrated combat networks · Hierarchical network · Degree distribution · Local world evolving

82.1 Introduction

Combat Analysis is an important area of military operations. With the development of information technology, combat have increasingly reduced their reliance on weapon platforms and force concentration. Through combat network, the

Y. Qi (✉) · Z. Liu
School of Electronic Engineering, Naval University of Engineering,
Jiefang Avenue. 717, Wuhan 430033, China
e-mail: yanbqi@163.com

K. Sun
Unit 91999, Qingdao 266000, China
e-mail: bill1302lzzk@163.com

W. Lu et al. (eds.), *Proceedings of the 2012 International Conference on Information Technology and Software Engineering*, Lecture Notes in Electrical Engineering 210, DOI: 10.1007/978-3-642-34528-9_82, © Springer-Verlag Berlin Heidelberg 2013

integration between various platforms is increasingly high while the distinction boundaries between different arms are becoming blurred. Integrated Combat Network (ICN) is a complex military network composed of command, sensor, firepower and communication platforms connected through information network. Then we will ask an important question: in order to reach the optimal operational effectiveness, what kind of topology should the network have?

Jeffery [1] first established a basic combat network model under antagonistic condition in 2004. He use adjacency matrix to represent and analyze his model. Though some researchers applied cares methods to command and control system modeling [2–4], combat system modeling [5, 6] and network analysis methods [7] research, as well as some of the complex network applications in the field of military networks Summary [8, 9] and other fields, These models either did not consider non antagonistic conditions, or their model did not match good with the reality.

Our objective herein is to fill this gap by developing an algorism model to generate a combat network model. First we proposed a command and control network based on hierarchy network, and then we used it as the initial network of the combat network to generate the combat network. Finally we verified the validity of the algorithm model by comparing the commands and control networks and combat networks under different command levels.

82.2 The Combat System of Systems Network Topological Model

82.2.1 Model Assumption

Considering the non antagonistic condition, make the following model assumptions according to the analysis of the true integrated combat network:

1. For the purpose of non antagonistic integrated combat network model, nodes are defined into four categories: decider node D, sensor node S, communication node C and influencer node I. suppose that the same type of nodes has the same function.
2. Assuming that all units on the battlefield are in control of the whole combat network, in other words, there is no isolated node in the network.
3. Each of the S and I must be directly or indirectly connected with D, indicating that the information is transmitted among the S, I and D. S or I can be linked to multiple or multi-level D, S can be linked to various D in its own combat Group, while I can only be linked to their own command center.
4. The link between two different sensors can be indirection; the link between two different deciders can also be indirection. For the convenience of calculation, it is assumed that each level of the command and control network has the same span value.

82.2.2 *Integrated Combat Network Generation Algorithm*

It was found that global preferential attachment mechanism was not suitable for combat network modeling for many the leaf nodes cannot connect directly to the central nodes. We propose a comprehensive network generation algorithm. First we build the command network as the initial network, and then we use an improved local world evolving construction method to make other nodes join into the entire network to generate the integrated combat network. As command network has modularity, local clustering, hierarchical and horizontal synergies characteristics, we propose a revised hierarchical network generation algorithm base on [10].

Assuming that the command module is hierarchical, the initial network is a global coupling network consisting of N nodes and the total number of node and command level is fixed and the command span K_i is varied, here $i \in [1, H]$ is accused of hierarchical label. The nodes which belong to the same Superior are fully connected; the nodes which belong to different Superiors have no connection. Detailed description of the algorithm is as follows:

Step 1: Let H represents the number of command and control network levels and $N_i (1 \leq i \leq H)$ represents the span size of level i.

Step 2: Let the number of the top-level nodes be $N_1 = K_1$ and all nodes of the top-level be interconnected, which consists the core modules of the command and control network. One of them performs as executive node, the others as spare Command and Control Center node.

Step 3: Let the command and control span of level i be K_i and each node of this level be connected to each node of level $i - 1$, and then all the nodes of this layer for pair wise connections, then i++;

Step 4: Repeat until $i = H$. Here we need some explanations: The first step determines the command level; the second step is based on the fact that usually there are more than two command centers but only one is on operation. The node relationship between each other is: if the level difference $\Delta H_{ij} = 1$, the relationship is command and the link is longitude; if $\Delta H_{ij} = 0$, the relationship is collaborative and the link is latitude; if $\Delta H_{ij} \geq 2$, they have no relationship. Let command span $N_i \equiv 3$, level $H = 4$, we can get a command and control network which is shown in Fig. 82.1.

Step 5: Nodes and links grow. Suppose that there are m_0 nodes and e_0 links in the initial network, its topology is like Fig. 82.1. Each time a new sensor, firepower or communication node and m links were added to the network;

Step 6: Connecting the new node to the network by Hierarchical world Priority. First select a level from the existing network randomly as the hierarchical world of the new node, and then link the new node to the command node and the subordinate node cluster of the command node. The connection probability is given by (82.1).

Fig. 82.1 Command and
control network topology

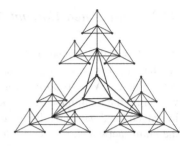

Table 82.1 Input data of the
combat network

Items	Value
Command level	4
Deciders	45
Span	3
Sensors	40
Influencers	50
Commutations	65

$$\Pi_L(k_i) = \Pi'(i \in LW) \frac{k_i}{\sum_j {}_L k_i} \equiv \frac{M}{m_0 + t} \frac{k_i}{\sum_j {}_L k_i} \qquad (82.1)$$

82.2.3 Simulation

Table 82.1 is combat scenario data for the model input. We use MATLAB7.1 to
program the algorithm, and then input the data and simulating; Fig. 82.2 shows the
visualization topology of the combat network.

82.3 Topological Analyses

82.3.1 Design of the Topological Parameter

We use the following characteristic parameters to analyze the integrated combat
network statistically from the angle of combat effectiveness.

1. The number of network nodes N.

   This implies the number of combat units on the battlefield, which indicates the
forces scale of both sides.

Fig. 82.2 Integrated combat network topology

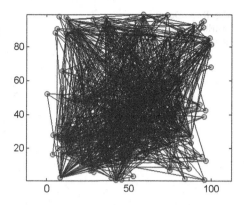

2. The number of links N_L.

This represents the amount of information exchange between network nodes.

3. Degree and degree distribution.

The degree of node i is the number of links connected to it, which reflect its importance in the whole network. $p(k)$ was defined as the probability that a node chosen uniformly at random has degree k. It reflects the macro statistical characteristics of the network system.

4. The average path length.

Average path length is also known as the characteristic path length, it is defined as the average distance between any two nodes in the network, and the expression is given by (82.2).

$$CPL = \frac{1}{\frac{1}{2}N(N+1)} \sum_{i \geq j} d_{ij} \tag{82.2}$$

5. Network clustering coefficient: node i clustering coefficient is defined as the actual presence of the number of edges connected to the node between the nodes and the total ratio of the number of edges that may exist, general expressions is given by (82.3).

$$C_i = 2E_i/(k_i(k_i - 1)) \tag{82.3}$$

The clustering coefficient C is the average of clustering coefficient of all nodes.

Fig. 82.3 Degree
distribution of command and
control network

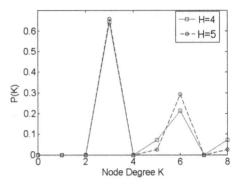

Fig. 82.4 Degree
distribution of the integrated
combat network when $H = 4$

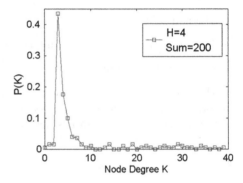

82.3.2 Comparable Analysis

First we compare the node degree distribution of the command and control network under different command level. Let command span $N_i \equiv 3$, the command level $H = 4$ and $H = 5$ respectively to analyze the node degree distribution law. The results can be seen from Fig. 82.3. The degree distribution curves of the two networks are essentially coincident; this indicates that under the condition of the same command span, when we increase the command level, the network topology doesn't change, verifying that the command network is hierarchical self-similar. This is also consistent with actual combat experience, thus improved the validation of our algorism model.

Second we compare the Degree distribution of integrated combat network under different command level. Also let $N_i \equiv 3$, the total number of nodes $Sum = 200$, $H = 4$ and $H = 5$ respectively. We can see in Figs. 82.4 and 82.5 that both the degree of distribution peak is substantially the same at about 0.43, but the peak in Fig. 82.5 moves slightly forward, indicating that the nodes who has a small degree in the network of $H = 5$ has a larger proportion than in the network of $H = 4$. On the tails of the two figures we can see that the largest degree value is up to 60 in Fig. 82.5, higher than 39 in the network of Fig. 82.4, indicating that when $H = 5$, the nodes of low degree has a larger proportion, and nodes of high degree has a

Fig. 82.5 Degree distribution of the integrated combat network when H = 5

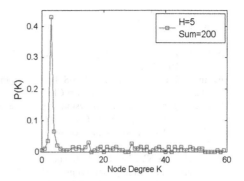

larger degree value, thus it has a more skewed degree distribution and a high invulnerability. This validates the fact that when the total number of nodes is fixed, a combat network which has higher command level has a more combat effectiveness.

82.4 Conclusions

In this paper we have proposed an integrated combat network topology generation algorithm. Starting with the existing combat model of cares, we consider the non antagonistic condition and study the characteristic parameters of the model based on the complex networks theory. By comparing the two command and control networks, two combat networks with different command level, we get the conclusion that under the same command span, the command and control network topology doesn't change when command level was increased, but the topology performance of the entire combat network is improved. This is in line with the characteristics of the actual combat network, verifying the validity of the algorithm model. Nonetheless, our algorism model itself has a limitation from its construction that we haven't consider the variation of command span. A future study might therefore use a new algorism model to analyze the combat model under the condition of varied command span.

References

1. Jeffrey RC (2004) An information age combat model. In: Paper for the 9th ICCRTS, ICCRTS 2004, Copenhagen
2. Ma XL, Sun KX, Wang HX (2010) Research on network topology of command and control organization based on theory of complex networks. Fire Control Command Control 35(2):69–71 (in Chinese)
3. Yu J, Wang W, Zhang GN (2011) Research on joint operations command system based on complex networks. Fire Control Command Control 36(2):5–10 (in Chinese)

4. Li JJ, Liu GG, Huang Q et al (2008) Analysis method and model of combat command architecture based on theory of complex networks. J Syst Simul 20(17):4712–4715 (in Chinese)
5. Di P, Li F, Hu B (2010) Complex network modeling and characteristics. J Naval Univ Eng 22(12):107–112 (in Chinese)
6. Jin WX (2010) SoS-Ops M&S based on the complex networks. Publishing House of Electronics Industry, Wuhan (in Chinese)
7. Li JJ (2009) Joint combat system network analysis methods and application research. Mil Oper Res Syst Eng 23(2):16–20 (in Chinese)
8. Hu XF (2010) A brief survey on war complex networks studies. Complex Syst Complex Sci 7(2–3):24–28 (in Chinese)
9. Wu Z, Hou XY, Xu T (2010) The application of complex network theory in the military domain. Natl Defence Sci Technol 31(5):1–6 (in Chinese)
10. Barabasi AL, Oltvai ZN (2004) Network biology: understanding the cell's functional organization. Nat Rev-Genet 5:101–114

Chapter 83
Study on Monitoring System for Pipeline Security Based on Multi-Sensor Data Fusion Technique

Jie Zhang

Abstract Based on multi-sensor data fusion technique, we designed monitoring system for pipeline security. The system acquires early warning signals by many sensors and processing modules when pipes are threatened. The non-stationary signal analysis method based on empirical mode decomposition is used to process the signals. The normalized kurtosis extracted from the decomposed results is composed of the feature vectors. Because there were separate sensors gathering modules in the monitoring system, we used the multi-sensor data fusion technique to fuse these individual recognition results for improving recognition rate. With software development platform, LabVIEW virtual instrumentation, this obtained the final decision. With the test, the system can achieve better detection performance than a system using the single sensor, so as to form a more complete, reliable pipeline real time detecting conclusion.

Keywords Pipeline · Multi-sensor data · Fusion technique · LabVIEW virtual instrumentation

83.1 Introduction

In the gas and oil pipeline operation, offenders often drill in the pipeline to steal oil and gas. If these things are not discovered and not restrained in time, that will not only seriously affect upstream normal production and downstream transmute

J. Zhang (✉)
Electronic Engineering Department, North China Institute of Aerospace Engineering,
Lang Fang City, China
e-mail: shaoyan8074@yahoo.com.cn

W. Lu et al. (eds.), *Proceedings of the 2012 International Conference on Information Technology and Software Engineering*, Lecture Notes in Electrical Engineering 210, DOI: 10.1007/978-3-642-34528-9_83, © Springer-Verlag Berlin Heidelberg 2013

production to cause serious economic losses, but also the leakage products of gas and oil pollutes the environment and generates the related secondary disasters [1]. Therefore, the safety monitoring system can report the scope and extent of the accident timely and accurately, can decrease the economic loss and reduce environmental pollution in the maximum. It is great significance to protect the security and reliability of production.

Regular leakage detection technology and existing mechanism is used in general only to find oil leakage or to call the police after these things occur. As a result, the questions, act of invading is realized timely when pipeline security is at stake, or pipelines are being destroyed but not cause damage, are pressing problems [2]. Based on multi-sensor data fusion technique, a new type of pipe safety monitoring system is designed. Many sensors and processing modules are used to acquire early warning signals when pipes are threatened. The nonstationary signal analysis method based on empirical mode decomposition is used to process the signals. The multi-sensor data fusion technique is used to fuse these individual recognition results for improving recognition rate. With LabVIEW virtual instrumentation, the security system is obtained. With the test, the leakage in the pipe can be found before destroy occurs. The gas and oil products and environmental pollution greatly reduced.

83.2 The Basic Principle of System

83.2.1 The Basic of Multi-Sensor Data Fusion

The basic principle of multi-sensor data fusion is like the process of human brain to comprehension and to handle with information. It is in full use of many sensors resources, by controlling and using these sensors and observational information reasonably, redundancy or complementary signals in time or space of these sensors are combined according to some guidelines to acquire consistency explanation or description of measured objects. With this, the performance of total information system is more superior to those of subset system. Redundancy data between sensors enhances system reliability and complementary data between sensors expands single performance. The essential difference between multi-sensor data fusion technique and classic signal processing is that multi-sensor information has more complex forms and can be in different information hierarchy.

83.2.2 The Process of Multi-Sensor Data Fusion

The process of multi-sensor data fusion is composed of many sensors, data preprocessing, data fusion centre and result output. The process can be seen in Fig. 83.1.

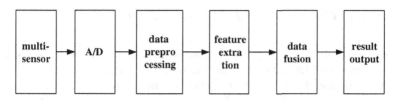

Fig. 83.1 Data fusion process

Those are converted to electrical signals, and then with A/D conversion, those are converted to digital quantities. The digital electrical signals are added with some interference noise inevitably because of the random environment factors. By preprocessing, the interference and noise can be filtered to get useful signals [3]. After this, the feature extraction of signals is done, data fusion of feature vectors is worked out, and the fusion output is gained.

83.2.2.1 Signal Acquisition

The pressure and the flux of pipelines changes when the leakage occurs. Criminals want to steal the crude oil, usually drilling in the oil pipeline first. When they use tools in the pipeline to drill, the special stress wave is generated by interaction friction between drilling tools and metal in the pipeline. By using corresponding sensors, the different elements of alarming signals can be received. With A/D conversion, the different signals can be converted to electrical signals, and then they are delivered to computer system.

83.2.2.2 Signal Preprocessing

Before data fusion, sensor output is in the preprocessing to filter intrinsic noise and to enhance Signal-to-Noise. The ways of signal preprocessing is mostly averaging, filtering, removing trend term and so on.

83.2.2.3 Feature Extraction

The feature extraction is done to original sensor information. The feature vectors can be gained according to specific pipeline measuring signals.

83.2.2.4 Data Fusion

The data fusion is a lot of, such as data correlation technique, estimated theory and recognition technique.

83.2.3 *The Gradation of Multi-Sensor Data Fusion*

In the process of multi-sensor data fusion, because of signals diversification, the fusion is done with definite gradations and steps according to data types and means of collection. Based on this, the gradation of multi-sensor data fusion must be introduced into. By means of level of abstraction, and means of circulation and transportation of data, data fusion is divided to high-level process and low-level process. The low-level process is composed of data preprocessing, target measuring, classification, identification and target tracking. The high-level process is composed of situation and threat assessment and extraction of total fusion [4]. The three basic frames of identification are data-level fusion, characteristic-level fusion and decision-making level fusion. Because of alarming pipeline signals are heterogeneous, such as pressure, flux and special stress wave, the fusion is only done in characteristic-level or decision-making level.

83.2.3.1 Characteristic Level Fusion

The representative feature is extracted from data, and then is fused to single feature vectors, and those are handled with by pattern identification.

83.2.3.2 Decision-making Level Fusion

In this method, the collecting information of every sensors exchanges to get preliminary results of measuring target. In the end, according to definite guideline and every judgemental reliability, the optimal decision is worked out. Some methods are majority vote method, bayesian method and generalized evidence theory. In the monitoring system for pipeline security, the D–S (Demp-ster-Shafer) evidence reasoning is used to fuse these individual recognition results for improving recognition rate. By receiving and analyzing the different elements of alarming pipeline signals, the characteristics, nature and location information of damage source can be obtained.

83.3 System Structure

In monitoring system, at the destruction pipeline area, high sensitivity sensors are fixed on the pipeline with certain distance to gather corresponding signals when pipeline is destroyed. The signal is enlarged, regularized, handled with synthetically by built-in Digital Signal Processor (DSP) to judge whether pipelines are threatened in the monitoring units.

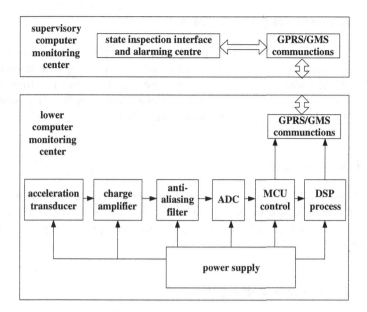

Fig. 83.2 System construction diagram

When pipelines are threatened by on-site unit' judgment, signal character of threatened events is uploaded to the monitoring units by GPRS communications network. In the junction center, the multi-sensor data fusion technique is used to fuse these individual recognition results for improving recognition rate and for different types of events different alarms are called. At the same time, in monitoring channeling interface, position and time of threatened events are recorded for management information and data analysis. System diagram is seen in Fig. 83.2.

83.3.1 System Hardware Design and Realization

83.3.1.1 High Sensitivity Sensors

The dissemination is decayed, considered the supply and low power consumption, high sensitivity sensors are selected to receive signals in this system.

83.3.1.2 Signal Conditioning Module

According to type of sensors, the corresponding charge amplifier is used; signal is anti-filtered when frequency analysis, the low pass filter is fixed whose cut-off frequency matches with frequency of measured signal. With a series of process, the signal is gathered availably in monitoring units obtained by sensors.

83.3.1.3 Digital Signal Processing Module

The collected signals is on time-frequency analysis to extract characters of signal in the time domain and frequency domain, and then first judgment is used integrated with the parameters of system. Once the signal is same as information in threatened events database, the infinite communications unit in the system is activated to upload information to monitoring centre.

83.3.1.4 Monitoring Module

The GPRS + GSM communication modules, audible alarming equipment and computer are composed of the system monitoring center. Based on the virtual instrument technology, safety controlling interface of pipes is designed to display real-time pipeline information.

83.3.1.5 Efficient Bulk Recharged

The system uses the bulk batteries and designs capacity for work half year, which all system is hidden in the ground.

83.3.2 System Software Design and Realization

83.3.2.1 Data Collection

Data collection module is the based part of system, that is realized by MPS-01020T multi-purposes card and data collected procedure. With software of upper monitor supposed and sub-VI and examples offered, according to the configuration of the various arguments system applications software is written.

83.3.2.2 Data Processing

The time domain analysis, the frequency domain analysis, multi-sensor data fusion is realized in data processing module. The true features and frequency distribution of effective signal is extracted from noised signal. The D-S (Demp-ster-Shafer) evidence reasoning is used to fuse these individual recognition results for improving recognition rate. By receiving and analyzing the different elements of alarming pipeline signals, the characteristics, nature and location information of damage source can be obtained. Analyze-Signal Processing sub-template, of LabVIEW, has many functions, those are signal generator, time domain analysis,

Table 83.1 Leakage detection rate with different fusion times

Fusion times	$\tau > 10\%$	$5\% < \tau < 10\%$	$\tau < 5\%$
t = 2	90.28	86.54	86.57
t = 4	93.21	90.85	89.27
t = 6	93.85	94.32	88.83

Table 83.2 Leakage detection results between wavelet transform and data fusion

Detecting method	$\tau > 10\%$	$5\% < \tau < 10\%$	$\tau < 5\%$
Wavelet transform	91.65	86.72	77.87
Data fusion	93.75	92.56	85.23

frequency domain analysis, the wavelet analysis, measuring functions, window functions of digital filter and other features [5].

83.3.2.3 Data Storage and System Configuration

The module is to store data collected, and by the computer screen it shows the signal at the various stages for the management for further enquiries, data processing, analysis and comparison, typed, etc. [6]. The module deploy major parameters of system, such as the argument, etc. [7].

83.4 Experiments

In testing process, the pipeline is 1.5 km from the base station, and buried deep to the area of about 1.5 m. In monitoring station, data acquisition module is capable of monitoring small changes of parameters such as the flow and negative pressure wave and promptly senting those to the data processing center. For the application effect validation of multi-sensor data fusion algorithm in leak detection, three kinds of leakage ($\tau > 10\%$, $5\% < \tau < 10\%$, $\tau < 5\%$) were compared, where τ is the ratio between pipeline leakage instantaneous flow and pipeline transient flow. Different fusion frequency leakage detection results (with fusion times respectively for t = 2, t = 4, t = 6) is shown in Table 83.1.

The application results between multi sensor with single sensor in pipeline leak detection in Table 83.2. Here takes the negative pressure wave data and the application of wavelet transform and data fusion in leakage detection method.

83.5 Conclusion

Based on multi-sensor data fusion technique, a new type of pipe safety monitoring system is designed. This method has a lot of advantages, those are response speed is fast, positioning accuracy is high, and the leakage loss of crude oil can be reduced effectively. Simple structure, reliable pump performance, high capacity of resisting disturbance, low rate of false alarm, and base station can meet the guardless demand in the field. Owing to pipeline without electricity and communication facilities and system needs to be buried in whole, therefore, the supply, communication issue, embedded software reliability and the measurement of weak signal in this system, are needed to be further improved.

Acknowledgments This work is under the support of the "Lang Fang Science and Technology Research and Development Program (2008011002)", authors hereby thank them for the financial supports.

References

1. Alleyne DN, Pavlakovic B, Lowe MJS, Cawley P (2001) Rapid long-range inspection of chemical plant pipework using guided waves. Insight 43:93–96
2. Lowe MJS, Alleyne DN, Cawley P (1997) Mode conversion of guided waves by defects in pipes. Rev Prog Quant Nondestr Eval 16:1261–1264
3. Smith D, Singh S (2006) Approaches to multisensor data fusion in target tracking. A survey. IEEE Trans Knowl Data Eng 18(2):1696–1710
4. Tafti AD, Sadati N (2008) Novel adaptive kalman filtering and fuzzy track fusion approach for real time applications. In: Proceedings of the 3rd IEEE conference on industrial electronics and application 15(5):120–125
5. Hongsheng L, Junwei Z, Yujun L (2005) Implementation of oil pipeline leak monitoring and location system based on virtual instruments. Comput Meas Control 13:415–417–433 (in Chinese)
6. Wolf Wayne (2001) Computer as components principles of embedded computing system design. Morgam Kau nann 2:34–35
7. Jia Li, Dianfu Yang (2002) Embedded remote monitoring system development. Process Autom Instrum 4:58–60 (in Chinese)

Chapter 84
Virtualization Technology Based on the FT Platform

Jing Wang, Qingbo WU, Yuejin Yan and Pan Dong

Abstract The FT processor, developed by national university of defense technology, composites system virtualization, network, security, floating-point operation union and accelerated memory on one chip. This study introduces the virtualization architecture of the FT platform and describes the virtualization mechanism of the FT platform from the CPU, Memory and IO point of view. The paper also evaluates the virtualization performance and presents our current research.

Keywords FT processor · FT platform · Virtualization · Hypervisor · Container

84.1 Introduction

System virtualization, the concept we usually use to refer virtualization, lies on the idea that using software to virtualize one or more virtual machines on a single physical machine. A virtual machine was originally defined by Popek and Goldberg in 1974 as "an efficient, isolated duplicate of a real machine" [1]. They explain these

J. Wang (✉) · Q. WU · Y. Yan · P. Dong
School of Computer Science, National University of Defense Technology, Sanyi Avenue, Kaifu, Changsha, Hunan, China
e-mail: wjjoyce115@hotmail.com

Q. WU
e-mail: wu.qingbo2008@gmail.com

Y. Yan
e-mail: yyj@nudt.edu.cn

P. Dong
e-mail: dongpan@kylin.com.cn

W. Lu et al. (eds.), *Proceedings of the 2012 International Conference on Information Technology and Software Engineering*, Lecture Notes in Electrical Engineering 210, DOI: 10.1007/978-3-642-34528-9_84, © Springer-Verlag Berlin Heidelberg 2013

notions through the idea of a virtual machine monitor (VMM). As a piece of software, a VMM has three essential characteristics: essentially identical, efficiency and resource control. By an "essentially identical" environment, the first characteristic, it means that any program running under the VMM should exhibit an identical effect with that if the program had been run on the original machine directly, with the possible exception of differences caused by the availability of system resources and timing dependencies. Efficiency, the second characteristic, means that the performance of software running in a virtual machine close to its direct running on the physical machine. Therefore, it demands that a statistically dominant subset of the virtual processor's instructions be executed directly by the real processor, with no software intervention by the VMM. The third characteristic, resource control, means that the VMM has complete control of resources (such as memory, peripherals and the like), including allocation, monitoring and recovery of resources.

It is not easy to meet the three characteristics above, because the instruction set of physical machine should correspondingly meet the following four conditions:

1. CPU needs to support multiple privilege level, and the instruction running in virtual machines should implement correctly in bottom privilege level.
2. The execution effect of non-privileged instruction (instructions that allow users to use directly) is not dependent on CPU's privilege level.
3. Sensitive instructions, which may influence system resource allocation, are privileged and not allowed to use directly by users.
4. Provide a memory protection mechanism, such as segment protection or page protection, to ensure the memory isolation between virtual machines.

As is known to all, the X86 architecture which occupies absolute monopoly position in personal computer field exists a virtualization flaws in its innate design, so that its support for virtualization is not enough. Especially, it doesn't meet the above third condition because its instruction set has 17 sensitive instructions (LGDT for example) that are not privileged. In other words, it may lead to crash when a virtual machine execution occur these instructions, thus affecting the stability of the whole physical machine. Domestic Loonson processor has the same problem. Although they have joined the Hardware-Assisted Virtualization strategy (such as Intel VT and AMD SVM) to make up for the loopholes, it is inferior compared to the overall virtualization strategy on FT platform.

The FT processor, which is developed by national university of defense technology, composites system virtualization, network, security, floating-point operation union and accelerated memory on one chip. The design of the FT processor is based on the sun4v virtualization architecture of Sun's server. The processor is compatible with SPARC V9 instruction set and supports 8 core *8 thread CMT (Chip Multi-threading) technology, which provides a strong hardware support for virtualization.

This paper will introduce the virtualization architecture of the FT platform and describes the virtualization mechanism of the FT platform from the CPU, Memory and IO point of view. Besides, we also evaluate the virtualization performance and present our current research.

84.2 Our Virtualization Mechanism

The virtualization architecture of FT platform blends in a Hardware Virtual Machine (HVM) design philosophy [2–5]. The key idea of this philosophy is allowing resources shared among virtual machines that execute different operating systems. The virtualization architecture is shown in Fig. 84.1. The HVM implements a layer called hypervisor as a VMM for the whole server, using it to create a number of virtual partitions and manage bottom hardware resources related to each partition. In contrast to X86's realization in software layer, the FT platform implement the hypervisor in the form of firmware as a part of the hardware platform. After system initialization, the hypervisor takes charge of system resources in an exclusive way, with super access and control authority; thus, it brings higher availability and avoid the 'single point of failure' phenomenon appeared in VMware/XEN/KVM's hypervisor.

The FT platform provides a complete hardware-level virtualization including CPU, memory, peripherals and the like. Below, we will analyze the current virtualization mechanism from the three aspects.

84.2.1 CPU Virtualization

The FT processor introduces an idea of "server on a chip", with 8 cores, and on-chip networking and security functionality. Every core on the chip can process 8 threads; theoretically, the processor could support 64 physical threads. Thus, we use hypervisor [6] to implement 64 virtual CPUs (VCPU). Each VCPU contains a

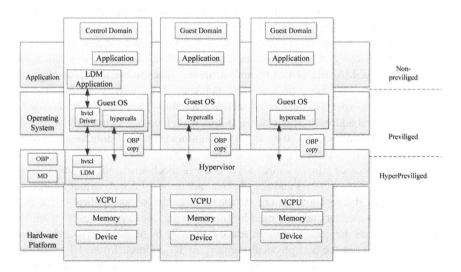

Fig. 84.1 Virtualization architecture of the FT platform

thread's state and software configuration information. Different from X86's four privilege levels (Ring 0–Ring 3), each VCPU in our platform has privilege at three levels: hyperprivileged, privileged and non-privileged. Applications that running in an operating system are under the non-privileged level, while the operating system kernel and partition software in the firmware get the privileged right. Only the hypervisor, core of the firmware, is running under the hyperprivileged level. Therefore, every operation of hardware access in the operating system should be achieved by the hypervisor.

The function of hardware partition in our microprocessor can assign dedicated VCPU for each operating system zone on the FT platform and guarantee that each partition gets no access to other partition's VCPU. We call such a virtual partition for logic domain; communications are allowed between two logical domains. All communications are realized and supervised by the hypervisor. In other words, each hardware partition in the FT processor is isolated which differs from X86's processor.

84.2.2 Memory Virtualization

Memory management of the FT platform is achieved by MMU and Hardware Address Translation layer (HAT); but our management of address space is very different from the X86's system. In the FT platform, the kernel's address space is separate from the user processes' address space. Address spaces are mapped by segments. They're created by the segment device driver and each segment driver manages mappings from a linear virtual address to memory pages of different equipment's. Layer segment manages the virtual memory as an abstract for a file. The segment driver calls for the HAT to create conversion relations for their address space and its physical pages.

Usually, the traditional operating system hypothesizes a sequential physical memory start from a zero page frame, and uses physical address to operate the page table and DMA. But in a hardware virtual machine environment, the physical memory assigned to logic domain may not start from zero or be continuous. Therefore, the hypervisor needs to increase another addressing mechanism to provide physical address virtualization. In this way, memory virtualization for our system includes two aspects: the memory assignment for logic domain and a page translation mechanism for logic domain.

The FT processor adopts a three-layer address mechanism: the virtual address (VA), the real address (RA) and the physical addresses (PA). A virtual address is used for a guest operation system (Guest OS) kernel or an application; a real address is the secondary address used by Guest OS and it can get access to memory or peripherals on the condition of an address transfer by MMU or Hypervisor; the physical address is the real physical memory or IO address. In the FT platform, applications are addressed by virtual address, while a guest operation system (Guest OS) can be addressed by both virtual address and real address. Only the hypervisor can use the physical address. Virtual address spaces that share the same piece of real addresses utilize context ID for identification. Guest OS is responsible for the virtual address management and creation of logic domain, while the hypervisor takes charge for management and creation of the server's real address space.

For address Translation, the FT processor implements a "Split Fire" memory management unit SFMMU, including a translation lookaside buffer (TLB) and its related register, as is shown in Fig. 84.2. These registers are used for management and control for the TLB, context/partition and Translation Storage Buffer (TSB). The TSB is a set of page table items resident in memory. Each item in the TLB and TSB has the same structure, including 64 bit label segment and 64 bit data segment. The former contains a context ID and a virtual page address, which are used to judge whether match the requested virtual page address or not; Data segment contains the RA or PA and length of page, which are used to calculate RA or PA mapped from the page to the real address space or physical address space. The TSB under control of the hypervisor can support both the VA to RA/PA mapping; while the TSB under control of Guest OS is unable to access and control PA, in this way, it can only support mapping from VA to RA.

In our system, an application must use a virtual address or a real address to access memory. Therefore, conversions between these two kinds of addresses are

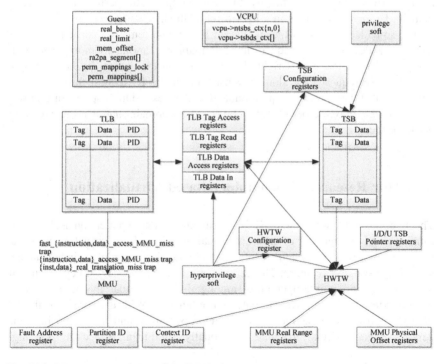

Fig. 84.2 Memory management of the FT platform

dynamically implemented by the processor on occasion of memory access. Exactly, this conversion is completed by the TSB. Modifications of the TSB could only be done by the hypervisor. The hypervisor allocates a different physical memory segment for each hardware partition to realize hardware virtualization partition. Except for address conversion, the TSB is also used by the hypervisor as a monitor to guarantee that each partition cannot get access to other partition's memory, thus to implement memory isolation between different hardware partition.

84.2.3 IO Virtualization

Following the start-up of the FT platform, the hypervisor will allocate the control permission of PCIE bus to the IO domain according to the description of logic domain MD. In fact, this progress is a mapping from PCIE configuration space, IO space and memory space to real address space of logic domain. An IO Domain can operate a device directly; on the other hand, if a non-IO Domain client software wants to acquire an IO service, the LDC (Logical Domain Channel) from Hypervisor will build a virtual IO service to implement a transfer between customer domain and the device through the IO Domain. Obviously, performance of this kind of virtual IO is not satisfactory. In order to improve the performance of the IO domain, FT platform brings in a hybrid IO method.

The so-called hybrid IO means that an IO domain remaps a particular limited space of a device to the request logic domain by sharing memory. The logic domain software has partial rights to write or read the device. In the hybrid IO approach, the request logic domain must transfer the IO request through the IO domain, but it can directly acquire data from devices' address space. In this way we ensure not only the peripherals' control to IO domain, but also the execution performance.

84.3 Our Research on Container-Based Virtualization

The hypervisor in our present FT platform implements application isolation by running them in a number of different virtual machines. Each virtual machine has relatively independent system resources and provides a complete independence support for system execution. This kind of strong independence strategy results in difficult application interoperability among different virtual machines, and, to certain extent, sacrifices system execution efficiency. With the spread of the FT platform, providing effective resources interoperability and a relative independence for each virtual machine are equally significant, especially for high

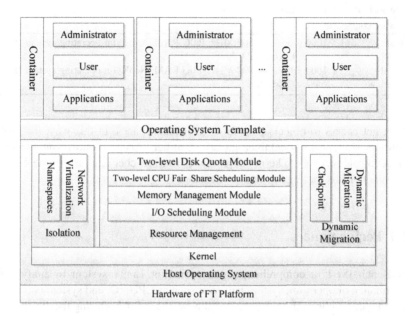

Fig. 84.3 Container-based virtualization framework

performance computing, Web services, database, and many other fields of application. Therefore, we introduce a container-based virtualization technology for the FT platform.

A container-based approach [7] strips out the system operation subject (processes) from resource subject and defines the container as an abstract operating system carrier to use system resources dynamically, realizing a set of reliable access control mechanism to ensure effective access control among shared resources. Figure 84.3 illustrates the framework of container-base virtualization approach on the FT platform. There are two compositions in this architecture: host platform and virtual platform. The hosting platform consists essentially of the shared OS images and a privileged host container. This is the container that a system administrator uses to manage other containers. The virtual platform is the view of system as seen by the guest container. Applications running in the guest containers work just as they would on a corresponding non-container-based OS image.

The isolation and resource management is crucial to realize container-based virtualization technology. To ensure system security isolation, the container-based approach directly involves the host operating system's internal objects (such as PID, UID, etc.). The basic techniques to securely use these objects include: separation of namespaces and access controls. The former means that global identifiers (e.g., process ids, user ids, etc.) live in completely different spaces, do not have pointers to objects in other spaces belonging to a different container, and thus

Table 84.1 Hardware configuration for test

Hardware configuration	
Virtual machine	FT 1000 processor, 800 MHz, 2.0G RAM, NeoKylin operating system (FT version), Kernel-2.6.32
Physical machine	FT 1000 processor, 800 MHz, 4.0G RAM, NeoKylin operating system (FT version), Kernel-2.6.32

cannot get access to objects outside of their namespaces. In this way, the global identifiers become per-container global identifier. Access control, on the other hand, controls accesses to kernel objects with runtime checks to determine whether the container has the appropriate permissions.

84.4 Performances

UnixBench [8] is a comprehensive test suite for Linux system to analyze the performance of operation system behavior such as CPU, IO, and system call and so on. In our experiment, the software version we use is 5.1.2. During the experiments we tested the system performance of virtual machine and physical machine separately. The test environment configuration parameters are shown in Table 84.1.

Table 84.2 provides the performance comparison of operating system running in the virtual machine and the physical machine. Numbers in the column represent scores from the test program; the higher the score, the better the performance. Column performance comparison calculates the score percentage of in virtual machine scores compared with physical machine (virtual machine/physical machine).

Table 84.2 Performance tested by Unixbench

Test item	Virtual machine	Physical machine	Performance comparison (%)
Dhrystone 2 using register variables	92.8	93.3	99.4
Double-precision Whetstone	28.5	27.7	102.9
Execl throughput	65.4	126.3	51.8
File copy 1024 bufsize 2000 max blocks	99.8	96.8	103.1
File copy 256 bufsize 500 max blocks	68.9	67.0	102.8
File copy 4096 bufsize 8000 maxblocks	163.6	179.0	91.4
Pipe throughput	72.5	65.5	110.7
Pipe-based context switching	32.5	37.9	85.8
Process creation	124.9	123.0	101.5
Shell scripts (1 concurrent)	105.3	117.3	89.8
Shell scripts (8 concurrent)	398.5	535.2	74.5
System call overhead	70.8	71.8	98.6
System benchmarks index score	86.3	94.7	91.1

As shown in Table 84.2, the Unixbench score of virtual machine accounts for 91.9 % of physical machine score. Some test programs thats which processes frequently, execl throughput, pipe-based context switching and shell scripts speaking for instance, behave a lower performance in the virtual machine, because the virtual address needs frequent switches and then the TLB needs to be cleared up and recreated. In other words, the TLB miss is the main restricting factor. While the other test program in a virtual machine gains a higher points than that of physical machine. Therefore, the test results indicate an excellent virtualization performance based on FT platform with a performance loss less than 9 %.

84.5 Conclusions

The container-based virtualization framework of the FT platform which is based on open source software OpenVZ, is currently in development.

References

1. Popek GJ, Goldberg RP (1974) Formal requirements for virtualizable third generation architecture, Commun ACM 412–421
2. Barney G (2006) The T1 hypervisor and the sun4v architecture API. http://opensparc. sunsource.net/conf/multicore/06/2006-03-23-multi-core-expo-Ashley.pdf
3. UltraSPARC Architecture (2007) Hyperprivileged, privileged, and non-privileged. Draft do. 91
4. UltraSPARC Virtual Machine Specification (2006) The sun4v architecture and hypervisor API specification, Revision 1.0
5. OpenSPARC T2 Core Microarchitecture Specification (2007) Revision 5
6. Hypervisor   [EB/OL].   http://www.ibm.com/developerworks/wikis/display/virtualization/ POWER5+Hypervisor
7. Soltesz S, Potzl H et al (2007) Container based operating system virtualization: a scalable, high-performance alternative to hypervisors. Eurosys 275–287
8. Linux benchmark suite homepage, http://lbs.sourceforge.net/

Chapter 85
Design and Implementation of Light Sensor Node Based on ZigBee

Junke Lv

Abstract ZigBee Home Automation (HA) network is a direction in the development of intelligent home system. Sensor nodes are very important devices of ZigBee HA network. After the detailed analysis about the characteristics of illuminance measurement cluster and the attribute report mechanism of ZigBee HA profile, this paper presents a design scheme concerning light sensor node in detail. The design scheme takes CC2530 System-on-Chip as the core control component, BH1620FVC as the measurement element and adopts modular structure to form light sensor. Based on Z-Stack protocol stack, the software achieves the function of light sensor node by creating a ZCL application. The analysis results of RF data packet are achieved through the software SmartRF^{TM} Packet Sniffer, which demonstrates that the scheme is effective. This design scheme is of universal property and can apply in the design of other measurement and sensing node devices.

Keywords ZigBee · Home automation · Light sensor · Z-stack

85.1 Introduction

ZigBee is a short-range wireless communication technology that is recently developed. It has following features: low power consumption, low cost, short delay and large network capacity. It is a wireless technology most possibly applied in

J. Lv (✉)
Xingzhi College, Zhejiang Normal University, 688 Yingbin Road,
Jinhua, Zhejiang, China
e-mail: ljk@zjnu.cn

W. Lu et al. (eds.), *Proceedings of the 2012 International Conference on Information Technology and Software Engineering*, Lecture Notes in Electrical Engineering 210, DOI: 10.1007/978-3-642-34528-9_85, © Springer-Verlag Berlin Heidelberg 2013

industrial monitoring, wireless sensing network and smart home [1]. ZigBee Alliance has established several application standards for various applications. ZigBee HA public application profile is the first public application established and released. ZigBee HA network is a direction in the development of intelligent home system.In HA public application profile, ZigBee Alliance has defined a series of devices among which the Sensor devices are very important devices. Z-Stack is an open source software platform which complies with ZigBee specification. It is very effective to apply Z-Stack to construct sensing node devices that comply with HA specification.

85.2 Analysis on Illuminance Measurement Cluster

Cluster is a related collection of attributes and commands, which together define an interface to specific functionality. A same cluster is adopted among different devices to implement the functions of a certain application. HA public application profile contains clusters of measurement and sensing functional domain. The following clusters are included: illuminance measurement, illuminance level sensing, temperature measurement, pressure measurement, flow measurement, relative humidity measurement, occupancy sensing, etc. [2]. Devices corresponding to these clusters are light sensor, temperature sensor, pressure sensor, flow sensor, occupancy sensor, etc. Illuminance measurement cluster is used for light sensor. These devices communicate with other devices through corresponding clusters mentioned above.

Throughout the ZigBee cluster library, a client/server model is employed. Typically, the entity that affects or manipulates the attributes of a cluster is referred to as the client of that cluster and an entity that stores those attributes of a cluster is referred to as the server of that cluster [3]. Client is used to describe the output cluster and it manipulates the attribute of server device by sending commands. Sever is used to describe the input cluster. It is an end of the storage attribute [4]. The server device responds to the commands sent from the client device and initiatively reports the attribute value. Obviously, light sensor devices function as servers while configuration tool or light controller and collector sever as client devices. The typical usage of illuminance measurement cluster is shown in Fig. 85.1.

Configuration tool or light controller configures the attributes of light sensor devices. The light sensor device generates notification information by means of sending attribute report and then sends the MeasuredValue attribute information to the collector which will execute corresponding action after receiving data. The attribute configuration and reporting between client and server can be carried out only under the same cluster condition.

We can see from ZIGBEE HOME AUTOMATION PUBLIC APPLICATION PROFILE that no illuminance measurement cluster specific commands are received and generated in the client or server cluster. Besides, this cluster must

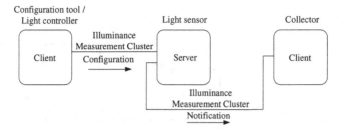

Fig. 85.1 Typical usage of illuminance measurement cluster

Fig. 85.2 Structure chart of hardware modularization of end device

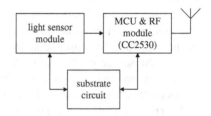

support the attribute reporting mechanism of ZigBee. In according with the attribute reporting mechanism, the client sends Configure Reporting command included minimum and maximum reporting interval and reportable change settings to the server; the server periodically or non-periodically sends illuminance measurement cluster attribute reporting using the Report Attributes command and according to the settings of the Configure Reporting command so as to deliver the results of MeasuredValue. Next, a design proposal concerning light sensor nodes is presented.

85.3 Design of Light Sensor Node

85.3.1 Design of Hardware

Light sensor node is a specific execution unit of wireless sensor networks. It provides hardware support for illuminance data collection. Subject to uncertainty of operating conditions at terminal node, battery-powered features, considerations must be made on the requirements for its low power consumption, miniaturization, modularity [5], low costs and standardization. The CC2530 is a true system-on-chip solution for IEEE 802.15.4, ZigBee and RF4CE applications. It enables robust network nodes to be built with very low total bill-of-material costs [6]. The light sensor device adopts a modular structure. It is composed of MCU and RF module, light sensor module, substrate circuit, etc. as shown in Fig. 85.2.

MCU and RF module consists of CC2530 chip, external crystal oscillator circuit and antenna, connecting to substrate by dual-in-line pin socket. The substrate

Fig. 85.3 Circuit schematic diagram of light sensor modular

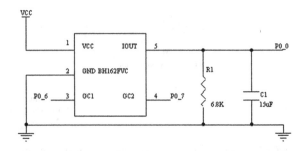

circuit mainly consists of peripheral circuits, including: keypad, LED, power circuit and external interface of P port. Keypad and LED are used for operating and indication of terminal node; external interface of P port leads to input pins connecting to light sensor module. Signal-collecting can be accomplished by P port. Light sensor module is made up of BH1620FVC which is a super-small analog current output ambient light sensor with excellent spectral sensitivity characteristic. The sensor modular circuit is shown in Fig. 85.3.

Where, GC1 and GC2 pins function as mode setting control ends. When GC1 = 0 and GC2 = 0, the sensor module is working in shutdown mode to save the electric energy of the system; when GC1 = 1 and GC2 = 0, the sensor module is working in a high gain mode with the measurable range of 0 ~ 1,000 lx, thus meeting the requirements of indoor luminous intensity measurement.

The hardware system is designed in modular form. Different sensor modules may be matched to flexibly and conveniently implement measurement or sensing with different types of physical quantity. Meanwhile, it has laid a foundation for the multi-type measurement of single nodes.

85.3.2 Design of Software

In the consideration of generality and convenience for development, terminal node software is designed by transplanting Z-Stack protocol stack owned by Texas Instruments. Z-Stack completely support IEEE 802.15.4 standard, complying with ZigBee 2007 specification. Supporting smart energy and HA profile, Z-Stack is the ZigBee solution on CC2530 chip in the system.

In Z-Stack, the functions of device node are implemented through creating the ZCL application of user. Following four modules should be created for a ZCL application of light sensor node:

1. Define zcl_LightSensor.h File

This header file should contain all the definitions needed for a new application. The application's endpoint and relevant events should be defined in this module.

2. Design zcl_LS_data.c File

This file mainly completes the definitions and declarations of some data:

(a) Declare all attributes of illuminance measurement cluster that are supported by the application. The attributes include MeasuredValue, MinMeasuredValue, MaxMeasuredValue, Tolerance, LightSensorType, etc.

(b) Construct attribute table zclLS_Attrs[MS_MAX_ATTRIBUTES]. This attribute table contains one entry of zclAttrRec_t type for each attribute supported by this endpoint. The attributes of illuminance measurement cluster must be included. The attribute table will be registered upon the initialization of endpoint.

(c) Create the input and output cluster ID tables. As for server device, the illuminance measurement cluster is only added in the input cluster. It usually does not appear in output cluster. The following sentences implement the addition of illuminance measurement cluster to the input cluster:

const cId_t zclLS_InClusterList [MAX_INCLUSTERS] =

```
{
ZCL_CLUSTER_ID_MS_ILLUMINANCE_MEASUREMENT
};
```

(d) Use SimpleDescriptionFormat_t type to define the application's simple descriptor table zclLS_SimpleDesc. In this descriptor, the input and output cluster ID tables of the application's endpoint of are designated.

3. Design zcl_LightSensor.c File

The file completes all functions required for the application, including:

(a) Add input cluster that is to be bound by application's endpoint to array bindingInClusters [] and use it for light sensor node to bind with other "complementary" nodes.

(b) Declare the application's command callback table zclLS_CmdCallbacks for the ZCL functional domains. The type of this table for the General functional domain is zclGeneral_AppCallbacks_t. Furthermore, create all the command callback functions to handle any incoming command from the ZCL clusters. These callback functions are used with the command callback tables mentioned above.

(c) Create Function zclLightSensor_Init(byte task_id). This Function is engaged in the initialization of this endpoint including registration of general cluster command callback function, attributes of application's attribute list, basic commands/response message events and key events.

(d) Establish Function uint16 zclLightSensor_event_loop(uint8 task_id, uint16 events). This function is used to receive and process the messages and key events put on the application task's queue. The light sensor node is also required to carry out REPORT_ATTRIBUTE_EVT about MeasuredValue. The following codes implement Report Attribute Event.

```
if (events & REPORT_ATTRIBUTE_EVT)

  {

  if (pMSReportCmd != NULL)
  {

  zcl_SendReportCmd (LIGHT_SENSOR_ENDPOINT, & zclLS_DstAddr,

  ZCL_CLUSTER_ID_MS_ILLUMINANCE_MEASUREMENT,
  pLSReportCmd,

      ZCL_FRAME_SERVER_CLIENT_DIR, 1, 0);
  osal_start_timerEx (zclLS_TaskID, REPORT_ATTRIBUTE_EVT, Report
  MinInterval*1000);

  }
  return (events ^ REPORT_ATTRIBUTE_EVT);

  }
```

(e) Implementation of analysis of Configure Reporting command and sending of Reporting Attribute command. In Z-Stack, the means of message is used to process the Configure Reporting command sent from the client. When the message received by the server is Configure Reporting command sent from the client, zclLS_ProcessInConfigReportCmd() is called in the function zclLS_ProcessIncomingMsg() to process the information received. Function zclLS_ProcessInConfigReportCmd() is used to implement the analysis of Configure Reporting Command and the sending of Reporting Attribute Command. The flow is shown in Fig. 85.4.

4. Design OSAL_LightSensor.c File

This file completes the addition of application's tasks.

(a) Add function zclLightSensor_event_loop at the end of array tasksArr[] which is a function pointer array. When the operating system calls function (tasksArr[idx]) (inx, events) to execute specific processing function, it will execute different functions in array tasksArr[] according to different idx values.

(b) Add zclLightSensor_Init(taskID) at the end of function osalInitTasks(). This initialization function can be called to implement the initialization of endpoint. If multiple application tasks are added, please make sure the addition order of the _event_loop task and _Init() of a same endpoint must be consistent to make sure the task_id value of _event_loop(uint8 task_id, uint16 events) is equal to the taskID value obtained from _Init(taskID).

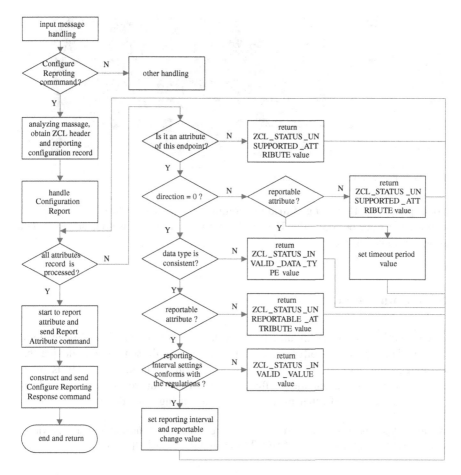

Fig. 85.4 Flow chart of analysis of configure reporting command and sending of report attribute command

85.4 Analysis of RF Data of Light Sensor Node

Under Z-Stack 2.3.0, the software programming of this light sensor is completed. CC2530 development suite is utilized to form a simple ZigBee network. After the coordinator conducts attribute configuration for light sensor node, the light sensor sends attribute value report on a regular basis. SmartRF^{TM} Packet Sniffer is utilized to capture packets. The result is shown in Fig. 85.5:

We can see from Fig. 85.5 that node with source address of $0 \times 796F$ sends attribute reporting data with Profile Id of 0×0104 (HA profile ID) and Cluster Id of 0×0400 (illuminance measurement cluster ID) to coordinator with the destination address of 0×0000 every certain time interval in the network with PAN ID of $0 \times 5B61$. After receiving Attribute Configure command, the light sensor

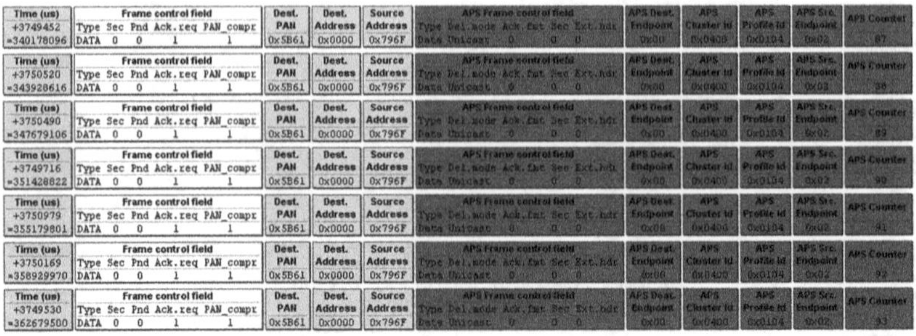

Fig. 85.5 Packet information diagram

node generates Reporting Attribute command and sends Attribute Report to coordinator node periodically. The result demonstrates that this node achieves data transmission accurately.

85.5 Conclusion

HA network of ZigBee Alliance is a development direction of intelligent home system. This article has offered a design scheme of light sensor node device that complies with ZigBee HA specification based on Z-Stack. The analysis result demonstrates that the scheme is effective. And it is of universal property and can apply in the design of other measurement and sensing node devices.

Acknowledgments This work is supported by scientific research fund of Zhejiang provincial education department (No. Y201016516). Thanks for the financial supports.

References

1. Qu L, Liu SD, Hu XB (2007) ZigBee technology and application. Beijing university of aeronautics and astronautics press, Beijing (in Chinese)
2. ZigBee Alliance (2010) ZigBee home automation public application profile. http://www. zigbee.org/Standards/ZigBeeHomeAutomation/Overview.aspx. Cited 08 Feb 2010
3. ZigBee Alliance (2008) ZigBee cluster library specification. http://www. zigbee.org/Standards/ Downloads.aspx. Cited 29 May 2008
4. Zhong YF, Liu YJ (2011) ZigBee wireless sensor network. Beijing university of posts and telecommunications press, Beijing (in Chinese)
5. Li CG, Liu WP, Hu CY, Hou Z (2010) Design and achievement of WSN node based on ARM and zigbee. J Comput Eng 36(17):135–137 (in Chinese)
6. Texas Instruments Incorporated (2009) A true system-on-chip solution for 2.4-GHz IEEE 802.15.4 and ZigBee applications. http://www.ti.com.cn/product/cn/cc2530,2011-02.247711. HTM. Cited 31 July 2009

Chapter 86
Functional Safety Assessment for the Carborne ATP of Train Control System

Yang Li and Dongqin Feng

Abstract Automatic train protection system (ATP) is the key safety-related system of train control system, so its safety reliability requirements are much higher. This article studies functional safety assessment (FSA) including function analysis, hazard identification, risk rank evaluation, risk reduction, and safety integrity verification during the development of carborne ATP. A risk rank evaluation method is proposed combining as low as reasonably practically (ALARP) criteria and risk matrix. Many safety defects of carborne ATP are found and then the design scheme is improved through FSA. The results of risk analysis software Isograph prove that carborne ATP meets the safety integrity level of its design goals at last.

Keywords Functional safety assessment · ATP · Risk assessment · Risk rank evaluation matrix · ALARP · Safety integrity level

86.1 Introduction

China train control system (CTCS) is researched and developed independently by China. ATP of CTCS is the key equipment to ensure train running safety. FSA is imperative to guarantee that ATP meets its high SIL besides implementing

Supported by 863 Program(NO.2012AA041102)

Y. Li · D. Feng (✉)
State Key Laboratory of Industrial Control Technology, Institute of Cyber-Systems and Control, Zhejiang University, NO. 38 ZheDa Road, Hangzhou, China
e-mail: dqfeng@iipc.zju.edu.cn

W. Lu et al. (eds.), *Proceedings of the 2012 International Conference on Information Technology and Software Engineering*, Lecture Notes in Electrical Engineering 210, DOI: 10.1007/978-3-642-34528-9_86, © Springer-Verlag Berlin Heidelberg 2013

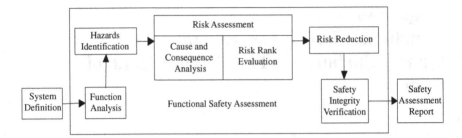

Fig. 86.1 FSA main contents for carborne ATP

technical specifications. FSA should be implemented during the whole life cycle of railway signal system [1] and the safety integrity requirements of it must be distributed reasonably [2].

This study deals with how to detect the risks of carborne ATP and risk rank evaluation matrix is proposed to define its risk ranks. We present risk control measures to reduce the unacceptable risks and verify the SIL of carborne ATP.

86.2 Functional Safety Assessment Method

FSA aims to ensure safety-related systems succeed their safe function. This paper only discusses FSA of carborne ATP in its design phase, standing to EN 50126, EN 50128, EN 50129, EN 50159 and IEC 61508 [3]. Its main contents are illustrated in Fig. 86.1.

86.3 Functional Safety Assessment for Carborne ATP

86.3.1 Design of Carborne ATP

This carborne ATP is designed for CTCS-1. It can be applied to diesel locomotive, electric locomotive and motor train unit. It is required to meet safety integrity level 4 (SIL4) under European railway safety standards.

The main functions of carborne ATP are preventing train overspeed and recording relevant data. External signals are conditioned to standard signal in Terminal board (TB) firstly, and then enter Mother board (MB). MB sends standard signals to each functional module. Controllers receive signals from functional modules and send safety command after operation. As shown in Fig. 86.2, internal modules of ATP mainly include controller (CON, 2-of-4 redundancy), digital interface unit (DIU, 2-of-2 redundancy), speed measure unit (SMU, 2-of-2 redundancy), signal interface unit (SIU, 2-of-2 redundancy), analog interface unit (AIU, 2-of-2 redundancy), MB and TB.I/O interfaces of ATP mainly include

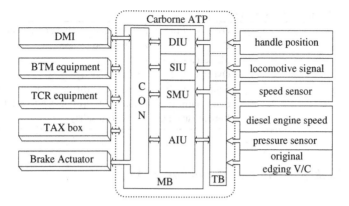

Fig. 86.2 Structure of carborne ATP

driver machine interface (DMI), balise transmission module (BTM), track circuit reader (TCR), TAX, brake actuator, handle position, locomotive signal, speed sensor, diesel engine speed, locomotive pressure sensor and original edging voltage/current.

86.3.2 Function Analysis of Carborne ATP

This section is to realize all sub-functions of main functions. Main functions are divided into several sub-functions with method of Function Fault Tree. Taking a main function failure as the top event, deduce the occurrence reasons layer upon layer, and find the basic cause of the top failure finally. According to its definition ATP has seven main functions as shown in Table 86.1. Sub-functions of slip escape protection are shown in Table 86.2.

86.3.3 Hazard Identification of Carborne ATP

Hazard identification is only for the main functions relevant to safety as shown in Table 86.1. After knowing all sub-functions and external interface, Hazard and Operability Study (HAZOP) should be promoted to find out their hazards of their failure modes and deduce failure modes of main functions. HAZOP results of slip escape protection are shown in Table 86.2. Failure modes of main functions are shown in Table 86.4.

Table 86.1 Main functions of ATP

1	Overspeed of railway line, locomotive and others protection
2	Rash advance railway light protection
3	Slip escape protection
4	Overspeed of dispatch protection
5	Send display information to DMI(irrelevant to safety)
6	Send alarm information to DMI(irrelevant to safety)
7	Record function(irrelevant to safety)

86.3.4 Risk Assessment of Carborne ATP

In order to assess the risks of hazards, cause and consequence analysis is promoted. Deduce possible causes and consequences of hazard; then determine risk rank with risk rank evaluation matrix.

86.3.4.1 Cause and Consequence Analysis of Hazard

HAZOP is commonly used to deduce the causes of hazards. The cause analysis results of slip escape protection are shown in Table 86.2. The next is to analyze the frequencies of hazards. As risk reduction measures have not been applied, the frequency of hazard is occurrence probability of its causes. Upon China Ministry of Railways requirements for railway signal system and characteristics of ATP, probability of cause hazard is divided into seven segments: A, B, C, D, E, F and G, as shown in Table 86.3. This research achieves probabilities of causes with the method of expert marking.

Consequence analysis aims to define severity of hazard effects, such as losses of lives and effects to surroundings. Severity is divided into seven levels: 1, 2, 3, 4, 5, 6 and 7, according to The Investigation and Handling Rules of Railway Traffic Hazard [4] and Emergency Rescue and Treatment Regulations of Railway Traffic Hazard [5], as shown in Table 86.3. This research achieves severity level of hazard with the method of expert marking.

86.3.4.2 Risk Rank Evaluation

Hazards possible frequency and severity of sub-functions have been obtained in cause and consequence analysis, and their risk ranks can be determined according to risk rank evaluation matrix. Risk rank of a main function is the highest one among its sub-functions', and then judge if risk ranks of main functions hazards can be tolerated according to ALARP criteria [6].

Risk rank evaluation matrix is formed with frequency segments and severity levels of hazards. As shown in Table 86.3, specific frequency and severity define a certain risk rank.

Table 86.2 HAZOP analysis results of slip escape protection

The main function	Subfunction	Hazard description	Causes of hazard	Final effects
Slip escape protection	1. Handle state detection	1.1 Unable to detect handle state	1. Handle fault 2. TB circuit fault 3. DIU fault	2A. Not slip escape protection timely
		1.2 Detection handle zero state for non zero state	Same as above	2A. Not slip escape protection timely
		1.3 Detection handle non zero state for zero state	Same as above	2B. Slip escape protection redundantly
	2. Run direction detection	Detection run direction opposite to actual	1. The speed sensor fault 2. SMU fault	2A. Not slip escape protection timely
	3. Speed detection	3.1 No speed	Same as above	2A. Not slip escape protection timely
		3.2 Detection exaggerated high speed	Same as above	2B. Slip escape protection redundantly
		3.3 Detection exaggerated low speed	Same as above	2A. Not slip escape protection timely
		3.4 Speed detection time partial long	1. System communication delay 2. Speed sampling cycle partial long	2A. Not slip escape protection timely
	4. Handle slip calculation	4.1 Error calculate not slip escape by handle state	CON fault	2A. Not slip escape protection timely
		4.2 Error calculate slip escape by handle state	CON fault	2B. Slip escape protection redundantly
	5. Phase slip calculation	5.1 Error calculate not phase slip	CON fault	2A. Not slip escape protection timely
		5.2 Error calculate phase slip	CON fault	2B. Not slip escape protection timely
	6. Brake command output	6.1 Brake command no output timely	CON fault	2A. Not slip escape protection timely
		6.2 Brake command output redundantly	CON fault	2B. Slip escape protection redundantly
	7. Safety relay output	7.1 not output timely	Relay failure	2A. Not slip escape protection timely
		7.2 Brake output redundantly	Relay failure	2B. Slip escape protection redundantly

Table 86.3 Risk rank evaluation matrix

Frequency segment	Severity level	1	2	3	4	5	6	7
	Mortality	0.1	0.3	1	3	10	30	50
	Frequency (time/year)	Risk rank						
A	$1 < f$	R1	R1	R1	R1	R1	R1	R1
B	$10^{-1} < f \leq 1$	R2	R1	R1	R1	R1	R1	R1
C	$10^{-2} < f \leq 10^{-1}$	R3	R2	R2	R1	R1	R1	R1
D	$10^{-3} < f \leq 10^{-2}$	R4	R3	R3	R2	R2	R1	R1
E	$10^{-4} < f \leq 10^{-3}$	R4	R4	R4	R3	R3	R2	R2
F	$10^{-5} < f \leq 10^{-4}$	R4	R4	R4	R4	R4	R3	R3
G	$f \leq 10^{-5}$	R4	R4	R4	R4	R4	R4	R4

Table 86.4 Results of risk rank evaluation and safety integrity verification

Main function failure modes	Risk rank evaluation before risk reduction			Safety integrity verification			
	Frequency segment	Severity level	Risk rank	Frequency	Risk rank	THR	SIL
Not overspeed protection timely	B	6	R1	6.82×10^{-5}	R4	7.781×10^{-9}	SIL4
Overspeed protection redundantly	B	1	R2				
Not rash advance protection timely	B	7	R1	6.82×10^{-5}	R4	7.781×10^{-9}	SIL4
Rash advance protection redundantly	B	2	R2				
Not slip escape protection timely	B	3	R1	6.82×10^{-5}	R4	7.781×10^{-9}	SIL4
Slip escape protection redundantly	B	1	R2				
Not dispatch protection timely	B	4	R1	6.82×10^{-5}	R4	7.781×10^{-9}	SIL4
Dispatch protection redundantly	B	1	R2				

According to ALARP criteria, risk ranks include four ranks: R1, R2, R3 and R4, from high to low. This research defines the values of risk ranks combining British standard Railway applications-Systematic allocation of safety integrity requirements [7] and characteristics of rail industry. Figure 86.3 illustrates the details of ALARP criteria.

The results of risk rank evaluation of main functions before risk reduction are shown in Table 86.4.

Fig. 86.3 ALARP criteria for risk acceptance

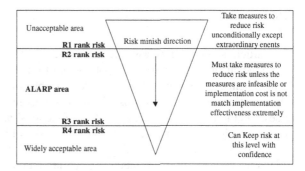

86.3.5 Risk Reduction of Carborne ATP

According to Table 86.4, risk ranks of the hazards numbered are all R1 or R2 which are intolerant. Risk control measures must be taken to reduce the risks. The process of risk reduction has two steps:

- Step 1. Classify intolerable risks by failure modes. Present the maximum possible improvement measures with method of expert brain storm, and then keep preliminary reasonable measures, abandon unreasonable obviously measures.
- Step 2. Analyze the measures kept in step 1 between cost and implementation effectiveness, and record the optimal control measures. Assess the risks of hazard again after taking the optimal measures. If the risks are still intolerable, repeat step 1 until improvement measures are infeasible or implementation cost is not match implementation effectiveness extremely. Or else these risks are considered as the tolerable risk.

Now all the risks of ATP are acceptable after risk reduction and then we can verify if their SIL meets the requirements.

86.3.6 Safety Integrity Verification of Carborne ATP

The safety integrity goal of carborne ATP is SIL4, namely tolerable hazard rate (THR) ranging from 10^{-9} to 10^{-8}/h. Carborne ATP SIL is the lowest one among all its main functions' SIL. In this study, SIL of carborne ATP is verified on the Fault Tree Analyze platform of Isograph. Firstly, taking a main function failure as the top event, build its fault tree model. Next the software estimates the THR of the top event with fault parameters of basic events.

THR of main functions are estimated by Isograph and then their SIL are confirmed according to their THR. The results of safety integrity verification are shown in Table 86.4. According to the verification results, SIL of carborne ATP is SIL4 meeting its safety requirement; its risk rank is R4 which is acceptable according with ALARP criteria.

86.4 Conclusion

This article studies safety assessment during the development of carborne ATP, while technical specifications are not enough to meet the higher safety reliability requirements. We promote function analysis obtaining sub-functions, and hazard identification to achieve hazards of sub-functions failure. Then we promote risk assessment to determine risk ranks of sub-functions' and main functions' hazards, and present improvement measures to reduce risks which are unacceptable. At last we verify SIL of carborne ATP by Isograph, and the results prove that carborne ATP developed with FSA meets its higher requirements of safety reliability absolutely, and that risk rank evaluation matrix proposed in this study can define risk ranks of hazards precisely.

It should be explained that this article only verifies the SIL of hazards whose risk ranks are R1, considering the others' SIL are approximated under the protection of ATP. Besides, all the fault parameters are achieved in the worst case, so the results are conservative.

References

1. Xiao N (2009) Research and application of risk assessment in railway signal system. Information science and technology institute of Southwest Jiaotong University (in Chinese)
2. Liu X (2011) The method to determine and distribution safety integrity requirements of railway signal system. Railway Commun Sig 47(4):55–57 (in Chinese)
3. IEC 61508 (2010) Functional safety of electrical/electronic/programmable electronic safety-related systems
4. Order of China Ministry of Railways (2007) The investigation and handling rules of railway traffic hazard, 30. (in Chinese)
5. Order of the State Council of China (2007) Emergency rescue and treatment regulations of railway traffic hazard, 501. (in Chinese)
6. Melchers RE (2001) On the ALARP approach to risk management. Reliab Eng Syst Saf 71:201–208
7. PD CLC/TR 50451 (2007) Railway applications-systematic allocation of safety integrity requirements

Chapter 87
Research of Personal Retrieval for Scientific Research Information Management System

Hu Liang and Xie Suping

Abstract This paper introduce the scientific research information classification of search results for solving some problem of information retrieval based on the personalized service. The classification based on K-NN algorithm improves the search accuracy of the specified topic information based on studying the sample. The test results presents the technologies can improve the search quality of specified field.

Keywords Information retrieval · Information classification · K-NN algorithm · Scientific research information system

87.1 Introduction

Traditional retrieval process of research and information management system is that firstly query based on user search terms, and then sort the results by relevance. Retrieve result sets usually are very much. The problems of polysemy, multi-word synonymous and differences in the existence of user habits, make the search results can't fully meet user needs, so to build the result set displaying form which facilitate the users to quickly find the information is a very important research direction. Statistical research shows that the average search term is short in length and rarely

H. Liang (✉)
College of Information Science, Beijing Language and Culture University, 3th Floor, Main building, 15 Xueyuan Road, Haidian, Beijing, China
e-mail: huliang@blcu.edu.cn

H. Liang · X. Suping
Computer and Information Management Center, Tsinghua University, Beijing, China
e-mail: xsp@cic.tsinghua.edu.cn

W. Lu et al. (eds.), *Proceedings of the 2012 International Conference on Information Technology and Software Engineering*, Lecture Notes in Electrical Engineering 210, DOI: 10.1007/978-3-642-34528-9_87, © Springer-Verlag Berlin Heidelberg 2013

use specialized query operators. Therefore, to consider the phrase relationship between the query words entered by the user or location adjacent relationship is very important to improve the retrieval results in the actual retrieval [1].

In order to improve the retrieval quality, this study organizes search results in a hierarchical manner, so that users can easily select a category, improve the efficiency and find research information faster. The realization process is that information retrieval module uses the K-NN algorithm to categorize search results based on the theme of information, and divides the search results into a simple hierarchy, which can both convenient the user searching and improve efficiency [2].

87.2 Concept and Classification of Scientific Research Information

In this study, the research information refers the primary data after once processing. It can identify the physical symbols to reflect the objective world, which includes documents, statements, figures and their data produced in research activities. The concept includes the meaning of two aspects: the objectivity and identification of the data. For example, we regard the rice as data, then the flour is the primary data after once processing and the noodle is data after secondary processing. Here the boundaries of the primary data is the processing degrees. Our standard is that information is new information which is just simply classified or graded physically, has not put too much labor, does not exceed the value of its own value-added, has no chemical changes and no other added information.

Research information classification is a very complex issue. In addition to the reasons of a wide variety and surprising number of information, the classification criteria is not unified and there exists many different habitual classification method [3–5]. The research information classification we study in this research refers to the process of automatically discriminating information categories according to the content of the information process under a given classification system [6–8]. Common research information classification is a multi-layer classification. Such as news, projects, institutions, achievements, database, legal policy is the class for the first level; achievements can be divided into papers, awards, patents and software copyrights, which is the second level; papers can also be divided into SCI, EI and ISTP, which is the third level; SCI can also be divided into the journal and Online version, which is the fourth level. It can also be divided continued. This is determined by the rich hierarchical nature of the objective world. The multi-layer classification for humankind is natural and easy to understand. However, it is quite complex for computers. Complex classification will increase the computational complexity and reduce efficiency. This study will simplify the research information into to three-layer tree. The first layer is the major categories of research information, such as news and achievements, the second layer is subclasses of the major class, such as papers, awarding, monographs, patents and software

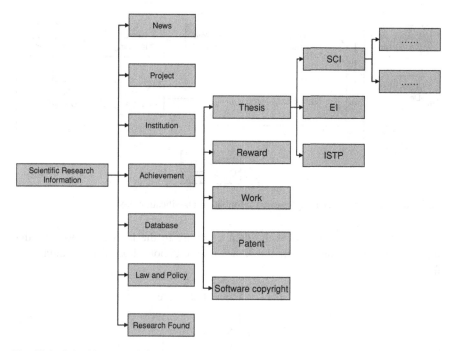

Fig. 87.1 Scientific research information classification

copyrights, the third layer is the theme of subclasses, such as the SCI, EI and ISTP papers and the national Natural science Foundation, the national Social science Foundation and the National 863 Program which is belonging to projects. The diagram is shown in Fig. 87.1.

87.3 Scientific Research Information Classification Based on K-NN

The scientific information classification model includes data collection, data processing and classifier. The data collection is used to solve the problem of data source, the data processing is used to process the original text and the classifier realize the function that sort the data according to the data property, as shown in Fig. 87.2.

The classifier construction is the key technology of the model. This study adopts the K-Nearest Neighbor (K-NN) algorithm to classify the text, because the K-NN algorithm is efficient and simple that probably realizes the same effect as the Bayesian Classification in theory. The K-NN algorithm computes the K nearest neighbor of the search items in the training sample set, finds the class of the most of the K nearest neighbor, and then classifies the search item as the class.

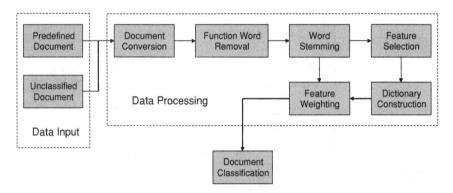

Fig. 87.2 Structure of scientific research information classification model

The idea of classification is the data x belong to the kind of sample set that includes the most samples in the k nearest neighbor of x. The formula is as following:

$$d_i(x) = k_i, i = 1, 2, \ldots, c$$

$$k(\overrightarrow{x}, \overrightarrow{y}) = \left[\sum_{i=1}^{n} (x_i - y_i)^2 \right]^{1/2} \tag{87.1}$$

The rule:

$$m = \max_{i=1,2\ldots,c} (d_i(x)), x \in \omega_m, \tag{87.2}$$

The algorithm flow is as following:

Step 1. Collect and process the data.

Step 2. Design the training set and testing set.

Step 3. Set the parameter k.

Step 4. Build a queue that is sorted by the distance and store the nearest training set.

Step 5. Get randomly the k items from the training set as the initial nearest items, compute the Euclidean distance of the k items from the training set, and then store these data in the queue.

Step 6. Traverse the training set, compute the distance L of the training set from the testing set, compare the L with the Lmax, and if L >= Lmax then visit the next item else store the current item in the queue.

Step 7. Find the class that includes the most items in the queue.

Step 8. Set the different parameter k, repeat the above steps, and finally get the parameter k of the lowest error.

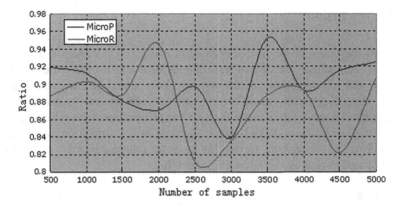

Fig. 87.3 Classification comparison of different samples

87.4 Performance Testing

For testing the performance, this study collects 108,238 data of scientific research, filters the distractions that will affect the classification, extracts the simple abstract of the information, establishes the semantic dictionary, builds the corresponding vector space model, and then builds the sample set for classification. The system computes the Euclidean distance between the search result set and the sample set by the K-NN algorithm and classify the search result. The testing adopts the MicroP and MicroR as the assessment index. The formula is as follows:

$$MicroP = \sum_{i=1}^{N} A_i \left/ \left(\sum_{i=1}^{N} A_i + \sum_{i=1}^{N} B_i \right) \right. \quad MicroR = \sum_{i=1}^{N} A_i \left/ \left(\sum_{i=1}^{N} A_i + \sum_{i=1}^{N} C_i \right) \right.$$

$$(87.3)$$

where MicroP is the average precision, MicroR is the average recall, N represents the total of the scientific research information classes, A_i represents the number of the correct data in the search results, B_i represents the number of the incorrect data in the search results, C_i represents the data that is correct but is not classified in the search result.

The search result is displayed according to the information levels in Fig. 87.3 and the system has higher average precision and recall, more than 80 %. As the chart shows, the dependency of the samples amount in our system is lower than the common system because the preprocessed data structure in the scientific research fields has more semantics that improve the classification performance.

87.5 Conclusion

Although the scientific research information management system can meet the basic demands of user, the cost of scientific research information retrieval doesn't decrease with the increasing of the acquired information quantity. For this, we build a personal retrieval system of scientific research information for improving the search efficiency and decrease the cost the information retrieval of user. The system provides not only the rich information retrieval, information customization and information comparison service but also the valued report for user through recording and analyzing the keywords search log.

Acknowledgments This author's work is supported by the Fundamental Research Funds for the Central Universities (XK201205) and National Key Technology R&D Program (2012BAH10F02 and 2012BAH10F02-A).

References

1. Rocchio J (1971) Relevance feedback in information retrieval. In: The SMART retrieval system: experiments in automatic document processing. Salton G Prentice-Hall, Englewood Cliffs, pp 313–323
2. Guo G, Wang H, Bell D (2003) K-NN model-based approach in classification. In: On the move to meaningful internet systems 2003: CoopIS, DOA, and ODBASE, LNCS 2888, Berlin, Springer, pp 986–996
3. Chen X (2006) The cluster analysis research in data mining. Comput Technol Dev 16(9):44–45 (in Chinese)
4. Yuan W, Gao M (2006) The study of personalization for search engine. Microelectron Comput 23(2):68–72 (in Chinese)
5. Liu J, Huang H (2006) The availability of search engine by classification and clustering. Railway Comput Appl 15(3):44–46 (in Chinese)
6. Dong H, Yu C (2005) Chinese ontology algorithm analysis of automatic acquisition and evaluation. Inf Theory Appl 28(4):415–418 (in Chinese)
7. Wang Z, Liu N (2009) Classification theory research for information retrieval based on pragmatic-philosophical view. J Libr Sci 31(11):1–4 (in Chinese)
8. Yao D, Zhao X, Wei Y (2010) The ontology study for information retrieval. Software Guide 8:42–43

Chapter 88
On the Average Diameter of Directed Double-Loop Networks

Sheng Li and Ying Li

Abstract In computer network system, the diameter always be represented the worst delay in the communication between any two nodes, it is an important factor to measure transmission efficiency, and the average diameter is the average delay between any two nodes, can it be measured transmission efficiency? The minimum distance diagram of a directed double-loop network is always an L-shaped tile, which can be described with four geometric parameters: a, b, p and q. According to the L-shaped tile, we proved that the average diameter of directed double-loop network was also up to four geometric parameters: a, b, p and q, we provided two algorithms to compute them, and simulated the relationship between average diameter and diameter. Simulation indicates that the distribution of the diameter and the average diameter are all axis-symmetrical figure; the average diameter approaches to a half of the diameter at the same point; the diameter does not always obtain minimum value, when the average diameter obtains minimum value. Finally, we verified that the lower bounds about the diameter and the average diameter of the directed double-loop networks provided by Wong and Coppersmith were correct, and confirmed Fang et al.'s suspicions. So we get that the average diameter also can be measured transmission efficiency in computer network system.

Keywords Directed double-loop networks · Diameter · Average diameter · L-shaped tile · Transmission efficiency · Shortest path

S. Li (✉)
Computer Department of Maanshan Vocational Technical College, Ma'anshan 243031 Anhui, China
e-mail: jsj818@yahoo.cn

Y. Li
Maanshan Teacher's College, Ma'anshan 243041 Anhui, China

W. Lu et al. (eds.), *Proceedings of the 2012 International Conference on Information Technology and Software Engineering*, Lecture Notes in Electrical Engineering 210, DOI: 10.1007/978-3-642-34528-9_88, © Springer-Verlag Berlin Heidelberg 2013

88.1 Introduction

Wong and Coppersmith [1] introduced the multiloop networks ML $(N; s_1,...,s_l)$ for organizing multimodule memory devices, double-loop network is one of them. Double-loop networks find applications in the design of communication networks, multimodule memory structures, data alignment in parallel memory systems, and supercomputer organizations.

A double-loop network is characterized by its number N of nodes and by its two chord lengths r and s. In this paper, we denote a double-loop network as $G(N; r, s)$, where the nodes are numbered 0–N−1, two chord lengths r and s satisfy $1 \leq r \neq s < N$, and each node $v(0 \leq v < N)$ is connected to each of the two nodes $v + r$ and $v + s$ by directed link, all node-index expressions are evaluated modulo N. Formally, a double-loop network $G(N; r, s)$ is a directed graph $G(V, E)$, where the vertex set of G is

$$V = \{0, 1, \ldots, N - 1\},$$

the edge set of G is

$$E = \{v \rightarrow v + r(\text{mod } N), v \rightarrow v + s(\text{mod } N)|v = 0, 1, \ldots, N - 1\}.$$

Figure 88.1 depicts a directed double-loop network $G(12; 1, 3)$ as an illustrative example.

Past research on double-loop network has dealt with both directed networks and undirected ones. Problems studied in connection with directed double-loop networks, are routing algorithms [2, 3], determination of graph diameter [4, 5], optimal assignment of chord lengths [6], construction of optimal double-loop networks [7–10], the wide diameter [11], and fault-tolerant [12, 13], etc. For more

Fig. 88.1 The directed double-loop network G(12;1,3)

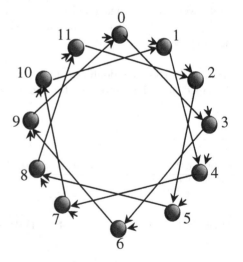

detail on properties of double-loop networks and their applications, we refer the reader to the survey papers written by Bermond et al. [14] and Hwang [15].

The current survey will minimize its overlapping with them by focusing on new methods. These include not just updated versions of important results, but also results being somewhat overlooked before, or simply results which we find a new way to organize. In this paper, we will research the average diameter of directed double-loop networks, which mention rarely in past literature. In Sect. 88.2, we will discuss the minimum distance diagram (MDD) of directed double-loop networks, it closely relate with the average diameter. In Sect. 88.3, we will get the formulas of average diameter by the L-shaped tiles. In Sect. 88.4, we will provide the algorithms to compute the average diameter rapidly. In Sect. 88.5, we try to get the relationship between the diameter and the average diameter by simulation results, and validate some important conclusions.

88.2 The Minimum Distance Diagram of Directed Double-Loop Networks

It is well known that a regular digraph is strongly connected if and only if it is connected. Since $G(N; r, s)$ is a 2-regular diagraph, we will substitute connectivity for strong connectivity throughout the chapter for brevity. When $G(N; r, s)$ is connected, it must satisfy Theorem 1 [15].

Theorem 1 $G(N, r, s)$ *is connected if and only if* $gcd(N; r, s) = 1$.

We can define a minimum distance diagram (MDD) in the rectangular coordinate system. Let node 0 occupy cell(0,0), and node v occupy cell(i, j) if and only if $ir + js \equiv v \pmod{N}$ and $i + j$ is the minimum among all (i, j) satisfying the congruence, where $\equiv$ means congruent modulo N. The MDD of $G(N; r, s)$ includes every node exactly once (in case of two shortest paths, the convention is to choose the cell with the smaller row index, i.e., the smaller j). Note that in a cell(i, j), i is the index of x-axis and j is the index of y-axis, respectively. Since $G(N; r, s)$ is node-symmetric, the minimum distance from u to v is same as from 0 to $v-u$. Wong and Coppersmith proved that the MDD of a directed double-loop network is always an L-shape (Fig. 88.2a) which can be characterized by four parameters a, b, p, q, we call it L-shaped tile, denote it by $L(N; a, b, p, q)$. Figure 88.2b gives the MDD of $G(10; 1, 3)$ in its L-shape.

The diameter of directed double-loop networks is the maximum of the shortest path between any two nodes; it represents the worst delay in the communication between two nodes. Research shows that the diameter can be obtained from the parameters of L-shaped title. For example, the diameter of $G(N; r, s)$, written as $D(N; r, s)$, is $\max\{a + b - p - 2, a + b - q - 2\}$.

Fig. 88.2 L-shaped tile

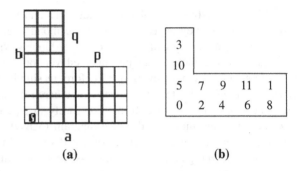

(a)   (b)

88.3 The Formula of Average Diameter

Fang and Tang [16] gave a survey on the average diameter of directed double-loop networks, but they did not obtain the formula of the average diameter, they only provided some simply results about it.

The average diameter of directed double-loop networks is the average distance between any two nodes, it represents the average delay in the communication, written as $\overline{D}(N; r, s)$. It should be noted that any distance function can be obtained from the L-shaped title directly, then how to get the function about the average diameter?

Because $G(N; r, s)$ is vertex-symmetric, The distance between node u and node v equals the distance node 0 to $u-v$, the average distance can be computed by node 0 to any other nodes. The L-shaped title of directed double-loop network can be seen as the area difference between a big rectangle (length is a, width is b, Fig. 88.3a) and a small one (length is p, width is q, Fig. 88.3b). The average diameter can be seen as the average of distance difference between node 0 to any other nodes in big rectangle and in small one. We denote the sum of distance from node 0 to any other nodes in big rectangle by φ, and denote the sum of distance from node 0 to any other nodes in small rectangle by γ. Because γ contains two parts, one is the sum of distance from node 0 to any other nodes in small rectangle, the other is the sum of distance between two nodes 0, and we denote them by δ and ζ, respectively, so we have:

$$\overline{D}(N; r, s) = \frac{1}{N}(\phi - \gamma) = \frac{1}{N}(\phi - \delta - \zeta).$$

Since, in big rectangle,

$$\begin{aligned}
\varphi &= [1 + 2 + 3 + \ldots + (a - 1)]b + [a + 2a + 3a + \ldots + (b - 1)a] \\
&= (1 + a - 1)\frac{a - 1}{2}b + a(1 + b - 1)\frac{b - 1}{2} \\
&= \frac{ab}{2}(a - 1) + \frac{ab}{2}(b - 1) \\
&= \frac{ab}{2}(a + b - 2).
\end{aligned}$$

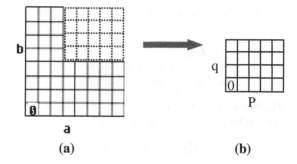

Fig. 88.3 The area difference between two rectangles

In the same way, in small rectangle,

$$\delta = \frac{pq}{2}(p + q - 2).$$

Because the distance between two nodes 0 is $a + b - p - q$, and the number of the path equals the number of the nodes in small rectangle, so we have:

$$\zeta = pq(a + b - p - q),$$

in conclusion, we have:

$$
\begin{aligned}
\overline{D}(N; r, s) &= \frac{1}{N}(\varphi - \delta - \zeta) \\
&= \frac{1}{N}\left[\frac{ab}{2}(a + b - 2) - \frac{pq}{2}(p + q - 2) - pq(a + b - p - q)\right] \\
&= \frac{1}{N}\left[\frac{ab}{2}(a + b - 2) - \frac{pq}{2}(2a + 2b - p - q - 2)\right].
\end{aligned}
$$

88.4 The Methods to Compute Four Parameters of L-Shaped Tile

According to the formula of average diameter, we must get four parameters of L-Shaped tile first if we want to compute $\overline{D}(N; r, s)$. In this chapter, we provide two algorithms to compute four parameters of L-Shaped tile.

88.4.1 Congruence Equation Algorithm

Lemma 1 *An L-shaped tile of directed double-loop network $G(N; r, s)$ is described by four parameters: a, b, p and q, then $(a - p)r + (b - q)s = 0 \ (mod \ N)$, $bs - pr = 0 \ (mod \ N)$, $ar - qs = 0 \ (mod \ N)$ [17].*

Definition 1 *For a given directed double-loop network $G(N; r, s)$, we define four parameters k_1, k_2, j_1, j_2 as follow:*

$$j_2 = Min\{j|ks = jr(mod \ N), j > k \geq 0, j = 1, 2, \ldots, N - 1\} \tag{88.1}$$

$$k_l = Min\{k|ks = jr(mod \ N), k \geq j \geq 0, k = 1, 2, \ldots, N - 1\} \tag{88.2}$$

$$j_1 = \{j|k_1s = jr(mod \ N), j \geq 0\} \tag{88.3}$$

$$k_2 = \{k|ks = j_2r(mod \ N), k \geq 0\} \tag{88.4}$$

Theorem 2 *For a given directed double-loop network $G(N; r, s)$, if four parameters k_1, k_2, j_1, j_2 define as **Definition 1**, if L-shaped tile of directed double-loop network $G(N; r, s)$ is described by four parameters: a, b, p and q, then $a = j_2$, $b = k_1$, $p = j_1$, $q = k_2$.*

Proof

Case 1: According to the definition of L-shaped tile of directed double-loop network $G(N; r, s)$, we get that a is the number of the cells on X-axis. From the congruence Eq. (88.1) $ks = jr(mod \ N)$, $j > k \geq 0$, we know that j_2 is the minimum step, in which there is same node on X-axis and Y-axis. Because there isn't same node in L-shaped tile, so there are $j_2 - 1$ steps on X-axis, one step links two nodes, so there are j_2 nodes on X-axis, so $a = j_2$.

Case 2: On the same principle, by congruence Eq. (88.2) we have $b = k_1$.

Case 3: By congruence Eq. (88.3), we have: $k_1s = j_1r(mod \ N)$, $j \geq 0$. It follows from **Lemma 1**, we have $bs - pr = 0 \ (mod \ N)$, so $bs = pr \ (mod \ N)$. Because $b = k_1$, so $p = j_1$.

Case 4: On the same principle, by congruence Eq. (88.4) we have $q = k_2$.

Algorithm

Input: $N, r, s, gcd(N; r, s) = 1, 1 \leq r < s < N$.

   Output: a, b, p, q.

Step1. Input N, r, s, for any $G(N; r, s)$, set $a\_found = FALSE$. $b\_found = -FALSE$. a_found is a flag to mark a has found, b_found is a flag to mark b has found, their default are FALSE.

Step2. Let array $aNode[N]$ and $bNode[N]$ store the nodes on a side and b side respectively. Let i vary from 1 to $N-1$, $aNode[i] = ir \ (mod \ N)$, $bNode[i] = is \ (mod \ N)$.

Step3. By **Definition 1 (1), (4)**, if $aNode[i] == bNode[j]$, and $i > j$, then we can get $a = i$. $q = j$. Use the same way, by **Definition 1 (2), (3)**, if $aNode[i] == bNode[j]$, and $j > i$, then we can get $b = j$, $p = i$.

Step4. Output a, b, p, q.

88.4.2 Inequality Algorithm

For any $G(N; r, s)$, assume $\gcd(N; r, s) = 1$, there are four steps in this algorithm as below.

Step 1. Set $l_{-1} = N$. Let $0 \leq l_0 \leq N$ be the integer satisfying

$$rl_0 + s \equiv 0 (\mod N),$$

and let l_i, q_i, $1 \leq i \leq m + 1$, be recursively defined by

$$l_{i-2} = q_i l_{i-1} + l_i, \, 0 \leq l_i \leq l_{i-1},$$

where m is chosen such that $l_{m + 1} = 0$.

Step 2. Define $U_{-1} = 0$, $U_0 = 1$ and

$$U_{i+1} = q_{i+1} U_i + U_{i-1}, \, 0 \leq i \leq m.$$

Note that l_i and U_i are decreasing in i. Hence

$$\frac{l_{m+1}}{U_{m+1}} < \frac{l_m}{U_m} < \ldots < \frac{l_0}{U_0} < \frac{l_{-1}}{U_{-1}}.$$

Step 3. Let u be the largest odd integer such that $\frac{l_u}{U_u} > 1$. Define

$$v = \left\lceil \frac{l_u - U_u}{l_{u+1} + U_u} \right\rceil - 1.$$

Step 4. $a = l_u - v l_{u + 1}$, $b = U_u + (v + 1) U_{u + 1}$, $p = l_u - (v + 1) \, l_{u + 1}$, $q = U_u - v U_{u + 1}$.

Example 2 For G(50;1,19), we have

$$50 = 1 \times 31 + 19, U_1 = 1 \times 1 + 0 = 1,$$
$$31 = 1 \times 19 + 12, U_2 = 1 \times 1 + 1 = 2,$$
$$19 = 1 \times 12 + 7, U_3 = 1 \times 2 + 1 = 3,$$
$$12 = 1 \times 7 + 5, U_4 = 1 \times 3 + 2 = 5,$$
$$7 = 1 \times 5 + 2, U_5 = 1 \times 5 + 3 = 8,$$
$$5 = 2 \times 2 + 1, U_6 = 2 \times 8 + 5 = 21,$$
$$2 = 2 \times 1 + 0, U_7 = 2 \times 21 + 8 = 50.$$

From $\frac{0}{50} < \frac{1}{21} < \frac{2}{8} < \frac{5}{5} < \frac{7}{3} < \frac{12}{2} < \frac{19}{1} < \frac{31}{1} < \frac{50}{0}$, we find $u = 3$.

Hence $v = 0$, $a = 7 - 0 \times 5 = 7$, $b = 3 + 1 \times 5 = 8$, $p = 7 - 1 \times 5 = 2$, $q = 3 - 0 \times 5 = 3$.

88.5 Relationship Between Diameter and Average Diameter

Wong and Coppersmith [1] proved

Theorem 3 $D(N; r, s) \geq \sqrt{3N} - 2$ and $\overline{D}(N; r, s) \geq \sqrt{25N/27} - 1$.

A directed double loop network is called tight optimal if its diameter achieves the lower bound $\lceil \sqrt{3N} \rceil - 2$, we denote the lower bound of $D(N; r, s)$ and $\overline{D}(N; r, s)$ as $l_b(D(N))$ and $l_b(\overline{D}(N))$, respectively. For any tight optimal double-loop $G(N; r, s)$, if its $\overline{D}(N; r, s)$ equals $l_b(\overline{D}(N))$, then $G(N; r, s)$ is called double optimal double-loop network. $D(N; r, s)$ is an important parameter to measure transmission efficiency of double-loop networks, then, can $\overline{D}(N; r, s)$ also be an important parameter as $D(N; r, s)$? What are relationships between $D(N; r, s)$ and $\overline{D}(N; r, s)$? Is there lower bound about the average diameter? These will describe in more detail by simulation examples below.

We realized the algorithm above with Java programming language. For any double-loop networks $(D(N; r, s))$ whose N is given random, r and s satisfy $1 \leq r \neq s < N$, their diameter and average diameter can be calculated and simulated respectively. For example, when $N = 50$, $r = 1$, s varies from 2 to $N - 1$, Simulation results are shown in Fig. 88.4. We substitute $D(N)$ and $\overline{D}(N)$ for $D(N; r, s)$ and $\overline{D}(N; r, s)$ for brevity. Many simulation experiments show that below:

1. For any directed double-loop network $G(N; r, s)$ whose N and r are fixed but s varies from $r + 1$ to $N - 1$, the distribution of their $D(N; r, s)$ and $\overline{D}(N; r, s)$ are all axis-symmetrical figure, and $\overline{D}(N; r, s)$ approaches to a half of $D(N; r, s)$ for the same N.
2. For any directed double-loop networks, there are lower bounds for $D(N; r, s)$ and $\overline{D}(N; r, s)$, whose values conform to Theorem 3.
3. $D(N; r, s)$ does not always obtain minimum value when $\overline{D}(N; r, s)$ obtains minimum value, but approaches to minimum value.

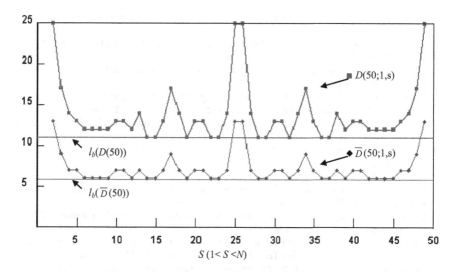

Fig. 88.4 The distribution about $D(50; 1, s)$ and $\overline{D}(50; 1, s)$

88.6 Conclusions

Simulation results show that $\overline{D}(N; r, s)$ and $D(N; r, s)$ can all be a good way to measure transmission efficiency. For double-loop networks, any distance function can be obtained from the L-shaped title directly, we provide a formula for $\overline{D}(N; r, s)$, and present two algorithms to compute it rapidly. Theorem 3 gives the lower bound of $\overline{D}(N; r, s)$ and $D(N; r, s)$, through the analysis of the experiment's result, we get that Theorem 3 is correct which confirmed Fang et al.'s [16] suspicions. According to the relationship between $\overline{D}(N; r, s)$ and $D(N; r, s)$, we find that there are many double optimal directed double-loop networks. Deriving these, formulas constitute a possible direction for further research, as it is the exploration of other topological and algorithmic properties of double loop networks.

Acknowledgments This research was supported by the Natural Science Foundation of Anhui Province under contract KJ2010A343. Sincere thanks are due to those anonymous referees who provided unusually detailed and very helpful comments that improved the accuracy and clarity of our presentation.

References

1. Wong CK, Coppersmith D (1974) A combinatorial problem related to multimodule organizations. J Assoc Comput 21(3):392–402
2. Xiebin Chen (2004) An optimal routing algorithm for double loop networks with restricted steps. Chin J Comput 27(5):596–600 (in Chinese)
3. Gómez D, Gutierrez J, Ibeas Á (2007) Optimal routing in double loop networks. Theoret Comput Sci 381(1–3):68–85

4. Cheng Y, Hwang FK (1988) Diameters of weighted double loop networks. J Algebra 9:401–410
5. Chen B, Xiao W, Parhami B (2005) Diameter formulas for a class of undirected double-loop networks. J Interconnect Netw 6(1):1–15 (in Chinese)
6. Chi-Feng Chan R, Chen C, Hong ZX (2002) A simple algorithm to find the steps of double-loop networks. Discrete Appl Math 121(1–3):61–72
7. Chen C, Lan JK, Tang WS (2006) An efficient algorithm to find a double-loop network that realizes a given L-shape. Theoret Comput Sci 359(1–3):69–76 (in Chinese)
8. Aguiló F, Fiol MA (1995) An efficient algorithm to find optimal double loop networks. Discrete Math 138(1–3):15–29
9. Li Q, Xu JM, Zhang ZL (1993) Infinite families of optimal double-loop networks. Sci China 23:979–992 (in Chinese)
10. Chen BX, Xiao WJ (2004) Optimal designs of directed double-loop networks//International symposium on computational and information sciences (CIS'04). Lecture notes in computer science (LNCS) 3314:19–24. Springer, Berlin (in Chinese)
11. Chen Y, Li Y, Wang J (2011) On the wide diameter of directed double-loop network. J Netw Comput Appl 34:692–696 (in Chinese)
12. Chen Y, Wang J, Li Y (2002) Fault-tolerant routing and diameters of directed double-loop networks. J Huazhong Univ Sci Technol (Nat Sci Ed) 38(2):12–15 (in Chinese)
13. Dharmasena, HP, Yan X (2005) An optimal fault-tolerant routing algorithm for weighted bidirectional double-loop networks. IEEE Trans Parallel Distrib Sys 16(9):841–852, 12
14. Bermond JC, Comellas F, Hsu DF (1995) Distributed loop computer networks: a survey. J Parallel Distrib Comput 24:2–10
15. Hwang FK (2001) A complementary survey on double-loop networks. Theoret Comput Sci 263:211–229
16. Fang MY, Tang HX (2009) A survey on the average diameter of double-loop networks with non-unit step. J Huazhong Univ Sci Technol (Nat Sci Ed) 37(6):12–15 (in Chinese)
17. Fiol MA, Yebra JL, Alegre I et al (1987) A discrete optimization problem in local networks and data alignment. IEEE Trans Comput 36:702–713

Chapter 89
A Real-Time SFM Method in Augmented Reality

Hongyan Quan and Maomao Wu

Abstract Augmented Reality technology is widely used. In this paper, we proposed an efficient method to obtain the satisfied SFM result in real-time using clustering method. First we use Lucas-Kanade method to calculate the optical flow of adjacent frames. We take advantage of the better movement features to calculate the fundamental matrix and use Sturm's method to restore the camera intrinsic parameters. We further use SVD decomposition to obtain the extrinsic parameters of adjacent frames. In order to obtain more accurate camera intrinsic and extrinsic parameters, we use clustering method to get the reconstruction result and further to amend the intrinsic and extrinsic parameters. Further experiments proved the validity of the method. The proposed method is a real-time algorithm and can meet the requirements of the study of augmented reality.

Keywords SFM · Intrinsic parameters · Extrinsic parameters · Reconstruction

89.1 Introduction

Fast and accurate reconstruction of the real scene is a key issue in augmented reality technology. At present, reconstruction technique is widely used in urban design [1], entertainment [2], health [3] and other fields. In the study of reconstruction, camera calibration is the key technology. At present there are different levels of bottleneck in the existing camera calibration method and this prevent it

H. Quan (✉) · M. Wu
East China Normal University Science, Building, B219, No. 3663,
Zhongshan North Road, Shanghai 200062, China
e-mail: ynyn888@163.com; hyquan@sei.ecnu.edu.cn

W. Lu et al. (eds.), *Proceedings of the 2012 International Conference on Information Technology and Software Engineering*, Lecture Notes in Electrical Engineering 210, DOI: 10.1007/978-3-642-34528-9_89, © Springer-Verlag Berlin Heidelberg 2013

from developing and application in practical. How to fast and accurate calculate the structure and motion parameters of the camera is the key problem to reconstruct and interact with the augmented reality environment.

Two existing calibration methods are traditional and self-calibration methods. The traditional calibration method is to use template of known structure as object to calibrate the camera. In order to obtain more accurate results it often takes optimal measures. Zhang [4] proposed a linear two-plane method. The traditional approach has the advantage of high precision, but can not be applied in practice because of using template. The common used self-calibration methods have Kruppa equations based method [5, 6] and absolute quadric based method [7, 8]. Structure from Motion (Abbreviated as SFM) is the method to obtain structure of the camera from object motion analysis [9, 10]. The main problems of existing SFM methods are how to obtain more accurate features of movement and how to calibrate in real-time [11, 12]. In order to overcome the shortcoming of existing SFM method, this paper proposes a real-time SFM method based on clustering, comparing with the existing methods, it has several characteristics:

- It can carry out the parameter recovering with a single video taking by an ordinary capturing device. It is very convenient for people to mixture the virtual object into the environment of augmented reality.
- It takes the measure of tracing good feature and this can keep the accuracy of the results.
- It takes the measure of clustering method to obtain more accurate reconstruction results and this can ensure the validity of the results.

This paper proceeds as follow. Sect. 89.2 describes the method of the real-time SFM. In Sect. 89.3 we will show experimental studies. The performance of running time is also discussed in Sect. 89.3, and it is concluded in Sect. 89.4.

89.2 Real-Time SFM Method

89.2.1 Selecting Good Features Using Continuity of Movement

For the continuity of movement, camera motion has small change between the adjacent frames. In the study, we regard the pixels of image as the particles. We first calculate the dense motion vector the particles and then track the particles to obtain more accurate movement vector. Lucas-Kanade (Abbreviation as LK) method [13] is a classical method of calculating motion vector. Due to its fast convergence speed and more accurate result, it has been widely used.

Because of the illumination and noise, these can bring calculation error to the movement results. In order to obtain more accurate result, we use the continuity of the movement to constrain. We always regard the current frame as a reference using continuity of the movement to obtain more accurate motion results from the successive frames until the number of feature point is less than the threshold.

If vec_{i_st} denotes the motion vector between the frame s and t of pixel I, the reference frame is c and $flag_p$ denotes flag of pixel p, the algorithm of selecting good features is described as following:

1. Initialization. For any adjacent two frames m and n set the flags of every pixel to 0 in the reference frame c.
2. If there still some pixels that want to deal with, go to step 3, else go to step 4.
3. For any untreated pixel $p(u, v)$ of frame c and its motion vector is $vec_{p\_cm} = (u_p, v_p)$, if it meets the following condition (just as shown in Fig. 89.1), we set $flag_p$ 1, else set $flag_p$ 0. Go to step 2:

$$vec_{p\_cm} + vec_{k\_mn} = vec_{p\_cn}$$

In which the pixel $k(u + u_p,\ v + v_p)$.

4. Count the amount of pixels which flag is 1 in the frame c and denote num_c.
5. If num_c is larger than num_{thres}, increase m and n, $vec_{p\_cm} = vec_{p\_cn}$, go to step 1, else go to step 6.

Which num_{thres} denotes the threshold of pixel number.

6. End of the algorithm.

In the study we set num_{thres} 100 and can obtain satisfied result.

It can be more accurate in camera parameter calculating to select good feature points using reference frame and the continuity of the particle movement. This can meet the real-time requirement of camera parameter recover.

89.2.2 Calculating Parameters of the Camera

Epipolar geometry is the geometry of the corresponding relations between the two images and it has nothing to do with the structure of the scene, only depends on the camera's internal and external parameters. According to the principle of epipolar geometry, two feature points of corresponding x_1 and x_2 should be distributed in the respective epipolar [14].Then have following result:

Fig. 89.1 Continuity of movement.
$vec_{p\_cm},\ vec_{k\_mn}$ and vec_{p_cn} are the motion vectors from different frames

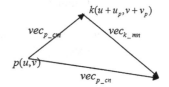

$$x_2^T F x_1 = 0 \qquad (89.1)$$

It is the epipolar constraining equation. In order to calculate fundamental matrix accurately, we use RANSAC algorithm to obtain more robust results.

We take the Sturm method [15] to restore the camera intrinsic parameters matrix K and further extrinsic parameters can be obtained by decomposition essential matrix. Assume the tracking of any feature point $p_i(i = 1, 2, \ldots, n)$ is $t_i = \{t_{ij}|j = 1, 2, \ldots, m\}$, the algorithm of reconstruction is describe as:

1. Calculating foundation matrix of every two adjacent frames using the RANSAC algorithm.
2. Restore the intrinsic parameters of every frame using Sturm method.
3. Essential matrix can be obtained from the intrinsic parameters matrix and the fundamental matrix by formula (89.2).

$$E = K'^T F K \qquad (89.2)$$

4. Extrinsic parameters can be obtained by decomposing the essential matrix with formula (89.3) [16]. And rotation matrix and translation matrix can be obtained.

$$E = U \Lambda V^T \qquad (89.3)$$

5. Calculate the 3D reconstruction result of the feature point p_i in the tracking t_{ij} and denote it as P$_{ij}$.
6. End of the algorithm.

89.2.3 Reconstruction Result Optimization using Clustering

For the calculating error of restoring intrinsic parameters and extrinsic parameters, it is inevitable containing the error in the 3D reconstructing results. In order to improve the accuracy of the 3D point, in the study we take advantage of the clustering measure to obtain the more accurate results.

The main idea of clustering is to cluster the point of having 3D similar positions to the same category according to certain rules. In the processing, we use the affinities of the 3D coordinates of reconstruction results to do the statistics.

For any track point q, the reconstruction results on its trajectory are denoted as $Q_i(i = 1, 2, \ldots, m)$. The number of its denote as m. The optimization algorithm of shortest distance clustering based method is described as follows:

1. Set initialization Category category $= 0$, and initialize each 3D point category for category.
2. The category increase. If all the 3D points are dealt with, go to step 6, otherwise go to next step.

3. Sign the 3D point $Q_j(x_j, y_j, z_j)$ $(j = 1, 2, \ldots, m)$ of 0 category for the current category. If exist another 3D point $Q_k(x_k, y_k, z_k)$ $(k \neq j)$. If Q_j and Q_k meet the condition (89.4), set $flag_{dis}$ of Q_j 1 and go to next step, else set $flag_{dis}$ 0 and go to step 3.

$$\sqrt{(x_j - x_k)^2 + (y_j - y_k)^2 + (z_j - z_k)^2} < threshold_{dis} \qquad (89.4)$$

which $threshold_{dis}$ denotes the threshold of the distance of 3D coordinate.

4. Direction testing of 3D reconstructing results of Q_j and Q_k. If they meet the direction consistency condition (89.5), set $flag_{dir}$ of Q_j 1 and go to next step, else set $flag_{dir}$ 0 and go to step 3.

$$\begin{cases} sign(x_j) \times signx(x_k) \geq 0 \\ sign(y_j) \times sign(y_k) \geq 0 \\ sign(z_j) \times sign(z_k) \geq 0 \end{cases} \qquad (89.5)$$

Which $sign(a)$ denote the direction sign of a.

5. If $flag_{dis}$ and $flag_{dir}$ of Q_j are both 1, set the category Q_k of the current value and go to step 2.
6. End of the algorithm.

In step 3, $threshold_{dis}$ is the threshold of distance between the clustering 3D point and the cluster center. In the experiment we set it 0.1. We think that the larger the cluster the more accurate the results. We judge the cluster by it size. If the number of cluster is meeting the condition (89.6), we think it is satisfied. We call it main categories. The results in the main categories we set it flag $flag_{3d}$ 1.

$$\frac{num_k}{n} > threshold_{sample} \qquad (89.6)$$

num_k denotes the samples of the kth class. n denotes the total number of the samples. In the study we set $threshold_{sample}$ 0.8 and obtain more satisfied results.

At last we use the coordinate's average of the 3D points which $flag_{3d}$ is 1.

After the shortest distance clustering, we obtain the more accurate results of 3D reconstruction. We further fix the intrinsic and extrinsic parameters of the cameras of every frame.

89.3 Experiments Results

The hardware environment is microcomputer with dual-core CPU and 4 GB Memory. The operating system is Windows 7.0. We use Visual C++ 6.0 development environment, OpenCV and OpenGL package to build the experimental platform. The verified experiment includes two parts. The first is the experiment is to verify the validity of the proposed method; the second is to verify and discuss its time performance.

89.3.1 Verifying the Validity of the Proposed Method

In this part of experiment, we use the video Mityatuz and 73v195t in DynTex database [17] to verify the validity of proposed method.

In the experiment, we statistics the feature number of the tracking points and the number of 3D points from clustering. We use the average re-projection error to verify the accuracy of the proposed method, and the re-projection is calculated by formula (89.7). Table 89.1 shows the statistic results. It can be seen from the results that the method proposed in this study is very valid. The re-projection error is smaller. Table 89.2 shows the results of internal and external camera parameters.

$$\varepsilon = \frac{\sum\limits_{i=1}^{n} \sqrt{d^2(m_i, \hat{m}_i) + d^2(m_i', \hat{m}_i')}}{n} \tag{89.7}$$

m_i and m_i' denotes the corresponding points on the two images. $\hat{m}_i$ and $\hat{m}_i'$ denotes the estimated value of m_i and m_i' with the use of the internal and external camera parameters respectively. n denotes the total number of the features. $d(m_i, \hat{m}_i)$ denotes the Euclidean distance between m_i and $\hat{m}_i$. $d(m_i', \hat{m}_i')$ denotes the Euclidean distance between m_i' and $\hat{m}_i'$.

89.3.2 Time Performance

In order to demonstrate the performance of the proposed method, we use the average reconstruction time of every frame to show the performance of the proposed method. In the study, we use the video Mityatuz [17], 73v195t [17], Temple [18] and Flower [18] to study.

Table 89.3 shows the experiment results of the time. From the results we can see that the proposed method in the paper is efficient. The Average Reconstruction Time of Every Frame is smaller. The reconstruction average time of every frame is

Table 89.1 Experiments of re-projection error

Video	The feature number	Number of the 3D points from clustering	The average re-projection error
Mityatuz	123	58	0.306722
73v195t	127	91	0.596488

Table 89.2 Results of internal and external camera parameters

Video	Frame	Focal length	Rotation matrix			Translation matrix		
Mityatuz	5	289.835	$\begin{bmatrix} -1.0004 \\ 0.0031 \\ -0.0001 \end{bmatrix}$	$\begin{matrix} -0.0032 \\ -0.9999 \\ 0.0010 \end{matrix}$	$\begin{matrix} -0.0001 \\ -0.0009 \\ -1.0000 \end{matrix}$	$\begin{bmatrix} 0.0038 \\ -0.0032 \\ 0.0195 \end{bmatrix}$		
73v195t	7	326.389	$\begin{bmatrix} 0.9863 \\ 0.0030 \\ -0.0496 \end{bmatrix}$	$\begin{matrix} -0.0029 \\ 0.9982 \\ 0.0372 \end{matrix}$	$\begin{matrix} -0.0495 \\ -0.0371 \\ 0.9981 \end{matrix}$	$\begin{bmatrix} -0.2412 \\ 1.9401 \\ -2.0892 \end{bmatrix}$		

Table 89.3 Experiments of re-projection error

Video	Frame number	Resolution	Average Reconstruction Time of Every Frame
Mityatuz	99	360 × 288	0.0780
73v195t	318	360 × 288	0.0657
Temple	120	576 × 352	0.1055
Flower	258	960 × 540	0.1695

0.104675 s. We can achieve a good performance which has 9.55338 frames per seconds. This can verify the time validity of the method.

89.4 Conclusion and Future Works

Currently, 3D reconstruction is a hot topic in computer vision field. SFM is an essential problem in such work. In this paper we study an efficient method of real-time SFM using clustering method to obtain the satisfied results. In order to preserve the better movement features, we trace several frames using continuity of the movement of video to constrain. We take advantage of the better movement features to calculate the fundamental matrix and use Sturm's method to restore the camera intrinsic parameters. We further use the method of SVD decomposition to obtain the extrinsic parameters of adjacent frames. In order to obtain more accurate camera intrinsic and extrinsic parameters, we use clustering method to obtain more accurate reconstruction result.

Further works include how we can apply the reconstruction method in the augmented reality scene construction in real-time.

References

1. Akbarzadeh A et al (2006) Towards urban 3D reconstruction from video. In: Abstracts of the 3rd international symposium on 3D data processing, visualization, and transmission, Chapel Hill, NC, 14–16 June2006
2. Carrano A, D'Angelo V et al (2010) An approach to projective reconstruction from multiple views. In: Abstracts of the signal processing, pattern recognition and applications, Innsbruck, Austria, 17–19 February 2010

3. Pollefeys M, Vergauwen M, Cornelis K, Verbiest F, Schouteden J, Tops J, Gool LV (2010) 3D acquisition of archaeological heritage from images. In: abstracts of CIPA conference, international archive of photogrammetry and remote sensing
4. Zhang ZY (2000) A flexible new technique for camera calibration. IEEE t pattern anal 22:1330–1334. doi:10.1109/34.888718
5. Li J (2011) Camera self-calibration method based on GA-PSO algorithm. Paper presented at the IEEE international conference on cloud computing and intelligence systems, Beijing, China, 15–17 (Sept 2011)
6. Li J, Boufama B (2008) Camera self-calibration from bivariate polynomials derived from Kruppa's equations. PR 41:2484–2492
7. Lotfi D, Marraki ME, Aboutajdine D (2011) A robust multistage algorithm for camera self-calibration dealing with varying intrinsic parameters. J Theor Appl Inf Technol 32:1–7
8. Chandraker MK, Agarwal S, Kahl F, Nistér D, Kriegman DJ (2007) Autocalibration via rank-constrained estimation of the absolute quadric. CVPR'07, Minneapolis, Minnesota
9. Dellaert F, Seitz S, Thorpe C, Thrun S (2002) Structure from motion without correspondence. Paper presented at IEEE computer society conference on computer vision and pattern recognition, 557–564 2002
10. Chen HT, Tien MC, Chen YW, Tsai WJ, Lee SY (2009) Physics-based ball tracking and 3D trajectory reconstruction with applications to shooting location estimation in basketball video. J Vis Commun Image R 20:204–216. doi:10.1016/j.jvcir.2008.11.008
11. Schweighofer G, Pinz A (2006) Robust pose estimation from a planar target. IEEE T Pattern Anal 28:2024–2030
12. Cornelis K, Pollefeys M, Van GL et al (2001) Tracking based structure and motion recovery for augmented video productions. In: Abstracts of the ACM symposium on virtual reality software and technology, NY, USA, 17–24
13. Lucas B, Kanade T (1981) An iterative image registration technique with an application to stereo vision. In: abstracts of the seventh international joint conference on artificial intelligence, Vancouver, Canada, p 674–679
14. Zhang ZY, Kanade T (1998) Determining the Epipolar geometry and its uncertainty: a review. Int J Comput Vision 27:161–195
15. Sturm P, Cheng ZL et al (2005) Focal length calibration from two views: method and analysis of singular cases. Comput Vis Image 99:58–95
16. Hartley RI (1992) Estimation of relative camera positions for uncalibrated cameras. Presented at the european conference on computer vision, p 579–587
17. Peteri R, Huskies M, Fazekas S (2006) DynTex:A comprehensive database of DynamicTextures. http://www.cwi.nl/projects/dyntex/. Accessed 2006
18. Zhang GF, Jia JY et al (2009) Consistent depth maps recovery from a video sequence. IEEE T Pattern Anal 31:974–988

Chapter 90
Delay-Dependent Observers for a Class of Discrete-Time Uncertain Nonlinear Systems

Dongmei Yan, Bing Kang, Ming Liu, Yan Sui and Xiaojun Zhang

Abstract This paper is concerned with the design problem of the systems with time-varying delay. Two Luenberger-like observers are proposed. One is delay observer which has internal time delay. This kind of observer is applicable when the time delay is known. The other is delay-free observer which does not use delayed information. This kind of observer is especially applicable when the time delay is not known explicitly. Delay-dependent conditions are derived for the existences of these two observers. A cone complementarity linearization algorithm is presented to calculate the observer gains. The effectiveness of the proposed methods is illustrated by numerical example.

Keywords Delay-dependent · Linear matrix inequality · Lipschitz nonlinear systems · Robust observer · Uncertain systems

90.1 Introduction

The theory of state observers for linear systems has been receiving many researchers' interests in the past decades. Rich literature have been published and different kind of observers have been proposed [1, 2]. When uncertainties appear

D. Yan (✉) · Y. Sui · X. Zhang
The Center of Network and Educational Technology, Jilin University, Changchun, China
e-mail: ydm@jluhp.edu.cn

B. Kang
College of Communications Engineering, Jilin University, Chang chun, China

M. Liu
The Agricultural Division, JiLin University, Changchun, China

W. Lu et al. (eds.), *Proceedings of the 2012 International Conference on Information Technology and Software Engineering*, Lecture Notes in Electrical Engineering 210, DOI: 10.1007/978-3-642-34528-9_90, © Springer-Verlag Berlin Heidelberg 2013

in the system model, a robust observer should be considered. Many results on this topic have been reported (see, e.g. [3–5] and references therein). On the other hand, time-delay is often encountered in many practical systems. For example, Trihh and Aldeen [6] proposed a reduced-order memoryless state observer for linear discrete-time delay systems by the augmentation approach. Boutayed [7] proposed a simple and useful approach to design observers for discrete-time systems with delays in the state and output vectors.

Recently, the design problem of observers for nonlinear systems, especially Lipschitz nonlinear systems, has received considerable attentions. The observer design problem of a class of Lipschitz nonlinear time-delay systems without parameter uncertainty was considered in [8]. A gain scheduled observer was designed and conditions for existence of the desired robust nonlinear observers were derived in [9]. However, the observer structure in [9] involves uncertainties. Observers for discrete-time Lipschitz nonlinear state delayed systems with or without parameter uncertainties were proposed in [10] where observer structures do not involve uncertainties and delay-independent conditions for existences of such observers were derived in terms of LMIs. Lu and Ho [11] also studied the robust H_∞ observer design problem for discrete-time Lipschitz nonlinear systems with time delay and parameter uncertainties using LMI approach. However, most of the above literature have assumed that the time delay is constant, which cannot be always the case in control systems. Furthermore, conditions for the existence of the desired observers in the above literature are all delay-independent, which are generally conservative than delay-dependent ones.

In this paper, the design problem of observers for a class of discrete time delay systems with Lipschitz nonlinear perturbations and norm-bounded uncertainties is considered. Two Luenberger-like observers are proposed and new delay-dependent existence conditions for these two observers are derived. Numerical examples are provided to illustrate the validity and less conservativeness of the proposed methods.

90.2 Problem Formulation and Reliminaries

Consider a discrete-time uncertain nonlinear system with time-varying delay:

$$x(k+1) = (A + \Delta A(k))x(k) + (A_d + \Delta A_d(k))x(k - \tau(k)) + Gg(x(k)) \quad (90.1)$$

$$x(k) = \varphi(k), \quad -\tau_{\max} \le k \le 0 \quad (90.2)$$

$$y(k) = (C + \Delta C(k))x(k) + Hh(x(k)) \quad (90.3)$$

where $x(k) \in \mathbb{R}^n$ is the state vector, $y(k) \in \mathbb{R}^m$ is the measurement output, $\varphi(k)$ is the initial condition sequence, $\tau(k)$ is an integer representing the time-varying bounded delay and satisfies $\tau_{\min} \le \tau(k) \le \tau_{\max}$, where $\tau_{\min} \ge 0$ and $\tau_{\max} > 0$ are constant scalars representing the minimum and maximum delays, respectively.

A, A_d, C, G and H are known real constant matrices with appropriate dimensions, $\Delta A(k)$, $\Delta A_d(k)$ and $\Delta C(k)$ denote parameter uncertainties and satisfy the following conditions:

$$[\Delta A(k) \quad \Delta A_d(k)] = D_1 F(k)[E_1 \quad E_2] \tag{90.4}$$

$$\Delta C(k) = D_2 F(k)E_1 \tag{90.5}$$

where D_1, D_2, E_1, E_2 are constant matrices with appropriate dimensions and $F(k)$ is an unknown time-varying matrix satisfying:

$$F^{\mathrm{T}}(k)F(k) \leq I, \quad \forall k \tag{90.6}$$

$g(\cdot) : \mathrm{R}^n \to \mathrm{R}^{n_g}$ and $h(\cdot) : \mathrm{R}^n \to \mathrm{R}^{n_h}$ are known nonlinear functions. $g(x(k))$ and $h(x(k))$ meet $g(0) = 0$ and the following global Lipschitz conditions:

$$\|g(x_1(k)) - g(x_2(k))\| \leq \|R_g(x_1(k) - x_2(k))\| \tag{90.7}$$

$$\|h(x_1(k)) - h(x_2(k))\| \leq \|R_h(x_1(k) - x_2(k))\| \tag{90.8}$$

for all x_1, $x_2 \in R^n$, where R_g and R_h are known constant matrices with appropriate dimensions.

The problem of interest is to design observers for the system (90.1–90.3) with norm-bounded uncertainties satisfying (90.4–90.6) and nonlinearities satisfying (90.7–90.8). The objectives are: (1) to design the following two different Luenberger-like observers to reconstruct the state $x(k)$ based on measurement output $y(k)$; (2) to provide an efficient procedure to compute the observer gains.

The two Luenberger-like observers have the following structures:

Delay observer: it has internal time delay, so it is applicable when the time delay is known.

$$\begin{aligned} \hat{x}(k + 1) = {}& A\hat{x}(k) + A_d\hat{x}(k - \tau(k)) + Gg(\hat{x}(k)) \\ & + L[y(k) - C\hat{x}(k) - Hh(\hat{x}(k))] \end{aligned} \tag{90.9}$$

Delay-free observer: it does not have internal time delay; hence, it is especially applicable when the time delay is not known explicitly.

$$\hat{x}(k + 1) = A\hat{x}(k) + Gg(\hat{x}(k)) + L[y(k) - C\hat{x}(k) - Hh(\hat{x}(k))] \tag{90.10}$$

The following lemma plays an important role in the development of main results in the following section.

Lemma 1 *Assume $x(k) \in R^n$, then for any matrices $Z > 0$, M, $N \in R^{n \times n}$, and a scalar function $h := h(k) \geq 0$, the following inequality holds:*

$$-\sum_{i=k-h}^{k-1} y^{\mathrm{T}}(i)Zy(i) \leq \xi^{\mathrm{T}}(k)\Lambda\xi(k) + h\xi^{\mathrm{T}}(k)\begin{bmatrix} M^{\mathrm{T}} \\ N^{\mathrm{T}} \end{bmatrix} Z^{-1}[M \quad N]\xi(k)$$

where $\Lambda = \begin{bmatrix} M^T + M & -M^T + N \\ * & -N^T - N \end{bmatrix}$, $\xi(k) = \begin{bmatrix} x(k) \\ x(k-h) \end{bmatrix}$, $y(i) = x(i+1) - x(i)$.

90.3 Main Results

In this section, two Luenberger-like observers are designed. Delay-dependent existence conditions for these two observers are derived. The observer gains can be obtained using an iterative algorithm.

90.3.1 Delay Observer Design

Before embarking on the delay observer design for uncertain Lipschitz nonlinear time-delay systems, results on the nominal observer design problem are presented.
    Consider the following nominal system:

$$x(k+1) = Ax(k) + A_d x(k - \tau(k)) + Gg(x(k)) \tag{90.11}$$

$$x(k) = \varphi(k), \quad -\tau_{\max} \le k \le 0 \tag{90.12}$$

$$y(k) = Cx(k) + Hh(x(k)) \tag{90.13}$$

    The observer described by (90.9) is adopted. Let the error vector be $e(k) = x(k) - \hat{x}(k)$, From (90.9), (90.11) and (90.13), it can be obtained that

$$e(k+1) = A_e e(k) + A_d e(k - \tau(k)) + G_e \begin{bmatrix} g(x(k)) - g(\hat{x}(k)) \\ h(x(k)) - h(\hat{x}(k)) \end{bmatrix} \tag{90.14}$$

where

$$A_e = A - LC, \quad G_e = [G \quad -LH] \tag{90.15}$$

    The following theorem gives a sufficient delay-dependent stability criterion for system (90.14).

Theorem 1 *For given $\tau_{\min}$ and $\tau_{\max}$, if there exist scalars $\varepsilon_i > 0$, $\lambda_i > 0$ ($i = 1, 2$), matrices $P > 0$, $Q > 0$, $S > 0$, $Z > 0$ and any matrices M_1, M_2, N_1 and N_2 such that*

$$\Psi = \begin{bmatrix} \Psi_{11} & \Psi_{12} \\ * & \Psi_{22} \end{bmatrix} < 0 \tag{90.16}$$

where

$$
\Psi_{11} = \begin{bmatrix}
\begin{array}{c} -P + (\tau_{max} - \tau_{min} + 1)Q \\ +(\varepsilon_1 + \tau_{max}\lambda_1)R_g^T R_g \\ +(\varepsilon_2 + \tau_{max}\lambda_2)R_h^T R_h \\ +M_1 + M_1^T + S \end{array} & -M_1^T + N_1 & 0 & 0 & 0 \\
* & -N_1^T - N_1 - Q + M_2^T + M_2 & -M_2^T + N_2 & 0 & 0 \\
* & * & -N_2^T - N_2 - S & 0 & 0 \\
* & * & * & -E & 0 \\
* & * & * & * & -\tau_{max}\Lambda
\end{bmatrix},
$$

$$
\Psi_{12} = \begin{bmatrix}
A_e^T P & (A_e - I)^T Z & \tau_{max} M_1^T & 0 \\
A_d^T P & A_d^T Z & \tau_{max} N_1^T & (\tau_{max} - \tau_{min})M_2^T \\
0 & 0 & 0 & (\tau_{max} - \tau_{min})N_2^T \\
G_e^T P & 0 & 0 & 0 \\
0 & G_e^T Z & 0 & 0
\end{bmatrix},
$$

$$
\Psi_{22} = \begin{bmatrix}
-P & 0 & 0 & 0 \\
* & -Z/\tau_{max} & 0 & 0 \\
* & * & -\tau_{max} Z & 0 \\
* & * & * & -(\tau_{max} - \tau_{min})Z
\end{bmatrix}, \quad \Lambda = \begin{bmatrix} \lambda_1 I & 0 \\ 0 & \lambda_2 I \end{bmatrix}, \quad E = \begin{bmatrix} \varepsilon_1 I & 0 \\ 0 & \varepsilon_2 I \end{bmatrix}
$$

then system (90.14) is asymptotically stable for any time-varying delay $\tau(k)$ satisfying $\tau_{min} \leq \tau(k) \leq \tau_{max}$. (Proof omitted)

On the basis of Theorem 1, robust observer design for uncertain Lipschitz nonlinear time-delay systems is considered in the following part.

Define the error vector as $e(k) = x(k) - \hat{x}(k)$. From (90.1), (90.3) and (90.9), it is easy to obtain

$$
\begin{aligned}
e(k+1) = (A - LC)e(k) + A_d e(k - \tau(k)) + (\Delta A(k) - L\Delta C(k))x(k) \\
+ \Delta A_d(k)x(k - \tau(k)) + G_e \xi(x(k), \hat{x}(k))
\end{aligned}
\tag{90.17}
$$

where G_e and $\xi(x(k), \hat{x}(k))$ are the same as those defined in (90.15).

Let the augmented state vector be $z(k) = \begin{bmatrix} x(k) \\ e(k) \end{bmatrix}$. Combining (90.1), (90.3), (90.9) and (90.17) yields

$$
z(k+1) = (A_z + \Delta A_z(k))z(k) + (A_{dz} + \Delta A_{dz}(k))z(k - \tau(k)) + G_z \xi_z(x(k), \hat{x}(k))
\tag{90.18}
$$

where

$$A_z = \begin{bmatrix} A & 0 \\ 0 & A - LC \end{bmatrix}, A_{dz} = \begin{bmatrix} A_d & 0 \\ 0 & A_d \end{bmatrix}, G_z = \begin{bmatrix} G & 0 \\ 0 & G_e \end{bmatrix},$$

$$\Delta A_z(k) = \hat{D}_1 F(k)\hat{E}_1, \ \Delta A_{dz}(k) = \hat{D}_2 F(k)\hat{E}_2, \ \xi_z(x(k), \hat{x}(k))$$
$$= \begin{bmatrix} g(x(k)) \\ g(x(k)) - g(\hat{x}(k)) \\ h(x(k)) - h(\hat{x}(k)) \end{bmatrix}.$$

The following theorem gives a sufficient delay-dependent stability condition for augmented system (90.18). And system (90.18) is asymptotically stable for all time-varying delay $\tau(k)$ satisfying $\tau_{\min} \leq \tau(k) \leq \tau_{\max}$.

90.3.2 Delay-Free Observer Design

The above delay observer contains delayed states, so it cannot be applicable when the delay is not available. The delay-free observer (90.10) does not need the exact value of the delay, so it is applicable especially when the delay is not known explicitly.

Define the error vector as $e(k) = x(k) - \hat{x}(k)$. From (90.1), (90.3) and (90.10), the error dynamics is described by:

$$e(k+1) = (A - LC)e(k) + A_d e(k - \tau(k)) + (\Delta A(k) - L\Delta C(k))x(k) \quad (90.19)$$
$$+ \Delta A_d(k)x(k - \tau(k)) + G_e \xi(x(k), \hat{x}(k))$$

Let the augmented vector be $\theta(k) = \begin{bmatrix} x(k) \\ e(k) \end{bmatrix}$. Combining (90.1), (90.3) with (90.19) yields

$$\theta(k+1) = (A_\theta + \Delta A_\theta(k))\theta(k) + (A_{d\theta} + \Delta A_{d\theta}(k))\theta(k - \tau(k)) + G_\theta \xi_\theta(x(k), \hat{x}(k)) \quad (90.20)$$

where

$$A_\theta = \begin{bmatrix} A & 0 \\ 0 & A - LC \end{bmatrix}, A_{d\theta} = \begin{bmatrix} A_d & 0 \\ A_d & 0 \end{bmatrix}, G_\theta = \begin{bmatrix} G & 0 \\ 0 & G_e \end{bmatrix}, \xi_\theta(x(k), \hat{x}(k))$$
$$= \begin{bmatrix} g(x(k)) \\ g(x(k)) - g(\hat{x}(k)) \\ h(x(k)) - h(\hat{x}(k)) \end{bmatrix},$$

$$\Delta A_\theta(k) = \bar{D}_1 F(k)\bar{E}_1, \ \Delta A_{d\theta}(k) = \bar{D}_2 F(k)\bar{E}_2$$

The following theorem gives a sufficient stability condition for augmented system (90.20).

Theorem 2 *For given $\tau_{\min}$ and $\tau_{\max}$, if there exist scalars $\varepsilon_i > 0$, $\lambda_i > 0$ ($i = 1, 2, 3$), $\delta_1 > 0$ and $\delta_2 > 0$, matrices $P > 0$, $Q > 0$, $S > 0$, $Z > 0$ and any matrices M_1, M_2, N_1 and N_2 such that*

$$\Phi = \begin{bmatrix} \Phi_{11} & \Phi_{12} \\ * & \Phi_{22} \end{bmatrix} < 0 \qquad (90.21)$$

Φ_{11}, Φ_{12}, Φ_{22} are omited. Then augmented system (90.20) is asymptotically stable for any time-varying delay $\tau(k)$ satisfying $\tau_{\min} \leq \tau(k) \leq \tau_{\max}$.

Remark 2 Condition (90.21) in Theorem 3 is either non-convex but they can be solved by the similar algorithm as that for Theorem 2.   □

90.4 Numerical Examples

In this section, a numerical example is given to illustrate the effectiveness of the proposed method.

Example Consider the following uncertain discrete-time Lipschitz nonlinear system with A, A_d, C, G, H, the uncertainties are assumed to be described by D_1, D_2, E_1, E_2, $F(k)$,

$$A = \begin{bmatrix} -0.08 & 0.3 & 0.1 \\ 0.2 & -0.1 & 0.2 \\ -0.2 & 0.2 & -0.1 \end{bmatrix}, \ A_d = \begin{bmatrix} 0.3 & 0.1 & 0 \\ -0.1 & 0 & -0.05 \\ 0.3 & 0.05 & 0.05 \end{bmatrix},$$

$$C = \begin{bmatrix} 0.1 & 0.2 & 0.1 \\ 0.1 & 0.2 & 0.1 \end{bmatrix}, \ G = \begin{bmatrix} 0.2 & 0.1 & -0.2 \\ 0.1 & 0.1 & 0 \\ 0.05 & 0.2 & -0.1 \end{bmatrix},$$

$$H = \begin{bmatrix} 0.1 & 0.1 & 0.1 \\ 0.2 & 0.1 & 0.2 \end{bmatrix}, \ D_1 = \begin{bmatrix} 0.05 & 0 \\ 0.2 & 0.1 \\ 0.1 & 0.1 \end{bmatrix},$$

$$D_2 = \begin{bmatrix} 0 & 0.2 \\ 0.1 & 0.1 \end{bmatrix}, \ E_1 = \begin{bmatrix} 0.1 & 0.1 & 0.1 \\ 0.1 & 0.1 & 0.2 \end{bmatrix},$$

$$E_2 = \begin{bmatrix} 0.1 & 0.2 & 0.2 \\ 0 & 0.1 & 0.1 \end{bmatrix}, \ F(k) = \begin{bmatrix} \sin(k) & 0 \\ 0 & \sin(k) \end{bmatrix}$$

□

Fig. 90.1 Estimation error of
delay observer

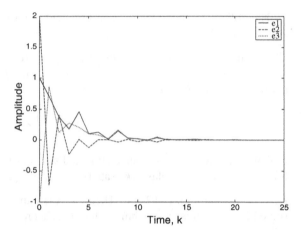

Fig. 90.2 Estimation error of
delay-free observer

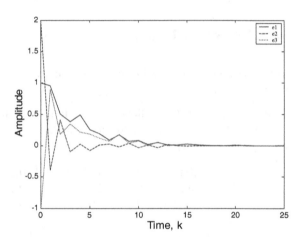

For simulation purpose, $g(x(k))$ and $h(x(k))$ are assumed to be

$$g(x(k)) = \begin{bmatrix} 0.2\sin(x_3(k)) \\ 0.2\sin(x_1(k)) \\ 0.2\sin(x_2(k)) \end{bmatrix}, \ h(x(k)) = \begin{bmatrix} 0.1\sin(x_2(k)) \\ 0.2\sin(x_1(k)) \\ 0.1\sin(x_3(k)) \end{bmatrix}$$

Assume $0 \le \tau(k) \le 4$. The aim is to design two observers with the structure (90.9) and (90.10) for the uncertain nonlinear time-delay system described by (90.1–90.3). Firstly, the delay observer is considered. Using the proposed algorithm, after 16 iterations, the observer gain is obtained as L_1; Secondly, the delay-free observer is considered. After 21 iterations, the observer gain is obtained as

$$L_1 = \begin{bmatrix} 0.0420 & 0.3674 \\ -0.1429 & 1.4533 \\ -1.6470 & 1.2910 \end{bmatrix}, \ L_2 = \begin{bmatrix} -0.2497 & 0.0966 \\ 1.0011 & -0.4147 \\ -0.9737 & 0.5439 \end{bmatrix}.$$

Assume the initial value $x(0) = \begin{bmatrix} 1 & 2 & -1 \end{bmatrix}^T$ and denote the error state $e_i = x_i - \hat{x}_i \, (i = 1, 2, 3)$. For the above two observer gains, the initial responses of error dynamics are shown in Figs. 90.1 and 90.2, respectively. The simulation results show that the methods proposed in this paper are valid and effective.

90.5 Conclusions

In this paper, the design problem of observers for a class of nonlinear uncertain time-delay systems has been studied. The nonlinearities are assumed to satisfy the global Lipschitz condition and the uncertainties are assumed to be time-varying but norm-bounded. Delay and delay-free observers have been designed for this class of nonlinear systems. Less conservative delay-dependent existence conditions for these two observers have been derived. Since the obtained conditions are non-convex, a cone complementarity linearization algorithm has been presented to calculate the observer gains. Numerical examples have illustrated the effectiveness of the proposed methods. It should be pointed that the proposed approach can be extended to Lipschitz nonlinear systems with multiple time delays. However, this paper has only considered the asymptotical stability of the error state system. The exponential stability problem would be a future research topic, and the methods in [8, 11] may be beneficial.

References

1. Luenberger DG (1971) An introduction to observers. IEEE Trans Autom Control 16:596–602
2. O'Reilly J (1983) Observers for linear systems. Academic, New York
3. Doyle JC, Stein G (1979) Robustness with observer. IEEE Trans Autom Control 24:607–611
4. Petersen IR (1985) A Riccati equation approach to the design of stabilizing controllers and observers for a class of uncertain systems. IEEE Trans Autom Control 30:904–907
5. Petersen IR, Hollot CV (1988) High-gain observers applied to problems in stabilization of uncertain linear systems, disturbance attenuation and H_∞ optimization. Int J Adapt Control Signal Process 2:347–369
6. Trinh H, Aldeen M (1997) A memoryless state observer for discrete time-delay systems. IEEE Trans Autom Control 42:1572–1577
7. Boutayeb M (2001) Observers design for linear time-delay systems. Syst Control Lett 44:103–109
8. Wang Z, Goodall DP, Burnham KJ (2002) On designing observers for time-delay systems with non-linear disturbances. Int J Control 75:803–811
9. Wang Z, Unbehauen H (2000) Nonlinear observers for a class of continuous-time uncertain state delayed systems. Int J Syst Sci 31:555–562
10. Xu S, Lu J, Zhou S, Yang C (2004) Design of observers for a class of discrete-time uncertain nonlinear systems with time delay. J Franklin Inst 341:295–308
11. Lu G, Ho DWC (2004) Robust $H\infty$ observer for nonlinear discrete systems with time delay and parameter uncertainties. IEE Proc Control Theory Appl 151:439–444

Chapter 91
Deconvolution Algorithm in the Retrieval of Raindrop Size Distribution from a VHF Profiling Radar

Yukun Du, Zheng Run and Runsheng Ge

Abstract The nonuniformity of atmospheric vertical movement broadens the returning signal spectrum of precipitation, and thus affects the retrieval of raindrop size distribution using profiling radar. The effect of the atmospheric vertical movement to the descending precipitation particles can be described by the convolution of the atmospheric vertical motion spectrum and the gravitation spectrum of precipitation particles. In order to reduce the influence of the nonuniformity of the atmospheric to the retrieval of raindrop size distribution, we need to deconvolute the precipitation spectrum from profiling radar. This study puts forward an improved iterative deconvolution method. Applying setting a corresponding threshold and using the smoothing approximation method to the received signal. The method effectively reduces the influence of the ill-posed problems of radar noise. Applying the gaussian fitting method to the atmospheric vertical motion returned radar signal, the constraint of the atmospheric spectrum width is reduced. Simulation results show that when the radar noise of the precipitation signal is less than effective signal at 120 db, no pretreatment is necessary, the deconvoluted signal spectrum is the same as the actual precipitation. The chapter also applied deconvolution method to the actual signal spectrum. The impact of the deconvolution to the retrieval of raindrop size distribution was also analyzed.

Y. Du (✉) · Z. Run · R. Ge
Chengdu University of Information Technology, Chengdu, China
e-mail: duyukun0725@yahoo.com.cn

Y. Du · Z. Run · R. Ge
Chinese Academy of Meteorological Sciences, Beijing, China

Y. Du · Z. Run · R. Ge
709-2 AMS, China Meteorological Administration, Zhongguancun South Street Num 46, Beijing 100-0081, China

W. Lu et al. (eds.), *Proceedings of the 2012 International Conference on Information Technology and Software Engineering*, Lecture Notes in Electrical Engineering 210, DOI: 10.1007/978-3-642-34528-9_91, © Springer-Verlag Berlin Heidelberg 2013

Keywords Phased array wind profiling radar · Signal processing · Deconvolution ·
Raindrop size distribution

91.1 Introduction

The retrieval of raindrop size distribution (DSD) in precipitation is crucial for
understanding rain forming processes in which the rainfall rate changes significantly [1]. However, for precipitation returned radar signal, the first thing is to
eliminate the impact of atmospheric vertical motion [2]. Atmospheric vertical
motion returned signal spectrum and precipitation returned signal spectrum in
Doppler spectrum measured by wind profiling radar accord with convolution
relationship. Wind profilers measure a Doppler spectrum that is a convolution
between the precipitation spectrum and a clear-air turbulent spectrum. In foreign,
there are many researches about the deconvolution of the measured spectrum,
such as the classic method Rajopadhyaya at. (1993) put forward Fourier transformation (FT) method and Schafer at. (2002) put forward interactive method.
Tikhonov through introducing the regular theory changed the ill-posed problems
to the well-posed problems [3]. But the solution calls for much computer
memory, and needs a lot of numerical iterative calculation. This chapter discusses a method based on Z transform [4], using the threshold value iteration
based on the polynomial long division method to realize the deconvolution. In
the iterative processes, the effect of noise accumulating can be reduced, and thus
can get more stable deconvoluted results. Due to the ill-posed problem can be
even more significant with the increasing of the number of iteration. For the
wide spectrum of clear-air turbulent, the method reduces the number of iteration
by sampling and smoothing. In addition, the method also has applicability and
the timeliness is improved.

91.2 Method and Theory

The power spectral density (S) measured by profiling radar is given by [5]:

$$
\begin{aligned}
S(w) &= G(w) + P(w); \\
P(w) &= G_0(w) * P_0(w + \bar{w});
\end{aligned}
\tag{91.1}
$$

$G(w)$ and $P(w)$ are the spectral densities dues to air motion and precipitation,
respectively; $G_0(w)$ is the normalized spectral density dues to clear-air vertical
motion, $P_0(w + \bar{w})$ is the translation of deconvolution spectrum. $\bar{w}$ is the Doppler

velocity dues to clear-air vertical motion. The asterisk represents the convolution operation.

The measured precipitation spectrum $S_{obs}(w) = G_0(w) * P(w + \bar{w})$ is the convolution of normalized clear-air vertical motion and the translation of deconvoluted spectrum.

The power spectral density is the data after sampling:

$$S_{obs}(w_i) = G_0(w_i) * P(w_i + \bar{w}_i) = \sum_{k=n_1}^{n_3} G_0(k) \cdot P(w_i + \bar{w}_i - k);$$

$$w_i = n_3, n_3 + 1, \ldots, n_3 + n_1 + n_2 - 1$$

(91.2)

n_1 is the starting point of atmospheric Doppler spectrum, n_2 is the end point of atmospheric Doppler spectrum, n_3 is the stating point of precipitation spectrum.

If $G_0(w_i)$ and $S_{obs}(w_i)$ are given, so the recursive method can be used to solve $P(w_i + \bar{w}_i)$, the process is as follows:

$$S_{obs}(n_3) = G_0(n_1) \cdot P(n_3 + \bar{w}_0); P(n_3 + \bar{w}_0) = S_{obs}(n_3)/G_0(n_1);$$
$$S_{obs}(n_3 + 1) = G_0(n_1) \cdot P(n_3 + \bar{w}_0 + 1) + G_0(n_1 + 1) \cdot P(n_3 + \bar{w}_0);$$
$$P(n_3 + \bar{w}_0 + 1) = [S_{obs}(n_3 + 1) - G_0(n_3 + \bar{w}_0 + 1) \cdot P(n_3 + \bar{w}_0)]/G_0(n_1);$$

...

$$S_{obs}(w_i) = G_0(n_1) \cdot P(w_i + \bar{w}_i) + \sum_{k=n_1+1}^{w_i} G_0(k) \cdot P(n_3 + \bar{w}_0);$$

$$P(w_i + \bar{w}_i) = \left[S_{obs}(w_i) - \sum_{k=n_l+1}^{w_i} G_0(k) \cdot P(n_3 + \bar{w}_0) \right]/G_0(n_1); i \geq 0$$

(91.3)

When $G_0(n_1)$ tend to 0, the derived $P(w_i + \bar{w}_i)$ tend to infinity, then the iterative algorithm is invalid. In measured Doppler velocity spectrum, the starting and ending points of the clear-air vertical motion spectrum are small scale measured values, and there are big measurement errors and quite a big noise interference, so through the traditional iteration algorithm, the result is usually not reliable. To avoid ill-posed problems and error accumulation, this study will improve the iterative method, it puts forward the algorithm:

Step 1. An appropriate nonnegative threshold aver_n will be set, according to the noise level;

Step 2. The starting point p_1 and the ending point p_2 will be pointed by searching the absolute value is not less than ave_n from the starting points of the clear-air vertical motion spectrum;

Step 3. $P(w_i + \bar{w}_i)$ will be iteratively derived, according to p_1 and p_2;

$$S_{obs}(n_3) = G_0(\mathrm{p}_1) \cdot P(n_3 + \bar{w}_0); P(n_3 + \bar{w}_0) = S_{obs}(n_3)/G_0(\mathrm{p}_1);$$
$$S_{obs}(n_3 + 1) = G_0(\mathrm{p}_1) \cdot P(n_3 + \bar{w}_0 + 1) + G_0(\mathrm{p}_1 + 1) \cdot P(n_3 + \bar{w}_0);$$
$$P(n_3 + \bar{w}_0 + 1) = [S_{obs}(n_3 + 1) - G_0(n_3 + \bar{w}_0 + 1) \cdot P(n_3 + \bar{w}_0)]/G_0(\mathrm{p}_1);$$

$$\dots$$

$$S_{obs}(w_i) = G_0(\mathrm{p}_1) \cdot P(w_i + \bar{w}_0) + \sum_{k=p_1+1}^{w_i} G_0(k) \cdot P(n_3 + \bar{w}_0);$$

$$P(w_i + \bar{w}_i) = [S_{obs}(w_i) - \sum_{k=p_1+1}^{w_i} G_0(k) \cdot P(n_3 + \bar{w}_0)]/G_0(n_1); i \geq 0$$

$$(91.4)$$

In the iterative process of the deconvolution, the high frequency detail information is restored and amplified, however, at the same time the high-frequency noise is also amplified, and that often submerges useful signals, but generally the noise can not create fixed peaks. With the increasing of iteration, the accumulation of noise would make derived spectrum gradually deviated from the real spectrum [6], < > denotes smooth approximation process, through this process it can weaken the noise influence constantly.

91.3 The Simulation Results and Analysis

In order to evaluate this algorithm, the output error $e_x(k)$, signal-to-noise ratio snr and RMS error e_{rms} will be used.

The output errors between the deconvoluted spectrum and input signal spectrum $e_x(k)$ is:

$$e_x(k) = \hat{x}(k) - x(k) \tag{91.5}$$

If the signal length for N, the total error among them is:

$$\partial_x = \sum_{k=0}^{N-1} |e_x(k)| \tag{91.6}$$

Therefore between $x(k)$ and $\hat{x}(k)$, the RMS error is:

$$e_{rms} = \left[\frac{1}{N} \sum_{k=0}^{N-1} e_x^2(k) \right]^{\frac{1}{2}} \tag{91.7}$$

To add the random noise $n(k)$.

Therefore:

$$SNR = 10 * \lg \left[\frac{\sum\limits_{k=0}^{N-1} x(k)}{\sum\limits_{k=0}^{N-1} n(k)} \right] \tag{91.8}$$

91.3.1 Noise Experimental

According to precipitation Doppler spectrum characteristics, simulating out precipitation spectrum (Gaussian Input), clear-air vertical motion spectrum (Gaussian Kernel), measured precipitation spectrum (Convolution Output). To add different size of random noise to measured precipitation spectrum, using the traditional and improving methods to deconvolute the measured precipitation spectrum, some results are shown in Fig. 91.1.

In different SNR case, the output errors $e_x(k)$, signal-to-noise ratio SNR and RMS error e_{rms} will be get, as is shown in Table 91 1.

The data in Table 91.1 shows that for a certain output signal the signal-to-noise ratio of the SNR, improved deconvolution algorithm produces errors less than traditional algorithm; the SNR is greater, the error is smaller; when $SNR > 10$ db, the improved method can get better results, but the traditional method can get better result when $SNR > 160$ db.

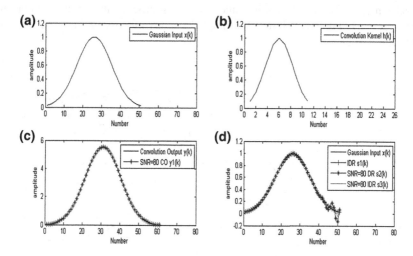

Fig. 91.1 Input $x(k)$, Gaussian Kernel $h(k)$ and Convolution Output $y(k)$, the results of tradition and improved deconvolution method

Table 91.1 The affect of noise in the two deconvolution methods

Output snr (db)	Traditional Method		Improved Method	
	∂_x	e_{rms}	∂_x	e_{rms}
10	4.2945e + 06	2.203e + 10	6.7883	1.126e + 02
40	1.8617e + 03	4.4777e + 03	2.3489	3.707e + 01
70	24.352	68.105	0.4253	0.917
100	3.764	12.2543	0.0242	0.0275
130	0.264	0.0846	0.0241	0.0232
160	0.187	0.0269	0.0224	0.0127
190	0.187	0.0269	0.0224	0.0127

91.3.2 Wide-Spectra Experiment

It is known that the Doppler spectral width ranges from 1 to 2 m/s [7], when the spectral width is wide, the corresponding clear-air vertical spectral points are more, then will degenerate precipitation spectral measurements and the occupied precision. In the study, the sampling and reconstruction method of the atmospheric spectrum will be used. It would keep full of information about the measured spectrum, and reduce atmosphere spectral points, improve the deconvoluted precision.

In the experiment of 91.6, the narrow gaussian spectrum width of 0.6 will be used to confirm the feasibility of the method. In this experiment, the gaussian spectrum width of 1.2 will be used. Figure 91.2 shows that the input gaussian spectrum, spectral width of 2.5 kernel function and the output of convoluted spectrum.

Deconvolute the convoluted spectrum, for the iterative points of the gaussian spectrum ($p = 21$), after sampling and smoothing process, the iterative points are cut to $\hat{p} = 11$, Fig. 91.3 shows the sampled deconvoluted spectrum and the convolution kernel.

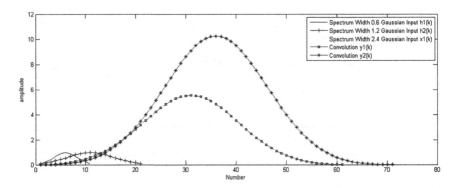

Fig. 91.2 Normalized gaussian convolution spectrum, respectively

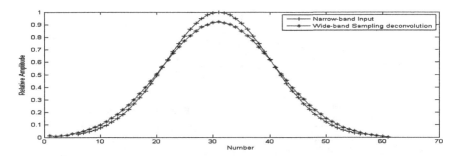

Fig. 91 3 The contrast figure of measured spectrum and deconvoluted spectrum

The sampled deconvoluted spectrum and original convoluted kernel have good uniformity. It is proved that the way to reduce the points is feasibility.

91.4 Radar Data Processing

Radar data processing flow chart is as follows:

First of all, the noise is dealt with using moving average. Then, the spectral peaks are recognized and the spectrum is reconstructed, therefore, the clear-air vertical peak St is identified. Compared with precipitation spectrum, clear-air vertical spectrum has a narrow spectral width and relatively small radial velocity. Assume that clear-air vertical spectrum to obey the gaussian distribution, as the following equation given [8]: (Fig. 91.4)

$$S_t(v) = \frac{1}{(2\pi\sigma^2)^{1/2}} \exp\left(-(w - \bar{w})^2/2\sigma^2\right) \qquad (91.9)$$

Gaussian parameters can be get with nonlinear squares fitting, including σ for spectral width, and the relatively accurate parameters of the initial estimates will be used. The two remaining gaussian distribution parameters: average speed $\bar{v}$ and σ in fitting process can be determined. As a result of these two parameters only have small range, so good initial estimation is easy to estimate.

When measured spectrum contain two distinct spectral peaks, with the section area max value method [9], the clear-air vertical motion spectral peak speed (V_{max}) is easy to be confirmed and the boundary (V_{min}) of clear-air vertical motion and precipitation spectrum will be positioned the at the minimum power spectral density. The range of spectrum is from ($V_{max} - V_{min}$) to ($V_{max} + V_{min}$). If G is well localized, fitting effects will be good.

Figure 91.5 With the section area max value method, spectrum is well identificated and gaussian fitting results are derived.

In the Fig. 91.5 the clear-air vertical motion spectrum is exactly recognized and is accurately fitted by gaussian function, with the fitting rate of 95 % or more.

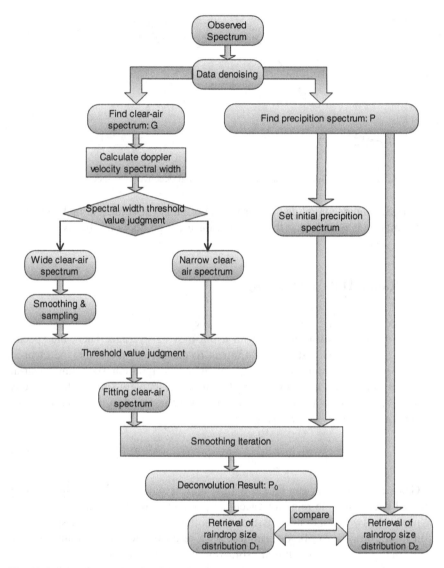

Fig. 91.4 Radar data processing flow chart

With the improved method to deconvolute the recognized precipitation and the atmospheric gaussian fitted spectrum, the result as shown uses solid line in Fig. 91.5, in order to further verify the deconvoluted result, we convolute the deconvoluted result and atmospheric gaussian fitted spectrum, the result as shown uses dotted line in Fig. 91.6.

In Fig. 91.6, convoluted spectra and original spectra fit well, it improves that the result of deconvolution is exactly. In the deconvoluted spectrum, some high

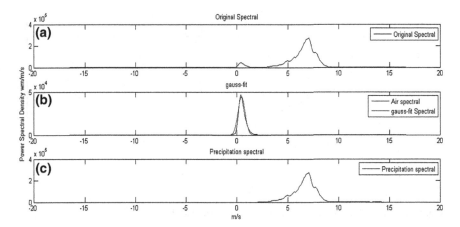

Fig. 91.5 Data processing results. **a** Twin peaks recognition reconstruction. **b** Gaussian fitting results. **c** Identified precipitation spectrum

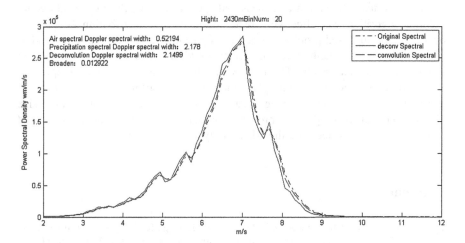

Fig. 91.6 The result of precipitation deconvolution

frequency information is to be amplified, and spectral peak details are more obvious.

The deconvoluted spectra and measured spectra will be delt with, by the retrieval of raindrop size distribution method. The results are shown in the Fig. 91.7

Figure 91.7 shows particle distribution details are more obvious, compared to the measured spectrum, the small particle number between the diameter of 0.5 and 1 mm and the large particle number between the diameter of 2.5 and 4 mm are decreased; meanwhile, the middle size particle number is increased; moreover, the derived raindrop size distribution curve is more neat.

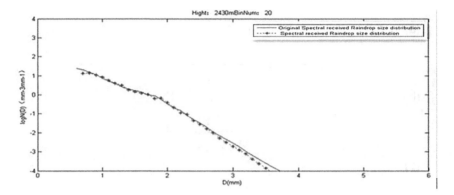

Fig. 91.7 The results of the retrieval of raindrop size distribution

91.5 Conclusion

In the numerical calculation and measured spectral results, the improved deconvolution algorithm can effectively recover the particle distribution details and overcome the ill-posed problem caused by the front-end signal amplitude tend to zero. In addition, the method can effectively inhibit the accumulating noise and get stable deconvoluted results. The method has universality for most precipitation spectrum and has good real-time performance; the algorithm can get more stable and more accurate results. It is significance to exact the retrieval of raindrop size distribution. But to complicated precipitation doppler spectrum, the deconvoluted results are easy distorted, and the adaptability would to improve.

References

1. Cifelli R, Williams CR, Rajopadhyaya DK, Avery SK, Gage KS, May PT (2000) Drop-size distribution characteristics in tropical mesoscale convective systems. J Appl Meteor 39:760–777
2. Gossard EE (1988) Measuring drop-size distributions in clouds with a clear-air-sensing Doppler radar. J Atmos Oceanic Technol 5:640–649
3. Tikhnov A, Arsenin V (1977) Solution of ill-posed problem. Wiley, New York
4. Hu GSH (2003) Digital signal processing, 2nd edn. Tsinghua University Press, Beijing, pp 438–440
5. Wakasugi K, Mizutani A, Matsuo M, Fukao S, Kato S (1986) A direct method for deriving drop-size distribution and vertical air velocities from VHF Doppler radar spectra[J]. Atmos. Oceanic Technol 3:623–629.
6. Wei WB, Jiang J (2005) An efficient threshold deconvolution algorithm. Acta Electronica Sinica 33:2531–2534
7. Strauch RG, Merritt DA et al (1984) The colorado wind-profiling network. J Atmos Oceanic 3(1):37–49
8. Kobayashi T, Adachi A (1986) Retrieval of arbitrarily shaped raindrop size distributions from wind profiler measurement. J Appl Atmos Oceanic Technol 433–442
9. Lucas C, Mackinnon AD, Vincent RA (2003) Raindrop size distribution retrievals from a VHF boundary layer profiler. J Atmos Oceanic

Chapter 92
Fast Cross-Layer Congestion Control Algorithm Based on Compressed Sensing in Wireless Sensor Network

Mingwei Li, Yuanwei Jing and Chengtie Li

Abstract A fast cross layer congestion control (FCLCC) algorithm based on compressed sensing technique is proposed in this paper. Firstly, the improved Homotopy method is applied for signal reconstruction, in which the high accuracy reconstruction signal is obtained only through Kth iteration (K is sparsity). The improved Homotopy method eliminates node-level congestion. Secondly, the channel selection algorithm is provided by eigenvalue-decomposition, which restrains link-level congestion. Simulation results demonstrate that FCLCC effectively relieve congestion, which is provided with good compression and reconstruction performance, and achieves high throughput and lower energy consumption. In addition, the congestion time is greatly cut down.

Keywords Wireless sensor network · Fast compressed sensing · Cross-layer · Congestion control · Eigenvalue-decomposition

92.1 Introduction

A wireless sensor network (WSN) consists of a large number of sensors with sensing, data processing, communications, and networking capabilities. Each node only communicates with several nodes nearby regions, in which data reach sink by multiple

M. Li (✉) · Y. Jing · C. Li
School of Information Science and Engineering, Northeastern University, Shenyang, China
e-mail: neuqlmw@hotmail.com

Y. Jing
e-mail: ywjing@mail.neu.edu.cn

C. Li
e-mail: lichengtie@126.com

W. Lu et al. (eds.), *Proceedings of the 2012 International Conference on Information Technology and Software Engineering*, Lecture Notes in Electrical Engineering 210, DOI: 10.1007/978-3-642-34528-9_92, © Springer-Verlag Berlin Heidelberg 2013

hops. When the rate of receiving packets is greater than relaying or scheduling, redundant packets should be buffered. When the packets queue is overflow, the redundant packets could be discard, which can result in node-level congestion in WSN. The second type is link-level congestion that is related to the wireless channels which are shared by several nodes. In this case, collisions could occur when multiple active sensor nodes try to seize the channel at the same time. Congestion not only increases packet delay, but also reduces throughput and wastes more energy. Most of the existing congestion control algorithms for WSN seldom concern the combined problem of data compression and channel allocation fairness. Thus, how to effectively reduce congestion in WSN is an important aspect in the study.

Compressed sensing [1] theory shows that a sparse or compressible signal $x \in \Re^{N \times 1}$ can be recovered from a relatively small number of linear measurements $y \in \Re^{M \times 1}$.

$$y = Ax + \varepsilon \tag{92.1}$$

where, sensing matrix $A \in \Re^{M \times N}$ is a known measurement matrix $(M \ll N)$, $\varepsilon \in \Re^{M \times 1}$ denotes possible measurement noise, which meats $(1 - \delta_S^2)\|Ax\|_2 \geq \zeta$ $(1 + \delta_S^2)\|\varepsilon\|_2$. CS theory considers that unknown signal x can be recovered from observation vector y with noise through optimization theory, if the following conditions be satisfied [2].

Condition 1: vector x is K sparse (x has K nonzero value), and $K \ll N$;
Condition 2: sensing matrix A meets restricted isometry property (RIP) [3, 4]. Sparse signal must keep consistent geometric properties in effect to measurement matrix, and make sure measurement matrix does not take two different-sparse signal mapping to the same sample set, that is $(1 - \delta_S)\|x\|_2^2 \leq \|Ax\|_2^2 \leq (1 + \delta_S)\|x\|_2^2$.

Congestion duration is shorter and network QoS is higher in WSN. In this paper, a fast cross-layer congestion control algorithm based on compressive sensing is presented. Firstly, transmission channel of measurement vector according to the eigenvalue decomposition principle is provided, and link level congestion is solved. Secondly, only K iterative times can get the reconstructed signal which reconstruction error is less than 0.1, accelerating the reconstruction speed, and improving the reconstruction accuracy. Fast reconstruction could greatly relieve node-level congestion.

92.2 Fast Reconstruction

Signal reconstruction research mainly concentrated on how to design of stable, low complexity, fewer observation number reconstruction algorithm to recover original signal accurately. However, if this technique is applied to the congestion in WSN, the realtime should be considered to fast relieve the congestion problem, reduce energy consumption, and improve the performance of network. This part will improve Homotopy algorithm [5], and accelerate reconstruction proceeding.

The original signal estimate is achieved from sensing data with noise by solving the following optimal problem [5]:

$$\min f_\lambda(x) = \min \left(\frac{1}{2} \|y - Ax\|_2^2 + \lambda \|x\|_1 \right) \tag{92.2}$$

The classic thought of convex optimization is that the necessary condition of minimum $f_\lambda(x)$ satisfy $\frac{\partial f}{\partial x} = 0$, calculating

$$\frac{\partial f}{\partial x} = -A^T(y - Ax) + \lambda \frac{\partial}{\partial x} \|x\|_1 \tag{92.3}$$

where, $\frac{\partial}{\partial x} \|x\|_1 = (\text{sgn}(x_1) \quad \text{sgn}(x_2) \quad \text{sgn}(x_3) \quad \cdots \quad \text{sgn}(x_{n-1}) \quad \text{sgn}(x_n))^T$, and mutual coherence meets $M(A) = \max_{i \neq j} |<a_i, a_j>| \leq \mu$.

Theorem 1 *From $x_0 = 0$, the signal estimate is obtained $x_l = x_{l-1} + \gamma_l d_l$ by K steps increasing or decreasing the number of sampling points. The parameters meet the following conditions:*

1) $K \leq \frac{\mu^{-1}+1}{2}$,
2) $d_l(I)$ satisfies: $\text{sgn}(d_l(I)) = -\text{sgn}(e_l(I))$,
3) $\diamondsuit e_l(i) = y(i) - Ax_l(i) = \sum_{j=1}^{k} \beta_l(i)a_j$,

$$c_l(i) = A^T e_l(i), \quad v_l = A^T d_l(I),$$

$$\gamma_l = \min\left\{ \min_{i \in I^c} \left\{ \frac{\lambda - c_l(i)}{1 - a_i^T v_l}, \frac{\lambda + c_l(i)}{1 + a_i^T v_l} \right\}, \min\left\{ -\frac{x_l(i)}{d_l(i)} \right\} \right\}.$$

Theorem 1 *Gives the signal reconstruction restriction that makes stable reconstruction signal got through K steps from the signal $x_0 = 0$.*

92.3 Channel Selection

Due to WSN communication channel with a finite number of good quality, in a addition need channel allocation according to the priority, also demand of all signals in the overall quality of communication, different signals sharing the same channel is allowed. This section proposes a channel selection strategy based on the eigenvalue decomposition, effectively avoiding the competition to the same channel lead to congestion.

Step1: Given a channel gain matrix $H \in \Re^{N_r \times N_t}$ and $r = rank(H) \leq \min(N_r, N_t)$, rank estimate r_t of r is determined, $H = U \sum V^T; U \in \Re^{N_r \times N_r}$ is orthogonal matrix of eigenvector of $HH^T; V \in \Re^{N_t \times N_t}$ is orthogonal matrix of eigenvector of $H^T H; \sum \in \Re^{N_r \times N_t}$ is diagonal matrix, $\sum = diag(\sigma_1, \sigma_2, \ldots, \sigma_r)$, where $\sigma_i = \sqrt{\lambda_i}; \lambda_i$ is the i eigenvalue of HH^T, and $\sigma_1 \geq \sigma_2 \geq \ldots \geq \sigma_r > 0, \sigma_1, \sigma_2, \ldots, \sigma_{r_t}$ are much larger than $\sigma_{r_t+1}, \ldots, \sigma_r$; so, a threshold can be set to determine $\sigma_i, i = 1, 2, \cdots, r_t$, as well remove the smaller $\sigma_j, j = r_t + 1, \cdots, r$ to save energy. In addition, $r_t \ll r$.

Step2: V is partitioned to $V = \begin{pmatrix} V_{11} & V_{12} \\ V_{21} & V_{22} \end{pmatrix}$, where $V_{11} \in \Re^{r_t \times r_t}, V_{12} \in \Re^{r_t \times (N_t - r_t)}, V_{21} \in \Re^{(N_t - r_t) \times r_t}, V_{22} \in \Re^{(N_t - r_t) \times (N_t - r_t)}$

Step3: E is determined through column pivoting decomposition

$$\begin{pmatrix} V_{11}^T & V_{21}^T \end{pmatrix} E = QR$$

where, Q is a unitary matrix, $R \in \Re^{r_t \times N_t}$ forms an upper triangular matrix with decreasing diagonal elements, and E is the permutation matrix. The positions of 1 in the first r_t columns of E correspond to the r_t ordered most significant transmitters.

The calculation of r_t refers to FCM method in Jing liang [6].

92.4 Reconstruction Error Analysis

This paper introduces the reconstruction performance index definition [7], $RQI = \dfrac{\|\hat{x} - x\|_2^2}{\|x\|_2^2}$. When RQI decreases, sample rate of original signal is deduced by feedback; When RQI increases, sample rate of original signal is increased, at the same time, the RQI is smaller, the reconstruction performance is better, and sample rate is higher. In this paper, reasonable satisfaction of RQI is $1 \times 10^{-7} \leq RQI \leq 4 \times 10^{-7}$ [8].

Theorem 2 *For a given $\zeta \in [10, \infty)$ and a vector $x \in \Re^{N \times 1}$, if $A \in \Re^{M \times N}$ is a structurally random matrix, and RQI has the restrictive condition $1 \times 10^{-7} \leq RQI \leq 4 \times 10^{-7}$, then estimate vector $\hat{x} \in \Re^{N \times 1}$ satisfy*

$$0 \leq \frac{\|\hat{x} - x\|_2}{\|x\|_2} \leq \frac{1}{\zeta(1 + \delta_S^2)} \tag{92.4}$$

Proof Signal reconstruction is actually solving [2]:

$$J = \frac{1}{2}\|A\hat{x} - y\|_2^2 + \lambda\|\hat{x}\|_1 \tag{92.5}$$

minimum. Substituting (92.1), it is got

$$J = \frac{1}{2}\|A(\hat{x} - x) - \varepsilon\|_2^2 + \lambda\|\hat{x}\|_1 \tag{92.6}$$

Supposing $z = \hat{x} - x$, if intend to get (92.6) minimum, only can J seek partial derivative for z, $\frac{\partial J}{\partial z} = z^T A^T A - \varepsilon^T A = 0$, get $z = (A^T A)^{-T} A^T \varepsilon$, then

$$
\begin{aligned}
\frac{\|\hat{x} - x\|_2}{\|x\|_2} &= \frac{\|(A^T A)^{-T} A^T \varepsilon\|_2}{\|x\|_2} \leq \frac{\|(A^T A)^{-T}\|_2 \|A^T \varepsilon\|_2}{\|x\|_2} = \frac{\|(A^T A x)^{-1}\|_2 \|(A x)^T \varepsilon\|_2}{\|x\|_2} \\
&\leq \frac{\|(A x)^T\|_2 \|\varepsilon\|_2}{\|A^T A x\|_2 \|x\|_2} \leq \frac{\|\varepsilon\|_2}{(1+\delta_S)\|A^T A x\|_2} \leq \frac{\|\varepsilon\|_2 \|x\|_2}{(1+\delta_S)\|x^T A^T A x\|_2} \\
&\leq \frac{\|\varepsilon\|_2}{(1+\delta_S)(1-\delta_S)\|A x\|_2}
\end{aligned}
\tag{92.7}
$$

Thanks to $(1 - \delta_S^2)\|Ax\|_2 \geq \zeta(1 + \delta_S^2)\|\varepsilon\|_2$, then (92.4)

$$\leq \frac{(1 - \delta_S^2)}{(1+\delta_S)(1-\delta_S)\zeta(1+\delta_S^2)} \leq \frac{1}{\zeta(1+\delta_S^2)} \leq 0.1$$

$\square$

Theorem 2 *Illustrates that sampling rate will be adjusted to the optimal based on the reconstruction of performance feedback, and effectively control congestion. The signal transmission quality does not be affected in this course because of the transmission data quantity decreases.*

92.5 Simulation

For the analysis of the proposed method in WSN performance, the MAC layer and the transport layer node queue overflow congestion and channel competition are considered in simulation experiment. In this paper, simulation software NS-2 is applied for testing throughput, packet loss rate and other performance indicators. Energy consumption is compared with the DCS-LO [9]. Simulations use 3 transmitters and 6 receivers, and the original packets length is N = 256, which each data is 8 bit. Observation data packet length is M after CS compression, and sparsity is $k = 50$. Node initial energy is 1 J, and the energy consumption of FCLCC algorithm is calculated by [5].

Figure 92.1 shows packet loss rate at the M values 32, 64,128, respectively. See from the picture, the simulation time at 5 s, the packet loss rate of different M

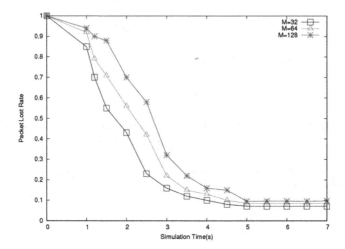

Fig. 92.1 Packet loss rate with simulation time

value can reach a very low, almost stable value, which illustrates algorithm can quickly relieve congestion. The packet loss rate of different M values is slightly different at the same time.

Figure 92.2 gives the network throughput comparison. The role of improving network performance in DCS-LO algorithm and the FCLCC algorithm within 4 s is roughly the equivalent, which demonstrates DCS-LO and FCLCC can effectively remove node level congestion. However, the FCLCC algorithm effectively intervened that case when network appeared in channel competition, which makes throughput stable, higher than the throughput of DCS-LO. The channel competition producing the link level congestion makes the throughput still in a state of oscillation in DCS-LO.

Figure 92.3 shows that the compression does not bring about the energy consumption. The variation of M value for the energy consumption has little effect, which is because energy of transmission account for the main part, in spite of data compression and reconstruction consume a part of energy. This proves adequately that compressed sensing theory applied to WSN congestion is rational.

Although the network packet loss rate and energy consumption are nearly the equivalent at different M value, M value should be compromise selection due to different MSE, so as to achieve the optimal effect.

92.6 Conclusion

This paper puts forward a congestion control algorithm based on compressed sensing theory applied to the transport layer node-level congestion and link layer

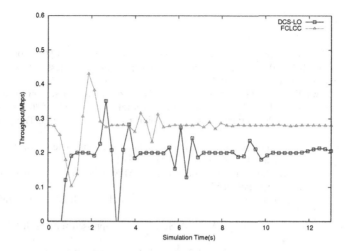

Fig. 92.2 Throughput with simulation time

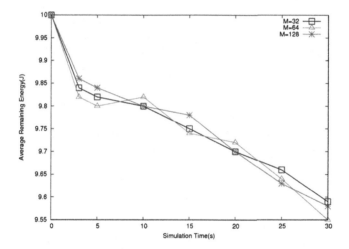

Fig. 92.3 Average remaining energy at different M value

link-level congestion in WSN. Channel allocation is achieved by eigenvalue decomposition method. That achieves the cross layer congestion control protocol design in WSN. Congestion control algorithm based on compressed sensing in WSN can improve network throughput, reduce network congestion time, and has maintained relatively low energy consumption.

References

1. Donoho D (2006) Compressed sensing. IEEE Trans Inf Theory 52(4):1289–1306
2. Chen W, Wassell IJ (2012) Energy-efficient signal acquisition in wireless sensor networks: a compressive sensing framework. IET Wireless Sens Syst 2(1):1–8
3. Haupt J, Bajwa WU, Raz G, Nowak R (2010) Toeplitz compressed sensing matrices with applications to sparse channel estimation. IEEE Trans Inf Theory 56(11):5862–5875
4. Davies ME, Eldar YC (2012) Rank awareness in joint sparse recovery. IEEE Trans Inf Theory 58(2):1135–1146
5. Donoho DL, Tsaig Y (2008) Fast solution of l_1-norm minimization problems when the solution may be sparse. IEEE Trans Inf Theory 54(11):4789–4812
6. Liang J, Liang QL (2009) Channel selection in virtual MIMO wireless sensor networks. IEEE Trans Veh Technol 58(5):2249–2257
7. Ward R (2009) Compressed sensing with cross validation. IEEE Trans Inf Theory 55(12):5773–5782
8. Malioutov D, Sanghavi S, Willsky A (2010) Sequential compressed sensing. IEEE J Sel Top Signal Process 4(2):435–444
9. Caione C, Brunelli D, Benini L (2012) Distributed compressive sampling for lifetime optimization in dense wireless sensor networks. IEEE Trans Ind Inf 8(1):30–40

Chapter 93
Design and Implementation of an Experiment Environment for Network Attacks and Defense

Songchang Jin, Shuqiang Yang, Songhe Jin, Hui Zhao and Xiang Wang

Abstract With the development of computer and network technology, large scale network attacks on the Internet occur frequently, such as worms, distributed denial of service, Trojan etc. The network attacks will have serious impacts on network service and network infrastructure. Data analyses on the variety of network attacks could help people to infer the type of network attacks. According to the association rules, people can take out proper disposal program timely to effectively reduce the harm caused by network attacks. Large scale complex network requires amount of money to support a large number of servers, switches, routers and other physical devices, and the network topology is not easy to change to support flexible network structure. This paper designs and implements an Experiment Environment for large scale network attacks and defense, and presents metrics and indices of effectiveness evaluation of large scale network attacks and defense. Experiments show that the system supports a variety of network attack experiments running at the same time, and can analyze network attack data effectively and predict the trend.

Keywords Network attack · Experiment environment · Emulab · Assessment analysis

S. Jin (✉) · S. Yang · H. Zhao · X. Wang
School of Computer, National University of Defense Technology, 410073 Changsha, China
e-mail: jsc04@126.com

S. Yang
e-mail: sqyang9999@126.com

S. Jin
School of Computer and Communication Engineering, ZhengZhou University of Light Industry, 450002 Zhengzhou, China

W. Lu et al. (eds.), *Proceedings of the 2012 International Conference on Information Technology and Software Engineering*, Lecture Notes in Electrical Engineering 210, DOI: 10.1007/978-3-642-34528-9_93, © Springer-Verlag Berlin Heidelberg 2013

93.1 Introduction

With the development of computer and network technology, large scale network attacks on the Internet occur frequently, and bring more and more serious harm to network services. Study on the characteristics of the various network attacks, can help people to discover network attacks timely. Before the arrival of more large-scale attacks, taking the proper coping strategies can effectively reduce the harm caused by the large-scale network attacks. The new antivirus or firewall software must be measured and then to release, so people can test its performance through network attacks.

Research on large scale network attacks and defense needs experiment environment. But it is difficult to carry out experiments on real network. The topology of the real network lacks flexibility; the Internet is an open network, network attacks may become uncontrollable, resulting in immeasurable loss; moreover, devices in laboratory are limited in types and quantity. So an emulation environment which can be used to simulate large scale network attacks and deploy different protection policies is needed to collect experiment data and analyze the efficiency of the policy enforcement.

This paper introduces an emulation environment based on Utah's Emulab software for large scale network attack experiment and the evaluation system.

93.2 Emulab Architecture

Emulab is a network testbed, giving researchers a wide range of environments in which to develop, debug, and evaluate their systems. The basic Emulab architecture consists of a set of experiment nodes, a set of switches that interconnect the nodes, and two control nodes, Boss and Users (see Fig. 93.1). The switches are used to interconnect the experiment nodes. The interconnections are physically separated into a dedicated control network and an experiment network for user-specified topologies [1]. Experiment nodes may be servers, personal computers or other devices, such as Intrusion Detection Systems, etc. Each experiment node has two or more network interfaces, one of which is connected to the dedicated control network and the other interfaces are connected to the experiment network.

The Boss server controls the testbed's operation including the ability to power cycle individual experiment nodes, while researchers log into the Users server to create and manage experiments and to store the data required or generated by their experiments. The testbed's switches are controlled using snmpit, a program that provides a high-level object interface to the individual SNMP MIB's of testbed switches. Other programs talk to the power controllers to power cycle nodes, load operating systems onto experiment nodes when requested, and interact with the database to reserve and assign nodes to experiments.

Fig. 93.1 Emulab
architecture

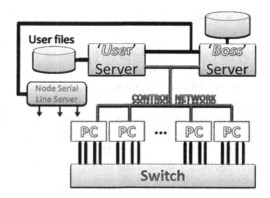

An Emulab experiment consists of a collection of nodes, software to run on the nodes, and an interconnection topology. An experiment is specified using a combination of a .ns [2] file and a web interface. The Emulab control software on the Boss server enables multiple, separate experiments to be simultaneously run on the testbed. The software isolates experiments by assigning each experiment to one or more unique Virtual Local Area Networks (VLANs) that connect together the experimental interfaces on each experiment node either using simulated band-width-limited and lossy links or using LANs. By using separate VLANs, an experiment's experimental traffic is isolated from other experiments. To prevent one experiment's network traffic from interfering with that of other experiments because of insufficient internal switch or inter-switch bandwidth, the assign program is responsible for mapping an experiment's link bandwidth requirements onto the available switch resources in a manner that ensures that the experiment's bandwidth demands match available inter- and intra-switch bandwidths. Note that unless an experiment is firewalled, all of the control network interfaces are on the same VLAN.

The Emulab process of swapping in a new experiment consists of several steps: mapping the researcher's desired network topology onto available nodes and switch resources, configuring VLANs on the switches to connect the experiment nodes into the researcher's desired network topology, installing an initial mini filesystem kernel and root filesystem onto the experiment nodes, and then loading and running the desired operating system and software.

In the Emulab trust and privilege hierarchy model, each researcher is a separate user of the testbed. Users working together are grouped into groups, and a project consists of a collection of related groups. Users may also belong to more than one project. Each testbed has its own complete (and independent) trust and privilege hierarchy.

93.3 System Design

In order to carry out large scale network attacks and defense experiments, analyze and evaluate effects of the attacks, we have implemented an experiment environment which has multiple-layer fidelity and supports large scale network security experiments by using Emulab based on Tianhe-1A supercomputer. The test shows that the environment could support researches on large scale network attacks. Figure 93.2 shows the function architecture of the system. There are three layers in the architecture: Resource layer, Emulation layer and User layer. The management module runs through these three layers.

93.3.1 Resource Layer

The resource layer is responsible for building resources pool, to achieve the underlying resource virtualization, such as CPUs, memories and so on. Taking into account the configuration of a single server in Tianhe-1A supercomputer (2 Intel Hexa Xeon X5670 6 core 2.93 GHz CPU, NVIDIA Tesla M2050 GPU, 32 GB Memory, 40 Gbps I/O), we start eight virtual machines on each server to expand the scale of the system. The flexibility of the virtual machines enables us to create various types of virtual machines According to the actual demand.

Figure 93.3 shows the resource layer of the system. We use 82 servers in Tianhe-1A supercomputer to build our system. One server acts as 'User Server'

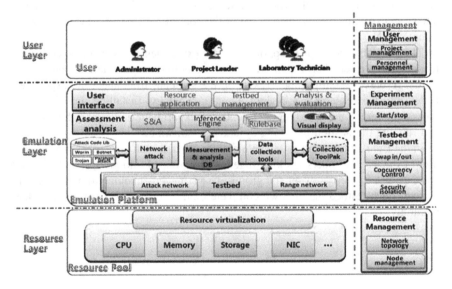

Fig. 93.2 Function architecture of the experiment environment

running FreeBSD operating system, one as 'Boss Server' running the same system as 'User Server' (see Sect. 93.2), the other as host nodes. Each host node runs into 64-bit NeoKylin operating system and starts 8 VMware virtual machines. So we get a total of 640 experiment nodes. All experiment nodes are virtual nodes, so we do not need serial line to control experiment node and 'Node serial line Server'.

93.3.2 Emulation Layer

The emulation layer is mainly composed of three parts: Testbed, Assessment analysis and User interface. It is the core component of the system.

93.3.2.1 Testbed

A testbed is a platform for experimentation of large development projects. Testbeds allow for rigorous, transparent, and replicable testing of scientific theories, computational tools, and new technologies [3]. The testbed in this paper is used to test and analyze variety of network attacks. The main function of the testbed is to provide the users a variety of network topologies and experimental resources. User can create many testbed in the system, each testbed is physical isolation.

Network topologies are divided into two categories: attack network and range network (defense network). According to the .ns file provided by the user, the system generates a corresponding physical network topology and environment, and maintains an attack code library. Users can upload their own attack code to the library. After a testbed is established, user can deploy attack code on attack network, select the appropriate measurement tools and data collection tools provided in the ToolPak and deployed them on defense network. All after that, starting experiment.

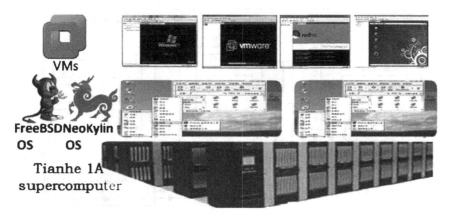

Fig. 93.3 System architecture

93.3.2.2 Assessment Analysis

The assessment analysis module carries out cluster analysis on the data received from the defense network, offers a variety of correlation analysis, filters false alarms, offers a visual way to display network attack events, and provides users with a detailed view of all aspects of the experimental network security (attack, flow, assets, vulnerability, etc.). Its main function is data integration and correlation analysis, integrating network attack data from various heterogeneous network and make them work in a common environment. So we developed data collection tools, correlation engine, rule base and data management and display tools. The architecture of the assessment analysis module is shows in Fig. 93.4.

The assessment analysis module consists of several sub-modules.

1. Various types of front-end probes are deployed on defense network to monitor host, service and network, to scan vulnerabilities and to collect data.
2. Data integration includes two types: Event data integration is used to integrate the security data get from front-end probes, such as Snort, Ntop, etc. Scan/ inventory data integration is used to integrate data from Nmap, nessus, ect.
3. Association analysis is the core of the assessment analysis. After preprocessing by security event analysis and network traffic analysis, the data is transmitted to association analysis. It analyzes the events and assets (Logical association), the event and the vulnerability of the events associated with other events (Cross-correlation), makes risk assessment based on association rules, and eventually generates the correct alarm and warning [4]. It converts these complex and abstract, large-scale network attacks into understandable warnings in an intelligent way for users.
4. DB, all the data from data integration and analysis are stored in the DB.
5. Security event display module presents the results of the evaluation of the network attack in such as chart, graphs and other.

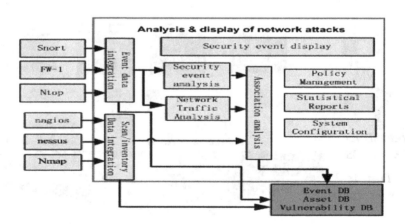

Fig. 93.4 Architecture of the assessment analysis module

The accuracy of the evaluation of the network attacks, to a large extent depends on the accuracy of the association rules.

93.3.2.3 User Interface

User interface provides an interactive interface for the users to access and control the testbeds.

Resource application is used for users to apply resource from the system to set up a testbed and the administrator respond to the users' requests.

Testbed management is used for all users including administrator and project leaders to manage the testbeds of their own.

Analysis and evaluation is used to present the evaluation results of the network attacks.

93.3.3 *Management*

Each layer of the system contains the management module. Resource management consists of network topology management and node management. Node management is responsible for maintaining the node type information, adding new nodes, deleting the existing nodes and so on.

Testbed management is responsible for swapping testbeds in/out, multi-testbeds' concurrency control and security isolation. Experiment management controls experiments to start or stop on the testbeds.

User management manages the users and projects in the system, including user registration and authorization, project application, authorization and cancellation, etc.

93.4 Experiment and Analysis

The system designed in this paper consists of 82 Tianhe-1A servers (see Sect. 93.3.1), 4 Cisco WS-C3750G-48TS-S gigabit switches including two control switches and two experiment switches, 1 PowerLeader PR2760T server acting as assessment analysis and web server.

With the system under normal operating conditions, we setup and swap in two testbeds. The network topologies created by netlab-client of the two testbeds are shown in Fig. 93.5. Figure 93.6 shows the active testbeds running in the system right now. We can see that the two testbeds just created appears in the head position of Fig. 93.6. So this means that the system supports several testbeds running at the same time.

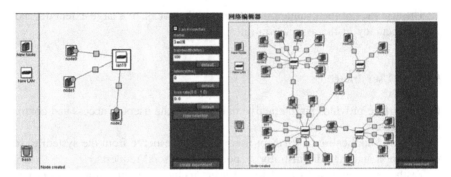

Fig. 93.5 Network topologies of two experiments in the system

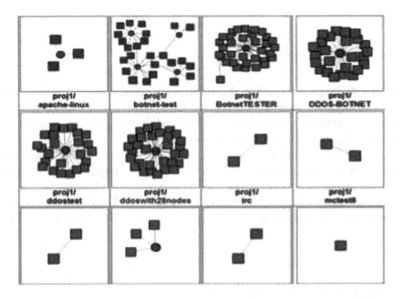

Fig. 93.6 Part of active testbed network topologies in the system

Table 93.1 Average time consuming (min) of creating and swapping in a multi-nodes testbed

OS\nodes	1	10	30	50	70	90
Windows XP SP2 32bit	11.4	12.1	12.4	12.9	13.2	13.8
Ubuntu 12.04 32bit	4.2	4.5	4.7	5.1	5.5	6.1
CentOS 5.5 64bit	4.8	5.0	5.4	5.9	6.2	6.9
FreeBSD 7.3 32bit	3.5	3.8	4.2	4.8	5.2	5.5

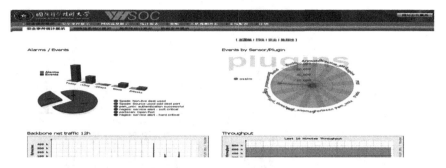

Fig. 93.7 Index page of the assessment analysis system

Fig. 93.8 Alarm/event 警报 / 事件

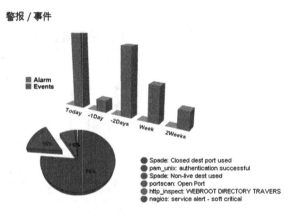

Fig. 93.9 Sensors event 传感器或者插件获取的事件

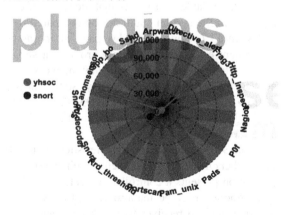

In the system, we create a lot of operating system images, including multi-version of 32 and 64 bit Windows Xp, Ubuntu, CentOS, Fedora and FreeBSD operating system. With Emulab's Frisbee [5], we can quickly establish a user-specified

Fig. 93.10 Network traffic within 12 h

12小时内网络流量

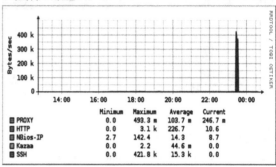

Fig. 93.11 Network traffic within 10 min

近10分钟网络流量

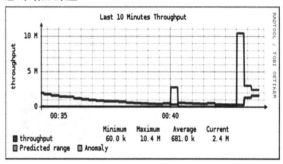

network topology and physical network operating system on experiment nodes. Table 93.1 lists the average time of creating and swapping in a multi-node test bed. Data in Table 93.1 shows that the speed of establishing a large-scale network topology in the system could not be compared by live network.

Figure 93.7 shows the index page of the assessment analysis system of the experiment environment. System provides a variety of views, including network attacks situation analysis and display, network attacks event analysis and display, network traffic analysis and display, statistical reports, security strategies, association rules. Figures 93.8, 93.9, 93.10, 93.11, 93.12, 93.13, 93.14 and 93.15 shows the data analysis results obtained from a number of network attacks during system operating normally.

Table 93.2 enumerates characteristics of traditional approaches. Each offers unique benefits, thus guaranteeing their continued importance [6]. For example, simulation presents a controlled, repeatable environment. However, its high level of abstraction may be inappropriate, for example, when studying the effects of interrupt-induced receiver live lock on a heavily-loaded system.

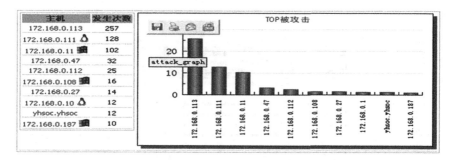

Fig. 93.12 Top 10 of the attacked nodes

Fig. 93.13 Top 10 of
attacking nodes

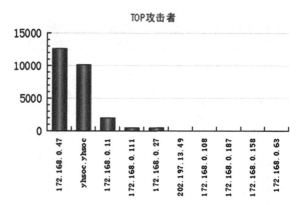

Fig. 93.14 Top 10 of ports
in using on attacked nodes

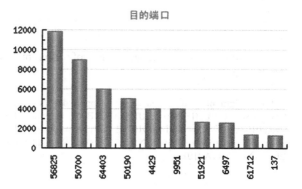

93.5 Conclusion

This paper introduces a design and implementation of an experiment environment
for large scale network attacks and defense, presents the method to measure the
impacts of network attack and defense. In the future, we will make good use of the

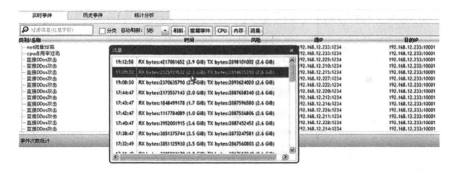

Fig. 93.15 Real-time attacks event and details from association analysis

Table 93.2 Characteristics of experimental platforms

Metric	Simulation.	Emulation	Live net	Our system
Ease of Use	√	ModelNet ?		√
Performance		√	√	√
Repeatability	√	√		√
Packet-level control	√			√
Coarse-grain control	√	√		√
Scalability	Varies	w/ModelNet	Varies	√
Param. Space Explor.	√	ModelNet ?		√
Reuse of models	√	ModelNet ?		√
Real links			√	√
Real routers			√	√
Real hosts	√		√	√
Real applications	√		√	√
Real users			√	

environment to study the efficiency of different network attacks and defense mechanisms.

Acknowledgments This research was supported by the Key Technologies R&D Program of China (No.2012BAH38B-04, 2012BAH38B06), National High-Tech R&D Program of China (No.2010AA012505, 2011AA010702, 2012AA01A401, 2012AA01A402), National Natural Science Foundation of China (No. 60933005), National Information Security 242 Program of China (No. 2011A010).

References

1. Cui Y, Liu L, Jin Q, Kuang X (2011) Implementation of an emulation environment for large scale network security experiments. In: Proceedings of the applied computing conference (ACC '11), 17–19 Nov 2011, Angers, France
2. The VINT Project (2011) The ns Manual, Nov 2011. http://www.isi.edu/nsnam/ns/ns-documentation.html

3. Testbed (2012) Wikipedia http://en.wikipedia.org/wiki/Testbed
4. Li Jingang (2010) Research and design of network security event correlation engine. Beijing University of Posts and Telecommunications, Beijing (in Chinese)
5. Hibler M, Stoller L, Lepreau J, Ricci R, Barb C (2003) Fast, scalable disk imaging with frisbee. In: Proceedings of the USENIX annual technical conference, June 2003
6. White B, Lepreau J, Stoller L et al (2002) An integrated experimental environment for distributed systems and networks. In: Proceedings of the 5th symposium on operating systems design and implementation, Boston, Massachusetts, 09–11 Dec 2002

Chapter 94
Study of Signal De-Noising in Wavelet Domain Based on a New Kind of Threshold Function

Jing-Bin Wang and Bao Wang

Abstract In this paper, on the basis of shortcomings of discontinuity in the hard threshold method and constant deviation in the soft threshold method, we propose an improved threshold function. The numerical experiments show that the proposed method can not only remove the noise effectively, but also have greater flexibility than the two classical threshold functions. We can get a higher SNR and better visual effects.

Keywords Wavelet transform · Threshold function · Dynamic threshold

94.1 Introduction

De-noising is a focus topic during all kinds of signal processing, which aims at removing the noise in the signal as much as possible and guaranteeing that the de-noised signal is the optimal estimate of the original signal. Many signal de-noising methods have been presented, such as Wiener filtering, Kalman filtering, the spectral subtraction, wavelet transform, etc. The methods of traditional de-noising are to filter out the noise frequency component of the signal through a filter. They are based on frequency domain, namely, separating the desired signal from the noise signal and ignoring the pulse signal, white noise, non-stationary process signal and so on. Compared with the traditional de-noising methods, wavelet transform analyses the noise signal both in time domain and frequency

J.-B. Wang (✉) · B. Wang
College of Mathematics and Computer Science, Fuzhou University, Fujian, Fuzhou, China
e-mail: wjbcc@263.net

B. Wang
e-mail: bpeng.jun@gmail.com

W. Lu et al. (eds.), *Proceedings of the 2012 International Conference on Information Technology and Software Engineering*, Lecture Notes in Electrical Engineering 210, DOI: 10.1007/978-3-642-34528-9_94, © Springer-Verlag Berlin Heidelberg 2013

domain and also realizes automatic zoom function [1], which has better performance on de-noising by distinguish the mutation part and noise from the signal
effectively. So far, wavelet de-noising can be divided into three categories:
wavelet transform modulus maximal based, threshold-based, wavelet coefficients
based. Where, the threshold-based method can obtain the best estimated value in
Besov space [2], which has been widely used in the field of signal de-noising.

94.2 Wavelet Threshold De-Noising

Assume the observation signal as follow.

$$f(\kappa) = s(\kappa) + n(\kappa) \tag{94.1}$$

Where f(k) is the signal with noise, s(k) is original signal, n(k) is the Gaussian
white noise with variance of σ^2 and obedience the distribution of $N(0, \sigma^2)$.
Applying discrete wavelet transform to the observation signal f(k).

$$\varpi f(j, k) = 2^{-j/2} \sum_{K=0}^{N-1} f(k)\psi(2^{-j} - k), j = 0, 1, \ldots, J - 1 \tag{94.2}$$

Where $\varpi f(j, k)$ is the wavelet coefficient and J denotes the number of layers
decomposed.

The main idea of wavelet threshold de-noising proposed by Donoho and
Johnstone [3] is described as follows. When $\varpi f(j, k)$ is less than a threshold, the
$\varpi f(j, k)$ is viewed caused by noise mainly and should be discarded. However, if
$\varpi f(j, k)$ is larger than the threshold, the $\varpi f(j, k)$ is viewed caused by signal mainly,
then $\varpi f(j, k)$ is reserved (hard threshold method) or shrunk to zero in accordance
with a fixed vector (soft threshold method), and obtain the signal de-noised by
wavelet reconstructed with the processed $\varpi f(j, k)$.

The process of wavelet threshold de-noise is described as follows [4].

Step1. Obtain a group of transformed wavelet coefficient by $\varpi f(j, k)$ by applying
the wavelet transform to the f(k).

Step2. To let the value of $|\varpi f(j, k) - \omega f(j, k)|$ as small as possible, we deal with
the $\varpi f(j, k)$ by hard or soft threshold function, which will generate a group
of estimated wavelet coefficient $\varpi f(j, k)$.

Step3. Obtain the estimated signal $\breve{f}(k)$ by re-constructing the signal with $\breve{\omega} f(j, k)$.

And the $\breve{f}(k)$ is the signal de-noised.

The hard threshold function and soft threshold function used by Donoho [3] are
described as Eqs. (94.3) and (94.4) respectively.

$$\breve{\omega} f(j, k) = \begin{cases} \omega f(j, k) & |\omega f(j, k)| \geq \lambda \\ 0 & otherwise \end{cases} \tag{94.3}$$

$$\breve{\omega}f(j,k) = \begin{cases} \text{sgn}(\omega f(j,k))(|\omega f(j,k)| - \lambda) & |\omega f(j,k)| \geq \lambda \\ 0 & otherwise \end{cases} \tag{94.4}$$

In which, $\omega f(j,k)$ is wavelet transform coefficient, $\breve{\omega}f(j,k)$ denotes the coefficient processed and λ denotes threshold. The image of hard threshold function and soft threshold function are shown as Figs. 94.1 and 94.2 respectively.

It is proved that the estimated signal obtained by wavelet threshold de-noise is effective on MSN by Donoho and Johnstone.

$$N^{-1}E\|\breve{\omega}f(j,k) - \omega f(j,k)\|^2 = N^{-1}\sum_{k=0}^{N-2}(\breve{\omega}f(j,k) - \omega f(j,k))^2 \tag{94.5}$$

94.3 A New Threshold Function

Though the hard and soft threshold functions are used widely and have achieved good performance, there are still many shortcomings [5]. During the method of hard threshold de-noise, the wavelet coefficient processed is discontinuous in the wavelet domain due to the discontinuity of $\breve{\omega}f(j,k)$ in the range of λ and $-\lambda$, which will lead to the results with big MSN [6]. While in soft threshold de-noise, the continuity of $\breve{\omega}f(j,k)$ is good in general and the de-noising effect is smooth, but the derivative of soft threshold function is discontinuous and difficult to get the higher derivative. Furthermore, when $\omega f(j,k) > \lambda$ and $\omega f(j,k) < -\lambda$, there is always constant deviation between the wavelet coefficient that processed by threshold function and the wavelet coefficient of original signal, which is not agree with the tendency that noise component gradually decrease with wavelet coefficient increase [7]. That will directly influence the approximation degree of re-constructed signal l, which is bound to bring the re-constructed signal inevitable deviation.

Fig. 94.1 The image of hard threshold function

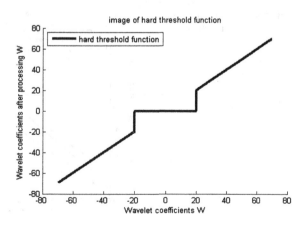

Fig. 94.2 The image of soft threshold function

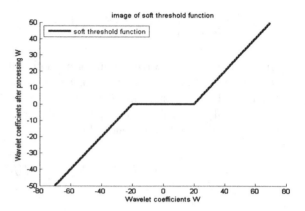

In order to overcome the lacks of the hard and soft threshold functions, we propose a new threshold function that is shown as follow.

$$\breve{\omega}(j,k) = \begin{cases} \operatorname{sgn}(\omega f(j,k))(|\omega f(j,k)| - \dfrac{\lambda}{(e^{(\frac{\omega f(j,k)}{\lambda})} - 1)^{\alpha}} & |\omega f(j,k)| \\ 0 & Otherwise \end{cases} \tag{94.6}$$

In which, $W(j,k)$ denotes the wavelet transform coefficient, $\overline{\omega}_{j,k}$ denotes the wavelet transform coefficient processed and a is the soft factor in the range from 0 to 1. Figure 94.3 has shown the image of our new threshold function.

As we can see from the Eq. (94.5), there are several characters of the new threshold function. Our new threshold is continuous and has higher derivative in case that $|\omega f(j,k)| > \lambda$, which makes the mathematical calculation simple. It also combines the hard threshold function with the soft threshold function. When $a = 0$, Eq. (94.5) is equivalent to soft threshold function. In case that

$$\alpha \neq 0 |\omega f(j,k)| > 0,$$

Fig. 94.3 The image of the new threshold function

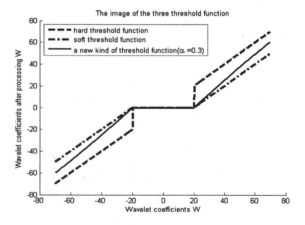

then

$$\frac{\breve{\omega}f(j,k)}{\omega f(j,k)} = \frac{\omega f(j,k)\frac{\lambda}{\left(e\left(\frac{A}{\lambda}-1\right)\right)^{\alpha}}}{\omega f(j,k)}$$

$$\breve{\omega}f(j,k) - \omega f(j,k) = -\frac{\lambda}{\left(e\left(\frac{|\omega f(j,k)|}{\lambda}\right)\right)}$$

Now, we can obtain

$$\lim_{\omega f(j,k)\to+\infty} \frac{\breve{\omega}f(j,k)}{\omega f(j,k)} = 1 \quad \text{and}$$

$$\lim_{\omega f(j,k)\to+\infty} \left(\breve{\omega}f(j,k) - \omega f(j,k)\right) = 0$$

Once

$$\alpha \neq 0|\omega f(j,k)| < 0$$

then

$$\lim_{\omega f(j,k)\to+\infty} \frac{\breve{\omega}f(j,k)}{\omega f(j,k)} = 1 \quad \text{and}$$

$$\lim_{\omega f(j,k)\to+\infty} \left(\breve{\omega}f(j,k) - \omega f(j,k)\right) = 0$$

So the new threshold function also takes $\breve{\omega}f(j,k) = \varpi f(j,k)$ as asymptote, and $\breve{\omega}f(j,k)$ sticks close to the hard threshold with the increment of $\varpi f(j,k)$. That overcomes the shortcoming of constant deviation between the soft threshold function $\varpi f(j,k)$ and $\breve{\omega}f(j,k)$.

Here, we analyze the impact of a on the threshold function.

$$\lim_{a\to+\infty} \breve{\omega}f(j,k) = \begin{cases} \text{sgn}(\omega f(j,k))(|\omega f(j,k)| - \frac{\lambda}{(e^{\frac{|\omega f(j,k)|}{\lambda}}-1)^{\alpha}} = \breve{\omega}f(j,k) - \lambda \times \text{sgn}(\omega f(j,k)) \\ 0 \end{cases}$$

As we can see, the new threshold function is close to the soft threshold function in case that $\alpha \to 0$ and the smaller a will lead to the tendency more obvious. When $a = 0$, the new threshold function is soft threshold function. When

$$\lim_{|\omega f(j,k)|\to+\infty} \frac{\lambda}{\left(e\left(\frac{|\omega f(j,k)|}{\lambda}-1\right)\right)^{\alpha}} = 0$$

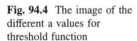

Fig. 94.4 The image of the different a values for threshold function

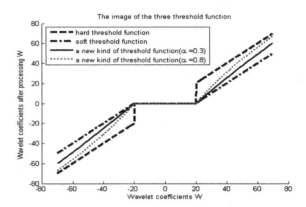

Equation (94.5) has indicated that $\widetilde{\varpi}f(j,k)$ is close to $\varpi f(j,k)$ with the increment of $|Wf(j,k)|$ in case that $\alpha \to 1$. In this case, new threshold function is close to hard threshold function and the larger a will lead to the tendency more obvious. So the parameter a can influence the tendency of the new threshold function, which reflects the flexibility of the new threshold function.

As we can see from Fig. 94.4, the new threshold function has combined the hard threshold function with soft threshold function. And the soft factor a can change the tendency of new threshold function, which also proves the flexibility of the new threshold function.

94.4 Experiment Results and Analysis

In order to study the effectiveness and superiority on de-noising of the new threshold function, we use the traditional methods based on global threshold value of the hard and soft threshold function to do a lot of de-noise tests (Figs. 94.5, 94.6, 94.7, 94.8, and 94.9).

In this paper, we use the signal-to-noise ratio (SNR) [8] and mean square error (MSE) to describe the performance of the de-noising, which can show the superiority of our algorithm intuitively. The formula of SNR is given as follow.

$$\text{SNR} = 10 \times \lg\left(\frac{power_{signal}}{power_{noise}}\right)$$

Where the $power_{signal} = \frac{1}{n}\sum_n f^2(n)$ denotes the power of original signal that with noise, $power_{noise} = \frac{1}{n}\sum_n \left[f(n) - \tilde{f}(n)\right]^2$ is the noise power, $f(n)$ is the original signal that with noise, $\tilde{f}(n)$ is the re-construct signal that de-noised by wavelet. And the formula of MSE is presented as follow.

Fig. 94.5 The original signal

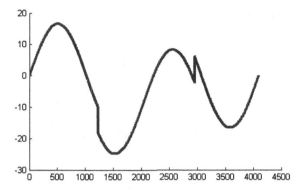

Fig. 94.6 The signal with noise

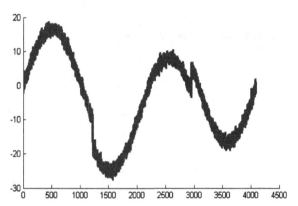

Fig. 94.7 Hard threshold function de-noising

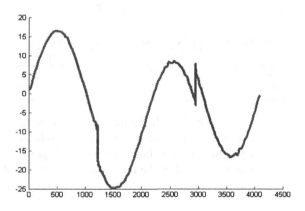

$$[\text{MSE}] = \frac{1}{N}\sum_{i=1}^{N}(X(i) - \tilde{X}(i))^2$$

Where $X(i)$ is the original signal and $\tilde{X}(i)$ represents the re-construct signal.

Table 94.1 shows the results of de-noised SNR and MSE for the three methods.

Fig. 94.8 Soft threshold
function de-noising

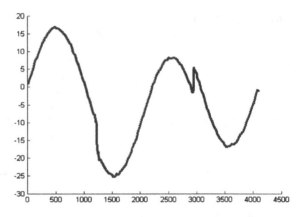

Fig. 94.9 The new threshold
function de-noising

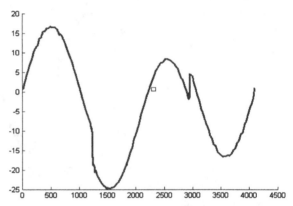

Table 94.1 SNR and MSE comparison of three methods

	Soft threshold	Hard threshold	New threshold
SNR/dB	22.9570	22.2068	23. 4827
MSE	0.0762	0. 0892	0.0589

As we can see from the results of the experiment, our new threshold function has better performance than the hard and soft threshold functions. Due to the flexibility of the new threshold function, our algorithm can obtain the smaller MSE and improve the SNR of the re-construct signal, also retain the important features of the original signal.

94.5 Conclusion

Due to the distributional characteristics of noise and useful signal frequency, and the time–frequency characteristics of the wavelet transform, the wavelet signal de-noise has more prominent advantages. According to the basic principles of the wavelet threshold de-noise and combining with the method of hard and soft threshold de-noise, we have proposed a new threshold function and the selection of threshold scheme. Our algorithm inherits the advantages and overcomes the lacks of the traditional methods, which has better flexibility and mathematical. As we can see from the experimental results, our algorithm can improve the SNR and MSE, and has better performance than the method of hard and soft threshold de-noise. Furthermore, the softening factor of our new threshold function is dynamic, which provides great convenience for the selection of threshold and improves the performance of de-noise. The value of softening factor a adopted in this paper is not the optimal, and the selection of best value of a is the problem that urgently need to be solved in the future.

References

1. Zhang DF (2007) Wavelet analysis and engineering application. National Defence Industry Press, Beijing (in Chinese)
2. Donoho DL, Johnstone IM (1994) Ideal spatial adaptation via wavelet shrinkage. Biometrika, UK, pp 613–627
3. Donoho DL (1995) Denoising by soft-thresholding. IEEE Trans Inform Theory 613–627
4. Wei-ren S, Luo XS, Hu N (2002) Wavelet-based multiresolution analysis to noise elimination. J Chongqing Univ 59–62 (in Chinese)
5. Liu ZA, Yang SY, Wang L (2009) An de-noising arithmetic based on max-modular of wavelet window. J Projectiles, Rockets, Missiles Guidance 2(1):286–289 (in Chinese)
6. Zhang Wei-qiang, Song Guo-xiang (2004) Signal de-noising in wavelet domain based on a new kind of thresholding function. J Xidian Univ 31(2):296–299. (in Chinese)
7. Wei XUE, Fu-hong G, Chen LZ, Sun XW (2008) Radar signal de-noising based on a new wavelet thresholding function. Comput simul 319–322 (in Chinese)
8. Gou XK, Guo TS (2010) Signal de-noising based on a new threshold function of wavelet. J Northwest Normal Univ 46(1):51–54 (in Chinese)

Chapter 95
Maneuvering Target Tracking Based on Adaptive Square Root Cubature Kalman Filter Algorithm

Sisi Wang and Lijun Wang

Abstract Concerning low accuracy even divergence of maneuvering target tracking due to inaccurate tracking model and statistical property, an adaptive Square Root Cubature Kalman Filter (SCKF) is proposed based on the standard SCKF and modified Sage-Husa estimator. The proposed algorithm can estimate the statistical parameters of unknown system noises online, and restrain the tracking error caused by unknown system noises effectively; hence it is applied to maneuvering target tracking. The simulation is preformed latterly and experimental results show that comparing with the standard SCKF algorithm, the adaptive SCKF can achieve better accuracy and stability for maneuvering target tracking while the system noises is unknown and time variation.

Keywords Adaptive SCKF · Maneuvering target tracking

95.1 Introduction

Maneuvering Target Tracking is a hotspot and difficulty in radar information processing fields at all times. In maneuvering tracking target applications, target dynamics are usually modeled in Cartesian coordinates. State vector encompass velocity and position components. The corresponding measurement information is expressed in polar coordinates, including distance components, bearing

S. Wang (✉) · L. Wang
School of Navigation, Guangdong Ocean University,
40#, East Jiefang Road, Zhanjiang, China
e-mail: mars32lin@sina.com

L. Wang
e-mail: 123wanglijun@163.com

W. Lu et al. (eds.), *Proceedings of the 2012 International Conference on Information Technology and Software Engineering*, Lecture Notes in Electrical Engineering 210, DOI: 10.1007/978-3-642-34528-9_95, © Springer-Verlag Berlin Heidelberg 2013

components even elevation components and so on. Hence, Maneuvering Target Tracking is considered as a typical multidimensional nonlinear estimate problem.

The square root cubature Kalman filter proposed in recent year is a effectively method for the multidimensional nonlinear estimate problem [1]. The simulation result proved that under the hypothesis of equal computational complexity, the SCKF algorithm has the higher accuracy than the common nonlinear estimate method such as Particle Filter and Unscented Kalman Filter and so on [2]. But while the filter is based on the inaccurate model and noise statistical property, it will lead to larger estimate error even divergence. And more often than not, the process noise is difficult represented its statistics property due to external interference, acceleration physical characteristic and manipulate and so on. Therefore, process noise is usually unknown and time variation.

To solve this problem, an adaptive SCKF (ASCKF) is proposed, which is a combination of modified Sage-Husa estimator and SCKF. The modified Sage-Husa suboptimal unbiased estimator which is embedded into the algorithm can recursively estimates and corrects the unknown noise on line, and thus the ASCKF can handle the recursive state filtering in the presence of unknown process noise covariance matrices.

95.2 Problem Statement

Considering general maneuvering target tracking problem in Cartesian coordinates, the discrete time state system and measurement equation can be described as following state space model.

$$x_k = f(x_{k-1}) + v_{k-1} \tag{95.1}$$

$$z_k = h(x_k) + \omega_k \tag{95.2}$$

Where $x_k \in \mathbb{R}^{n_x}$ is the state of the dynamic system at discrete time k; $f: \mathbb{R}^{n_x} \times \mathbb{R}^{n_x} \to \mathbb{R}^{n_x}$ and $h: \mathbb{R}^{n_x} \times \mathbb{R}^{n_x} \to \mathbb{R}^{n_x}$ are some known functions; $z_k \in R^{n_x}$ is the measurement; $\{v_{k-1}\}$ and $\{\omega_k\}$ are independent process and measurement Gaussian noise sequences with means q and r and covariances Q_{k-1} and R_k, respectively.

95.3 Square Root Cubature Kalman Filter

The symmetry and positive definiteness of error covariance matrix are often lost in update cycle of the CKF, and then Ienkaran Arasaratnam proposed the SCKF algorithm base on the standard CKF. The rest of this section is devoted to describe the SCKF in detail.

Firstly, according the 3rd spherical-radial rule, a total of $m = 2n$ cubature points is chosen, where n is the state-vector dimension, and cubature points set $\{w_i, \xi_i\}$ are used to piecewise linear approach the probability distribution function of state vector.
Where

$$w_i = \frac{1}{m}, \xi_i = \sqrt{\frac{m}{2}}[l]_i, \quad i = 1, 2, \ldots, m; \; m = 2n, \tag{95.3}$$

where $[l] \in \mathbb{R}^n$ is a generator, and $[l]_i$ denote the ith element of $[l]$. For example, $[l] \in \mathbb{R}^2$ represents the following set of points:

$$\left\{ \begin{pmatrix} 1 \\ 0 \end{pmatrix}, \begin{pmatrix} 0 \\ 1 \end{pmatrix}, \begin{pmatrix} -1 \\ 0 \end{pmatrix}, \begin{pmatrix} 0 \\ -1 \end{pmatrix} \right\} \tag{95.4}$$

Assume the posterior density probability $p(x_{k-1}|z_{1:k-1}) \sim N(x_{k-1}; x_{k-1/k-1}, P_{k-1/k-1})$ at time $k - 1$ is known, the Cholesky factor $S_{k-1/k-1}$ of error covariance is available as $S_{k-1/k-1} = chol\{P_{k-1/k-1}\}$. The SCKF algorithm can be represented as following three steps:

Step 1: Initialization

$$x_{0|0}, \; S_{0|0} = chol(P_{0|0})$$

Step 2: Time update

$$X_{i,k-1|k-1} = S_{k-1|k-1}\xi_i + \hat{x}_{k-1|k-1}$$
$$X_{i,k|k-1}^* = f(X_{i,k-1|k-1}, u_{k-1})$$
$$\hat{x}_{k|k-1} = \frac{1}{m}\sum_{i=1}^{m} X_{i,k|k-1}^* + q \tag{95.5}$$
$$S_{k|k-1} = Tria\left(\left[\chi_{k|k-1}^*, S_{Q,k-1}\right]\right)$$

Where

$$Q_{k-1} = S_{Q,k-1}S_{Q,k-1}^T$$
$$\chi_{k|k-1}^* = \frac{1}{\sqrt{m}}\left[X_{1,k|k-1}^* - \hat{x}_{k|k-1} \quad X_{2,k|k-1}^* - \hat{x}_{k|k-1} \quad \cdots \quad X_{m,k|k-1}^* - \hat{x}_{k|k-1}\right]$$

$$\tag{95.6}$$

Step 3: Measurement update

$$X_{i,k|k-1} = S_{k|k-1}\xi_i + \hat{x}_{k|k-1}$$
$$Z_{i,k|k-1} = h(X_{i,k|k-1}, u_k)$$

$$\hat{z}_{k|k-1} = \frac{1}{m}\sum_{i=1}^{m} Z_{i,k|k-1}$$

$$S_{zz,k|k-1} = \textbf{\textit{Tria}}\left(\left[\begin{array}{cc} \zeta_{k|k-1} & S_{R,k} \end{array}\right]\right)$$

$$P_{xz,k/k-1} = \chi_{k/k-1}\chi_{k/k-1}^{T} \qquad (95.7)$$

$$W_k = \left(P_{xz,k|k-1}/S_{zz,k|k-1}^{T}\right)/S_{zz,k|k-1}$$

$$\hat{x}_{k|k} = \hat{x}_{k|k-1} + W_k\left(z_k - \hat{z}_{k|k-1}\right)$$

Where

$$R_k = S_{R,k}S_{R,k}^{T}$$

$$\zeta_{k|k-1} = \frac{1}{\sqrt{m}}\left[\begin{array}{cccc} Z_{1,k|k-1} - \hat{z}_{k|k-1} & Z_{2,k|k-1} - \hat{z}_{k|k-1} & \cdots & Z_{3,k|k-1} - \hat{z}_{k|k-1} \end{array}\right]$$

$$\chi_{k|k-1} = \frac{1}{\sqrt{m}}\left[\begin{array}{cccc} X_{1,k|k-1} - \hat{x}_{k|k-1} & X_{2,k|k-1} - \hat{x}_{k|k-1} & \cdots & X_{m,k|k-1} - \hat{x}_{k|k-1} \end{array}\right]$$

$$(95.8)$$

95.4 Adaptive SCKF

Many adaptive filtering methods are proposed to avoid divergence due to inaccurate and time variation of the noise statistics property [3, 4]. Among them, the Sage-Husa noise estimator is used most extensively [5]. However, it should be noted that adaptive filtering algorithms which the Sage-Husa estimator is applied can not estimate process and measurement noise simultaneously. And while the process noise is time variation, the latest data need be centered on, so the sub-optimal modified Sage-Husa estimator is introduced.

Accordingly, aim at Maneuvering Target Tracking, an ASCKF which combines the advantages of modified Sage-Husa noise estimator and SCKF can recursively estimate the unknown process noise, and then filter in the nonlinear system. The detailed SCKF algorithm is described as below formulas.

Step 1: Initialization

$$x_{0|0}, S_{0|0} = chol\left(P_{0|0}\right), \hat{q}_0 = q_0, \hat{Q}_0 = Q_0 \qquad (95.9)$$

Step 2: Time update
Compute from (95.5–95.6) on the basis of given $S_{k-1|k-1}$, $\hat{x}_{k-1|k-1}$, $\hat{q}_{k-1}$ and $\hat{Q}_{k-1}$.

Step 3: Measurement update
Compute from (95.7–95.8), the updated state estimate $\hat{x}_{k|k}$ is achieved.

$$\hat{q}_k = (1 - d_k)\hat{q}_{k-1} + d_{k-1}\left[\hat{x}_{k|k} - \Phi_k\hat{x}_{k-1|k-1}\right]$$
$$\hat{Q}_k = (1 - d_k)\hat{Q}_{k-1} + d_k\left[W_k\tilde{z}_k\tilde{z}_k^T W_k^T + P_{k|k} - \Phi_k P_{k-1|k-1}\Phi_k^T\right]$$

(95.10)

Where

$$\tilde{z}_k = z_k - h\left(x_{k/k-1}\right), \quad d_k = \frac{1 - b}{1 - b^{k+1}}, \quad 0.95 < b < 0.99 \tag{95.11}$$

b is a forgetting factor, $\tilde{z}_k$ is predict residual and Φ_k is transition matrix.

95.5 Numerical Simulations

95.5.1 Simulation Scenario

Consider Maneuvering Target Tracking scenario where a target executes maneuvering turn in a horizontal plane at a constant, but unknown turn rate Ω, and the process noise is unknown. So the kinematics of the turning motion can be modeled by the following nonlinear process equation [6]. Where the state of target is $x = [x \quad \dot{x} \quad y \quad \dot{y}]$, x and y denote position, and $\dot{x}$ and $\dot{y}$ denote velocities in the x and y directions, respectively; T is the time-interval between two consecutive measurements

$$x_k = \begin{bmatrix} 1 & \frac{\sin \Omega T}{\Omega} & 0 & -\left(\frac{1-\cos \Omega T}{\Omega}\right) \\ 0 & \cos \Omega T & 0 & -\sin \Omega T \\ 0 & \frac{1-\cos \Omega T}{\Omega} & 1 & \frac{\sin \Omega T}{\Omega} \\ 0 & \sin \Omega T & 0 & \cos \Omega T \end{bmatrix} x_{k-1} + v_k \tag{95.12}$$

Where the process noise with an unknown covariance as below equations.

$$\mathbf{v_k} \sim N(0, Q_{k-1}), Q_{k-1} = diag[\eta\mathbf{M} \quad \eta\mathbf{M}], \mathbf{M} = \begin{bmatrix} T^3/3 & T^2/2 \\ T^2/2 & T \end{bmatrix} \tag{95.13}$$

The scalar parameter η is related to process noise intensities. A radar is fixed at the origin of the target and equipped to measure the range r, and bearing θ. Hence, we write the measurement equation.

$$\begin{pmatrix} r_k \\ \theta_k \end{pmatrix} = \begin{bmatrix} \sqrt{x_k^2 + y_k^2} \\ \tan^{-1}\left(\frac{x_k}{y_k}\right) \end{bmatrix} + \mathbf{w_k} \tag{95.14}$$

Where the measurement noise $\mathbf{w_k} \sim N(0, R)$, with $R = diag\left[\sigma_r^2 \quad \sigma_\theta^2\right]$
Data

$$T = 1s, \Omega = -3° s^{-1}, \sigma_r = 10m, \sigma_\theta = \sqrt{10}\,mrad$$

True initial state

$$\mathbf{x_0} = \begin{bmatrix} 1,000 \text{ m} & 300 \text{ ms}^{-1} & 1,000 \text{ m} & 0 \text{ ms}^{-1} \end{bmatrix}$$

And the associated covariance

$$\mathbf{P_{0/0}} = diag \begin{bmatrix} 100 \text{ m}^2 & 10 \text{ m}^2\text{s}^{-2} & 100 \text{ m}^2 & 10 \text{ m}^2\text{s}^{-2} \end{bmatrix}$$

Initial state estimate

$$\hat{\mathbf{x}}_{0/0} \sim N(x_0, P_{0/0})$$

Q_k is designed to change according to

$$Q_k = \begin{cases} diag[\eta_2 M & \eta_2 M] \, 1 \leq k \leq 30 \\ diag[\eta_3 M & \eta_3 M] \, 31 \leq k \leq 70 \\ diag[\eta_4 M & \eta_4 M] \, 71 \leq k \leq 100 \end{cases} \tag{95.15}$$

Where $\eta_2 = 10$, $\eta_3 = 40$, $\eta_4 = 90$, $M = \begin{bmatrix} T^3/3 & T^2/2 \\ T^2/2 & T \end{bmatrix}$

95.5.2 Simulation Results and Analysis

250 independent Monte Carlo runs were made, where the prior process noise covariance $Q_0 = [\eta_1 M \quad \eta_1 M]$, and $\eta_1 = 0.1 \text{ m}^2\text{s}^{-3}$ for a more accurate simulation. All the filters are initialized with the same condition in each run. The total number of scans per run is 100.

To compare various nonlinear filters performances, the root-mean square error (RMSE) of the position and velocity is used.

$$RMSE_{pos(k)} = \sqrt{\frac{1}{N} \sum_{n=1}^{N} \left((x_k^n - \hat{x}_k^n)^2 + (y_k^n - \hat{y}_k^n)^2 \right)} \tag{95.16}$$

Where $(x_k^n y_k^n)$ and $(\hat{x}_k^n \hat{y}_k^n)$ are the true and estimated positions at the n-th Monte Carlo run. Similarly to the RMSE in position, the formula of RMSE in velocity is also written. Figure 95.1 shows the trajectory of target and Fig. 95.2 compares the RMSEs in position and velocity, respectively. In addition, the mean of target state estimation RMSE is also shown in Table 95.1.

In Fig. 95.2, the RMSEs of various filters are compared across 100 time steps, and By reason of the large estimation errors caused by unknown and time variant process noise, it can be seen from the Fig. 95.2 that RMSE of standard SCKF sometimes deviate from the true RMSE very large, especially when the true process noise is much different from the prior knowledge. Note that as expected in

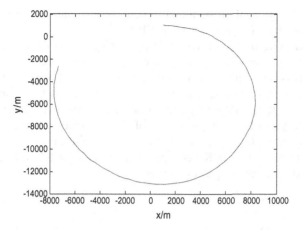

Fig. 95.1 Target trajectory

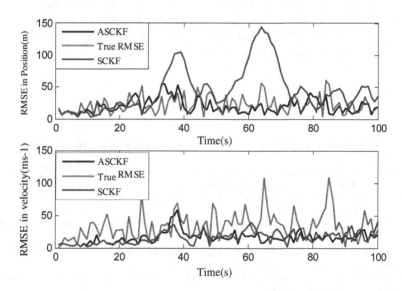

Fig. 95.2 True RMSE Vs. filter-estimated RMSEs (ASCKF and SCKF for the scenario)

Table 95.1 Mean of target state estimation RMSE

Parameters	SCKF	ASCKF	Unit
Position	44.9961	20.1605	m
Velocity	16.2349	15.8687	ms^{-1}

Table 95.1, the mean of position and velocity estimations RMSE of ASCKF are obviously less than that of standard SCKF.

The better performance of the ASCKF in maneuvering target tracking due to the modified Sage-Husa estimator which can estimate the unknown process noises online, whereas the standard SCKF depends on the fixed prior knowledge about the process noises. In new algorithm, the estimated noise statistics are recursively used by the ASCKF to adaptively compensate the influence caused by inaccurate a priori knowledge and changing statistics of system noise.

Ultimately, according to the contents mentioned above, it can be concluded that the performance of the ASCKF is superior to standard SCKF in the condition of unknown and time variant process noises.

95.6 Conclusions

In this paper, a new ASCKF, which is based on modified Sage-Husa estimator, has been proposed for maneuvering target tracking with unknown and time variant system noise statistics. The simulation results show that the proposed filter provides better performance in tracking accuracy than the CKF.

References

1. Arasara J, Haykin S (2009) Cubature Kalman filter. IEEE Trans Autom control 54(6): 1254–1260
2. Haykin S (2009). Cubature filters: new generation of nonlinear filters that will impact the literature. Available via internet. http://soma.mcmaster.ca
3. Hu C, Chen W, Chen Y, Liu D (2003) Adaptive Kalman filtering for vehicle navigation. J Global Positioning Syst 2(1):2–47
4. Mohamed AH, Schwarz KP (1999) Adaptive Kalman filtering for INS/GPS. J Geodesy 73:193–203
5. Sage A, Husa GW (1969) Adaptive filtering with unknown prior statistics. In: Proceedings of joint automatic control conference, USA, pp 760–769
6. Bar Shalom Y, Li XR, Kirubarajan T (2001) Estimation with applications to tracking and navigation. Wiley, New York

Chapter 96
A Novel Spatial Indexing Mechanism Leveraging Dynamic Quad-Tree Regional Division

Jine Tang, Zhangbing Zhou, Ke Ning, Yunchuan Sun and Qun Wang

Abstract Spatial index and query are enabling techniques for achieving the vision of the Internet of Things. Quad-tree is a commonly used spatial index which divides the space region into four small regions recursively, until the number of data objects in each region is within the given range. Based on the research of the traditional quad-tree, a new spatial index based on dynamic quad-tree regional division is proposed in this paper. The new index structure will first carry on clustering in the queried region, and then chose the biggest covering area to carry on quad-tree partition for each cluster. The construction of the tree is simple and it can save the storage space and improve the query efficiency by dynamically moving one quad-tree to index data objects in the entire region.

Keywords Spatial indexing · Quad-tree · Virtual quad-trees

J. Tang (✉) · Z. Zhou · Q. Wang
China University of Geosciences, Beijing, China
e-mail: tangjine2008@163.com

Z. Zhou
e-mail: zbzhou@cugb.edu.cn

Q. Wang
e-mail: qunw@cugb.edu.cn

K. Ning
Kingdee Research, Kingdee International Software Group Company Limited, Shenzhen, China
e-mail: ke_ning@kingdee.com

Y. Sun
Beijing Normal University, Beijing, China
e-mail: yunch@bnu.edu.cn

Z. Zhou
Institut-MinesTELECOM /TELECOM SudParis, Cedex, France

W. Lu et al. (eds.), *Proceedings of the 2012 International Conference on Information Technology and Software Engineering*, Lecture Notes in Electrical Engineering 210, DOI: 10.1007/978-3-642-34528-9_96, © Springer-Verlag Berlin Heidelberg 2013

96.1 Introduction

Things (or objects) in the Internet of Things can be modeled as Minimum Bounding Rectangles (MBRs) in spatial database. In recent years, there will produce a lot of spatial data when carry on study of the geographical resources, the global environment and geographical information, at the same time, the remote sensing, satellite image processing, human body medical imaging applications also provide a number of spatial databases [1, 2] information.

Many multi-dimensional indexing techniques have been proposed, and quad-tree is one of the earliest used data structures dealing with the high dimensional data [3] which is the expansion of the binary search tree in multi-dimensional space, and has been widely used in geographic information systems.

However, the traditional quad-tree will continue to divide layer by layer and form much larger quad-tree after dividing the whole region, since the spatial distribution of the data objects is chaotic. After region division by using quad-tree, many division regions do not store data objects, but some regions store a lot of data objects, so there will appear many empty division regions, leading to the low spatial region utilization.

Based on the research of the above quad-tree making full use of the advantages of a single quad-tree, a new spatial index based on dynamic quad-tree regional division is proposed in this paper, which will carry on dynamic movement to the quad-tree having been formed, and take shape of multiple virtual quad-trees the same as the initial formed quad-tree, until the regions which the initial formed quad-tree has traversed completely covering all the queried objects. The new index tree can avoid conducting division to the regions without data objects.

This paper is organized as follows. Section 96.2 proposes the corresponding construction method of our new spatial index tree. In Sect. 96.3 we introduce the corresponding operation algorithms. Section 96.4 gives experimental results. In Sect. 96.5 we discuss related works. A conclusion is given in Sect. 96.6.

96.2 Spatial Indexing Mechanism Based on Dynamic Quad-Tree Regional Division

The new index structure will carry on division to the region which the data objects occupy. In order to effectively achieve the regions which are mainly occupied by the data objects, we adopt the cluster algorithm [4] to get the dense regions. We first use the center and the four corners of the MBR of the whole queried region as the initial cluster centers, then the whole region will be formed into five clusters. The clustering result is shown in Fig. 96.1. The five clusters are the regions mainly occupied by data objects. The following work is to carry on quad-tree region division towards the clustered regions. The idea of quad-tree region division is as follows:

- In order to make every clustered region be indexed by the quad-tree, we chose the MBR of a certain cluster which has the biggest covering area as the initial region.
- Through using the MBR with the same size of the above selected MBR to contain the cluster whose cluster center is the center of the whole region, we will carry on quad-tree region division on the region with regional center as the cluster center.
- After carrying on quad-tree region division on the region with regional center as the cluster center, there will form the first two quad-trees whose data objects are completely included in the division region or the data objects intersect with the division lines. The next is to dynamically move the first two achieved quad-tree regions to the remaining four clusters with the four corners of the whole region as the clustering centers.
- Due to using the biggest covering region as the division region of the five clusters, the data objects in each cluster will be indexed by the quad-tree. Finally, there will form eight virtual quad-trees in the whole division process.

When carry on dividing on each cluster, there are two cases to be processed:

- For the data objects completely contained in the sub-division region, these data objects will be stored in the corresponding child node.
- For the data objects not completely contained in the sub-division regions, namely the data objects crossing the division lines, we adopt the same quad-tree to store the data objects. The four child nodes are respectively the four crossing lines segmented by the center of the division region. For example, the data objects crossing the division line a are stored in the nodes corresponding to a. If the data objects occupy the whole division region, they are stored in the root node of the corresponding quad-tree.

Figure 96.1 shows the data objects clustering division region. In the process of division, first, through selecting the center and the four corners of the Q as the initial clustering centers, the whole region Q is formed into five clusters. Second, since the covering region of Q_1 is biggest, we chose the MBR of Q_1 as the initial division region for quad-tree index. Similar to [5], the four sub-division regions are marked as 0, 1, 2, 3, and the location of each region is $Q_{1.0}$, $Q_{1.1}$, $Q_{1.2}$, $Q_{1.3}$. At the same time, we mark the four sub-crossing line as 0, 1, 2, 3, and the location of each crossing is $L_{1.0}$, $L_{1.1}$, $L_{1.2}$, $L_{1.3}$.

Figure 96.2 corresponds to the quad-tree division on Q_1. The initial quad-tree index structure is shown in Figs. 96.3 and 96.4. Figures 96.3 and 96.4 are the index structures corresponding to Q_1 region. Due to P_1 crossing the division line, P_1 will be stored in the quad-tree based on the division lines. P_2, P_3 and P_4 are stored in the corresponding quad-tree child nodes. In our paper, the index structure of the other four clustering regions will adopt the structure the same as the Q_1 quad-tree index structures, which only needs to change the identifiers of each corresponding division region and the data objects stored in each corresponding sub-division region.

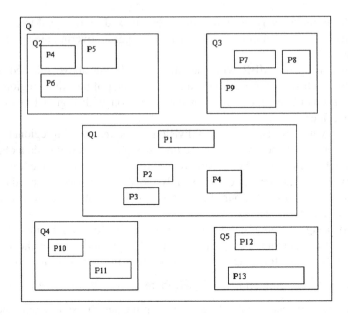

Fig. 96.1 Data objects clustering division region

Fig. 96.2 The quad-tree division on Q_1

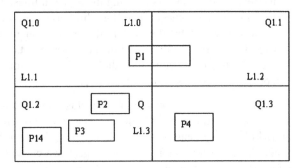

Fig. 96.3 The initial quad-tree index structure about the data objects completely contained in each sub-division region

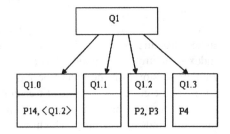

Fig. 96.4 The initial quad-
tree index structure about the
data objects crossing the
division line

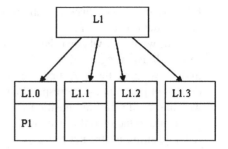

96.3 Operation of Index Structure and Algorithm Description

96.3.1 Insertion Algorithm

Input: Identification oid of the data object to be inserted.
   Output: The leaf node with oid.

- First, by using the clustering algorithm, we can achieve the cluster which the inserted data object is in. If the sub-division region or the division line sub-region of the above cluster can contain the inserted data object, then the data object is inserted in the corresponding node;
- If the corresponding sub-region is full, then we find the under-flow sibling node of the node where the inserted data object is. The data object is inserted into the above under-flow sibling node, and at the same time the actual position of the inserted data object is stored following the inserting position of the inserted data object;
- If the sibling node of the node where the inserted data object is are all full, then we find the under-flow sibling node in the neighbor cluster of the cluster where the inserted data object is. If it still can not find the under-flow sibling node, then we continue to find in the other clusters according to the distance between the cluster where the inserted data object is and the other clusters until the inserting position is found.

It can be seen from Fig. 96.2, data object P_{14} is the inserted object. P_{14} belongs to the cluster Q_1 through clustering. Because $Q_{1.2}$ is the region containing P_{14}, P_{14} should be inserted into the $Q_{1.2}$ node. If the maximum number of the data objects stored in the node is two, then inserting P_{14} will result in the overflow of $Q_{1.2}$. Due to $Q_{1.0}$ empty and there existing space to store another data object, we insert the P_{14} into $Q_{1.0}$. The actual inserting position is stored following the inserted data object.

96.3.2 Deletion Algorithm

Input: Identification oid of the data object to be deleted;

- First, by using the clustering algorithm, we can achieve the cluster which the deleted data object is in.
- Second, we check whether the deleted data object is completely contained in a certain division region. If the deleted object is completely contained in a certain division region, then we find the corresponding division region node. If the deleted object is in the corresponding node, then we delete the data object. Else, we continue to find the position of the deleted data object in the sibling node according to the record of storage position in the child node of the sibling node. If the position is still not found, we continue to search in the child node of the sibling node in the neighbor cluster or some farther clusters according to the distance between the cluster where the deleted cluster is and the remaining clusters. When the position is found, we delete the data object.

96.3.3 Query Algorithm

Input: Query window
   Output: The leaf nodes of the data objects falling into the query window.

- First, by using the clustering algorithm, we can achieve the cluster which the query window is in.
- The second is to find the division regions which intersect with the query window in the above cluster. Then, the data objects falling in the crossing regions are returned. We also should find the identifiers of the crossing regions in the sibling nodes in the same cluster and the other different clusters. In the end, the data objects whose actual positions are among the crossing region are returned.

96.4 Test Results and Performance Analysis

In order to verify the performance of new proposed spatial index, we use the new spatial index a based on dynamic quad-tree regional division proposed in this paper and the traditional quad-tree region division index b to carry on comparison. The experimental environment is Microsoft Windows, using C++ language to program, and experimental data comes from the real two-dimensional spatial data in a city, and we randomly generate the data objects to be queried in the above geographical space. Figures 96.5, 96.6 and 96.7 show the comparisons of the CPU time spent on insertion, deletion, query algorithm under the same conditions by using the two indexes.

   The experimental results show that the new index structure is superior to the traditional quad-tree index structure in the insertion, deletion and query performance. The pointer-based traditional quad-tree occupies too much memory. The data structure of pointer-based quad-tree needs relatively a large amount of storage space and it leads to frequent disk accesses when following pointers. Furthermore,

Fig. 96.5 The CPU running time comparison when carry on inserting

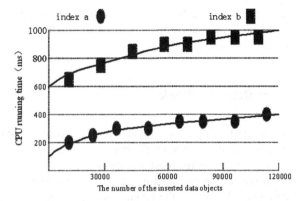

Fig. 96.6 The CPU running time comparison when carry on deleting

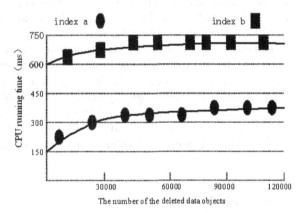

Fig. 96.7 The CPU running time comparison in the different query regions

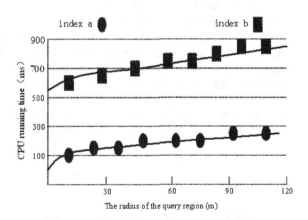

the traditional quad-tree is formed through dividing the whole region layer by layer until the data object number of all the leaf nodes does not exceed the maximum value.

96.5 Related Work

In the past thirty years, quite a few techniques have been developed for constructing spatial index [6–9], but there is no consensus on the best spatial index structure suitable for most situations.

In the traditional quad-tree [3], when the number of data objects in a certain division is lager than the given value, we need continue to divide downward until the number of data objects is not lager than the given value. QR-tree [10, 11] is another spatial index structure to index the data objects crossing the division line. It constructs the horizontal and vertical nodes to store the data objects intersecting with the horizontal and vertical division lines.

Our paper divides the crossing lines into four parts. It can avoid searching the data objects on the whole division line for some relatively smaller query window. Due to using two initial quad-trees to dynamically form the other virtual quad-trees, the number of the pointers is greatly reduced compared to the original quad-tree. Dynamic quad-tree always moves towards the regions having data objects, and it can avoid invalid region division and the index, make the new index be capable of adapting to the regional distribution of any shape, so the usability is wide.

96.6 Conclusion

To support spatial query in the context of the Internet of Things, a new spatial index based on dynamic quad-tree regional division is proposed in this paper, which uses one dynamic quad-tree to carry on traversing in the regions having data objects, and forms multiple virtual quad-trees the same as the initial quad-tree. Experiments results show that the new index structure can adapt the changing spatial area and has a wide application in practice.

Acknowledgments This work was supported by the Fundamental Research Funds for the Central Universities (China University of Geosciences at Beijing).

References

1. Zhang Z, Zhang J, Yang J, Yang Y (2007) A new approach to create spatial index with r-tree. In: Proceedings of the sixth international conference on machine learning and, cybernetics, pp 2645–2648

2. Mouratidis K, Papadias D, Papadimitriou S (2008) Tree-based partition querying: a methodology for computing medoids in large spatial datasets. Very Large DataBases J, pp 923–945
3. Finkel RA, Bentley JL (1974) Quad trees: a data structure for retrieval of composite keys. Acta Informatica 4(1):1–9
4. Pettinger D, Fatta GD (2010) Space partitioning for scalable k-means. In: Ninth international conference on machine learning and applications 2010:319–324
5. Xie S, Feng X, Wang J, Zhou L (2009) A new quad-tree structure and construct algorithm based on dominant attribute storage. Wuhan University of Technology, Wuhan
6. Liu R, An X, Gao X (2009) Spatial index structure based on r-tree. J Comput Eng 35:32–34
7. Floriani LD, Fellegara R, Magillo P (2010) Spatial indexing on tetrahedral meshes. In: The 18th ACM SIGSPATIAL international conference on advances in geographic, information systems, pp 506–509
8. Barroso M, Reyes N, Paredes R (2010) Enlarging nodes to improve dynamic spatial approximation trees. In: Proceedings of the third international conference on similarity search and applications, pp 41–48
9. Weiss K, Fellegara R, Floriani LD, Velloso M (2011) The pr-star octree: a spatio-topological data structure for tetrahedral meshes. In: The 19th ACM SIGSPATIAL international conference on advances in geographic, information systems, pp 92–101
10. Fu Y, Hu Z, Zhou GWD (2003) Qr-tree: a hybrid spatial index structure. In: Proceedings of the second international conference on machine learning and, cybernetics, pp 459–463
11. Zhang D, Chen Z (2005) The spatial index studies based on improved qr-tree. Heilongjiang Academy of Engineering, pp 18–20

Chapter 97
Feature Evaluation of Radar Emitter Signal Based on Logistic Regression Model

Yanli Deng, Weidong Jin and Zhibin Yu

Abstract This paper presents a method which is feature evaluation of radar emitter signal based on logistic regression model by different features of radar emitter signal. Firstly, using the signal to noise ratio and the correct recognition rate of features establishes mathematical models. Secondly, it is posed the evaluation measures and evaluation criterions. Lastly, feature evaluation of radar emitter signal is completed. The simulation results show that the method is effective and feasible. It provides a new idea and approach for the feature evaluation of the radar emitter signal.

Keywords Radar emitter · Feature evaluation · Logistic regression model

97.1 Introduction

With the rapid development of electronic science and technology, complex radars become more and more widely. In such a dense electromagnetic environment and the highly complex radar signal's waveform, using the traditional five parameters

The work is supported by the national natural science foundation(No. 60971103), Author: Yanli Deng (1985-), she is a Ph.D student in Southwest Jiaotong University, Sichuan. She studies the radar emitter signal processing. (E-mail:dyllovebeijing@163.com). Wei-dong JIN (1959-), he is a Ph.D supervisor. And he studies intelligent information processing, system simulation and optimization theory and optimal control.

Y. Deng (✉) · W. Jin · Z. Yu
School of Electrical Engineering, Southwest Jiao tong University,
Cheng du Sichuan 610031, China

W. Lu et al. (eds.), *Proceedings of the 2012 International Conference on Information Technology and Software Engineering*, Lecture Notes in Electrical Engineering 210, DOI: 10.1007/978-3-642-34528-9_97, © Springer-Verlag Berlin Heidelberg 2013

(pulse width, pulse repetition frequency, time of arrival, the carrier frequency and angle of arrival) which discern the radar emitter can not meet demand. The majority of scholars are continuous to explore and study the so called "the sixth parameter" in order to compensate for the lack of five parameters for radar emitter recognition. A lot of articles based on the analysis of the radar intra-pulse feature are emerging, such as derived features [1], resemblance coefficient [2], the entropy features [3], time–frequency atomic features [4], ride frequency features [5] and complex features [6].

Facing so many intra-pulse features, numerous articles only discuss the ability of feature classification according to the correct recognition rate. Or using the correct recognition rate under different SNR determines noise immunity of features. When the differences of ability to identification of features are too large, using the correct recognition rates evaluates feature easily. But when the differences of ability to identification of features are too small, using the correct recognition rates evaluates feature hardly. Feature evaluation for radar emitter can be studied from multiple angles [7]. The complexity of time and the complexity of space for radar emitter have been studied before. In this paper, it proposes a evaluation method based on logistic regression model for radar emitter from the noise sensitivity and correct recognition rate stability. First, using SNR and the correct recognition rate creates a mathematical model. Then, according to model, the relevant statistic values are calculated. Finally, it is the experimental simulation. The simulation results show that the method is effective and feasible. It provides a valuable reference for the feature evaluation of the radar emitter signal.

97.2 Logistic Regression Model

The recognition results of the radar intra-pulse features are marked as Y. It can be seen as a binary discrete variable. $Y = 1$ represents the correct identification and $Y = 0$ indicates an error identification. The linear regression is not suitable to deal with the relationship between the binary discrete variable and independent variable X because of the non-normal errors, error variance of nonconstant and regression function restriction. But logistic regression model can handle well. The correct recognition rates of radar emitter are affected by various factors. Noise or human disturbance factors are most prominent. Therefore, the signal to noise ratio and correct recognition rate of radar emitter in the only case of noise factor as follows [8, 9]

$$p = \frac{\exp(a + bx)}{1 + \exp(a + bx)} \tag{97.1}$$

Where a and b are regression intercept and regression coefficient respectively. x is signal to noise ratio and p is correct recognition rate of radar emitter signal in the x. Equation (97.1) can be written by derivation as follows:

$$\ln(\frac{p}{1-p}) = a + bx \tag{97.2}$$

It can be gotten the logistic regression model of signal to noise ratio and correct recognition rate by the least squares method estimating parameters a and b.

The correct recognition rate p is a point estimation. In order to discuss the stability of the correct recognition rate p, it is necessary to solve the confidence interval of the correct recognition rate p. Parameter estimation $\hat{a}$ and $\hat{b}$ have been gotten by the least squares method as follows:

$$E = \ln\left(\frac{p}{1-p}\right) = \hat{a} + \hat{b}x \tag{97.3}$$

The variance of E is calculated as follows:

$$Var(E) = Var(\hat{a}) + x^2 Var(\hat{a}) + 2xCov(\hat{a}, \hat{b}) \tag{97.4}$$

Where $Var(\hat{a})$ is variance of $\hat{a}$, $Cov(\hat{a}, \hat{b})$ is covariance. The confidence interval of E under the $1 - \alpha$ as follows:

$$\left[E - z_{\frac{\alpha}{2}}\sqrt{Var(E)}, E + z_{\frac{\alpha}{2}}\sqrt{Var(E)}\right] \tag{97.5}$$

E_{low} and E_{up} are marked as lower bound and upper bound respectively in (97.5). Confidence interval of correct recognition rate p as follows:

$$\left[\frac{\exp(E_{low})}{1 + \exp(E_{low})}, \frac{\exp(E_{up})}{1 + \exp(E_{up})}\right] \tag{97.6}$$

97.3 Feature Evaluation

The correct recognition rates of different intra-pulse features are distinct in the same SNR. It will be obtained different logistic regression model and parameter owing to distinct intra-pulse feature. Divergence of model reflects the ability of identification in distinct intra-pulse feature.

In the Eq. (97.1), the abscissa is SNR and the vertical axis the correct recognition rate for the radar emitter. Regardless of SNR is what value, the vertical axis of curve is always the bigger the better. An area between the curve and the abscissa is regarded a evaluation criteria using Eq. (97.1).The size of the area under the curve is determined by the curve shape and SNR range. To facilitate comparison, the area should be normalized.

97.3.1 Normalized Area

Based on the above considerations, the normalized area is looked upon a measure of the ability to identification of impulse feature. SL is defined as follows:

$$SL = \frac{S}{|\Delta X|} \tag{97.7}$$

Where S is the area of the curve in the interval ΔX, ΔX is SNR range size. The magnitude of SL reflects the ability to identification of different feature on logistic regression model. When the SL is the greater, it shows the ability to identification of the feature the better. The difference of SL values bigger demonstrates the difference of ability to identification for radar emitter the bigger.

97.3.2 Strong Deviation Interval

Using Eq. (97.1) is more or less error between the predict value of correct recognition rate and the true value. When the predict value and true value are large difference, SL is different to truly reflect the ability to identification for radar emitter. Therefore, the interval of including true curve replaces in front of curve.

Using Eq. (97.6) can calculate the confidence interval of correct recognition rate with respect to same confidence level in the range of SNR. This confidence interval is the true value falling into the band in the confidence level of $1 - \alpha$. According to Eq. (97.7), it can get upper and lower bound of the normalized area. So strong deviation interval has been gained as follows:

$$SL_{minus;} = (1 - \alpha)SL_{low} \quad SL_+ = (1 - \alpha)SL_{up} + \alpha \tag{97.8}$$

Equation (97.8) ensure the normalized area of true curve fall into the range of this deviation for greater than or equal to the probability of $1 - \alpha$.

The strong deviation interval can be used as a measure that assesses the ability to identification of different intra-pulse feature. The lower bound of strong deviation interval means the ability to identification of feature the worst and the upper bound of strong deviation interval means the ability to identification of feature the best. When the lower bound of deviation within one feature is greater than the upper bound of deviation within another feature, the former's ability to identification is better than the latter. If strong deviation intervals of two features are in the confidence level of $1 - \alpha_1$ and $1 - \alpha_2$ respectively, conclusions have a credibility of $(1 - \alpha_1)(1 - \alpha_2)$.

Based on logistic regression for feature evaluation model Specific steps are as follows:

Fig. 97.1 The correct recognition rates of two features when SNR are under every 2 db in the range of 0 ∼ 10 db

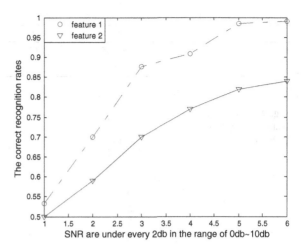

(a) According to Eq. (97.2), using the least squares method estimates the parameters in order to get logistic regression model. SNR affects the correct recognition rate whether significantly taking advantage of Wald test.
(b) Calculate the normalized area SL and estimate the ability to identification of different intra-pulse feature good or bad.
(c) Calculate strong deviation interval and compare the degree of ability to identification.

97.4 Simulation

Example 97.1 The correct recognition rates of two features for radar emitter are shown in Fig. 97.1 when SNR are under every 2 db in the range of 0 ∼ 10 db.

The ability to identification of feature 1 is better than feature 2 in Fig. 97.1. The following step verifies feasibility of feature evaluation based on logistic model.

Based on the above steps, (x, p) is test sample. x is SNR and p is the correct recognition rate. Using the function lsqcurvefit and function trapz estimates parameter and solves the area. $(1 - \alpha) = 0.95$. The calculated values are shown in Table 97.1.

SL value of feature 1 is greater than the SL value of feature 2 from Table 97.1. It shows that the ability to identification of feature 1 is superior to feature 2's. The lower bound of deviation of feature 1 is equivalent to feature 2's. It shows that the ability to identification of feature 1 is more obvious than feature 2's. True values of two features fall into the interval in Figs. 97.2 and 97.3 respectively. The simulation results are consistent with the logic analysis. It show that the method is effective and feasible.

Table 97.1 The calculated value of two different features

	a	b	SL	SL_	sl_+
Feature 1	0.07	0.456	2.355	0.771	0.896
Feature 2	0.17	0.159	0.985	0.623	0.774

Fig. 97.2 True value of feature 1 falls into the interval

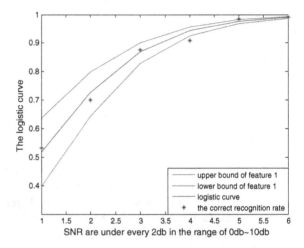

Fig. 97.3 True value of feature 2 falls into the interval

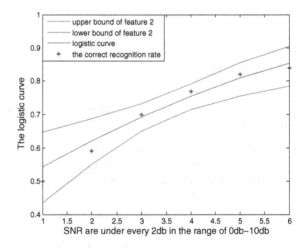

Example 97.2 Reference [3] proposes the entropy feature (referred to feature 1), Ref. [1] extracts the feature using instantaneous autocorrelation method (referred to feature 2), Ref. [5] extracts ridge frequency feature (referred to feature 3) and Ref. [4] extracts the feature using time–frequency atom method. The correct recognition rates of four features for radar emitter are shown in Fig. 97.4 when SNR are under every 2 db in the range of 0 ~ 10 db.

Fig. 97.4 The correct recognition rates of four features when SNR are under every 2 db in the range of 0 ~ 10 db

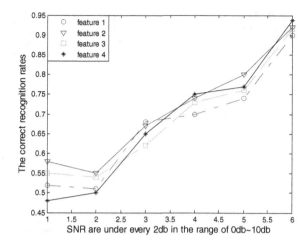

Table 97.2 The calculated value of four different features

	a	b	SL	SL−	sl+
Feature 1	−0.12	0.194	0.796	0.599	0.754
Feature 2	0.63	−0.038	0.596	0.498	0.703
Feature 3	−0.11	0.201	0.81	0.602	0.763
Feature 4	−0.25	0.241	0.879	0.612	0.766

The correct recognition rates of four features are very close in Fig. 97.4. It is unable to determine which features better. The calculated parameter values based on above evaluation method are shown Table 97.2.

The SL value of feature 2 is the least in the Table 97.2. It shows that the ability to identification of feature 2 is the worst. But the SL value of feature 4 is the maximum in the Table 97.2. It shows that the ability to identification of feature 4 is the best. The deviation's lower bound of feature 4 is less than the deviation's upper bound of feature 3. It illustrates that the ability to identification of feature 4 is a little obvious than feature 3's. True values of feature 3 and feature 4 fall into the interval in Fig. 97.5. In Fig. 97.5, the intervals of two features have a large number of overlapping. It is once again to see that difference of feature 3 and feature 4 are not too large.

In summary, feature 4 is the best while feature 2 is the worst. Feature 3 is slightly weaker than feature 4.

97.5 Conclusion

In this paper, it proposes an evaluation method based on logistic model for radar emitter. Using the SNR and the correct recognition rate of features establishes logistic model. It puts forward the normalized area under the logistic function curve as measure and evaluation criteria. Finally feature evaluation of radar

Fig. 97.5 The deviation
interval of feature 3 and
feature 4

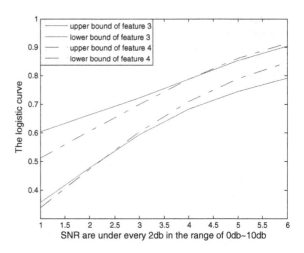

emitter signal is completed. To further study the degree of feature's performance, strong deviation interval has been posed. The simulation results show that the method is effective and feasible.

References

1. Yunwei Pu (2007) Deinter leaving Models and algorithms for advanced radar emitter signals. Southwest Jiaotong University, Cheng Du (in Chinese)
2. Gexiang Z, Laizhao H, Weidong J (2005) Ressemblance coeffcient based intra-pulse feature extraction approach for radar emitter signals. Chin J Electron 14(2):337–341 (in Chinese)
3. Gexiang Z, Laizhao H, Weidong J (2004) Entropy feature extraction approach of radar emitter signals. In: Proceedings of international conference on intelligent mechatronics and automation, Chengdu, China, pp 621–625 (in Chinese)
4. Ming Z, Yunwei P, Weidong J, Laizhao H (2007) Feature extraction of radar emitter signal based on time-frequency atomic methods. J Radio Sci 22(3):458–462 (in Chinese)
5. Zhibin Yu (2010) Study on radar emitter signal identification based on intra-pulse features. Southwest Jiaotong University, Cheng Du (in Chinese)
6. Gexiang Z, Weidong J, Laizhao H (2004) Radar emitter signal recognition based on complexity features. J Southwest Jiaotong Univ 12(2):116–122 (in Chinese)
7. Skrypnyk I (2011) Irrelevant features, class separability and complexity of classification problems. In: IEEE International conference on tools with artificial intelligence, pp 998–1003
8. Jun H (2005) Research on high resolution MMW ATR algorithm performance evaluation. National Defense University, Chang Sha
9. John N, Willimanm W, Michael HK (1990) Linear regression model of application. China Statistics Press, Bei Jing

Chapter 98
A Novel Algorithm of Modulation Classification and Carrier Frequency Estimation for MFSK

Cai Qiaolian, Ren Chunhui, Gao Chiyang and Hou Yinming

Abstract The properties of the spectra of squared MFSK signal is investigated and applied for modulation order classification and carrier frequency estimation in this paper. The proposed algorithm employed the equal distance place of sub-carrier frequencies of squared signal as the key feature. We only need some knowledge of received signals that can be obtained through the front-end processing. The algorithm is appropriated for both continuous phase displacement frequency shift keying (CPFSK) signals and the discontinuous phase displacement frequency keying (DPFSK) signals. Simulation results show that this algorithm has good performance in classification with a success rate higher than 96 % when the SNR is greater than or equal to 1 dB. In addition, the proposed algorithm performs well in the estimation of carrier frequency.

Keywords FSK · Spectrum · Classification · Carrier frequency estimation

98.1 Introduction

Automatic modulation classification (AMC) has become the key technology in military and civilian communication system, such as signal recognition, signal detection, spectral analysis and electronic rescue. The Feature Based (FB) methods

C. Qiaolian (✉) · R. Chunhui · H. Yinming
School of Electronic Engineering, Qingshuihe Campus, UESTC, Chengdu City, China
e-mail: qiaolian_cai@163.com

R. Chunhui
e-mail: chuiren@uestc.edu.cn

G. Chiyang
China Software Testing Center 13/F, CCID Plazza, 66 Zizhuyuan Road, Beijing 10048, People Republic of China

W. Lu et al. (eds.), *Proceedings of the 2012 International Conference on Information Technology and Software Engineering*, Lecture Notes in Electrical Engineering 210, DOI: 10.1007/978-3-642-34528-9_98, © Springer-Verlag Berlin Heidelberg 2013

need less prior information and low computational complexity have practical value in AMC. Different features such as wavelet coefficients [1–3], the time frequency representation [4], and spectrum feature [5–8] have been considered in the development of AMC algorithms.

For MFSK signals, the spectrum feature is obvious. DFT is used to identify the number of sub-carrier frequencies of MFSK signal in [7]. The algorithm in [7] requires the DFT resolution to be as high as possible. However, it may increase the complex of computation and lower the accuracy of the classification. To solve this problem, an improved approach adding some rules is proposed in [8]. The above approaches are suitable to the discontinuous phase displacement frequency keying (DPFSK) signals. But to the continuous phase frequency shift keying (CPFSK) especially the signals with modulation index less than 1, they are not suitable. In this paper, we proposed a new algorithm based on the spectrum of squared signals which is suitable to both DPFSK and CPFSK. Furthermore, we can estimate the carrier frequency of MFSK basing on the proposed algorithm.

98.2 The Signal Models and Spectrum of Signals

The received signal can be expressed as

$$s(t) = x(t) + w(t) \tag{98.1}$$

where $x(t)$ is the transmitted signal that is independent of zero-mean additive white Gaussian noise $w(t)$.

The time-domain signal model of DPFSK can be described as

$$x_{DCFSK}(t) = \sum_n \cos(2\pi(f_c + f_n)t + \varphi_c)p(t - nT) \tag{98.2}$$

where $p(t)$ is a standard rectangular pulse of duration T, T is the symbol period, N is the total number of samples. f_c is the carrier frequency, φ_c is the initial carrier phase which is zero generally. f_n is the modulated frequency of the nth transmitted symbol and it takes one of M equal likely possible values

$$f_n = f_d \times (2m - 1 - M), m = 1, 2, \ldots, M \tag{98.3}$$

As described in Eq. (98.3), f_n is independent identically distributed, where f_d is the frequency deviation.

The power spectral density (PSD) of received DPFSK can be expressed as

$$P_{DPFSK}(f) = \frac{T}{16} \sum_{i=1}^{M} \left\{ \sin c^2[\pi(f - f_i)T] + \delta(f - f_i) \right\}, f \geq 0 \tag{98.4}$$

Where $f_i = f_c + f_n$. Equation (98.4) implies that DPFSK have M spectrum peaks when f is greater than or equal to 0.

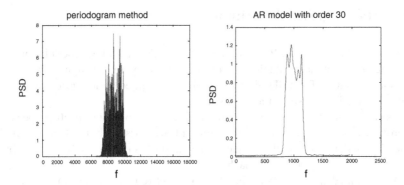

Fig. 98.1 The PSD of 8CPFSK applied periodogram method

The time-domain signal model of CPFSK can be described as

$$x(t) = \cos(2\pi f_c t + \phi(t; \mathbf{I}) + \varphi_c) \tag{98.5}$$

$$\phi(t; \mathbf{I}) = 2\pi f_d T \sum_{k=-\infty}^{n-1} I_k + 2\pi f_d (t - nT) I_n$$

$$= \theta_n + 2\pi h I_n q(t - nT) \tag{98.6}$$

$$h = 2f_d T, \theta_n = \pi h \sum_{k=-\infty}^{n-1} I_k, q(t) = \begin{cases} 0 & t < 0 \\ t/2T & 0 \le t \le T \\ 1/2 & t > T \end{cases}$$

Where $\{I_n\}$ is the amplitude sequence, which is obtained from digital base band sequence $\{a_n\}$ mapping to the amplitude level. The carrier phase within the time interval $nT \le t \le (n+1)T$ is determined by the accumulation in Eq. (98.6). h is called modulation index.

The PSD of baseband CPFSK can be shown as follow [9]

$$P'_{CPFSK}(f) = T \left[\frac{1}{M} \sum_{n=1}^{M} A_n^2(f) + \sum_{n=1}^{M} \sum_{m=1}^{M} B_{nm}(f) A_m(f) \right] \tag{98.7}$$

$$A_n(f) = \sin c(\pi(fT - \tfrac{1}{2}(2n - 1 - M)h))$$
$$B_{nm}(f) = \frac{\cos(2\pi fT - \alpha_{nm}) - \psi \cos \alpha_{nm}}{1 + \psi^2 - 2\psi \cos 2\pi fT}$$
$$\alpha_{nm} = \pi h(m + n - 1 - M), \psi \equiv \frac{\sin M\pi h}{M \sin \pi h}$$

Then the PSD of CPFSK denoted by P_{CPFSK} with carrier frequency is the result that the PSD of baseband signal is carried to the carrier frequency.

The spectrum of CPFSK is relatively smooth when $h < 1$ and it is broader when $h > 1$. Especially, it has M spectrum peaks when $h = 1$. It can be seen that the PSD of CPFSK has obvious spectrum peaks when h is closer to 1 or greater.

98.3 The Proposed Algorithm

The approaches based on PSD in [5, 8] plays well for DPFSK modulation order classification, but not suitable to CPFSK, especially to the signals with small modulation index. The PSD of 8CPFSK ($h < 1$) applied periodogram method [8] and AR model [5] is shown as Fig. 98.1.

As shown in Fig. 98.1, the PSD of CPFSK has no obvious spectrum peaks, so approaches in [5, 8] are not practical for CPFSK classification when modulation index is smaller than 1 or the frequency deviation is small. To solve this problem, we proposed an improved approach.

As described in Sect. 98.2, the PSD of CPFSK has obvious spectrum peaks when h is closer to 1 or greater. While in practice, $h \geq 0.5$ is satisfied generally. Then the classification can be performed based on spectrum of squared MFSK signals. In communication theory, each sub-carrier frequency of the squared MFSK signal is twice of the original MFSK signal's. Accordingly, modulation index is twice of the original signal's, so the modulation index of squared signal satisfies the condition $h \geq 1$. In this way we can obtain obvious spectrum peaks easier based on the squared signal.

We use AR model to estimate the PSD of squared signal as in [5] other than periodogram method. Because periodogram method will cause many false spectrum peaks surrounding a spectrum peak corresponding to a true sub-carrier frequency, and AR model has better frequency resolution and smoother wave forms.

The following analysis is based on CPFSK signal which is also applicable to DPFSK signal. The sub-carrier frequencies, frequency deviation f_d and PSD referred below pertain to the squared signal except especial statement.

The proposed algorithm can be performed by the following 4 steps.

Step 1: Calculate the PSD of the squared signal with AR model, adjust the magnitudes of the PSD, normalize the PSD magnitudes and set a threshold.

The PSD magnitudes corresponding to the sub-carrier frequencies should be local peaks greater than the threshold. These peaks are defined as true peaks.

We could normalize the magnitudes of PSD and set a threshold to extract the spectral peaks as in [8]. However, if the threshold is too large, some of the true spectral peaks will be missed in many cases. If the threshold is too small, the false spectral peaks out of the frequency range of the true peaks are extracted in low SNR. We propose an improved approach, that is to adjust the magnitudes of PSD with the method that sets the minimum magnitude value of PSD to 0, and then normalize the max value as 1 such as Eq. (98.8).

$$P'(f) = P(f) - \min(P(f)) \ , \ \ P'(f) = P'(f)/\max(P'(f)), f \geq 0 \qquad (98.8)$$

In this way, we can choose a small threshold, such as $\mu = 0.35$, without worrying about the bad algorithm performance in low SNR. Figure 98.2 shows the PSD of the squared signal before and after adjusting the magnitude with the small threshold when SNR is 3 dB.

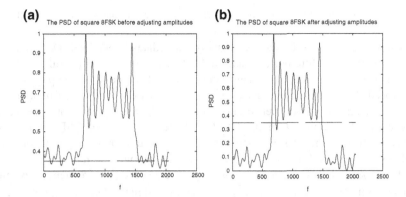

Fig. 98.2 The normalized PSD of 8FSK before and after adjusting the magnitudes, **a** the normalized PSD before adjusting the magnitudes**b** the normalized PSD after adjusting the magnitudes

As comparison in Fig. 98.2, after adjusting the PSD magnitudes, the spectral peaks greater than the threshold which can be set as a small value are in the frequency range of the true peaks although in low SNR.

Step 2: The PSD magnitude corresponding to a sub-carrier frequency should be a local peak. The frequencies of the local peaks greater than the threshold will be acquired in this step as the sub-carrier frequency candidates (SFC) $\{f_i\}, i = 1, \ldots K$.

Step 3: Extract the equally distance frequencies in $\{f_i\}$, because the sub-carrier frequencies are equally spaced with the frequency deviation f_d.

To reduce false spectrum peaks, a low order, i.e., 30, of AR model is used in this paper. But the resolution is reduced at the same time making that the true sub-carrier frequencies in $\{f_i\}$ aren't spaced accurately equally. The frequencies satisfying equal distance roughly can be extracted according to the following procedures.

Procedure 1: Each neighboring frequencies distance in $\{f_i\}$ can be calculated as $\{\Delta f_i\}, i = 1, \ldots, K - 1$. The set $\{\Delta f_i\}$ may be separated into several non-overlapping subsets. The mean value of each set can be treated as one candidate frequency deviation f_d.

The estimator of f_d is formed as in [8]: (1) Calculate $\{\Delta f_i\}, i = 1, \ldots, K - 1$, where $\Delta f_i < \Delta f_{i+1}$. (2) Calculate $\{\Delta f_{i+1} - \Delta f_i\}$ resulting in distinct values that are denoted by $\{\Delta^2 f_j\}, j = 1, \ldots, K - 2$. (3) Find the largest one in $\{\Delta^2 f_j\}$ that not greater than the mean value of $\{\Delta^2 f_j\}$, denote it by $\Delta^2 f_T$. (4) Separate $\{\Delta f_i\}$ into several subsets such that the range of $\{\Delta f_i\}$ in each subset is not larger than a small value χ where $\chi = \min(\Delta^2 f_T, \Delta f_1/8, \xi/4)$. We can define the valve ξ as $\xi = \widehat{R_s}/\beta$, where $\widehat{R_s}$ is the calculated value of the symbol rate with the method referred in [10] and β is a positive parameter that is greater than or equal to 4. And then count the element number of each subset denoted by V_l. (6) Calculate the mean value $\overline{f}_{d,l}$

of the Δf_i in each subset and each $\bar{f}_{d,l}$ can be treated as one candidate frequency deviation.

Up to now, a set of frequency deviation candidates is obtained. Different candidates will result different estimate of M. It needs to establish a rule to decide which candidate is acceptable. This can be finished according to the procedure 2.

Procedure 2: For the equal distance of sub-carrier frequencies, we can define the custom candidate sub-carrier frequencies (CCSF) with a candidate frequency deviation. The number T of the CCSF and the number K of CSF are possibly different. If a candidate frequency deviation is the true frequency deviation, the difference between T and K should match the condition $(T - K) < V_l$.

The procedure 2 is realized by searching the true candidate frequency deviation from the sets $\{\bar{f}_{d,l}, V_l\}$. Each set $\{\bar{f}_{d,l}, V_l\}$ is arranged in the descending order of V_l, and if V_l is the same value, the set will be arranged in the ascending order of $\bar{f}_{d,l}$. Now the arranged sets should be searched from the one with the greatest V_l to the one with the smallest V_l. If the current candidate frequency deviation is the true one, searching will be stopped.

The frequency of the normalized PSD magnitude being 1 is denoted by f_{top}. The frequencies which is made up by adding some frequencies to the left and the right of f_{top} with the current frequency distance $\bar{f}_{d,l}$, is defined as the custom candidate frequencies (CCSF). The distance between the leftmost frequency of CCSF and the minimum frequency of CSF is not greater than $0.5\bar{f}_{d,l}$ and it is the same between the rightmost frequency of CCSF and the maximum frequency of CSF. If the condition $(T - K) < V_l$ is matched, this $\bar{f}_{d,l}$ is the rough estimate of the frequency deviation which is denoted by f_{Ro} and searching is stopped.

To enhance the performance of the algorithm, the candidates in $\{f_i\}$ will be retained if the frequency distance Δf_i between the neighboring candidate frequencies satisfies the condition $abs(\Delta f_i - f_{Ro}) \leq 0.5f_{Ro}$. Between the frequencies that the frequency distance Δf_i meets the condition $(\Delta f_i - f_{Ro}) > 0.5f_{Ro}$ we will insert some frequencies with the frequency deviation f_{Ro}. Then the number of the retained and inserted frequencies denoted by $\{f_r\}, r = 1, 2, \ldots, \widehat{M}$ is the rough estimate of M.

Step 4: Set a dynamic threshold and search the possibility missing spectrum peaks.

The frequencies of the spectrum peak magnitudes lower than the fix threshold will be missing especially the ones out of the frequency range of $\{f_r\}$. In order to enhance the robustness of the algorithm, dynamic threshold is applied in this paper.

If the value of $\widehat{M}$ is less than 2, based on the initial threshold set as 0.35, the threshold will be decreased as $\mu \Leftarrow \lambda \times \mu$, where λ is a positive constant that is less than 1. Here the value of λ is 0.95. Then the classifier will repeat from the beginning. Because M is a power of 2, if the value of $\widehat{M}$ is not a power of 2, the threshold will continue being adjusted. If frequencies of the spectrum peak magnitudes out of the frequency range of $\{f_r\}$ are searched with the new threshold, which are expressed by $\{f_{m\_add}\}, m = 1, 2, \ldots D$, each frequency in $\{f_{m\_add}\}$ satisfying the equal distance condition will be added to the old $\{f_r\}$ as a new $\{f_r\}$ and the new value of $\widehat{M}$ is $\widehat{M} + 1$. The threshold is adaptively adjusted until any of the following conditions is conformed. The stop conditions include: (a) The estimate value $\widehat{M}$ is a power of 2. (b) The adjusted threshold meets the lower limit, i.e., $\mu = 0.25$.

The estimated value of M is decided by $\widetilde{M} = 2^{round(\log 2(\widehat{M}))}$. Thus when $\widehat{M}=2$, the signal is 2FSK; when $2 < \widehat{M} \leq 5$, $\widetilde{M}=4$, the signal is 4FSK; when $6 \leq \widehat{M} \leq 11$, $\widetilde{M}= 8$, the signal is 8FSK; when $12 \leq \widehat{M} \leq 22$, $\widetilde{M}= 16$, the signal is 16FSK.

Because the frequencies in $\{f_r\}$ isn't the sub-carrier frequencies correctly sometimes, it needs a more accurate frequency deviation to estimate sub-carrier frequencies correctly. If $\widehat{M}$ is a power of 2, the classifier will calculate the frequency distances of each neighboring frequencies in $\{f_r\}$, and then the mean value of the frequency distances is the estimate of frequency deviation. The value of the estimate is denoted by $\widehat{f_d}$. If $\widehat{M}$ isn't a power of 2, the estimate is calculated by $\widehat{f_d} = (f_{max} - f_{min})/(\widehat{M} - 1)$. In this way, the sub-carrier frequencies can be estimated by f_{max}, f_{min} and $\widehat{f_d}$ which are denoted by $\{\widehat{f_m}\}, m = 1, 2, \ldots, \widetilde{M}$. Then the carrier frequency of original signal can be estimated by

$$\widehat{f_c} = \frac{1}{\widetilde{M}} \sum_{i=1}^{\widetilde{M}} (\widehat{f_i} + \widehat{f_{\widetilde{M}-i-1}})/4 \tag{98.9}$$

The proposed algorithm process is shown in Fig. 98.3.

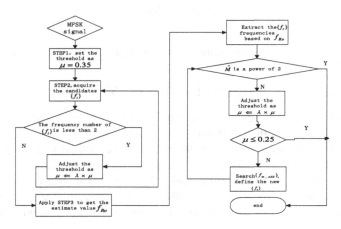

Fig. 98.3 The process of the proposed algorithm

Fig. 98.4 The classification performance

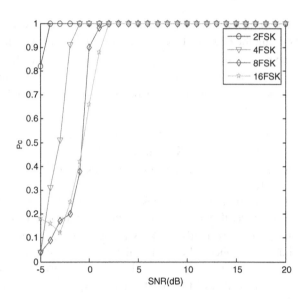

98.4 Simulation Results

We consider 2FSK, 4FSK, 8FSK and 16FSK as candidate modulation. In the Nyquist theorem, the sample frequency must be greater than twice of the highest sub-frequency of the squared signal. In simulations, for each signal, the normalized sample frequency f_s is 1; the carrier frequency f_c is 0.01; the symbol rate R_s is 0.04; the modulate index h is 0.53; the number of symbols is 500; the order of AR model is 30.

The correct classification rate P_c is defined as $P_c = \frac{1}{N_s} \sum_{n=1}^{N_s} e_n$, where $e_n =$

$\begin{cases} 1 & \widetilde{M}_n = M \\ 0 & otherwise \end{cases}$ and N_s is the number of Monte Carlo experience. The normalized

mean squared err (NMSE) is defined as $NMSE = \frac{1}{N_s} \sum_{i=1}^{N_s} \left(1 - \frac{\hat{f}_{ci}}{f_c}\right)^2$, where $\hat{f}_{ci}$ is the ith estimate of the carrier frequency. The simulation results versus SNR are shown as Figs. 98.3 and 98.4 with 100 Monte Carlo experiences.

It can be easily noticed from Fig. 98.4 that a higher SNR is required to achieve a certain classification performance for higher order modulations. For example, recognition of 2FSK, 4FSK, 8FSK and 16FSK signals with a probability of 1 requires -4 dB, -1 dB, 2 dB and 2 dB SNR respectively. This algorithm has good performance in classification with a success rate higher than 96 % when the SNR is greater than or equal to 1 dB.

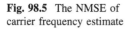

Fig. 98.5 The NMSE of
carrier frequency estimate

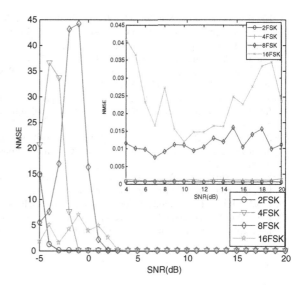

As shown in Fig. 98.5, the accuracy of carrier frequency is better for 2FSK and
4FSK than that for 8FSK and 16FSK. The stable carrier frequency estimate
NMSEs of 2FSK and 4FSK are around 10^{-3}, when the ones of 8FSK and 16FSK
are around 10^{-2} in the higher SNR.

98.5 Conclusion

We proposed a squared signal PSD based algorithm to recognize the modulation
order both of DCFSK and CPFSK and estimate the carrier frequency. First, the
algorithm calculated the PSD of the squared FSK signal with AR model and then
adjusted the PSD amplitudes in order to set a small initial threshold that can
estimate the frequency deviation accurately. The adaptive threshold has a significant role in enhancing the robust of the algorithm performance.

References

1. Maliatsos K, Vassaki S, Constantinou P (2007) Interclass and Intraclass modulation
   recognition using the Wavelet Transform. Pers Indoor Mobile Radio Commun PIMC 2007:1–
   5
2. Ho KM, Vaz C, Daut DG (2009) A wavelet-based method for classification of binary
   digitally modulated signals. Sarnoff Symposium, 2009 IEEE, pp 1–5
3. Ho KM, Vaz C, Daut DG (2010) Automatic classification of amplitude, frequency, and phase
   shift keyed signals in the wavelet domain. Sarnoff symposium, 2010 IEEE, pp 815–819

4. Haq KN, Mansour A, Nordholm S (2010) Classification of digital modulated signals based on time frequency representation. International conference on signal processing and communication systems, ICSPCS 2010, pp 1–5
5. Liu YY, Liang GL, Xu XK, Li XX (2008) The methods of recognition for common used M-ary digital modulations. International conference on wireless communications, networking and mobile computing, WiCOM 2008,pp 1–4
6. Zhou M, Feng QY (2011) A new feature parameter for MFSK/MPSK recognition. International conference on intelligence science and information engineering, ISIE 2011, pp 21–23
7. Zaihe Y, Shi YQ,Su W (2003) M-ary frequency shift keying signal classification based-on discrete Fourier transform. IEEE Military Communications Conference, MILCOM 2003, pp 1167–1172
8. Zaihe Y, Shi YQ, Su W (2004) A practical classification algorithm for M-ary frequency shift keying signals. IEEE Military Communications Conference, MILCOM 2004, pp 1123–1128
9. Proakis JG Digital communications, 4th edn. Publish House of Electronics Industry, Beijing
10. Lei XM (2010) A research on the recognition and parameter estimation of the signal with low intercept probability. University of Electronic Science and Technology. (in Chinese)

Chapter 99
The Design and Analysis of Regional Positioning System Based on Zigbee Technology

Jianpeng Lu and Hongbin Gao

Abstract As regional positioning widely applied in wireless sensor networks, it attracts more and more attention in recent years. A regional positioning system based on ZigBee technology is designed and implemented in this paper, and different distance measurement methods is employed to explore the relationship between RSSI and distance. Through experiment, we can obtain the empirical value of RF parameter A and transmission parameter N. Using the least squares method for error correction in the distance fitting method. Error analysis for the positioning result shows the system can achieves a positioning accuracy of 2.21 m.

Keywords ZigBee · Positing system · RSSI · Curve fitting · AN ranging

99.1 Introduction

With the continuous development of wireless sensor networks, emerging wireless business has been driven, and the information of node location has drawn much attention. Although the global positioning system based on satellite communications can effectively solve the large number of military and civilian positioning, but its accuracy is limited to the positioning of the larger scope's things and can do nothing to the positioning of the regional things. There exist obviously inadequate

J. Lu (✉) · H. Gao
School of Information Science and Engineering, Hebei University of Science
and Technology, Shijiazhuang, Hebei 050018, China
e-mail: jianpeng061@163.com

H. Gao
e-mail: gao_hb@hebust.edu.cn

W. Lu et al. (eds.), *Proceedings of the 2012 International Conference on Information Technology and Software Engineering*, Lecture Notes in Electrical Engineering 210, DOI: 10.1007/978-3-642-34528-9_99, © Springer-Verlag Berlin Heidelberg 2013

of the positioning accuracy of the internal things such as office buildings, ware-houses, the hall, for all this, proposing a new technique that can effective posi-tioning the region's things is very necessary. At present, the common technology is infrared technology, ultrasound technology, ZigBee/IEEE802.15.4 on behalf of the wireless network technology and RFID technology [1, 2].

ZigBee is an emerging short-range, low-rate wireless network technology. It has its own radio standards. It can coordinate communication between the thou-sands of tiny sensors, and can be very good for node positioning [3].

99.2 Basic Framework

RSSI (Received Signal Strength Indication) ranging method: Knowing the trans-mit power and the received power, and then we can calculate the propagation loss, using the theoretical or empirical signal propagation model transform the propa-gation loss into distance.

The positioning system is divided into the structures of the underlying hardware and the design of the upper software: lower layer using a mesh network topology is divided into three types of node: coordinator node, positioned node and reference node. According to the position of reference nodes and the distance between positioned node and reference node, we can calculate the absolute position of the positioned node. The overall structure is shown as Fig. 99.1 [4].

Node position calculated use triangular positioning algorithm: Known three reference nodes A, B, C and its respective coordinate(x_a, y_a), (x_b, y_b), (x_c, y_c), and their distance to the unknown node D is d_a, d_b, d_c, assuming the coordinates of node D is(x, y), [5] there are formula (99.1):

Fig. 99.1 Hardware structure of system

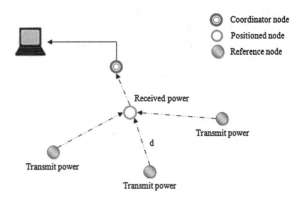

$$
\begin{cases}
\sqrt{(x-x_a)^2 + (y-y_a)^2} = d_a \\
\sqrt{(x-x_b)^2 + (y-y_b)^2} = d_b \\
\sqrt{(x-x_c)^2 + (y-y_c)^2} = d_c
\end{cases}
\tag{99.1}
$$

Thus we can calculate the (x, y) as formula (99.2):

$$
\begin{bmatrix} x \\ y \end{bmatrix} =
\begin{bmatrix} 2(x_a - x_c) & 2(y_a - y_c) \\ 2(x_b - x_c) & 2(y_b - y_c) \end{bmatrix}^{-1}
\begin{bmatrix} x_a^2 - x_c^2 + y_a^2 - y_c^2 + d_c^2 - d_a^2 \\ x_b^2 - x_c^2 + y_b^2 - y_c^2 + d_c^2 - d_b^2 \end{bmatrix}
\tag{99.2}
$$

99.3 Data Measuring and Processing

99.3.1 Data Acquisition

In the system, data collected by the reference node send to the positioned node, and then forwarded to coordinator node, finally send to the host computer through the serial port. Part of the sent data is shown as Fig. 99.2:

The data's format received in the system as follows: flag (1 byte), the IEEE address (8 bytes), short network address (2 bytes), the RSSI (1 byte) and LQI (1 byte), data length (1 byte).

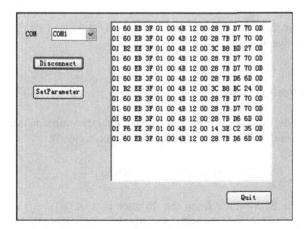

Fig. 99.2 Received data by the serial port

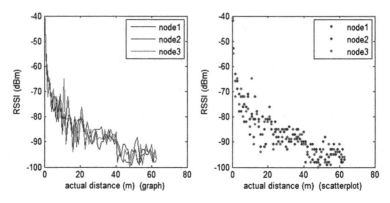

Fig. 99.3 Relationship between RSSI and distance

99.3.2 Relationship Between RSSI and Distance

In the experiment RSSI value were collected from three nodes at intervals of 1 m. The relationship between the received signal strength and the distance is shown as Fig. 99.3:

The data between RSSI and distance received from different node is different, but overall is consistent. It affected by the component selection and welding technology, but the impact is smaller. In experiment the transmit power is set to 0 dBm when measuring the distance to about 60 m the data is instability. In the experiment using CC2530 chip as hardware, according to CC2530 Data Sheet, the RF receiving sensitivity is 97 dBm, so there will be irregularities measured about 60 m.

99.3.3 RSSI Ranging

In this paper, the theoretical Shadowing model and fitting are two ways to calculate the distance between nodes [6, 7].

99.3.3.1 Shadowing Model

Theoretical model commonly used in wireless signal transmission is the Shadowing model. The formula of the model is shown as formula (99.3):

$$RSSI(d) = RSSI(d_0) - 10n \log(\frac{d}{d_0}) + \zeta \qquad (99.3)$$

$RSSI(d)$ is the RSSI value when the distance is d; $RSSI(d_0)$ is the RSSI value when the distance is d_0; n is the path loss exponent, indicating that the growth rate of path loss with distance; d is the actual distance; d_0 is the reference distance; ζ is

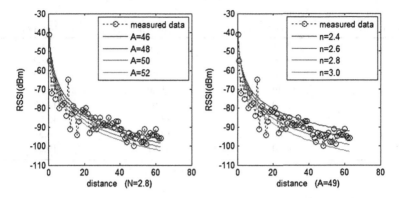

Fig. 99.4 Analysis of the value A and n

the shadowing factor in dB. It is a normal random variable with mean 0 and standard deviation σdB (dB).

In the statistics not include the impact of shadowing factor on the RSSI and d_0 value is 1 m. We can get the formula (99.4):

$$RSSI(d) = RSSI(d_0) - 10n\log(d) \tag{99.4}$$

Used in this experimental model is the formula (99.5):

$$RSSI(d) = -(10n\log(d) + A) \tag{99.5}$$

RF parameter A is defined to be the absolute value of the average energy when 1 m distant from the transmitter, that is, the received signal strength when 1 m from the launch node; n is the signal propagation constant, relating to the signal transmission environment; d is distance.

The calculation of RSSI and distance impacted by A and n. Analysis of the value A and n in matlab shown as Fig. 99.4:

Left graph is when n = 2.8 and take different values of A. It can be seen that with the increase of A the curve down; Right graph is when A = 49 and take different values of n. It can be seen that with the increase of n the curve down. Different value of A and n were measured in the actual calculation, and finally choosing a smaller error value. After measurement experiment, when A = 53, n = 2.5 the calculated distance is closest to the actual distance.

99.3.3.2 Curve Fitting

Fitting is a function that known a number of discrete function value {f1, f2, ..., fn}, by adjusting the number of undetermined coefficients f (λ1, λ2, ..., λn) makes the function and the known points with minimum differences (least squares) [8].

The intensive collection in the experiment will affect the fitting effect. In order to achieve a better fitting result we use data means value to handle the data

Table 99.1 Data of collection and calculation

Serial number	Actual distance	AN calculate	Fitting calculate
1	1.71	1.20	1.08
2	3.20	2.79	2.62
3	5.60	5.25	4.59
4	10.40	11.80	9.58
5	12.80	10.96	11.98
6	15.00	13.18	15.03
7	17.00	17.92	16.57
8	20.00	18.91	19.23

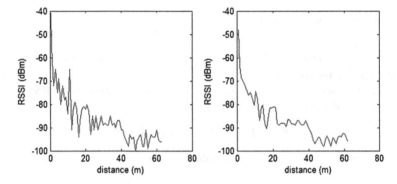

Fig. 99.5 Node means value before and after

Fig. 99.6 Data fitting

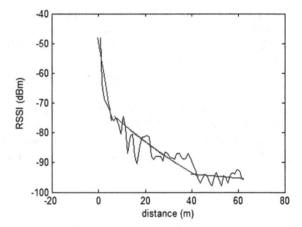

Fig. 99.7 Location of
reference node

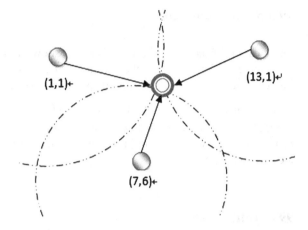

(1,1)

(13,1)

(7,6)

Table 99.2 Calculated and actual position

ID	Actual location	AN calculate	Error	Fitting calculate	Error
1	(2.50,5.00)	(2.67,5.34)	0.38	(2.39,5.46)	0.47
2	(4.80,4.00)	(4.16,3.74)	0.33	(4.70,3.07)	0.94
3	(6.00,4.00)	(6.50,4.90)	1.03	(6.25,4.43)	0.50
4	(8.00,4.60)	(6.80,4.90)	1.24	(8.57,5.43)	1.00
5	(10.00,3.40)	(7.96,2.56)	2.21	(10.2,1.83)	1.58
6	(9.00,5.00)	(8.32,3.82)	1.36	(8.18,5.76)	1.12

collected in the experiment so that the resulting curve is relatively smooth, as shown in Fig. 99.5:

Left graph is the data of the collected RSSI and distance and the right graph is the curve that we get several means of the data. We use least squares to fit the obtained data as shown in Fig. 99.6:

According to the trend of the curve we piecewise fit the collected data in matlab. The curve is the collected data of RSSI and distances. The polygonal line is the fitting curve.

99.3.4 Experiment

In the experiment we use two methods to calculate the real-time data as shown in Table 99.1:

Through the above data the AN measurement data has a small error within a short distance. Compared to the AN measurements the fitting measurement error is larger within a short distance. The overall measurement can meet the basic requirements t to calculate the distance in positioning.

99.3.5 Location Calculate

The location of the reference node is set as shown in Fig. 99.7:

In the experience we use the formula (99.2) to calculate the node's position. The calculated and actual position of positioned node is shown in Table 99.2:

The formula used in error calculation is $\sqrt{(x - x_1)^2 + (y - y_1)^2}$. (x, y) is actual distance, (x_1, y_1) is calculated distance. It can be seen from the experimental data that the error is within 2.3 m, and can meeting the requirements of the regional location accuracy.

99.4 Conclusion

Shadowing model and curve fitting are two ways to analyze the RSSI ranging. Through comparing the different A and n values a suitable value to the experimental environment is found. It can be seen from the experiment the data can meet the positioning requirements. Some errors also exist in the data, so the algorithms used in this paper need further research to improve the positioning accuracy.

References

1. Wang W, Lin JG (2011) Indoor location technique based on RFID. China Instrum 2:54–57
2. Zhang TT (2011) The research and application of wireless sensor network technology. Sci Technol Inf 2:231
3. Zhong LW, Duan CY (2009) ZigBee2007/PRO protocol stack experiments and practice. Beijing University of Aeronaut Astronaut Press
4. Farahani S (2008) ZigBee wireless networks and transceivers. Newnes
5. Liu XP (2009) Research and application on RSSI localization algorithm for wireless sensor network. Northwestern University (2009)
6. Quan HL, Chen YF, Quan rong LU (2008) Novel path loss model for propagation of radio waves. China's Electron J Sci Res Inst 1:40–43
7. Ming ZHU, Zhang HQ (2010) Wireless location technology for indoor environments based on RSSI. Mod Electron Tech 17:45–48
8. Wang Z, Jiang F, Zhao HM (2011) Distance measurement technology in RSSI of the ZigBee network. J Test and Meas Technol 4:301–304

Chapter 100
A New Watermarking Algorithm for Geo-Spatial Data

Haojun Fu, Changqing Zhu, Jianfeng Yuan and Hui Xu

Abstract This paper researched the security of geo-spatial data in the circulation process. Firstly, the deficiencies of current watermarking algorithms were analyzed through researching existing algorithms. Secondly, the data encryption algorithm idea was introduced into the design of the watermark based on the circulate characteristics of geo-spatial data, watermark information was embedded into the data bytes through the mapping function with additive rule, based on which a new byte stream-based geo-spatial data watermarking algorithm was proposed. Finally, experiments on the proposed watermarking algorithm were performed. The experiments showed that the proposed algorithm was robust and could effectively reduce the availability of geo-spatial data, problems such as the geo-spatial data of access controlling and flow direction tracking in the circulation process were well solved in this paper.

Keywords Geo-spatial data · Digital watermarking · Byte stream · Robustness

H. Fu (✉)
Institute of Surveying and Mapping, Information Engineer University,
Zhengzhou 450052, China
e-mail: fhjun121@163.com

C. Zhu · H. Xu
Key Laboratory of Virtual Geographic Environment of Ministry of Education,
Nanjing Normal University, Nanjing 210046, China
e-mail: zcq88@263.net

H. Xu
e-mail: xh901128@126.com

H. Fu · J. Yuan
Geographic information applications, The 61363 Troop, Xi'an, China
e-mail: dolpinpaper1314@126.com

W. Lu et al. (eds.), *Proceedings of the 2012 International Conference on Information Technology and Software Engineering*, Lecture Notes in Electrical Engineering 210, DOI: 10.1007/978-3-642-34528-9_100, © Springer-Verlag Berlin Heidelberg 2013

100.1 Introduction

With the ever-accelerating pace of the geo-spatial data digitalization and network process, the spreading and circulation between different regions become much more convenient, which leads to a growing phenomenon of the illegal use of geo-spatial data and copyright disputes. Therefore, how to effectively manage the data using limits of authority in the circulation process, protect data copyright even track data using condition, has become important problems that should be solved urgently in the current geo-spatial information security.

Digital watermarking technology, a new developed frontier research fields on information security [1, 2], has a close relationship with the data encryption technology. Data encryption technology is a kind of active control data mechanism, which makes illegal users unable to use the data normally by controlling the access of data files. However, the data encryption technology is unable to track the encrypt data flow direction, the data will turn into plaintext completely once the encryption technology got cracked, which will no longer be able to prevent the illegal distribution and use of data. Comparatively speaking, digital watermarking technology is a kind of passive control data mechanism, which shows promising potential in the field of copyright protection [3–7], utilization tracking and content authentication of geo-spatial data, but it is unable to control the proper geo-spatial data access permissions actively for it cannot control the geo-spatial data using limits of authority in the circulation process. Therefore, a new byte stream-based geo-spatial data watermarking algorithm absorbing the data encryption algorithm idea was proposed on the basis of the analysis of the current geo-spatial data security technology in this paper, which achieve the geo-spatial data of access controlling and flow direction tracking in the circulation process.

100.2 A Byte Stream-Based Digital Watermarking Algorithm

The design purpose of this algorithm is to protect the geo-spatial data copyright, identify the geo-spatial data flow direction and control the proper geo-spatial data access permissions actively in the circulation process. Therefore, the algorithm should not only contains both the copyright protection and flow direction tracking function of the traditional digital watermarking technology but also be able to reduce the data accuracy and quality, even encrypt the data. In this algorithm, meaningless watermark information will be adopted, the generation of which can be referenced to Ref. [3]. The randomness of meaningless watermark information enhances the confidentiality of the watermarking algorithm so that the illegal users can hardly understand or decrypt data without random seed points. The following explains mainly the byte stream-based digital watermarking algorithm from two aspects—watermark information embedding and watermark information detection and removal.

100.2.1 Watermark Information Embedding

Owing to the diversity of the geo-spatial data sources, the watermarking algorithm of different data types should be determined by the specific data characteristics. However, the geo-spatial data are organized in the unit of byte stream in the computer carrier access and circulation process, so as to achieve effective control of geo-spatial data in the circulation process and make the algorithm universal, it should always take the byte stream as watermarking embedding unit as general consideration in the design of the watermarking algorithm, while take a bit as a unit when embedding a single watermark bit as individual consideration. In this way, the practicality of the algorithm are enhanced—the watermark information can be embedded into the type stream fast in real time and the data accuracy and quality can be reduced even the data can be encrypted while embedding watermark information. The same time, the pretreatment of the byte can be carried out before watermark information embedding, such as bit inversion and scrambling operation which can enhance the unreliability of geo-spatial data containing watermarking to unauthorized users. The watermarking embedding algorithms of geo-spatial data in circulation status are determined by the following aspects: (1) the embedding position of watermark information in the byte stream; (2) the watermark information embedding rules; (3) the watermark information embedding method. The three aspects above determine the robustness and removability of the watermarking embedding algorithm and the influence degree on the data quality.

The embedding position of watermark information in the byte stream is directly related to the information locating when the watermark information is detected and removed. In the geo-spatial data access and circulation process, the byte stream may be transferred asynchronously, certain part of byte information in the byte stream may got lost, the information may be changed as a result of that the byte stream got attacked, etc. Therefore, the embedding position of watermark information should not be determined by the location of the byte stream in the geo-spatial data or the position of the bytes in the byte stream but the byte itself, namely, keep the byte itself the same pace with the embedding position of watermark bit to make it possible to detect watermarking blind. To this end, the mathematical mapping idea is introduced to build data-watermarking synchronization function, that is, establish a function which is able to locate the embedding position of watermark bit according to the byte itself, namely establish a multi-to-one mapping relationship which is constructed between the byte and the watermark bit to determine the embedding position of watermark bit. Due to the unique multi-to-one characteristic of the mapping function, the algorithm is able to keep the byte the same pace with the watermark bit in the geo-spatial data circulation process. The concrete algorithm is as follows:

Assuming the watermark message length as N, transfer all the bytes in the geo-spatial data byte stream into corresponding values then map the values to $[1, N]$ by bytes based on the multi-to-one mapping ideology. Set the mapping function as $f(x)$, the corresponding value of the byte itself as x, the watermark information as

$W = \{w[i], \ i = 0, 1, \ldots, N - 1\}$ $(1 \leq f(x) \leq N)$. In order to establish a multi-to-one mapping relationship, the watermark bit is embedded with the corresponded byte value repeatedly. Taking the robustness of the algorithm into account, $f(x)$ should be able to map x to $[1, N]$ as uniformly as possible, and the change of x caused by watermarking should not affect the number of bytes corresponding with the embedding position of watermark bit obviously. Thus, the watermark bit corresponding to the byte value won't be affected by watermarking.

The watermark information embedding rules should be taken into consideration when the embedding position of watermark information has been determined. The watermarking embedding rules of current algorithms can be divided into two types—additive rules and multiplicative rules. Considering that the watermark information need to be removed to ensure the normal use of legal users during the late using stage of geo-spatial data, additive embedding rules are used in this algorithm in order to make sure that the geo-spatial data still be able to be used after the embedding watermarking removed, the watermark information is embedded with the byte value in a strong intensity, so that the data accuracy and quality can be effectively reduced even make the data unavailable while embedding watermark information.

There are two kinds of watermark information embedding methods: transformation domain-based and space domain-based. Space domain-based watermark information embedding method is used in this paper in order to isolate the interaction relationship between the bytes in the byte stream, enhance the independence of every single watermark bit in the watermarking blind detection, and eliminate the dependence between the watermark bits.

100.2.2 Watermark Information Detection and Removal

Watermarking detection is actually a reverse process of watermarking embedding. There is need to discriminate the watermark information extracted according to the maximum membership degree law in the final. The removal of the watermark information as a data reduction process is crucial. Firstly, the binary watermark information detected by the watermarking detecting algorithm is meaningless so that certain relevant detection should be done with the original binary watermark information corresponded with the random value which was stored in the watermarking relationship mapping table [8], then judge that if there is watermark information in the under-tested data according to the correlation coefficient. When watermark information was detected in geo-spatial data, it can be removed to restore the data to the original status by authorized users then the users are able to get the correct geo-spatial data; otherwise, the users who access to the geo-spatial data illegally have no privilege to remove the watermark information that is why the geo-spatial data is still untrusted, thereby the illegal users are unable to use the correct geo-spatial data. Meanwhile, the copyright of the data can be protected and the outflow source of illegal data can be tracked according to the data flow direction, the copyright or some other relevant information extracted from the watermark information.

100.3 Experiment Analysis

For the purpose of verifying the watermarking algorithm proposed in this paper, the watermarking embedding proposal above has been realized in the VC++6.0 programming environment. Furthermore, an experiment has been carried out on a size of 808 × 758 color raster map to analyze the performance in reducing the data quality and accuracy and the robustness of the algorithm.

100.3.1 Robustness Analysis

Raster map embedded with the watermark information based on the algorithm proposed in this paper is unusable, that is the users are unable to use the data file normally. In order to verify the robustness of the proposed watermarking algorithm, common data attacks on the raster map containing watermark information should be taken. Geo-spatial data are organized in the unit of byte stream in the computer carrier access and circulation process, so that the attack object here should be the byte stream which comprises the raster map and there is need to detect the watermarking in the attacked data. Considering the geo-spatial data feature showed in the access and circulation process, data processing methods such as increasing bytes randomly, deleting bytes randomly, cutting byte segments, shifting bytes and some others are chosen as watermarking attack methods. The watermarking detection results under different attack intensities are shown in Table 100.1, detection results are described by correlation coefficient, the "$\sqrt{}$" in the brackets behind the correlation coefficient shows that watermark information

Table 100.1 Detection results of attacking experiments

Attack intensities	Attacks intensities/Detecting result					
Increasing bytes randomly	Byte increasing (%)	10	20	30	40	50
	Detecting result	0.99 ($\sqrt{}$)	0.90 ($\sqrt{}$)	0.84 ($\sqrt{}$)	0.73 ($\sqrt{}$)	0.64 ($\sqrt{}$)
Deleting bytes randomly	Byte deleting (%)	10	20	30	40	50
	Detecting result	1.0 ($\sqrt{}$)	0.93 ($\sqrt{}$)	0.85 ($\sqrt{}$)	0.79 ($\sqrt{}$)	0.66 ($\sqrt{}$)
Modifying byte randomly	Byte modifying (%)	10	20	30	40	50
	Detecting result	0.89 ($\sqrt{}$)	0.78 ($\sqrt{}$)	0.69 ($\sqrt{}$)	0.64 ($\sqrt{}$)	0.41 (x)
Cutting byte segments	Byte cutting (%)	10	20	30	40	50
	Detecting result	1.0 ($\sqrt{}$)	1.0 ($\sqrt{}$)	1.0 ($\sqrt{}$)	0.98 ($\sqrt{}$)	0.91 ($\sqrt{}$)
Shifting bytes	Shifted byte (bit)	1	3	5	6	7
	Detecting result	1.0 ($\sqrt{}$)	1.0 ($\sqrt{}$)	1.0 ($\sqrt{}$)	1.0 ($\sqrt{}$)	1.0 ($\sqrt{}$)

Table 100.2 Statistics of data error

Error size	Watermarked raster map data		Raster map data after remove watermark information	
	Number of data	Percentage	Number of data	Percentage
0	863,768	47.01	1,837,392	100.00
0 ~ 50	53,843	2.93	0	0.00
50 ~ 100	75,840	4.13	0	0.00
100 ~ 200	255,318	13.90	0	0.00
>200	588,623	32.04	0	0.00

was detected, on the contrary, the "x" in the brackets behind the correlation coefficient shows that watermark information was not detected.

As can be seen from the attack test above, the algorithm proposed in this paper is not only able to reduce data accuracy and quality but also robust against byte missing, byte shifting, byte addition, byte deletion and some other common attacks in the geo-spatial data circulation process, which can effectively protect the copyright of geo-spatial data in the circulation process and track the outflow source of data. However, it can be inferred from Table 100.2, the ability of the algorithm to resist attack like modifying the byte value in a strong intensity is relatively weak.

100.3.2 Error Analysis

The corresponding appropriate pretreatment (such as inverse scrambling, parameter correct of relevant data, etc.) on the watermarked raster map should be done before carry out the error comparison with the original raster map; meanwhile, the removal of the watermark information which is concerned with the credibility of the geo-spatial data used by different users is also an important step, for this reason that there is need to take an error comparison between the watermarked raster map and the original one. The comparison result is shown in Table 100.2.

It can be inferred from the error comparison result, the data change caused by watermark information is big enough to result in a great degree reduction of the raster map data quality and accuracy. Before the watermarked data was used, the map data is incredible even unavailable if the watermarking has not been removed so that the copyright interest of the data owners can be protected effectively. In addition, there is no different between the watermarked raster map which the watermarking have been removed and the original one, accordingly, the usage of the geo-graphic data will not be influenced, then the using equity of the legitimate users can be protected effectively.

100.4 Conclusions

Considering the circulate characteristics of geo-spatial data showed in the computer carrier access and circulation process, the data encryption algorithm idea was introduced into the design of the watermark, watermark information was embedded into the data bytes through the mapping function by quantitative methods, based on which a new byte stream-based geo-spatial data watermarking algorithm was achieved. Experiments on the availability of watermarked data and the robustness of proposed algorithm were performed, and the experiment results showed that the proposed algorithm was not only able to reduce data accuracy and quality but also robust against byte missing, byte shifting, byte addition, byte deletion and some other common attacks, which can achieve effective control of geo-spatial data in the circulation process. The algorithm proposed in this paper is valuable for the transmission security of geo-spatial data in the circulation process, especially under network environment.

References

1. Sun SH, Lu ZM, Niu XM (2004) Digital watermarking technique and application. Science press, Beijing, pp 10–14 (in Chinese)
2. Zhong H, Zhang XH, Jiao LC (2006) Digital watermarking and image authentication—algorithm and application. Science press, Beijing, pp 1–2 (in Chinese)
3. Yang CS, Zhu CQ (2011) Research on watermarking algorithm robust to geometrical transform for vector geo-spatial data based on invariant function. Acta Geodaetica et Cartographica Sinica 2(40):256–261 (in Chinese)
4. Zhu CQ, Wang ZW, Long Y, Yang CS (2009) An adaptive watermarking algorithm for DEM based on DFT. In: Proceedings 24th international cartographic conference, Chile (in Chinese)
5. Fu H, Zhu CQ, Miao J, Hu QY (2011) Multipurpose watermarking algorithm for digital raster map based on wavelet transformation. Acta Geodaetica et Cartographica Sinica 3(40):397–400 (in Chinese)
6. Li Y, Xu L (2003) A blind watermarking of vector graphics images. In: 5th international conference on computational intelligence and multimedia applications (ICCIMA'03), p 424
7. Voigt M, Yang B, Busch C (2004) Reversible watermarking of 2D-vector data. In: Proceedings of the 2004 multimedia and security workshop on multimedia and security. Magdeburg, Germany, pp 160–165
8. Yang CS, Zhu CQ, Wang QS (2008) Research on the autocorrelation detection in digital watermarking technology. In: Proceeding of the 21st international congress for photogrammetry and remote sensing, vol XXXVII. Part B4-1, Commission IV, pp 127–130

Chapter 101
An Investigation of Stationary Signal Blind Separation Using Adaptive Algorithms

Xu Peng-fei and Jia Yin-jie

Abstract Blind source separation attempts to recover unknown independent sources from a given set of observed mixtures. The two adaptive algorithms -EASI and RLS are introduced in this chapter, the separation simulation of a set of stationary signals is constructed. Through the comparative analysis, the result shows that the RLS algorithm has better convergence speed and the steady state performance than the EASI algorithm for blind signal separation of stationary signal. In the RLS algorithm, we need select appropriate forgetting factor to meet the demand of convergence speed and steady state performance.

Keywords Blind signal separation · Stationary signal · Adaptive algorithm · Forgetting factor

101.1 Introduction

The seminal work on blind source separation (BSS) was first introduced by Jutten and Herault [1] in 1985, the problem is to extract the underlying source signals from a set of mixtures, where the mixing matrix is unknown. In other words, BSS seeks to recover original source signals from their mixtures without any prior

X. Peng-fei (✉)
Faculty of Computer and Communication Engineering,
Huaian College of Information Technology, 223003 Huai'an, Jiangsu, China
e-mail: jionad@126.com

J. Yin-jie
Faculty of Computer Engineering, HuaiYin Institute of Technology, 223003 Huai'an,
Jiangsu, China
e-mail: jiayinjie@126.com

W. Lu et al. (eds.), *Proceedings of the 2012 International Conference on Information Technology and Software Engineering*, Lecture Notes in Electrical Engineering 210, DOI: 10.1007/978-3-642-34528-9_101, © Springer-Verlag Berlin Heidelberg 2013

information on the sources or the parameters of the mixtures. This situation is common in acoustics, radio, medical signal, image processing, hyper spectral imaging and other areas.

The problem of basic linear BSS can be expressed algebraically as follows (assumes that the signal is continuous signal):

The vector of the source signals is:

$$S(t) = [s_1(t), s_2(t)\ldots s_n(t)]^T \tag{101.1}$$

where the source signals are assumed to be statistically independent.

The vector of the observed or mixed signals is:

$$X(t) = AS(t) = [x_1(t), x_2(t)\ldots x_m(t)]^T \tag{101.2}$$

where A is a non-singular (m × n) mixing matrix.

The problem can be formulated as the computation of an unmixing or separating matrix W, whose output y is given below.

$$Y(t) = WX(t) \tag{101.3}$$

$Y(t)$ is a scaled and permuted version of the original source signals.

In short, the basic goal of BSS is to find the separating matrix W, without knowing the mixing matrix A.

To evaluate the performance of the BSS algorithms, we use the cross-talking error as the performance index [2]:

$$E = \sum_{i=1}^{n} \left(\sum_{j=1}^{n} \frac{|c_{ij}|}{\max_k |c_{ik}|} - 1 \right) + \sum_{j=1}^{n} \left(\sum_{i=1}^{n} \frac{|c_{ij}|}{\max_k |c_{kj}|} - 1 \right) \tag{101.4}$$

where C $(C = WA)$ is the matrix with elements of c_{ij}. The lowest bound of E is 0, in general, the smaller the value of E is, the better the separation effect is. Simply put, ECT is the difference degree of the separating matrix and the mixing matrix.

The chapter is organized as follows. In Sect. 101.2, The EASI algorithm and RLS algorithm are introduced briefly. Section 101.3 gives some experimental results to demonstrate the effectiveness of the algorithms for stationary signal with conclusion and discussion on future works given in the last section.

101.2 Overview of the Adaptive Algorithm

101.2.1 The EASI Algorithm

Equivariant variant Adaptive Separation via Independence (EASI) algorithm [3] is a basic algorithm among various on-line ICA (Independent Component Analysis) algorithms. We focus on the EASI algorithm because it has simple parallel structure which are based on the idea of serial updating: This specific form of matrix updates systematically yields algorithms with a simple structure for both real and complex mixtures. Most importantly, the performance of an EASI algorithm does not depend on the mixing matrix. In particular, convergence rates, stability conditions, and interference rejection levels depend only on the (normalized) distributions of the source signals. Closed form expressions of these quantities are given via an asymptotic performance analysis. Iterative formula is as follows:

$$W(0) = I \tag{101.5}$$

$$y(t) = W(t)S(t) \tag{101.6}$$

$$g(y(t)) = y^3(t) \tag{101.7}$$

$$W(t+1) = W(t) + u[y(t)y^T(t) - I + g(y(t))y^T(t) - y(t)g^T(y(t))]W(t) \tag{101.8}$$

where g is nonlinear transform function which is used to signal nonlinear extraction according to the nature of the signal selection. In general, for sub Gaussian signal we select $g(y) = y^3$, for super Gaussian signal we select $g(y) = y - \tanh(y)$. In this simulation we select $g(y) = y^3$, Communication signal belongs to sub Gaussian signal. Here, the selecting of step size u is very important to the convergence of the algorithm, which is related to the effect of blind signal separation directly.

The larger step size is, the worse steady state performance is. if the step size is too small, the convergence rate is very slow.

101.2.2 The General Gradient RLS Algorithm

Recursive least squares (RLS) [4] algorithm is based on nonlinear principle component analysis (PCA) algorithms. Here the gradient descent algorithm use the general gradient. The steps are as follows.

We first perform pre-whitening treatment on the data:

$$\bar{x}(t) = x(t) - E\{x(t)\} \tag{101.9}$$

$$v(t) = E\{\bar{x}(t)\bar{x}^T(t)\}^{-1/2}x(t) \tag{101.10}$$

Then, the following cost function is optimized:

$$J(W) = \sum_{i-1}^{t} \beta^{t-i} \|v(i) - W(i)g(W^T(i-1)v(i))\|^2 \tag{101.11}$$

where β is forgetting factor, g is nonlinear transform function.

So we can get the iterative formulas are as follows:

$$y(t) = W(t-1)v(t) \tag{101.12}$$

$$z(t) = g(y(t)) \tag{101.13}$$

$$h(t) = P(t-1)z(t) \tag{101.14}$$

$$m(t) = \frac{h(t)}{\beta + z^T(t)h(t)} \tag{101.15}$$

$$P(t) = \frac{1}{\beta} Tri[P(t-1) - m(t)h^T(t)] \tag{101.16}$$

$$W(t) = W(t-1) + m(t)[v^T(t) - z^T(t)W(t-1)] \tag{101.17}$$

where $Tri[]$ is a tri-diagonal matrix. $W(0)$ and $P(0)$ both are an unit matrix, g can select different value for different signal.

101.2.3 The Natural Gradient RLS Algorithm

Natural gradient-based RLS algorithm is an improved version of the general gradient -based RLS, that is, the natural gradient is used instead of the general gradient. Generally speaking, the use of natural gradient descent can speed up the convergence, it is expected to achieve better results. Natural gradient concept was first put forward by literature [6], then was corrected by literature [7].

The iterative formula of natural gradient-based RLS algorithms are as follows (continued by formulas 101.12 and 101.13):

$$Q(t) = \frac{P(t-1)}{\beta + z^T(t)P(t-1)y(t)} \tag{101.18}$$

$$P(t) = \frac{1}{\beta}[P(t-1) - Q(t)y(t)z^T(t)P(t-1)] \tag{101.19}$$

$$W(t) = W(t-1) + [P(t)z(t)v^T(t) - Q(t)y(t)z^T(t)W(t-1)] \tag{101.20}$$

where W(0) and P(0) both are unit matrix, g can select different value for different signal. It is proved that [8]: Natural gradient is an extension of general gradient in Riemann space, while the general gradient is special performance of the natural gradient in the uniform Euclidean space. The parameter structure of the contrast function is considered in the natural gradient, so it is more suitable for blind signal separation than general gradient. In a word, compared with the former RLS, the convergence speed of natural gradient algorithm increased significantly, while the calculated load is acceptable.

Fig. 101.1 The ECT curve of the three algorithms

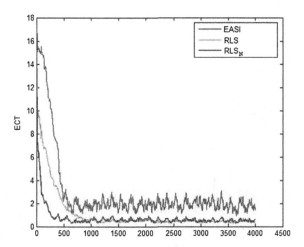

101.3 Experimental Results

In order to verify the effectiveness of the above algorithm for adaptive BSS of stationary signal, the research simulations are done by matlab software. The first line of Fig. 101.3 are the five unknown communication signals, they are arranged in the order from left to right, using five receiving antennas. The mixing matrix A is a 5×5 random matrix, the elements of A are randomly generated subject to the uniform distribution in [0, 1] such that A is nonsingular. The mixed signals $X(t)$ are sampled at 10 kHz (sample number is 4,000). All signals here are sub Gaussian stationary signals, therefore the separation algorithm parameter selection are as follows:

The EASI algorithm parameters: nonlinear transform function $g(y) = y^3$, step size $u = 0.006$; The RLS algorithm parameters: nonlinear transform function $g(y) = \tanh(y)$, forgetting factor $\beta = 0.99$.

Simulation curve by the ECT is shown below, in Fig. 101.1, from the top to the bottom, there are three curves named *EASI*, *RLS* (general gradient RLS) and RLS_N (natural gradient RLS). In terms of convergence speed and steady state performance, the RLS algorithm is better than EASI algorithm. Further, we also can see, the natural gradient algorithm has faster convergence speed than general gradient algorithm, while the steady state performance slightly inferior to the latter. The ECT of three algorithms respectively is 1.903, 0.43861 and 0.52693.

The following experiment shows the effect of different forgetting factors β on natural gradient RLS algorithm. For natural gradient RLS algorithm, here we use four different forgetting factors(beta = 0.91, beta = 0.97, beta = 0.99, beta = 1.0) and perform twenty randomized experiments, finally we can get the convergence curves below:

From Fig. 101.2, forgetting factor is used to balance the convergence speed and the steady state performance. The bigger the β is, the better the steady state

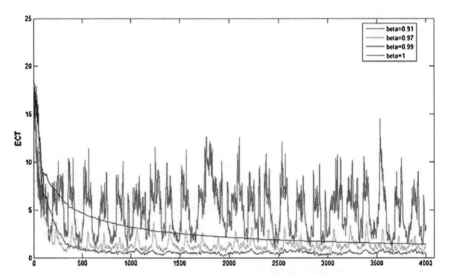

Fig. 101.2 The convergence curve of natural gradient RLS under different forgetting factors

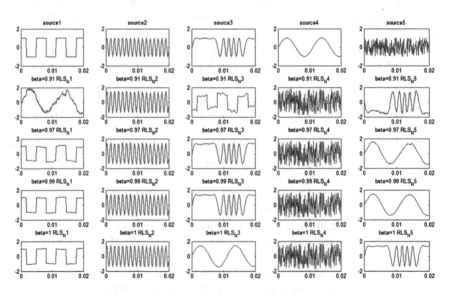

Fig. 101.3 The separation waveform of natural gradient RLS under different parameter

performance is. When $\beta = 0.91$, $\beta = 0.97$, $\beta = 0.99$, $\beta = 1.0$, their ECT respectively are 2.7921,1.0332, 0.42692,0.42159. But the bigger the β is, the slower the convergence speed is. Such as black curve (beta = 1.0).

Vice versa, The smaller the β is, the faster the convergence speed is, the worse the steady state performance is. Such as red curve (beta = 0.91).

At last, the waveform of the recover signal by natural gradient RLS under different forgetting factors shows in Fig. 101.3:

From Fig. 101.3, we can intuitively get a conclusion: when $\beta = 1.0$, 0.99, the separation effect is significantly better than the situation when $\beta = 0.97$, $\beta = 0.91$.

So, the selecting of forgetting factor is very important, small forgetting factor can accelerate the convergence, but the steady state performance is poor. The big forgetting factor has the opposite effect, the performance is stable but the convergence is slow. Therefore we need select appropriate forgetting factor to meet the demand of convergence speed and steady state performance. Here, combined Fig. 101.2 with Fig. 101.3, $\beta = 0.99$ is more appropriate.

101.4 Conclusions and Future Work

For blind signal separation of stationary signal, the RLS algorithm has better convergence speed and the steady state performance than the EASI algorithm. The natural gradient algorithm has faster convergence speed than general gradient algorithm, while the steady state performance slightly inferior to the latter. At the same time, we need select appropriate forgetting factor to meet the demand of convergence speed and steady state performance by the natural gradient algorithm. But for speech signals, because of nonstationarity of the signal, the RLS algorithm is more dependent on signal statistics than EASI, the separation performance of RLS algorithm is worse than EASI, this point deserves further analysis.

References

1. Herault J, Jutten C (1986) Space or time adaptive signal processing by neural network models[A]. In: Neural networks for computing: Denker J S. AIP conference processings 151 [C]. American Institute of Physics, New York
2. Yang HH, Amari S (1997) Adaptive on-line learning algorithm for blind separation: maximum entropy and minimum mutual information[J]. Neural Comput 9(5):1457–1482
3. Cardoso JF, Laheld B (1996) Equivariant adaptive source separation. IEEE Trans Signal Process 44(12):3017–3029
4. Pajunen P, Karhunen J (1998) Least-squares methods for blind source separation based on nonlinear PCA. Int J Neural Syst 8:601–612
5. Zhu XL, Zhang XD (2002) Adaptive RLS algorithm for blind source separation using a natural gradient [J]. IEEE Signal Process Lett 9(12):432–435
6. Cichocki A, Unbehauen R, Moszczynski L (1994) A new online adaptive learning algorithm for blind separation of source signals [A]. In: Proceedings ISANN[C]. Taiwan, pp 406–411
7. Amari S, Cichocki A, Yang HH (1996) A new learning algorithm for blind signal separation[A]. In: Advances in neural information processing systems[C]. Cambridge, pp 757–763
8. Amari S (1998) Natural gradient works efficiently in learning. Neural Comput 10:251–276

Chapter 102
A Novel Algorithm for 2D-Radar and Infrared Sensor Track Association

Jiesheng Liu, Dengyi Zhang, Hanbao Wu and Biyin Zhang

Abstract Aiming at the problem of track association for the 2D radar and the infrared sensors, this paper has proposed an algorithm of track association for dissimilar platforms and dissimilar sensors based on multiple hypothesis to associate, which adequately utilizes the complementary of 2D radar and infrared sensors by getting similarity of the key factor with multiple hypothesis and judging the checkup condition. The experiment shows that the method proposed is available and practical.

Keywords Dissimilar-sensor · Multiple hypothesis method · Similarity · Track association

102.1 Introduction

In the target tracking, the system composed of the radar and the infrared sensor is a kind of typical dissimilar multi-sensor fusion system. How to sufficiently exploit the complementation of two sensors, i.e. the infrared and radar, so that each can shine more brilliantly in other's company, is the current hotspot of research among experts both abroad and in China. However, dissimilar sensor information fusion is facing many difficulties, among which the main one is that there is no unified mathematical tool and method [1–5]. Therefore, for the information fusion of dissimilar sensor, we can only discuss the fusion methods aiming at the specific

J. Liu · D. Zhang
Computer School, Wuhan University, Wuhan 430072, China

H. Wu · B. Zhang (✉)
Wuhan Digital Engineering Institute, Wuhan 430074, China
e-mail: succor_com@sina.com

W. Lu et al. (eds.), *Proceedings of the 2012 International Conference on Information Technology and Software Engineering*, Lecture Notes in Electrical Engineering 210, DOI: 10.1007/978-3-642-34528-9_102, © Springer-Verlag Berlin Heidelberg 2013

objects. Generally speaking, the track association is the primary step of the fusion system. The traditional association algorithm mainly solves the track association on sensor of the same type and same dimension. When sensors are not in the same platform, it is often necessary for these methods to convert track information from different sensors into the same measurement space. Because the information detected by the 2D radar and the infrared sensor can't completely express the position of one target in the three-dimensional space: with the former lacking the elevation information and the coordinate conversion error being too large, and the latter not possessing any range information and unable to carry out coordinate conversion. Thus a drop in the fusion performance occurs.

Aiming at the track association problem on dissimilar platforms, i.e. 2D radar and infrared sensor, this paper has proposed one track association algorithm of dissimilar platforms and dissimilar sensors based on the multiple hypothesis method. This method has obtained the 2D common vector of the sensor measurement space via the pure bearing cross-positioning under the hypothesized condition. And by combining with the high-precision infrared angular measurement and the radar distance measurement information, estimates on the corresponding vertical vector are obtained respectively. Finally according to the similarity of the vertical vectors, we can judge if the tracks are actually associated or not. The simulated experiments have shown that this method is simple and effective.

102.2 Problem Description

Assume that the target dynamic model in the Cartesian coordinate system is:

$$x_{k+1} = F_k x_k + w_k \qquad (102.1)$$

where x_k is status vector, F_k is status transition matrix, and w_k is process noise. Assume that w_k is the Gauss white noise with zero mean value, then

$$E(w_k) = 0 \quad D(w_k) = E(w_k w_j^T) = Q\delta_{kj} \qquad (102.2)$$

where δ_{kj} is Kronecker delta function. Assume that 2D radar is sensor 1, infrared is sensor 2, and the measurement model of the sensor is:

$$z_k^{(i)} = H_k^{(i)} x_k + v_k^i \quad i = 1, 2, k = 0, 1, \cdots \qquad (102.3)$$

When $i = 1$, the measurement corresponding to 2D radar $z_k^{(1)} = [\alpha_1, d_1]^T + v_k^1$, where d_1, α_1 respectively stand for the distance and elevation.

When $i = 2$, the measurement corresponding to the infrared $z_k^{(2)} = [\beta_2, \alpha_2]^T + v_k^2$, where β_2, α_2 respectively stand for the bearing and elevation. And the relationships between the distance, bearing, elevation and status variable are respectively:

$$d = \sqrt{x^2 + y^2 + z^2} \tag{102.4}$$

$$\alpha = \arctan(x/y) \tag{102.5}$$

$$\beta = \arctan\left(z \Big/ \sqrt{x^2 + y^2}\right) \tag{102.6}$$

where x, y, z is the three-dimensional coordinate in the Cartesian coordinate system, and $v_k^{(i)}$ is the Gauss white noise with zero mean value, and covariance $R^{(i)}$ is known. And for the 2D radar, its covariance matrix is:

$$\text{cov}(v_k^{(1)}) = R^{(1)} = \begin{bmatrix} \sigma_{\alpha_1}^2 & \\ & \sigma_{d_1}^2 \end{bmatrix} \tag{102.7}$$

For the infrared sensor, its covariance matrix is:

$$\text{cov}(v_k^{(2)}) = R^{(2)} = \begin{bmatrix} \sigma_{\beta_2}^2 & \\ & \sigma_{\alpha_2}^2 \end{bmatrix} \tag{102.8}$$

where $\sigma_{d_1}^2, \sigma_{\alpha_1}^2$ are respectively the variances of distance and bearing measurement noise for 2D radar, and $\sigma_{\alpha_2}^2, \sigma_{\beta_2}^2$ are respectively the variances of the bearing and elevation measurement noise for infrared sensor.

Because this paper mainly discusses the track association problem of 2D radar and infrared sensor, the time alignment problem is ignored. Assume that the detection information on the target for the two sensors have the same time, i.e. the target detection information for the 2D radar is $(d_1, \alpha_1, \sigma_{d_1}^2, \sigma_{\alpha_1}^2, t)$, and the platform position is (x_1, y_1), while the target detection information for the infrared sensor is $(\alpha_2, \beta_2, \sigma_{\alpha_2}^2, \sigma_{\beta_2}^2, t)$, and its platform position is (x_2, y_2). For the track association problem on dissimilar platforms, i.e. 2D radar and infrared sensor, we have to judge that if the detection information $(d_1, \alpha_1, \sigma_{d_1}^2, \sigma_{\alpha_1}^2, t)$ from 2D radar and the detection information $(\alpha_2, \beta_2, \sigma_{\alpha_2}^2, \sigma_{\beta_2}^2, t)$ from the infrared sensor represent the same target.

102.3 Track Association Method

This paper will discuss on the track association method for dissimilar platforms, i.e. 2D radar and infrared sensor, which mainly consists of 5 steps as follows:

1. Establish the association hypothesis condition: assume that the detection data from 2D radar and the infrared sensor are the same target.
2. Firstly perform the cross-positioning on the bearings detected by two sensors, so that the common 2D vector (x, y) of the measurement space for two sensors can be obtained.

3. By combining the distance measurement information of the 2D radar and angular measurement information of infrared sensor, the corresponding vertical vector estimates z_1, z_2 can be obtained.
4. Then perform the similarity d_k^2 calculation on the vertical estimates z_1, z_2.
5. Finally judge according to similarity d_k^2 of z_1, z_2: given the threshold χ_α^2, and if $d_k^2 < \chi_\alpha^2$, the hypothesis condition becomes tenable, and the association is successful. Otherwise, the hypothesis condition is not tenable, and the association fails.

102.3.1 Establish Association Hypothesis Condition

The 2D radar detection track corresponds to one target, and the infrared sensor detection track corresponds to one target. And thus the association problem between one 2D radar track and one infrared sensor track will be converted into one multiple hypothesis checkup problem as follows:

H_0 assume that 2D track is associated with the infrared track, and $d_k^2 < \chi_\alpha^2$, then the 2D track is successfully associated with the infrared track.

H_1 assume that 2D track is associated with infrared track, with $d_k^2 \geq \chi_\alpha^2$, then 2D track fails to associate with infrared track.

102.3.2 Calculate Common 2D Vector

Firstly perform the cross-positioning of the bearing information detected by the 2D and infrared sensor as shown in the following diagram: (Fig. 102.1)

Because of the assumed hypothesis, 2D radar detection target and the infrared sensor detection target are the same target, the bearing cross-position is (x, y), which can build up an equation as follows:

Fig. 102.1 2D radar and infrared sensor space position distribution

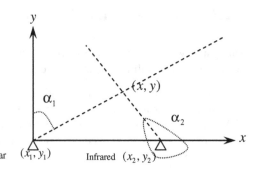

$$\tan \alpha_1 = \frac{x - x_1}{y - y_1} \quad \tan \alpha_2 = \frac{x - x_2}{y - y_2} \tag{102.1}$$

By using the simultaneity of formula (102.1), we can calculate and obtain the x and y values.

102.3.3 Estimates on Vertical Vectors

Assume that the cross-position of 2D radar and infrared sensor detection target is (x, y, z). In Sect. 102.3.1, by using the bearing information of two sensors, the 2D vector (x, y) of the measurement space is obtained. In this section, by using the distance measurement information of 2D radar and the angular measurement information of infrared sensor, the corresponding vertical sector z can be obtained respectively.

The equation can be built up as follows:

$$d_1 = \sqrt{(x - x_1)^2 + (y - y_1)^2 + z_1^2} \tag{102.2}$$

$$\tan \beta_2 = \frac{z_2}{\sqrt{(x - x_2)^2 + (y - y_2)^2}} \tag{102.3}$$

From formula (102.2), we can calculate the vertical vector z_1 of 2D radar detection target, and from formula (102.3), we can calculate the vertical vector z_2 of infrared sensor detection target.

102.3.4 Similarity Calculation

From Sects. 102.3.1 and 102.3.2, we can obtain that the 2D radar detected target information is (x, y, z_1), and the infrared sensor detected target information is (x, y, z_2), where (x, y) is obtained by calculating two sensors' bearing information together. The z_1, z_2 are obtained by respectively using the 2D radar distance measurement information and infrared sensor angular measurement information. Finally the similarity between z_1 and z_2 is used to judge and determine if the tracks are associated.

From the simultaneity of (102.1), (102.2) and (102.3), we can obtain the function of z_1, z_2, i.e.

$$z_1 = f_1(\alpha_1, \alpha_2, d_1) \tag{102.4}$$

$$z_2 = f_2(\alpha_1, \alpha_2, \beta_2) \tag{102.5}$$

Record $z = z_1 - z_2$, then:

$$z = f_1(\alpha_1, \alpha_2, d_1) - f_2(\alpha_1, \alpha_2, \beta_2) = f_3(\alpha_1, \alpha_2, \beta_2, d_1) \quad (102.6)$$

The location error of Δz is [6]:

$$\Delta z = \frac{\partial f}{\partial \alpha_1} \rho_{\alpha_1} + \frac{\partial f}{\partial \alpha_2} \rho_{\alpha_2} + \frac{\partial f}{\partial \beta_2} \rho_{\beta_2} + \frac{\partial f}{\partial d_1} \rho_{d_1} \quad (102.7)$$

This location error is a random variable of zero mean value, whose variance is:

$$\sigma_z^2 = \left(\frac{\partial f}{\partial \alpha_1}\right)^2 \sigma_{\alpha_1}^2 + \left(\frac{\partial f}{\partial \alpha_2}\right)^2 \sigma_{\alpha_2}^2 + \left(\frac{\partial f}{\partial \beta_2}\right)^2 \sigma_{\beta_2}^2 + \left(\frac{\partial f}{\partial d_1}\right)^2 \sigma_{d_1}^2 \quad (102.8)$$

Are z_1, z_2 associated? According to the idea of "nearest neighbor", given that the statistic distance is defined as the weighting norm of innovation vector $d_k^2 = \tilde{z}_k^T S_k^{-1} \tilde{z}_k$, where $\tilde{z}_k$ stands for residual error vector, S_k is the innovation covariance matrix, and d_k^2 is the norm of the residual error vector, we can understand the statistic distance between two detected tracks. d_k^2 is a normalized random variable. When $\tilde{z}_k$ is of normal distribution, $d_k^2 = x$ will conform with the probability density function of χ^2 distribution whose freedom is M

$$f(x) = \frac{x^{\frac{M}{2}-1}}{2^{\frac{M}{2}} \Gamma\left(\frac{M}{2}\right)} \exp\left(-\frac{x}{2}\right) \quad (102.9)$$

where in M is the number of measurement dimensions. Actually, it converts the problem to judge if two tracks fall into association threshold or wave threshold into a statistic checkup problem.

102.3.5 Hypothesis Judgment

Through the d_k^2 calculation, we know that according to χ^2 checkup:

If the random variable d_k^2 is less than the threshold value χ_α^2, the association will be considered as successful, and checkup H_0 will be accepted.

If the random variable d_k^2 is more than the threshold value χ_α^2, the association will be considered as failed, and checkup H_1 will be accepted.

Therefore the size of the wave threshold and the fall-into probability are associated. From the above expressions we can see that, the boundary of the wave threshold is corresponding to χ_α^2, whose size mainly depends on the measurement noise. The threshold point χ_α^2 can be found in the χ^2 distribution table according to freedom M and the given fall-into probability P.

If there is an ambiguity problem in the association process of two sensors, the minimum value d_k^2 should be taken so as to be corresponding to it [7].

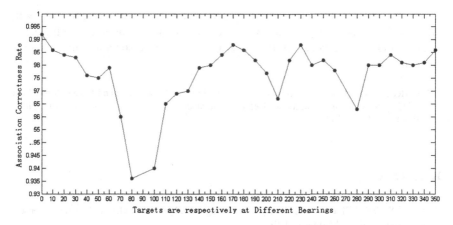

Fig. 102.2 Association correctness rate of 2D radar and infrared sensor detection data when targets are respectively at different bearings

102.4 Experiments

Assume that the coordinates $R_1(0,0)$ of infrared sensor is at the origin point O, and the 2D radar coordinates is $R_2(-50\text{ km}, 0)$. The target distance from the origin point O is 20 km, and the bearings are respectively $0°$, $10°$, $20°$, $30° \cdots 340°$, $350°$, with an elevation of $23.5°$, which are dispersed in a circle. The 2D radar distance error root mean square $\sigma_{d_1} = 80$ m, and the bearing error root mean square $\sigma_{\alpha_1} = 0.30°$. The infrared sensor bearing error root mean square $\sigma_{\alpha_2} = 0.15°$, and the elevation error root mean square $\sigma_{\beta_2} = 0.15°$, as shown in Fig. 4.1:

Here we set the freedom $M = 1$, and give the confidence level $P = 0.99$, and then we will obtain the following simulation results:

Notes: because when the bearings are 90 and 270°, the target will be in alignment with two sensors (their bearing lines coincide). It is impossible to carry out cross-positioning, and so there is no calculation result.

102.5 Conclusion

From the above numerical experiment we can see that the track association correctness rates for dissimilar platforms such as 2D radar and infrared sensor when they are detecting targets at different bearings are all more than 93.0 %, which will meet the actual requirements. This approach has fully exploited the complementation of 2D radar and infrared sensor, and effectively realized the track association between dissimilar platforms, i.e. 2D radar and infrared sensor, possessing certain technical application significance.

Besides, theoretically when the target projection position is intersecting with the connecting line of two sensors or their extension line, the approach in the paper is not valid, which needs further research on other association methods for compensation.

Acknowledgments The work was supported by National Defense Advanced Research of China (No. 51306030201) and National Defense Advanced Research Foundation of China (No. 9140A06040111CB39).

References

1. Huang X, Yang H et al (2006) Fused tracking based on 2D radar and infrared sensor. Firepower Command Control 31(9):54–57
2. Wang G Research on distributed detection, tracking and dissimilar sensor data association and guidance. Beihang University Ph. D dissertations. www.intsci.ac.cn/research/caiqs04.ppt
3. Han H, Han C, Zhu H et al (2004) Dissimilar multi-sensor data association algorithm based on fuzzy clustering. Xi'an Jiaotong Univ Acad J 38(4):388–391
4. Wu J et al (2007) Synchronized track fusion of radar and infrared. Ship Electron Eng 27(4):113–114
5. Wang G, Mao S, He Y (2001) Target tracking algorithm of infrared sensor. Firepower Command Control 26(2):5–9
6. He Y, Xiu J et al. (2006) Radar data processing with application. Electronics Industry Press, pp 70–71
7. Hu S (1994) Fuzzy mathematics and its application. Shichuan University Press, Chengdu

Chapter 103
Interference Management by Component Carrier Selection for Carrier Aggregation System in Heterogeneous Networks

Changyin Sun, Lan Huang, Jing Jiang and Guangyue Lu

Abstract In this paper, interference management scheme by component carrier selection for carrier aggregation system in Heterogeneous Networks is proposed. The scheme dynamically selects the component carrier with the cooperation of the other cells to deal with the interference between Macro cell and Pico cell. The component carrier selection criterion is based on the combination of cell specific Harmonic Mean rate metric and carrier specific PF metric. As a result, the selection algorithm will reserve some portion of carriers on which the Macro nodes are partially deactivated allowing the users of Pico cell to be served with lower interference. Additionally, greedy binary carrier search algorithm is used to reduce the search complexity of the algorithm. Simulation results show that the proposed scheme can achieve high cell-average and cell-edge throughput for the Pico cell in the Heterogeneous Networks.

Keywords Heterogeneous network · Inter-cell interference management · Greedy search · Carrier aggregation

103.1 Introduction

The new deployment strategy using heterogeneous networks is a key aspect of LTE Advanced. In this network topology, the system consists of regular (planned) placement of macro base-stations that typically transmit at high power level,

C. Sun (✉) · L. Huang · J. Jiang · G. Lu
Department of Communication and Information Engineering, University of Xi'an Posts and Telecommunications, Xi'an, People's Republic of China
e-mail: changyin_sun@163.com

W. Lu et al. (eds.), *Proceedings of the 2012 International Conference on Information Technology and Software Engineering*, Lecture Notes in Electrical Engineering 210, DOI: 10.1007/978-3-642-34528-9_103, © Springer-Verlag Berlin Heidelberg 2013

overlaid with several pico base-stations, femto base-stations and relay base-stations, which transmit at substantially lower power levels and are typically deployed in a relatively unplanned manner. These deployments can improve the overall capacity and the cell edge user performance [1, 2].

With HetNet deployment in same spectrum, users can experience severe interference. This effect is due to the often geographically random low power node deployment as well as the near-far problem arising from the imbalance in path-gains and transmission powers between the macro cell and low power nodes. 3GPP-LTE has devoted significant standardization effort towards devising inter-cell interference coordination (ICIC) schemes for minimizing interference, culminating in the so-called "enhanced" ICIC (eICIC) in LTE-Advanced.

The eICIC is critical to heterogeneous network deployment, without it the range extension concept [3] loses its advantage and efficiency. A basic ICIC technique involves resource coordination amongst interfering base-stations, where the resource partitioning can be performed in time-domain, frequency domain, or spatial domain. The main frequency domain ICIC scheme used in LTE-Advanced is carrier aggregation, which basically enables an LTE-Advanced user equipment (UE) to be connected to several carriers simultaneously.

Carrier aggregation (CA) not only allows resource allocation across carriers but also allows scheduler based fast switching between carriers without time consuming handovers, which means that a node can schedule its control information on a carrier and its data information on another carrier. So by scheduling control and data information for both Macro and Pico layers on different component carriers, interference on control and data can be avoided. It is also possible to schedule center Pico-eNodeB(eNB) user data information on the same carrier that the Macro layer schedules its users, as the interference from the Macro layer on center Pico-eNB users can be tolerated, while Pico-eNB users in the range extension areas are still scheduled in the other carrier where the Macro-eNB users are not scheduled.

For frequency domain ICIC, the partition of the available spectrum into separate component carriers (CC) can be static or semi-static based on factors such as users distribution among the different network layers, but it is inefficient in term of spectrum reuse. Among previous work, many studies have been done on component resource allocation [4–6], carriers selection [7, 8] and interference coordination [9]. However, the problems addressed in prior literature were either for homogeneous network, or for Home eNB in heterogeneous network with carrier aggregation. So in this paper, interference management by component carrier selection for carrier aggregation system in Heterogeneous Networks is addressed. The Heterogeneous Networks consists of complementing the Macro layer with low power nodes such as Pico base stations, and Range Extension is assumed to extend the coverage area of the Pico nodes. The proposed scheme deals with the inter cell interference by dynamically selecting the component carrier with the cooperation of the other cells. Based on the selection criterion which is the combination of cell specific Harmonic Mean rate metric and modified carrier specific Proportional Fair (PF) metric, the selection algorithm will reserve some portion of carriers on which the Macro nodes are partially deactivated allowing the users of Pico cell to be served with lower interference. The performance of the proposed approach is confirmed by simulation results compared with the performance of the baseline approach.

103.2 System Model

We consider Hetnet network with the following features: (1) a certain number of Pico-eNBs are deployed throughout one Macro cell layout; (2) The Pico-eNBs are randomly distributed; (3) The users are randomly distributed throughout the cell area.

Now consider the downlink transmission with M users and K carriers in the network. For simplicity, the transmit power is kept the same in each carrier per cell q, i.e. P_q. Let the binary matrix $\mathbf{A} = \{a_{k,m} | a_{k,m} \subset \{0,1\}\}_{K \times M}$ describes the carrier selection among the users, where $a_{k,m} = 1$ denotes that carrier k is assigned to user m, otherwise, $a_{k,m} = 0$.

Now, denote by $S_{i,k,m}$ the channel power gain to the selected mobile user m in cell q from the cell i base station, in resource slot t. The channel gains are assumed to be constant over each such resource slot, i.e. we have a block fading scenario. Note that the gains $S_{q,k,m}$ correspond to the desired communication links, whereas the gains $S_{i,k,m}$ for $i \neq q$ correspond to the unwanted interference links. Assuming the transmitted symbols to be independent random variables with zero mean and a variance P_q, the signal to noise-plus-interference ratio (SINR) for each user is given by:

$$SINR_{q,k,m} = \frac{P_q \, S_{q,k,m}}{\sum_{i=1,i\neq q}^{N} a_{k,i} \, P_i \, S_{i,k,m} + N_0} \tag{103.1}$$

where N_0 is the variance of the independent zero-mean AWGN.

Then the achievable data rate for user m is given by:

$$R_{q,m} = \sum_{k=1}^{K} \log_2(1 + SINR_{q,k,m}) = \sum_{k=1}^{K} R_{q,k,m} \tag{103.2}$$

The interference management by component carrier selection is to find the CC selection matrix $\mathbf{A}$ such that the objective function is optimized. Assuming at time slot t, only one user is scheduled at each base-station q, and the goal is to achieve the highest system throughput (sum rate) while ensuring proportional fairness among different users, the following objective function needs to be maximized:

$$\mathbf{A} = \arg \max_{\mathbf{A}, m \subset U_q} \sum_{q=1}^{N} \sum_{k=1}^{k} w_m \, a_{k,q} \, R_{q,k,m} \tag{103.3}$$

where w_m be the proportional fair weight, and U_q be the users set that served by base station q.

Solving the exhaustive resource allocation problem (103.3) for large networks presents the system designer with an exponentially complex task. In this paper we study approach with reduced search complexity.

103.3 Interference Management by Component Carrier Selection

To achieve a good performance versus complexity compromise, we assume the total number of N cells in the network are clustered into groups of G ≪ N cells. For a given cluster Q, the interference from the remaining N–Q cells will simply contribute as noise, i.e. the sum throughput of the cells in Q is given as:

$$R_Q = \sum_{q \subset Q} \sum_{k=1}^{K} \log_2 (1 + SINR_{q,k,m}) \qquad (103.4)$$

where $SINR_{q,k,m}$ is given by:

$$SINR_{q,k,m} = \frac{P_q S_{q,k,m}}{\sum_{i \subset Q, i \neq q} a_{k,i} P_i S_{i,k,m} + I_N + N_0} \qquad (103.5)$$

and I_N is the interference from cells in the network which are not part of the cluster. We assume here that I_N can be estimated, so the idea is to do optimization only locally within each cluster. Hence, the following problem is solved for each cluster Q,

$$\mathbf{A} = \arg \max_{\mathbf{A}, m \subset U_q} \sum_{q \subset Q} \sum_{k=1}^{K} w_m \, a_{k,q} \, R_{q,k,m} \qquad (103.6)$$

103.3.1 Step1: Cell Specific CC Selection

As mentioned above, Pico cell users in the range extension area suffer from a high level of interference, so they will have less priority than Macro cell users when CC is selected by (103.6) co-operatively. This causes insufficient offloading from the Macro to Pico cells and cell splitting gain due to HetNet becomes limited. For this issue, cell specific CC selection based on Harmonic Mean rate metric is proposed here. Since we are concerned with best effort packet data service, we will use the long term average throughput achieved by the users as the quality of service metric and require that all users within different cell are treated the same way. To this end, we define the Harmonic Mean rate R_{hm} as following:

$$\frac{1}{R_{hm,q}} = \frac{1}{U_q} \sum_{i \in U_q} \frac{1}{\bar{R}_{q,i}} \tag{103.7}$$

where U_q be the set of users that are served by the base-station q, and $\bar{R}_{q,i}$ is the aggregated rate of user i across the K carrier.

For any pair of base-station q and q' in the homogeneous network, we expect $R_{hm,q} \approx R_{hm,q'}$. For pico cell q' and Macro cell q in Hetnet network, we have $R_{hm,q'} = R_{hm,q} / N_L$, where N_L is a offloading factor which reflect the difference of transmit power, associated users and interference between Micro cell and pico cell. Then we define cell specific Harmonic Mean rate selection metric for cell q:

$$w_{m \subset U_q} \propto \beta_q = 1 / R_{hm,q} \tag{103.8}$$

Based on the Harmonic Mean rate metric (103.8), the Pico cell users will be treated with fairness when they have chances to be allocated on CCs.

103.3.2 Step2: Proposed Carrier Selection and User Scheduling Algorithm

The joint CC selection and user scheduling problem (103.6) requires high computation and feedback overheads in the optimal algorithm, so we decouple this problem into two sub-optimum problems: intra-cell user scheduling (103.9) and inter-cell CC selection (103.10).

$$m^* = \arg \max_{m \subset U_q} \{\lambda_k (\breve{\mathbf{a}}_{k,q})\} \tag{103.9}$$

$$\mathbf{A} = \arg \max_{\mathbf{A}} \sum_{k=1}^{K} \hat{\lambda}_k (\mathbf{a}_k) \tag{103.10}$$

where $\hat{\lambda}_k (\mathbf{a}_k) = \sum_{q=1}^{Q} a_{k,q} \beta_q \alpha_{k,q} R_{q,k,m} (\breve{\mathbf{a}}_{k,q})$, β_q is defined in (103.8), $\alpha_{k,q}$ is the carrier specific PF weight factor. $\lambda_k (\breve{\mathbf{a}}_{k,q})$ is defined as:

$$\lambda_k (\breve{\mathbf{a}}_{k,q}) = \alpha_{k,q} \times w_{m,k} \times R_{q,k,m} (\breve{\mathbf{a}}_{k,q}) \tag{103.11}$$

where $\breve{\mathbf{a}}_{k,q} = [a_{k,1}, a_{k,2}, \cdots, a_{k,q-1}, a_{k,q+1}, \cdots, a_{k,Q}]$ and $w_{m,k}$ is the intra-cell PF weight.

Let $\mathbf{A} = [\mathbf{a}_1^T \, \mathbf{a}_2^T \cdots \mathbf{a}_K^T]^T$, then the optimum $\mathbf{A}$ is solved by the greedy search algorithm as following:

Greedy Search Algorithm

1: $\boldsymbol{\alpha} = [1]^{1 \times Q}$, $\mathbf{e} = [1]^{1 \times Q}$

2: **for** $k = 1$ to K **do**

3:      $\mathbf{a}(1) = [1]^{1 \times Q}$, $\bar{\lambda}(0) = 0$.

4:      **for** $l = 1$ to Q **do**

5:          **for** $q = 1$ to Q **do**

6:              $m^* = \arg \max\limits_{m \subset U_q} \{\lambda_k(\mathbf{a}(l), m)\}$, $[\mathbf{U}(l)]_q = m^*$.

7:          **End for** q.

8:          $\bar{\lambda}(l) = \hat{\lambda}_k(\mathbf{a}(l), \mathbf{U}(l))$

9:          **if** $\bar{\lambda}(l) \langle \bar{\lambda}(l-1)$,**break**

10:         **else** $q^* = \arg \min\limits_{q} \beta_q \times [\boldsymbol{\alpha}]_q \times R_{q,k,[\mathbf{U}(l)]_q}$,

11:             $[\mathbf{a}(l+1)]_{q^*} = 0$, $[\mathbf{U}(l+1)]_{q^*} = 0$, $\mathbf{a}^*(k) = \mathbf{a}(l)$.

12;   **End if**

13   **End for** l

14      $\boldsymbol{\alpha} = \boldsymbol{\alpha} + \mathbf{e} - \mathbf{a}^*(k)$.

15: **End for** k

16: $\mathbf{A} = [\mathbf{a}^*(1)^T, \mathbf{a}^*(2)^T, \cdots, \mathbf{a}^*(K)^T]^T$

The greedy search algorithm is motivated by the observation that the capacity gains may be achieved by deactivating cells which do not offer enough capacity to outweigh the degradation caused through interference to the system [9]. Among the cells, the candidate cell to be deactivated will probably be the cell which has a low channel gain but high interference from other cells, this candidate cell is fund in the 10th step and the resulted gain is verified in the 9th step.

It is obvious that fairness issues will arise when increasing the system capacity by deactivating some cells, so carrier specific weighting vector $\boldsymbol{\alpha}$ is introduced. The carrier specific weighting vector $\boldsymbol{\alpha}$ is updated in the 14th step:

$$\alpha_{q,k} = \sum_{j=1}^{k-1} \bar{a}_{j,q} + 1 \qquad (103.12)$$

where $\bar{a}_{j,q} = 1 - a_{j,q}$. The carrier specific weighting means that if one cell has been deactivated on carrier k−1, then it will have a increased scheduling metric on carrier k. As a result, the cell-edge user in one cell will be assigned a carrier that is non-overlapping with the carrier assigned to the user in its interfering cell, since this interfering user may not be scheduled on the interfering carrier for interference avoidance.

Table 103.1 Simulation parameters

Parameters	Value
Carrier bandwidth	10 MHz
Number of carriers	2
Path loss model pico (dB)	$140.7 + 36.7\log10(R)$, R in km
Path loss model macro (dB)	$128.1 + 37.6\log10(R)$, R in km
Shadowing standard deviation (dB)	8
Transmit power macro (dBm)	43
Transmit power pico (dBm)	30

103.4 Simulation Results

In this section, simulation results are presented to evaluate the performances of the proposed scheme in a carrier aggregation system. For comparison purpose, the performances of the network with all the carriers fully reused between macro and pico station are also given. For simplicity, we only consider one cell network which contains 1 Macro-eNB and 2 Pico-eNBs. The radius of Macro cell is 500 m. And the total number of users is 40. The range extension threshold is 10 dB. Detailed simulation parameters including channel model and system assumptions are summarized in Table 103.1. Most of them are set according to the LTE-A simulation assumptions.

First, uniform user distribution across the macro cell is assumed, and the cumulative distribution function (CDF) performances of the networks with the new CC selection scheme ('New' in the figure) and baseline scheme are compared in top subplot of Fig. 103.1. Compared to the baseline scheme where carriers are fully reused ('Reuse 1') between Macro cell and pico cell, the throughput performance of the pico cell with the proposed scheme is significantly improved when proposed metric is applied. On the contrary, the throughput performance of the Macro cell with the proposed scheme is a little worse than that of the baseline scheme.

Next, the CDF performances with none uniform user distribution are compared in bottom subplot of Fig. 103.1. In this case, the users are dropped at the dense area which overlay both the pico cells and is about 12 % of the Macro cell area. Similar to the uniform conditions, there is consistent throughput improvement with the proposed interference management scheme for pico cell, so the pico cell achieves high cell-average and cell-edge throughput.

In general, the results suggest that the interference management scheme might decide to leave certain set of resources in the Macro cell especially to serve users that are at cell edges in the pico cell. However, the interference reduction gain for edge of the cell users is offset by losing out the opportunity to transmit to another user in the Macro cell.

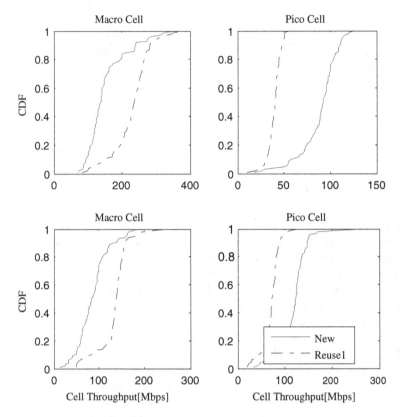

Fig. 103.1 CDF performances of micro cell and pico cells

103.5 Conclusion

In this paper, we consider interference management by component carrier selection for carrier aggregation system in Heterogeneous Networks. The proposed scheme dynamically selects the component carrier with the cooperation of the other cells. The selection criterion is based on the cell and carriers specific PF metric. Additionally, greedy binary carrier search algorithm is used to reduce the search complexity of the algorithm. Simulation results for different user distribution conditions suggest that the interference management might decide to leave certain set of resources in the Macro cell especially to serve users that are at cell edges in the pico cell. As a result, the pico cell achieves high cell-average and cell-edge throughput.

Acknowledgments This work was supported in part by National Natural Science Foundation of China (Grant No. 61102047); The Natural Science Foundation of Shaanxi Province(2011JQ8027); The Natural Science Foundation of Education department of Shaanxi Province(11JK1021).

References

1. Qualcomm Incorporated (2010) LTE advanced: heterogeneous networks. Feb [Online] Availalbe: http://www.qualcomm.co.kr/documents/lteadvanced-heterogeneous-networks
2. Lindbom L, Love R, Krishnamurthy S, Yao C, Miki N, Chandrasekhar V (2011) Enhanced Inter-cell interference coordination for heterogeneous networks in LTE. Advanced a survey, CoRR abs/1112.1344
3. 3GPP Tdoc R4-083018, Qualcomm Europe (2008) Range expansion for efficient support of heterogeneous networks (Aug)
4. Wang Y, Pedersen KI, Sorensen TB, Mogensen PE (2010) Carrier load balancing and packet scheduling for multi-carrier systems. IEEE Trans Wireless Comm, 9(5):1780–1789
5. Songsong S, Chunyan F, Caili G (2009) A resource scheduling algorithm based on user grouping for LTE. Advanced system with carrier aggregation In: proceeding symp computer network and multimedia technology (CNMT) Jan
6. Wang H, Rosa C, Pedersen K (2011) Performance analysis of downlink inter-band carrier aggregation in LTE. Advanced in proceeding. IEEE ICC, June
7. Garcia L, Pedersen K, Mogensen P (2009) Autonomous component carrier selection for local area uncoordinated deployment of LTE. Advanced in proceeding IEEE VTC 2009 Fall, Sept
8. Yuan Y, Li A, Gao X, Kayama H (2011) A new autonomous component carrier selection scheme for home eNB in LTE. A system, In: proceeding IEEE VTC 2011 Spring, May
9. Kiani S, Gesbert D, Kirkebo J, Gjendemsjo A, Oien G (2006) A simple greedy scheme for multicell capacity maximization. In: Proceeding international telecommunications symposium, pp 435–440, Sept

Printed in the United States
By Bookmasters